Organic Electronics for Electrochromic Materials and Devices

Organic Electronics for Electrochromic Materials and Devices

Hong Meng

WILEY-VCH

Author

Prof. Hong Meng
Peking University
Shenzhen Graduate School
Building G 306
Lishui Road, Nanshan Disctrict
518055 Shenzhen
China

Cover Design: Wiley
Cover Image: © Andrew Goodsell/Shutterstock

Library of Congress Card No.:
applied for

British Library Cataloguing-in-Publication Data
A catalogue record for this book is available from the British Library.

Bibliographic information published by the Deutsche Nationalbibliothek
The Deutsche Nationalbibliothek lists this publication in the Deutsche Nationalbibliografie; detailed bibliographic data are available on the Internet at <http://dnb.d-nb.de>.

Print ISBN: 978-3-527-34871-8
ePDF ISBN: 978-3-527-83061-9
ePub ISBN: 978-3-527-83062-6
oBook ISBN: 978-3-527-83063-3

Typesetting SPi Global, Chennai, India
Printing and Binding
CPI Group (UK) Ltd, Croydon CR0 4YY

C110994_120321

Contents

Preface

In recent years, with the development of artificial intelligence, more and more industries strive to be "smart." As a new generation of display technology, organic electrochromic (OEC) devices offer numerous advantages such as flexibility, full colors, wide origins of materials, fast switching time, low driving voltage, and simple configuration. In addition, these devices possess "smart" characteristics of multi-stimulation and multi-response. Therefore, the OEC industry is emerging as a potential display competitor in the field of electronic information.

This book covers major topics related to the phenomenon of electrochromism, including the history of organic electrochromism, fundamental principles, different types of electrochromic materials, development of device structures, multifunctional devices, their characterizations and applications, and future prospects of OEC technology. It also spotlights recent research progress reported by academic institutes and enterprises, and discusses the existing challenges in further development of this area.

This book provides a comprehensive review of OEC materials and devices, and can be used as a teaching reference for undergraduate and graduate students as well as teachers in the fields of organic chemistry and polymer science etc. Also, this book can be adopted as a comprehensive reference for researchers engaged in the development of OEC technology enterprise in the field of electrochromism.

Shenzhen, PR China *Hong Meng*
11 November 2020

About the Author

Prof. Dr. Hong Meng obtained his BS in Chemistry at Sichuan University in 1988 and MS degree in Organic Chemistry from Peking University in 1995. He then studied Polymer Science and Engineering and acquired his second MS degree from the National University of Singapore in 1997. After working at the Institute of Materials Science and Engineering (IMRE) in Singapore for two years, he went to the United States in 1999 and completed his PhD under the supervision of Prof. Fred Wudl at the University of California, Los Angeles (UCLA), in 2002. Prof. Dr. Meng worked as a research consultant at Bell Labs, Lucent Technologies, with Dr. Zhenan Bao for one year. He then joined DuPont Experimental Station, Central Research and Development, as a senior research chemist in 2002. In 2012, he moved back to China and worked as the CEO of a laser printing industry at Leputai Technology Company. In 2014, he joined the School of Advanced Materials at Peking University, Shenzhen as a chair professor. He has been engaged in the research and development of solid-state organic synthesis, organic semiconductor device engineering, organic electronics, and other relevant fields, especially organic light-emitting diodes, organic electrochromics, organic thin-film transistors, organic conductive polymers, and nanotechnology. He has published more than 200 articles in internationally renowned journals, participated in the writing of three book chapters and co-edited two books in the field of organic optoelectronic technology. He has obtained more than 150 patents for inventions in the United States and China, among which, several patented products have been commercialized.

1 Introduction

1.1 General Introduction

Electrochromism is the phenomenon that describes the optical (absorbance/ transmittance/reflectance) change in material via a redox process induced by an external voltage or current [1]. Usually the electrochromic (EC) materials exhibit color change between a colored state and colorless state or between two colors, even multicolor. In nature, its origin is from the change of occupation number of material's internal electronic states. As the core of EC technology, the EC materials have built up many categories during decades of development, for example, according to the coloration type, it could be classified as anodically coloring materials (coloration upon oxidation) or cathodically coloring materials (coloration upon reduction) [2]. Based on the light absorption region in the solar radiation, which consists of these three parts: ultraviolet (UV), visible (Vis), and near-infrared radiation (NIR) lights (Figure 1.1), it could be divided into visible EC materials (wavelength: 380–780 nm), which can be seen by the human eye and therefore are suitable for smart window and indicator applications, and NIR EC materials (wavelength: 780–2500 nm), which have great potential for thermal regulation technologies and even in national defense-related applications [3]. And on the basis of materials species, there are mainly inorganic, organic, and hybrid EC materials [4, 5] (https://commons.wikimedia.org/wiki/File.Solar_spectrum_en.svg). Inorganic EC materials are transition metal oxides (TMOs) (WO_3, NiO, TiO_2, and Prussian Blue [PB]), organic EC materials including small molecules (e.g. viologen), conjugated polymers (e.g. poly(pyrrole), poly(thiophene), and poly(carbazole)) and aromatic polymers (e.g. polyimides [PIs] and polyamides [PAs]), organic–inorganic hybrid materials referring to metallo-supermolecular polymers, and metal–organic framework (MOF). Among them, inorganic materials exhibit excellent long-term stability compared with organic ones; however, considering the structure variety, flexibility, and low-cost solution processability, organic EC materials are superior to inorganic materials. The organic–inorganic hybrid materials are designed to combine advantages of both organic and inorganic materials.

EC materials exhibit color changes during the redox process; therefore the electrochromic devices (ECDs) generally consist of three elements: electrodes, EC materials, and electrolyte solution. The electrodes offer a constant supply of electric

Organic Electronics for Electrochromic Materials and Devices, First Edition. Hong Meng.

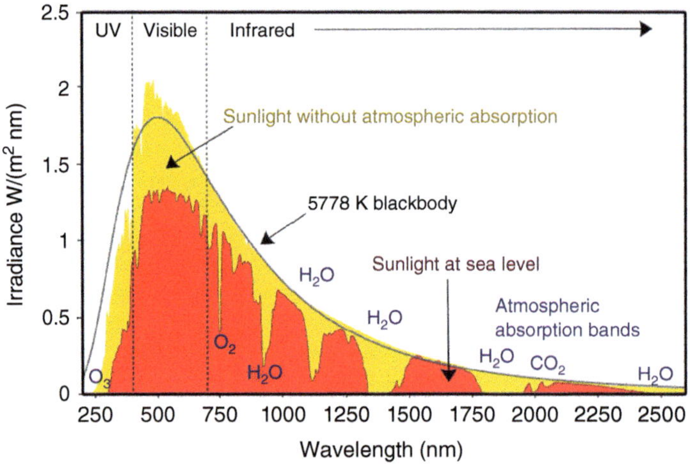

Figure 1.1 Solar irradiance spectrum above atmosphere and at the surface of the Earth. Source: Nick84: https://commons.wikimedia.org/wiki/File:Solar_spectrum_en.svg, Licensed under CC BY-SA 3.0.

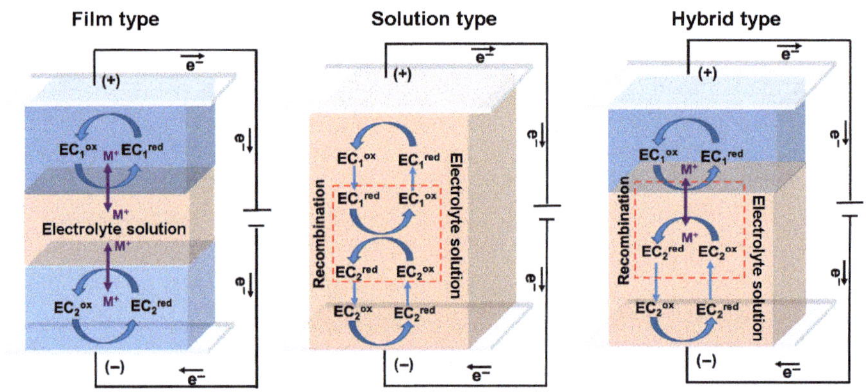

Figure 1.2 The scheme of three types of electrochromic devices.

current, and ions are conducted by the electrolyte solution. Then the EC materials undergo electrochemical oxidation and/or reduction, which results in changes in the optical bandgap and colors. As shown in Figure 1.2, a typical ECD has five layers: two transparent conducting oxide (TCO) layers, EC layer, ion-conducting layer (electrolyte solution), ion storage layer. Particularly, the ion storage layer acts as the "counter electrode" to store ions and keep electric charge balance. And according to the exact state of EC materials, there are three types of ECD: film type (I), solution type (II), and hybrid type (III). The Type I ECD is the most common; many kinds of EC materials are suitable for this type including TMOs, conjugated/non-conjugated polymers, metallo-supermolecular polymers, and MOF/covalent organic framework (COF) materials, which using spin-coating, spray-coating, and dip-coating processes to form uniform films; these films won't dissolute in electrolyte solutions. Type II ECD requires that the EC materials have good solubility in electrolyte solutions. Therefore many organic small molecules such as viologen, terephthalate derivatives, and isophthalate derivatives are appropriate for this type of device. Meanwhile the fabrication method for this type of device is the most simple one. It just needs to dissolve the electrolyte and EC material in a specific solvent and inject into the prepared conducting cell. Type III ECD uses film-type EC materials as working electrode and solution-type EC materials as ion storage layer.

1.2 The History of Electrochromic Materials

The word "electrochromism" was invented by John R. Platt in 1960 [6], in analogy to "thermochromism" and "photochromism." However, the EC phenomenon could be traced to the nineteenth century, as early as 1815. Berzelius observed the color change of pure tungsten trioxide (WO_3) during the reduction when warmed under a flow of dry hydrogen gas. Then from 1913 to 1957, some patents described the earliest form of ECD based on WO_3 [7, 8]. Therefore the origins of electrochromism are the nineteenth and twentieth centuries. After then, electrochromism technology began to undergo rapid development, especially the exploration of many classes of EC materials. As showed in the technology roadmap (Figure 1.3), we summarized several generations of EC materials during long-term development.

The first-generation EC material is TMOs (e.g. WO_3, NiO, and PB). Among them, WO_3 plays an important role in the electrochromism field; as the first founded EC material, it has already realized commercialization in smart windows application. PB was discovered as a dye by Diesbach in 1704, and then the electrochemical behavior and EC performance of PB was firstly reported by Neff at 1978 [9]. Benefitted from the structure stability and reversible redox process of those inorganic TMOs, the electrochromism based on the thin-film TMOs are widely investigated, including

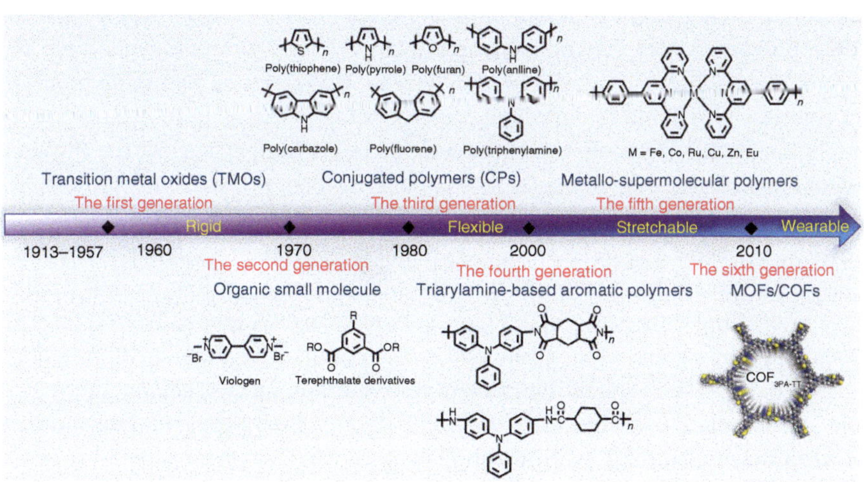

Figure 1.3 The roadmap of EC materials development.

the development of new TMOs materials, introduction of new nanostructures, and different element doping.

Following the first-generation TMO EC materials, organic small molecule EC materials have emerged since 1970. Among them, viologen as the most representative small molecule was first discovered by Michaelis and Hill in 1932 [10], and because of the violet on the reduction, these 1,1′-disubstituted-4,4′-bipyridine compounds were named "viologen." Then in 1973, Shoot made a new flat alphanumeric display using heptyl viologen; this can be regarded as the beginning of the use of viologen for electrochromism [11]. After a century's development, viologen already has been successfully commercialized. Besides the viologen, other small molecules EC materials such as terephthalate derivatives, isophthalate derivatives, methyl ketone derivatives, and some dye molecules have also attracted much attentions from scientists due to their simple structure and low cost.

The third-generation EC materials are conjugated polymers. In 1983, Francis Garnier and coworkers firstly characterized the EC properties of a series of five-membered heterocyclic polymers including poly(pyrrole), poly(thiophene), poly(3-methylthiophene), poly(3,4-dimethylthiophene), and poly(2,2′-dithiophene). Since then, conjugated polymers were given rise to the rapid emerge as a new class of electrochromism [12]. Five years later, Berthold Schreck observed the electrochromism phenomenon of poly(carbazole), which showed a color change from pale yellowish to green together with the conductivity enhancement [13]. To date, the conjugated polymer EC system has been well developed, from better understandings on mechanisms to completed color pallette with soluble or electro-deposited polymers, and even full-color display samples or roll-to-roll fabricated flexible devices.

Later, in early 2000, triarylamine (TA)-based aromatic polymers especially the PIs and PAs have drawn considerable attention from the research community as the fourth-generation EC materials. The correlation between electrochemical properties and chemical structures of different aromatic PIs was firstly described in 1990. Ten years later, Zhiyuan Wang and coworkers [14] reported the first EC behavior of poly(ether naphthalimide)s, which showed stepwise coloration process, from colorless to red and to dark blue corresponding to the neutral, radical anion, and dianion species, respectively. However, due to the high rigidity of the PIs/PAs backbone and strong intermolecular interactions, the poor processability limited the development of PIs or PAs EC materials. Therefore the TA groups were introduced to the PIs/PAs backbone to improve the solubility of aromatic polymers. The first TA-based polyamide PA was synthesized in 1990 [15], and the first aromatic polyimides integrating interesting EC properties containing TA groups were disclosed in 2005 [16]. Since then, Liou, Hsiao and, other groups have developed numerous TA-based EC PIs/PAs. Most of the PIs/PAs were solution processible and thermally stable with excellent adhesion with indium tin oxide-coated glass electrode and had good electrochemical stability. Now the TPA-based PIs/PAs are considered as great anodic EC materials due to proper oxidation potentials, electrochemical stability, and thin-film formability.

Benefiting from the bloom and revolution of organic polymers, metallo-supermolecular polymers were developed by incorporating metal centers into synthetic polymer chains, as the fifth-generation EC materials. The first metal-containing polymer, poly(vinyl-ferrocene), was reported in as early as 1955 [17]. However, due to the insolubility of those macromolecules and the limitation of characteristic technologies in the early years, the metallo-supermolecular polymers haven't been rapidly developed until the mid-1990s [18]. Since then, metallo-supermolecular polymers began to be widely explored in EC field with the advantages of beneficial properties of both organic and inorganic materials. Especially, because the transition metal complexes often exhibit well-defined redox events and intense charge transfer transitions in the NIR region, metallo-supermolecular polymers often show potentials in NIR EC application.

More recently, with the active researches on the crystalline and porous MOFs and COFs, the sixth-generation MOFs/COFs EC materials have emerged. In 2013, the first EC properties of MOFs using naphthalene diimide (NDI) as organic linker were reported by Professor M. Dinca's group [19]. And the first COFs EC material using the TPA as building block was revealed by Yuwu Zhong and Dong Wang and coworkers in 2019 [20]. All in all, some essential features of MOFs/COFs give them advantages in EC, including designable and precise molecular structure, simple self-assembly synthesis, and porous structure that facilitate the electrolyte ions transport. However, these new EC materials haven't been fully revealed; many efforts should be taken to improve the device performance of MOFs/COFs-based electrochromism.

1.3 The Key Parameters of Electrochromism

In order to elucidate the EC properties of EC materials, in situ UV–Vis–NIR spectroelectrochemistry (SEC) measurements were performed on a spectrophotometer, combining with an electrochemical workstation to apply and control the potential in the SEC cell. This SEC spectrum dynamically records the absorption change of EC materials during different applied voltage, which reflects the color change during the whole redox process. As an example, the SEC of a black-to-transmissive EC material is shown in Figure 1.4. Both absorbance model and transmittance model could be used to carry out the SEC measurement.

1.3.1 Electrochromic Contrast

EC contrast ($\Delta\%T$) is measured as the percent transmittance change of the EC material at a specific wavelength [22] it is a primary parameter for characterizing the EC materials. It is calculated from the difference of light transmission (TM) in the bleached and colored state TM_b and TM_c at the specified wavelength. And the transmittance values are generally recorded upon the application of square-wave potential steps to the electroactive film placed in the beam of a spectrophotometer. Each color has a characteristic wavelength as shown in Figure 1.5b, such as the

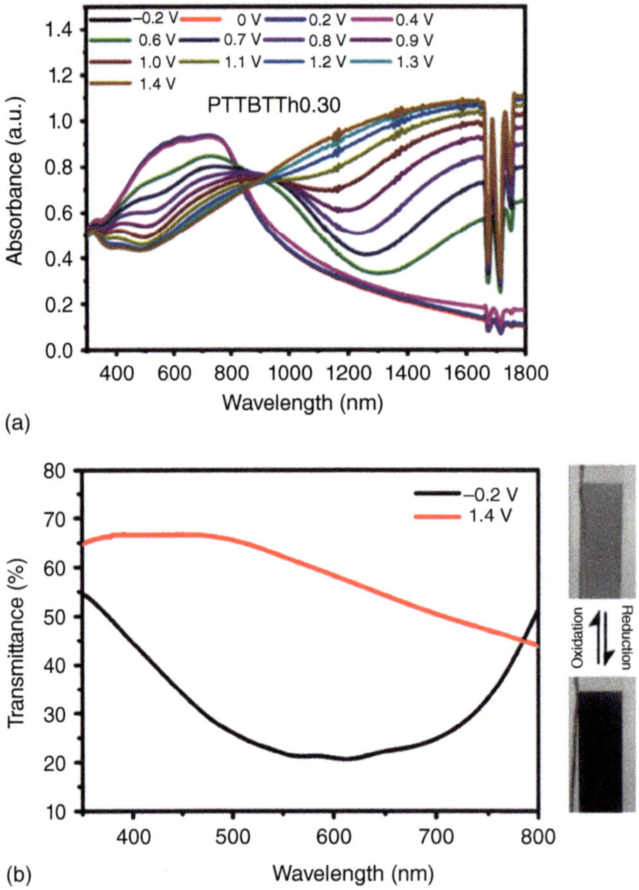

Figure 1.4 The spectroelectrochemistry (SEC) of a black-to-transmissive EC material. (a) Absorbance model and (b) transmittance model. Source: Reproduced by permission Li et al. [21]. © 2018, Royal Society of Chemistry.

wavelength of blue color is ranging from 455 to 490 nm. Therefore, in most cases, the contrast in the characteristic wavelength is chosen to evaluate the degrees of color change. Usually the absorption of this characteristic wavelength also reaches its maximum value (λ_{max}). Moreover, there is evidence that human eyes are most sensitive to green light with a wavelength of 555 nm [25]. It's also recommended to calculate the contrast in 555 nm for comparison in different publications. Specifically, a contrast test example is shown in Figure 1.5a, the TM_b and TM_c of λ_{max} 425 nm are 1% and 99%, respectively; therefore the $\Delta\%T$ is calculated as 98%. For applications such as smart windows, in which the difference between the bleached and colored states is expected to be the highest, the $\Delta\%T$ should be higher than 80%. Many inorganic EC materials, organic small molecules, and PEDOT series polymers, which have a high transmittance in the bleached state, can achieve this index. Especially, some reported small molecule EC materials even show $\Delta\%T$ exceed 95% [23].

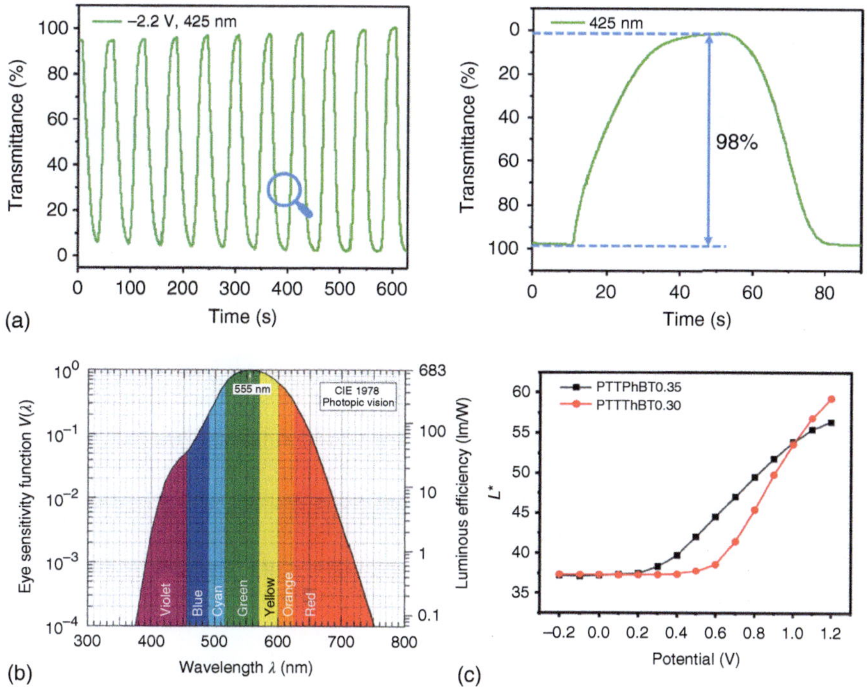

Figure 1.5 (a) The electrochromic contrast of a small molecule EC material. Source: Jiang et al. [23], (b) sensitivity function of the human eye $V(\lambda)$ and luminous efficacy vs wavelength. Source: Fred Schubert [24]. © 2006, Cambridge University Press. (c) The change of the lightness values from the neutral to the oxidized states. Source: Li et al. [21]. © 2018, Royal Society of Chemistry.

Meanwhile, for some broad absorption or color-to-colorless EC materials, measurements on the relative luminance change ($\Delta\%Y$) during an EC switch are more realistic for exhibiting the overall EC contrast, which conveys more information on the perception of transmittance to the human eye. As an example, a luminance change curve during the redox process of black-to-transmissive EC materials is shown in Figure 1.5c. The lightness L^* value (from 0 (black) to 100 (white)) of 37.5 (black state) increases to 60 (bleach state); therefore the $\Delta\%Y$ is calculated as 22.5%.

Except for the aforementioned method for electrochemical contrast measurements, a photopically weighted value called photopic contrast was proposed by Javier Padilla et al. [26]. The photopic contrast also reflects an overall contrast during the whole visible region, which is more consistent with the real application condition. It can be calculated using the following equation:

$$T_{\text{photopic}} = \frac{\int_{\lambda_{\min}}^{\lambda_{\max}} T(\lambda)S(\lambda)P(\lambda)\mathrm{d}\lambda}{\int_{\lambda_{\min}}^{\lambda_{\max}} S(\lambda)P(\lambda)\mathrm{d}\lambda}$$

where T_{photopic} is the photopic transmittance, $T(\lambda)$ is the spectral transmittance of the device, $S(\lambda)$ is the normalized spectral emittance of a 6000 K blackbody, and $P(\lambda)$

is the normalized spectral response of the eye. λ_{min} and λ_{max} define the considered range of wavelengths.

1.3.2 Switching Time

In the context of electrochromism, the switching time (t) can be defined as the time needed for EC materials to switch from one redox state to the other. It is generally followed by a square wave potential step method coupled with optical spectroscopy. Switching time depends on several parameters, such as the ability of the electrolyte to conduct ions as well as the ease of intercalation and deintercalation of ions across the EC active layer, the electrical resistances of electrolytes, and the transparent conducting films. Usually the liquid electrolyte has a lower resistance than the solid electrolyte; therefore the half device and the liquid electrolyte ECD will exhibit a rapid switching than solid ECD. Meanwhile, the large area ECD such as the large smart windows will show a lower switching compared with the small laboratory samples due to the larger electrical resistances of transparent conducting films. However, fast switching is not required in all applications, such as the switchable window technologies; the obvious color change process will increase the fun of user experience. Conversely, the sub-second magnitude rapid switching is particularly desired for display applications.

Usually the switching times are evaluated at the λ_{max} or 555 nm together with the EC contrast. Therefore there are two kinds of switching time. One is electrochemical switching time, as shown in Figure 1.6a, which is the time required for the current density to change by 90% or 95% between two constant voltages. Meanwhile the switching time of oxidization ($t_{oxidization}$) and reduction ($t_{reduction}$) process can be estimated from this curve. The other is optical switching time (Figure 1.6b), which defines the time needed for the transmittance to change by 90% or 95%. Correspondingly, the coloration switching time ($t_{coloration}$) and bleaching switching time ($t_{bleaching}$) are recorded in this measurement. It is worth noting that the pulse length of potential step has influence in transmittance. A shorter potential step will achieve a smaller contrast, and longer potential will allow EC materials to reach stationary transmittance value in both coloration and bleaching state. But after a certain length, continuing increase the pulse length won't boost the contrast. Therefore the pulse length that just reached the highest contrast are applied to switching time as well as contrast tests.

However, the aforementioned method of switching time is an experiential measurement, which has a difference in varied research groups, such as the different percentage of transmittance change (90% or 95%). Therefore, it is difficult to compare switching time data between different research groups. In recent years, Javier Padilla and coworkers proposed a standard method for calculating EC switching times. They fitted the contrast values as a function of pulse length to the following exponential increase function:

$$\Delta TM(t) = \Delta \mathrm{TM}_{max}(1 - \mathrm{e}^{-t/\tau})$$

where ΔT_{max} represents the full-switch contrast obtained for long pulse lengths and τ is the time constant. If switching time t is equal to τ, 63.2% of the maximum

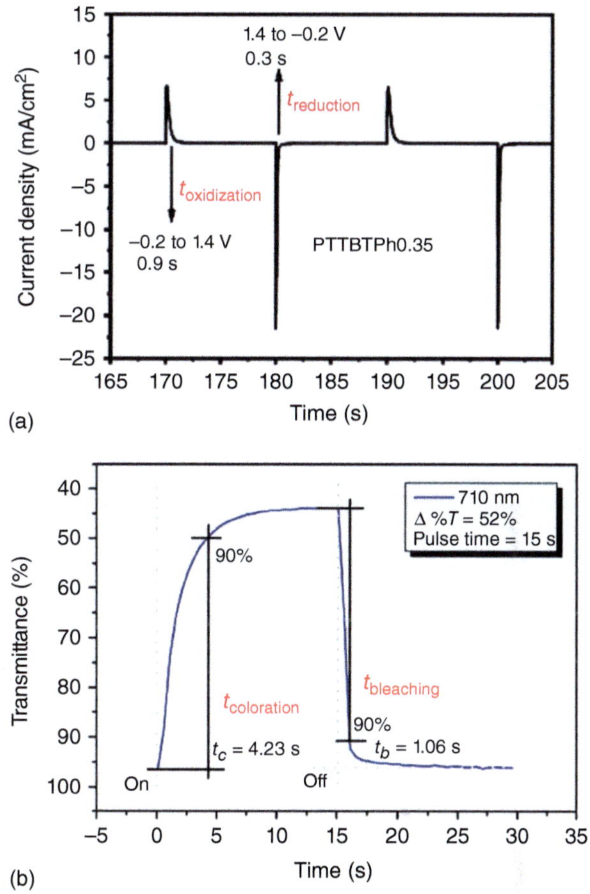

Figure 1.6 The switching time of EC materials. (a) Electrochemical switching time. Source: Li et al. [21]. © 2018, Royal Society of Chemistry (b) Optical switching time. Source: Hsiao et al. [27].

transmittance change is reached. At a time of 2.3τ, 90% ΔTM_{max} is switched, identically 95% and 99% of ΔTM_{max} corresponding to 3τ and 4.6τ.

Therefore, for new EC materials, the same chronoabsorptometric responses [28] should be measured and fitted to the aforementioned function. From the fittings, the max values of ΔTM_{max} (the contrast corresponding to a full switch), the time constant τ, and the corresponding regression coefficient r^2 will be obtained. Afterwards the switching time $t_{90\%}$ or $t_{95\%}$ will be easily calculated. This method allows an easy direct comparison between different reported values.

1.3.3 Coloration Efficiency

Coloration efficiency (CE) plays a fundamental role in the evaluation of the efficiency of charge utilization during the EC processes. It relates the optical absorbance change of an EC material at a given wavelength (ΔA) to the density of injected/ejected electrochemical charge necessary to induce a full switch (Q_d). The higher CE value indicates a large transmittance change with a small amount of charge, which makes more effective use of the injected charge. CE value can be

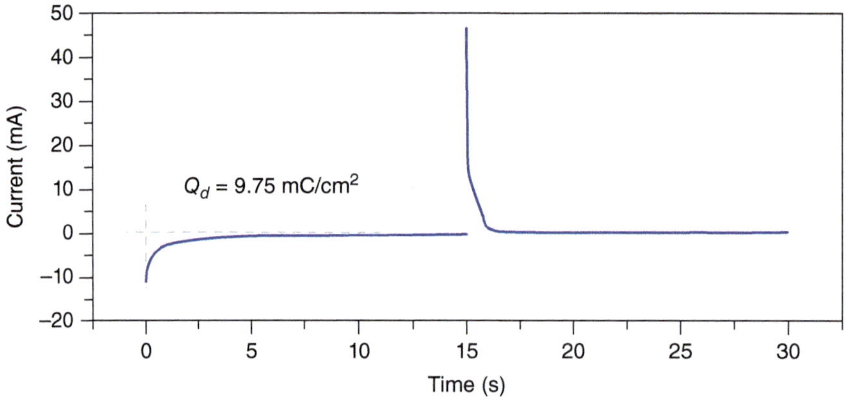

Figure 1.7 The calculation of Q_d. Source: Hsiao et al. [27].

calculated using the following equation:

$$\mathrm{CE} = \frac{\Delta A}{Q_d} = \frac{\log\left(\frac{T_{\mathrm{ox}}}{T_{\mathrm{neut}}}\right)}{Q_d}$$

where T_{ox} and T_{neut} are the transmittances in the oxidized and neutral states, respectively, and Q_d represents the injected/ejected charge per unit area, which could be obtained from the integral area of the current density curve during voltage switching (Figure 1.7).

Very similar to the previous parameters contrast and switching time, the CE values are also different depending on the selective wavelength, as shown in Figure 1.8. Several kinds of CE value of the same EC materials are exhibited: in most reported literature, the CE of characteristic wavelength reaching the maximum contrast ($\lambda_{\max}$) is calculated, which is the maximum CE value ($\mathrm{CE}_{\max}$). Also, the CE value at 555 nm

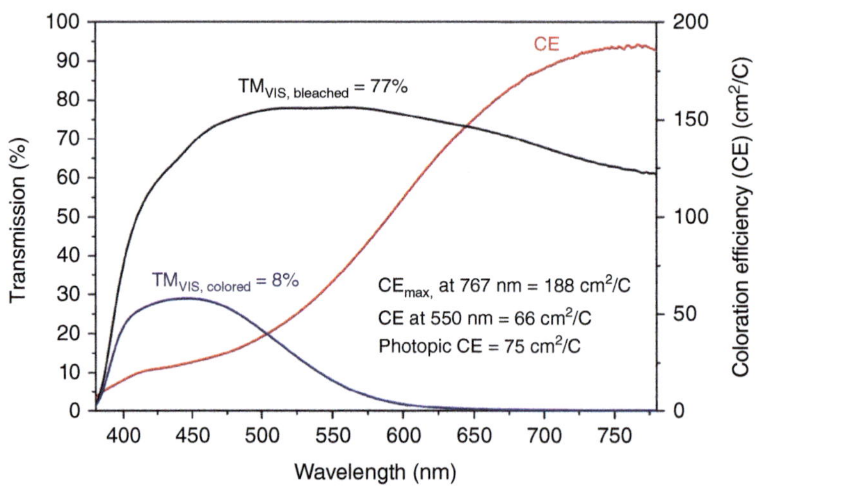

Figure 1.8 Different types of CE value of the same EC materials. Source: Kraft [25].

is calculated for comparison in different publishes because of the highest sensitivity of human eyes at 555 nm as well as the photopic CE based on the photopic transmittance mentioned in the contrast part, which considered the light transmission over the wavelength range between 380 and 780 nm normalized with the spectral sensitivity of the human eye. Therefore, more plot of CE values should be obtained rather than single-valued CE values, to give more information about the performance of EC materials.

In addition, when insightfully considering the injected/ejected charge Q_d, we can find it in fact to consist of three part: faradaic charge Q_F associated with doping/de-doping, capacitive charge Q_C due to the capacitive nature of the ECD, and parasitic charge Q_P associated with electrolyte/impurity reactions. Among them, the faradaic charge is the source of redox activity leading to chromic change actually. Therefore, Fabretto et al. reported a new technique for measuring CE by extracting the faradaic charge from the total charge and calculated the only faradaic charge-based CE value [29]. As we discussed, the total charge flow is simply the addition of the three individual charge flows and is given by

$$Q_d = Q_F + Q_C + Q_P$$

where the parasitic current was a small component (approximately <2%) compared with the other two and therefore can be ignored. Then the time–evolution total current flow can be described as following:

$$I_d(t) = I_F(t) + I_C(t) = \frac{n\mathrm{FAC}_0 D^{1/2}}{\pi} t^{-1/2} + I_0 \mathrm{e}^{-t/\mathrm{RC}} = k\,t^{-1/2} + I_0 \mathrm{e}^{-t/\mathrm{RC}}$$

where n is the number of electrons transferred per molecule, F is the Faraday constant (96 500 C/mol), A is the electrode area (cm^2), C_0 is the concentration of species in the bulk solution (mol/cm^3), D is the apparent diffusion coefficient (cm^2/s), t is time in seconds, I_0 is the maximum current flow at $t = 0$, R is the cell resistance, and C is the double layer capacitance. Then fitting the experimental data to this equation and substituting the constant k, at last, a plot of the time–evolution faradaic current will be obtained, and the corresponding faradic-corrected CEs can be calculated. Usually, the faradic-corrected CEs are larger than the uncorrected results, because the total charge ingress/egress (i.e. Q_d) is larger than the faradic charge (i.e. QF).

1.3.4 Optical Memory

The optical or EC memory (also called open-circuit memory) of an EC material can be defined as the propensity of the material to retain its redox/colored state upon removing the external bias. Usually, the memory effect are often observed in film-state EC materials such as conjugated polymers, which well adhered onto the electrode, and hence restrict the movement of the electrons. In contrast, some solution-based ECDs (e.g. viologens) will exhibit a self-erasing effect, which means the colored state disappeared rapidly in the absence of applied voltage because the electrons diffuse freely in this type of device. The memory effect is useful for the energy-saving devices and also can be applied for data storage. Figure 1.9 shows

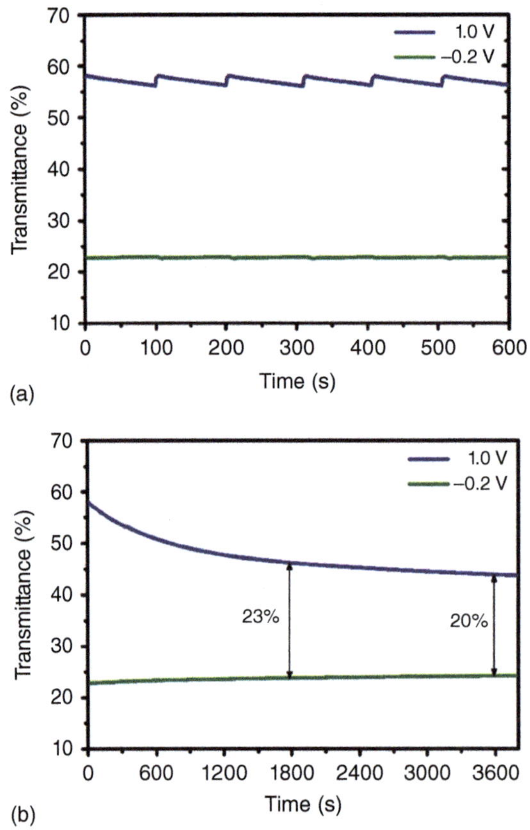

Figure 1.9 Open-circuit memory tests of PBOTT-BTD spray coated on an ITO-coated glass slide in 0.1 M TBAPF6/ACN at 423 nm: (a) short- and (b) long-term performance. Source: Li et al. [21].

an optical memory test. The short-term memory was investigated by applying a potential pulse for 2 seconds prior to forming the open-circuit state for 100 seconds; the transmittance change at 423 nm was monitored simultaneously (Figure 1.9a). Then a long-term memory is also studied by applying a potential for two seconds and removing the bias for one hour (Figure 1.9b). The EC conjugated polymers remain in the initial transmittance contrast well in the absence of an applied voltage, which exhibits a good optical memory.

1.3.5 Stability

In most cases of laboratory study, researchers record the number of redox cycles that an EC material stand without significant loss in the performance as the electrochemical stability, irreversible oxidation or reduction at extreme potentials, side reactions with water or oxygen, and heat release in the system during switches may cause the degradation of electrochemical stability. Usually, the charge density Q_d recorded under electrochemical cycling is up to 10^4–10^6, as shown in Figure 1.10a. The charge density of a Ti-doped V_2O_5 EC film haven't changed through 2×10^6 cycles; meanwhile the transmittance change at a certain wavelength during continuous cycling is also important to describe the stability

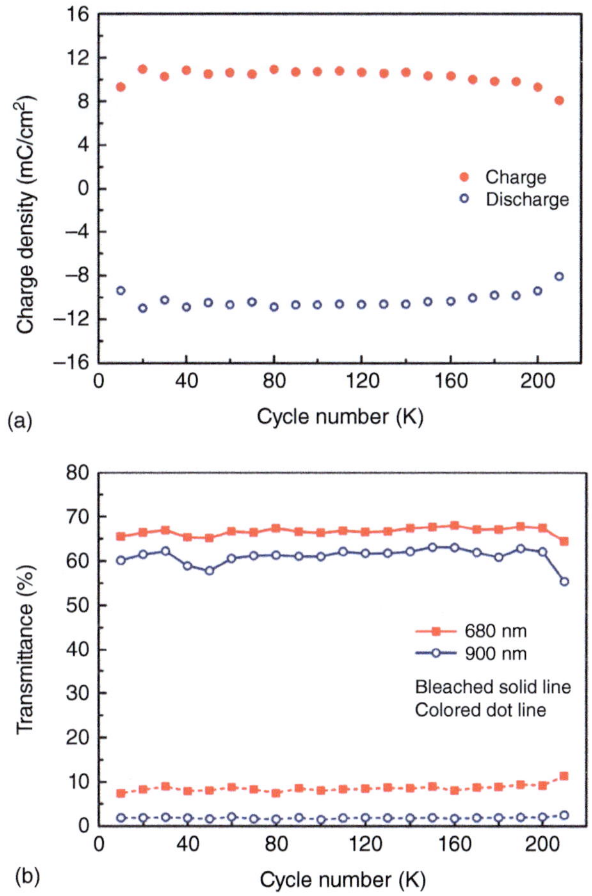

Figure 1.10 Charge density (a) and transmittance (b) variation curves of ECD with the cycle number K : 1000. Source: Wei et al. [30].

of an EC material. Such as shown in Figure 1.10b, the transmittance of the ECD remains stable through 200 000 cycles. Actually, the CE change after numerous cycles also can be used to evaluate the long-term stability of EC materials, because it contains information of both transmittance and charge density.

However, if we consider the real application of ECD in building windows, there are more strict conditions for durability and reliability. For instance, a lifetime over 20 years with more than 10^6 switching cycles is necessary. Extreme weather conditions such as temperatures below −20 °C and above +40 °C are huge challenge for both EC materials and electrolytes, as well as other degradation factors such as high solar irradiation levels, fast temperature changes, uneven temperature distribution and additional stresses, rain, humidity, mechanical shock, and drying. Therefore, in 1998, Carl M. Lampert proposed a standard test guideline for industry application of EC [31], as shown in Figure 1.11. Recently, the International Organization for Standardization (ISO) also has launched an international standard: Glass in building – Electrochromic glazings – accelerated aging test and requirements (ISO 18543) for EC use in buildings.

EC color/bleach cycling	1. 25 000 cycles at 25 °C 2. 10 000 cycles at – 20 °C 3. 25 000–100 000[b] cycles with UV[a] at 65 °C
UV[a] stability	Prolonged cycles (12 h colored, 12 h bleach) for a total exposure of 6000 MJ/m^2, UV intensity integrated between 300 and 400 nm
Heat storage	500 h at 90 °C
Low temperature storage	1000 h at 20 °C to – 30 °C
Humidity/temperature storage	1000 h at 70 °C, 90% humidity
Thermal cycling	85 °C for 4 h, followed by – 40° C for 4 h and followed by 37 °C for 16 h at 100% R.H. (4 h of ramp between each condition) (repeat four times)
Thermal shock	One hour colored and one hour bleached. Test in a UV chamber, with UV[a] lamp on during coloration. Spray with 25 °C water in bleached state (24 cycles)

[a]Black panel temperature during the bright periods of 60 °C, and 25 °C in the dark periods.
[b]Asahi has used up to 100 000 cycles.

Figure 1.11 Recommended testing guidelines for EC windows for exterior architectural applications. Source: Lampert et al. [31].

1.4 Conclusion

In this chapter, a broad overview of electrochromism, EC materials, device structure, development history, and key parameters of electrochromism have been introduced briefly. More detailed descriptions of each area will be discussed in Chapters 2–15. In summary, research in EC technologies has achieved significant breakthroughs over the decades. Many generations of EC materials have been developed, ranging from traditional metal oxides to more recent organic polymers, small molecules, and hybrid materials. Moreover, benefit from the ECD design and structural optimization, flexible substrate-based devices were fabricated with the low-price roll-to-roll process, which makes the EC technology have large scope applications, such as smart windows for reducing building energy consumption, self-powered EC window using organic photovoltaic cells as power supplement, car rear-view mirrors for greater safety, and smart sunglasses for better UV-radiation protection. Many of these technologies and applications have been commercialized and are available on the market. With the concerted efforts of researchers and engineers, we believe that the new EC materials and advanced technologies will constantly develop and more advanced ECD with low manufacturing cost will be exploited to realize practical applications.

References

1 Fletcher, S. (2015). The definition of electrochromism. *Journal of Solid State Electrochemistry* 19 (11): 3305–3308.

2 Camurlu, P. (2014). Polypyrrole derivatives for electrochromic applications. *RSC Advances* 4 (99): 55832–55845.

3 Wang, Z., Wang, X., Cong, S. et al. (2020). Fusing electrochromic technology with other advanced technologies: a new roadmap for future development. *Materials Science & Engineering R: Reports* 140.

4 Wu, W., Wang, M., Ma, J. et al. (2018). Electrochromic metal oxides: recent progress and prospect. *Advanced Electronic Materials* 4 (8).
5 Mortimer, R.J. (2011). Electrochromic materials. *Annual Review of Materials Research* 41 (1): 241–268.
6 Platt, J.R. (1961). Electrochromism, a possible change of color producible in dyes by an electric field. *The Journal of Chemical Physics* 34 (3): 862–863.
7 Hutchison, M.R. (1913). Electrographic display apparatus and method. US Patent 1,068,774, filed 29 July 1913.
8 Lehovec K. (1957). Photon modulation in semiconductors. US Patent 2,776,367, filed 1 January 1957.
9 Neff, V.D. (1978). Electrochemical oxidation and reduction of thin-films of prussian blue. *Journal of the Electrochemical Society* 125 (6): 886–887.
10 Michaelis, L. and Hill, E.S. (1933). The viologen indicators. *The Journal of General Physiology* 16 (6): 859–873.
11 Schoot, C.J., Ponjee, J.J., van Dam, H.T. et al. (1973). New electrochromic memory display. *Applied Physics Letters* 23 (2): 64–65.
12 Garnier, F., Tourillon, G., Garzard, M., and Dubois, J.C. (1983). Organic conducting polymers derived from substituted thiophenes as electrochromic material. *Journal of Electroanalytical Chemistry* 148: 299–303.
13 Mengoli, G., Musiani, M.M., Schreck, B., and Zecchin, S. (1988). Electrochemical synthesis and properties of polycarbazole films in protic acid media. *Journal of Electroanalytical Chemistry and Interfacial Electrochemistry* 246 (1): 73–86.
14 Zheng, H.B., Lu, W., and Wang, Z.Y. (2001). Electrochemical and electrochromic properties of poly(ether naphthalimide)s and related model compounds. *Polymer* 42 (8): 3745–3750.
15 Oishi, Y., Takado, H., Yoneyama, M. et al. (1990). Preparation and properties of new aromatic polyamides from 4,4′-diaminotriphenylamine and aromatic dicarboxylic acids. *Journal of Polymer Science Part A: Polymer Chemistry* 28 (7): 1763–1769.
16 Cheng, S.-H., Hsiao, S.-H., Su, T.-H., and Liou, G.-S. (2005). Novel aromatic poly(amine-imide)s bearing a pendent triphenylamine group: synthesis, thermal, photophysical, electrochemical, and electrochromic characteristics. *Macromolecules* 38 (2): 307–316.
17 Arimoto, F.S. and Haven, A.C. (1955). Derivatives of dicyclopentadienyliron. *Journal of the American Chemical Society* 77 (23): 6295–6297.
18 Whittell, G.R. and Manners, I. (2007). Metallopolymers: new multifunctional materials. *Advanced Materials* 19 (21): 3439–3468.
19 Wade, C.R., Li, M., and Dincă, M. (2013). Facile deposition of multicolored electrochromic metal–organic framework thin films. *Angewandte Chemie International Edition* 52 (50): 13377–13381.
20 Hao, Q., Li, Z.-J., Lu, C. et al. (2019). Oriented two-dimensional covalent organic framework films for near-infrared electrochromic application. *Journal of the American Chemical Society* 141 (50): 19831–19838.
21 Li, W., Ning, J., Yin, Y. et al. (2018). Thieno[3,2-*b*]thiophene-based conjugated copolymers for solution-processable neutral black electrochromism. *Polymer Chemistry* 9 (47): 5608–5616.

22 Beaujuge, P.M. and Reynolds, J.R. (2010). Color control in π-conjugated organic polymers for use in electrochromic devices. *Chemical Reviews* 110 (1): 268–320.

23 Jiang, M., Sun, Y., Ning, J. et al. (2020). Diphenyl sulfone based multicolored cathodically coloring electrochromic materials with high contrast. *Organic Electronics* 83: 105741.

24 Fred Schubert, E., Chapter 16. *Human Eye Sensitivity and Photometric Quantities in Light-Emitting Diodes*, 2e. Cambridge University Press.

25 Kraft, A. (2018). Electrochromism: a fascinating branch of electrochemistry. *ChemTexts* 5 (1): 1–18.

26 Padilla, J., Seshadri, V., Filloramo, J. et al. (2007). High contrast solid-state electrochromic devices from substituted 3,4-propylenedioxythiophenes using the dual conjugated polymer approach. *Synthetic Metals* 157 (6–7): 261–268.

27 Hsiao, S.-H. and Lin, S.-W. (2016). Electrochemical synthesis of electrochromic polycarbazole films from *N*-phenyl-3,6-bis(*N*-carbazolyl)carbazoles. *Polymer Chemistry* 7 (1): 198–211.

28 Hassab, S., Shen, D.E., Österholm, A.M. et al. (2018). A new standard method to calculate electrochromic switching time. *Solar Energy Materials and Solar Cells* 185: 54–60. https://doi.org/10.1016/j.solmat.2018.04.031.

29 Fabretto, M., Vaithianathan, T., Hall, C. et al. (2008). Faradaic charge corrected colouration efficiency measurements for electrochromic devices. *Electrochimica Acta* 53 (5): 2250–2257.

30 Wei, Y., Zhou, J., Zheng, J., and Xu, C. (2015). Improved stability of electrochromic devices using Ti-doped V_2O_5 film. *Electrochimica Acta* 166: 277–284.

31 Lampert, C.M., Agrawal, A., Baertlien, C., and Nagai, J. (1999). Durability evaluation of electrochromic devices – an industry perspective. *Solar Energy Materials and Solar Cells* 56 (3): 449–463.

2

Advances in Polymer Electrolytes for Electrochromic Applications

2.1 Introduction

The ionic conduction medium between the electrodes and electrochromic (EC) materials is an electrolyte, which is one of the most essential active components in electrochromic device (ECD). Electrolyte provides an indispensable role as the prime ionic conduction medium between the electrodes while preventing electron conduction between the two electrodes during EC operation. The important electrolyte properties greatly affecting the EC performance are the electrolyte ionic conductivity, ion dissociation, transport rate of ion through bulk and interface, and thermal stability [1]. Electrolytes were initially reported in the early 1970s, including ceramic, glass, crystalline, and polymer electrolytes (PEs). PE was first introduced by Fenton et al. in 1973 [2] and widely applied since 1980s [3]. In the past decades, PEs attracted much attention from all over the world's researchers due to their promising applications in electrochemical storage/conversion devices.

In general, electrolytes can be classified into PEs, liquid electrolytes, ceramic electrolytes, and solid inorganic electrolytes [4–6]. Briefly, PE is a membrane composed of a dissolution of salts in a polymer matrix with high molecular weight [7]. PE is widely applied in electrochemical devices such as solid-state batteries and rechargeable batteries, ECDs, supercapacitors, fuel cells, dye-sensitized solar cells, and EC windows. Technologically, PEs evolved from polymer, liquid ionic conductor and solid-state ionic conductor. PEs can be prepared by dissolving metal salts in polar polymer hosts, which could be used to replace the liquid ionic solution. Liquid electrolytes possessing high ionic conductivity have several inevitable drawbacks such as the possibility of electrolyte leakage, low chemical stability, hydrostatic pressure considerations, and difficulty in assured sealing and are unsafe for practical applications especially in scaling-up processes [8]. Comparing with liquid electrolytes, PE has several prominent advantages, such as high ionic conductivity, safety, flexibility, wide electrochemical windows, and so on [9, 10].

Organic Electronics for Electrochromic Materials and Devices, First Edition. Hong Meng.

2.2 Requirements of Polymer Electrolytes in Electrochromic Applications

Factors that lead to the use of PE in an EC devices are as follows (Figure 2.1).

- High optical transparency
  High optical transparent PEs are highly desirable for EC applications. High transmittance enhances the transparency of an EC device in the bleached state. Electrolytes cannot affect the optical shifts during colored and bleached states.
- Flexible and mechanically stretchable
  PE owing to flexible and mechanically stretchable advantages can be better applied in ECDs, especially in flexible and stretchable devices. The polymeric matrix is of vital importance. The PE requires to keep mechanical stability during bending and stretching states.
- Multifunctional separator
  There is a direct contact between the two electrodes without any separator, which will lead to a short circuit. PE can act as a separator layer in an EC device, making these devices light and compact without the need for additional separators. Moreover, PEs demand intimate binding effect, making favorable electrical contact with the electrodes and good adherence to EC layers.
- Wide range of working temperature
  PEs inherently exhibit the ability to operate over a wide range of temperature as compared to the other counterparts.
- Cation coordination ability

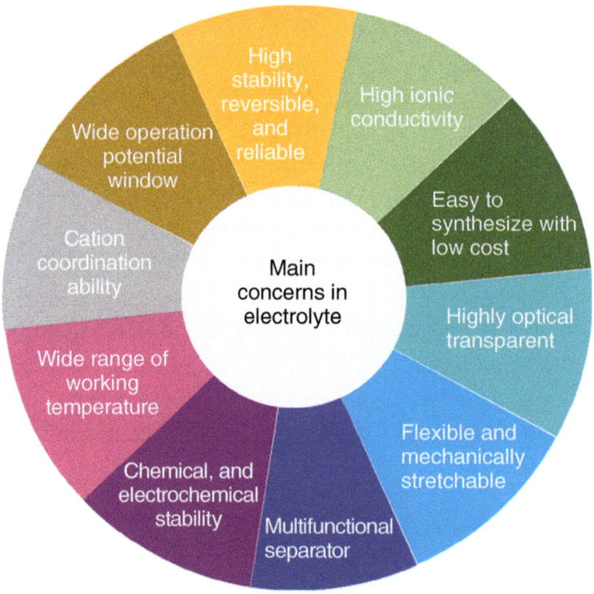

Figure 2.1 Main concerns in electrolyte.

Polymers in electrolytes with benign cation coordination ability can better make for coordination bonding interactions with the metal cations.

- Wide operation potential window
  Stable electrochemical window is the range of working potential in an EC device without breakdown of electrolyte itself. The maximum potential is determined by the potential of the oxidation reaction. Meanwhile, the minimum potential is determined by the potential of the reduction reaction. To ensure the long-time and cycling stability of the ECD, the working potential window of a device should be considered in the electrochemical operating range of the electrolyte.
- Thermal, photo, chemical, and electrochemical stability
  An ideal PE should possess good thermal and photo stability. During electro chromic processes, the device may release heat, which may result in degradation of the EC devices. Similarly, an ideal electrolyte with photo stability prevents device from being destroyed via prolonged exposure to air. It is necessary to keep good electrochemical stability in the voltage range for ECDs. Undesired interaction effects between the EC layer and electrolyte should not exist in the ECD during EC processes so that devices can keep reversible and long lifetime performance for practical application. This chemical stability must be ensured both during the deposition and during the cycling processes. Currently, small molecule EC materials-based ECDs have appeared in lots of researches. However, this kind of ECD mostly can be operated under higher potential. Therefore, PEs with wide potential window are demanded.
- High ionic conductivity ($>10^{-4}$ S/cm) with low electronic conductivity
  PEs should be a good ionic conductor and electronically insulating ($\sigma_e < 10^{-12}$ S/cm) so that ion transport could be facilitated to minimize self-discharge. In EC process, the essence of color change is the transfer of ions into and out of an EC film. The electrolyte should also maintain its high ionic conductivity even after thousands of cycles. The flexibility of polymer matrix chains in the amorphous phase allows ions to be transported frequently. This easy transport of ions is hindered in the crystalline phase, where the material is densely packed and there is not enough space to allow rapid transport of ions [11]. The ion transference number is important in the characterization of PEs. A large transference number can reduce concentration polarization of electrolytes during charge–discharge steps [12].
- Easy to synthesize with low cost
  The price of PE cannot be the factor of resulting in hindering practical application. Inexpensive, easily available PEs are highly necessary for the successful commercialization and implementation of EC devices. It is worth mentioning that the thickness of PE film should be controlled and easy to operate.
- Safe and environmentally friendly
  Safe and environmentally friendly are highly desirable for practical application of the ECDs. So far, ECDs have been widely used in people's daily lives, such as EC glasses, car mirrors, and cockpit glasses. A nontoxic, environmental, and recyclable electrolyte is a top priority in an EC device, which cannot jeopardize our health.

2.3 Types of Polymer Electrolytes

PEs are formed by dispersing a salt into a neutral polymer matrix. The composition of the matrix plays a great role in mechanical strength, electrolyte/active material contact, flexibility, and processability in ECDs. PEs applied for ECDs can be mainly classified into gel polymer electrolytes (GPEs), self-healing PEs, cross-linking polymer electrolytes (CPEs), ceramic PEs, ionic liquid (IL) PEs, and gelatin-based PEs.

2.3.1 Gel Polymer Electrolytes (GPEs)

GPE also known as plasticized PE was first introduced by Feuillade and Perche in 1975 [13]. GPEs are usually synthesized by incorporating a larger quantity of liquid plasticizer and/or solvents into a polymer matrix to form a stable gel with the polymer host structure, having relatively higher ambient-temperature ionic conductivity [14]. Generally, GPE consists of an ionically conducting medium polymer such as poly(ethylene oxide) (PEO) and a metal salt (lithium) swollen with a suitable solvent. GPE could exhibit fast ion diffusive nature and cohesive properties comparable to solids [15]. The first one is known for its high conductivity with the H^+ donors originating from, e.g. sulfuric (H_2SO_4) or phosphoric (H_3PO_4) acid [16]. In the second group mobile Li^+ species are provided by dissolution of lithium perchlorates ($LiClO_4$) [17], triflates ($LiCF_3SO_3$) [18], fluorophosphates ($LiPF_6$) [19], or fluoroborates ($LiBF_4$) [20] in protic solvents (propylene, acetonitrile [ACN], ethylene carbonates, etc.). As was reported, binary or ternary solvents, such as EC + PC and DMC + PC + EC, were also employed [21, 22]. These types of electrolytes are characterized by a higher ambient ionic conductivity but poor mechanical properties. Although most PGEs have high ionic conductivity of 10^{-3} S/cm at room temperature, poor mechanical properties and considerable viscosity inevitably result in internal short circuit and cell leakage. UV and heat irradiation are other potential factors for degradation of electrolytes. Plasticizers such as PC, EC, diethyl carbonate (DEC), and dimethyl carbonate (DMC) can be applied in developing the physical characteristics of the overall blend. Plasticizers improve the ionic conductivity by increasing the amorphous phase content, dissociating ion aggregates, increasing ionic mobility within the gel electrolytes, or lowering the glass transition temperature (T_g) of the system [23]. The balance between increasing ionic conductivity and decreasing mechanical strength needs to be maintained while plasticization occurs. Gel electrolytes have been used in many applications but most of recent published papers are about EC applications. In the GPE, the immobilized solvent in the polymer matrix has strong effects on the ionic conductivity. So far, PEO, poly methyl methacrylate (PMMA), poly(vinylidene fluoride) (PVDF), poly(vinylidene fluoride-*co*-hexafluoropropylene) (PVDF-HFP), and poly acrylonitrile (PAN)-based gel electrolytes have aroused researchers' interests [24, 25]. Recently, some of gel electrolytes have aroused researchers' interests, such as poly ethylene oxide (PEO), polyethylene glycol (PEG) [26], boronate esters [27], sodium polystyrene sulfonate [28], waterborne polyurethane, [29] polyester and PEO [30], polyvinyl alcohol (PVA), PAN, PMMA, poly(ethyl methacrylate) (PEMA) and poly(2-ethoxyethyl methacrylate) (PEO EMA), [31] poly vinyl chloride (PVC), poly(vinyl sulfones) (PVS), and PVDF [32].

2.3.1.1 PEO-/PEG-Based Electrolytes

PEO (with a molecular mass above 20 000 g/mol) or PEG (with a molecular mass below 20 000 g/mol) simply refers to an oligomer or polymer of ethylene oxide. PEO-based electrolyte is the earliest and most widely studied matrix for the formation of PE. Since Armand proposed a PE based on polyethylene oxide (PEO) for lithium batteries in 1978, research in this field has been extensively carried out [33]. Polyethylene oxide (PEO) has a polar ether group and significant segment mobility. As a wide range of salt polymers, especially high lithium ion stable polymers, it has excellent compatibility [34]. But because of high crystallinity, low melting point, limited operating temperature range, low hydroxyl ion migration number, poor interface characteristics, and other shortcomings, it fails to achieve the desired effect. In PEO-based electrolytes, the most common method of increasing ion dissociation is the use of low lattice energy salts and to add fillers, which may prevent formation of polymer crystals, resulting in fast ion transportation via an interaction between the fillers and electrolyte [35]. PEO–H_3PO_4 and $(PEO)_8$–$LiClO_4$ were studied for electrochromism in devices [36]. However, the low Li^+ conductivity inhibits its applicability. Later, PEO–$Li_SO_3CF_3$ and PEO–LiN $(SO_2CF_3)_2$ were reported to reach higher ion conductivities [37, 38]. PEO used as electrolyte in the gel form with an ionic conductivity of 2 mS/cm was reported in WO_3 film ECD, which exhibited superior EC performance and memory characteristics [39]. In addition, PEO-based GPE plasticized with ethylene carbonate/propylene carbonate or *N*-butyl-3methylpyridinium trifluoromethane sulfonylimide (PTFSI) has been studied by Desai et al. for the ECD [40]. Ionic conductivity can be significantly enhanced and the phase separation of the PEO and plasticizer was inhibited. Yang et al. studied the performance of the EC device prepared using poly(2,5-dimethoxyaniline) (PDMA) and tungsten oxide (WO_3) as electrode materials and PEO-$LiClO_4$ as GPE plasticized with polycarbonate [41]. Similarly various other hybrid electrolytes based upon PEO were successfully applied in EC applications [42].

2.3.1.2 PMMA-Based Polymer Electrolytes

Due to high degree of crystallization, low ionic conductivity at ambient temperature exists in PEO-based PE [43]. PMMA along with PVDF PEs have recently caused great attention in EC application. PMMA including amorphous phase and flexible backbone could increase ionic conductivity. As was reported, PMMA can provide a high transparency, excellent environmental stability, good gelatinizing properties, high solvent retention ability, and excellent compatibility with the liquid electrolytes [44, 45]. Moreover, PMMA shows good interfacial stability toward electrodes and high solvation ability to form complexation between polymer and salt. Highly ionic conductive PMMA exists as a gel and is mainly used for electrochromism [45]. Anderson et al. majored on ECDs with films of W oxide and vanadium pentoxide and an intervening layer of a PPG–PMMA–$LiClO_4$ [46, 47]. Reynold and coworkers first reported a polymeric ECD using poly(3,4-proplenedioxythiophene) ($PProDOTMe_2$) and poly[3,6-bis(2-(3,4-ethylenedioxy)-thienyl)-*N*-methylcarbazole] (PBEDOT-*N*-MeCz) as the cathodically and anodically coloring polymers, respectively [48]. In this device, the EC films were electrosynthesized in poly (3,4-ethylenedioxythiophene) (PEDOT)-PSS electrodes at the oxidation potential of the monomer, and the electrolytes were prepared by using PMMA as the novel polymer matrix. The GPE having composition of 70 : 20 : 7 : 3 (ACN: PC: PMMA: $TBAPF_6$ [tetrabutylammonium hexafluorophosphate]) was applied in the ECDs [48]. For the durability of the device, GPE was evaporated at the edges during the sealing process. The ECD exhibited a large transmittance change ($\Delta\%T$) of 51% at the wavelength of 540 nm and only 5% contrast loss after 32 000 switches. Beaupre et al. reported a flexible EC cell by using PMMA-LiClO4 electrolyte, which was plasticized with propylene carbonate to form a highly transparent and conductive gel. Sonmez et al. has explored a highly transparent and conductive gel with $LiClO_4$ plasticized with PC, having composition of PC: PMMA (MW: 350.000): $LiClO_4$ (70 : 20 : 7 : 3), which was applied in EC device [49]. ACN was also added as a high vapor pressure solvent to make the gel ingredients blend easily. Oral et al. uses similar electrolyte of LiClO4: ACN: PMMA: PC in the ratio of 3 : 70 : 7 : 20 to study the EC properties [49]. Tung et al. studied the EC properties of the device of PEDOT-Prussian Blue (PB) using PMMA as GPE [50]. The device exhibited high coloration efficiency and good long-term cycling stability. Recently, Yang et al. studied a new type of GPE composed of free-standing aramid nanofibers, which was used to fabricate all-solid-state near-infrared (NIR) ECDs for NIR sheltering applications [51]. This new type of GPE showed excellent mechanical and heat endurance compared with currently available GPEs. Kim et al. presented a novel ECD-based photonic device, which can modulate IR light intensity in a planar optical waveguide ECD by using PMMA gel electrolyte consisting of 5% (w/w) PMMA, 4% (w/w) phenothiazine, 0.1 m $LiClO_4$, and $11.25 \approx 10^{-3}$ m ferrocene [52]. The results confirm a new approach to consider ECD-based optical modulators for the development of planar photonic-integrated circuits and systems.

2.3.1.3 PVDF-Based Polymer Electrolytes

PVDF is another popular host material for electrolytes and has recently been widely used in ECDs. PVDF offers many advantages, such as good thermomechanical properties, fairly high permittivity, high hydrophobicity, thermal and chemical stabilities, and chemical resistance. As a semi-crystalline polymer, the PVDF crystalline phase provides thermal stability, while the amorphous phase provides the flexibility required for ECDs. PVDF can be soluble in high boiling point and commercial solvents, such as *N*-methyl-2-pyrrolidone (NMP) and dimethylformamide (DMF). However, comparing with lithium salt, PVDF, which is expected to have a low donor number (DN), is insoluble and cannot be used effectively in polymer salt complexes. For EC application, PVDF-based PEs have been identified as interesting candidates and their study is in progress. As was reported by Fabrettos group, an ECD was prepared by using PVDF for studying the coloration efficiency [53]. P(VDF-TrFE) as GPE was also studied as potential PE in ECD with polyaniline as EC materials [1]. The gel electrolyte device reached an average ionic conductivity of 2.84×10^{-5} S/cm and shows stable and reversible light modulation up to 65% in gel state. It was found that the gel-state device was affected by the number of free ions, while the movement of ions in the electrolyte bulk and the modulation of light in the semisolid device are indicated by the electrolyte/EC material interface.

The ionic conductivity of PVDF PEs can be enhanced by incorporating substantial amounts of plasticizers or combining with IL. Jia et al. have studied 1-butyl-3-methylimidazolium hexafluorophosphate-loaded $SCCO_2$-treated electrospun P(VDF-HFP) membrane as an electrolyte in EC device [54].

Recently, Reynold's group reported paper-based ECDs consisting of PEDOT:PSS electrodes and [EMI][TFSI]/PVDF-HFP ion gel electrolyte layer (Figures 2.2 and 2.3) [55]. The ECDs incorporating an ion gel electrolyte were demonstrated where a magenta-to-colorless device achieves a color contrast (ΔE^*) of 56, attributing to a highly color-neutral bleached state of the extracellular protein (ECP) ($a^* = -0.5$, $b^* = 2.9$). It was found that the gel-state device was affected by the number of free ions, while the movement of ions in the electrolyte bulk and the modulation of light in the semisolid device were indicated by the electrolyte/EC material interface.

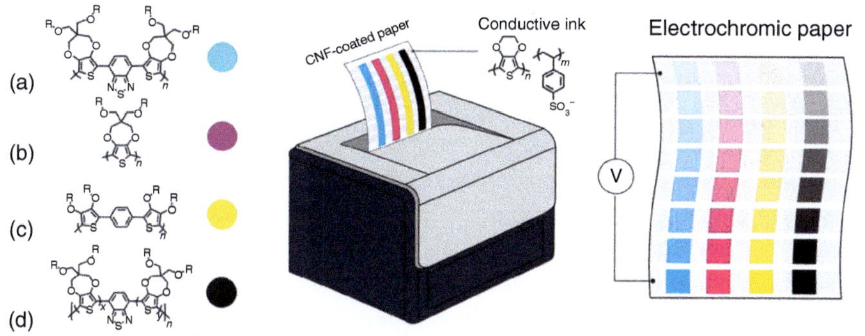

Figure 2.2 Schematic showing printing of colored-to-clear electrochromic paper incorporating cellulose nanofiber (CNF)-coated paper substrates and PEDOT:PSS electrodes as well as the repeat-unit structures of (a) ECP-Cyan, (b) ECP-Magenta, (c) ECP-Yellow, and (d) ECP-Black. R = ethylhexyl. Source: Lang et al. [55].

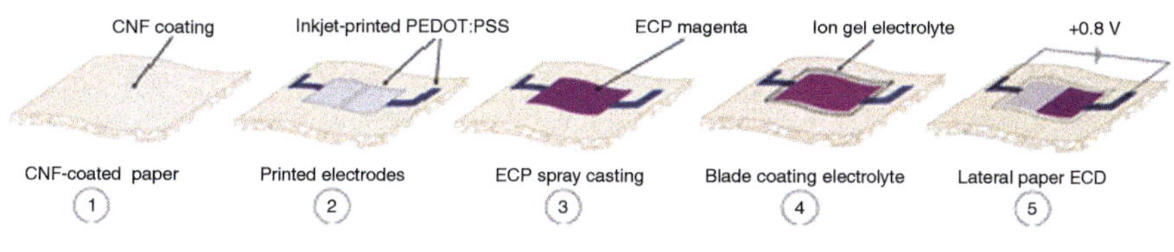

Figure 2.3 A fabrication process for lateral paper ECDs showing inkjet-printed PEDOT:PSS electrodes, deposition of ECPs, and [EMI][TFSI]/PVDF-HFP ion gel electrolyte layer. Devices are operated by applying a 0.8 V bias across the two lateral pixels. Source: Lang et al. [55].

2.3.2 Self-Healing Polymer Electrolytes

A self-healing material is a material that possesses abilities to partially repair damage and restore mechanical properties during its service lifetime [56]. Self-healing materials are particularly attractive in the field of smart materials because of their inherent ability to repair cracks, thereby avoiding such risks, reducing waste, and improving equipment life, durability and reliability. It is believed that the use of an EC polymer film with self-healing properties would be an effective approach to overcome the risk of the scar generation in ECDs. Until now, self-healing polymers based on the Diels–Alder (DA) reaction [57] is an effective method for the implementation of intrinsic self-healing into functional materials, which have been explored for various electronic and optoelectronic applications like conducting films [58], organic light-emitting devices [OLEDs] [59], supercapacitors, etc. [60] Self-healing can come up from the microscopic to macroscopic level. Some properties in engineering materials may degrade over long time because of environmental conditions, fatigue or operation damage. Self-healing materials can address this kind of degradation. The ability to self-heal upon breakdown or damage depends on various kinds of physical interactions such as hydrogen bonding [61], π–π stacking [62], ion dipole interaction [63], or chemical interactions such as disulfide bond [64, 65] and imine bond. [66] They heal autonomously or in response to a multitude of externally applied stimuli such as pH change, light [67], electricity, temperature [68], and pressure. Geubelle and coworkers first reported a completely autonomous man-made self-healing material [69]. Later, an epoxy system containing microcapsules covered with (liquid thermosetting) monomers was explored [70]. If a microcrack occurs in the system, the microcapsule breaks and the crack will be filled by the monomer. This novel self-healing system could be performed well in polymers and polymer coatings [71].

Zhang and coworkers [72] prepared a dual physically cross-linked hydrogel containing the hydrogen bonding between poly(acrylamide-*co*-acrylic acid) and PVA, hence producing self-healing abilities in hydrogel. As was reported, ion gels with self-healing performance, which composed of polymer networks and electrolyte solutions, have attracted great attention [73]. Different strategies and approaches to devise self-healing materials in ECs have been investigated. For example, multifunctional EC ion gels were enabled by a blend of active methyl viologen dication in polystyrene-block-poly(methyl methacrylate)-blockpolystyrene (PS-PMMA-PS) triblock copolymer with ILs [74]. Poly[styrene-ran-1-(4-vinylbenzyl)-3-methylimidazolium hexafluorophosphate] (P[S-r-VBMI] [PF6]) random copolymer (Figure 2.4) [75], poly(styrene-ran-methyl methacrylate) (PS-rPMMA) random copolymer [76], (poly(methyl methacrylate)-*b*-polystyrene)$_6$ star-shaped block copolymer (Figure 2.5) [77], and poly(vinylidene fluoride-*co*-hexafluoropropylene) (PVDF-*co*-HFP) [78], etc. matrices have been developed for achieving self-healing solid PEs. The ion conductivity of PVDF-*co*-HFP-based polyelectrolytes could be improved by using various PVDF-*co*-HFP-based polymers and amorphous polymer electrolytes such as PEG [78], PMMA [79] and poly(vinyl acetate) (PVAC) [80, 81]. In addition, materials containing IL/polymer blends

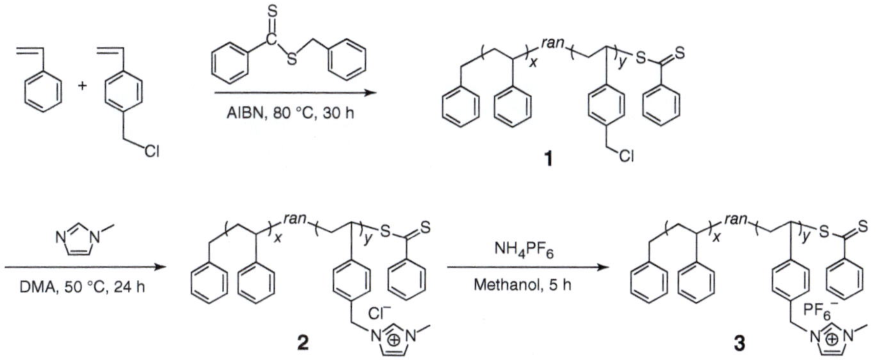

Figure 2.4 Synthetic routes for P[S-r-VBMI][PF_6]. Source: Seo and Moon [75].

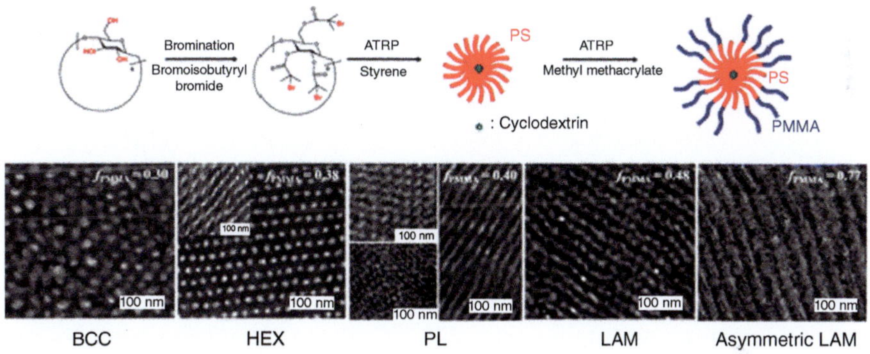

Figure 2.5 Synthetic routes for $(PS\text{-}b\text{-}PMMA)_{18}$. Source: Jang et al. [77].

for self-healable EC have been reported in Leong's group [82]. The interaction of polymer blends or the inclusion of reaction sites and addition of ILs may induce rapid healing at lower temperatures, attributing to the increased fluidity and inhibition of the glass transition temperatures. Moreover, self-healing polymers based on DA reaction [83] is another effective method for self-healing functional materials. A polymer *N*-(4-aminophenyl)-*N*-phenylbenzene-1,4-diamine maleimide (DATPFMA) was the first report on a bifunctional self-healing EC polymer, which exhibited superior cyclic stability and colors variations [57]. The *N*,*N*′-bis(4-aminophenyl)-*N*,N_0-diphenyl-1,4-phenylene diamine used as core and electro-active unit for synthesizing the polymer poly(pentafluorophenyl methacrylate) (PPFMA) provided more electrochemical active sites to enhance the EC performance (Figure 2.6) [84]. This DA polymer can also repair cracks by retro-DA and DA reaction for the self-repair capability [57].

However, developing a material that simultaneously exhibits excellent EC properties and good self-healing behavior at low temperature is still an unmet challenge, such as large optical contrast, high coloration efficiency, and long-term, stability, and fast healing process.

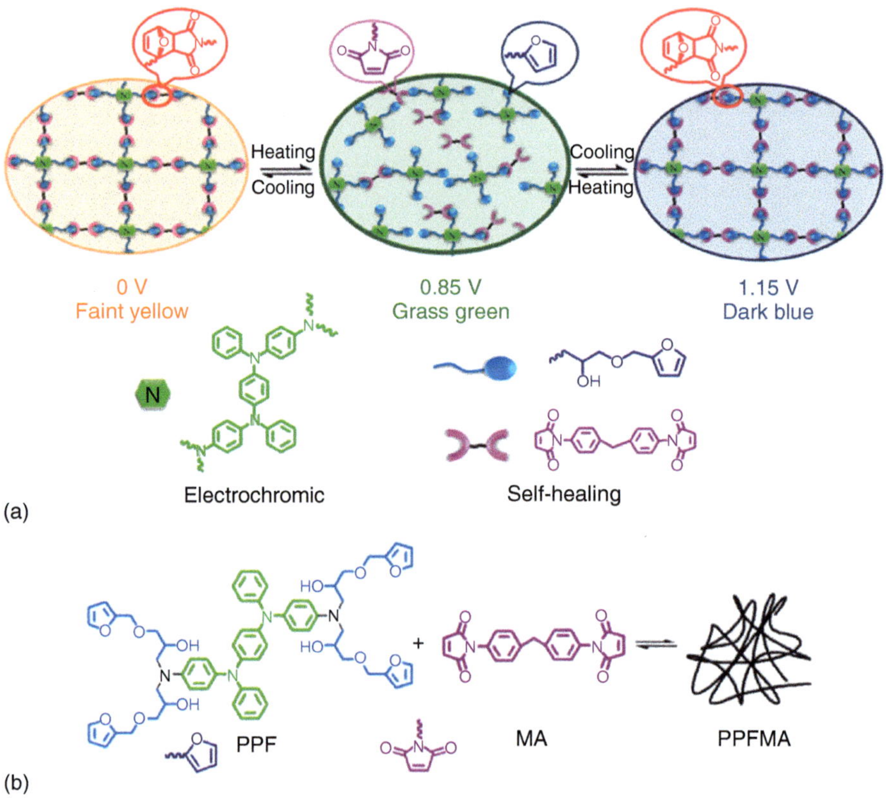

Figure 2.6 (a) The mechanism of self-healing process for electrochromic PPFMA film and (b) the retro-DA and DA reaction for self-healing polymer PPFMA. Source: Zheng et al. [84].

2.3.3 Cross-linking Polymer Electrolytes (CPEs)

The high ionic conductivity of PEs at low temperature is related to the amorphous nature of the polymer matrix (Figure 2.7). The cross-linked PEs show good ionic conductively and amorphous feature [85]. Cross-linking was reported to improve the dimensional stability and increase the dynamic storage modulus of the PE. This technique was used to modify positively various PE properties [86, 87], while brittleness, low elasticity, and processability inevitably occur in cross-linked polymer [88]. In the study of lithium salts-based electrolytes, PEO, PMMA, PVC, etc. were commonly organized into cross-linked PEs. Matsui et al. have prepared a cross-linked PE of poly(ethylene oxide) 2-(2-methoxy ethoxy) ethyl glycidyl ether with and without allyl glycidyl ether P(ethylene oxide [EO]/2-(2-methoxyethoxy)ethyl glycidyl ether [MEEGE]/allyl glycidyl ether [AGE]) complexed in $LiN(CF_3SO_3)_2$ salts [89]. Kuratomi et al. studied a cross-linked copolymer of ethylene oxide and propylene oxide with $LiBF_4$ or $LiN(CF_3SO_2)_2$ salts [90]. The concentration and anions in lithium salts are key parameters in determining electrochemical performance. As was studied, good ionic conductivity and mechanical strength have been exhibited in a cross-link of high molecular weight poly(oxy ethylene)s PEs [91]. Lee et al.

Figure 2.7 Pictorial model of a preparation of a cross-linked polymer.

have also reported a cross-linked composite PE by polymerizing alkyl monomer and polyethylene glycol dimethylcrylate (PEGDMA) in $LiPF_6$/EC [92]. The studies indicate that ionic conductivity and flexibility of the PEs are dependent on the monomer content.

In fact, while many conventional ECDs using the EC media of cross-linked polymer gels are exposed to the dynamic range of real-world temperatures, cross-linked polymer gels may not be optically viable for commercial use due to visual irregularities and/or defects. The patent in 2003 reported a self-healing cross-linked polymer gel electrolyte used as EC medium, which could solve these detriments and complications in ECDs [92].

Another intriguing field for the possible application of ionic liquid polymers (ILPs) is using polymerized ionic liquids (PILs) in ECDs. Among different IL electrolytes, cross-linked IL PEs have been explored recently. Yan and coworkers reported thermo- and electro-dual responsive ILPs electrolytes, which were synthesized through the co-polymerization of an ionic liquids monomer (ILM), 3-butyl-1-vinyl-imidazolium bromide ([BVIm][Br]) and *N*-isopropylacrylamide (NIPAM) using diallyl-viologen as both the cross-linking agent and the EC material, respectively (Figure 2.8) [93]. The electro- and thermo-responsive behaviors can be found simultaneously under electrical and temperature stimuli, which demonstrates the application of the dual-response smart window [93].

2.3.4 Ceramic Polymer Electrolytes

Differing from liquid and gel electrolytes, salt in solid-like PEs is dissolved directly into the solid medium. It is usually a relatively high dielectric constant polymer (PEO, PMMA, PAN, polyphosphazenes, siloxanes, etc.) and a salt with low lattice energy.

Multiple advantages of using solid PEs in electrochemical cells are as follows:

(1) Non-volatility.
(2) No decomposition at the electrodes.
(3) No possibility of leaks.
(4) Decreased cell price (such as PEO, PMMA, etc.).
(5) Flexibility.
(6) Lowering the cell weight – solid-based cells do not need heavy steel casing.
(7) Safety.

Many studies have been addressed to incorporate inert oxide ceramics particles into PE, in order to improve the mechanical properties, reduce polymer crystallinity, and thus solve the problem of low ionic conductivity of solid polymeric

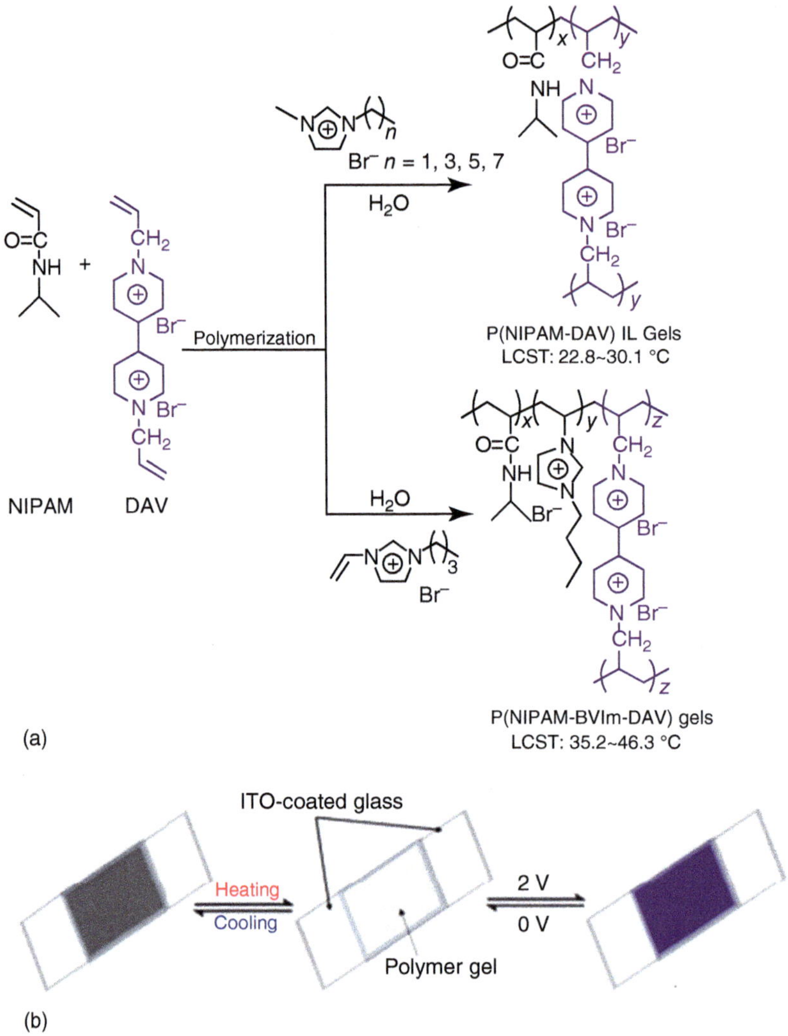

Figure 2.8 (a) General synthetic routes for P(NIPAM-fv [DAV]) IL gels and P(NIPAMBVIm-DAV) gels and (b) a schematic mechanism of thermochromic and electrochromic devices based on the prepared polymer gels. Source: Chen et al. [93].

electrolyte (SPE). For example, various composites had been mixed into polymers (see in Figure 2.9) [94], including inert ceramic fillers [95, 96], fast ion-conductive ceramics [97, 98], lithium salts [99], IL [100], etc.

By combining the advantages of these two materials, i.e. ceramic filler and polymer material, new hybrid composite materials with high dielectric constants can be fabricated. Different types of inert ceramics had been incorporated into the polymer by blending the polymers with fillers, such as SiO_2 [101], Al_2O_3 [102], TiO_2 [103], zeolite, etc. In 1982, Weston and Steele firstly prepared a composite by mixing PEO with Al_2O_3 [104], proving that PEO after addition of fillers exhibited an improvement

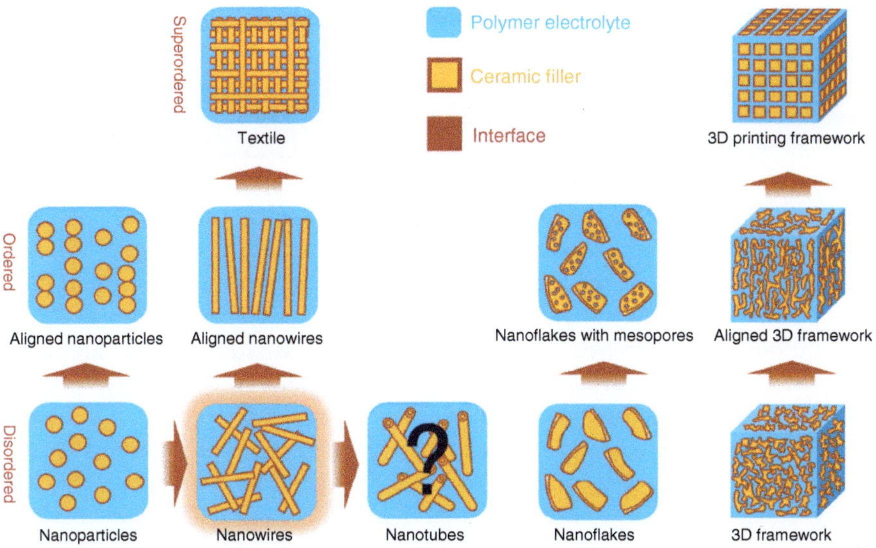

Figure 2.9 The evolutions of morphology of ceramic filler in SCEs. Source: Li et al. [94].

of mechanical properties and ionic conductivity even in a temperature exceeding 100 °C. The idea of adding ceramic fillers into polymer complex is to provide a stable solid matrix [104]. Composite polymeric materials containing fine ceramic particles are considered as heterogeneously disordered systems [105]. Later, Capuano et al. (1991) [106] studied the effects of the doping amount and particle size of $LiAlO_2$ powder on the conductivity of solid electrolyte. The conductivity achieved the highest after the doping amount of $LiAlO_2$ was around 10 wt.%. In other words, particle size of the inert ceramic material influenced to the conductivity of the SPE and which increased as the size is <10 μm [107]. As was reported, Al_2O_3 can also effectively decrease the crystallinity and glass transition temperature of PEO, proving that the decrease of polymer crystallinity promotes the improvement of ionic conductivity [102]. Liang et al. [108] reported that PEO-PMMA-based composite electrolyte exhibited an improvement of the ionic conductivity from 6.71×10^{-7} to 9.39×10^{-7} S/cm, in which PEO-PMMA and nano Al_2O_3 were used as a host matrix and filler, respectively [108]. Moreover, SiO_2 is also a common inert ceramic filler material used in synthesizing SPE. A composite of a PEO matrix and SiO_2 fillers was reported and reached an ionic conductivity of 2×10^{-4} S/cm at ambient temperature, [109] with 2.5 wt.% filler loadings. In addition, SiO_2 can be designed as a three-dimensional skeleton doped into polymers. In recent years, PEO-Silica aerogel was reported to achieve high ionic conductivity of 6×10^{-4} S/cm and high modulus of 0.43 GPa [110]. The mechanical properties and ionic conductivity has been enhanced by controlling powder dispersion. In addition, some studies have integrated a variety of inorganic ceramics into the polymer, and the ionic conductivity has also been improved.

Fast ion conductor ceramics, also known as active inorganic electrolytes, exhibit a high ionic conductivity of up to 10^{-2} S/cm at 25 °C. Composite of fast ion conductor ceramics with polymer can strengthen interfacial contact. Since 2009, polymer-in-ceramic (PIC) approach was introduced into the field of composite polymeric electrolytes [111]. It consists of introducing polymer salt complexes into porous ceramic structures. The studied systems exhibited excellent mechanical properties, high conductivities, and good stability vs Li metal electrodes under prolonged storage [111]. As was reported recently, composite electrolytes from "ceramic-in-polymer" (CIP) to PIC with different sizes of garnet particles are investigated for their effectiveness in dendrite suppression [112]. With a large specific surface area, nanoscale garnet ceramic fillers improve the transition rate of ions [113]. The PIC electrolyte with 80 vol% 5 μm $Li_{6.4}La_3Zr_{1.4}Ta_{0.6}O_{12}$ (LLZTO) (PIC-5 μm) shows the highest tensile strength of 12.7 MPa [113]. The morphology, distribution, and three-dimensional structure of ceramic filler particles may affect the ionic conductivity of polymer composite electrolyte [114]. Cui et al. explored the different morphologies of LLZO to increase ionic transport [115]. These LLZO nanowires show an ionic conductivity of 6.05×10^{-5} S/cm at 30 °C. A polymer–ceramic composite electrolyte of PEO and $Li_{0.33}La_{0.557}TiO_3$ nanowires has been studied and exhibited the lithium-ion conductivity of 2.4×10^{-4} S/cm at 25 °C [116]. Furthermore, glass–ceramic electrolytes are another solid ceramic electrolytes in which ions transfer through the ceramic phase by means of vacancies or interstitials within the lattice. Typical examples of this are glass/glass–ceramic Li_2S-P_2S_5 and ceramic thio-lithium superionic conductor (LISICON) $Li_{4-x}Ge_{1-x}P_xS_4$ ($0 < x < 1$), which are the most promising electrolytes to increase ion conductivity up to 10^{-5} S/cm [117].

The effects of ceramic fillers on the ionic conductivity of solid polymer systems have been discussed for quite many years. The published papers show that increasing the conductivity is the key parameter. Polymer-ceramic composites offer superior mechanical properties and great conductivities. However, exploring ceramics materials is still a challenge.

2.3.5 Ionic Liquid Polymer Electrolytes

ILs are molten salt liquid at room temperature (<100 °C) and consisted of dissociated ions [118]. IL electrolytes have been widely applied in electrochemical application due to non-volatility, thermal stability, non-flammability, high ionic conductivity, and a wide electrochemical stability window [119]. Recently, the incorporation of room-temperature ILs into PEs to prepare solvent-free PEs has been considered as a promising approach to improve the performance of ECDs. For example, room temperature ionic liquid (RTIL) (1-butyl-3-methyl-imidazolium hexafluorophosphate [BMIM] PF_6) [120] was reported to increase the optical contrast and coloring efficiency of the ECD. Zhou et al. have studied an RTIL non-volatile polymer electrolyte based on poly(propylene carbonate) as the host for ECD application [121]. In this work, 1-butyl-3-methylimidazolium tetrafluoroborate ($BMIM^+BF_4^-$) with high ionic conductivity, good electrochemical stability, and low viscosity was added into poly (propylene carbonate) (PPC)/$LiClO_4$ SPE to form novel PEs with improvement in ionic conductivity, thermal and electrochemical properties, and safety [121].

Recently, combining RTIL and ceramic filler into PEs have been explored. Murali and coworker [122] studied zinc ion conducting gel polymer electrolytes (ZCGPEs) made of poly(vinyl chloride) (PVC)/PEMA blend, zinc triflate [$Zn(OTf)_2$] salt, 1-ethyl-3-methylimidazolium bis(trifluoromethylsulfonyl) imide (EMIMTFSI) IL, and silica (SiO_2), which exhibited the highest ionic conductivity.

PILs are polyelectrolytes consisting of a polymeric backbone and IL components in monomer repeating units. Variability and designability of ILs enrich the performance, which attract great attention in the field of polymer material [123]. Mecerreyes in 2011 reviewed the synthetic aspects of ILs, the synthetic methods of PILs, and their physicochemical properties [124]. Yuan et al. also reported the applications of PILs and ILs [125]. ILs can be widely applied in diverse field through adjusting ionic conductivity, hydrophilicity, hydrophobicity, chemical durability, and thermodynamic stability. Marcilla et al. [126] have reported a new PE based on ILs and PILs in ECDs having configuration PEDOT/electrolyte/PEDOT. The electrolyte was prepared by mixing 1-butyl-3-methylimidazolium-bis(trifluoromethanesulfonimide) [bmim][Tf_2N], 1-butyl-3-methylimidazolium tetrafluoroborate [bmim][BF_4], and 1-butyl-3-methylimidazolium bromide [bmim][Br], with poly(1-vinylethyl-imidazolium) bearing similar counter anions [Tf_2N], [BF_4] and [Br]. By using this electrolyte, the cycle life was significantly increased. Hydrophobic ILs ($P(EO)_{10}$ lithium bis (trifluoromethanesulfonyl) imide (LiTFSI) + 0.96PYR_{14}TFSI) electrolyte was reported to improve the performance of the all-solid-state ECD [127]. Puguan and Kim reported a switchable single-molecule ECD derived from a viologen-tethered triazolium-based poly(ionic liquid) (Figure 2.10) [128].

Yuan's group reported a flexible and voltage-switchable polymer Velcro constructed using host–guest recognition between poly(ionic liquid) strips. [129] The reversible phase transition behavior of PILs can be activated by temperature, electrochemical redox reactions, and chemical redox reactions. Vancso and coworkers

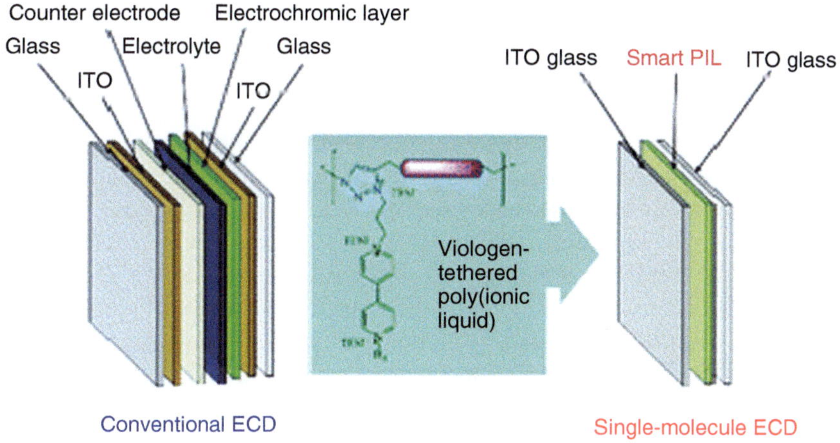

Figure 2.10 A switchable single-molecule electrochromic device derived from a viologen-tethered triazolium-based poly(ionic liquid). Source: Puguan and Kim [128].

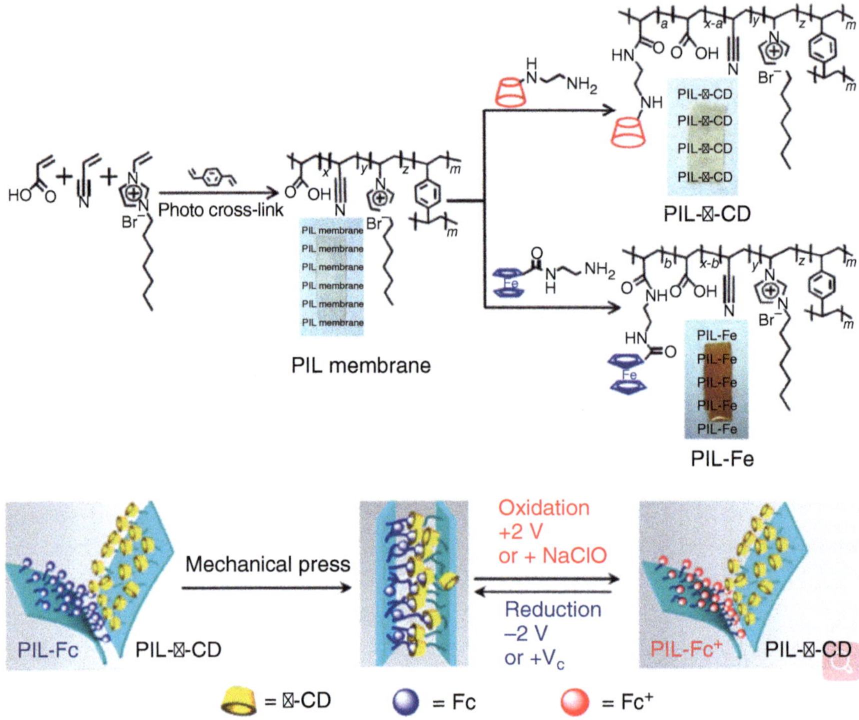

Figure 2.11 Schematic representation of the synthesis of PIL-b-CD and PIL-Fc membranes and the hook-and-loop strategy for adhesion based on b-CD and Fc-modified PIL membrane surfaces. Source: Guo et al. [129].

prepared a redox-active organometallic PIL [130]. The prepared polymer Velcro based on PILs exhibited strong adhesion in air and in aqueous solutions (including acidic and basic water and artificial seawater) and could be unfastened and fastened by mechanical, chemical and electrochemical means (Figure 2.11).

2.3.6 Gelatin-Based Polymer Electrolytes

Natural polymers can become a promising substitute for synthetic polymers and cause great attention from researchers due to their biodegradability, eco-friendliness, low production cost, and good physical and chemical properties. Among natural polymers, polysaccharides and proteins are best candidate due to their abundance in environment. Gelatin can be used as PE in EC applications. Gelatin is a polypeptide mainly composed of proline, hydroxyproline, and glycine. Avellaneda et al. [131] have reported a solid gelatin PE prepared by dissolving gelatin, $LiClO_4$, and glycerol into water for EC applications. The ionic conductivity for this electrolyte reached from 1.53×10^{-5} S/cm at room temperature to 4.95×10^{-4} S/cm at 80 °C. ECDs with the configuration K-glass/Nb_2O_5:Mo EC-layer/gelatin-based electrolyte/$(CeO_2)_x(TiO_2)_{1-x}$ ion-storage (IS) layer/K-glass have been assembled and reported to increase the ionic conductivity and good long-time cyclic stability

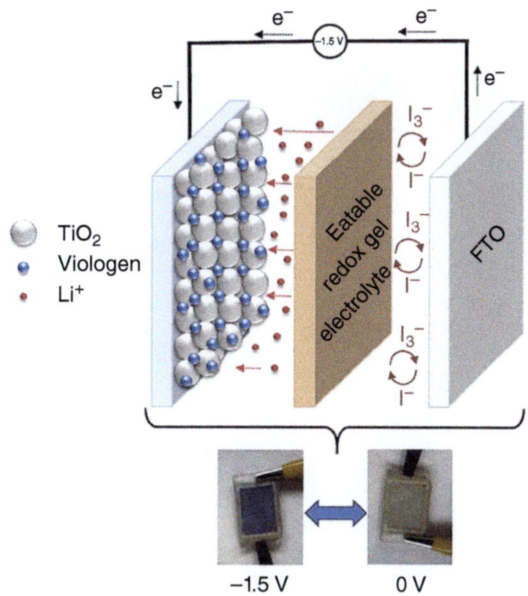

Figure 2.12 An ECD structure containing a gel electrolyte composition combining lithium iodide LiI in 1-butyl-3-methylimidazolium iodide (BMII) ionic liquid, triiodide I_3^-/I^- redox mediator, and biodegradable gelatin. Source: Reproduced with permission of Danine et al. [134].

[132]. Silva et al. explored a solid-state ECD including gelatin-based electrolytes [133]. This prototype ECD exhibited good optical density. The ECD display incorporating gelatine I and gelatine II samples presented in the visible region an average transmittance above 68% in the bleached state [133]. A novel gel electrolyte composition combining lithium iodide LiI in 1-butyl-3-methylimidazolium iodide (BMII) IL, triiodide I_3^-/I^- redox mediator, and biodegradable gelatin is proposed for ECDs (Figure 2.12) [134]. Fast switching times and high cycling stability, up to 20 000 cycles, are recorded in this ECD.

Moreover, gelatin-based electrolyte can be used in reflective ECD. Reflective ECD with glass/Cr/PB/electrolyte/CeO_2–TiO_2/ITO/glass configuration was reported to reveal a change from 8% to 15% at 450 nm for bleached and colored states.

Tihan et al. have used a new DNA–$LiClO_4$-based solid PE in an ECD (Figure 2.13) [135]. The novelty of the present research consists in the use of DNA membrane with different $LiClO_4$ ratios in order to achieve new EC windows with good performances [135].

2.4 Conclusion and Future Outlook

The field of PE is steadily broadening, as more and more various application are presented. Although lithium-ion-based electrolytes have been commercialized, such as applied in pull down window shades on Boeing 787 Dreamliner, smart glass in Ferrari 575 M Superamerica, and Flexity 2 light rail vehicles, some disadvantages such as poor safety, easy leakage, and unstable electrochemical performance limit its further development and wider applications. In this chapter, we have provided fundamental understanding of requirements for EC applications. In the meantime,

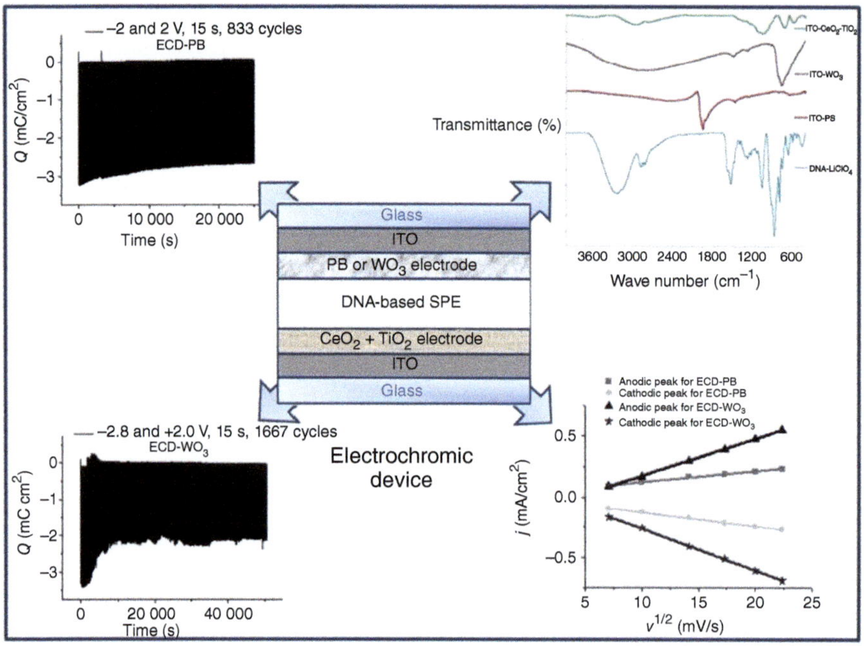

Figure 2.13 An electrochromic device containing DNA-based electrolyte. Source: Tihan et al. [135].

recent progresses on polymer-based electrolytes were summarized, including PEO/PVDF/PMMA gel polymer, self-healing polymer, cross-linking polymer, ceramic polymer, IL polymer, and gelatin PEs. The use of composite PEs has been identified as a promising method to improve the performance of electrolytes.

Although extensive researches have been devoted to the polymer-based electrolytes, some basic issues still need to be addressed before commercialization. Above all, ionic conductivity is still on lower orders of magnitude than lithium-based liquid electrolytes. Temperature is one of the most dominant parameters for determining the ionic conductivity. In particular, ionic conductivities in polymer-based solid electrolytes drop dramatically at lower temperatures. Moreover, the mechanism of interfacial interaction between electrolyte and an EC layer is still unknown.

For now, the GPEs, typically based on PEO/PVDF/PMMA, have also been extensively explored, but they exhibit too low ionic conductivity at room temperature ($\approx 10^{-5}$ S/cm) and poor resistance to oxidation at relatively high voltage. Poor mechanical performance limits the IL electrolyte from increasing the ionic conductivity. Safety issues cannot be neglected. Developing a self-healing electrolyte that simultaneously exhibits excellent EC properties and good self-healing behavior at low temperature is still an unmet challenge. The cross-linked composite PE

Table 2.1 Polymer hosts generally studied with the examples of gel polymer electrolyte complexes and their respective highest ionic conductivities.

Polymer electrolyte	EC materials	Ionic conductivity (S/cm)	Electrochemical stability window (V)	Optical modulation (%)	Stability	References
p(trimethylenecarbonate [TMC])/PEO	Prussian Blue/PEDOT	1.33×10^{-6}	$-1.5 \sim 1$	$8 \sim 30$	Full color switch (>600 s)	[138]
PEO/PVDF	TiO_2	6.98×10^{-6}	—	90	—	[15]
PEO/$LiClO_4$	PEDOT:PSS	4.2×10^{-4}	$-3.0 \sim 3.0$	22	No significant change	[139]
P(VDF-TrFE)/PEO	PEDOT:PSS	6.79×10^{-4}	$-1.0 \sim 0.5$	60	—	[140]
PEMA/poly (vinylidene fluoride) (PVdF)-HFP	WO_3//CeO_2–TiO_2	1.46×10^{-6}	$-1.3 \sim -0.8$	81	—	[141]
PVdF-HFP/silane-functionalized ZrO_2	PANI:DBSA	1.78×10^{-3}	$-4.0 \sim 4.0$	70	—	[142]
PVDF-HFP/POEI-ISF_4	Nanofibers	8.62×10^{-3}	$0 \sim 1.2$	68.7	95.5% of ΔT	[143]
PVDF-*co*-HFP/[EMI][TFSI]	1,1′-bis(3-fluoro-4-(trifluoromethyl) phenyl)-4,4′-bipyridinium bis(trifluoromethylsulfonyl)imide (TFMFPhV(TFSI)$_2$)	6.7×10^{-3}	$-0.35 \sim 0$	~73	~14% decrease of ΔT after 24 h	[144]
IL/PVDF-HFP	PEDOT:PSS	1.13×10^{-3}	$-1.4 \sim 1.4$	~24	—	[54]
TMPD/HV(BF_4)$_2$/SN/PVDF-HFP	TMPD/HV(BF_4)$_2$/succinonitrile	1×10^{-3}	$0 \sim 0.9$	60.1	74.7% of the original ΔT after 2000 cycles	[145]
[Emim]BF_4/PMMA	WO_3	2.9×10^{-3}(RT)	$-3.0 \sim 1.5$	88	—	[146]
Poly(ether ether ketone) membrane	IR-VEDs	6.8×10^{-3}	$-1.0 \sim 1.0$	47	Good cycling stability	[147]
PVDF-HFP:PMMA/$LiClO_4$	—	2.83×10^{-4}	—	—	—	[148]

(Continued)

Table 2.1 (Continued)

Polymer electrolyte	EC materials	Ionic conductivity (S/cm)	Electrochemical stability window (V)	Optical modulation (%)	Stability	References
PMMA/SN-PC	WO_3//brain natriuretic peptide (PBNPs)	1.46×10^{-3}	$-2.5 \sim 2.0$	52.4	44.5% after 2250 cycles	[149]
Poly(ε-caprolactone) (PCL)/SiO_2	PEDOT: PSS	5.2×10^{-3}	$-4.0 \sim 1.0$	30.6	Good stability (100 cycles)	[150]
Polyvinyl butyral (PVB)-based GPEFs	WO_3//$Ni_{1-x}O$	4.0×10^{-5}	$-2.0 \sim 2.0$	65.8	Fading of 4.8% (16 h)	[151]
PAN–DMF–LiTFSI	WO_3//CeO_2–TiO_2	2.54×10^{-4}	$-2.0 \sim 1.0$	29.1	—	[152]
PADA	LiNiO//WO_3	1.33×10^{-2} (RT)	$-2.30 \sim 2.30$	61	58.4% after 50 cycles	[153]
PVDF based	P(PVK-*co*-EDOT)	—	$-0.9 \sim 0.5$	39.1	—	[154]
2-APPG /PMDA/ICS	WO_3 film	1.01×10^{-3}	$-3.5 \sim 2.5$	38.2	450 cycles	[155]

Table 2.2 Examples of self-healing and cross-linking polymer electrolyte complexes and their respective highest ionic conductivities.

Electrolyte	Ionic conductivity (S/cm)	Electrochemical window (V)	Application	References
B-PVA/KCl/graphene oxide (GO)	47.5×10^{-3}	$0 \sim 1.0$	Flexible energy-storage device	[156]
SiO_2–UPy	8.0×10^{-5}	$0 \sim 6.0$	Lithium-ion batteries	[157]
P(AA-VIm-VSN)	1.26×10^{-4}	$-1.2 \sim 0.3$	Flexible electrochromic devices	[158]
[P(PO/EM)]/lithium trifluoro(perfluoro-tert-butyloxyl)borate (LiTFPFB)	1.55×10^{-4} (70 °C)	$3 \sim 4.2$	Lithium metal batteries	[159]
Zn–metal-organic gel (MOG)	—	$0 \sim 2.1$	Flexible supercapacitor	[160]
Poly(ethylene-co-acrylic lithium (fluoro sulfonyl)imide) (PEALiFSI)	5.84×10^{-4}	$2.5 \sim 4.0$	Solid-state Li-ion batteries	[161]
Li + -PEO	—	$-4 \sim 4$	Micro-super-capacitors	[162]

could effectively encapsulate the electrolyte solution without solvent leakage and exhibit favorable interfacial characteristics. However, low ionic conductivity at room temperature still need to be improved. The difficulty of ceramic PEs is how to construct good dispersion and enhance the interaction between filler and polymer, which restricts the further improvement of ionic conductivity. Gelatin-based PEs have proved to be one of the alternative binders, but its rheological behavior in the tested conditions inhibited such material to be used at industry level.

To overcome these challenges, several future research directions are suggested as follows.

(1) Improving electrolyte's electrochemical stable potential window (ESPW) value. The ESPWs are dependent on the cation and anion of the conducting salt and the solvent. It should be noted that in most cases, an increase in electrolyte ESPWs may result in the deterioration of other properties such as ionic conductivity and viscosity of electrolytes. The improvement of ESPWs could be achieved by exploring new organic solvents and new conducting salt or by optimizing/modifying the commonly used organic electrolytes. However, it is difficult to achieve all the requirements, such as high ESPW, high ionic conductivity, high physicochemical stability, low viscosity, and eco-friendliness. Some tradeoffs may be reasonable in solving the practical problems.

Table 2.3 Comparison of the Li^+ conductivities of some reported solid polymer electrolytes.

Solid polymer electrolytes	Type of electrolytes	Anionic center	Ionic conductivity (S/cm) at RT	References
P(MEO-MALi)	Random copolymer	$-CO_2^-$	1.5×10^{-7}	[163]
PAE_8-*co*-E_3SO_3Li	Random copolymer	$-SO_3^-$	2.0×10^{-7}	[164]
	Random copolymer	$-(C_2O_4)_2B^-$	1.9×10^{-7}	[165]
POE	Random copolymer	$-CF(CF_3)\ SO_3^-$	1×10^{-7}	[166]
P(LiSMOE$_n$)s	Random copolymer	$-C_6H_5\cdot SO_3^-$	1.5×10^{-7}	[167]
Polyphosphazenes	Random copolymer	$-SO_2N^-SO_2CF_3$	3.0×10^{-6}	[168]
Li[poly(styrenesulfonyl (trifluoromethylsulfonyl)imide) (PSTFSI)-*co*-methoxy poly(ethylene glycol) acrylate (MPEGA)]	Random copolymer	$-SO_2N^-SO_2CF_3$	7.7×10^{-6}	[169]
Poly(2-oxo-1-difuluoroethylene sulfonylimide) (LiPI)-polyether	Blended copolymer	$-CO_2CF_2SO_2N^-$	1×10^{-6}	[170]
P(LiSMOE$_n$)	Blended copolymer	$-CO_2^-$	1.5×10^{-7}	[167]
Poly(5-oxo-3-oxy-4-trifluoro-methyl-1,2,4-pentafluoropentylene sulfonylimide lithium) (LiPPI)	Blended copolymer	$-COCF(CF_3)O\ (CF_2)_2SO_2N^-$	1×10^{-5}	[171]
Poly (estercarbonate) (PEC) based	Blended copolymer	$-SO_2N^-SO_2CF_3$	1×10^{-9}	[172]
LiTFSI	Blended copolymer	$-SO_2(CF_2)_4SO_2N^-$	1×10^{-6}	[173]
LiPSFSI/PEO	Blended copolymer	$-SO_2N^-SO_2F$	1×10^{-8}	[174]
poly[(4-styrenesulfonyl) (trifluoromethyl(S-trifluoro-methylsulfonylimino)sulfonyl) imide] (PSsTFSI$^-$)/PEO	Blended copolymer	$-SO_2N^-\ SO(=NSO_2CF_3)CF_3$	1×10^{-8}	[174]
PEO based	Block copolymer	$-CO_2^-$	1×10^{-7}	[175]
poly (ethylene glycol) macromer (PEGM) based	Block copolymer	$-SO_2N^-SO_2CF_3$	2.3×10^{-6}	[176]
Polystyrene-PEO based	Block copolymer	$-SO_2N^-SO_2CF_3$	1.3×10^{-5}	[177]
PEO-PSLiTFSI	Block copolymer	$-SO_2N^-SO_2CF_3$	3.0×10^{-8}	[170]
Methacrylic-PEO	Block copolymer	$-SO_2N^-SO_2CF_3$	3.4×10^{-8}	[177]
PEALiFSI	Block copolymer	$-CON^-SO_2CF_3$	5.84×10^{-4}	[178]
PEO-PLSS	Graft copolymer	$-SO_3^-$	7.0×10^{-8}	[179]
LiBOPEG$_{600}$	Homopolymer	$-OP^-(C_2O_4)_2$	1.6×10^{-6}	[180]

Table 2.4 Composition and performance of ionic liquid electrolytes.

Year	Electrolyte	Ionic conductivity (S/cm) at RT	References
2005	PVA–KOH	1.0×10^{-2}	[181]
2005	PEGDA–PVdF–EC–DMC–ethylmethyl carbonate (EMC)–$LiPF_6$	1.5×10^{-3}	[182]
2010	$[PPyr_{11}TFSI]_{28\%}LiTFSI_{12\%}[Pyr_{14}TFSI]_{60\%}$	1.6×10^{-6}	[183]
2010	$PEO_{20}LiTFSI_1[Pyr_{1.2O1}TFSI]_{1.5}$	7.0×10^{-5}	[184]
2010	PEO_{25}–LiTf–IL	3.0×10^{-4}	[185]
2011	$PEO_{20}LiTFSI_1[S_{2.2.2}TFSI]_1$	5.0×10^{-4}	[186]
2012	$PEO_{20}LiTFSI_1[Pip_{13}TFSI]_{1.27}$	2.1×10^{-4}	[187]
2013	poly(methyl methacrylate–acrylonitrile–vinyl acetate) [P(MMA–AN–VAc)]-*N*-methyl-*N*-butyl pyrrolidinium bis(trifluoromethansulfonyl)imide	1.2×10^{-3}	[188]
2013	$PEO_{20}LiTFSI_2[Pyr_{14}TFSI]_6$	5.0×10^{-4}	[189]
2013	Chitosan–PEO–NH_4NO_3–EC	2.06×10^{-3}	[190]
2014	Poly(ε-caprolactone)–NH_4SCN–EC	3.8×10^{-5}	[191]
2014	LiTFSI in P14TFS LiTFSI/P14TFSI/PEO	0.4×10^{-3}	[192]
2015	$PEO_{20}LiTFSI_2[Pyr_{1.2}O_1TFSI]_4$	2.5×10^{-4}	[193]
2017	PEO-1-butyl-3-methylimidazolium bis(trifluoromethylsulfonyl)imide	2.2×10^{-4}	[194]
2017	PEO-(EMimFSI)	1.3×10^{-3}	[195]
2017	Oligoethylene glycol/*N*-butyl imidazole	4.8×10^{-4}	[196]
2018	PVDF–HFP(EMITFSI)	0.76×10^{-3}	[197]
2018	PVDF–HFP–(SiO_2PPTFSI)	0.64×10^{-3}	[198]
2019	PIL:[PVEIm][TFSI]	5.92×10^{-4} (60 °C)	[199]
2019	PEO TBPIIP	2.51×10^{-3}	[200]
2019	PEO–EMimFSI	3.6×10^{-4}	[201]

(2) Increasing the working temperature range for ES operation. Some additives like ethylene glycol may be explored to decrease the power temperature limit in aqueous electrolytes. Regarding organic electrolytes, the development of new organic solvent mixtures could possibly obtain electrolytes with a wider operating temperature range.

(3) Optimizing the interaction effects between the electrolyte and electrode materials to improve the EC performance. In situ characterization methods such as fourier transform IR (FTIR) and Raman microscopy can be used to characterize the electrolytes in operating condition. The interface between an EC layer

and electrolyte can be observed under conventional transmission electron microscopy (TEM) [136, 137].

(4) Combining theoretical and experimental investigations to analyze, guide, and design electrolyte. The corresponding ionic mechanism can be simulated and interpreted by material calculations. It is necessary to fundamentally understand the mechanisms of electrolyte ion dynamics by using both theoretical modeling and experimental approaches.

(5) Development of standard methods to evaluate the performance of electrolytes. It is necessary and important to develop proper and standardized methods to evaluate the performance of different electrolytes and their associated ECs.

Finally, struggling to find suitable composite electrolytes with high conductivities during wide range of temperature always deserves more study. Commercial electrolytes require high ionic conductivity at room temperature, safety, and easy processing.

References

1 Nguyen, C.A., Xiong, S., Ma, J. et al. (2009). *The Journal of Physical Chemistry B* 113: 8006.

2 Fenton, D.E., Parker, J.M., and Wright, P.V. (1973). *Polymer* 14: 589.

3 Shriver, D. and Bruce, P. (1995). *Solid State Electrochemistry*, 95. Cambridge: Cambridge University Press.

4 Choudhury, N., Sampath, S., and Shukla, A. (2009). *Energy & Environmental Science* 2: 55.

5 Agrawal, R. and Pandey, G. (2008). *Journal of Physics D: Applied Physics* 41: 223001.

6 Meyer, W.H. (1998). *Advanced Materials* 10: 439.

7 Ramesh, S. and Lu, S. (2012). *Journal of Applied Polymer Science* 126: 484.

8 Wu, T.-Y., Li, W.-B., Kuo, C.-W. et al. (2013). *International Journal of Electrochemical Science*: 10720.

9 Sa'adun, N.N., Subramaniam, R., and Kasi, R. (2014). *The Scientific World Journal*: 2014.

10 Susan, M.A.B.H., Kaneko, T., Noda, A., and Watanabe, M. (2005). *Journal of the American Chemical Society* 127: 4976.

11 Thakur, V.K., Ding, G., Ma, J. et al. (2012). *Advanced Materials* 24: 4071.

12 Sekhon, S., Arora, N., and Singh, H.P. (2003). *Solid State Ionics* 160: 301.

13 Feuillade, G. and Perche, P. (1975). *Journal of Applied Electrochemistry* 5: 63.

14 Song, J., Wang, Y., and Wan, C.C. (1999). *Journal of Power Sources* 77: 183.

15 Chen, P., Liang, X., Wang, J. et al. (2017). *Journal of Sol-Gel Science and Technology* 81: 850.

16 Azens, A., Avendano, E., Backholm, J. et al. (2005). *Materials Science and Engineering B* 119: 214.

17 Naji, A., Ghanbaja, J., Willmann, P. et al. (1996). *Journal of Power Sources* 62: 141.

18 Fey, G., Hsieh, M., Jaw, H., and Lee, T. (1993). *Journal of Power Sources* 44: 673.

19 Takami, N., Ohsaki, T., and Inada, K. (1992). *Journal of the Electrochemical Society* 139: 1849.

20 Kumar, B., Schaffer, J.D., Nookala, M., and Scanlon, L.G. Jr., (1994). *Journal of Power Sources* 47: 63.

21 Imoto, K., Takahashi, K., Yamaguchi, T. et al. (2003). *Solar Energy Materials and Solar Cells* 79: 459.

22 Agnihotry, S., Pradeep, P., and Sekhon, S. (1999). *Electrochimica Acta* 44: 3121.

23 Cowie, J.M.G. and Arrighi, V. (2007). *Polymers: Chemistry and Physics of Modern Materials*. CRC press.

24 Wang, Y., Chen, K.S., Mishler, J. et al. (2011). *Applied Energy* 88: 981.

25 Miller, M. and Bazylak, A. (2011). *Journal of Power Sources* 196: 601.

26 Beaujuge, P.M. and Reynolds, J.R. (2010). *Chemical Reviews* 110: 268.

27 Gaupp, C.L., Zong, K., Schottland, P. et al. (2000). *Macromolecules* 33: 1132.

28 Pennarun, P.-Y. and Jannasch, P. (2005). *Solid State Ionics* 176: 1103.

29 Pennarun, P.-Y., Papaefthimiou, S., Yianoulis, P., and Jannasch, P. (2007). *Solar Energy Materials and Solar Cells* 91: 330.

30 Boehme, J.L., Mudigonda, D.S., and Ferraris, J.P. (2001). *Chemistry of Materials* 13: 4469.

31 Yang, C.-H., Huang, L.-R., Chih, Y.-K., and Chung, S.-L. (2007). *The Journal of Physical Chemistry C* 111: 3786.

32 Reiter, J., Krejza, O., and Sedlaříková, M. (2009). *Solar Energy Materials and Solar Cells* 93: 249.

33 C. Sherwood, F. Price, R. Stein, "Effect of shear on the crystallization kinetics of poly (ethylene oxide) and poly (e-caprolactone) melts", presented at *Journal of Polymer Science, Polymer Symposia*, 1978.

34 Armand, M., Chabagno, J., and Duclot, M. (1978). Second International Meeting on Solid Electrolytes, presented at Extended Abstracts (20–22 September), St. Andrews, Scotland.

35 Pitawala, H., Dissanayake, M., and Seneviratne, V. (2007). *Solid State Ionics* 178: 885.

36 Pedone, D., Armand, M., and Deroo, D. (1988). *Solid State Ionics* 28: 1729.

37 Visco, S.J., Liu, M., Doeff, M.M. et al. (1993). *Solid State Ionics* 60: 175.

38 Baudry, P., Aegerter, M.A., Deroo, D., and Valla, B. (1991). *Journal of the Electrochemical Society* 138: 460.

39 Su, L., Wang, H., and Lu, Z. (1998). *Materials Chemistry and Physics* 56: 266.

40 Desai, S., Shepherd, R.L., Innis, P.C. et al. (2011). *Electrochimica Acta* 56: 4408.

41 Yang, C.-H., Chong, L.-W., Huang, L.-M. et al. (2005). *Materials Chemistry and Physics* 91: 154.

42 Nishikitani, Y., Uchida, S., Asano, T. et al. (2008). *The Journal of Physical Chemistry C* 112: 4372.

43 Mendoza, N., Paraguay-Delgado, F., Hechavarría, L. et al. (2011). *Solar Energy Materials and Solar Cells* 95: 2478.

44 Panero, S., Scrosati, B., Baret, M. et al. (1995). *Solar Energy Materials and Solar Cells* 39: 239.

45 Vasilopoulou, M., Raptis, I., Argitis, P. et al. (2006). *Microelectronic Engineering* 83: 1414.

46 Y. Zhou, P. Gu, J. Tang, "Electrochromic device with a polymer ionic conductor", presented at *Optical Materials Technology for Energy Efficiency and Solar Energy Conversion XII*, 1993.

47 Andersson, A.M., Granqvist, C.G., and Stevens, J.R. (1989). *Applied Optics* 28: 3295.

48 Argun, A., Cirpan, A., and Reynolds, J. (2003). *Advanced Materials* 15: 1338.

49 Oral, A., Koyuncu, S., and Kaya, İ. (2009). *Synthetic Metals* 159: 1620.

50 Tung, T.-S. and Ho, K.-C. (2006). *Solar Energy Materials and Solar Cells* 90: 521.

51 Yang, X., Cong, S., Li, J. et al. (2019). *Solar Energy Materials and Solar Cells* 200: 109952.

52 Kim, J.T., Song, J., Ryu, H., and Ah, C.S. (2020). *Advanced Optical Materials*: 1901464.

53 Fabretto, M., Vaithianathan, T., Hall, C. et al. (2008). *Electrochimica Acta* 53: 2250.

54 Jia, P., Yee, W.A., Xu, J. et al. (2011). *Journal of Membrane Science* 376: 283.

55 Lang, A.W., Österholm, A.M., and Reynolds, J.R. (2019). *Advanced Functional Materials* 29: 1903487.

56 Çelik, E. (2013). *Journal of Materials*: 2013.

57 Zheng, R., Zhang, J., Jia, C. et al. (2017). *Polymer Chemistry* 8: 6981.

58 Li, J., Qi, S., Liang, J. et al. (2015). *ACS Applied Materials & Interfaces* 7: 14140.

59 Liang, J., Li, L., Niu, X. et al. (2013). *Nature Photonics* 7: 817.

60 Sun, H., You, X., Jiang, Y. et al. (2014). *Angewandte Chemie International Edition* 53: 9526.

61 Ko, J., Kim, Y.-J., and Kim, Y.S. (2016). *ACS Applied Materials & Interfaces* 8: 23854.

62 Han, L., Lu, X., Wang, M. et al. (2017). *Small* 13: 1601916.

63 Cao, Y., Morrissey, T.G., Acome, E. et al. (2017). *Advanced Materials* 29: 1605099.

64 Canadell, J., Goossens, H., and Klumperman, B. (2011). *Macromolecules* 44: 2536.

65 Yoon, J.A., Kamada, J., Koynov, K. et al. (2012). *Macromolecules* 45: 142.

66 Chao, A., Negulescu, I., and Zhang, D. (2016). *Macromolecules* 49: 6277.

67 Wei, Z., Yang, J.H., Zhou, J. et al. (2014). *Chemical Society Reviews* 43: 8114.

68 Xu, Z., Zhao, Y., Wang, X., and Lin, T. (2013). *Chemical Communications* 49: 6755.

69 White, S.R., Sottos, N.R., Geubelle, P.H. et al. (2001). *Nature* 409: 794.

70 Caruso, M.M., Delafuente, D.A., Ho, V. et al. (2007). *Macromolecules* 40: 8830.

71 Trask, R., Williams, G., and Bond, I. (2007). *Journal of the Royal Society Interface* 4: 363.

72 Gong, Z., Zhang, G., Zeng, X. et al. (2016). *ACS Applied Materials & Interfaces* 8: 24030.

73 Moon, H.C., Lodge, T.P., and Frisbie, C.D. (2015). *Chemistry of Materials* 27: 1420.

74 Michaelis, L. and Hill, E.S. (1933). *The Journal of General Physiology* 16: 859.

75 Seo, D.G. and Moon, H.C. (2018). *Advanced Functional Materials* 28: 1706948.

76 Zhou, N.C., Xu, C., Burghardt, W.R. et al. (2006). *Macromolecules* 39: 2373.

77 Jang, S., Moon, H.C., Kwak, J. et al. (2014). *Macromolecules* 47: 5295.

78 Alsalhy, Q.F., Rashid, K.T., Ibrahim, S.S. et al. (2013). *Journal of Applied Polymer Science* 129: 3304.

79 Saikia, B.J. and Dolui, S.K. (2016). *Journal of Polymer Science Part A: Polymer Chemistry* 54: 1842.

80 Ulaganathan, M., Nithya, R., Rajendran, S., and Raghu, S. (2012). *Solid State Ionics* 218: 7.

81 Ponmani, S., Kalaiselvimary, J., and Prabhu, M.R. (2018). *Journal of Solid State Electrochemistry* 22: 2605.

82 Ko, J., Surendran, A., Febriansyah, B., and Leong, W.L. (2019). *Organic Electronics* 71: 199.

83 Kavitha, A.A. and Singha, N.K. (2009). *ACS Applied Materials & Interfaces* 1: 1427.

84 Zheng, R., Fan, Y., Wang, Y. et al. (2018). *Electrochimica Acta* 286: 296.

85 Nishimoto, A., Agehara, K., Furuya, N. et al. (1999). *Macromolecules* 32: 1541.

86 Kang, Y., Cheong, K., Noh, K.-A. et al. (2003). *Journal of Power Sources* 119: 432.

87 Kono, M., Hayashi, E., and Watanabe, M. (1998). *Journal of the Electrochemical Society* 145: 1521.

88 Wen, Z., Itoh, T., Uno, T. et al. (2003). *Solid State Ionics* 160: 141.

89 Matsui, S., Muranaga, T., Higobashi, H. et al. (2001). *Journal of Power Sources* 97: 772.

90 Kuratomi, J., Iguchi, T., Bando, T. et al. (2001). *Journal of Power Sources* 97: 801.

91 Matoba, Y., Ikeda, Y., and Kohjiya, S. (2002). *Solid State Ionics* 147: 403.

92 Lee, K.-H., Kim, K.-H., and Lim, H.S. (2001). *Journal of the Electrochemical Society* 148: A1148.

93 Chen, F., Ren, Y., Guo, J., and Yan, F. (2017). *Chemical Communications* 53: 1595.

94 Li, S., Zhang, S.Q., Shen, L. et al. (2020). *Advanced Science* 7: 1903088.

95 Aravindan, V. and Vickraman, P. (2008). *Journal of Applied Polymer Science* 108: 1314.

96 Liu, W., Lin, D., Sun, J. et al. (2016). *ACS Nano* 10: 11407.

97 Ling, S.-G., Peng, J.-Y., Yang, Q. et al. (2018). *Chinese Physics B* 27: 038201.

98 Keller, M., Appetecchi, G.B., Kim, G.-T. et al. (2017). *Journal of Power Sources* 353: 287.

99 Do, J.-S., Chang, C.-P., and Lee, T.-J. (1996). *Solid State Ionics* 89: 291.

100 Subianto, S., Mistry, M.K., Choudhury, N.R. et al. (2009). *ACS Applied Materials & Interfaces* 1: 1173.

101 Ketabi, S. and Lian, K. (2013). *Electrochimica Acta* 103: 174.

102 Tambelli, C., Bloise, A., Rosario, A. et al. (2002). *Electrochimica Acta* 47: 1677.

103 Pal, P. and Ghosh, A. (2018). *Electrochimica Acta* 260: 157.

104 Weston, J. and Steele, B. (1982). *Solid State Ionics* 7: 75.

105 Kontos, G., Soulintzis, A., Karahaliou, P. et al. (2007). *Express Polymer Letters* 1: 781.

106 Capuano, F., Croce, F., and Scrosati, B. (1991). *Journal of the Electrochemical Society* 138: 1918.

107 Chandrasekhar, V. (1998). *Blockcopolymers-Polyelectrolytes-Biodegradation*, 139. Springer.

108 Liang, B., Tang, S., Jiang, Q. et al. (2015). *Electrochimica Acta* 169: 334.

109 Nan, C.-W., Fan, L., Lin, Y., and Cai, Q. (2003). *Physical Review Letters* 91: 266104.

110 Lin, D., Yuen, P.Y., Liu, Y. et al. (2018). *Advanced Materials* 30: 1802661.

111 Syzdek, J., Armand, M., Gizowska, M. et al. (2009). *Journal of Power Sources* 194: 66.

112 Chen, L., Li, Y., Li, S.-P. et al. (2018). *Nano Energy* 46: 176.

113 Kumar, B. and Scanlon, L.G. (2000). *Journal of Electroceramics* 5: 127.

114 Yao, P., Yu, H., Ding, Z. et al. (2019). *Frontiers in Chemistry* 7: 522.

115 Liu, W., Lin, D., Sun, J., Zhou, G., and Cui, Y. (2016). *ACS Nano* 10: 11407–11413.

116 Zhu, P., Yan, C., Dirican, M. et al. (2018). *Journal of Materials Chemistry A* 6: 4279.

117 Zhao, Y., Wu, C., Peng, G. et al. (2016). *Journal of Power Sources* 301: 47.

118 Čolović, M., Jerman, I., Gaberšček, M., and Orel, B. (2011). *Solar Energy Materials and Solar Cells* 95: 3472.

119 Lewandowski, A. and Świderska-Mocek, A. (2009). *Journal of Power Sources* 194: 601.

120 Ma, L., Li, Y., Yu, X. et al. (2009). *Solar Energy Materials and Solar Cells* 93: 564.

121 Zhou, D., Zhou, R., Chen, C. et al. (2013). *The Journal of Physical Chemistry B* 117: 7783.

122 Candhadai Murali, S.P. and Samuel, A.S. (2019). *Journal of Applied Polymer Science* 136: 47654.

123 MacFarlane, D.R., Forsyth, M., Howlett, P.C. et al. (2016). *Nature Reviews Materials* 1: 1.

124 Mecerreyes, D. (2011). *Progress in Polymer Science* 36: 1629.

125 Yuan, J., Mecerreyes, D., and Antonietti, M. (2013). *Progress in Polymer Science* 38: 1009.

126 Marcilla, R., Alcaide, F., Sardon, H. et al. (2006). *Electrochemistry Communications* 8: 482.

127 Shin, J.-H., Henderson, W.A., Tizzani, C. et al. (2006). *Journal of the Electrochemical Society* 153: A1649.

128 Puguan, J.M.C. and Kim, H. (2019). *Journal of Materials Chemistry A* 7: 21668.

129 Guo, J., Yuan, C., Guo, M. et al. (2014). *Chemical Science* 5: 3261.

130 Sui, X., Hempenius, M.A., and Vancso, G.J. (2012). *Journal of the American Chemical Society* 134: 4023.

131 Avellaneda, C.O., Vieira, D.F., Al-Kahlout, A. et al. (2008). *Solar Energy Materials and Solar Cells* 92: 228.

132 Avellaneda, C.O., Vieira, D.F., Al-Kahlout, A. et al. (2007). *Electrochimica Acta* 53: 1648.

133 Silva, M., Barbosa, P., Rodrigues, L. et al. (2010). *Optical Materials* 32: 719.

134 Danine, A., Manceriu, L., Fargues, A., and Rougier, A. (2017). *Electrochimica Acta* 258: 200.

135 Tihan, G.T., Mindroiu, M., Rau, I. et al. (2020). *Materials Chemistry and Physics* 241: 122349.

136 Zachman, M.J., Tu, Z., Choudhury, S. et al. (2018). *Nature* 560: 345.

137 Wang, X., Zhang, M., Alvarado, J. et al. (2017). *Nano Letters* 17: 7606.

138 Barbosa, P.C., Rodrigues, L.C., Silva, M.M. et al. (2010). *Electrochimica Acta* 55: 1495.

139 Eren, E. (2019). *Journal of the Turkish Chemical Society Section A* 5: 1413.

140 Nguyen, C.A., Xiong, S., Ma, J. et al. (2011). *Physical Chemistry Chemical Physics* 13: 13319.

141 Sim, L.N., Majid, S.R., and Arof, A.K. (2012). *Solid State Ionics* 209: 15.

142 Puguan, W., J. Chung, and H. Kim. *Electrochimica Acta*, 2016, 196, 236

143 Yu, H.F., Kao, S.Y., Lu, H.C. et al. (2018). *Solar Energy Materials & Solar Cells* 177: 32.

144 Oh, H., Seo, D.G., Yun, T.Y. et al. (2017). *Organic Electronics* 51: 490.

145 Chang, T.H., Hu, C.W., Kao, S.Y. et al. (2015). *Solar Energy Materials & Solar Cells* 143: 606.

146 Tang, Q., Li, H., Yue, Y. et al. (2017). *Materials and Design* 118: 279.

147 Li, X., Zhang, L., Wang, B. et al. (2020). *Electrochimica Acta* 332: 135357.

148 Gohel, K. and Kanchan, D.K. (2018). *Journal of Advanced Dielectrics* 8: 1850005.

149 Wang, J.Y., Wang, M.C., and Jan, D.J. (2017). *Solar Energy Materials & Solar Cells* 160: 476.

150 Liu, X., Li, K., Hou, C. et al. (2019). *Materials Science and Engineering B* 241: 36.

151 Zhang, F., Dong, G., Liu, J. et al. (2017). *Ionics* 23: 1879.

152 Sim, L.N., Sentanin, F.C., Pawlicka, A. et al. (2017). *Electrochimica Acta* 229: 22.

153 Chen, W., Zhu, C., Guo, L. et al. (2019). *Journal of Materials Chemistry C* 7: 3744.

154 Yao, M., Hu, H., Wang, N. et al. (2020). *Journal of Colloid and Interface Science.*

155 Zhou, J., Wang, J., Li, H., and Shen, F. (2018). *Organic Electronics* 62: 516.

156 Peng, H., Lv, Y., Wei, G. et al. (2019). *Journal of Power Sources* 431: 210.

157 Zhou, B., Jo, Y.H., Wang, R. et al. (2019). *Journal of Materials Chemistry A* 7: *10354*.

158 Wang, Y., Zheng, R., Luo, J. et al. (2019). *Electrochimica Acta* 320: *134489*.

159 Wang, Q., Cui, Z., Zhou, Q. et al. (2020). *Energy Storage Materials* 25: 756.

160 Karan, C.K., Mallick, S., Raj, C.R., and Bhattacharjee, M. (2019). *Chemistry - A European Journal* 25: 14775.

161 Ahmed, F., Choi, I., Rahman, M.M. et al. (2019). *ACS Applied Materials & Interfaces* 11: *34930.*

162 Shi, Y., Wang, Y., Gu, Y. et al. (2019). *Chemical Engineering Journal: 123645.*

163 Kobayashi, N., Uchiyama, M., and Tsuchida, E. (1985). *Solid State Ionics* 17: 307–311.

164 Sun, X.G., Hou, J., and Kerr, J.B. (2005). *Electrochimica Acta* 50: 1139–1147.

165 Sun, X.G., Reeder, C.L., and Kerr, J.B. (2004). *Macromolecules* 37: 2219–2227.

166 Benrabah, D., Sylla, S., Sanchez, J.Y., and Armand, M. (1995). *Journal of Power Sources* 54: 456–460.

167 Xu, W., Siow, K.S., Gao, Z., and Lee, S.Y. (1998). *Chemistry of Materials* 10: 1951–1957.

168 Allcock, H.R., Welna, D.T., and Maher, A.E. (2006). *Solid State Ionics* 177: 741–747.

169 Gupta, H., Balo, L., Singh, V.K. et al. (2017). *Solid State Ionics* 309: 192–199.

170 Watanabe, M., Suzuki, Y., and Nishimoto, A. (2000). *Electrochimica Acta* 45: 1187–1192.

171 Watanabe, M., Tokuda, H., and Muto, S. (2001). *Electrochimica Acta* 46: 1487–1491.

172 Fenton, D.E., Parker, J.M., and Wright, P.V. (1973). *Polymer* 14: 589.

173 Geiculescu, O.E., Yang, J., Zhou, S. et al. (2004). *Journal of The Electrochemical Society* 151: A1363.

174 Ma, Q., Zhang, H., Zhou, C. et al. (2016). *Angewandte Chemie, International Edition* 55: 2521–2525.

175 Ryu, S.W., Trapa, P.E., Olugebefola, S.C. et al. (2005). *Journal of the Electrochemical Society* 152: A158–A163.

176 Porcarelli, L., Shaplov, A.S., Salsamendi, M. et al. (2016). *ACS Applied Materials & Interfaces* 8: 10350–10359.

177 Bouchet, R., Maria, S., and Meziane, R. (2013). *Nature Materials* 12: 452–457.

178 Ahmed, F., Choi, I., and Rahman, M.M. (2019). *ACS Applied Materials & Interfaces* 11: 34930–34938.

179 Park, C.H., Sun, Y.K., and Kim, D.W. (2004). *Electrochimica Acta* 50: 375–378.

180 Sun, X.G. and Angell, C.A. (2004). *Solid State Ionics* 175: 743–746.

181 Yang, C.C., Hsu, S.T., and Chien, W.C. (2005). *Journal of Power Sources* 152: 303–310.

182 Yang, C.M., Ju, J.B., Lee, J.K. et al. (2005). *Electrochimica Acta* 50: 1813–1819.

183 Appetecchi, G.B., Kim, G.T., Montanino, M. et al. (2010). *Journal of Power Sources* 195: 3668–3675.

184 Abitelli, E., Ferrari, S., Quartarone, E. et al. (2010). *Electrochimica Acta* 55: 5478–5484.

185 Pandey, G.P., Kumar, Y., and Hashmi, S.A. (2010). *Indian Journal of Chemistry* 49: 743–751.

186 Fisher, A.S., Khalid, M.B., Widstrom, M. et al. (2011). *Journal of Power Sources* 196: 9767–9773.

187 Yongxin, A., Xinqun, C., Pengjian, Z. et al. (2012). *Journal of Solid State Electrochemistry* 16: 383–389.

188 Liao, Y., Sun, C., Hu, S., and Li, W. (2013). *Electrochimica Acta* 89: 461–468.

189 Joost, M., Kim, G.T., Winter, M., and Passerini, S. (2013). *Electrochimica Acta* 113: 181–185.

190 Ithnin, R., Illias, H.A., and Kadir, M.F.Z. (2013). *Optical Materials* 35: 1834–1841.

191 Woo, H.J., Liew, C.W., Majid, S.R., and Arof, A.K. (2014). *High Performance Polymers* 26: 637–640.

192 Costa, C.M., Ribelles, J.G., Lanceros-Méndez, S. et al. (2014). *Journal of Power Sources* 245: 779.

193 de Vries, H., Jeong, S., and Passerini, S. (2015). *RSC Advances* 5: 13598–13606.

194 Huo, H., Zhao, N., and Sun, J. (2017). *Journal of Power Sources* 372: 1–7.

195 Song, S., Kotobuki, M., Zheng, F. et al. (2017). *Journal of Materials Chemistry A* 5: 6424–6431.

196 Yang, G., Chanthad, C., Oh, H. et al. (2017). *Journal of Materials Chemistry A* 5: 18012–18019.

197 Guo, Q., Han, Y., Wang, H. et al. (2018). *Journal of Physical Chemistry C* 122: 10334–10342.

198 Cheng, Y., Zhang, L., Xu, S. et al. (2018). *Journal of Materials Chemistry A* 6: 18479–18487.

199 Ma, F., Zhang, Z., Yan, W. et al. (2019). *ACS Sustainable Chemistry & Engineering* 7: 4675–4683.

200 Xie, Z., Wu, Z., An, X. et al. (2019). *Journal of Membrane Science* 586: 122–129.

201 Song, S., Wu, Y., Tang, W. et al. (2019). *ACS Sustainable Chemistry & Engineering* 7: 7163–7170.

3

Electrochromic Small Molecules

3.1 Background of Small Molecule Electrochromic

The first reports on electrochromism was published in the 1930s. In the 1960s, when Plant studied organic dyes, it was found that their color changed when voltage was applied, and after this researchers deepen the study on electrochromism. The first electrochromic device (ECD) was fabricated by S. K Deb by WO_3 film in 1969. At the same time, he also proposed the mechanism of "oxygen vacancy color center."

Since the 1970s, a series of small molecule electrochromic (EC) materials mainly including inorganic molecules have come to most people's attention gradually. At the same time, the mechanism of electrochromic materials has been studied since then and more EC materials were found. Early EC materials are mainly transition metal oxides (such as W, Mo, V, Nb, Ti, Ir, Rh, Ni, Co, etc.) and Prussian Blue. It is usually divided into a reduced coloring state and an oxidation coloring state, and some materials may appear in different colors in both the oxidized state and the reduced state.

In 1987, Japanese researchers have reported EC properties of a series of organic materials with simple molecular structure, such as dimethyl terephthalate [1]. The effects of temperature, time, and voltage on the properties of the materials were studied.

Common organic small molecule EC materials include viologen, metal organic chelate (such as peptide cyanine) with more research, and other esters, ketones, acids, and other derivatives, which have been studied in recent years. Viologen is introduced in Chapter 6. This chapter mainly introduces other types of small molecule organic electrochromic (OEC) materials. Compared with inorganic small molecule and organic polymer EC materials, organic small molecule EC materials have the following advantages: (i) high purity and easy purification; (ii) rich, long-lasting, and bright color; (iii) simple synthetic route and broad prospects for industrialization; (iv) film can be formed by both solution method and electrochemical method; (v) they can be used at both cathode and anode; (vi) can be treated with aqueous solution; (vii) and structure and properties adjustability by molecule design. However, small molecule EC materials still have some undeniable defects,

Organic Electronics for Electrochromic Materials and Devices, First Edition. Hong Meng.

Figure 3.1 Commonly utilized device structure for small molecule electrochromic materials.

such as high discoloration voltage, poor cyclicity, and high sensitivity to water and oxygen.

Typical EC devices generally have five layers, including transparent conductive oxide (TCO) layer, ion storage (IS) layer, ionic conductive layer (electrolyte), EC layer, and another layer of TCO, which can be simplified into four or three layers under special situations. The EC device of soluble small molecule EC materials are usually fabricated with sandwich structures as shown in Figure 3.1. Using indium tin oxide (ITO) glass as electrode, electrolyte and EC material can be dissolved in a specific solvent and injected between two layers of glass. In this chapter, we will introduce the types, characteristics, and applications of small molecule EC materials in detail according to the material classification.

3.2 Technology Development of Small Molecule Electrochromic Materials

In Figure 3.2, the types of small molecule EC materials are used as abscissa and the development process as ordinate. The general development process of small molecule EC materials is classified and ordered, which are mainly divided into violene–cyanine hybrids, terephthalate/isophthalate derivatives, methyl ketone derivatives, fluoran dye derivatives, and triphenylamine (TPA) dye derivatives. With gradually deepening of researching, not only the types of small molecule EC materials are gradually enriched but also their stability is improved in the process, from the study of Kenichi Nakamura et al. on the electronic properties of phthalic acid in 1987 to the preparation and research of bistable EC materials and modification of electrodes in recent years. The practicability and stability of small molecular EC materials are gradually improving with a better industrial prospect.

3.3 Violene–Cyanine Hybrids (AIE PL OEC)

The violene–cyanine hybrid material is an EC material with high color intensity and high thermodynamic stability. The first report was published in 1999, where Siegfried Hünig et al. reported several violene/anthocyanin structures. As shown in Figure 3.3, in these series the function of the cyanine structure is to provide color; the oxidation levels are −4, −2, 0, 2, and 4 [5].

Violene–cyanine hybrids materials are classified into closed-shell and open-shell materials. Free radical reactions are usually associated with low activation energy. The open-shell system decomposes faster than similar closed-shell system, so

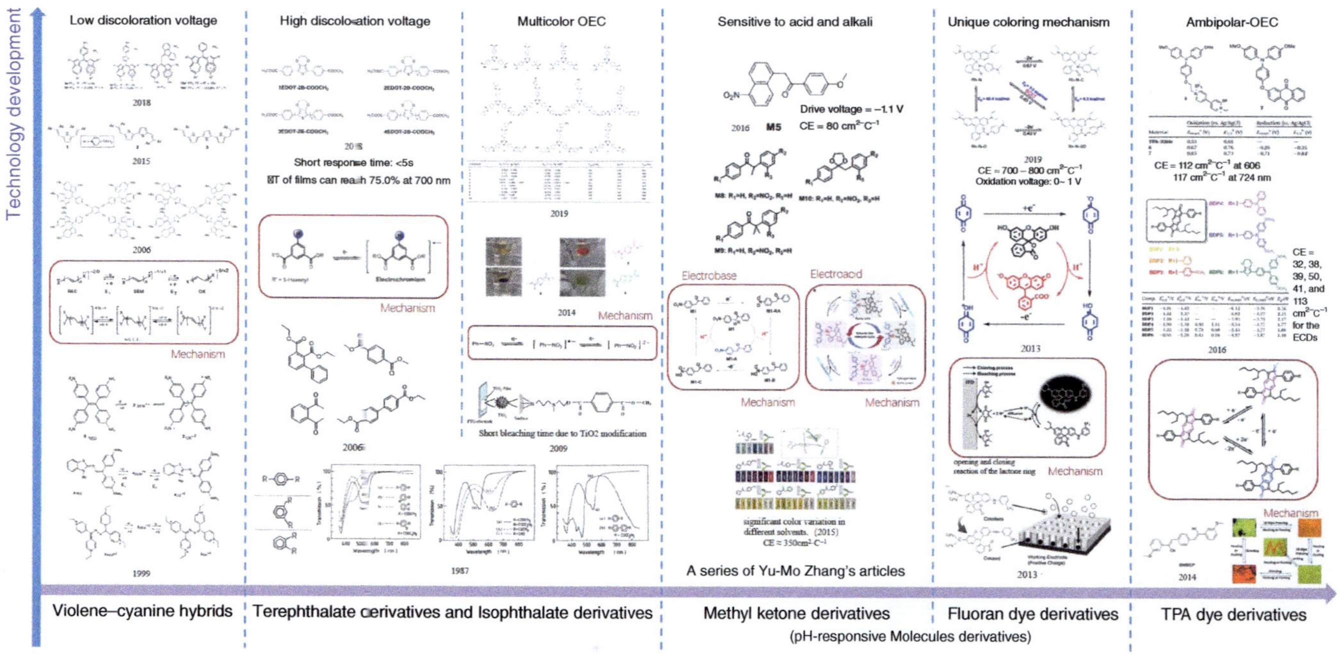

Figure 3.2 Technology development of small molecule electrochromic materials. Source: (C2) Reproduced with permission of Xu et al. [2] from Copyright © Royal Society of Chemistry 2014, (D2) Reproduced with permission of Zhang et al. [3] from Copyright © Springer Nature 2019, (F2) Reproduced with permission of Sun et al. [4] from Copyright © Royal Society of Chemistry 2014.

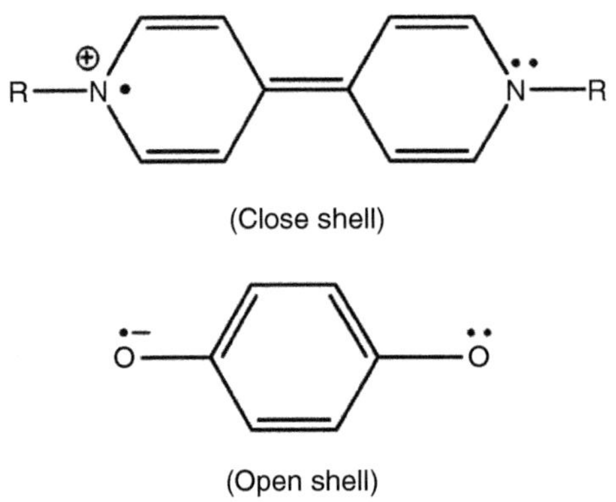

Figure 3.3 General structure for violene-type electrochromics.

the stability can be improved by using the free radical structure of closed-shell molecules. In the electric field, the structure of violene–cyanine hybrid material changes as shown in Figure 3.3.

By changing the substituents, including phenothiazine, phenylzine, phenoxathiazine, thianthracene, aromatic dicarboxylic acid ester, etc. or replacing the C atom of intermediate vinyl group with N atom, the structure and properties of the material can be adjusted. According to this principle, researchers can design EC small molecules with specific color and potential range.

The ECD remains strong fluorescence of film in its yellow neutral state, and most importantly, its fluorescence can be switched on/off by voltage along with its color. These materials show synchronous fluorescence switching and EC switching of the ECD (Figure 3.4).

In 2006 and 2018, Shunji Ito et al. reported two series of cyanine–cyanine hybrid materials. One series is the tetracations composed of di(1-azulenyl)-methylium units based on a new structural principle of a cyanine–cyanine hybrid for the design of

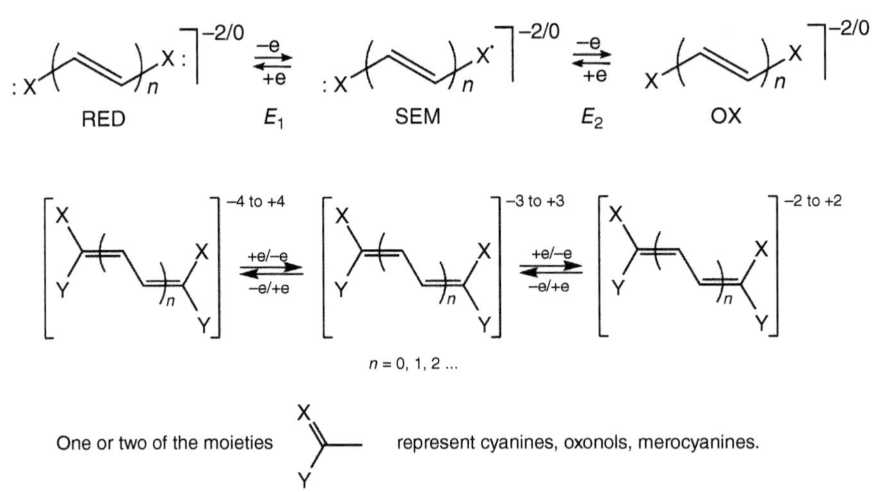

Figure 3.4 Electrochromic mechanism of violene–cyanine hybrid material. Source: Siegfried Hünig et al. [6].

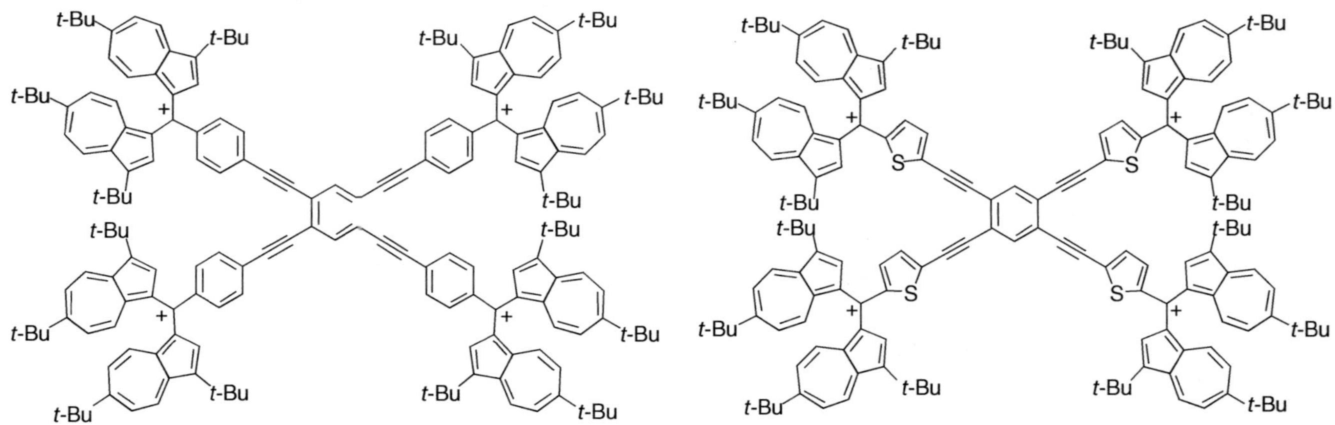

Figure 3.5 Structure of cyanine–cyanine hybrid materials. Source: Shunji Ito et al. [7]. © 2006, American Chemical Society.

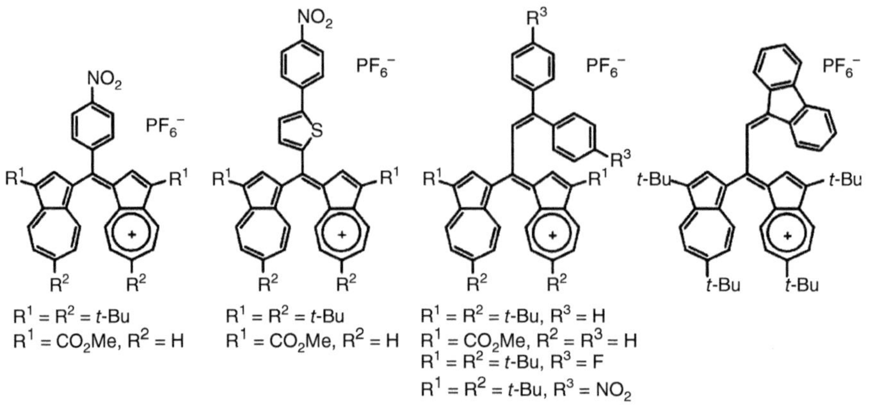

Figure 3.6 Structure of cyanine–cyanine hybrid materials. Source: Ito et al. [8]. Licensed under CC-BY-4.0.

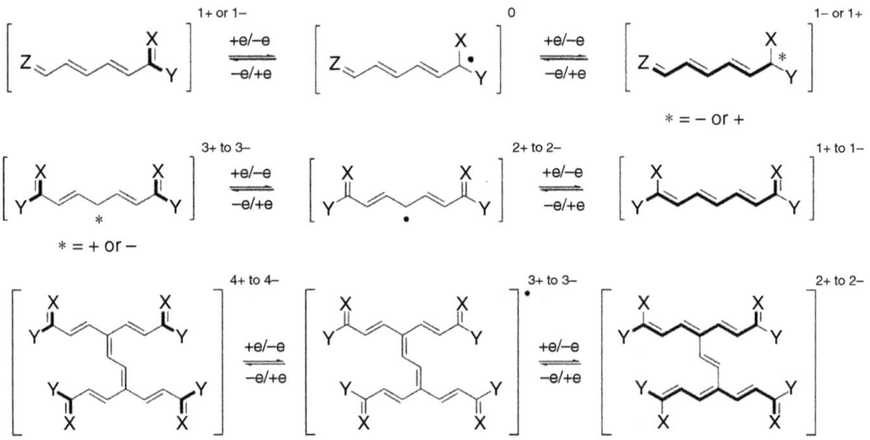

Figure 3.7 Structure of the cyanine–cyanine hybrid with a cyanine unit at one/both terminal. Source: Shunji Ito et al. [7]. © 2006, American Chemical Society.

EC materials with two color changes in 2006 (Figure 3.5). And the other one is two types of cations composed of di(1-azulenyl)methylium units based on the new structural principle of cyanine–cyanine (C–C) hybrid in 2018 (Figure 3.6). This kind of material has a lower driving voltage, usually less than 1 V. The discoloration mechanism is shown in Figure 3.7. However, the electrochemical behavior was not ideal for the presumed C–C hybrid. Therefore, the effects of substituents and intermolecular forces on the stability of EC materials should be considered in the design of such structures.

In 2016, Yusuke Ishigaki et al. designed new violene/cyanine hybrid-type electron donors, which have two 4-methoxyphenyl groups each on the two cyanine parts and undergo reversible two-stage one-electron oxidation (as shown in Figure 3.8). These materials were analyzed by X-ray, and the results revealed that the violene system consisting of thiophene unit(s) and the ethenyl groups is nearly planar.

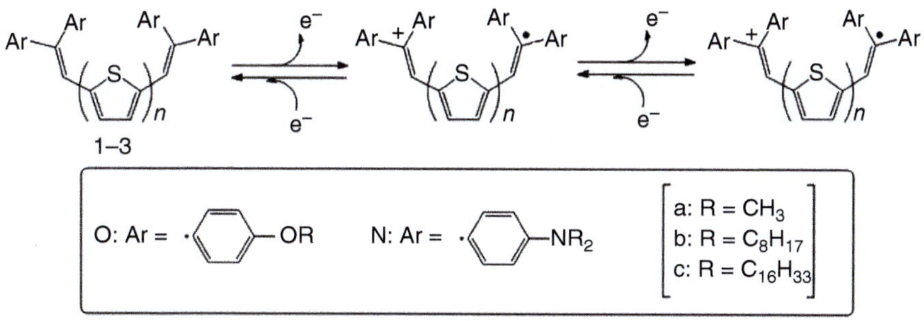

Figure 3.8 Mechanism of violene–cyanine hybrids.

And they proposed that a short contact between thiophene sulfur atom (S) and the ipso carbon (C_{ipso}) of the aryl group is highly likely to be associated with electronic interaction between σ* (C–S) and π (C_{ipso}), resulting in conformational stability for the coplanarly extended geometries, which are suitable for electronic delocalization over the extended π-systems [9]. They also proposed that the number of thiophene units can be used as a variable to modify and tune the spectral properties

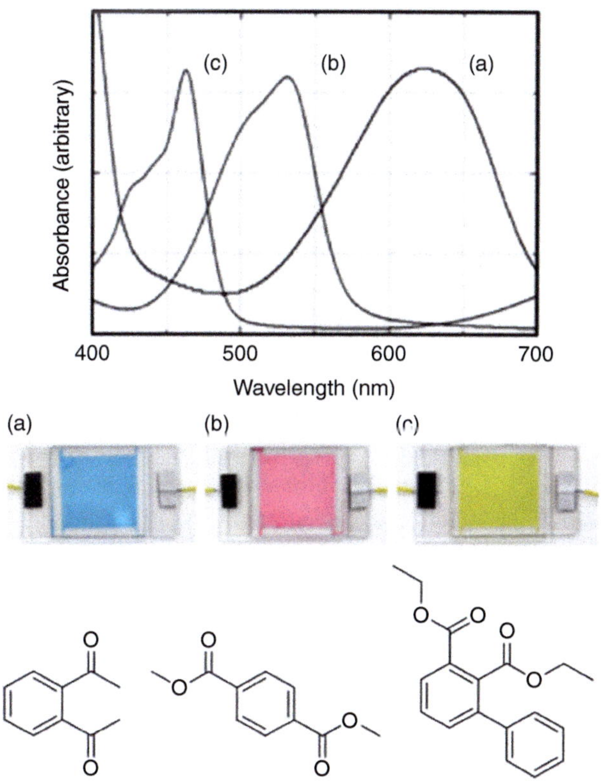

Figure 3.9 Structure, absorption spectra, and images of colored cells. Source: Reproduced with permission of Kobayashi et al. [10] from Copyright © Elsevier 2008.

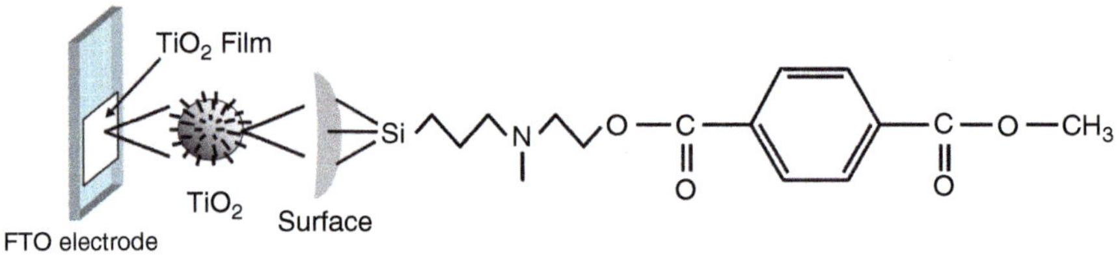

Figure 3.10 Structure of the TP film. Source: Tanaka et al. [11].

(UV–Vis–near-infrared [NIR], **fluorescence spectrumetry** [FL]) as well as the **lowest unoccupied molecular orbital** (LUMO)-level without changing the **highest occupied molecular orbital** (HOMO)-level (Figures 3.9 and 3.10).

Violene–cyanine hybrid materials are able to adjust the electrochemical and photochemical properties by adjusting the oxidation level. They have high color intensity and high thermodynamic stability. For molecules with similar structure, the free radical structure with closed-shell structure can improve the stability of materials. However, the influence of substituents and intermolecular forces on the stability of EC materials should be considered in molecular design. Otherwise it's difficult to get satisfactory materials and properties.

3.4 Terephthalate Derivatives (Multicolor OEC)

Phthalates and their derivatives are a kind of EC materials with various colors and high adjustability. They usually show transparent color in undoped state and turn to different colors including tricolor in the effect of external electric field. Therefore, it can meet the current market demand for multicolor EC materials. Because of its good solubility, it can combine flexible substrates to produce flexible electronic products such as electronic paper.

In 1987, Kenichi Nakamura et al. have studied the EC properties of phthalic acid. In 2008, Norihisa Kobayashi et al. synthesized a series of materials, which show various colors including tricolors (cyan, magenta, and yellow) using acetylbenzene (Figure 3.11), dimethyl terephthalate, and diethyl diphthalate as raw materials [10]. In the experiment they found that the stability of this material is greatly affected by oxygen, water, and applied voltage. So it is necessary to dehydrate, deoxidize, and do sealing treatment well while fabricating devices.

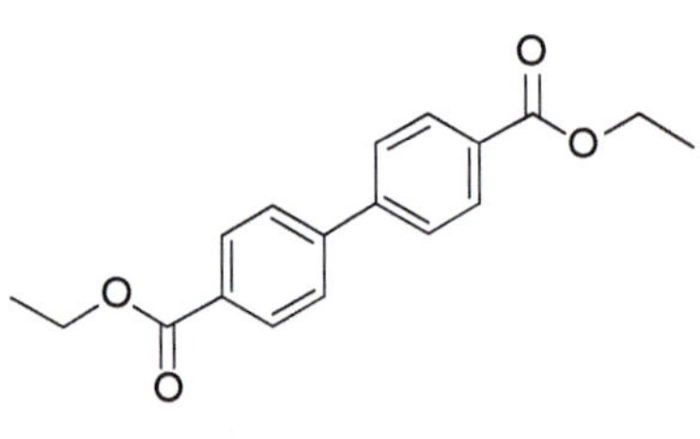

Figure 3.11 Structure of biphenyl ester compounds with larger conjugated system.

(a)

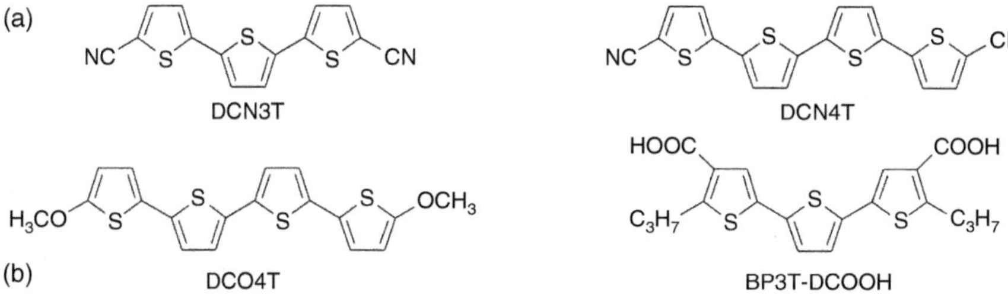

(b)

Oligothiophene derivative	$E_{ox}{}^{a}$/V (vs Ag/Ag$^+$)	$E_{red}{}^{b}$/V (vs Ag/Ag$^+$)
DCN3T	1.16	−1.5
DCN4T	1.22	−0.93
DMO4T	1.23	−1.31
BP3T-DCOOH	1.93	−0.83

(c) [a] The first oxidation potential, [b] the first reduction potential.

Figure 3.12 Structures of oligothiophene derivatives, 5,5″-dicyano-2,2′: 5′,2″-terthiophene (DCN3T), 5,5‴-dicyano-2,2′:5′,2″:5″,2‴-tetrathiophene (DCN4T), 5,5‴-dimethoxy-2,2′:5′,2″:5″,2‴-tetrathiophene (DMO4T), and 4,4″-bicarboxyl-5,5″-bipropyl-2,2′: 5′,2″-terthiophene (BP3T-DCOOH) (a) and their electrochromism properties (b). Source: La et al. [12].

It is well known that viologen is an EC material with higher commercialization level at present. How to achieve a faster response as viologen system when using phthalate derivatives (PDs) is a valuable research issue (Figures 3.12 and 3.13).

In 2009, Satomi Tanaka et al. proposed a strategy to solve this problem. PDs can be modified onto electrodes [11]. On the base of the experimental research, the **dimethyl terephthalate (**DMT) solution still has good switching performance after 150 switches using the modified electrode. These materials have the advantages of simple synthesis and low cost but also have the disadvantages of high drive voltage and low stability. In order to improve the stability, ferrocene can be added to the electrolyte solution as an electron donor (i.e. anti-redox material). Such materials and electrolyte are usually sandwiched between ITO electrodes in the fabrication of devices. While fabricating, sealing measures should be taken to prevent the influence of water and oxygen on the stability and color of the material.

The discoloration mechanism of diester is realized by two gains and losses electronic stages. Alkoxyl and benzene ring can produce synergistic electron effect. If changing the core of the molecule to biphenyl group, a larger conjugated system can be formed because the two phenyl rings of biphenyl ester compounds are in the same plane. Besides, by using NiO-modified electrodes, it can achieve high coloration efficiency and long-term switching stability.

On the basis of small molecules, some researchers have also studied the polymer EC materials based on benzene/diphenyl ester derivatives. In 2011, Yuichi Watanabe et al. studied the EC properties of the cross-linked phthalate polymer (polyethylene

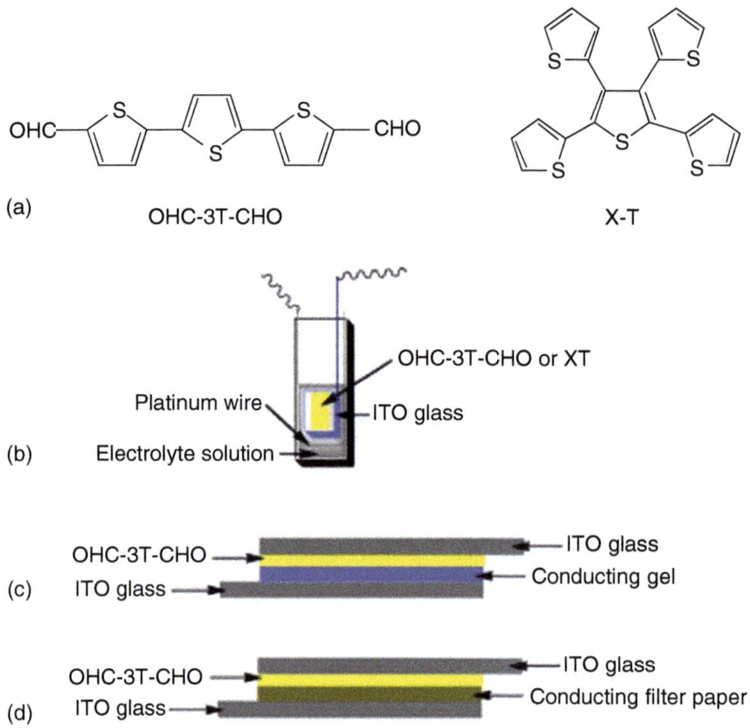

Figure 3.13 (a) Molecular structures of OHC-3T-CHO and X-T. (b–d) Schematic diagram of electrochromic devices. Source: Yin et al. [13].

terephthalate [PET] derivative)-modified electrode. By electrochemical doping process, PET showed a color change from transparent to magenta. Compared with monomer DMT, the electrodes modified by PET film have better coloring and memory properties and switching stability. It also proves that small molecule EC materials still have many defects, which need further study and improvement.

3,4-ethoxylene dioxy thiophene (EDOT) is a commonly used EC monomer material. Comparing with the extensive research on EC materials of polythiophene derivatives, EDOT oligomers studies on oligomers of EDOT are relatively less popular. Oligomers containing EDOT and an aromatic group films were observed that they exhibited reversible color changes upon electrochemical doping and dedoping.

In 2008, La Ming et al. reported the EC properties of four oligothiophenes. Oligothiophenes with well-defined structures have recently received a great deal of attention not only as model compounds for conducting polythiophenes but also as a new class of functional π-electron systems. They have potential application to field-effect transistors, photovoltaic devices, and organic electroluminescent devices. All of the materials exhibited two anodic oxidation and one cathodic waves. These materials undergo reversible clear color changes on electrochemical doping and dedoping. The cyclic voltammetry (CV) test showed the electrochemical properties of the

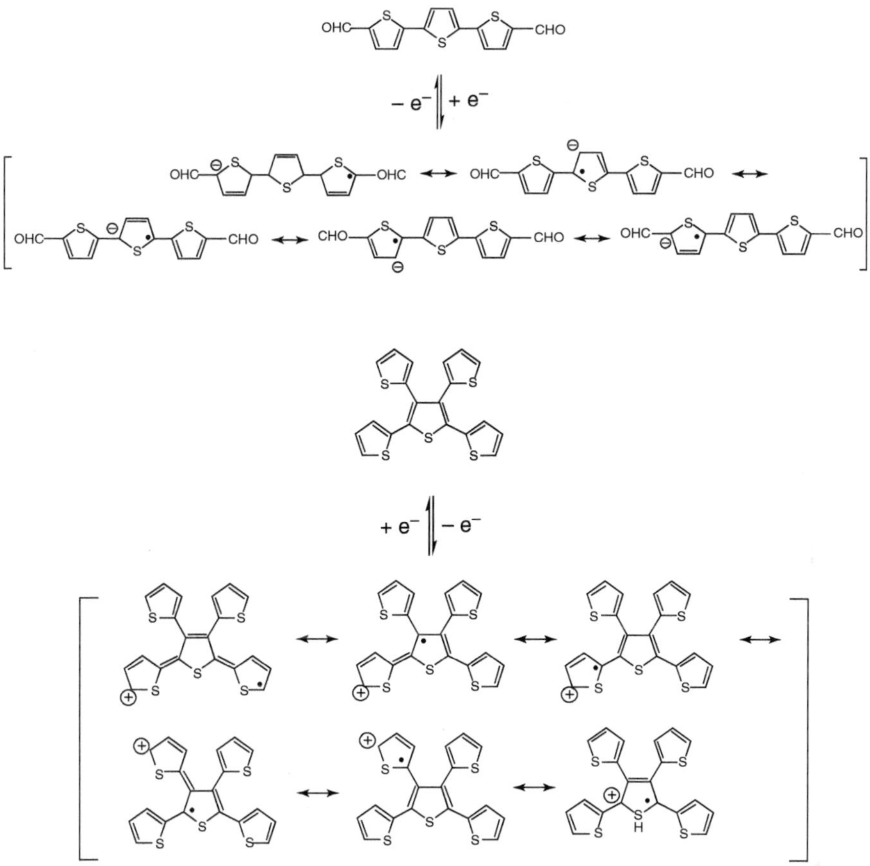

Figure 3.14 Mechanism of oligothiophene derivatives. Source: Yin et al. [13].

oligothiophene derivatives as shown in Figure 3.14. But the mechanism was not well described in their report [12].

In 2010, Binbin Yin et al. synthesized oligothiophene derivatives, 5,5″-biformyl-2,2′:5,2″-terthiophene (OHC-3T-CHO) and 2,3,4,5-tetrathiopenyl-thiophene (X-T) (Figure 3.14), and studied their EC properties as cathode coloring materials [13]. They designed and fabricated three ECDs. In different devices, the material will show slightly different colors after redox process. The ITO/OHC-3T-CHO/electrolyte solution/ITO device exhibited reversible, clear color change from yellow to bright bluish green on electrochemical doping and dedoping, while ITO/OHC-3T-CHO/conducting gel/ITO device and ITO/OHC-3T-CHO/conducting filter paper/ITO device exhibited color change from yellow to green. The ITO/X-T/electrolyte solution/ITO device showed reversible, clear color change between yellowish orange and silver gray on electrochemical doping and dedoping (Figure 3.15).

The EC mechanism of these two materials was analyzed and characterized (Figure 3.16). OHC-3T-CHO is a cathode coloring material. In the process of

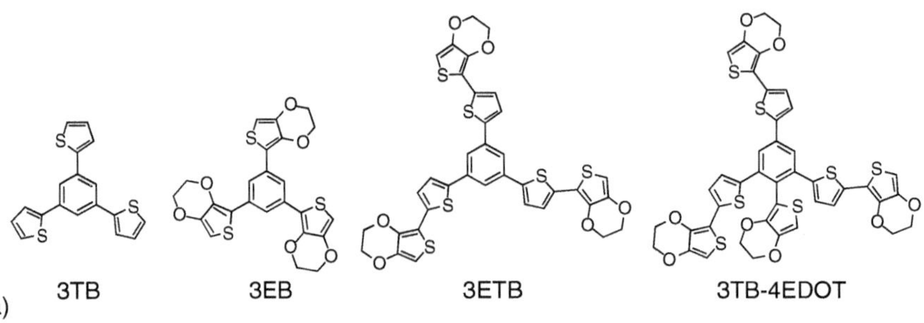

(a)

Table 1. Photoelectric properties data of 3TB, 3EB, 3ETB, and 3TB-4EDOT.

Compound	λ_{max} (nm) (absorption)	λ_{edge} (nm)	λ_{max} (nm) (emission)	$E_g^{opt\,a)}$ (eV)	E_{ox}^{onset} (V) vs SCE	E_{red}^{onset} (V) vs SCE	$E_g^{electro\,b)}$ (eV)
3TB	293	330	370	3.76	1.5	−0.85	2.35
3EB	302	345	378	3.54	0.85	−0.75	1.60
3ETB	369	426	423	2.91	0.65	−0.80	1.45
3TB-4EDOT	367	426	448	2.91	0.66	−0.75	1.41

a) E_g^{opt} (eV) = 1240/l_{edge}; E_g^{opt}: the optical band gap.

b) $E_g^{electro}$ (eV): the electrochemical band gap.

(b)

Figure 3.15 (a) Molecular structures of 3TB, 3EB, 3ETB, and 3TB-4EDOT. (b) Photoelectric properties data of 3TB, 3EB, 3ETB, and 3TB-4EDOTs. Source: Zeng et al. [14].

electrochemical doping, OHC-3T-CHO obtains an electron and forms anionic free radicals corresponding to the 630 nm new absorption band. In electrochemical denucleation, the anionic radical of OHC-3T-CHO loses an electron and changes the neutral OHC-3T-CHO, corresponding to yellow. When + 4.6 V is applied between ITO electrode and platinum wire electrode in electrolyte solution, one electron is lost in electrochemical doping, and cationic free radicals corresponding to new absorption bands at 412, 520, and 553 nm are formed. When 4.6 V is applied between ITO electrode and platinum wire electrode in electrolyte solution, the cation radical of X-T obtains an electron and changes the neutral X-T, which corresponds to the yellow orange color.

Faster response time was achieved using these two materials, but due to the lack of time testing equipment, the accurate response time has not been tested yet in this work, but this material provides a strategy to reduce the response time of EC materials.

In 2017, Jinming Zeng et al. studied a series of star oligomers containing thiophene or EDOT (Figure 3.17). Studies have shown that, compared with organic ester of small molecular weight, these oligomers exhibit lower doping voltage and significantly less number of cycles. But the color of these oligomers was not as rich as that of ester materials. Oligomers are formed in the process of electrochemical doping.

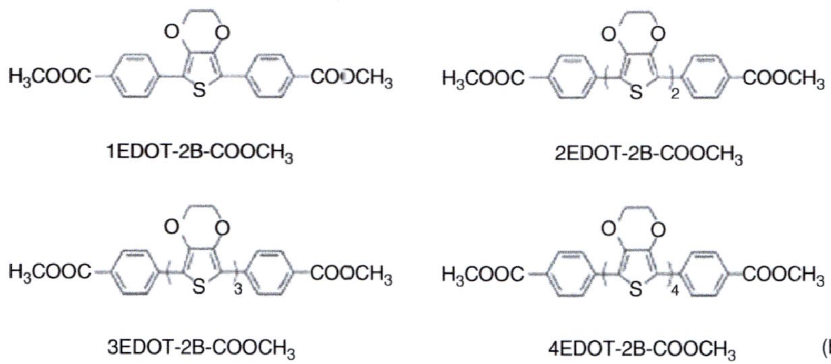

(a)

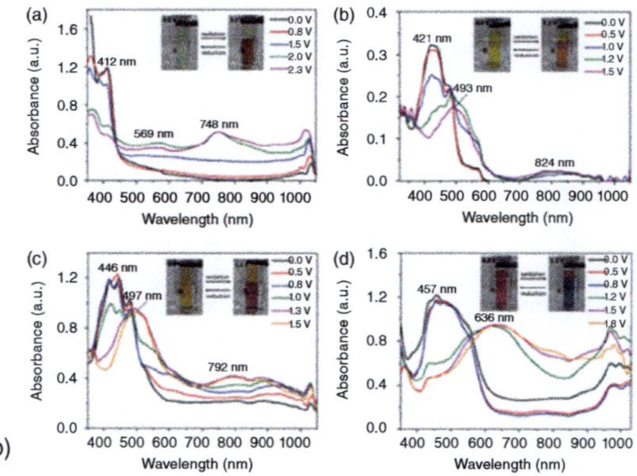

(b)

Compounds	λ_{max} (nm)		λ_{odg} (nm)		E_g^{opt} (eV)	E_{ox}^{onset} (V)	E_{red}^{onset} (V)	E_{HOMO} (eV)	E_{LUMO} (eV)	E_g^{elec} (eV)
	Sol	Film	Sol	Film						
1EDOT-2B-COOCH$_3$	374	391	425	442	2.81	1.25	−1.14	−5.65	−3.26	2.39
2EDOT-2B-COOCH$_3$	428	438	482	508	2.44	0.87	−0.69	−5.27	−3.71	1.56
3EDOT-2B-COOCH$_3$	443	451	525	562	2.21	0.30	−0.75	−4.70	−3.65	1.05
4EDOT-2B-COOCH$_3$	485	493	562	596	2.08	0.37	−0.75	−4.77	−3.65	1.12

(c)

Figure 3.16 (a) Molecular structures of 1EDOT-2B-COOCH$_3$, 2EDOT-2B-COOCH$_3$, EDOT-2B-COOCH$_3$, and 4EDOT-2B-COOCH$_3$ (b) and (c) photoelectric properties of the oligomers. Source: Reproduced with permission of Wan et al. [15] from Copyright © John Wiley and Sons 2018.

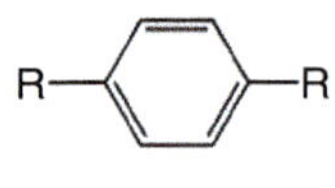	$COOCH_3$ $COOC_2H_5$ CHO $COCH_3$	Dimethyl terephthalate (*p*-DMP) Diethyl terephthalate (*p*-DEP) Terephthalate aldehyde (*p*-PA) *p*-Diacetyl benzene (*p*-DAB)
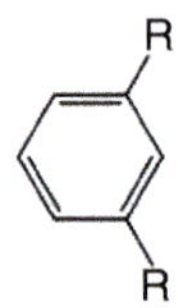	$COOCH_3$ CHO $COCH_3$	Dimethyl *m*-phthalate (*m*-DMP) Isophthalaldehyde (*m*-PA) *m*-Diacetylbenzene (*m*-DAB)
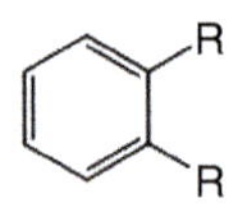	$COOCH_3$ $COOC_2H_5$	Dimethyl phthalate (*o*-DMP) Diethyl phthalate (*o*-DMP)

Figure 3.17 Structure of common isophthalate derivatives.

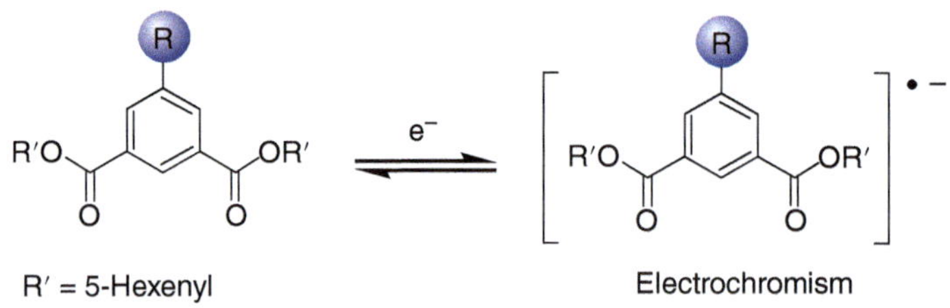

(a)

Compounds R′ =	Redox potential[a] vs Ag/Ag$^+$(V)	Applied DC voltage[b] (V)
H, M1	−2.68 (rev)[c]	5.05
Ph, M2	−2.55 (rev)	5.14
4-Fluorophenyl, M3	−2.58 (rev)	5.20
4-Methoxyphenyl, M4	−2.62 (rev)	5.54
4,4-Biphenyl, M5	−2.57 (rev)	4.38
2-Thienyl, M6	−2.63 (rev)	5.15
CH_3, M7	−2.72 (rev)	5.32
OM_e, M8	−2.61 (rev)	5.21
NO_2, M9	−1.88, −2.71, −2.81 (rev)	4.17, 5.19
Nitrobenzene	−2.04, −3.08	4.38, 5.27

(b)

Figure 3.18 (a) Electrochromism of 5-substituted isophthalate derivatives and (b) redox potentials of compounds in cyclic voltammetry and applied voltage for color display. Significant (a) represents redox potential vs. Ag/Ag$^+$ (reference electrode) determined in a conventional three electrodes cell by using 0.2 M tetrabutylammonium hexafluorophosphate as the supporting electrolyte in γ-butyrolactone, ITO glass as the working electrode, and platinum as the counter electrode, (b) represents applied DC voltage for color displays from sandwich-type ITO electrode cells, 0.2 M tetrabutylammonium hexafluorophosphate as supporting electrolyte in γ-butyrolactone, (c) represents reversible peak. Source: Sharmoukh et al. [16].

The films of these oligomers exhibit reversible color changes under electrochemical doping and dedoping with a shorter discoloration/bleaching time. Compared with ordinary linear oligomers, star oligomer has higher ionic radical stability. But they also found that except P3ETB, these oligomer films could not change back to their intrinsic colors upon electrochemical dedoping. Only the P3ETB film changed color reversibly from orange to light green, which showed a short response time and good electrochemical stability. The doping and dedoping times of the P3ETB film were 1.2 and 0.9 seconds, respectively [14].

In 2018, Zhijun Wan et al. published a research on flexible multicolor ECDs driven by low voltage. Four EC materials containing 1–4 EDOT units were synthesized and blocked with methyl benzoate (Figure 3.18a) [15]. Some flexible organic ECDs are fabricated using these oligomers and the EC properties were studied. These oligomers exhibit reversible color changes after electrochemical doping and dedoping. The device structure is ITO–PET plastic slide/active layer/conducting gel/ITO-PET. The maximum absorption peaks of these oligomer films were located at 391, 438, 451, and 493 nm, respectively. With increasing EDOT component, a red shift was observed in the electronic absorption spectra of solution and thin films. This was due to the increase of conjugated length and the broadening of backbone planarity. In addition, the red shift of the thin films was more pronounced than that in solution due to molecule aggregation causing increased conjugation. Among the four compounds, 4EDOT-2b-$COOCH_3$ showed the highest optical contrast, which is 75.2% at 700 nm and response time is less than five seconds.

For all oligomers, the electrochemical bandgaps were less than the optical bandgaps because the electrochemical determination results in the formation of charge carriers. According to the CV curves, the EC mechanism was speculated that the oxidative coloring of 1EDOT-2B-$COOCH_3$ was attributed to the formation of cationic radical, while the oxidative coloring of the other oligomers was attributed to the formation of bipolaron. In the bipolaron structure, the positive charges are predominantly localized in benzene. These four oligomers can show clear and reversible color changes under the applied electric field. If several or four of them are mixed together and integrated on one device, the overlapped part will display a color, which is close to black.

3.4.1 Conclusion

Terephthalate derivatives is a series of materials, which is easy to synthesize, and they can show various colors by adjusting the molecule structure. They are very suitable for commercialization because of their simple synthesis route and purification. However, there are some crucial problems needing further research such as low stability, low cycle number, and water-oxygen sensitivity. It is proved that the increase of conjugate structure area is good for increasing the stability in the published literature. But there should be a balance of the conjugation area and the processing performance because the increasing conjugation structure will decrease the solubility of materials.

3.5 Isophthalate Derivatives

Isophthalate derivatives are a kind of EC materials with simple preparation process and low cost, but they also have some disadvantages such as low stability and cycle times. Therefore, besides studying the properties of materials themselves, strategy of improving the coloring ability of such materials has become a research hot spot.

In 1987, Kenichi Nakamura et al. studied the EC properties of phthalic acid derivatives but did not carefully study the mechanism. But their research laid a foundation for the research of small molecule EC materials.

Compared to viologen and phenothiazines, less attention has been paid to the 1,4-terephthalate derivatives (Figure 3.20). Because functional groups can be easily introduced at the 5-position of these materials and there is no report on the electrochromism of 1,3-isophthalates before this report, in 2008, W. Sharmoukh et al. prepared different 5-substituted isophthalates and observed that the functional groups of 5-position isophthalates had a significant effect on the observed color (Figure 3.21) [16]. Among those materials, electrochemically active nitro group induced polychromatic display, which showed that the use of more kinds of 5-substituted isophthalate derivatives could control further EC performance. Among all the compounds, 5-(2-thiophenyl)-phthalate is pale yellow, while other compounds are neutral and colorless.

CV was performed on each sample to understand the electrochemical behavior using 0.2 M tetrabutylammonium hexafluorophosphate (TBAPF6) as electrolyte and *G*-butyrolactone as solvent. The results showed that these compounds exhibited reversible reduction peaks in the range of 2.5–2.90 V (compared with Ag/Ag^+). For these compounds, it can be reasonably inferred that the longer the length of the conjugated system, the smaller the HOMO–LUMO level and the redshift of the maximum absorption peak. Based on this conjecture, it is inferred that the color can be controlled by introducing a functional group at five sites. The observed color trend of reduced particles is proved by the calculation of time-dependent density functional theory (TDDFT).

As can be seen from the Figure 3.19, **molecular orbitals** (MOs) between LUMO and LUMO + 10 do not participate in the absorption process, which means that the transition of MOs from HOMO to LUMO and LUMO + 10 does not participate in the absorption process. Interestingly, the electron density is mainly distributed on two

Figure 3.19 Calculated HOMO, LUMO, and LUMO + 10 of the reduced species of M1. Source: Sharmoukh et al. [16].

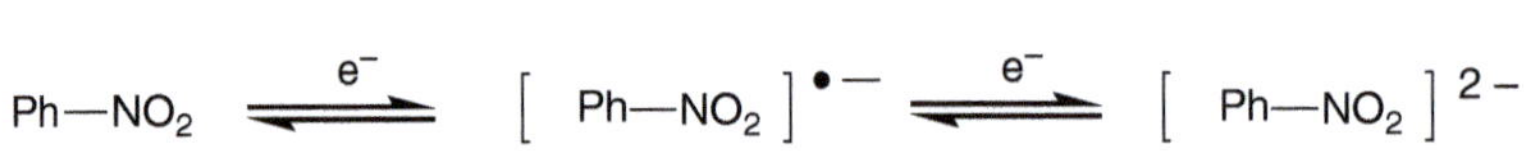

Figure 3.20 Known reduction process of nitrobenzene. Source: Sharmoukh et al. [16].

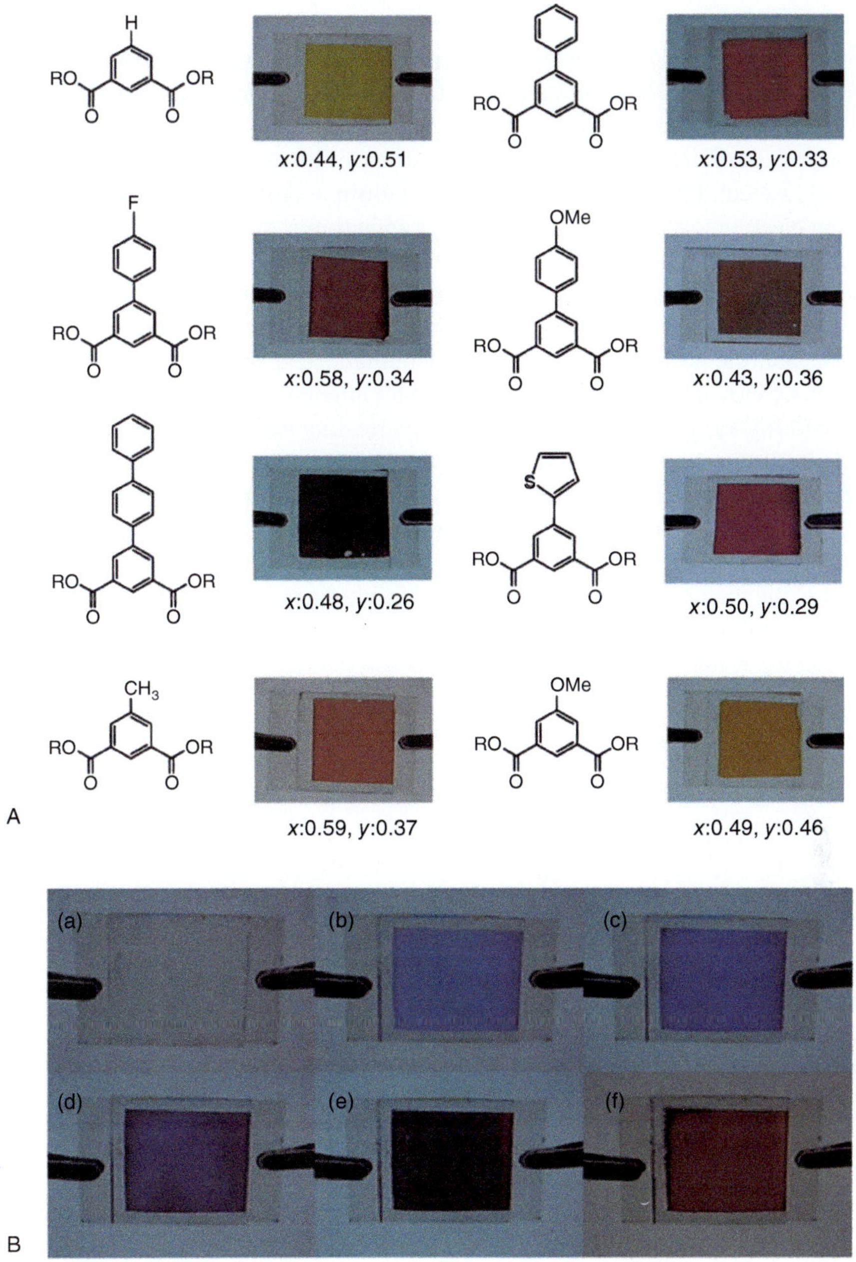

Figure 3.21 (A) Electrochromism of 5-substituted isophthalate derivatives (with the computed coordinates in the CIE chromaticity diagram) and (B) multicolor display of 5-nitro-isophthalate derivative with a supply of DC voltage: before voltage application (a), colors of the first reduced species (b, c), and the second reduced species (d–f). Source: Reproduced with permission of Sharmoukh et al. [16] from Copyright © Royal Society of Chemistry 2008.

esters of MOs between LUMO and LUMO + 10. This may be responsible for the quite pure colors for the 5-substituted isophthalate derivatives investigated. So what happens if an electrochemical active group is introduced? The authors considered the aromatic nitro compounds and prepared quasi solid-state devices using poly(vinyl pyrrolidone) gel electrolytes.

Anionic radicals of aromatic nitro compounds have been a very important electrochemical research subject. The reduction mechanism of aromatic nitro compounds has been electrochemically characterized. As shown in the following, the electrochemical process of nitro compounds consists of two electrons forming two reducing species in one step. The first reducing species is an anionic radical, which can be further reduced to form two ions. Considering the multistep reduction process of nitrobenzene, the author prepared 5-nitroisophthalic acid M9 and studied its EC behavior. As shown in (Figure 3.22), the CV shows the multistep reduction process (relative to Ag/Ag^+) at 1.88, 2.71, and 2.81 V. As mentioned in the literature, in the case of nitrobenzene, two reduction peaks were observed in almost the same mode, and their transfer potentials were 2.04 and 3.08 V (relative to Ag/Ag^+), respectively, which indicated that the reduction of nitrobenzene required higher energy than that of M9. It should be noted that M9 has two additional electron-withdrawing esters. Therefore, the first two reduction peaks of M9 can be attributed to the formation of anionic radicals and diions, which are similar to the principle of nitrobenzene.

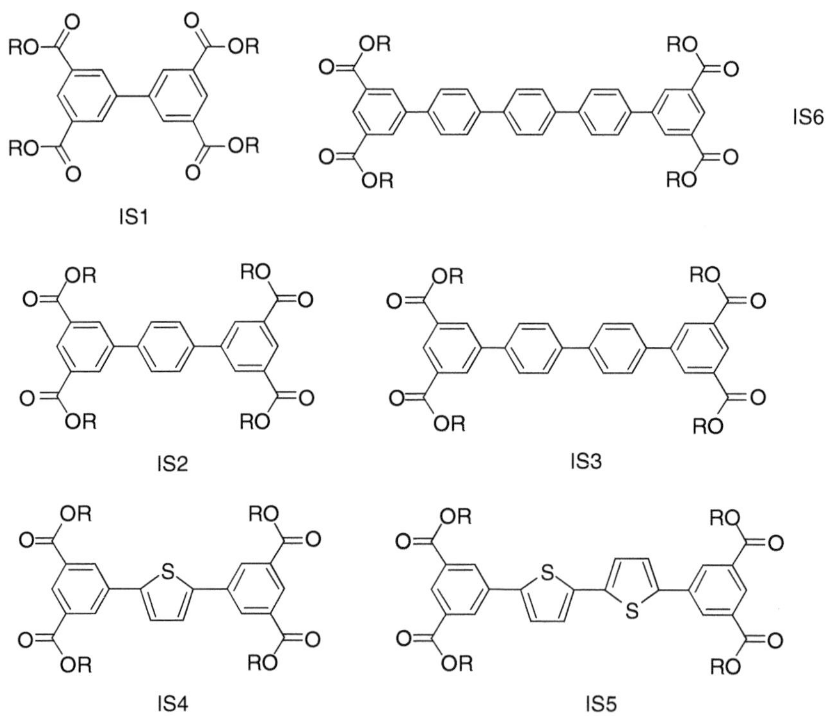

Figure 3.22 Structures of Bis-isophthalate derivative IS1-6. Source: Sharmoukh et al. [17].

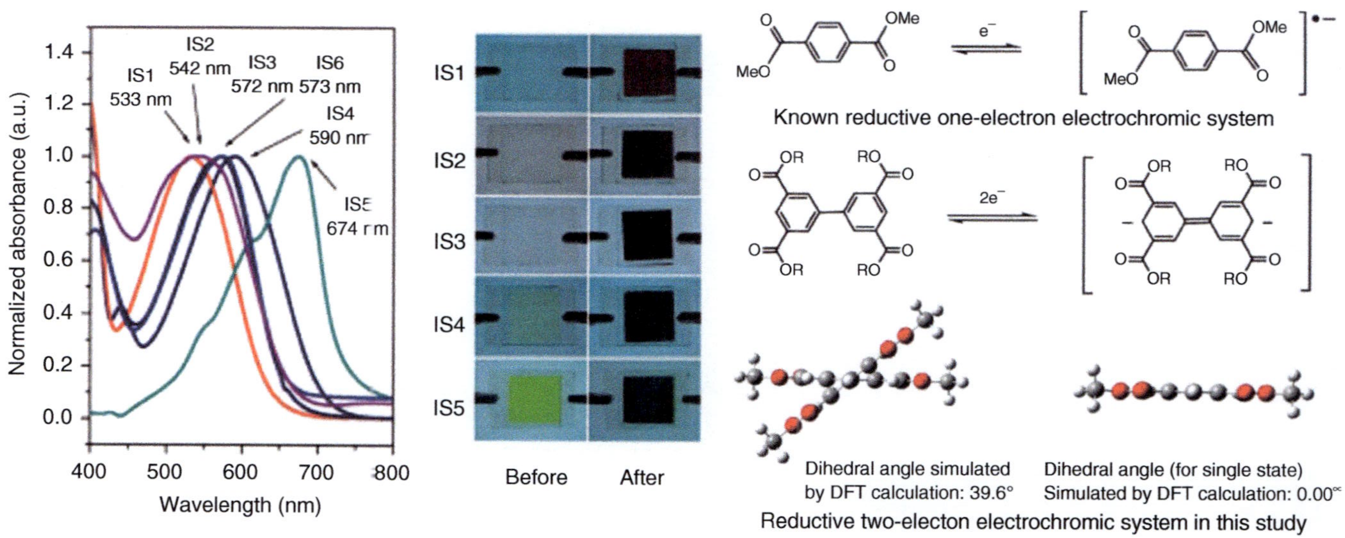

Figure 3.23 UV–visible spectra and photos of displayed colors before and after voltage supply. Source: Reproduced with permission of Sharmoukh et al. [17] from Copyright © American Chemical Society 2008.

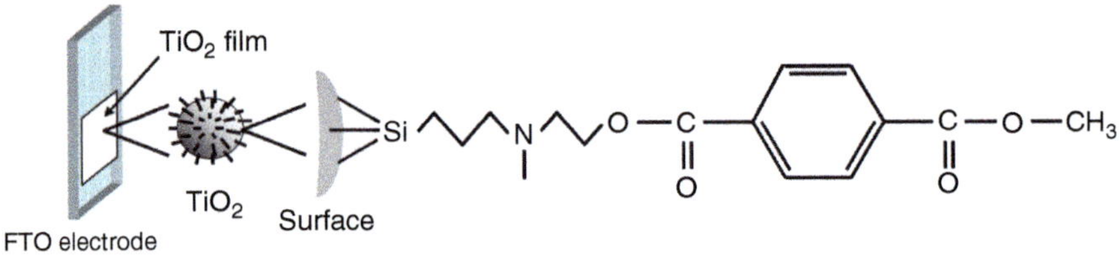

Figure 3.24 Structure of the TP film. Source: Tanaka et al. [11].

The current-voltage curves of M9 and nitrobenzene were obtained by comparing their electrochemical measurements, and two redox peaks were observed. M9 will show bright blue and red appearing at 4.17 and 5.19 V, respectively (Figure 3.23). Simple nitrobenzene showed no detectable color change in the range of 0–6 V, which indicated that two esters played an important role in EC reaction. The authors speculate that the blue and red species are anionic radicals and anions, respectively, but further research and calculation methods are still needed. This study shows that the EC properties can be further controlled by using more 5-substituted isophthalate derivatives. This discovery opens not only the academic door for the study of various spectroelectrochemical methods of nitro radicals and diions but also the door for the development of multicolor display devices.

In 2008, W. Sharmoukh et al. designed and prepared a new EC system based on diester isophthalate (Figure 3.24). New EC systems are more bistable and can be tailored to control the displayed colors. They showed electrochromism behavior by two-electron reduction. The displayed colors showed greatly enhanced bistability and were dependent on the length of the conjugated bridge between the two isophthalate groups. We believe that the connection of two EC systems with a conjugated bridge is a good basic strategy to obtain new bistable EC systems. We believe that this method can be applied to the design of more diverse EC systems.

If two isophthalate groups were connected by a conjugated bridge, the generated radicals in each part could be coupled to form a more stable quinoid dianion, like Chichibabin's hydrocarbon. In the case of the singlet diradical form, it was theoretically anticipated that the dihedral angle (39.6°) between two arenes was decreased to be coplanar via two-electron reduction (Figure 3.25).

The electrochemical properties of materials were studied by CV. It was found that with the increase of the length of the biphenyl bridge, the two separated reduction peaks in the CV curve gradually approached and coincided. The second reduction required higher energy than the first reduction. Then the ECDs were fabricated, whose color was characterized by UV–Vis absorption spectra. It was found that if thiophene group was introduced into the bridge, the absorption peak can be red-shifted (Figure 3.25). Among compounds with thiophene units in the bridge, IS4 is dark blue. Thiophene is widely used in organic electronic systems because of the electron transfer properties. IS4 showed two separate reduction peaks in CV, indicating that there was a simple electronic communication between the two isophthalic acid, groups and the electronic communication is still good while thiophene ring increased [17].

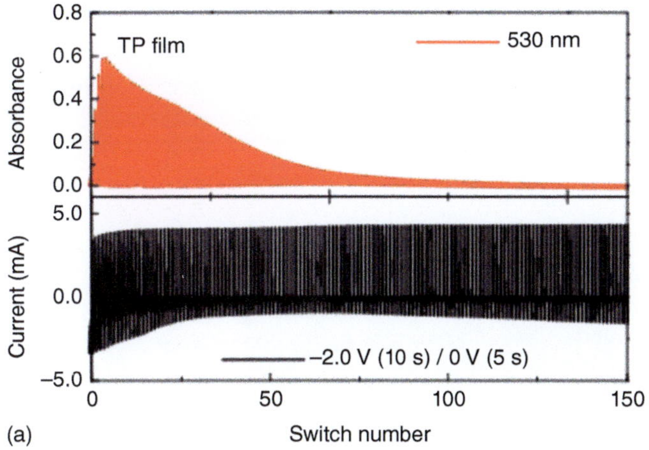

Figure 3.25 Optical (at 530 nm) (top) and electrical current (bottom) responses by stepping the potential for TP film (a) and DMT solution (b). Source: Tanaka et al. [11].

So it indicates that the color and the electronic communication capability is highly dependent on the length of the conjugate bridge. As the length of conjugate bridge increases, the memory effect decreases. This result can help to explain the difference of properties between different structural materials.

How to improve the stability of small molecule EC materials has been a key problem for researchers. If we want to commercialize small molecule EC materials, their stability must be well improved, and the influence of water and oxygen on the materials must be reduced.

Therefore in 2009, Satomi Tanaka et al. studied the EC properties of PDs (TP). By electrochemical reaction, the material can be reversibly changed from colorless and transparent to three different colors. However, the stability of ester small molecules is not as good as that of polymer materials. In order to improve the ability of keeping colored under open-circuit conditions, the terephthalate derivative/TiO_2-modified

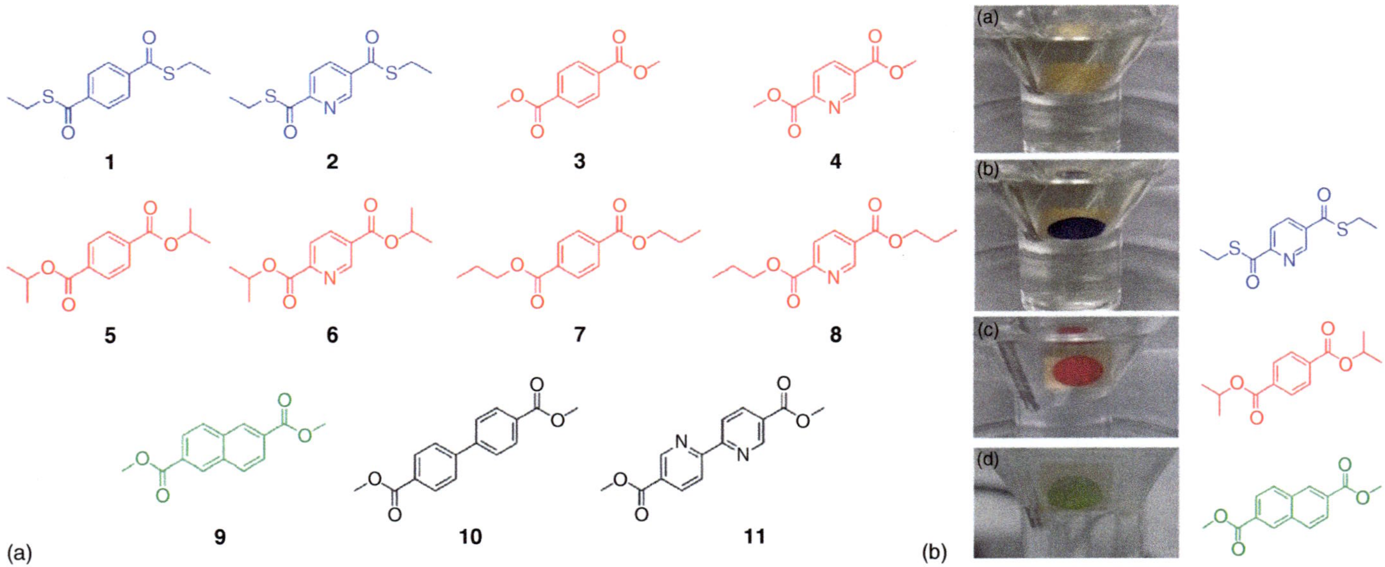

Figure 3.26 (a) Structure of eleven esters synthesized and (b) photographs obtained of a gold micro-mesh electrode in CH_3CN: (a) no applied potential, (b) 10 mM 2, (c) 10 mM 5, and (d) 15 mM9. Source: Reproduced with permission of Xu and Webster [2] from Copyright © Royal Society of Chemistry 2014.

electrode was prepared (Figure 3.26). The method is to modify TiO_2 with PDs [11]. The physically dispersible dissolution of the modified TiO_2 is carried out in a certain order to improve its chemical properties in the solution. By electrochemical reduction, the material exhibits bright magenta with absorption peaks at 530 nm. The color of fuchsin derives from anionic radicals produced by the electrochemical reaction of DMT with phthalate analogues.

The modified electrode has better short-circuit bleaching effect and open-circuit memory effect than the material itself. Cyclic tests showed that DMT solution needed longer bleaching time than TP membrane (Figure 3.27). Electrochemical cycling tests showed that DMT solution needed longer bleaching time than TP film. Because the colored anion radicals of DMT had to diffuse to the electrode to be oxidized, it took a long time (36 seconds) to bleach DMT solution.

The bleaching reaction of TP membrane includes two processes. The first process is due to the oxidation of TP in 0–2 seconds, which results in the disappearance of magenta color. TP should transfer electrons to TIO_2 and then transfer electrons to the electrode. The second process is due to the disappearance of blue–black during two to four seconds, at which time the titanium dioxide that is still in the reduction state should be oxidized.

The modification effect of phthalic acid derivatives on electrodes is similar to that of viologen system. By fourier transform infrared spectrometer (FTIR), it is found that there are significant differences between C=O ester Tensile vibration at 1717 cm^{-1} and C–O ester Tensile vibration at 1273 cm^{-1}. These peaks decrease after switching experiments, indicating that some C=O bonds and C—O bonds disappear on the surface of the electrode. On the other hand, the peak value of benzene did not change at 1473 cm^{-1}. Therefore, ester bonds that are not on the electrode side of the TP benzene ring will break when the potential changes, and this property still needs more experiments to prove. However, it can be inferred that the modified electrode has great potential in the application of color electronic paper.

In fact, the anionic radicals of para-diesters are more stable than the ortho- or meta-substituted forms because of the higher electron density in the aromatic rings, resulting in less nucleophilicity of the OR or/or-SH groups and thus less readily reacting with water or oxygen molecules. In 2014, Xiuhui Xu et al. studied the EC properties of 11 kinds of oxygen/thiodiesters. And three primary color materials were screened out (Figure 3.28) [2].

According to the following electron reduction equation (Figure 3.29f), the nine materials shown in Figure 3.29 can be reduced in acetonitrile solution without oxidation. But not all compounds can achieve chemical reversibility. Because some ions are chemically unstable, which prevents them from being oxidized and reduced to their original compounds. It has been reported that for compounds with long-term stability of free radical anions formed by single electron reduction, free radicals eventually decompose with molecular oxygen (or water) to form carboxylate anions. Voltammetry shows that phenyl ester is more difficult to reduce than similar pyridyl ester derivatives. And with the increase of the length of alkyl chain on ester, the reduction potential becomes slightly negative. When the concentration is 10 mm, the response time of the microgrid electrode in the forward (coloring)

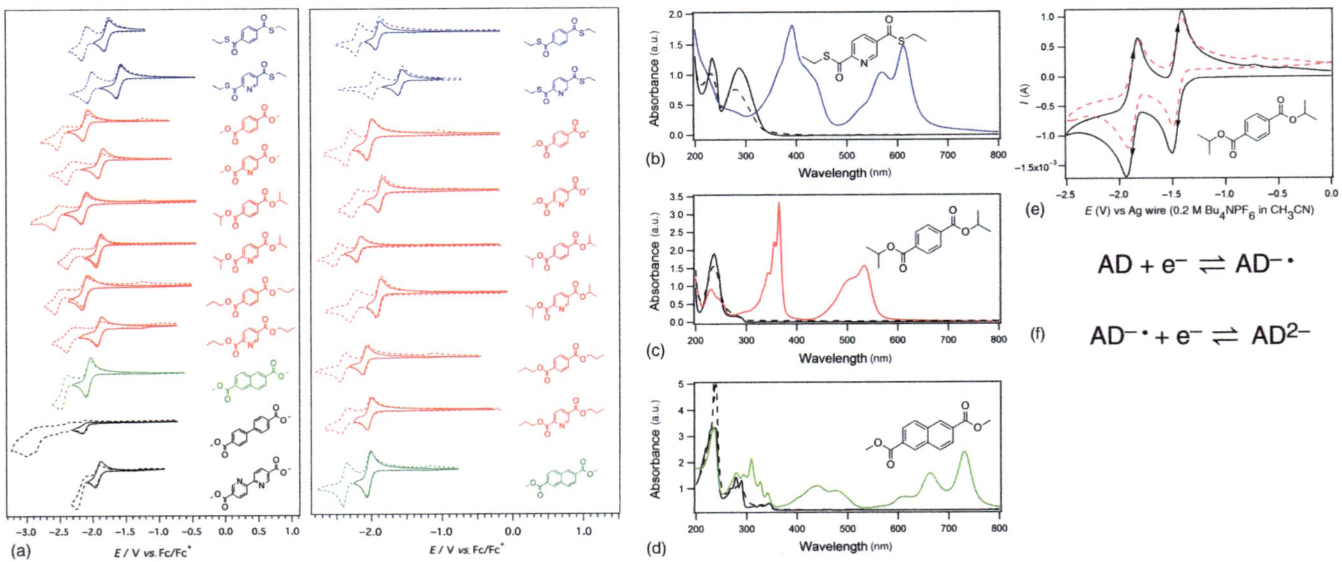

Figure 3.27 (a) Cyclic voltammograms of diesters and (b) in situ electrochemical UV–vis spectra of (b) 2, (c) 5, (d) 9, and (e) cyclic voltammograms recorded at a scan rate of 0.1 V s1 with a 1 mm diameter planar glass carbon (GC) electrode. (f) The reduction reactions of similar electrochromic esters. Source: Xu and Webster [2].

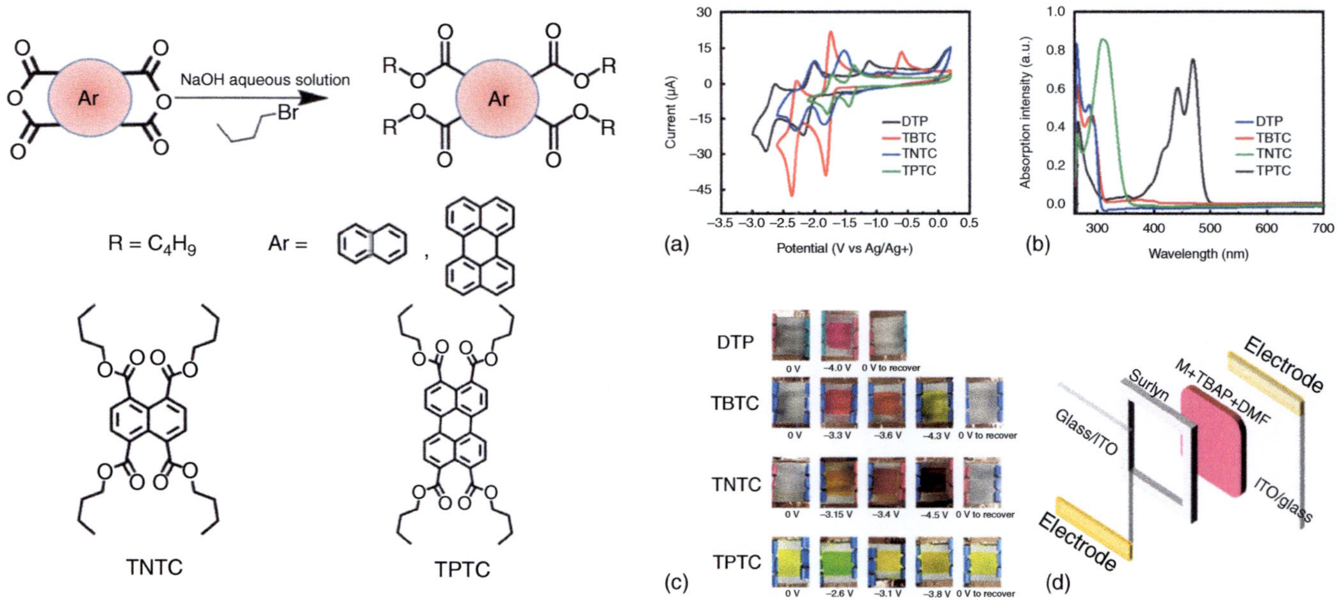

Figure 3.28 Structure, synthesis, and properties of EC materials based on phthalates. Source: Reproduced with permission of Pengfei Zhang [18].

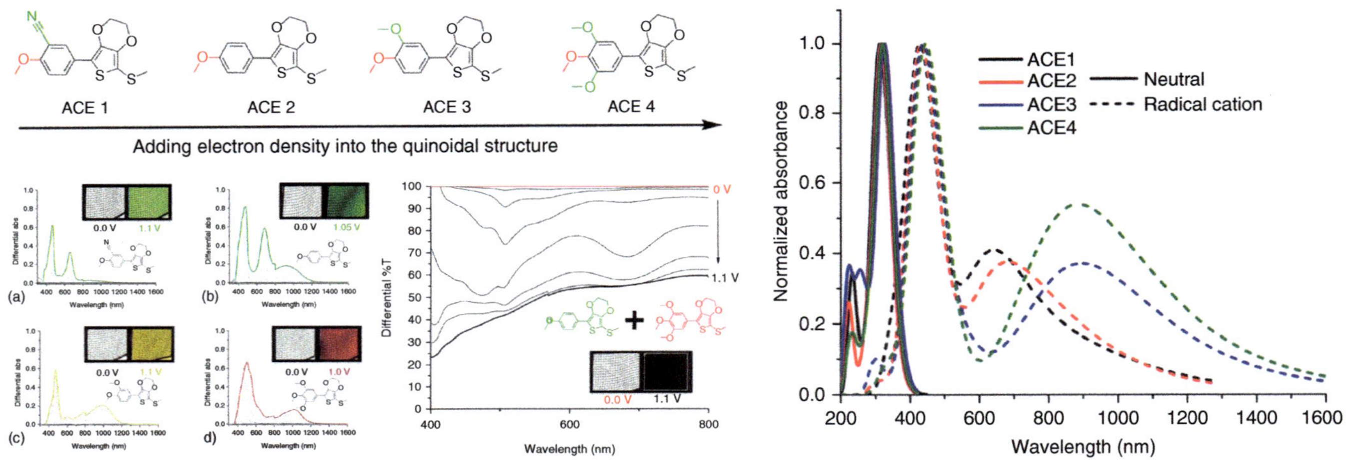

Figure 3.29 Structure and UV spectra of conjugated anodically coloring electrochromic molecules. Source: Christiansen et al. [19].

step is about two seconds. However, although the response in a short time scale (a few seconds) is highly reversible, the lifetime of the colored state in a long time is strongly affected by trace pollutants. Therefore, in order to maintain the coloring state, the solution around the microgrid must be completely free of oxygen and trace water. In order to solve this problem, it is necessary to synthesize compounds with lower reduction potential so as to reduce the reactivity of the reduced state to trace water and oxygen.

In 2019, Pengfei Zhang et al. designed and prepared a series of new EC materials based on phthalates, an inexpensive class of benzoate materials with relatively simple molecular structure. Despite being good candidates for EC devices due to clear color changes when reduced, devices fabricated using PDs suffer from a high driving voltage. They propose a facile strategy of replacing the benzene core with polyaromatic esters to enlarge the conjugated area and to synthesize PDs to lower the driving voltage in EC devices. Additionally, devices show good memory effect (hundreds of seconds), the ability to undergo multiple color changes and enhanced stability, depending on the size of the conjugated bridge between the two sides of the molecule. The fabricated EC devices based on polyaromatic esters demonstrate a low driving voltage (2.6 V). We have shown that the number of aromatic ester rings in the conjugated area is very critical to obtain different colors in this new class of EC materials. This series of EC materials has promising potential for future industry applications due to the vivid color change upon electrochemical stimulation at a low driving voltage [18].

In the field of anodically coloring OECs, there is a knowledge gap on how absorption in the charged state is controlled. In 2019, Dylan T. Christiansen et al. established a new paradigm for the design of conjugated anodically coloring EC molecules [19]. By simply adjusting the position and type of substituents, the color of the material can be controlled, and the mixture of the two substances can form a reversible discoloration from colorless to black. The experimental results show that the energy levels of free radical cations can be controlled by the electron-rich/electron-deficient part which cross-conjugates with chromophores, and finally the absorption and color can be controlled.

In recent years, azobenzene and its derivatives have been widely used in the fields of photoresponsive materials, supramolecular structures, and nonlinear optics due to their unique photophysical properties. Photoisomerization of azobenzene and its derivatives can be produced under appropriate wavelength irradiation. The more thermodynamically stable the trans-isomers are, the more unstable the homoisomers are. The color, refractive index, dielectric constant, and dipole moment of the isomers have significant changes, so azobenzene and its derivatives have great potential as functional materials. But in fact, the application of azobenzene in EC is rarely reported. Its ECD has the advantages of fast switching time, high contrast, excellent optical storage performance, and high redox stability.

In 2012, Li-hua He et al. studied the EC properties of azobenzene-4,4′-dialkyl dicarboxylate derivatives, which have good reversible photoisomerization properties as well as EC properties [20]. The prepared ECD device can change color from colorless to magenta between 0.0 and 3.0 V. In addition, ECD has fast switching

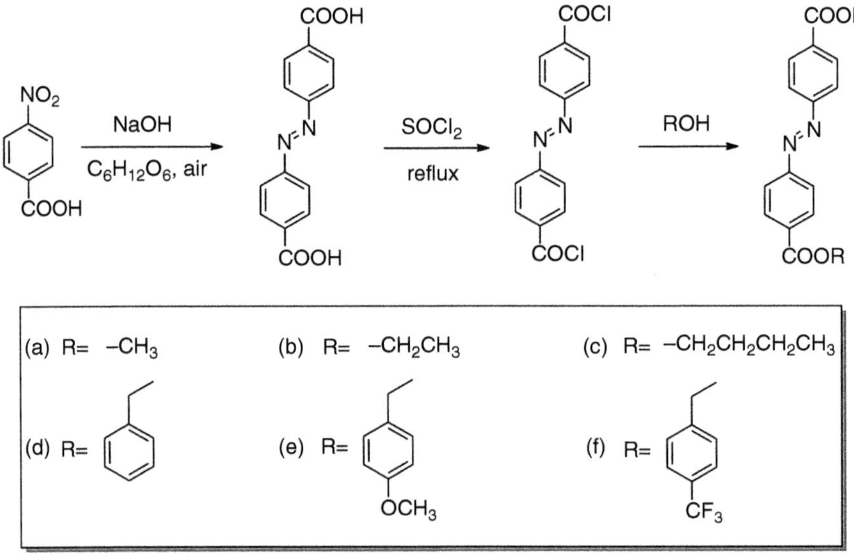

Figure 3.30 Synthesis and structures of azobenzene-4,4′-dicarboxylic acid dialkyl ester derivatives (ADDEDs). Source: He et al. [20].

$$\text{ADDED} + \text{Fc(II)} \rightleftharpoons \text{ADDED}^{\cdot -} + \text{Fc(II)}$$

Figure 3.31 Electrochromic mechanism of ADDEDs. Source: He et al. [20].

time, reasonable contrast, satisfactory optical memory, and redox stability. New redox azobenzene derivatives are expected to be used in full-color electronic display devices, electronic paper, intelligent windows, optical storage devices, dual stimulus response systems, and other fields.

This material also uses ITO glass sandwich device structure. The material and electrolyte are dissolved in **N,N-dimethylformamide (**DMF) and injected into ECD. The properties of these compounds are measured by CV. It can be seen from the figure that only compound C shows a quasi-reversible single-electron reduction, indicating that compound C has only one pair of redox peaks. The possible reason for this phenomenon is that the second reduction requires higher energy than the first reduction. Other materials in the same series show two pairs of separated redox peaks, which indicates that two quasi-reversible redox peaks can be observed due to the substituent effect. The first reduction above 1.0 V can be considered as the reduction of adduct, and the second redox pair can be attributed to the reduction of adduct (Figure 3.30).

The corresponding EC mechanism is as follows (Figure 3.31).

The B/C/D of this series of materials has excellent stability and optical transparency in nearly 100 cycles. However, due to the influence of substituents, A and E are relatively unstable, and the optical transparency is relatively low.

Substituents usually have a significant effect on small molecule EC materials. In 2017, Jun-fei Long et al. studied a series of pyridinium-based EC materials, which have 2 color states. A series of terpene pyridines were designed and synthesized by introducing esters into the phenyl ring of 2,4-bis(4-pyridyl)-4-benzylpyridine derivatives [21]. It can be found that the types of carboxylic acid and nitrogen substituents have a significant effect on their EC properties. These materials have the advantages of high color rendering efficiency, reasonable contrast, good optical storage performance, and good redox stability and have broad application prospects. The discoloration mechanism of this series of pyridinium is similar to the discoloration mechanism of viologen. Viologen usually shows only one reversible color change, and free radical cations can be produced in the electrochemical reduction process. Some pyridinium can only observe one reversible color change because the reversibility and stability of free radical cation reduction to neutral molecule are poor in liquid dielectrics. Therefore, Vin Uales et al. reported a bi-ECD based on viologen with polyvinyl alcohol gel as polyelectrolyte, in which polyelectrolyte can stabilize free radical cation reduction to medium state, which provides a better strategy for the preparation of bi-color ECDs with such materials.

Devices were fabricated with ITO sandwich structures and volt-ampere characteristic curve was measured by three-electrode system. It is found that carboxylic acid has a significant influence on the electrochemical process. Two pairs of redox processes can be observed obviously because aromatic esters have strong conjugation and stabilize neutral molecules by resonance, which facilitate the formation of molecules with one or two positive charges (Figure 3.32).

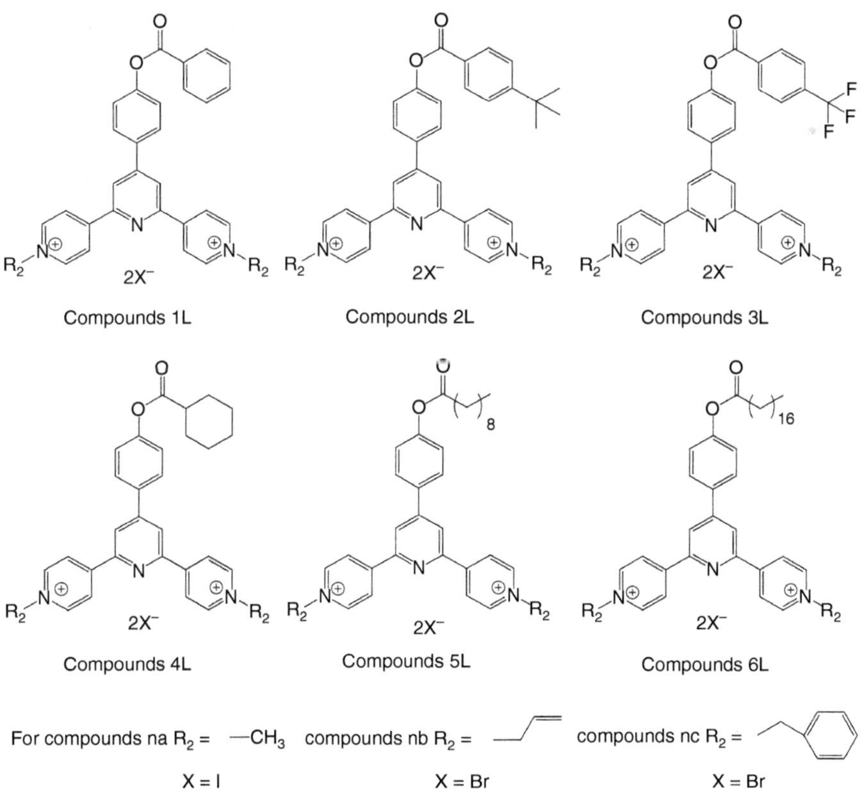

Figure 3.32 Structure of pyridinium electrochromic materials. Source: Long et al. [21].

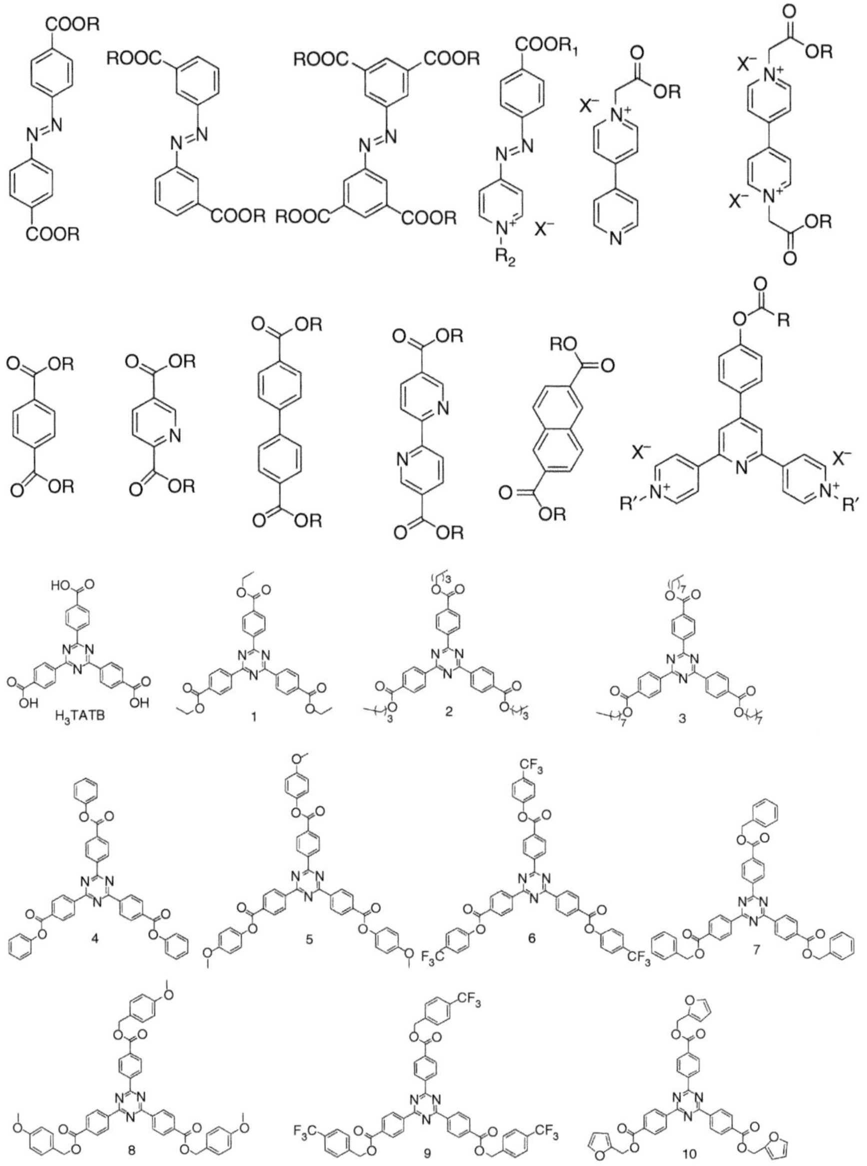

Figure 3.33 Molecular structures of electrochromic 2,4,6-triphenyl-1,3,5-triazine-based esters with electron donor–acceptor structure. Source: Zhang et al. [22].

In 2019, Weijing Zhang et al. studied the EC properties of 2,4,6-triphenyl-1,3,5-triazinyl esters with electron donor–acceptor (D–A) structure. These materials have 2,4,6-triphenyl-1,3,5-triazine nuclei and three ester side arms. Thus the structure of electron donor–pi–receptor is formed (as the compounds **1–10** shown in Figure 3.33). The ECDs of some compounds (**1**, **2**, **4**, **5**, and **10**) show strong discoloration, fast switching time (<2 seconds), high optical contrast (>60%), good switching stability, high optical density (>1.0), and high coloring efficiency (>1000 cm^2 C^{-1}). The coloring state of these devices depends largely on the conjugate structure of the tripod elements. Short chain length (compounds **1** and **2**) and electron donor groups (compounds **4**, **5**, **8**, and **10**) are beneficial to EC properties, while long chain length (compounds **3**) and electron acceptor groups (compounds **6** and **9**) are unfavorable to EC performance.

Two reduction peaks can be observed in compounds with alkyl chains, indicating that there are two single-electron reductions, but the chemical reversibility of the first cathode peak can be attributed to the chemical instability of the monoanion species, while the diion species have better chemical reversibility. Irreversible anodic peaks can be observed in compound **6** with electron-withdrawing groups, which may be due to the oxidation of some unstable intermediates, suggesting that electron-withdrawing groups are not conducive to the electrochemical properties of triazine esters. In the measurements of the EC switching characteristics of compounds, it is also proved that the electron-withdrawing group can reduce the EC properties of materials, while the electron-donating group can improve their stability (Figure 3.33) [22].

3.5.1 Conclusion

Isophthalate derivatives have similarity with Terephthalate derivatives, the coloration reactions are usually driven by ester group. Researchers found that functional groups can be easily introduced at the 5-position of these materials, but different functional groups have different effects on the color of materials. Interestingly, if an electrochemical active group nitro is introduced at 5-position, it induced polychromatic display. These materials have advantages that their structure is simple and adjustable, showing bright color. But it is also necessary to find a better way to solve the problem of low stability of this material.

3.6 Methyl Ketone Derivatives

Multi-response molecular switch is an intelligent molecule, which can control molecular structure through activation or inactivation mechanism. It can be applied to multifunctional display, intelligent camouflage, multifunctional sensors, composite materials, optical, electrical, thermal, acid base, and other external stimuli. It has become an important field of logic gate operation of color switch. It has become an important field of logic gate operation of color switch and multimode data storage. In designing multi-response color switches, the usual strategy is to combine and integrate the known switch units with conjugate linkers and non-conjugate spacers. However, this method usually faces the disadvantages of complex technology, precise bandgap matching, and unstable performance.

Another method is to use multi-response switch units, such as azo and bis and thioether and spiropyran, which respond to two or more different stimuli. The key of this strategy is to develop a new multi-response color switch unit.

Yu-Mo Zhang et al. have done a lot of work in the field of EC materials for methyl ketone and its derivatives. In 2013, they designed methyl ketone as a switching unit for electroaddressing molecular color switches. CV, X-ray photoelectron spectroscopy (XPS), infrared (IR) spectroscopy, and in situ ultraviolet–visible

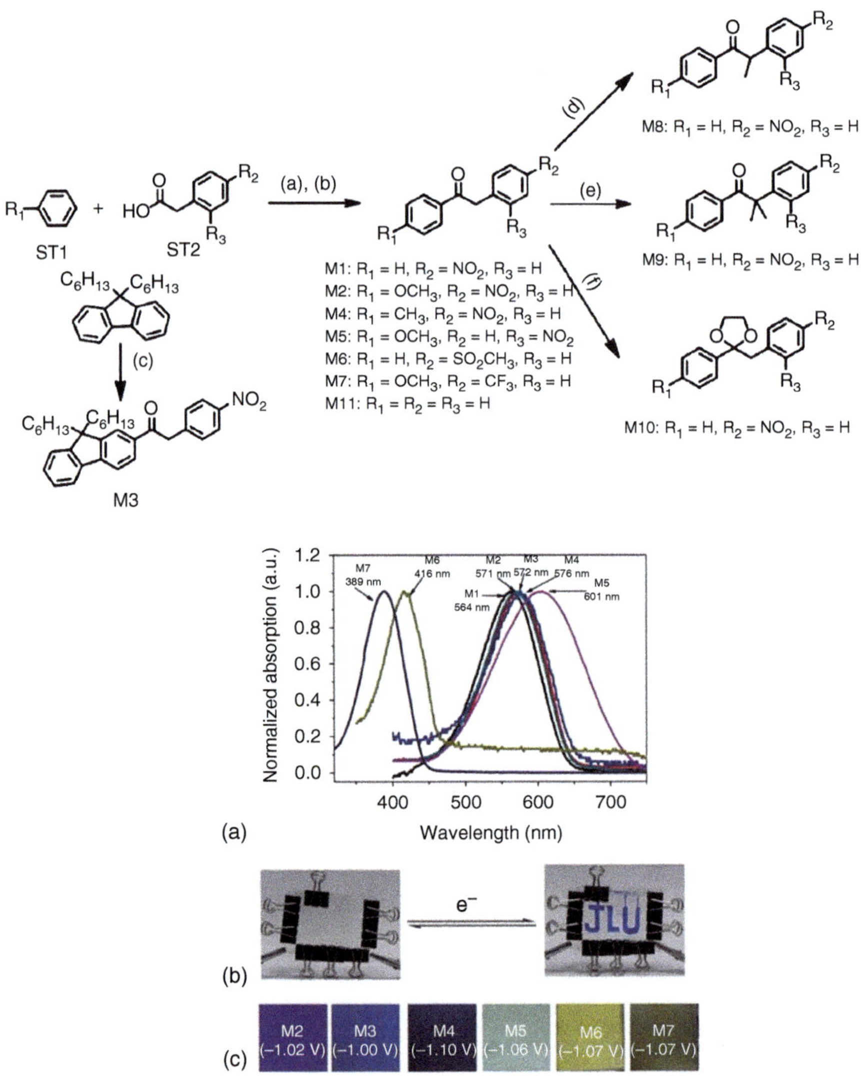

Figure 3.34 Synthetic scheme for methyl ketone and (A) absorption spectra of M1–M7 in acetonitrile after the voltage is supplied. (B) A real object illustration of an electrically driven display based on M1. (C) The real color of M2–M7 switched electrically in their ITO devices, with the switch threshold (from CV) provided in brackets. Source: Reproduced with permission of Zhang et al. [23] from Copyright © Royal Society of Chemistry 2013.

spectroscopy (UV–VIS) were used to demonstrate the mechanism of acid/base (free radical ion) induced interstitial electron transfer in color switch molecules (Figure 3.34) [23].

In 2015, they synthesized a series of new compounds and studied them by characterization and density functional theory (DFT). It was found that the material would change color by regulating alkalinity, solvation, and applied electric field and that the multicolor ECDs (green, blue, and magenta) were durable and had high color rendering efficiency, fast switching time (50 ms), and superior reversibility. This study is a successful attempt to integrate basochromism, solvatochromism, and electrochromism into the electronic display. This integration not only brings a new way for the structure design of toner but also stimulates further development in many other properties such as conductivity and redox properties through this method. Thus, it accelerates the application of multipurpose in data recording, ultrathin flexible display, optical communication, and sensing elements (Figure 3.35) [3].

In 2016, Yu-Mo Zhang et al. developed a novel methyl ketone bridge multi-stimulus response M5 device for visible and near-IR absorption ECDs. The basic and EC properties of methyl ketone M5 containing nitronaphthalene were studied in detail by using spectroelectrochemical method and DFT calculation method. Colorless M5 can be converted to green–yellow under the stimulation of alkali and electricity and has a wide range of visible and near-IR absorption (400–500 nm and 550–800 nm, respectively). Figure 3.36 explains the relationship between the structure and the absorption bands. The absorption bands of M5 at 692 nm are distributed to 55 intramolecular charge transfer (ICT) bands, ranging from HOMO, which is mainly located in the oxygen anion part, to LUMO, which is mainly located in the nitro group. The absorption band at 432 nm is endowed with a π–π^* transition from Homo to Lumo + 2, which is distributed on the conjugated system.

The switching mechanism of M5 can be extrapolated by the test of electrochemical workstation with ultraviolet spectroscopy (as shown in the following). The intermediate M5-enolate is generated by the proton transfer from the methylene of M5 to the reductive product of M5, such as radical anion of nitro, radical anion of carbonyl, and dianion of nitro. Thus, the EC mechanism of M5 is proton-coupled electron transfer (PCET), as shown in as shown in Figure 3.37. In this diagram, radical anion/dianion of nitro or carbonyl can induce the proton transfer of methylene, similar to chemical base. Thus, the radical anion/dianion is named electrobase, and the mechanism of proton transfer is defined as electrobase mechanism. Therefore, free radical anion/diion is called electroalkali, and the mechanism of proton transfer is defined as electroalkali mechanism. The multi-stimulus color switch can be successfully applied to ECDs and has the advantages of short switching time and good reversibility (Figure 3.37) [24].

This new switching method and molecular structure operation will stimulate and accelerate the further development of a wide range of switching materials and their applications in ultrathin flexible displays. This work also verified that the combination of the ketyl radical system with the isophthalate EC system is a successful strategy to obtain a multi-EC system.

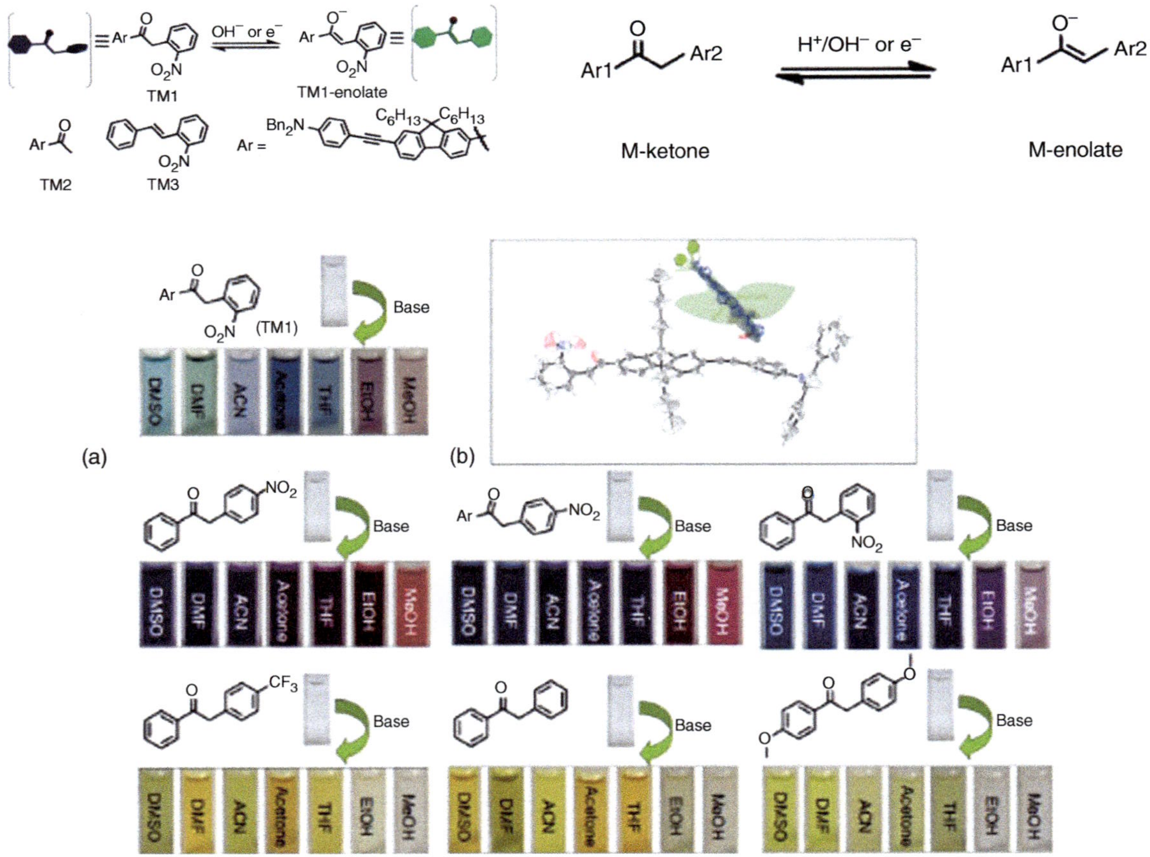

Figure 3.35 The structures of seven methyl ketone molecules and the colors of their enolates in different solvents and the appearance of TM1 molecule in crystal. Source: Reproduced with permission of Zhang et al. [3] from Copyright © Springer Nature 2019.

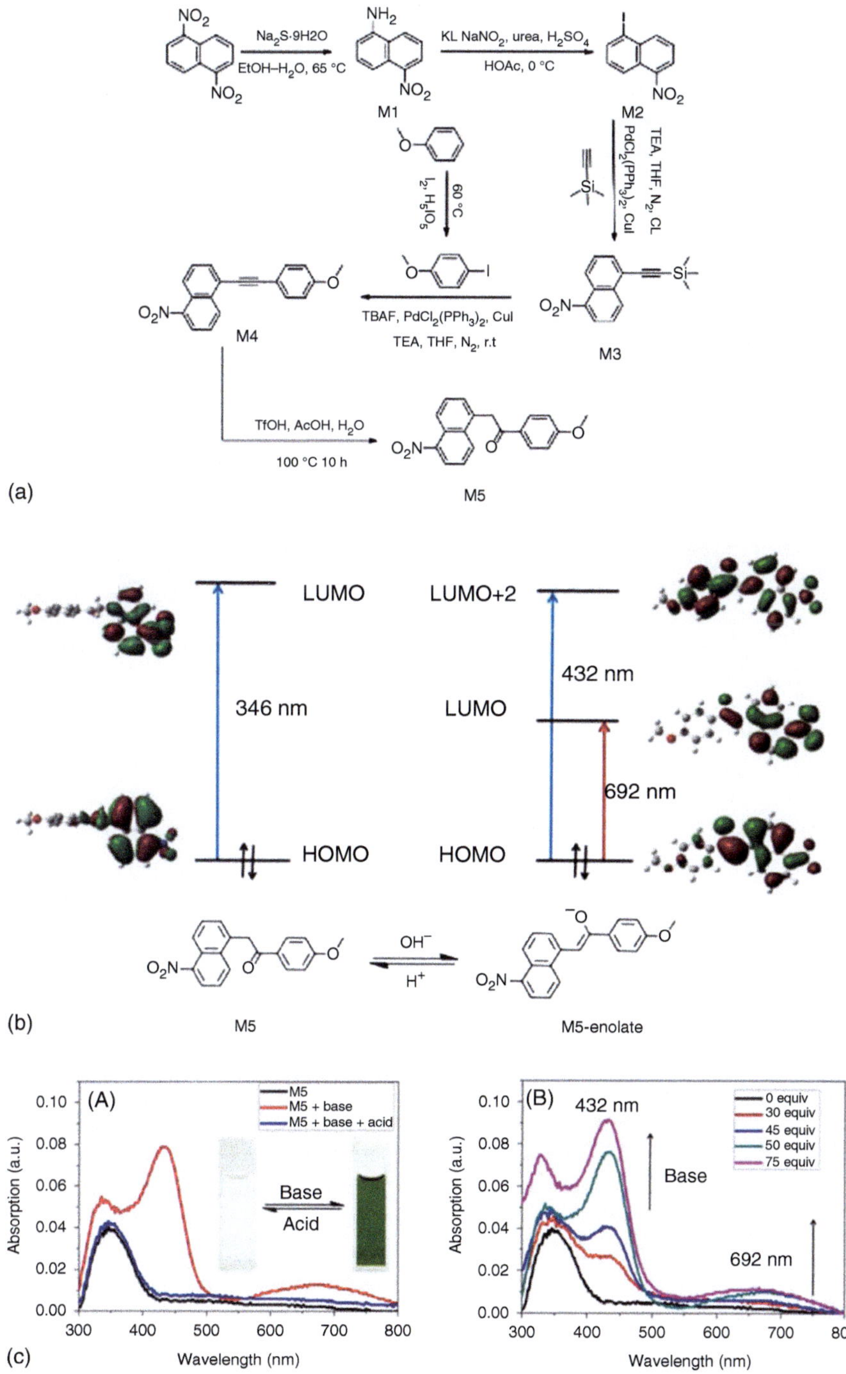

Figure 3.36 (a) The synthetic route of M5; (b) The HOMO (bottom) and LUMO or LUMO+2 (top) orbits of M5 and M5-enolate; (c) (A) absorption spectra of M5 (1.0×10^{-5} M, black curve), treated with potassium tert-butoxide (red curve) and then neutralized with CH_3COOH (blue curve) in DMF. (B) Absorption spectra of M5 (1.0×10^{-5} M) in the presence of potassium *tert*-butoxide (0, 30, 45, 50, and 75 equiv) in DMF. Source: Reproduced with permission of Zhang et al. [24] from Copyright © Royal Society of Chemistry 2016.

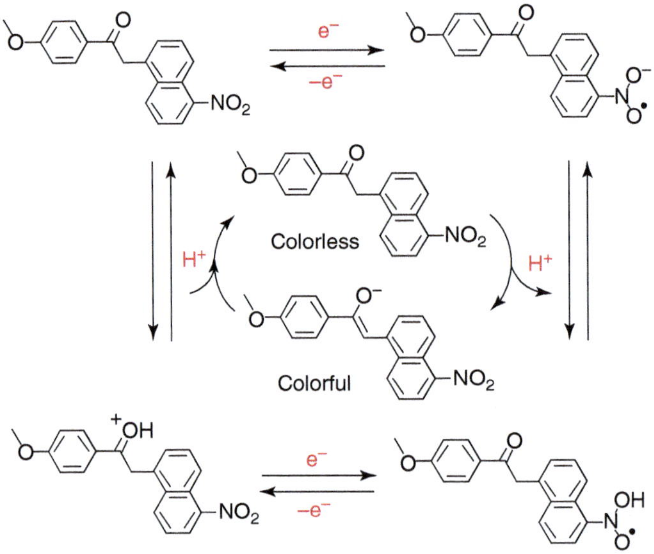

Figure 3.37 The electrochromic mechanism of M5 with a one-electron reduction mechanism. Source: Zhang et al. [24].

3.6.1 Conclusion

Methyl ketone derivatives have a very broad application prospect as multi-responsive molecules. The EC mechanism of this material is PCET. Nitro or carbonyl radical anion/diion can induce proton transfer of methylene, similar to chemical base. This mechanism is also defined as electro alkali mechanism. This kind of material has the advantages of short switching time and good reversibility. It can activate the response in many ways.

3.7 Diphenylacetylenes

EC materials, such as cyanobiphenyls, are the typical liquid crystal (LC) component. In 1981, Nakamura's group were the first to find an electrochromic liquid crystal (ECLC), composed of LC molecular cyano-alkyl-biphenyls (nCB) and cyano-alkoxybiphenyls (nCOB) and doped with tetra-alky-ammonium iodide $[(C_nH_{2n+1})N]^+I^-$ electrolytes. In 2005, Nicoletta et al. developed an electrochromic polymer-dispersed liquid crystal (EC-PDLC) with ethyl viologen as the EC material, being able to change both transmittance and color upon the application of alternating current (AC) and direct current (DC) field in a single cell. They also used commercially available materials to improve preparation process.

Actually the ECLC cell has some disadvantages such as long switching time, insufficient ability, and poor electro-optical performance. Researchers has tried to use the nonaqueous aprotic electrolytes to improve the performance of EC. And ionic liquids (ILs), due to their negligible volatility, wide potential window, and good conductivity, have been widely used in electrochemistry for several years.

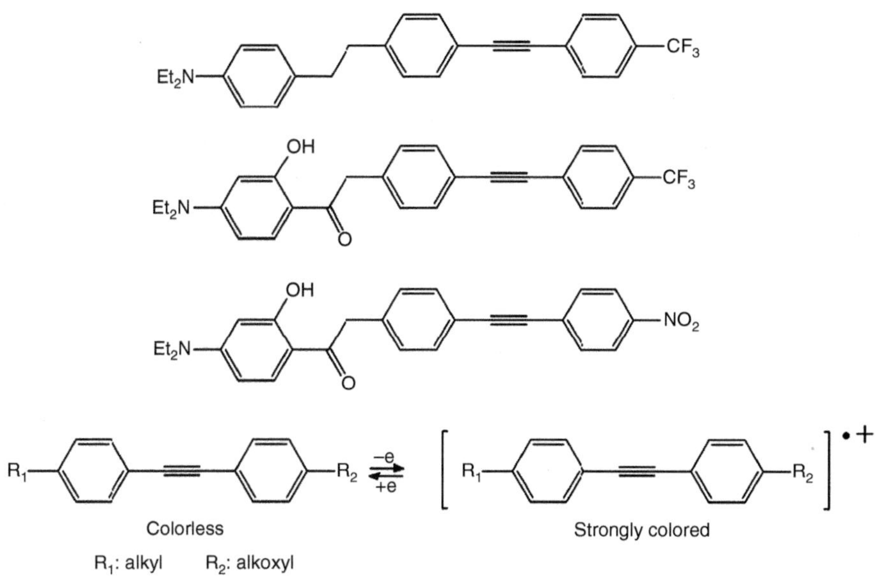

Figure 3.38 The structure, electrochromic mechanism, and photographs of the four states of an EC-PDLC cell of the substituted diphenylacetylene species.

In 2016, Zemin He et al. reported an ECLC material composed of only LC and IL [25]. The LC containing the substituted diphenylacetylene serves as EC material to realize transmittance and color change under the DC field, while the IL with the designable cation and anion served as electrolyte. By choosing an appropriate IL electrolytes, the electro-optic performance of ECLC containing 1-ethoxy-4-[2-(4-pentylphenyl)ethynyl]-benzene (PEB) and IL was evaluated. Compared with other PEB-IL, PEB-[Bmim][NTf_2] with [Bmim][NTf_2] electrolyte shows a satisfactory transmittance at low operating voltage. The light scattering of the ECLC cell is also improved because of the lower EC potential.

The combination of electrically controlled transmittance of PDLC films with the color control of ECDs can be achieved by adding electroactive molecules in a PDLC film, which would trigger potential application in large-area displays, smart windows, and electro-optical shutters (Figure 3.38).

3.8 Fluoran Dye Derivatives

Unlike most other EC materials that change color by electron transfer reaction, the color change reaction of fluoran dyes is based on the opening and closing reaction of the lactone ring, and therefore this can be activated in two ways as shown in Figure 3.39a. The first one is the redox reaction by Lewis acid base (electron transfer with anode) and the other is the redox reaction by Brønsted–Lowry acid base (proton transfer with proton donor). All the reported papers for ECDs based on fluoran dye used the first method for EC reaction.

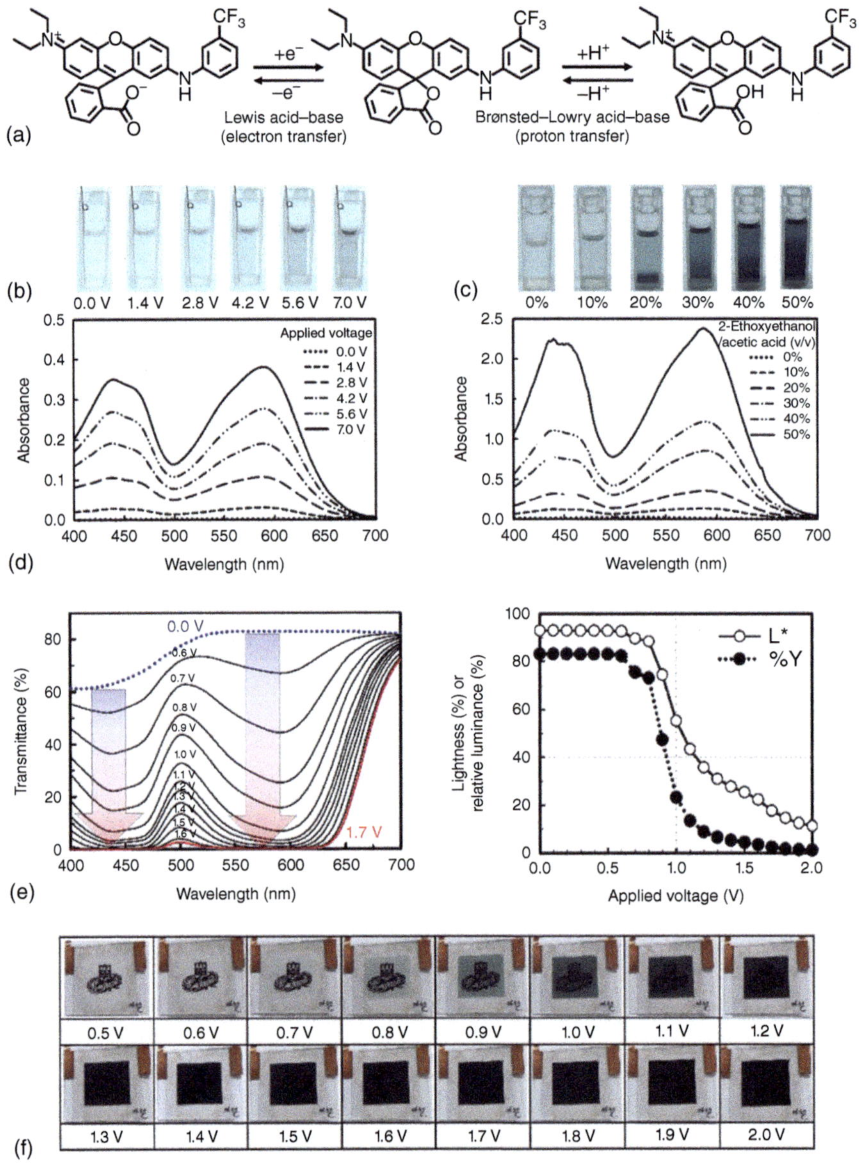

Figure 3.39 (a) Lactone ring-opening–closing reaction of fluoran dye. (b) The change in absorption spectrum of 1×10^{-6} M 6′-(diethylamino)-2′-[[3-(trifluoromethyl)phenyl]amino] spiro[isobenzofuran-1(3H),9′-[9H]xanthene]-3-one (DATFMF) and 1×10^{-6} M tetra-*n*-butylammonium tetrafluoroborate (TBTF). (c) The change in absorption spectrum of 1×10^{-6} M DATFMF solution dissolved in 2-ethoxyethanol depending on the amount of added acetic acid. (d) Optical transmittance changes of the ECD with respect to applied voltage from 0.0 to 1.7 V. (e) Lightness (L*) and percent relative luminance (%Y) changes of the ECD as a function of applied voltage. (f) Actual photographs of the ECD with respective applied voltage. Source: Reproduced with permission of Kim et al. [26] from Copyright © John Wiley and Sons 2018.

In 2018, Gyeong Woo Kim et al. reported a high-performance EC light shutter based on fluoran dye [26]. The EC reactions of the fluoran dye that transmit or absorb visible light depending on its redox state are activated by the protons generated from hydroquinone derivatives. This EC light shutter can be used with optical see through-head mounted display (OST-HMD) as an augmented reality (AR) display system, which will improve the visibility of the displayed images in strong ambient light conditions (Figure 3.39).

In order to be appropriate for use as an optical shutter, EC dye and its related composition materials in ECD must be highly transparent in bleached state and large absorbance in all visible wavelength region at colored state. The fluoran derivatives are usually transparent because central carbon of spirostructure linking xanthene and lactone unit through sp^3 hybrid orbitals breaks conjugation length, resulting in a large bandgap. On the other hand, when the fluoran derivative is oxidized and the lactone ring is opened, the conjugation length becomes longer as the sp^3 hybrid orbital of central carbon is converted to sp^2 hybrid orbital, which reduces the bandgap and shifts absorption wavelength to red region. Therefore, the fluoran derivatives are usually applied as thermochromic dyes and EC materials.

Referring to what was said at the beginning of the chapter, there are two methods to active fluoran dyes. If a proton generator that makes protons by redox reactions at the electrode of ECD is applied to the fluoran-based ECD system, then it may cause an EC reaction of fluoran dye by the second method. After experimentation, it is found that hydroquinone oxidizes generate two protons and convert to benzoquinone by losing two electrons from the electrode, so hydroquinone is likely to act as a proton generator. And the lactone ring-opening reaction is activated by the protons generated from oxidation of **dimethoxy benzoquinone** (DMHQ) at the anode, i.e. DMHQ changes the lactone ring-opening reaction by electron transfer (ECD 1) to the lactone ring-opening reaction by proton transfer (ECD 2 and 3) (Figure 3.40).

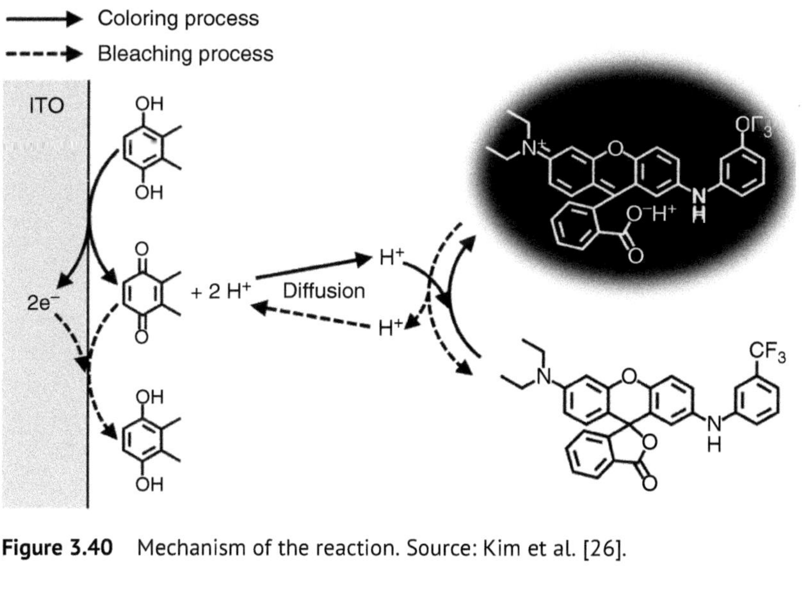

Figure 3.40 Mechanism of the reaction. Source: Kim et al. [26].

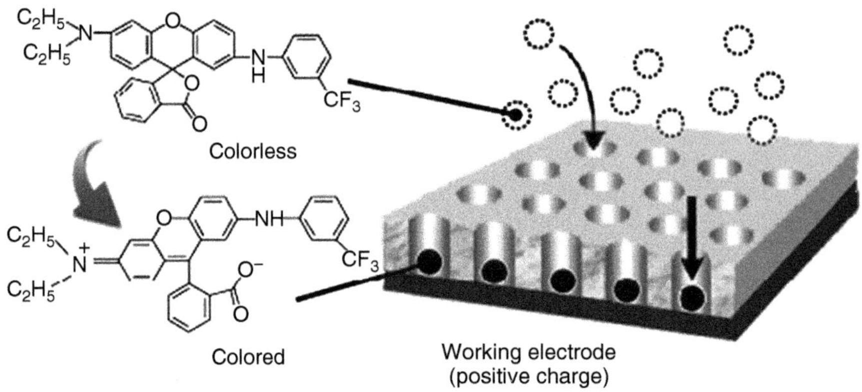

Figure 3.41 Mechanism of imaging using a leuco dye on a mesoporous TiO_2 electrode with vertical pores. Source: Weng et al. [27].

They fabricated an OST-HMD using this material and test its performance. The OST-HMD with EC optical shutter demonstrates a significant improvement in ambient contrast ratio and color gamut. The proposed AR display system will maximize the usability by broadening the usage range of AR displays due to the ambient light-independent visibility.

Actually in 2010, to realize high scanning speed and high display quality, Wu Weng et al. have demonstrated a high-speed and high-resolution EC passive-matrix display using the fluoran dyes as mentioned earlier with a mesoporous TiO_2 electrode with vertical pores [27]. The vertical pores of the electrode can support effective diffusion of fluoran dyes perpendicular to the electrode and can prevent the diffusion of the dye around the electrode. It provides a more transparent background of display, which is very desirable for future E-paper applications (Figure 3.41).

In the study of EC properties of fluorescent dyes, the free radical ions of electroreduction of *p*-benzoquinone (BQ) were introduced into the molecules and their properties were tested. In 2013, Yu-Mo Zhang et al. did this work and reported that the devices show excellent switchability and reversibility with no degradation during our extended switching test of 1500 cycles [28]. In previous studies, the authors found that the radical anion of the nitro molecule can capture an acidic proton from a methylene ketone group and trigger the molecular enolization. Thus, they conjecture that all radical anions could be used as "electro-acid/base" for switching appropriate pH-sensitive molecules.

A molecule is designed as shown in Figure 3.42. The color change of the molecule can be proved to be driven only by an applied electric field. Fluorescein is easy to degrade, thus a compact overlayer NiO film can be electrodeposited onto the surface of counter ITO electrode to avoid the undesired oxidation and improve the response time and reversibility of this device. The coloring mechanism of this material is shown as follows.

With the development of electrofluorescent discoloring devices, development of desirable RGB color-tunable electrofluorochromism (EFC) devices has great significance for accelerating practical optoelectronic applications. One strategy is

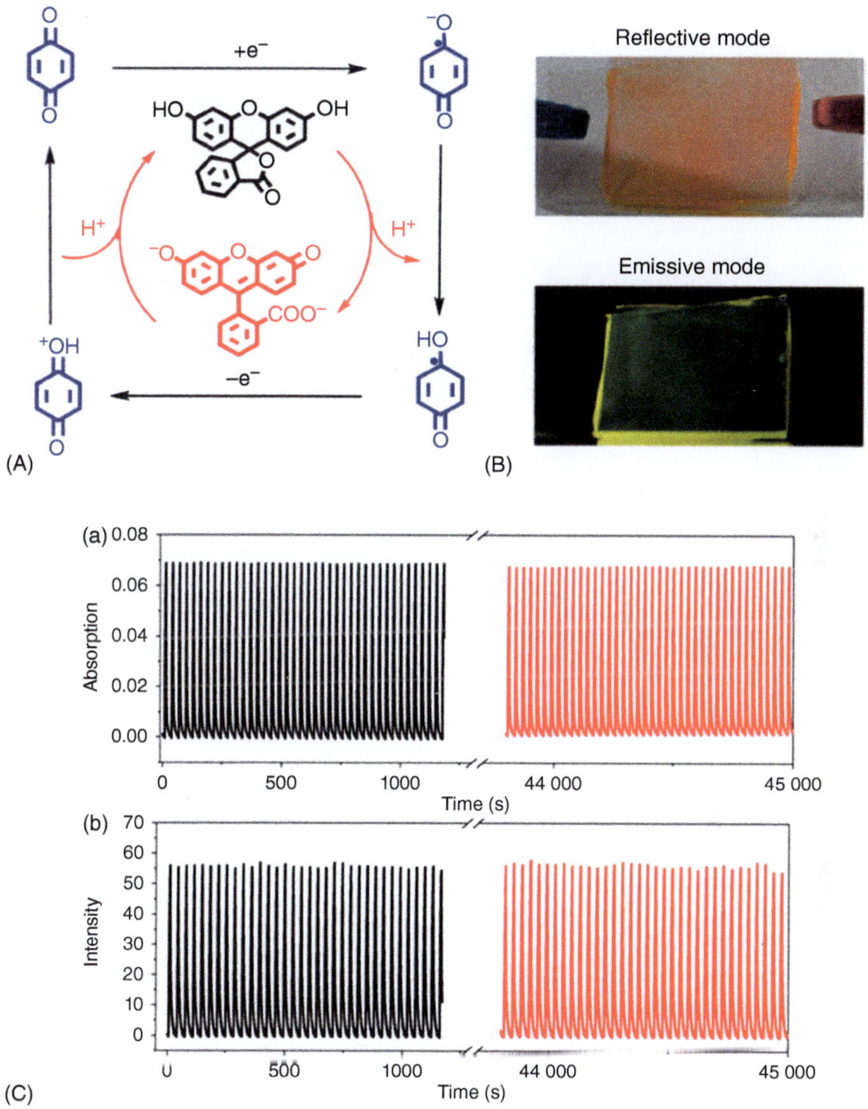

Figure 3.42 (A) The switch mechanism of fluorescein with electro-reduced *p*-benzoquinone as in situ base, (B) photographs of solid thin film on reflective mode and emissive mode (e.g. 365 nm) under −1.5 V, (C) absorbance (515 nm) and emission (535 nm) variation with the switch cycles by alternative −1.5 V (5 seconds) and 1.0 V (25 seconds) using activated NiO electrode in polymethylmethacrylate (PMMA) film. Source: Reproduced with permission of Zhang et al. [28] from Copyright © Royal Society of Chemistry 2014.

separating the redox part and the fluorophore and make use of proton transfer to control the emission to even the quenching of photoluminescence by the radical anion/cation, which was reported by Xiaojun Wang et al. [29] A pH-sensitive luminescent material was synthesized as shown in the following. And RGB color-tunable EFC devices (red, green, and blue) with a "turn-on" mode were newly fabricated

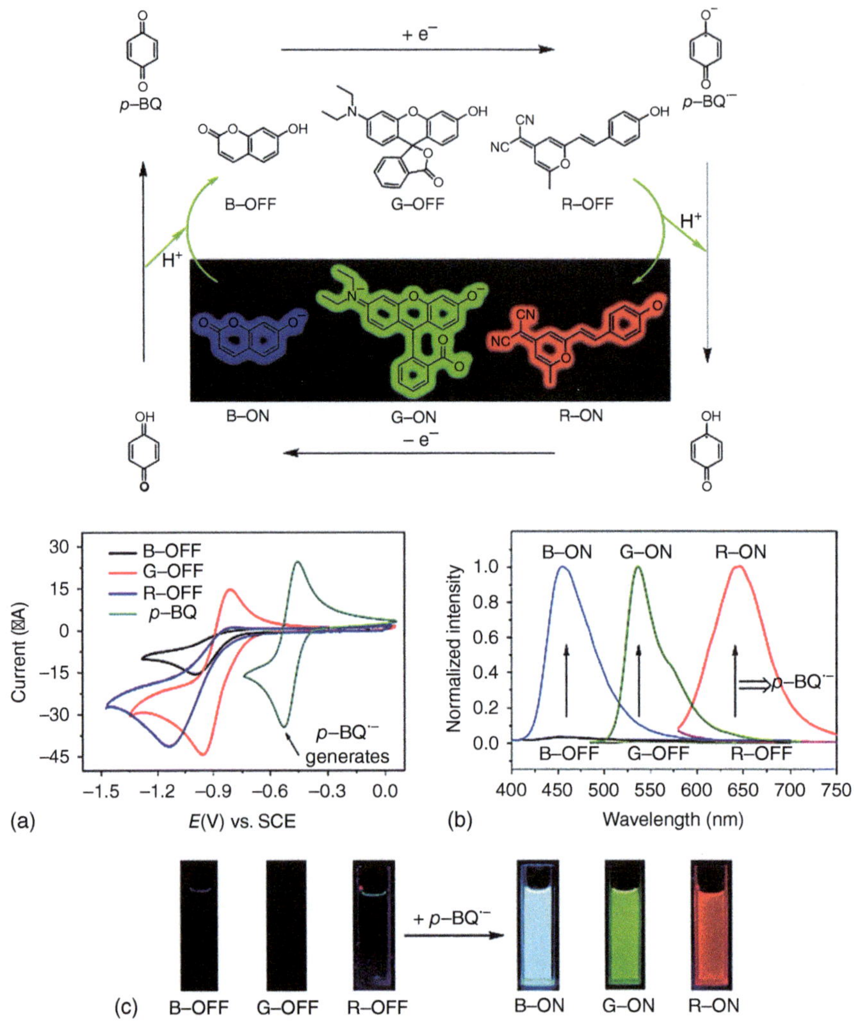

Figure 3.43 (a) The fluorescence switching mechanism induced by *p*-BQ as an electro base, (b) the cyclic voltammograms of the B, G, and R molecules and *p*-BQ, (c) the fluorescence spectra, and (d) the pictures under UV light of the B, G, and R molecules. Source: Wang et al. [29]. © 2017, Royal Society of Chemistry.

based on the same electro-base mechanism. Such "turn-on" color-tunable EFC devices show high color purity, relatively quick cycling speeds along with high fluorescence contrast ratios, and excellent durability (Figure 3.43).

They also proposed a biomimetic system on reversible bond-coupled electron transfer (BCET) and investigated in a switchable Rh–N molecule with redox active subunits at the same time. Rh–N molecule was synthesized, which was designed by ingenious interconnecting color-switchable rhodamine with a redox active phenylenediamine. According to experiment result and calculations, the following electron transfer mechanisms are proposed (Figure 3.44) [30].

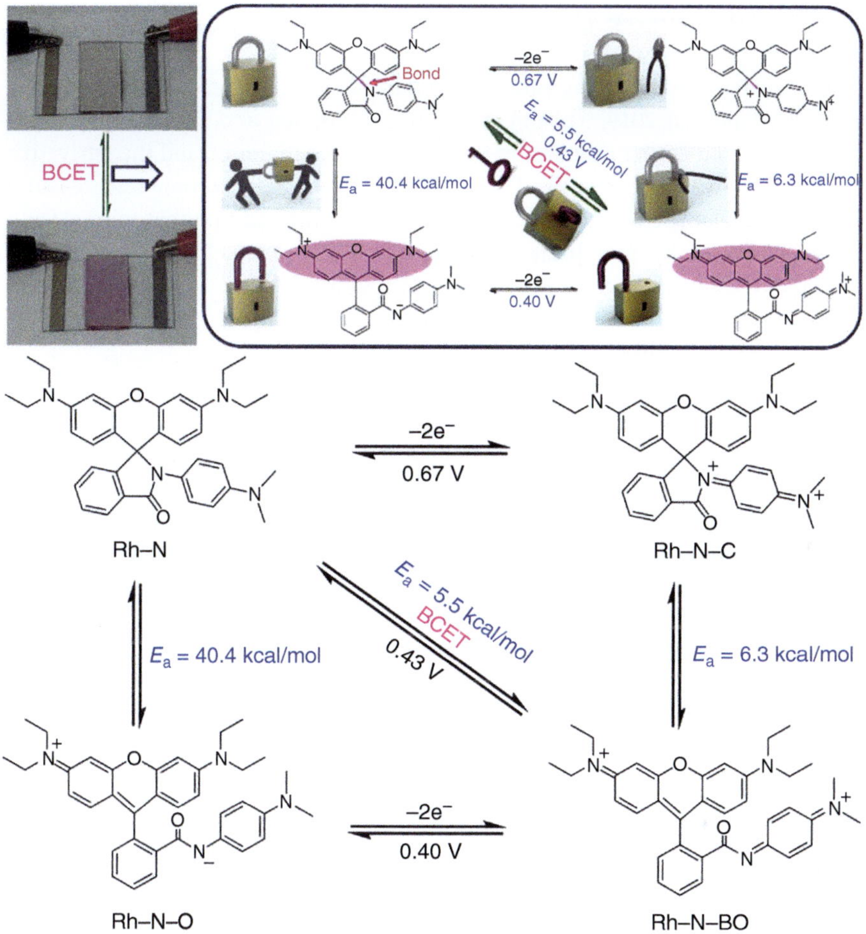

Figure 3.44 Bonding alteration process of Rh–N via bond-coupled electron transfer. Source: Reproduced with permission of Wang et al. [30]. © 2017, Royal Society of Chemistry.

In this BCET system, an organic cation takes away the positive charge from the oxidative part right after its C—N bond cleavage. This amazing process has similar characteristic with PCET and metal ion-coupled electron transfer (MCET) transferring proton and metal ion along with electron transfer. These delightful similarities motivate us to highly recommend a unified concept, cation-coupled electron transfer (CCET), for simplifying those existing concepts on similar couplings (such as PCET, MCET, and BCET). It is well known that, in versatile biological process, very low activation energy is required to effectively achieve multistep complex reaction while involving ion (such as proton, metal ion, electron, small organic ion, etc.) transfer, bond cleavage/formation, structural alteration, and configuration inversion, etc. It is also well known that ion introducing or eliminating will greatly alter electron density on active side of the molecule or molecular aggregate under enzyme catalysis.

3.9 PH-Responsive Molecule Derivatives

In recent years, improving the bistable properties of EC materials has become a research hot spot. In 2019, Yuyang Wang et al. designed and synthesized a new material, interconnecting rhodamine B with phenylenediamine, and the derivative of urea, urea–N, was applied to the devices. Urea–N electro-acid proton releases fully reversible BCET in Rh–N molecule. This work demonstrated that this colored state is actively stabilized by the supramolecular interaction between Urea–N and Rh–M. Meanwhile, the faster coloring response of "Urea–N + Rh–M" also illustrates that the proton transfer between electro-acid and pH-sensitive dye can be accelerated by the intermolecular hydrogen bonding (Figure 3.45).

In the same years, they also proposed to use PCET and BCET technology to prepare ideal bistable EC materials. The PCET mechanism enhances the bistable properties of EC materials by adjusting supramolecular interactions, broadening color selection with available pH indicators, and avoiding high-energy intermediates, which may reduce the material's stability. In order to further improve the properties of the materials, we should also explore a method that can make the bistable retention time longer. The key problem is to improve the coupling efficiency of proton transfer and electron transfer between molecules.

The devices realized using materials based on these concepts show wide color tunability and good EC properties, including on-switching times lower than 1.5 seconds, off-switching times lower than 7.0 seconds, 70% transmittance change, coloration efficiency of 1240 $cm^2\ C^{-1}$, bistability lasting for more than 52 hours, and reversibility for more than 12 000 cycles.

ECDs of copolymer RHMAs containing five layers were fabricated to demonstrate their potential use in display applications. Different colors could be obtained under the same voltage treatments. They also studied the bistability, which is the ability to maintain colorless and colored states without the constant application of a voltage, for our optimized devices.

The colorless state of RHMA-M1 was maintained for more than 30 days without voltage. The colored state was stable for about 1h without voltage after a stimulation of +1.7 V for 1.5 seconds, with a transmittance change higher than 70%. This kind of material has excellent cycle stability and can maintain a color decay rate of less than 7.8% after 12 000 cycles (Figure 3.46) [31].

As mentioned in Section 3.8, the new molecule, which was BQ, introduced a radical ion from electro-reduced BQ as a new source of convenient in situ base for color. Emission switching of fluorescein is also a pH-responsive molecules derivative with better performance, which show excellent switchability and reversibility with no degradation during switching test of 1500 cycles. During the experiment, they found that only when BQ and fluorescence coexist in the solution can the electric field switch the emission of fluorescein at 535 nm. While preparing the device, NiO was coated on an ITO electrode to prevent the unwanted oxidation. In this way, the problem of degradation of fluorescein was solved successfully.

The chromism mechanism was conjectured in Figure 3.47. (i) BQ was first reduced to BQ, (ii) double BQ grabbed two protons from the nearby fluorescein molecule to

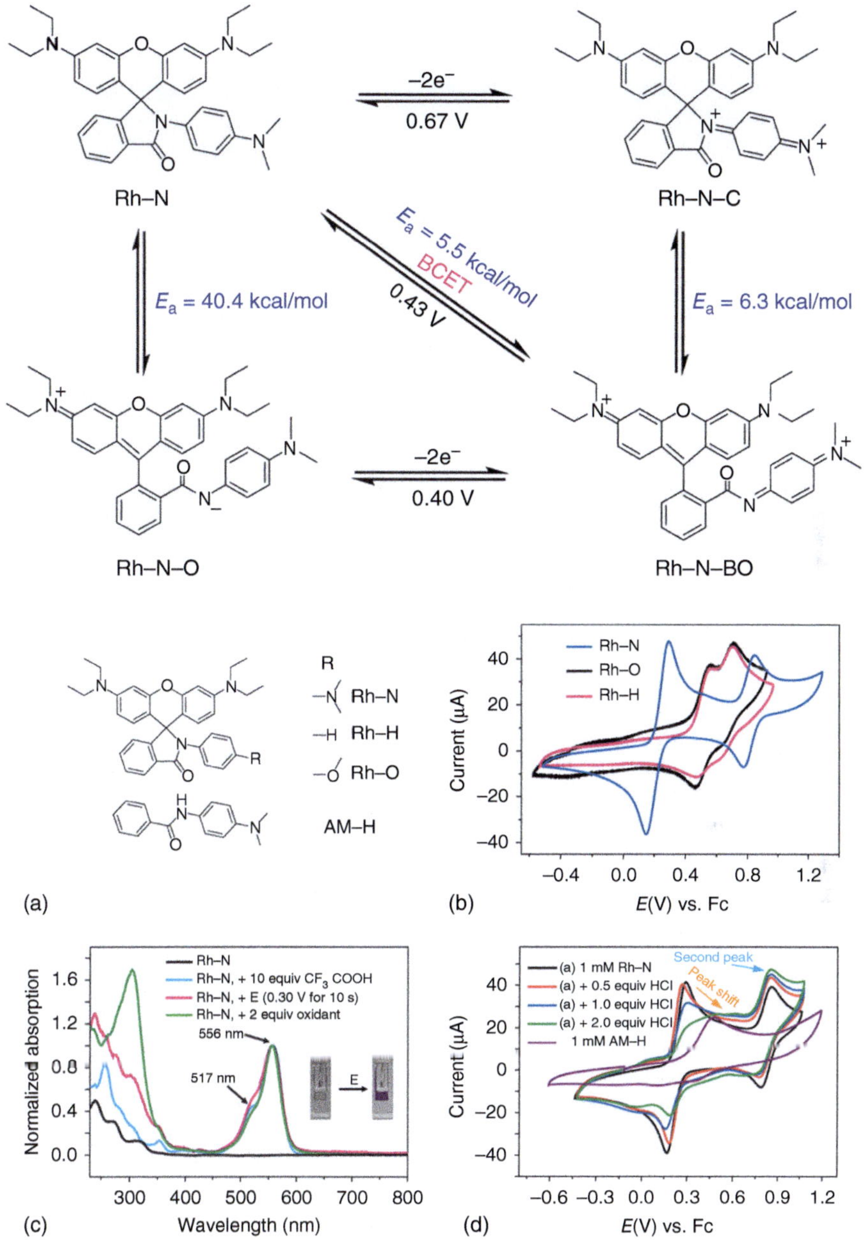

Figure 3.45 (a) Bonding alteration process of Rh−N via bond-coupled electron transfer. (b) Structure of Rh−N, Rh−H, Rh−O, and AM−H. (c) The cyclic voltammograms of Rh−O, Rh−H, and Rh−N. (d) The electrochemical UV–vis spectra before and after applied voltage for Rh−N (1.0×10^{-4} M) with 0.1 mol/l TBAPF6 in CH_3CN comparing with the Rh−N (1.0×10^{-5} M) added with 10 equiv CF_3COOH and 2 equiv tris(*p*-bromophenyl)aminium hexachloroantimonate. (d) The cyclic voltammograms of Rh−N (1.0×10^{-3} M) with different equivalent HCl and cyclic voltammogram of AM-H (1.0×10^{-3} M) in 0.1 M TBAPF6/CH_3CN. Source: Wang et al. [30]. © 2017, Royal Society of Chemistry.

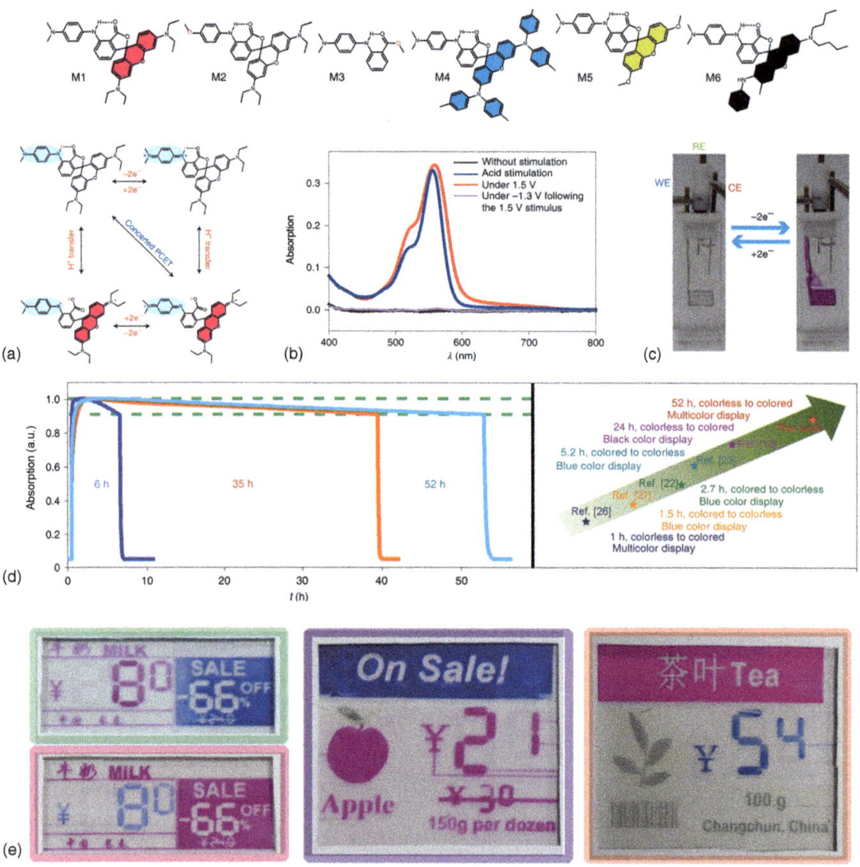

Figure 3.46 (a) Molecular structures of RHMA-M1–RHMA-M6 (the colored shadings represent different color chromophores for the molecules). (b) The proposed mechanism (dashed line, hydrogen bonding between H and O; blue shading, *p*-phenylenediamine; pink shading, fluoran). (c) Absorption spectra of the RHMA-M1 solution (0.4 mg ml^{-1} in THF) without stimulation. (treated with CF_3COOH, at +1.5 V vs saturated calomel electrode (SCE) and at −1.3 V vs SCE using air as a reference) and photos of the RHMA-M1 solution before and after treatment at +1.5 V vs SCE (WE, working electrode; CE, counter electrode; RE, reference electrode). (d) The absorption of devices with conducting layer thicknesses of 250 μm (blue), 348 μm (orange) and 425 μm (cyan). On the right, a comparison is made with the retention properties of different reported electrochromic materials. (e) Multicolour electronic shelf label (ESL) prototypes (the colored outlines represent the rim of ESLs). Source: (a, b, e) Wang et al. [31]. © 2019, Springer Nature.

generate BQH and the di-anionic form of fluorescein (the colorant), and (iii) BQH was oxidized to BQH$^+$ at reversed bias and gave the proton back to the di-anionic form of fluorescein first and its mono-anion next and finally transformed back to neutral fluorescein again with the reversal in color.

This work demonstrated the switching of both the color and the emission of fluorescein in a prototype ITO device by using electro-reduced radical ion of BQ as a convenient electrobase in situ as a new source of convenient in situ base for color/emission switching of fluorescein in thin film devices [32].

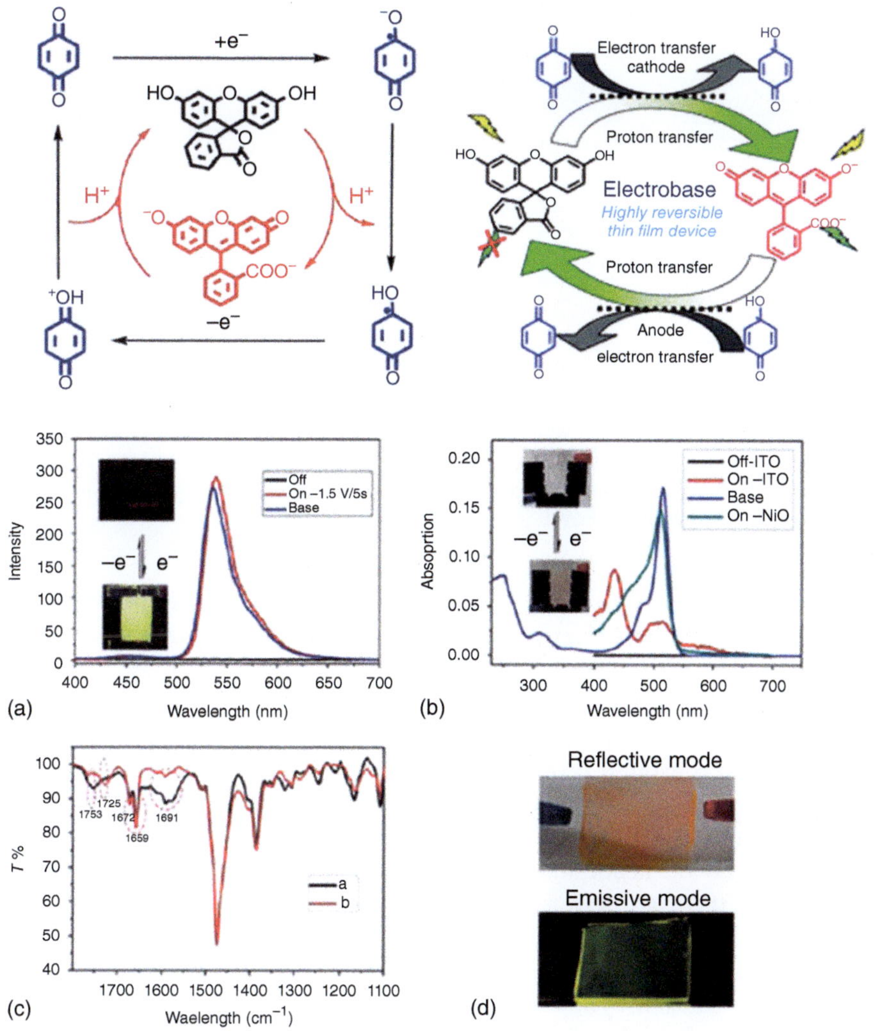

Figure 3.47 The switch mechanism of fluorescein with electro-reduced *p*-benzoquinone as in situ base and (a) emission spectra of the ITO cell sandwiched with the mixture solution of BQ–fluorescein, (b) absorption spectra of the cell, (c) IR spectra of BQ–fluorescein solution before and after reduction in acetonitrile, and (d) photographs of solid thin film on reflective mode and emissive mode under 1.5 V. Source: Reproduced with permission of Zhang et al. [32].

3.10 TPA Dye Derivatives

Because of its excellent electrochemical properties, TPA and its derivatives can be introduced into the structure of EC materials to achieve more interesting and excellent performance and even to achieve dual response characteristics.

In 2014, Jingwei Sun et al. designed and synthesized a D–A cruciform conjugated luminophore DMCS-TPA, which has a dual response function of

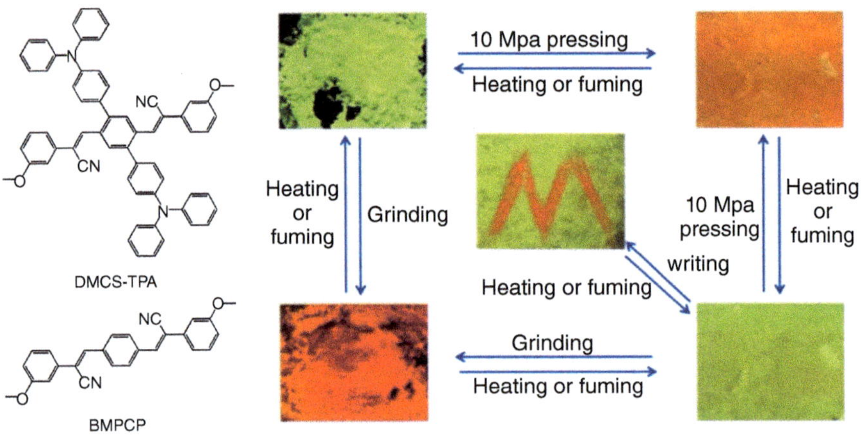

Figure 3.48 (a) Structure of DMCS-TPA and BMBCP and (b) photographs of DMCS-TPA under 365 nm UV light upon different treatments. Source: Reproduced with permission of Sun et al. [4] from Copyright © Royal Society of Chemistry 2014.

aggregation-induced emission (AIE) effect and high contrast mechanochromic (MC) behavior [4] and the liner molecule BMBCP without TPA was prepared for comparation. DMCS-TPA shows both AIE effect and high contrast MC behavior with a remarkable spectral of 87 nm and exhibits EC properties with multicolor showing. The results suggest that incorporation of electroactive moieties into luminophores to constitute D–A cruciform conjugated structures is a promising design strategy for preparing dual functional materials combining MC and EC properties (Figure 3.48).

Compared with EC anode material, the number of studies on cathode material is less, so there are few reports about this material at present. Recently, organic/polymeric semiconductors with D–A structures have been popularly exploited as EC materials. The ICT interactions in the D–A molecules are capable of achieving a low bandgap; by applying positive and negative potentials, multicolored electrochromism can be facilely realized in the ECDs based on the D–A materials (Figure 3.49).

In 2016, Yuan Ling et al. designed and synthesized a series of ambipolar EC materials composed of electron-withdrawing benzodipyrrolidone (BDP) moieties [33]. The authors chose BDP as an electron defect unit to construct low-bandgap D–A structural molecule. By introducing suitable electron donors, this material can realize different ICT interactions between electron donors and acceptors and make the target chromophores have different colors covering the whole visible range. Multicolor electrochromism at negative potential is caused by a reversible two-step reduction involved in the electrochemical process of BDP moieties. Upon accepting two electrons, the BDP unit with a quinoid structure turns into a dianion with a different aromatic resonance structure. Therefore, by simply etching the ITO electrodes, reversible multicolored electrochromism can also be achieved in single-layered ECDs at a fixed potential owing to the ambipolar characteristics

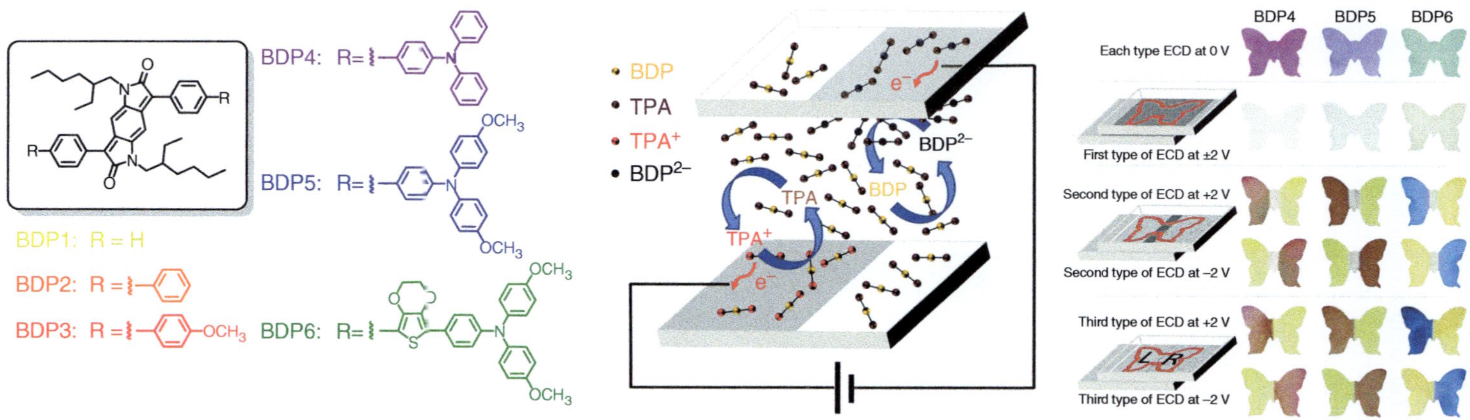

Figure 3.49 (a) Chemical structures of ambipolar EC materials BDP1–BDP6, (b) working principle of the second- and third-type ECDs based on ambipolar BDPs, and (c) photographs of the ECDs based on ambipolar BDP4, BDP5, and BDP6 at different potentials. Source: Ling et al. [33].

of the BDPs ambipolar electrochromes and present a new path to design novel ambipolar materials for efficient multicolored ECDs.

In 2017, De-Cheng Huang et al. combined TPA with viologen structure, proposing a facile "linkage" approach to merging ambipolar EC behaviors using a single small molecule. Based on this principle, they synthesized two bipolar EC materials containing ether bonds, TPA-VIO and TPA-OAQ. By incorporating cathodic EC moieties into TPA as a charge trapping layer, the resultant ambipolar ECDs not only reduced the driving voltage and switching time in the EC procedure compared with the devices using TPA only but also showed the ability to merge the colors

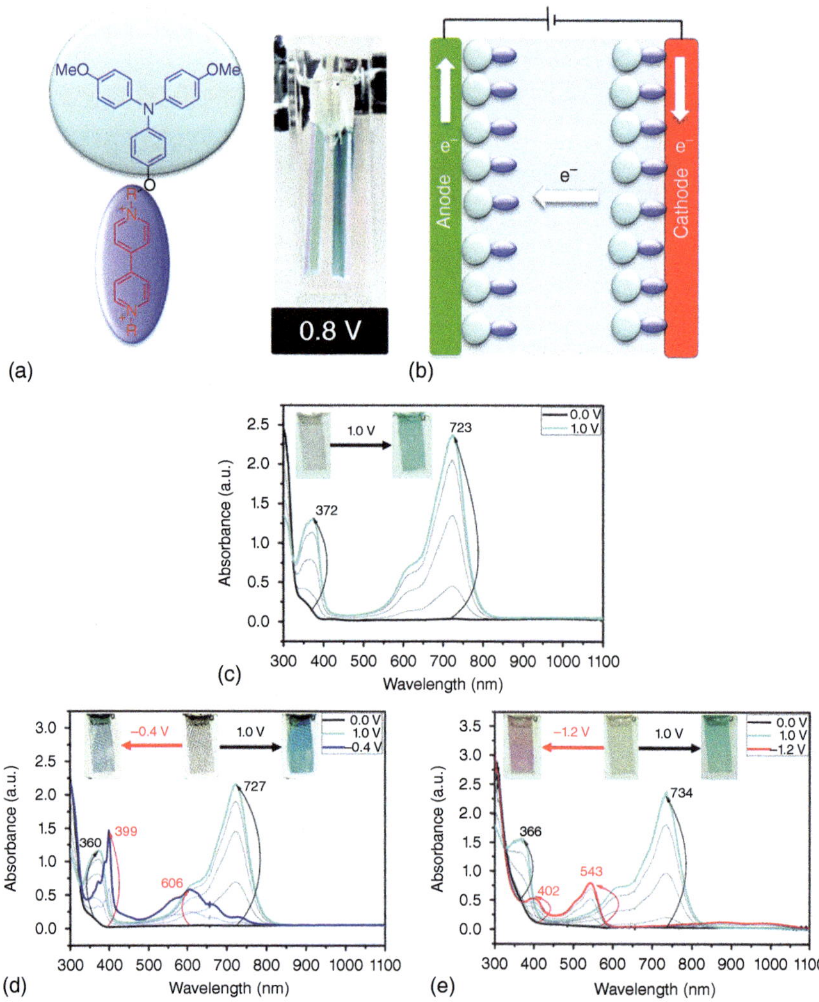

Figure 3.50 (a) Schematic diagram of color change for ambipolar materials, (b) structures of ambipolar TPA derivatives, (c) electrochromic behavior and corresponding photographs, and (d) CV diagrams for ambipolar materials. Source: Reproduced with permission of Huang et al. [34] from Copyright © Royal Society of Chemistry 2017.

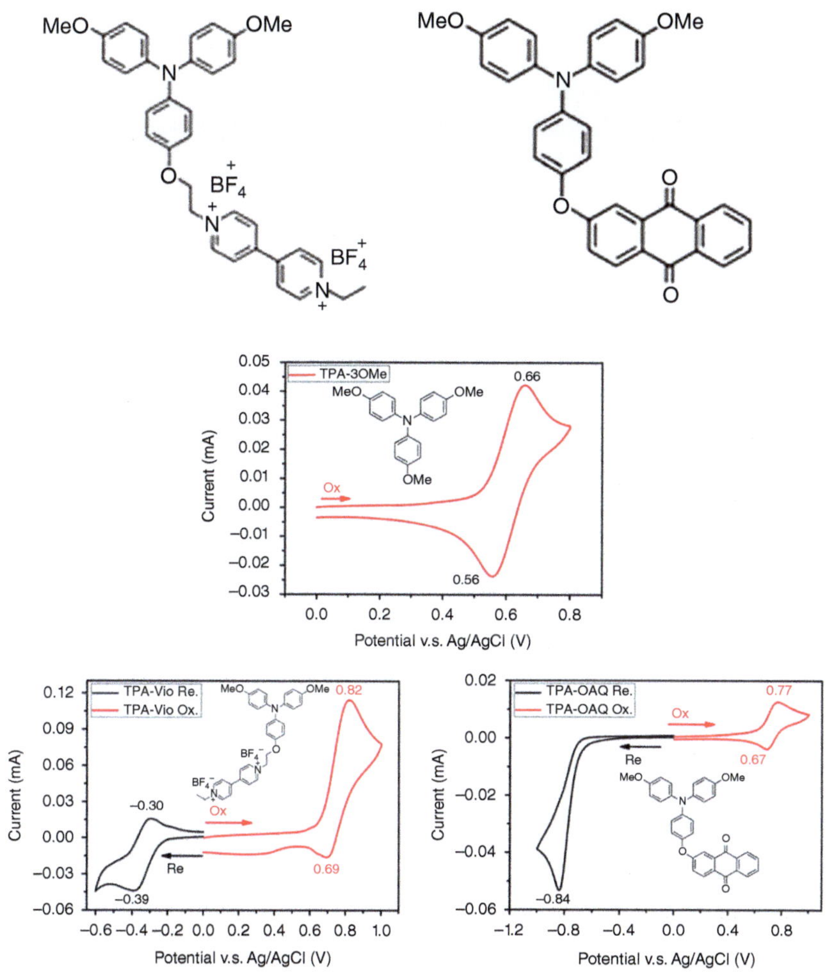

Figure 3.50 (*Continued*)

contributed by the anodic TPA and cathodic EC moieties. This enhanced the electrochemical stability and EC performance of the obtained ECDs containing ambipolar EC materials. These results demonstrated conclusively that the incorporation of suitable counter EC moieties into TPA derivatives is a facile and feasible approach to fabricating novel ECDs (Figure 3.50) [34].

3.11 Hydrocarbon Derivatives-NIR-OEC

In addition to EC materials in the visible range, materials with strong absorption/emission of IR light have also been paid more attention in recent years. Organic molecules that strongly absorb/emit NIR light have many potential applications, including security inks, bioimaging, photodynamic therapy, photoimmunotherapy,

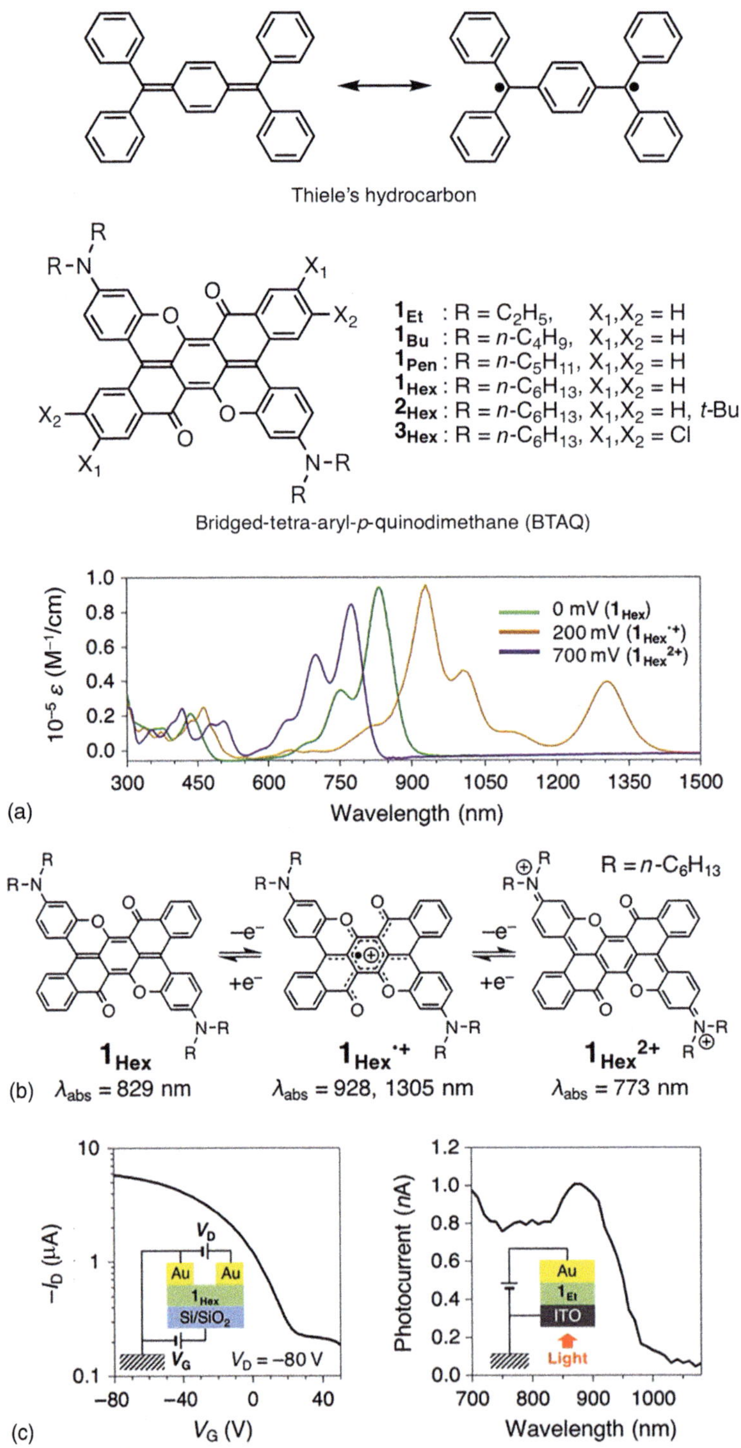

Figure 3.51 (a) Structures of Thiele's hydrocarbon and bridged tetra-aryl-*p*-quinodimethanes (BTAQs) and (b) absorption spectra during stepwise electrochemical oxidation of 1Hex in CH_2Cl_2 with 0.1 M *n*-Bu_4NClO_4 in air. (c) Twostep equilibrium of 1Hex, (d) transfer characteristics of a transistor made with 1Hex, and (e) near-infrared photocurrent in a film of 1Et at an applied voltage of 1 V. Schematic device structures are shown in insets of panels (c) and (d). Source: Okamoto et al. [30].

and photovoltaic devices. Although various types of NIR absorbing/emitting organic molecules have been developed so far, most of them suffer from relatively low stability and high synthetic cost.

Based on fluorinated colorless dyes, a kind of quinolone compound, called bridged tetra-aryl-*p*-quinodimethanes (BTAQs), was synthesized [35]. They were proved to have closed shell structures in the ground state. The quinoid structure of the BTAQs showed no biradical character. Instead, they exhibited two-step near-IR electrochromism, which shows strong absorption bands in the 800–900 nm region. The near-IR absorption peaks are related to the surrounding substituents. The absorption peaks of tetrachloro derivatives (3hex) are the longest (874 nm) (Figure 3.51).

According to the result of DFT calculations, they found that the absorption band corresponds to the HOMO → LUMO transition and is attributable to the $\pi \rightarrow \pi^*$ transition, because the HOMO and LUMO are delocalized over the π-skeleton. The peak position of the NIR absorption band of BTAQs was dependent upon solvent polarity. Because they do not show biradical character in the ground state, BTAQs are relatively stable. BTAQs is a new class of organic molecules, which have potential applications in various technologies utilizing NIR light.

3.12 Conclusions and Perspective

In this chapter, we mainly summarized the recent development of EC small molecules, which draw less attention but are promising materials. We introduce the EC materials with different structure, design strategy, and mechanisms. This kind of materials have the advantages of bright color, high contrast, good solubility, easy processing, and so on. They are also more flexible to adapt to different substrate materials and achieve more functions, which are currently impossible to achieve. At the same time, its relatively simple synthesis steps enable it to have great advantages in industrialization. However, these small molecular compounds are also faced with the defects of serious influence of water and oxygen on their stability and high driving voltage. Therefore, researchers should pay more attention to improve the stability of these materials. In this respect, researchers have made remarkable achievements, such as the discovery of bistable materials, electrode modification methods, and so on. We believe that these materials will play an indispensable role in the huge market of EC materials in the future.

References

1 Kenichi Nakamura, Y.O., Sekikawa, T., and Sugimoto, M. (1987). Electrochromic characteristics of organic materials with a simple molecular structure. *Japanese Journal of Applied Physics* 26 (6): 931–935.

2 Xu, X. and Webster, R.D. (2014). Primary coloured electrochromism of aromatic oxygen and sulfur diesters. *RSC Advances* 4 (35): 18100.

3 Zhang, Y.-M., Wang, X., Zhang, W. et al. (2015). A single-molecule multicolor electrochromic device generated through medium engineering. *Light: Science & Applications* 4 (2): e249–e249.

4 Sun, J., Lv, X., Wang, P. et al. (2014). A donor–acceptor cruciform π-system: high contrast mechanochromic properties and multicolour electrochromic behavior. *Journal of Materials Chemistry C* 2 (27): 5365.

5 Siegfried Hünig, K.M., Wenner, H. et al. (1999). Violene/cyanine hybrids: a general structure for electrochromic systems. *Chemistry A European Journal* 5 (7): 1969–1973.

6 Siegfried Hünig, K.M., Wenner, H. et al. (2000). Violene/cyanine hybrids as electrochromics part 2: tetrakis(4-dimethylaminophenyl)ethene and its derivatives. *Chemistry A European Journal* 6 (14): 2618–2632.

7 Shunji Ito, K.A., Kawakami, J., Tajiri, A. et al. (2006). Synthesis, stabilities, and redox behavior of mono-, di-, and tetracations composed of di(1-azulenyl)methylium units connected to a benzene ring by phenyl- and 2-thienylacetylene spacers. A concept of a cyanine-cyanine hybrid as a stabilized electrochromic system. *Journal of Organic Chemistry* 72 (1): 162–172.

8 Ito, S., R, S., Mizushima, A. et al. (2018). Cyaninecyanine hybrid structure as a stabilized polyelectrochromic system: synthesis, stabilities, and redox behavior of di(1-azulenyl)methylium units connected with electron-accepting π-electron systems. *Arkivoc* (2): 145–169.

9 Ishigaki, Y. et al. (2017). Bis(diarylethenyl)thiophenes, -bithiophenes, and -terthiophenes: a new series of electrochromic systems that exhibit a fluorescence response. *Canadian Journal of Chemistry* 95 (3): 243–252.

10 Kobayashi, N., Miura, S., Nishimura, M., and Urano, H. (2008). Organic electrochromism for a new color electronic paper. *Solar Energy Materials and Solar Cells* 92 (2): 136–139.

11 Tanaka, S., Watanabe, Y., Nagashima, T., and Kobayashi, N. (2009). Phthalate-derivative/TiO_2-modified electrode for electrochromic application. *Solar Energy Materials and Solar Cells* 93 (12): 2098–2101.

12 La, M., Liu, M., M., Liu, P. et al. (2008). Electrochromic properties based on oligothiophene derivatives. *Chinese Journal of Chemistry* 26 (8): 1523–1526.

13 Yin, B., Jiang, C., Wang, Y. et al. (2010). Synthesis and electrochromic properties of oligothiophene derivatives. *Synthetic Metals* 160 (5–6): 432–435.

14 Zeng, J., Zhang, X., Zhu, X., and Liu, P. (2017). Synthesis and electrochromic properties of star-shaped oligomers with phenyl cores. *Chemistry, an Asian Journal* 12 (17): 2202–2206.

15 Wan, Z., Zeng, J., Li, H. et al. (2018). Multicolored, low-voltage-driven, flexible organic electrochromic devices based on oligomers. *Macromolecular Rapid Communications* 39 (10): e1700886.

16 Sharmoukh, W., Ko, K.C., Ko, J.H. et al. (2008). 5-Substituted isophthalate-based organic electrochromic materials. *Journal of Materials Chemistry* 18 (37): 4408.

17 Sharmoukh, W., Ko, K.C., Park, S.Y. et al. (2008). Molecular design and preparation of bis-isophthalate electrochromic systems having controllable color and bistability. *Organic Letters* 10 (23): 5365–5368.

18 Pengfei Zhang, X.X. et al. (2019). Multi-colour electrochromic materials based on polyaromatic esters with low driving voltage. *Journal of Materials Chemistry C* 7: 9467–9473.

19 Christiansen, D.T., Tomlinson, A.L., and Reynolds, J.R. (2019). New design paradigm for color control in anodically coloring electrochromic molecules. *Journal of the American Chemical Society* 141 (9): 3859–3862.

20 He, L.-h., Wang, G.-m., Tang, Q. et al. (2014). Synthesis and characterization of novel electrochromic and photoresponsive materials based on azobenzene-4,4′-dicarboxylic acid dialkyl ester. *Journal of Materials Chemistry C* 2 (38): 8162–8169.

21 Long, J.-F., Tang, Q., and Lv, Z. (2017). Synthesis and characterization of dual-colored electrochromic materials based on 4′-(4-alkyl ester)-4,2′:6′,4″-terpyridinium derivatives. *Electrochimica Acta* 2017 (248): 1–10.

22 Zhang, W., Zhu, C., Huang, Z. et al. (2019). Electrochromic 2,4,6-triphenyl-1,3,5-triazine based esters with electron donor-acceptor structure. *Organic Electronics* 67: 302–310.

23 Zhang, Y.-M., Li, M., Li, W. et al. (2013). A new class of "electro-acid/base"-induced reversible methyl ketone colour switches. *Journal of Materials Chemistry C* 1 (34): 5309.

24 Zhang, Y.-M., Xie, F., Li, W. et al. (2016). A methyl ketone bridged molecule as a multi-stimuli-responsive color switch for electrochromic devices. *Journal of Materials Chemistry C* 4 (21): 4662–4667.

25 He, Z., Yuan, X., Zhao, Y. et al. (2016). A greener electrochromic liquid crystal based on ionic liquid electrolytes. *Liquid Crystals* 43 (8): 1110–1119.

26 Kim, G.W., Kim, Y.C., Ko, I.J. et al. (2018). High-performance electrochromic optical shutter based on fluoran dye for visibility enhancement of augmented reality display. *Advanced Optical Materials* 6 (11): 1701382.

27 Weng, W., Higuchi, T., Suzuki, M. et al. (2010). A high-speed passive-matrix electrochromic display using a mesoporous TiO_2 electrode with vertical porosity. *Angewandte Chemie (International Ed. in English)* 49 (23): 3956–3959.

28 Zhang, Y.M. et al. (2014). Highly durable colour/emission switching of fluorescein in a thin film device using "electro-acid/base" as in situ stimuli. *Chemical Communications* 50 (12): 1420–1422.

29 Wang, X., Li, W., Li, W. et al. (2017). An RGB color-tunable turn-on electrofluorochromic device and its potential for information encryption. *Chem Commun. (Camb)* 53 (81): 11209–11212.

30 Wang, X., Wang, S., Gu, C. et al. (2017). Reversible bond/cation-coupled electron transfer on phenylenediamine-based rhodamine B and its application on electrochromism. *ACS Applied Materials & Interfaces* 9 (23): 20196–20204.

31 Wang, Y., Wang, S., Wang, X. et al. (2019). A multicolour bistable electronic shelf label based on intramolecular proton-coupled electron transfer. *Nature Materials* 18 (12): 1335–1342.

32 Zhang, Y.M., Li, W., Wang, X. et al. (2014). Highly durable colour/emission switching of fluorescein in a thin film device using "electro-acid/base" as in situ stimuli. *Chem Commun. (Camb)* 50 (12): 1420–1422.

33 Ling, Y., Xiang, C., and Zhou, G. (2017). Multicolored electrochromism from benzodipyrrolidone-based ambipolar electrochromes at a fixed potential. *Journal of Materials Chemistry C* 5 (2): 290–300.

34 Huang, D.C. et al. (2017). Preparation and optoelectronic behaviours of novel electrochromic devices based on triphenylamine-containing ambipolar materials. *Journal of Materials Chemistry C* 5 (36): 9370–9375.

35 Okamoto, Y., Tanioka, M., Muranaka, A. et al. (2018). Stable Thiele's hydrocarbon derivatives exhibiting near-infrared absorption/emission and two-step electrochromism. *Journal of the American Chemical Society* 140 (51): 17857–17861.

4

Viologen OEC

4.1 The Introduction of OEC and Viologen

4.1.1 General Introduction

Organic electrochromic materials (OECs), which exhibit reversible change of color upon the application of electrochemical potential, have broad promising application in energy-saving windows, full color passive displays, auto-dimming mirrors, and wearable devices for military and/or health monitoring purpose [1–3] due to their cost-saving, easy fabrication, coloration efficiency, and high contrast properties. Therefore, a series of organic semiconductors have been applied to OEC, which attracted worldwide attention of researchers.

In general, we collectively refer to 1,1′-disubstituted-4,4′-bipyridine compounds as viologen, which was first coined by Michaelis in 1933 [4]. Since the 1970s, a variety of novel OEC have been subsequently reported, such as conducted polymers, conjugated small molecules, covalent organic framework, etc. Among them, viologen has its merits suitable for electrochromic device (ECD) application with low driving voltage, high contrast, flexible switching, color mixing capability, and simple device setup. It is also the first OEC material that has been successfully commercialized by Gentex Corporation [5]. There are many book chapters and reviews that have comprehensive introduction on viologen; a part of great works were selected in the following Table 4.1.

4.1.2 Research History of Viologen

As shown in Scheme 4.1, in the nineteenth century, as the electrochromic (EC) properties of viologen had been discovered, the researchers started to tune the exhibited color by simply adding different N-substituents, counter anion modification, and polymerization strategy. In the 2000s, since EC materials exhibit the function of adjusting transmittance, which has attracted wide attention, researchers began to focus on improving the performance of ECDs. The stability and efficiency have been significantly improved by electrode modification, for example, choose the nanocrystalline electrode. More advanced structure sprang up, such as porous materials and supramolecular materials; with that comes the new function, for

Organic Electronics for Electrochromic Materials and Devices, First Edition. Hong Meng.

Table 4.1 Recent review and book chapter about viologen-based EC materials and devices.

Book chapter/review	Year	Author	Highlight
The electrochemistry of the viologens	1981	Bird and Kuhn [6]	Redox ability of viologen
The bipyridines	1984	Summers [7]	Synthesis and properties
Chemistry of viologens	1991	Sliwa et al. [8]	Chemical mechanism of viologen
The viologens: physicochemical properties, synthesis, and applications of the salts of 4,4′-bipyridine	1998	Monk [9]	Fundamental research and application
The viologens	2007	Monk et al. [10]	Basic structures and properties
Electrochromic materials and devices based on viologens	2015	Monk et al. [11]	Comprehensive review, from materials to ECD device
Viologens-based electrochromic materials and devices	2018	Yu and coworkers. [12]	Synthesis of viologen-based electrochromes with co-redox species and their ECD performance
Design strategies and redox-dependent applications of insoluble viologen-based covalent organic polymers	2019	Škorjanc et al. [13]	Covalent organic polymers
Viologen-inspired functional materials: synthetic strategies and applications	2019	Ding et al. [14]	Viologen-like materials and their applications

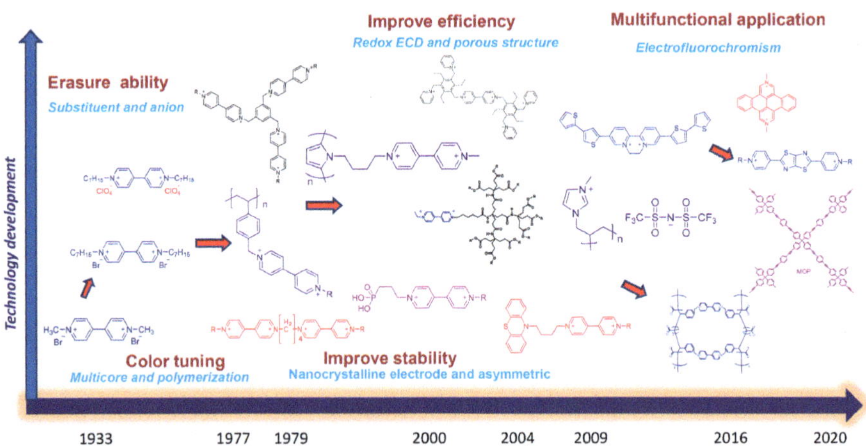

Scheme 4.1 Development routes of viologen.

example, electrofluorochromic materials. All in all, the study of viologen as an EC material has been less than 100 years since its discovery; we can divide the detailed study of this material into four stages as follows.

4.1.2.1 First Stage (1930s–1970s)

The first stage is the development of viologens, from the discovery to the electrochemistry and electrochromism researches. In 1933, Michaelis firstly discovered that 1,1′-dimethyl-4,4′-bipyridilium (MV) exhibited violet in the reduced state, so it was termed "viologen" [4]. After that, a series of theoretical researches allude to the intermediate state structure of viologens [15–18]. At the same time, they did fundamental research in color tuning by synthesizing different N-substituents. In 1973, Shoot made a new flat alphanumeric display using heptyl viologen. This can be regarded as the initial application of viologen for electrochromism [19].

4.1.2.2 Second Stage (1970s–2000s)

Since the existence of comproportionation reactions and the diffusion of radical cation, the stability and switching speed of viologen had been greatly damaged. In this stage, basically it is the deeper research of viologens including the optimization in device performance. Here are several typical methods.

Anionic modification: In 1977, Bruinink tried non-halogen large anions (ClO^{4-}, BF^{4-}) in order to avoid the recrystallization at the radical cation state. In addition, he applied SnO_2 as the electrode to obtain good inert transparency and higher overvoltage [20].

Polymerization: Polymers with viologen and viologen-like structures can exhibit unique properties, such as higher stability, excellent mechanical properties, and higher solvent processability. In 1979, Hirotsugu Sato used the radical film formed by the deposition of viologen on the electrode to achieve the ECDs. Since that, people took the advantage of the polymer effect to obtain longer memory period and lower potential ECDs [21]. Then in 1988, researchers have begun to attach pendant viologen on a conductive polymer backbone. Shu and Wrighton synthesized a pyrrole-containing viologen derivative and electrochemically completed the polymerization to synthesize a side chain-type viologen polymer with polypyrrole as the main chain [22].

In addition, the mechanism of radical cation of viologen was developed by means of modern spectroscopic technology. In 1985, Gaudiello et al. first systematically expounded the dimerization of viologens by electron spin resonance spectrometer (ESR) spectra [23]. After that, in 1992 Monk et al. used in situ electrochemical electron paramagnetic resonance spectrometer (EPR) and UV–vis spectroscopy to investigate the mechanism of the bipyridylium comproportionation reaction [24].

4.1.2.3 Third Stage (2000s–2010s)

At this stage, scientists started to optimize viologen-based device by electrode modification. More and more extended viologen structures have emerged during the early 2000, such as dendrimer [25], supramolecular cycle rings [26], unsymmetrical viologen [27], etc.

In 1999, Bonhote first applied nanocrystalline metal oxide as EC electrodes. A series of viologen molecules substituted with anchoring groups as a monolayer adsorbed on the surface of the nanocrystalline TiO_2 electrode had been synthesized for realizing higher conductivity, fast electron exchange, and optical amplification purposes. In this study, a transmissive device was fabricated by adding Prussian blue to the anode, and a reflective device was fabricated using Zn on the anode and a white reflector between the counter electrode and working electrode. This research laid an important foundation for obtaining high-performance ECDs [28]. In 2004, Moller et al. has successfully implemented coloration and switching times of ECD utilizing inkjet printing technology and nanocrystalline semiconductor electrode modified with a self-assembled monolayer of viologen [29].

In 2005, Jee-Hyun Ryu et al. proposed a new concept of reflective display by incorporating viologen moiety onto the surface of monodisperse polymeric microspheres. The ECD system featured on inherent white background yet exhibited undesirable response time [30]. In the follow-up study, he concentrated on enhancing the switching speed by introducing porous structure into the ECD. The response time of the device decreased to 270 ms since the high specific surface area can increase the electrolyte contents and thus effectively improve the ionic conductivity [31]. At the same time, optical companies such as Gentex and Essilor began to officially launch EC commercial devices based on viologen [5, 32–35].

4.1.2.4 Fourth Stage (2010s–Present)

At this stage, researchers have begun to use porous materials to achieve higher ion conductivity and contrast and have also used novel materials such as metal organic frameworks and conjugated organic frameworks for related exploratory research. The device structure is constantly optimized, and multifunctional devices have appeared, creating opportunities for a wider range of applications.

In 2010, Marina Freitag and Galoppini formed a host–guest complexes system by encapsulating the nonfunctionalized viologens V into cucur[7]bituril (CB[7]), which acts as a molecular container, anchoring the V^{2+}@CB[7] onto the surface of TiO_2 films. This strategy effectively prevents the side reaction of viologen and improves the cycle life [36].

In 2011, Xi Tu et al. incorporated cathodic material viologen with a series of anodic materials to select the most suitable cathode–anode composite for ECD application. As a result, the device structure was simplified, rendering multicolor (colorless-wine-blue) at different voltages [37].

In order to ameliorate the durability of ECDs, directly combining the electrochromes with polymer electrolyte or electrode was the most universal method. In this way, the switching properties and coloration efficiency of the ECD could be greatly damaged. In 2014, Eunhee Hwang et al. designed an electrolyte-free and flexible ECD using graphene quantum dot (GQD)-viologen nanocomposites. Thanks to the high ionic conductivity and heat resistance of graphene, the device exhibited excellent operational stability, desirable switching performance, and thermal stability [38].

For future advanced device, further developments of multifunctional materials are in great demand. In 2015, Christian Reus et al. extended π-system with an electron-deficient phosphole core; an EC system that can be reduced with a simple element of metal was obtained [39]. Afterwards, in 2018, Guoping Li et al. replaced phosphorus with other oxygen elements. Due to their high visible-light absorption and optimal energy levels, it can be applied as the photocatalysts of visible-light-driven hydrogen evolution [40]. In 2017, Alexis N Woodward designed a dual function device by adding thiazolo-[5,4-*d*]thiazole (TTz) heterocycle between the two pyridines. In this way, the device exhibits strong blue fluorescence and vibrant and reversible EC properties [41]. In Section 4.1.3, we will discuss the redox electrochemistry and problems that existed in viologen.

4.1.3 Electrochromism and Electrochemistry of Viologens and Their Device

Firstly, the structure of the viologen itself is as follows. As Figure 4.1 shows, in the redox process, there are three oxidation states in the viologen molecule. The dication salt is the most stable form among the three states, and therefore it can be purchased or synthesized in the laboratory. Generally, the dipyridilium salts are colorless unless the existence of counter ion or the charge-donating species. For example, if we choose the feeble absorbance anions such as chloride or sulfate, the dication salt is colorless, but if we change the strong absorbance anions like iodide, it will show scarlet [42]. The one electron-reduced state is called radical cation. This kind of radical is also stable since the radical electron on the nitrogen delocalize the π-framework of the entire two-ring bipyridyl nucleus. Hence the color of radical cation can be kept for a long period of time in the solution. Owing to the intramolecular photo-effected electronic excitation process, it shows intensive color change. The third redox form is neutral compound or di-reduced viologen. This compound is simply viewed as a reactive amine; however, it doesn't show vibrant color since no obvious optical charge transfer is accessible [43–46]. The potential needed to initiate the reduction depends both on the substituents at nitrogen and on the bipyridyl nucleus.

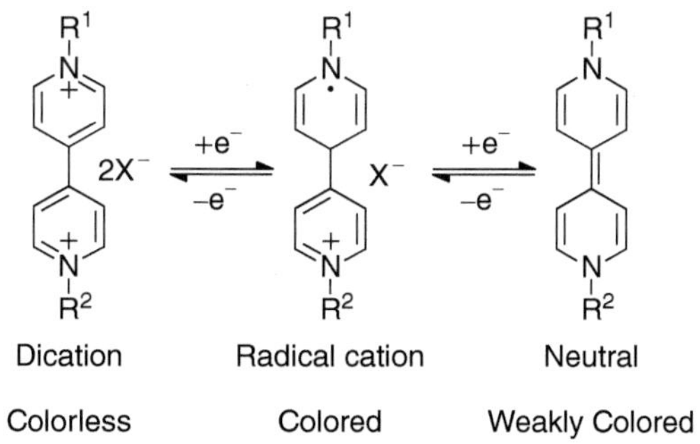

Figure 4.1 Viologens in three forms: R^1 and R^2 refer to the substituent at nitrogen.

4.2 Different Structures of Viologen-Based Electrochromic Materials

4.2.1 Synthesis of Viologens

To date, some typical methods have been developed to synthesize different viologen and their derivatives, which are briefly introduced as follows.

4.2.1.1 Direct Substitution Reaction

Direct substitution reaction for pyridine and the corresponding halogenated hydrocarbons is the simplest and most common method of synthesizing such a compound. The mechanism is mainly based on Menshutkin reaction [48], the reaction and the steps are simple, and it is easy to control different counter anion category [49].

4.2.1.2 Zincke Reaction

The Zincke reaction has also been widely applied for the synthesis of the conjugated viologens compound. This reaction provides an excellent sample for the synthesis of substituted structures with conjugated rings [50].

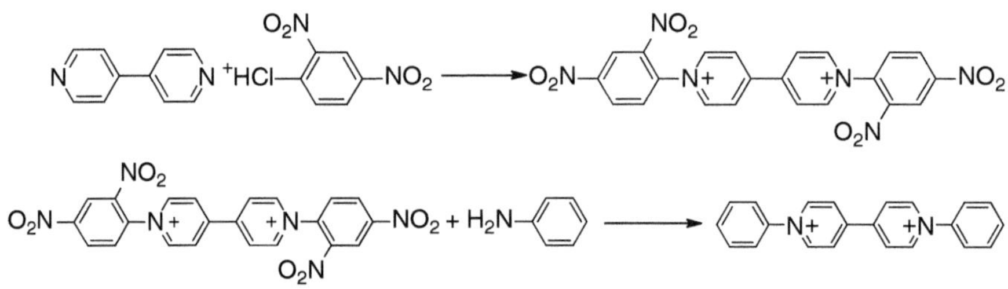

4.2.1.3 Methods for Synthesizing Bipyridine

(1) The reactions of oxonium ion salt and primary amino provide different structures of viologen.

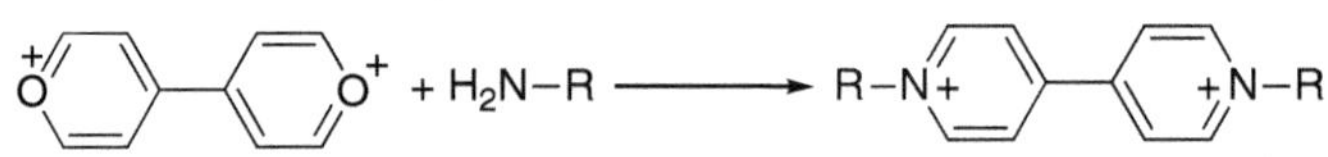

(2) Pyridine molecules can form viologen by self-coupling (usually add sodium).

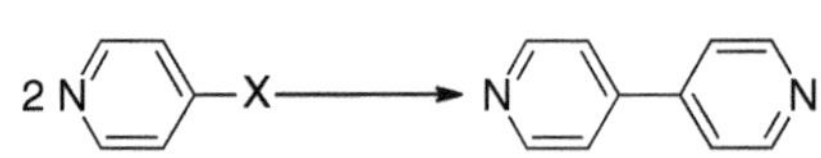

(3) Conjugation-extended viologen molecules can be synthesized by Stille coupling reaction [51].

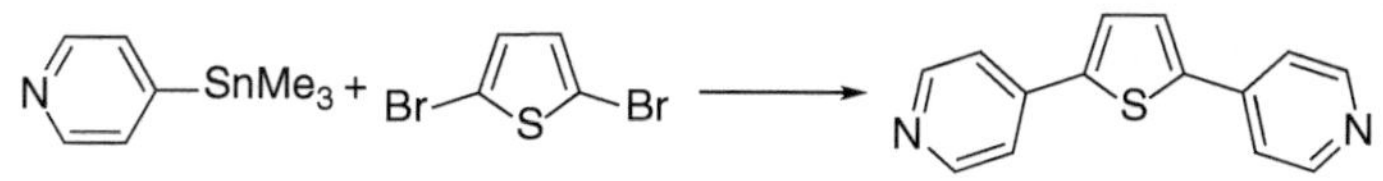

4.2.2 The 1,1′-Substituted Viologen

4.2.2.1 Simple Alkyl

As a commercial material, the solubility of viologens is a critical factor. Researchers found that the solubility of free radical species in aqueous solution is lower than that of ionic species, and its solubility is significantly affected by the size of the cation and the counter anion species. Therefore, they studied the solubility, stability, film-forming properties, and absorption spectra of viologens with simple alkyl chains from C_1 to C_{13}.

The R groups protect the anion from the solvent molecules. In general, the larger the radius of anions, the smaller the electrostatic repulsion between the dipyridinium ions. Consequently, this kind of balanced state can make the viologen system more stable. With all things considered, researchers found C_7 has the best EC performance. In addition, polyfluoroalkyl substitutions have also been reported [52].

From the aforementioned Table 4.2, we can draw a rough conclusion. The viologen radical cations substituted by simple alkyl groups usually appear blue-violet, while the radical cations substituted by aryl groups are green [55]. As the dimerization rate increases with the length of alky substituents, the formation of dimers will make the color shift to red.

4.2.2.2 Acid Group

Since 1999, Bonhote et al. [28] firstly used TiO_2 nanocrystalline electrode and anchor group substituted viologens to achieve higher conductivity, sharp color exchange,

Table 4.2 Optical and colorimetric properties of selected viologen radical cations.

R	Anion	Solvent	Color	ε_{max}
Methyl	Cl^-	H_2O	Blue	13 700
Ethyl	ClO_4^-	DMF	Blue	12 200
Heptyl[a)]	Br^-	H_2O	Mauve	26 000
Octyl[a)]	Br^-	H_2O	Crimson	28 900
Benzyl	Cl^-	H_2O	Mauve	17 200
4-Cyanophenyl	BF_4^-	PC	Green/black	83 300

a) Radical cation dimer.
Source: Ho et al. [12]; Monk et al. [47] ; Van Dam and Ponjee [53]; Xu et al. [54].

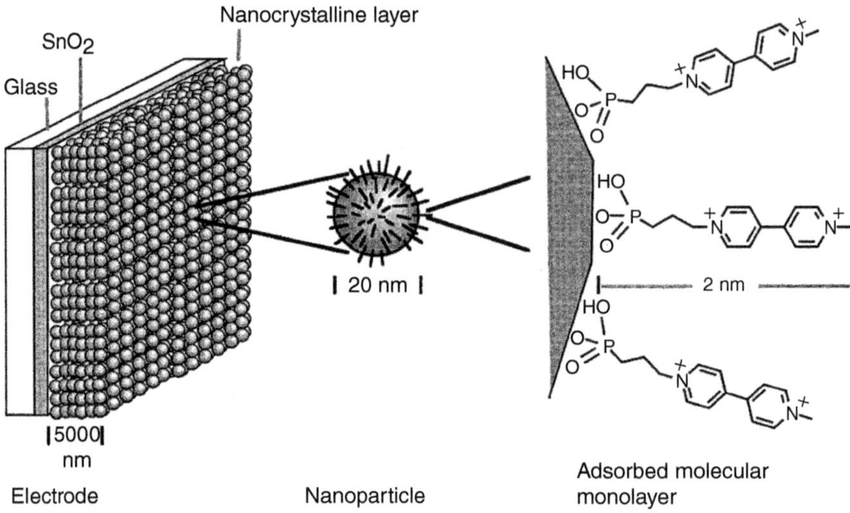

Figure 4.2 Adsorption of viologen on the surface of nanocrystalline electrode. Source: Bonhote et al. 28. Copyright (1999) Elsevier.

and fast electron transfer ECD, as shown in Figure 4.2. The mechanism of this experiment is that the highly specific surface area of the nanocrystal electrode can absorb a single layer of ions, thereby obtaining such a derivatized highly porous film, and the linear structure of the nanocrystal can intensify the light absorption, thereby obtaining the effect of optical amplification. The adsorption behavior of viologen with different structures on the surface of TiO_2 was studied by absorption spectra, which also can verify the presence of dimerization. This work provides a direction for researchers to realize the higher-performance ECD.

Furthermore, since the viologen dication salt is reduced at the cathode under controllable diffusion rate, the researchers also made a follow-up research on the effect of the thickness of the TiO_2 film on the transmittance and the coloration of the device [56].

4.2.2.3 Ester and Nitrogen Heterocycle

In general, the ester groups possess EC effect and they usually exhibit red color. The researchers have synthesized species, which have different colors at different voltages by combining ester-containing species with viologen. The species appear red at 2.0 V and blue at 0.8 V, though their optical contrast is less than 43% (Figure 4.3) [57].

Nitrogen heterocycle species (such as phenothiazine, indole, morpholine, pyrrole, and carbazole) are important anodic materials in electrochromism. Furthermore their drive voltage is about 2 V, which is higher than usual viologen

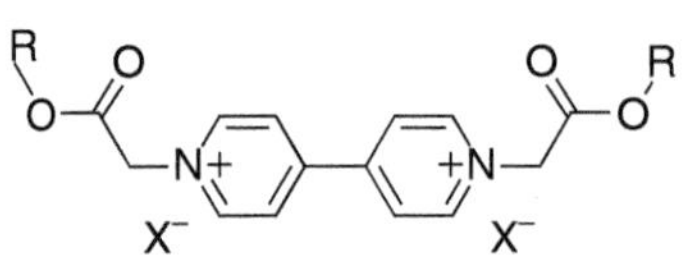

Figure 4.3 Structure of ester viologen.

discoloration voltage by 1 V. The researchers applied this property as an opportunity to implement a cathode–anode coupling material, which also can simplify the device structure [37].

4.2.2.4 Asymmetric Substitution

Studies showed that symmetric viologen has the trend to form dimer, which highly impairs the performance of ECD. A feasible way to effectively inhibit the formation of dimer is asymmetry strategy. Moon and coworkers [58] found the asymmetric viologen-based ECD (benzyl heptyl viologen dication, BHV^{2+}) had higher contrast ratio, effective coloration, and remarkable coloring/bleaching stability because of the asymmetric substitution.

Additionally, the asymmetric structure may come to neutral and gray color, since the neutral color such as gray is more suitable as a filter option for its wild adaptability in various environments. Devices that intrinsically exhibit colorless to gray/dark green/black color are popular in e-books and smart windows for privacy purpose. Alesanco et al. [58, 60] found the asymmetric viologens offer more neutral-color state and maintain satisfactory bleached state (i.e. $\%T_b \approx 70\%$) than their corresponding symmetric viologens (Figures 4.4 and 4.5). After all, the insufficient response speed and lack of colorless state remain challenges to obtain high-performance ECDs.

4.2.3 Conjugate Ring System Expansion

4.2.3.1 Thiazolothiazole (TTz) Unit

It is noteworthy to mention that by adding fluorescent building block, the EC materials may have bifunctional characteristics. Walter and coworkers [41] used a rigid thiazolothiazole (TTz) structure as the core, as shown in Figure 4.6, and prepared a series of methyl-, octyl-, and benzyl-substituted viologen derivatives. The introduction of the TTz structure gave the viologen derivative an outstanding

Potential (V)	$\%T_b$	$\%T_c$	$\Delta\%T^{(b)}$	$x^{(b)}$	$y^{(b)}$	$Y^{(b)}$	$L*^{(c)}$	$a*^{(c)}$	$b*^{(c)}$	Color(d)
off		–	–	0.309	0.326	78.4	91	0	–2	
0.5		59.7	17.9	0.307	0.337	64.0	94	–6	2	
–0.6		41.5	36.0	0.317	0.353	44.1	72	–7	8	
–0.7	77.5	31.4	46.1	0.326	0.366	33.1	64	–7	11	
–0.8		23.1	54.4	0.335	0.380	24.2	56	–8	15	
–0.9		16.6	60.9	0.316	0.370	18.8	50	–10	9	
–1.2		0.9	76.6	0.234	0.559	3.3	21	–38	19	

Figure 4.4 Color coordinates of asymmetric viologens device. [a]Transmittance change at $\lambda = 600$ nm: ($\%T_b - \%T_c$), where T_b and T_c are the percentage of transmittance at bleached (off) and colored states, respectively. [b]Color coordinates (D65): xyY 1931 (x = blue/red; y = blue/green). [c]Color coordinates (D65): L*a*b* 1976 (a* = (−)/red (+); b* = blue (−)/yellow (+)) dColor interpretation of the corresponding color coordinates (L*a*b) acquired through a color converter software (included to ease their interpretation). Source: Adapted with permission from Alesanco et al. 60. Copyright (2016) American Chemical Society.

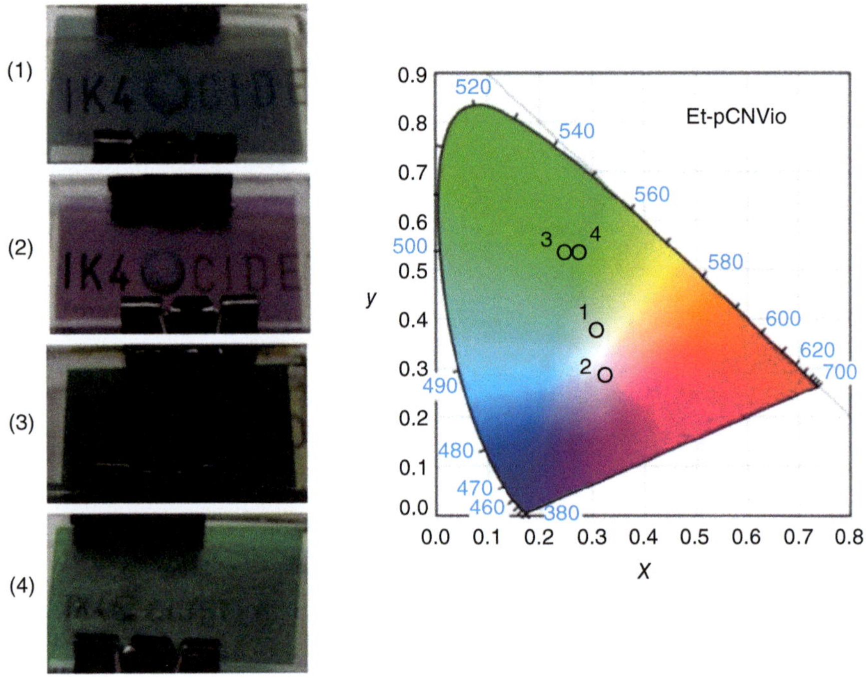

Figure 4.5 Different devices and color coordinates. Source: Adapted with permission from Alesanco et al. 60. Copyright (2016) American Chemical Society.

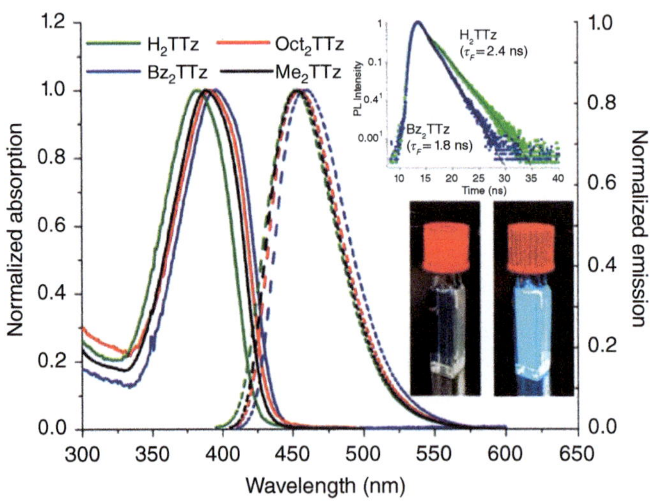

Figure 4.6 Normalized absorption and emission spectra of H_2TTz^{2+}, Me_2TTz^{2+}, Oct_2TTz^{2+}, and Bz_2TTz^{2+}. Source: Adapted with permission from Woodward et al. 41. Copyright (2017) American Chemical Society.

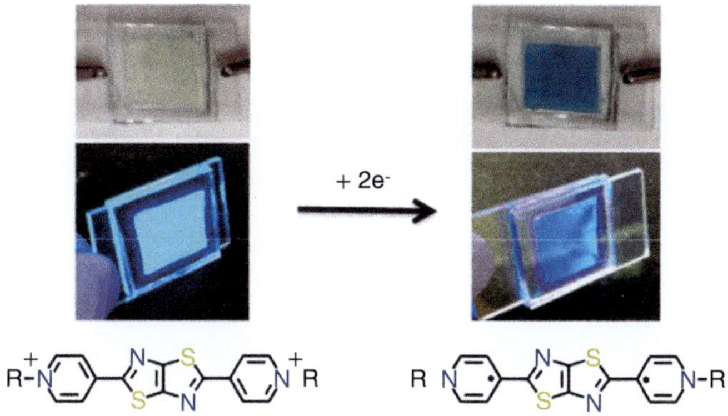

Figure 4.7 Fluorescence quenching. Source: Adapted with permission from Woodward et al. 41. Copyright (2017) American Chemical Society.

low reduction potential (−0.46 ~ −0.58 V vs saturated calomel electrode [SCE]) and high fluorescence quantum yield ($\Phi F = 0.96$). After applying a −0.7 V bias voltage in the Bz_2TTz^{2+} derivative solution for 25 minutes, fluorescence quenching can be observed (Figure 4.7). This dual-function interaction that affects each other's performance is considered to have promising prospects in sensor and biometric applications.

4.2.3.2 Perylenediimide (PDI) Unit

Perylenediimide (PDI) is a kind of EC material which exhibits red in the neutral state, blue in the semi-reduced state, and purple in the reduced state. Moon et al. [59] reported the synthesis of PDI-based viologen, and they synthesized PDI-2 V^{2+} multilayer film, which exhibited multiple-color electrochromism ranging from red to blue. The multilayer film is prepared by the layer by layer (LBL) method, which alternatively coated the designated substrates with aqueous polyanion and polycation solutions.

The results in Figure 4.8 show that the PDI-2 V ECD has multiple intense colors ranging from red to purple, and the intensity of color is proportional to the film layers. As shown in Figure 4.9, as the scanning rate gradually increases from 5 to 200 mV/s, the peak current and the scan rate have a linear relationship, proving that the PDI-2 V^{2+} multilayer film exhibits good stability under multiple cycles.

4.2.3.3 PBEDOTPh

The P(BEDOT-Ph) is an applicable polymer backbone to realize richer color of viologen since it has different EC colors and its redox potential don't overlap. The resulting viologen-bearing conductive polymer demonstrated five distinct colors, which is more plentiful than that of the individual materials (Figure 4.10) [61].

4.2.3.4 Heteroatoms Bridged

A series of chalcogen-bridged and phosphorus-bridged [39, 40, 62] viologens have been reported, as the chalcogen and phosphorus cores are electron deficient,

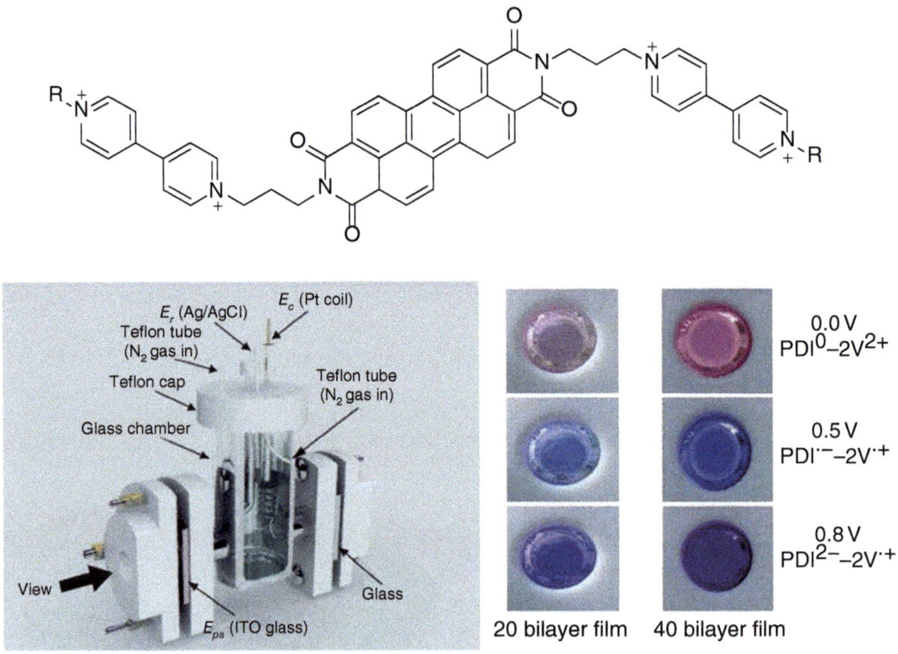

Figure 4.8 Multilayer PDI-MV device. Source: Adapted with permission from Kim et al. [59]. Copyright (2012) Royal Society of Chemistry.

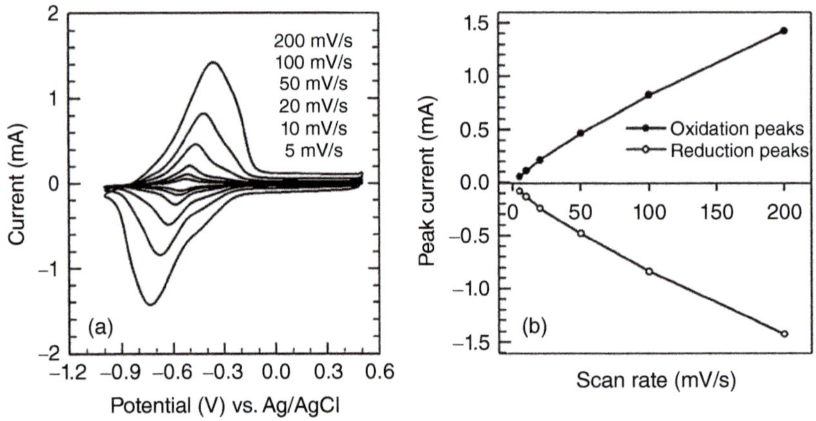

Figure 4.9 Cyclic voltammograms of a 10 bilayer film at various scan rates and plot of peak current vs. scan rate.

offering better ECD performance. Baumgartner and coworkers [62–64] synthesized a series different N-substituents phosphoryl-bridged viologen (Scheme 4.2). Research demonstrated that the combination of the π extended conjugation and the electron-deficient core considerably has lower highest occupied molecular orbital–lowest unoccupied molecular orbital (HOMO–LUMO) bandgap. Among

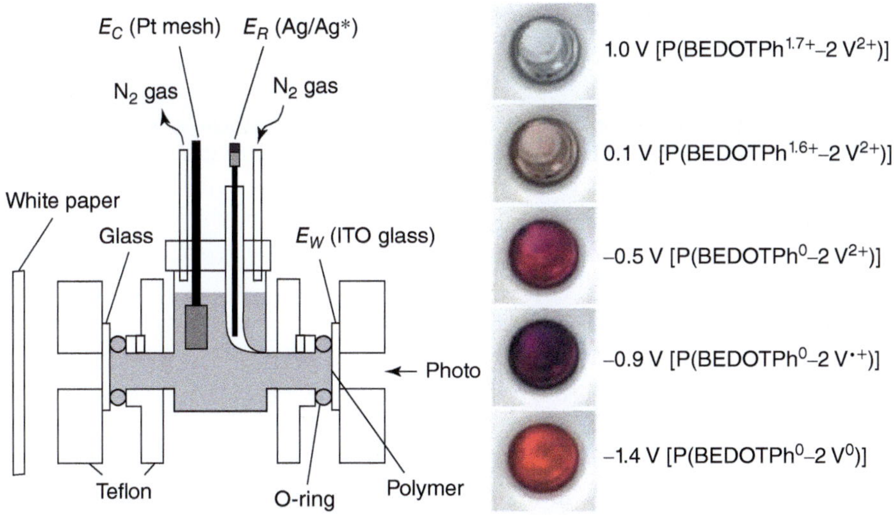

Figure 4.10 The structure of photographs device. Source: Adapted with permission from Ko et al. 61. Copyright (2012) Wiley.

them, the N-arylation substituents has the lowest HOMO–LUMO bandgap. Interestingly, ECD experiments verified that the energy bandgap modification does not change the color of the viologen (Table 4.3).

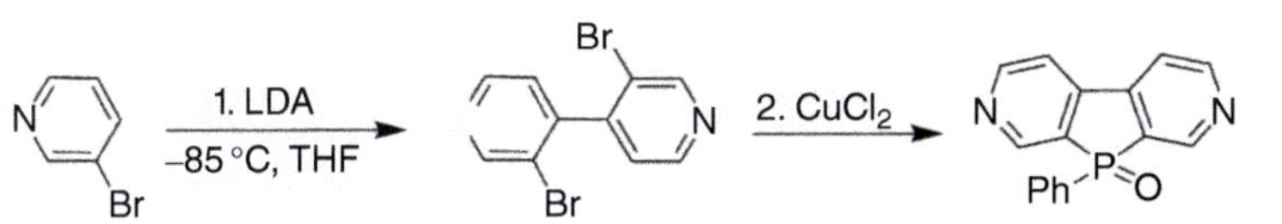

Scheme 4.2 The synthesis route of phosphoryl-bridged pyridiniums. Source: Based on Durben and Baumgartner [64].

He and coworkers [40] synthesized a series of chalcogenoviologens with heavy atoms (S, Se, and Te), which have narrow HOMO–LUMO bandgaps. Cyclic voltammetry (CV) experiments showed these materials have low reduction potential ascribe to the introduction of heteroatoms. This was attributed to a more rigid and conjugated structure. The results showed the Se–BnV^{2+} and Te–BnV^{2+} have desired electron-accepting characteristics and strong UV–vis absorption, providing a new strategy to explore visible light-driven hydrogen evolution. Lichtenberger and coworkers [65] did detailed research to investigate the intramolecular electron transfer in bipyridinium disulfides dication. Results showed the bipyridinium moiety provides a low-energy pathway for the reduction of S—S bond. Besides, they found the first reduction potential depends on the dihedral between the two heterocycles. The planarity of the structure promotes the electron delocalization, which gives us a way to lower the reduction potential by effective molecular modification.

Table 4.3 Optical and electronic properties of the chalcogen-bridged viologens and already known viologens.

Compound	λ_{max} (nm) (ε [M^{-1} cm^{-1}])	E_{red} (V)	E_g[a] (eV) (calculated)[b]	E_{LUMO}[c] (eV) (calculated)	E_{HOMO} (eV) (calculated)
2a	270 (2608) 345 (5443)	−2.69, −2.07	3.50 (4.50)	−2.73 (−2.07)	−6.23 (−6.57)
3a	272 (6424) 388 (3060)	−1.12, −0.67	3.06 (3.81)	−4.13 (−3.92)	−7.19 (−7.73)
5a	271 (12079) 395 (5670)	−1.05, −0.60	3.00 (3.40)	−4.20 (−3.88)	−7.20 (−7.283)
2b	267 (3370) 355 (4002)	−2.60, −2.08	3.38 (4.31)	−2.72 (−2.09)	−6.10 (−6.40)
3b	271 (8824) 408 (2460)	−1.12, −0.69	2.81 (3.54)	−4.11 (−3.90)	−6.92 (−7.44)
5b	273(11782) 422 (3851)	−2.64, −1.05, −0.63	2.78 (3.48)	−4.17 (−3.86)	−6.95 (−7.281)
2c	280 (8291) 385 (4917)	−2.56, −2.11	3.08 (3.95)	−2.69 (−2.07)	−5.77 (−6.02)
3c	272 (12 343) 479 (3380)	−1.18, −0.77	2.38 (3.22)	−4.03 (−3.82)	−6.41 (−7.04)
5c	278 (16 300) 491 (3680)	−2.66, −1.10, −0.68	2.33 (3.17)	−4.12 (−3.80)	−6.45 (−6.97)
MeV	275 (12 773)	−1.20, −0.81	3.97	−3.99	−7.96
BnV	279(13458)	−1.08, −0.74	3.78	−4.06	−7.84
P-BnV	257 (10 700) 298 (8270)	−0.97, −0.53	3.5	−4.3	−7.8

a) Energy gap values were calculated from the absorption spectra.
b) Theoretical calculations have been carried out by using the GAUSSIAN09 suite of programs.
c) Energy levels vs vacuum level were calculated from CV data and from the optically determined energy gap.
Source: Li et al. 40. Copyright (2018) Wiley.

4.2.3.5 Bithiophene Bridged

Materials in which photoluminescence is modulated by redox processes are called electrofluorochromic materials. Bithiophene-bridged viologens are typical EC fluorescent materials, which quench fluorescence by electrochemical switching.

In 2014, the Beneduci group [66] introduced the bithiophene group between 4,4′-bipyridines and long alkyl chain substituents onto the N-quaternary ammonium salt. The resulting compound exhibited a unique ionic liquid (IL) crystal property and coupled EC and fluorescent properties (Figure 4.11). In the redox state, it has up to 80% fluorescence contrast and short switching time (<3.5 seconds). At the same time, the application of potential bias on the isotropic liquid will induce the crystal structure change into the smectic phase and is reflected as a significant

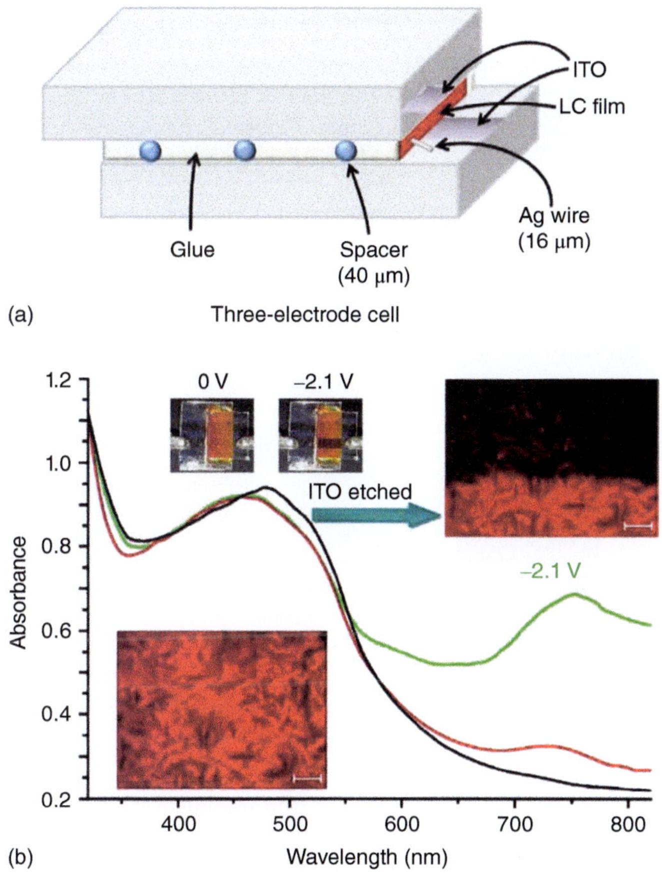

Figure 4.11 (a) Three-electrode liquid crystalline cell. (b) Optical absorption and images of the SmA phase under different potential. Source: Adapted with permission from Beneduci et al. 66. Copyright (2014) Springer.

red shift in the fluorescence spectrum. This completely reversible phenomenon allows color adjustment of fluorescence.

However, the fluorescence conversion of this material needs to be completed at high temperatures (>100 °C). The required high-temperature working environment limits their direct application in room temperature equipment. In the same year, the Benedduci group [67] refined the thienoviologen, which is prepared into a polymer gel. By changing the percentage of fluorophores in the gel, the modulation of the fluorescence intensity in the visible light range at room temperature can be effectively achieved (Figure 4.12).

4.2.4 Viologen-Based Polymer

In 1977, the researchers started to utilize the polymerization strategy, because usually, the viologen-based ECD belongs to solution-type ECD. The response time of ECD is easily affected by the diffusion of viologen. Besides, the solution-type ECD

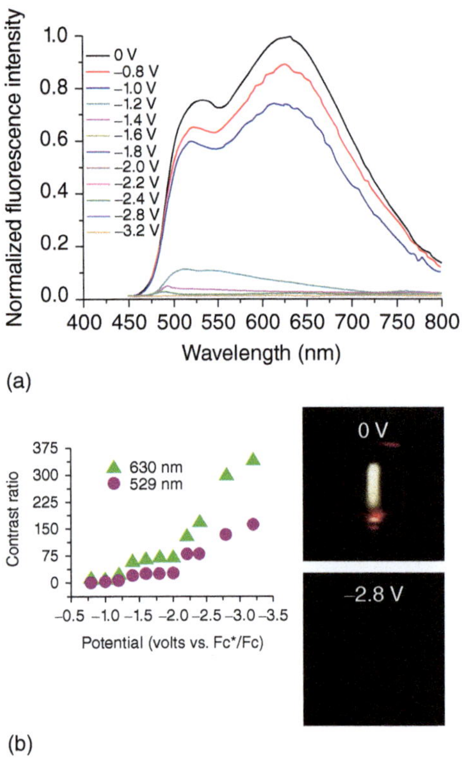

Figure 4.12 (a) Fluorescence spectra of the applied voltage of the gel of 10% (w/w) of thienoviologen. (b) Fluorescence contrast ratio at 630 and 529 nm and images under applied voltage. Source: Adapted with permission from Beneduci et al. 67. Copyright (2014) Wiley.

also pose a risk of leakage. Therefore, one solution is to immobilize the viologen film on the electrode by electropolymerization.

There are three advantages of viologen-based polymer:

1) Polymer-based ECD is more safe and has better performance.
2) Polymer is easy to volume-produce.
3) Viologen-based polymer integrates viologen and polymer characteristics, which are expected to represent richer multicolor with low driving voltage.

From the perspective of the position, polyviologens can be classified into main-chain and side-chain polymers.

4.2.4.1 Viologen in the Side Chain

Compared with main-chain polyviologens, it is more popular to attach viologen to conductive polymer backbone through suitable chemical bond. It is noteworthy that Zincke reaction is the most common way to introduce viologen to polymer backbone [68, 69]. As shown in Figure 4.13, the polymer backbone basically are polyethene [70, 71] (**P1**), polyphenylene [69] (**P2**), polypyrrole [68] (**P3**), polyimide [72] (**P4**), polycarbazole [73] (**P5**), polythiophene, and their derivatives [74, 75] (**P6**, **P7**).

Both poly(3,4-ethylenedioxythiophene) (PEDOT) and viologen are cathodic coloring materials (i.e. represented as transparent in oxidized state and coloring in the reduced state), which are expected to obtain synergic electrochromism. Lee and

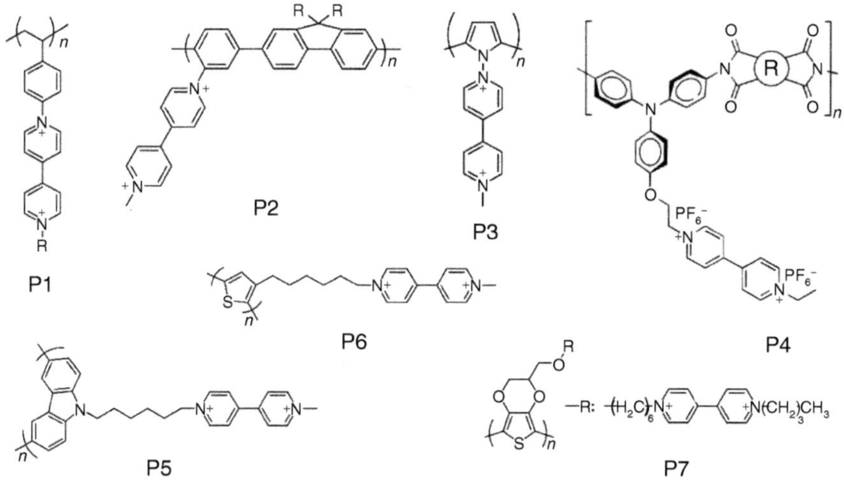

Figure 4.13 Side-chain viologen-based polymers.

coworkers [75] verified that hypothesis; they firstly incorporated the viologen unit into the ethylenedioxythiophene (EDOT) monomer, and then the PEDOT-V^{2+} film was formed by electropolymerization. The results demonstrate the contrast ratio of PEDOT-V^{2+} was enhanced ($\Delta T = 65\%$ at 610 nm) compared with the single PEDOT polymer.

People have been looking forward to solve the aggregation and dimerization of viologen and improve the efficiency of electron transport. As shown in Figure 4.14 and Figure 4.15, researchers found that the connection of viologen at the end of hyperbranched glycerol branches can achieve this goal [76]. And in this process, the researchers found that the high-molecular-weight viologen

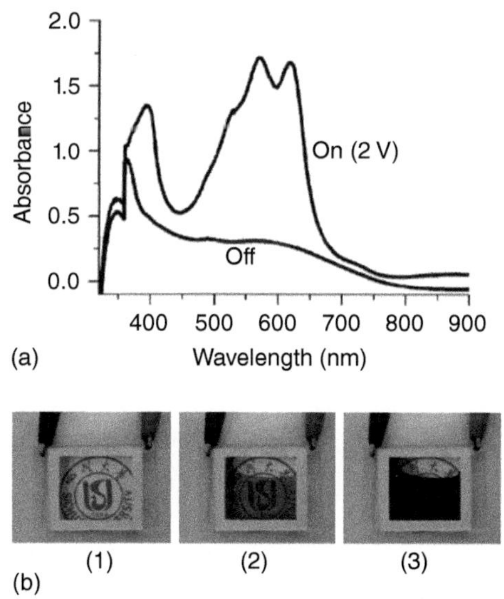

Figure 4.14 Spectral absorption of the device. Source: Adapted with permission from Cao et al. 76. Copyright (2009) Royal Society of Chemistry.

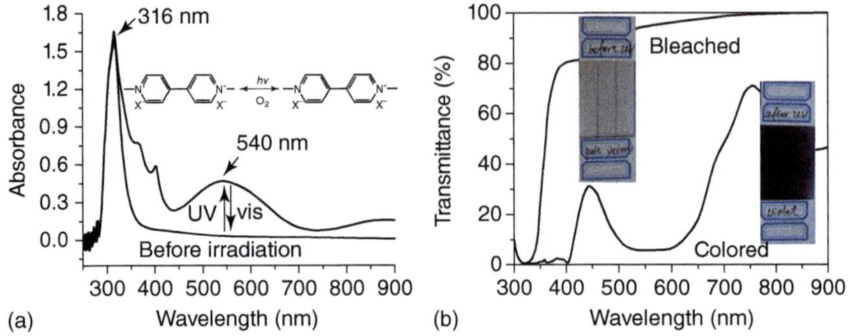

Figure 4.15 Reversible photochromism spectrum. Source: Adapted with permission from Cao et al. 76. Copyright (2009) Royal Society of Chemistry.

molecules exhibit photochromic properties in UV light or even in natural light, but the low-molecular-weight viologens do not have photochromism, and this discoloration process can be bleached by O_2.

4.2.4.2 Viologen in the Main Chain

Main-chain polyviologens usually synthesized through Menshutkin reaction need halogenated olefin or electroactive monomer. Oyaizu and coworkers [77] synthesized the polyviologen, which combines polyethylene glycol with 3,3′-dimethyl-4,4′-bipyridine via Menshutkin reaction, as shown in Figure 4.16. The introduction of methyl substituents in bipyridine results in the change of conformation and ways of the π–π stacking, which makes this polymer possess blue color and fast response time. What's more, results showed that smudging in the EC display could be effectively inhibited due to this modification, which promotes intensive commercialization of polyviologens.

Khokhlov et al. [78] developed cathode–anode complementary EC conjugated polyviologens containing triarylamines (TAAs). The polymers exhibit various advantages: (i) containing a high concentration of electrochromes in a polymer matrix, (ii) improving the solubility and glass-transition temperatures, and (iii) the redox potentials of TAA and viologen did not overlap, providing the enhancement of contrast and multi-electrochromism.

Viologen is a kind of redox building block, which is popular to put in the conjugated or stimuli-responsive polymer backbone, expecting to obtain better ECD performance and multifunctional characteristics.

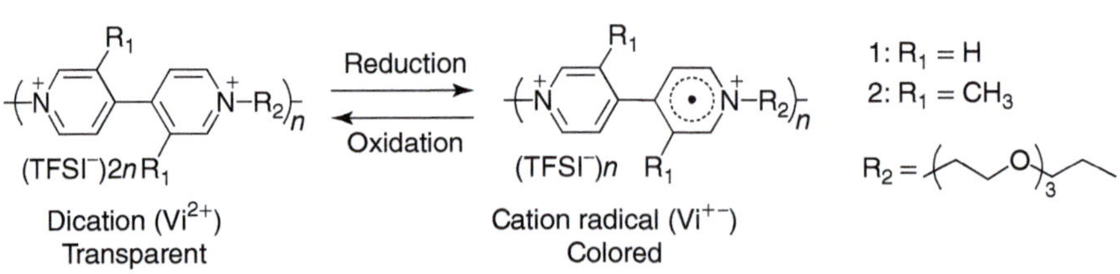

Figure 4.16 Synthesis of polyviologens via Menshutkin reaction. Source: Sato et al. 77, https://www.mdpi.com/2073-4360/9/3/86, licensed under CC BY 4.0.

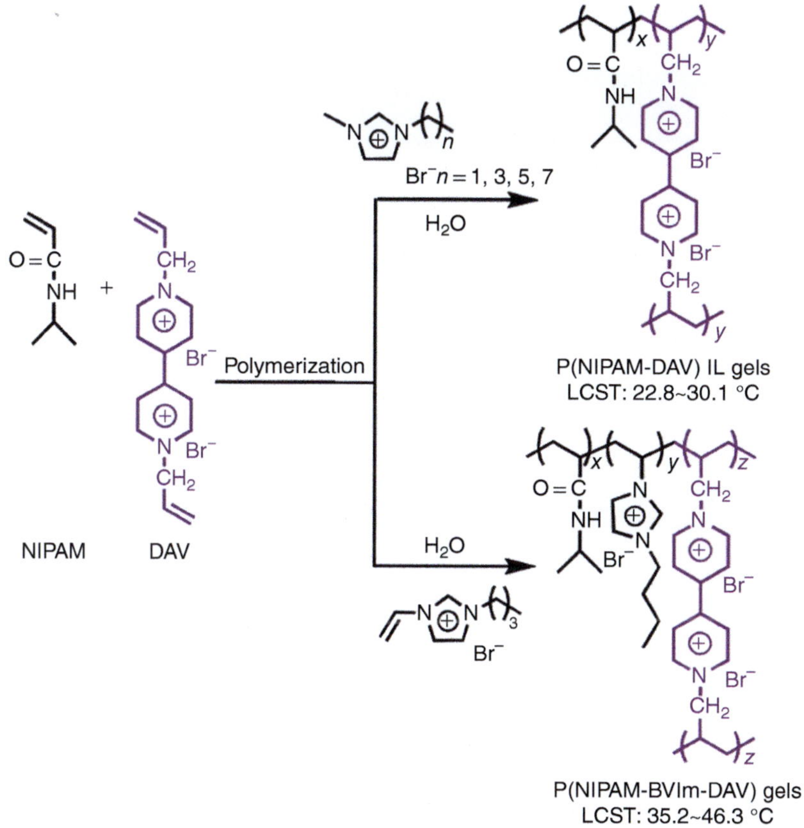

Figure 4.17 Synthesis of P(NIPAM-DAV) IL gels and P(NIPAM-BVIm-DAV) gels. Source: Chen et al. 81. Copyright (2016) Royal Society of Chemistry.

Jang and coworkers [79] synthesized viologen containing poly(2-isopropyl-2-oxazoline) (PiPOx-v) via one-pot cationic ring-opening polymerization. The PiPOx is known as a thermosensitive polymer. A simple sandwich-type ECD was fabricated, and the results showed the ECD exhibits four kinds of color various at different potential and temperature.

A polythiophene with a viologen-like moiety in the main chain was synthesized via Stille coupling reaction and Menshutkin reaction [80]. This conjugated polymer containing both p- and n-dopable units exhibits high electrostability.

As shown in Figure 4.17, researchers have designed Poly(Nisopropylacrylamide-diallyl-viologen) (P[NIAPM-DAV]) IL gel-based thermochromic and electrochromic devices (TEDs) [81]. The smart window system based on the thermochromic and EC dual responsive poly(ionic liquid) electrolytes shows both tunable transparency and EC properties. More importantly, the thermo-sensitive and EC behavior can be observed simultaneously (Figure 4.18). Although this polymer is not very ideal in terms of the response time (3 ~ 5 seconds) and cycle stability (100 cycles), it would have great potential in smart windows and advanced sensors.

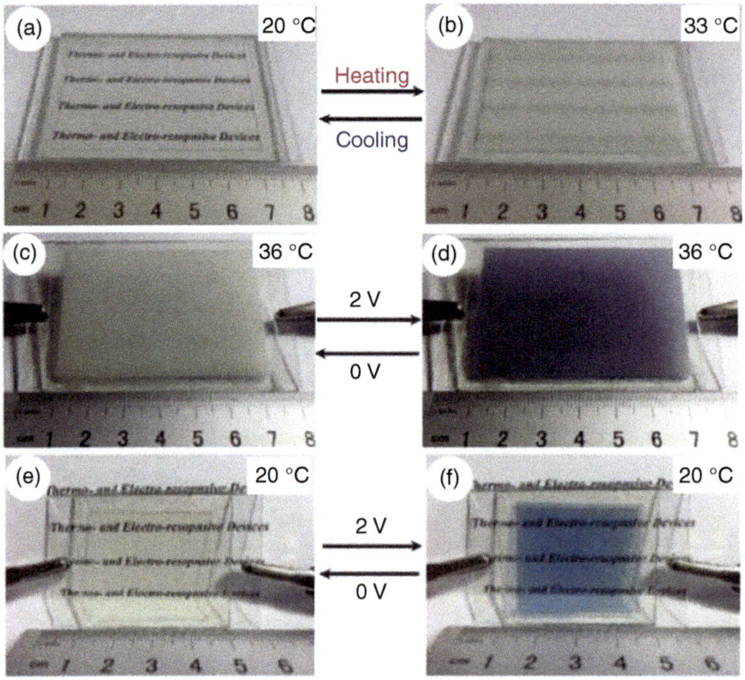

Figure 4.18 Thermochromic and electrochromic effect. Source: Adapted with permission from Chen et al. 81. Copyright (2016) Royal Society of Chemistry.

4.3 Viologen Electrochromic Device

4.3.1 Device Structure

4.3.1.1 Five-Layer Classic Structure

Generally speaking, mostly OEC devices come from this infrastructure. As shown in Figure 4.19, the five-layer device contains the following fundamental components: transparent conductor, EC layer, ion conductor layer, ion storage layer, and another transparent layer.

Transparent conductor layer: The ECD needs at least one transparent conductive electrode (TCE), which can simultaneously transmit light and conduct electrical current. Metal oxide has been known as the most common TCE used in ECD [82], such as indium tin oxide (ITO) films, doped oxide films (SnO_2: F, Cd_2SnO_4), etc. Shortly after, materials based on carbon have been rapidly emerging [83], such as carbon nanotubes and single- and multilayer graphene. These materials have

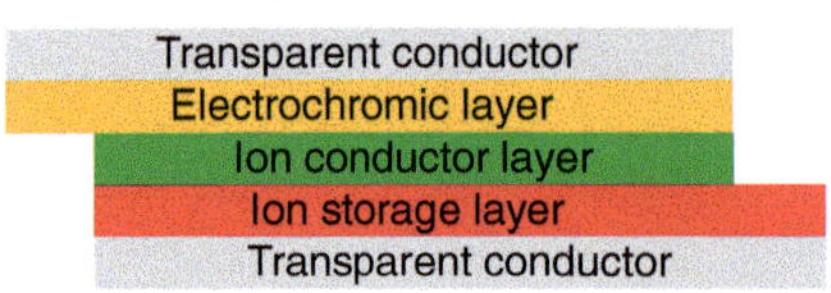

Figure 4.19 Illustration of five-layer classic structure.

the potential to replace the metal oxide attribute to the high electron mobility and high transmittance. For the all-polymer ECDs, another option is PEDOT: poly(sodium-p-styrenesulfonate) (PSS) [84], as the conductive transparent polymer spin coated onto the transparent plastic film.

Electrochromic layer: This is the most important part of the device. This layer usually requires EC materials with high transparency in the neutral state and a high transmittance change when applied potential. For small molecular viologens, they usually dissolve in an organic solvent. For viologen-based polymers, the traditional approach is to deposit a thin film on the electrode substrate by electropolymerization.

Ion conductor layer: Generally, the ion conductor layer can also be called "electrolyte," which should be a transparent electronic insulator with high ionic conductivity. For small molecular viologens, the electrolyte used generally dissolved in non-aqueous solution, such as polycarbonate (PC), acetonitrile (ACN), and N,N-dimethylformamide (DMF). The risk of using this kind of solvent is leakage issues. Gradually, the use of electrolyte solutions has changed into electrolyte gels and polymers in recent years.

Ion storage layer: This layer usually requires a strong ability of storing electrons to stabilize the intermediate during the redox process, prevent irreversible transformation, and improve the cycle stability of the device.

The five-layer structure is only a theoretical structure. In order to save costs and simplify the production process in practical experiments, researchers usually use materials with multiple functions to simplify with a layer of multilayer functions.

4.3.1.2 Simple Sandwich Structure

The simplest form can be a three-layer structure (Figure 4.20). Though it is the simplest form, it can contain multiple color species at the same time. Separate color or mixed color development can be achieved at different driving voltages [85].

Because of its simple structure, this requires that the constituent materials in the same layer independently bear different functions. We can sort them into four substances according to function: viologen, redox mediator, electrolyte, and other filler. This section will be introduced in 4.3 later..

4.3.1.3 Cathodic Anode Separation Structure

This structure can enhance the change of light transmittance by the double EC layer, but it demands that two different EC materials have a certain voltage relationship and color relationship. The anode material is typically phenazine, Prussian blue [86]. The ratio of cathodic and anodic coloring materials can be controlled by chemical bonds (Figure 4.21) [87].

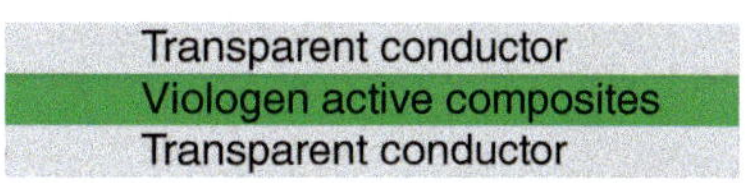

Figure 4.20 Illustration of simple sandwich structure.

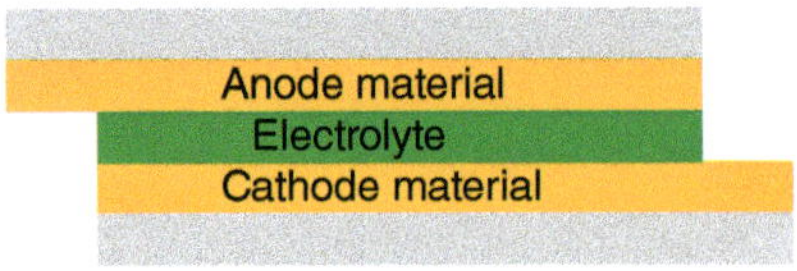

Figure 4.21 Illustration of cathode–anode separation structure.

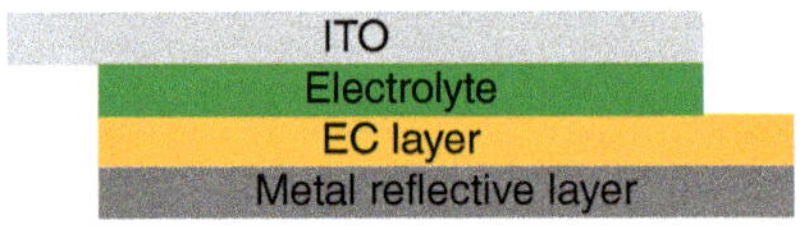

Figure 4.22 Illustration of reflective device structure.

4.3.1.4 Reflective Device Structure

Usually the composition of the layers but the reflective layer is identical to the aforementioned transmissive device. The conductive reflective layer is preferably silver, iridium, aluminum, nickel or platinum, or zinc oxide (Figure 4.22).

4.3.2 Electrolyte

The electrolyte layer that has the function to transfer ions is an important site where the EC reactions occur. The EC matrices can be classified into liquid and gel. Considering simplicity and stability, the liquid matrix is commonly used in the laboratory, while the gel matrix is mainly used in commercial production. The electrolyte layer is located between the EC materials and the electrode, so the contact area of those two materials can decide the ion conductivity.

Studies have shown that [79, 86, 88] bonding the EC materials with electrolytes gel can control the diffusion rate and local concentration, reduce the rate of redox reactions on the electrode surface. What's more, this kind of immobilization strategy can be also applied in IL and polymer IL. Researchers found that [89] the presence of polymer IL can reduce the rectified current density of the viologen-based ECDs and provide additional ions to electrolyte, which can improve the ion conductivity. In the case of constant absorbance, a lower measured current density is obtained, thereby reducing the driving voltage and enhancing the cycle stability.

4.3.3 Redox Mediator

Due to the presence of free radical intermediates in the chromogenic process of the viologen dication, this highly active species is prone to irreversible transformation and thus impairs the lifetime of the device, and it is easy to produce aggregates during the oxidation and re-reduction process, which can also harm the reversible process. It is known that the redox mediators generally have fast, stable, and reversible redox performance while maintaining high transparency. The most widely used substances are ferrocene, phenazine, 2,2,6,6-tetramethylpiperidine-1-oxyl (TEMPO), GQDs, and reduced graphene oxide (RCD) (Figure 4.23) [37, 38, 90–93].

It is worth mentioning that, when the GQDs anion is utilized as the redox mediator, the electrolyte-free device can be realized [38]. The influence on the mechanical bending or the temperature can be greatly decreased since the GQDs has excellent thermostability and flexibility. As shown in Figure 4.24, the homogeneous MV^{2+}

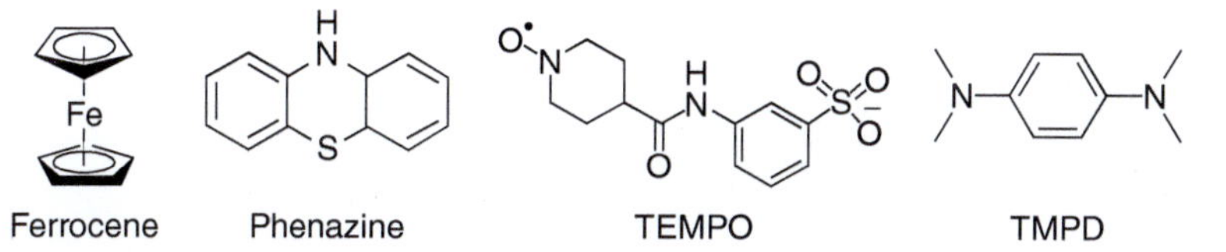

Figure 4.23 The structure of common redox mediators.

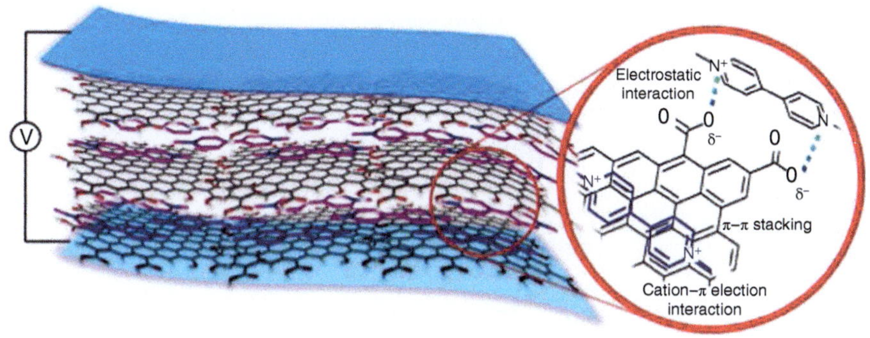

Figure 4.24 Combination of GQDs anion and viologen. Source: Hwang et al. 38. Copyright (2014) Wiley.

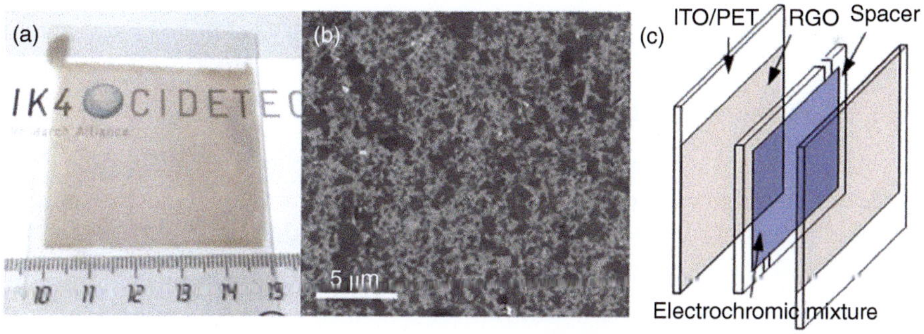

Figure 4.25 Photograph of RGO film in ethyl viologen device. Source: Adapted with permission from Palenzuela et al. 92. Copyright (2014) American Chemical Society.

and graphene quantum dots (GQDs) composition is caused by the interaction of electrostatic, π–π stacking, and cation–π electrons, which results in a stable switching performance of ECD. The resulting device showed excellent operational stability over 3000 seconds when switched between 0 and −2.8 V, with faster switching rates, in comparison to the common MV^{2+} device (Figure 4.25).

The solution-phase graphene oxide/reduced graphene oxide (GO/RGO) has been demonstrated to be a catalyst facilitating the $MV^{2+}/MV^{+\cdot}$ redox reaction and consequent color switching [92], making the ECD have lower operational voltages.

4.3.4 Conductive Medium

Isocyanate: Isocyanates are usually cross-linked by the action of a hydroxy compound to form a polymer network.

Ionic liquids: The IL can act not only as an electrolyte layer in ECD but also as an electrolyte medium for electropolymerization of EC conjugated polymers. The introduction of IL/polymer IL [89] provides an opportunity for the fabrication of low-energy-driving ECD. Imidazole-based ionic liquids have the most types and are currently the most widely used ionic liquids. Wang et al. [94]reported 1-ethyl-3-methylimidazolium tetrafluoroborate ([Emim]BF_4)-doped poly(methyl methacrylate) (PMMA)-based gel electrolyte for EC application. The results showed that the ionic conductivity greatly improved (2.9×10^{-3} S/cm at 20 °C) and the anodic potential effectively decreased.

Microporous organic polymer (MOP) gel: As shown in Figure 4.26, this kind of electrolyte [95] can effectively separate the anion and cation in the electrolyte system by forming a cross-linked network structure and prevent the formation of electron transfer complexes, which can effectively improve the operational stability of ECD.

Coumarin: As shown in Figure 4.27, a functional gel based on a copolymer of coumarin and polyacrylate was synthesized. The gel can self-heal by giving 12 hours of light [96]. In this way, phase separation or deposition and electrolyte leakage can be suppressed. Moreover, self-healing can be achieved by light irradiation when there are microcracks caused by external impact. Thereby this kind of functional gel has a significant effect on prolonging the service life of ECD.

4.3.5 Problems with Viologen Compound

However, there are still many problems (such as insufficient switching speed, low device stability, poor color contrast, and undesirable side reaction) and mechanisms that need to be solved and discovered. In Section 4.3.5., the very limited reports in this area will be introduced in detail.

4.3.5.1 Dimerization

Due to the presence of free radical intermediates, the problem of dimerization and irreversible transformation of intermediates is produced in mechanism. Through the aforementioned question, the first thing that has received widespread attention is the radical cation dimerization.

The mechanism of the generation of dimer, which is accepted by most people, is the comproportionation reaction in eq. (4.1). Once the radical cation is reduced to neutral state (V^0), V^0 could react with V^{2+} to obtain the dimerized viologen radical cation $(V^{+\cdot})_2$ and then disassociate to monometric $V^{+\cdot}$ species. The disassociation constant in organic solution is sufficiently large, so the dimer is scarcely formed in the organic electrolyte ascribe to the weak solvation energy. As for the aqueous solution, the equation was proved to be quasi-reversible [24]. According to the following

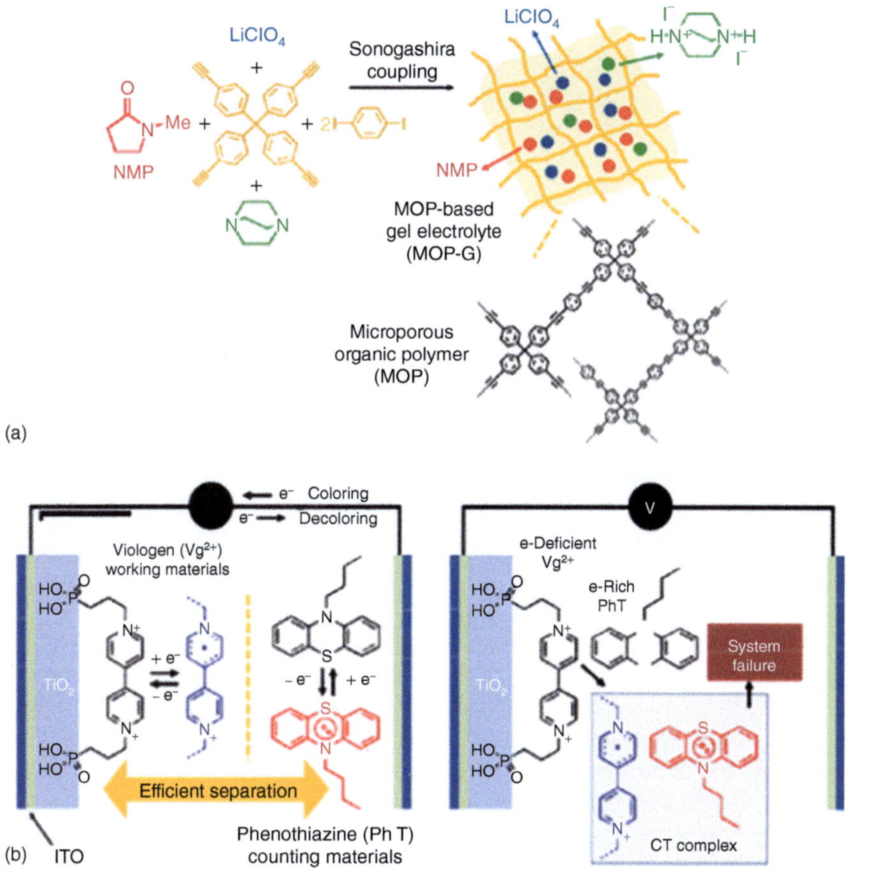

Figure 4.26 (a) Formation of gel electrolytes based on MOP chemistry and (b) operation illustrations of electrochromic cells with (left) gel and (right) liquid electrolytes. Source: Ko et al. 95. Copyright (2018) Royal Society of Chemistry.

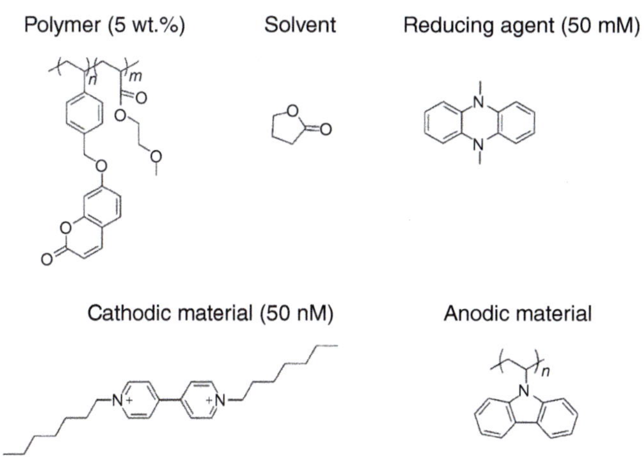

Figure 4.27 Self-healing electrolyte gel.

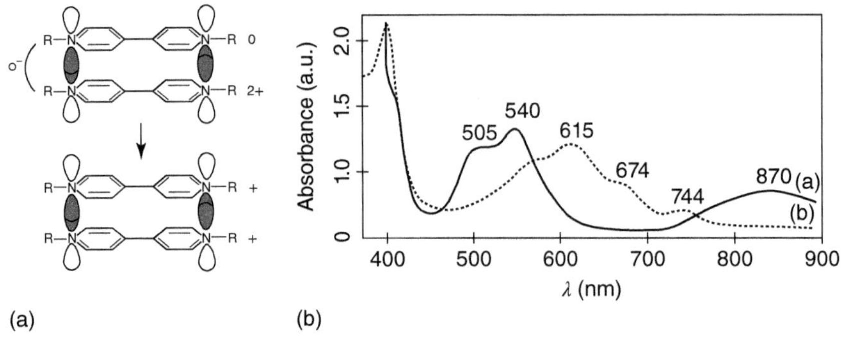

Figure 4.28 (A) Dimerization behavior of radical cation. (B) Absorbance spectrum of dimerized heptyl viologen radical cation (solid line) and monometric radical cation (dotted line). Source: Monk 97. Copyright (1997) Elsevier.

research [97], the dimer of viologen exhibits a poor write–erase property because of the overlapping of π-electrons between the two radical cation, imparting irreversible comproportionation, which has bad influence on the device.

$$V^{2+} + V^0 \rightarrow (V^{+\cdot})_2 \rightarrow 2V^{+\cdot} \tag{4.1}$$

ESR and UV–vis spectroscopy had been applied to verify whether the dimer is produced. Figure 4.28 shows the dimerized radical cation transform to the monomer radical cation owing to the intramolecular electron transfer and the evidence for the presence of methyl viologen radical cation dimer. With the potential of 1.0 V, there was only an absorbance peak at 620 nm. When increasing the potential to 3 V, the λ_{max} was blue shifting to around 540 nm and had a new absorbance at 874 nm. The color changed from blue to purple-red since the dimer wasn't totally transformed.

Here are some conclusions we gathered:

(1) It has been investigated that for the correlation between the degree of dimerization and alky chain length on bipyridinium, the color of radical cation solution changes from blue-purple to deeper red with the increasing CH_2 units, suggesting positive correlation [47].
(2) The bigger size anion of viologen can form an electrostatic shielding effect on the cation and can also reduce the degree of dimerization of the viologen molecule.
(3) The adsorption of TiO_2 nanocrystal electrodes by phosphoric acid or carboxylic acid group is utilized to reduce the dimerization by the spatial effect, and the lifetime and stability of the device can also be improved. In order to enhance the electron transmission between the viologen molecule and the electrode, researchers take the TiO_2 electrodes to adsorb phosphate groups to achieve the purpose. They synthesized a series of different viologen molecules and measure their absorption spectra to investigate their dimerization ratio [6, 23].
(4) The degree of dimerization can be controlled by different combinations of polymer gels, using the same compound to give both blue and violet colors [98, 99].

4.3.5.2 Aggregation and Solubility

In addition to the problem of dimerization, viologen usually took the method of directly contacting the surface of the electrode to form film on the device, so its solubility and diffusion rate in the solution also have negative influence on its EC properties.

(1) Earlier studies have suggested that the dicationic and radical cation states of viologens are very soluble, yet, after many electrochemical cycles, uncontrolled crystallization will occur. Once the viologen crystallization occurs, it will have influence on its write–erase ability or cycling stability [100].
(2) When the cathode has a large number of electrons passing through, $PV^{+}\cdot$ will begin to form agglomerates, which would hinder the recombination during the bleaching stage.
(3) It is noted that alternation of the counter anion of viologen is another important method to modulate the thermodynamic properties of viologens. Changing the type of counter ions (such as Br^- to PF_6^-) can easily change the solubility characteristics of the viologen compound, which is also widely used in the purification and recrystallization process of viologen [31].Among them, tetracyanoquinodimethane (TCNQ) anion has attracted researchers' attention as its special green color [101].

4.3.5.3 Response Time

The insufficient response speed is another demerit we should pay attention to. For the commercial video images application, the switching speed needs to decrease to less than one second. However, the low ion conductivity in the electrolyte is certainly the main causes.

Firstly, the immobilization of EC layer on electrode is the common method researchers thought.

In terms of dynamics, the reaction rate can be increased by expanding the contact area of the reactants. So Jee-Hyun Ryu et al. [31] concentrated on enhancing the switching speed by introducing porous structure into the ECD. The response time of the device decreased to 270 ms since the high specific surface area can increase the electrolyte contents and thus effectively improve the ionic conductivity.

4.3.5.4 Driving Voltage

In general, the viologen-based ECDs has a driving voltage of 1–4 V. In terms of saving energy, researchers expect to get the lower drive voltage. A redox mediator means reversible colorless oxygen-reducing species such as ferrocene will be very helpful [90]. Sheng-Yuan Kao et al. adopted TEMPO as the redox mediator and then got a very low driving voltage of 0.4 V, which is the lowest driving voltage currently reported and also improves the cycle stability [91].

4.3.5.5 Conclusion

In this chapter, we briefly introduce the development of the viologen compound and its color display mechanism and unique advantages as well as a series of problems.

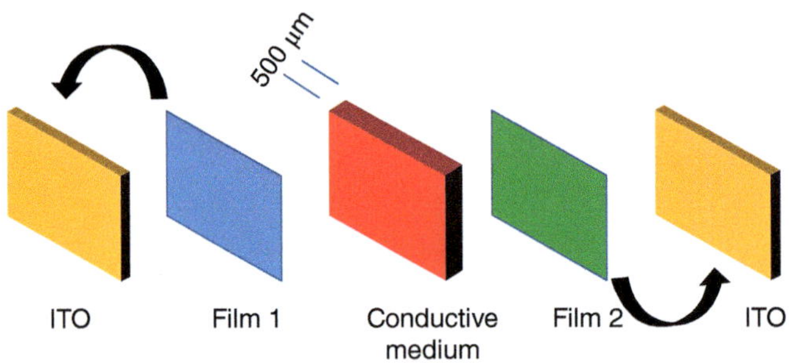

No.	Cathodic material	Polymer matrix	Solvent	Catalyst
Film 1	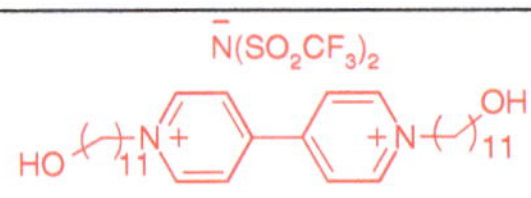 9.8 wt.%	PMMA 7.9 wt.%	PC 79%	DBTDA 1% 110 μl

No.	Redox mediator	Polymer matrix	Solvent	Catalyst
Film 2	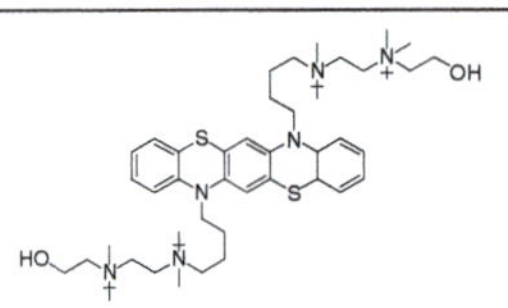 18.5 wt.%	PMMA 7.1 wt.%	PC 71%	DBTDA 1% 110 μl

Figure 4.29 The representative structure and components of ECD. Source: Byker [5]; Byker [33].

In conclusion, research in OEC viologen has achieved significant breakthroughs over the decades. It is not difficult to find that the researcher's center of gravity shifts from the structural construction of the viologen itself to the construction of the viologen OEC device. Different electrolyte and different additives have been adopted to achieve higher stability and faster response time and lower drive voltage. Although the technology in this field is quite mature, we can find that the field still has great potential waiting to be discovered.

4.3.6 Examples of Viologen-Based ECD

As a reference, we give several formulations used in ECD (Figure 4.29 and Table 4.4).

4.4 Companies Operating in the Field of Viologen Electrochromism

4.4.1 Gentex

Founded in 1974, Gentex is a high-tech company headquartered in Zeeland, Michigan. It is a listed company on the National Association of Securities Dealers

Table 4.4 Representative of composition of ECD.

No.	Solvent	Measurement data	Viologen I	Viologen II	Reducing agent	Conductive electrolyte	Other filler
1	PC 82.3 wt.%	Potential: 0.9 V ΔT: 16%–75%	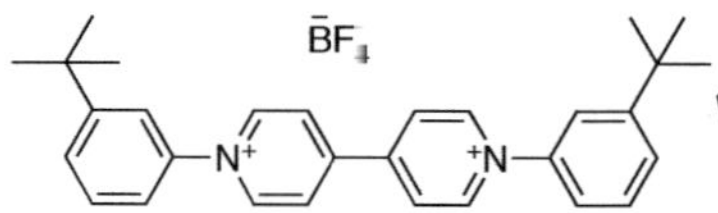 3.5 wt.%	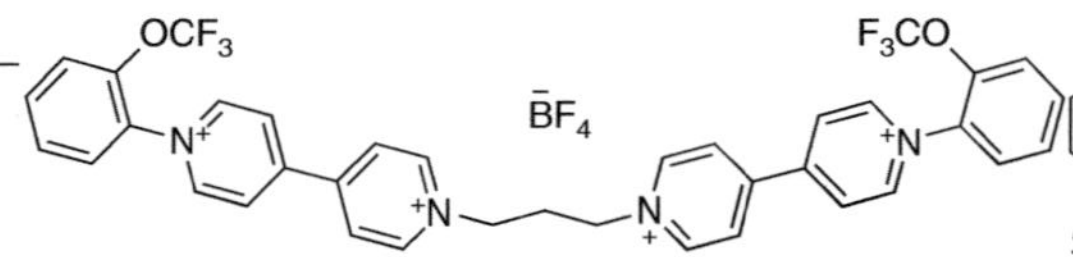 2.0 wt.%	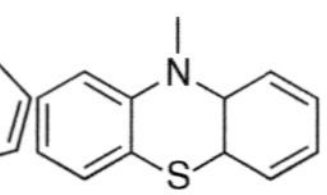 5.9 wt.%	TBA–BF_4 5.5 wt.%	Berry red 0.8 wt.%
2	PC 88.8 wt.%	Potential: 0.9 V	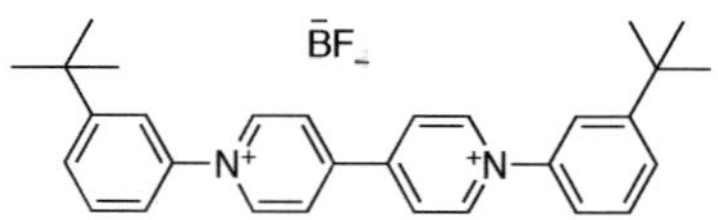 4.3 wt.%		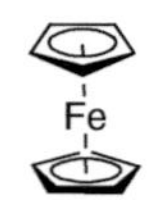 1.2 wt%	TBA–BF_4^- 4.3 wt.%	Red 2 BLSE 1.4 wt.%

Source: Aiken et al. [34]; Biver et al. [34, 35].

Automated Quotations (NASDAQ) market in the United States and is the world's leading supplier of automatic anti-glare rearview mirrors for both inside and outside the car. Additionally, it has more than 105 patents of viologen. Hence, it is absolutely a leader in the EC industry.

For its product orientation, Gentex's devices use gel electrolytes, which require excellent thermal and light stability. That's why we often find substances such as UV absorbers in Gentex's patents.

Looking at the patents of Gentex, its general development direction are the following:

(1) To control dimerization through film formation and asymmetric molecules.
(2) Increase solubility by ion-pairing to improve stability.
(3) Thermal curing to form a cross-linked gel system.
(4) Self-healing cross-linked gel.

The idea that Gentex offers is a self-healing gel that uses uniform painting and device testing methods and is heated by the excess of isocyanate-copolymer polyol (or terpolymer polyol) and tin reagents.

4.4.2 Essilor

Essilor originated in Paris, France in 1849. In 1959, Essilor created resin lenses and in the same year invented the world's first progressive lens, named "Wanli Road." It has more than 5600 patents to date.

Since Essilor is an eyewear company, it focuses on the transparent EC system. We can find photochromic substances in their systems. And the color of the system is usually brown, gray, or other common colors of sunglasses.

4.4.3 Haoruo

Ningbo Haoruo Electronic Technology Co., Ltd. is located in Ningbo, Zhejiang, China. The company is a manufacturer engaged in the research and production of EC glass (also known as EC glass) in China.

Haoruo has similar business scope to Gentex. Therefore, it also has its own unique and comprehensive system in the field of viologen.

It synthesizes a series of 3,3′-substituted viologen compounds by a simple sodium coupling method, and a light self-healing gel is prepared by the optical activity of the coumarin group [96]. It is a potential emerging Chinese company.

4.5 Conclusion

As the only EC material that has been successfully commercialized, the viologen system has its unique advantages. Firstly, from the perspective of structure, the 4,4′-bipyridine compound is facile in synthesis and stable in properties. Although some derivatives are toxic, their toxicity is relatively weakened as the substituent alkyl chain grows. What's more, the remarkable electron-trapping properties make the viologen-based ECD have low-energy driving and long service life. Finally, the viologen system has less color change, but complex colors and multiple color changes can be achieved by dyes or other anode materials. In the recent decades, researches in viologen have been intensively vibrant. From the perspective of material, the structures based on viologen have been expanded to polymer, supramolecular, and metal complex. The structure diversity leads to richer color and unexpected properties. From the perspective of the ECD, the device structure tends to be simplified and flexible. The electrolytes we used before had low-boiling point and thus came to leakage risks. Now we try to use the IL, polymer electrolyte, or gel. In this way, the leakage problem is solved and also have the advantage of offering higher switching speed. As an in-depth research in electrochemistry and electrochromism, it is believed that the problems that existed in the device would be successively solved.

References

1 Lampert, C.M. (1994). Towards large-area photovoltaic nanocells: experiences learned from smart window technology. *Solar Energy Materials and Solar Cells* 32 (3): 307–321.

2 Wang, M., Xing, X., Perepichka, I.F. et al. (2019). Electrochromic smart windows can achieve an absolute private state through thermochromically engineered electrolyte. *Advanced Energy Materials* 9 (21): 1900433.

3 Grätzel, M. (2001). Ultrafast colour displays. *Nature* 409 (6820): 575–576.

4 Michaelis, L. and Hill, E.S. (1933). The viologen indicators. *The Journal of General Physiology* 16 (6): 859–873.

5 Byker, H. J. (1998). Variable reflectance automobile mirror. Google Patents, US 5751467, filed 23 April 1992 and issued 12 May1998.

6 Bird, C.L. and Kuhn, A.T. (1981). Electrochemistry of the viologens. *Chemical Society Reviews* 10 (1): 49.

7 Summers, L.A. (1984). The bipyridines. *Advances in Heterocyclic Chemistry* 35 (10): 281–374.

8 Sliwa, W., Bachowska, B., and Zelichowicz, N. (1991). Chemistry of viologens. *Heterocycles (Sendai)* 32 (11): 2241–2273.

9 Monk, P.M.S. (1998). *The Viologens: Physicochemical Properties, Synthesis and Applications of the Salts of 4,4′-Bipyridine*. Wiley.

10 Monk, P., Mortimer, R., and Rosseinsky, D. (2007). *Electrochromism and Electrochromic Devices*. Cambridge University Press.

11 Monk, P.M.S., Rosseinsky, D.R., and Mortimer, R.J. (2015). *Electrochromic Materials and Devices*. John Wiley & Sons, Ltd.

12 Ho, K.-C., Lu, H.-C., and Yu, H.-F. (2018). Viologens-based electrochromic materials and devices. In: *Electrochromic Smart Materials*, 372–405.

13 Škorjanc, T., Shetty, D., Olson, M.A., and Trabolsi, A. (2019). Design strategies and redox-dependent applications of insoluble viologen-based covalent organic polymers. *ACS Applied Materials & Interfaces* 11 (7): 6705–6716.
14 Ding, J., Zheng, C., Wang, L. et al. (2019). Viologen-inspired functional materials: synthetic strategies and applications. *Journal of Materials Chemistry A* 7 (41): 23337–23360.
15 Michaelis, L. (1935). Semiquinones, the intermediate steps of reversible organic oxidation–reduction. *Chemical Reviews* 16 (2): 243–286.
16 Kosower, E.M. and Cotter, J.L. (1964). Stable free radicals. II. The reduction of 1-methyl-4-cyanopyridinium ion to methylviologen cation radical. *Journal of the American Chemical Society* 86 (24): 5524–5527.
17 Leest, R. (1973). The coulometric determination of oxygen with the electrolytically generated viologen radical-cation. *Journal of Electroanalytical Chemistry and Interfacial Electrochemistry* 43 (2): 251–255.
18 Bard, A.J., Ledwith, A., and Shine, H. (1976). Formation, properties and reactions of cation radicals in solution. In: *Advances in Physical Organic Chemistry*, vol. 13, 155–278. Elsevier.
19 Schoot, C.J., Ponjee, J.J., van Dam, H.T. et al. (1973). New electrochromic memory display. *Applied Physics Letters* 23 (2): 64–65.
20 Bruinink, J., Kregting, C., and Ponjee, J. (1977). Modified viologens with improved electrochemical properties for display applications. *Journal of the Electrochemical Society* 124 (12): 1854–1858.
21 Sato, H. and Tamamura, T. (1979). Polymer effect in electrochromic behavior of oligomeric viologens. *Journal of Applied Polymer Science* 24 (10): 2075–2085.
22 Shu, C.F. and Wrighton, M.S. (1988). Synthesis and charge-transport properties of polymers derived from the oxidation of 1-hydro-1′-(6-[pyrrol-1-yl]hexyl)-4,4′-bipyridinium bis (hexafluorophosphate) and demonstration of a pH-sensitive microelectrochemical transistor derived from the redox properties of a conventional redox center. *The Journal of Physical Chemistry* 92 (18): 5221–5229.
23 Gaudiello, J.G., Ghosh, P.K., and Bard, A.J. (1985). Polymer films on electrodes. 17. The application of simultaneous electrochemical and electron spin resonance techniques for the study of two viologen-based chemically modified electrodes. *Journal of the American Chemical Society* 107 (11): 3027–3032.
24 Monk, P.M.S., Fairweather, R.D., Ingram, M.D., and Duffy, J.A. (1992). Evidence for the product of the viologen comproportionation reaction being a spin-paired radical cation dimer. *Journal of the Chemical Society, Perkin Transactions* 2 (11).
25 Heinen, S. and Walder, L. (2000). Generation-dependent intramolecular CT complexation in a dendrimer electron sponge consisting of a viologen skeleton. *Angewandte Chemie International Edition* 39 (4): 806–809.
26 Balzani, V., Bandmann, H., Ceroni, P. et al. (2006). Host–guest complexes between an aromatic molecular tweezer and symmetric and unsymmetric dendrimers with a 4,4′-bipyridinium core. *Journal of the American Chemical Society* 128 (2): 637–648.

27 Foster, J.A. and Steed, J.W. (2010). Exploiting cavities in supramolecular gels. *Angewandte Chemie International Edition* 49 (38): 6718–6724.

28 Bonhote, P., Gogniat, E., Campus, F. et al. (1999). Nanocrystalline electrochromic displays. *Displays* 20 (3): 137–144.

29 Möller, M., Asaftei, S., Corr, D. et al. (2004). Switchable electrochromic images based on a combined top–down bottom–up approach. *Advanced Materials* 16 (17): 1558–1562.

30 Ryu, J.H., Shin, D.O., and Suh, K.D. (2005). Preparation of a reflective-type electrochromic device based on monodisperse, micrometer-size-range polymeric microspheres and viologen pendants. *Journal of Polymer Science Part A: Polymer Chemistry* 43 (24): 6562–6572.

31 Ryu, J.H., Lee, J.H., Han, S.J., and Suh, K.D. (2006). Novel electrochromic displays using monodisperse viologen-modified porous polymeric microspheres. *Macromolecular Rapid Communications* 27 (14): 1156–1161.

32 Guarr, T. F., Roberts, K. E., Lin, R., et al. (2004), Controlled diffusion coefficient electrochromic materials for use in electrochromic mediums and associated electrochromic devices. Google Patents, US 6710906B2, filed 13 November 2001, issued 23 March 23 2004.

33 Byker, H. J. (1994). Electrochromic devices with bipyridinium salt solutions. Google Patents, US 5336448, filed 25 June 25 1991, issued date 09 August 9 1994.

34 Aiken, S., Gabbutt, C. D., Heron, B. M., et al. 2017. Electrochromic single and two-core viologens and optical articles containing them. Google Patents, WO 2015/040033 A3, World Intellectual Property Organization, filed 16 September 2014, issued 26 March 2015.

35 Biver, C., Archambeau, S., Berit-Debat, F. 2017. Electrochromic composition. Google Patents, US 9841652 B2, filed September 16 2014, issued 12 December 12 2017.

36 Freitag, M. and Galoppini, E. (2010). Cucurbituril complexes of viologens bound to TiO_2 films. *Langmuir* 26 (11): 8262–8269.

37 Tu, X., Fu, X., Jiang, Q. et al. (2011). The synthesis and electrochemical properties of cathodic–anodic composite electrochromic materials. *Dyes and Pigments* 88 (1): 39–43.

38 Hwang, E., Seo, S., Bak, S. et al. (2014). An electrolyte-free flexible electrochromic device using electrostatically strong graphene quantum dot–viologen nanocomposites. *Advanced Materials* 26 (30): 5129–5136.

39 Reus, C., Stolar, M., Vanderkley, J. et al. (2015). A convenient N-arylation route for electron-deficient pyridines: the case of π-extended electrochromic phosphaviologens. *Journal of the American Chemical Society* 137 (36): 11710–11717.

40 Li, G., Xu, L., Zhang, W. et al. (2018). Narrow-bandgap chalcogenoviologens for electrochromism and visible-light-driven hydrogen evolution. *Angewandte Chemie International Edition* 57 (18): 4897–4901.

41 Woodward, A.N., Kolesar, J.M., Hall, S.R. et al. (2017). Thiazolothiazole fluorophores exhibiting strong fluorescence and viologen-like reversible electrochromism. *Journal of the American Chemical Society* 139 (25): 8467–8473.

42 Rosseinsky, D.R. and Monk, P.M. (1993). Comproportionation in propylene carbonate of substituted bipyridyliums. Rates and equilibria by rotating ring–disc electrode and associated electrochemical studies. *Journal of the Chemical Society, Faraday Transactions* 89 (2): 219–222.

43 Ito, M., Sasaki, H., and Takahashi, M. (1987). Infrared spectra and dimer structure of reduced viologen compounds. *Journal of Physical Chemistry* 91 (15): 3932–3934.

44 Mohammad, M. (1987). Methyl viologen neutral MV. 1. Preparation and some properties. *The Journal of Organic Chemistry* 52 (13): 2779–2782.

45 Bockman, T. and Kochi, J. (1990). Isolation and oxidation-reduction of methylviologen cation radicals. Novel disproportionation in charge-transfer salts by X-ray crystallography. *The Journal of Organic Chemistry* 55 (13): 4127–4135.

46 Porter, W.W. and Vaid, T.P. (2005). Isolation and characterization of phenyl viologen as a radical cation and neutral molecule. *The Journal of Organic Chemistry* 70 (13): 5028–5035.

47 Monk, P.M., Hodgkinson, N.M., and Ramzan, S.A. (1999). Spin pairing ('dimerisation') of the viologen radical cation: kinetics and equilibria. *Dyes and Pigments* 43 (3): 207–217.

48 Menschutkin, N. (1890). Beiträge zur Kenntnis der Affinitätskoeffizienten der alkylhaloide und der organischen amine. *Zeitschrift für Physikalische Chemie* 5 (1): 589–600.

49 Singh, R.P. and Shreeve, J.M. (2003). Syntheses of the first N-mono- and *N,N′*-dipolyfluoroalkyl-4,4′-bipyridinium compounds. *Chemical Communications (Camb)* (12): 1366–1367.

50 Oh, H., Seo, D.G., Yun, T.Y. et al. (2017). Novel viologen derivatives for electrochromic ion gels showing a green-colored state with improved stability. *Organic Electronics* 51: 490–495.

51 Takahashi, K. and Nihira, T. (1992). One-pot synthesis and redox properties of conjugation-extended 4,4′-bipyridines and related compounds. new ligands consisting of a heterocyclic three-ring assembly. *Bulletin of the Chemical Society of Japan* 65 (7): 1855–1859.

52 Singh, R.P. and Shreeve, J.n.M. (2003). Bridged tetraquaternary salts from *N,N′*-polyfluoroalkyl-4,4′-bipyridine. *Inorganic Chemistry* 42 (23): 7416–7421.

53 Van Dam, H. and Ponjee, J. (1974). Electrochemically generated colored films of insoluble viologen radical compounds. *Journal of the Electrochemical Society* 121 (12): 1555.

54 Xu, J.W., Toppare, L., Li, Y. et al. (2019). *Electrochromic Smart Materials: Fabrication and Applications*. Royal Society of Chemistry.

55 Alesanco, Y., Viñuales, A., Ugalde, J. et al. (2018). Consecutive anchoring of symmetric viologens: electrochromic devices providing colorless to neutral-color switching. *Solar Energy Materials and Solar Cells* 177: 110–119.

56 Nakajima, R., Yamada, Y., Komatsu, T. et al. (2012). Electrochromic properties of ITO nanoparticles/viologen composite film electrodes. *RSC Advances* 2 (10): 4377–4381.

57 Zhu, C.-R., Long, J.-F., Tang, Q. et al. (2018). Multi-colored electrochromic devices based on mixed mono- and bi-substituted 4,4′-bipyridine derivatives containing an ester group. *Journal of Applied Electrochemistry* 48 (10): 1121–1129.

58 Kim, M., Kim, Y.M., and Moon, H.C. (2020). Asymmetric molecular modification of viologens for highly stable electrochromic devices. *RSC Advances* 10 (1): 394–401.

59 Kim, S., Shim, N., Lee, H., and Moon, B. (2012). Synthesis of a perylenediimide-viologen dyad (PDI-2V) and its electrochromism in a layer-by-layer self-assembled multilayer film with PEDOT:PSS. *Journal of Materials Chemistry* 22 (27): 13558.

60 Alesanco, Y., Vinuales, A., Cabanero, G. et al. (2016). Colorless to Neutral Color Electrochromic Devices Based on Asymmetric Viologens. *ACS Applied Materials & Interfaces* 8 (43): 29619–29627.

61 Ko, H.C., Kim, S., Lee, H., and Moon, B. (2005). Multicolored electrochromism of a poly{1,4-bis[2-(3,4-ethylenedioxy)thienyl]benzene} derivative bearing viologen functional groups. *Advanced Functional Materials* 15 (6): 905–909.

62 Stolar, M., Borau-Garcia, J., Toonen, M., and Baumgartner, T. (2015). Synthesis and tunability of highly electron-accepting, N-benzylated phosphaviologens. *Journal of American Chemical Society* 137 (9): 3366–3371.

63 Reus, C., Stolar, M., Vanderkley, J. et al. (2015). A convenient N-arylation route for electron-deficient pyridines: the case of pi-extended electrochromic phosphaviologens. *Journal of the American Chemical Society* 137 (36): 11710–11717.

64 Durben, S. and Baumgartner, T. (2011). 3,7-Diazadibenzophosphole oxide: a phosphorus-bridged viologen analogue with significantly lowered reduction threshold. *Angewandte Chemie International Edition* 50 (34): 7948–7952.

65 Hall, G.B., Kottani, R., Felton, G.A. et al. (2014). Intramolecular electron transfer in bipyridinium disulfides. *Journal of the American Chemical Society* 136 (10): 4012–4018.

66 Beneduci, A., Cospito, S., La Deda, M. et al. (2014). Electrofluorochromism in π conjugated ionic liquid crystals. *Nature Communications* 5 (1): 1–8.

67 Beneduci, A., Cospito, S., Deda, M.L., and Chidichimo, G. (2015). Highly fluorescent thienoviologen-based polymer gels for single layer electrofluorochromic devices. *Advanced Functional Materials* 25 (8): 1240–1247.

68 Yamaguchi, I. and Asano, T. (2011). Uncatalyzed synthesis of polypyrrole with viologen side groups and its chemical properties. *Journal of Materials Science* 46 (13): 4582–4587.

69 Yamaguchi, I., Mizoguchi, N., and Sato, M. (2009). Self-doped polyphenylenes containing electron-accepting viologen side group. *Macromolecules* 42 (13): 4416–4425.

70 Sampanthar, J., Neoh, K., Ng, S. et al. (2000). Flexible smart window via surface graft copolymerization of viologen on polyethylene. *Advanced Materials* 12 (20): 1536–1539.

71 Janecek, E.R., McKee, J.R., Tan, C.S. et al. (2015). Hybrid supramolecular and colloidal hydrogels that bridge multiple length scales. *Angewandte Chemie International Edition* 54 (18): 5383–5388.

72 Yen, H.J., Tsai, C.L., Chen, S.H., and Liou, G.S. (2017). Electrochromism and nonvolatile memory device derived from triphenylamine-based polyimides with pendant viologen units. *Macromolecular Rapid Communications* 38 (9).

73 Lim, J.Y., Ko, H.C., and Lee, H. (2006). Single-and dual-type electrochromic devices based on polycarbazole derivative bearing pendent viologen. *Synthetic Metals* 156 (9–10): 695–698.

74 Ko, H.C., Park, S.-a., Paik, W.-k., and Lee, H. (2002). Electrochemistry and electrochromism of the polythiophene derivative with viologen pendant. *Synthetic Metals* 132 (1): 15–20.

75 Ko, H.C., Kang, M., Moon, B., and Lee, H. (2004). Enhancement of electrochromic contrast of poly(3,4-ethylenedioxythiophene) by incorporating a pendant viologen. *Advanced Materials* 16 (19): 1712–1716.

76 Cao, L.-C., Mou, M., and Wang, Y. (2009). Hyperbranched and viologen-functionalized polyglycerols: preparation, photo- and electrochromic performance. *Journal of Materials Chemistry* 19 (21): 3412–3418.

77 Sato, K., Mizukami, R., Mizuma, T. et al. (2017). Synthesis of dimethyl-substituted polyviologen and control of charge transport in electrodes for high-resolution electrochromic displays. *Polymers (Basel)* 9 (3): 90036.

78 Keshtov, M.L., Udum, Y.A., Toppare, L. et al. (2013). Synthesis of aromatic poly(pyridinium salt)s and their electrochromic properties. *Materials Chemistry and Physics* 139 (2–3): 936–943.

79 Nam, J., Jung, Y., Joe, J., and Jang, W.-D. (2018). Dual stimuli-responsive viologen-containing poly(2-isopropyl-2-oxazoline) and its multi-modal electrochromic phase transition. *Polymer Chemistry* 9 (26): 3662–3666.

80 Krompiec, M., Grudzka, I., Filapek, M. et al. (2011). An electrochromic diquat-quaterthiophene alternating copolymer: a polythiophene with a viologen-like moiety in the main chain. *Electrochimica Acta* 56 (24): 8108–8114.

81 Chen, F., Ren, Y., Guo, J., and Yan, F. (2017). Thermo- and electro-dual responsive poly(ionic liquid) electrolyte based smart windows. *Chemical Communications* 53 (10): 1595–1598.

82 Ellmer, K. (2012). Past achievements and future challenges in the development of optically transparent electrodes. *Nature Photonics* 6 (12): 809–817.

83 Bonaccorso, F., Sun, Z., Hasan, T., and Ferrari, A.C. (2010). Graphene photonics and optoelectronics. *Nature Photonics* 4 (9): 611–622.

84 Argun, A.A., Cirpan, A., and Reynolds, J.R. (2003). The First Truly All-Polymer Electrochromic Devices. *Advanced Materials* 15 (16): 1338–1341.

85 Mishra, S., Pandey, H., Yogi, P. et al. (2017). Interfacial redox centers as origin of color switching in organic electrochromic device. *Optical Materials* 66: 65–71.

86 Kuo, T.-H., Hsu, C.-Y., Lee, K.-M., and Ho, K.-C. (2009). All-solid-state electrochromic device based on poly(butyl viologen), Prussian blue, and succinonitrile. *Solar Energy Materials and Solar Cells* 93 (10): 1755–1760.

87 Michaelis, A., Berneth, H., Haarer, D. et al. (2001). Electrochromic dye system for smart window applications. *Advanced Materials* 13 (23): 1825–1828.

88 Alesanco, Y., Vinuales, A., Rodriguez, J., and Tena-Zaera, R. (2018). All-in-one gel-based electrochromic devices: strengths and recent developments. *Materials (Basel)* 11 (3): 11030414.

89 Lu, H.C., Kao, S.Y., Yu, H.F. et al. (2016). Achieving low-energy driven viologens-based electrochromic devices utilizing polymeric ionic liquids. *ACS Applied Materials & Interfaces* 8 (44): 30351–30361.

90 Chen, P.-Y., Chen, C.-S., and Yeh, T.-H. (2014). Organic multiviologen electrochromic cells for a color electronic display application. *Journal of Applied Polymer Science* 131 (13): 40933.

91 Kao, S.-Y., Kawahara, Y., Nakatsuji, S.i., and Ho, K.-C. (2015). Achieving a large contrast, low driving voltage, and high stability electrochromic device with a viologen chromophore. *Journal of Materials Chemistry C* 3 (14): 3266–3272.

92 Palenzuela, J., Vinuales, A., Odriozola, I. et al. (2014). Flexible viologen electrochromic devices with low operational voltages using reduced graphene oxide electrodes. *ACS Applied Materials & Interfaces* 6 (16): 14562–14567.

93 Chang, T.-H., Hu, C.-W., Kao, S.-Y. et al. (2015). An all-organic solid-state electrochromic device containing poly(vinylidene fluoride-*co*-hexafluoropropylene), succinonitrile, and ionic liquid. *Solar Energy Materials and Solar Cells* 143: 606–612.

94 Tang, Q., Li, H., Yue, Y. et al. (2017). 1-Ethyl-3-methylimidazolium tetrafluoroborate-doped high ionic conductivity gel electrolytes with reduced anodic reaction potentials for electrochromic devices. *Materials & Design* 118: 279–285.

95 Ko, J.H., Lee, H., Choi, J. et al. (2019). Microporous organic polymer-induced gel electrolytes for enhanced operation stability of electrochromic devices. *Polymer Chemistry* 10 (4): 455–459.

96 Cao, Z.-H.; Hu, S.-S. (2018), Under a kind of illumination can gelation and selfreparing electrochromic solutions and its application. CN 108251099 A, filed date 29 December 2016, issued 06 July 2018.

97 Monk, P.M. (1997). The effect of ferrocyanide on the performance of heptyl viologen-based electrochromic display devices. *Journal of Electroanalytical Chemistry* 432 (1–2): 175–179.

98 Chang, T.-H., Lu, H.-C., Lee, M.-H. et al. (2018). Multi-color electrochromic devices based on phenyl and heptyl viologens immobilized with UV-cured polymer electrolyte. *Solar Energy Materials and Solar Cells* 177: 75–81.

99 Zhao, S., Huang, W., Guan, Z. et al. (2019). A novel bis(dihydroxypropyl) viologen-based all-in-one electrochromic device with high cycling stability and coloration efficiency. *Electrochimica Acta* 298: 533–540.

100 Mortimer, R.J. (1999). Organic electrochromic materials. *Electrochimica Acta* 44 (18): 2971–2981.

101 Wang, G., Fu, X., Deng, J. et al. (2013). Electrochromic and spectroelectrochemical properties of novel 4,4′-bipyridilium–TCNQ anion radical complexes. *Chemical Physics Letters* 579: 105–110.

5

Metallohexacyanates

5.1 Background

In the electrochromic (EC) field, research on the development of non-lithium technologies has attracted more and more attention, due to the high cost and low abundance of Li. Recently, practically unlimited elements, such as Na, Mg, and Ca, are being actively investigated for charge transfer [1–3]. As a result, different materials are required for cathodes, anodes, and electrolytes in the non-lithium electrochromic devices (ECDs), in order to provide the necessary structural, electronic, and ion diffusion properties. To solve this problem, a variety of research directions are carried out, such as modifying the existing Li-based framework and using it for different guest atoms, synthesizing conceptually new non-Li-based materials, and studying materials that have never been used for EC applications. Due to their unique structure and electronic and ionic properties, these new materials have very broad application prospects in non-lithium technologies. Prussian blue (PB) and its analogs are good examples of the existing Li-based materials with a well-defined face-centered cubic structure, which may unexpectedly appear "new life" in the field of EC industry [4].

Prussian blue (PB) was the first purely synthetic pigment and is extensively used as a pigment in oil paintings and other cultural objects. Due to the peculiarities of its atomic structure and ionic and electronic properties, PB is recognized as belonging to a very important family of functional materials: the transition metal hexacyanoferrates [5–7]. It has been proposed for use in various fields, e.g. electrocatalysis [8], fuel cells [9], electrochromic devices ECDs [10, 11], spintronics [12], sensors [13, 14], photochargeable devices [15], and hydrogen storage devices [16]. Today, PB is a typical EC material, which has been widely studied for many years. Early studies concentrated on the preparation of PB films by electrochemical reduction of solutions containing iron(III) and hexacyanoferrate(III) ions, such as galvanostatic, potentiodynamic, and potentiostatic methods. To realize industrial production, traditional chemical synthesis of PB has attracted more and more attention. ECDs, where PB was employed as the sole electrochromic material, usually showed unsatisfactory contrast and stability. Thus, in most reports, PB is adopted in modified electrodes for ECDs and multifunctional devices.

Organic Electronics for Electrochromic Materials and Devices, First Edition. Hong Meng.

5.2 Technology Development of PB

PB was firstly discovered by a paint maker Diesbach around 1706. When he was trying to make a red lake of cochineal, a product showing a strong reddish blue color was accidentally obtained when using a potassium sample contaminated by iron. The product was named PB and then widely used in the paint industry. The reaction process can be described as follows: (i) potassium hexacyanoferrate was obtained by calcining the mixture of grass ash and cow blood, and (ii) ion hexacyanoferrate was prepared by reacting potassium hexacyanoferrate with $FeCl_3$. The EC properties of PB was firstly reported by Neff in 1978 [17]. After that, research on the electrochemical mechanism and ECDs of PB has been gradually carried out. Its technology development is summarized in Figure 5.1. With the rapid development of nanotechnology, recent research on PB electrochromism is concerned on the ECDs utilizing nanomaterials of PB or its analogs [18, 19]. Besides, PB ECDs with self-powered ability, sensor, catalysis, or other performance, which show multifunctional applications, have attracted more and more attention.

5.3 Crystal Structure

The crystal structure of PB and its analogs has been widely studied by a variety of modern characterization and analytical techniques, including UV-vis, Raman, XRD, XPS, SEM, TEM, TGA, and so on. It is demonstrated that PB is the classic example of face-centered cubic structure with a mixed-valence iron(II)–iron(III) inorganic complex, in which alternating Fe^{II} and Fe^{III} centers are bridged by cyanide ions in a $Fe^{II}—C{\equiv}N—Fe^{III}$ fashion. This situation leads to the formation of an infinite three-dimensional network, presenting the idealized $Fe^{III}{}_4[Fe^{II}(CN)_6]_3{\cdot}xH_2O$ formula [4]. PB is classified as both soluble and insoluble, depending on the absence or presence of a monovalent ion in the lattice. $KFe^{III}[Fe^{II}(CN)_6]$ is known

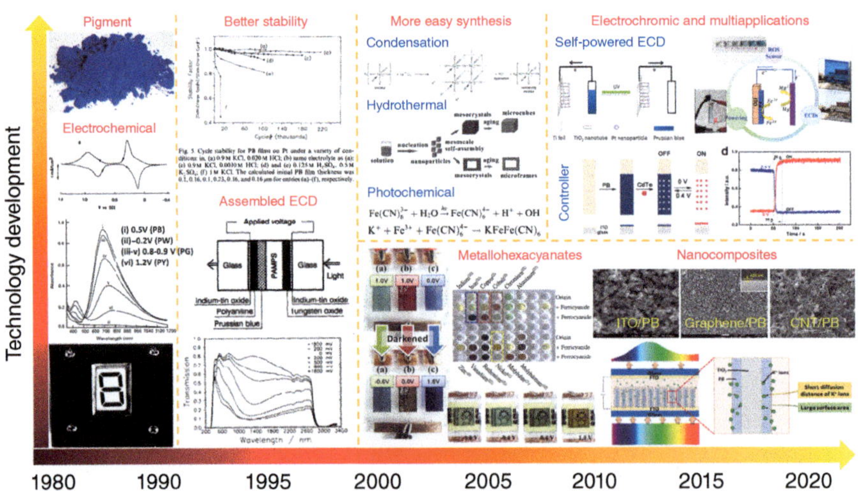

Figure 5.1 Technology development of PB.

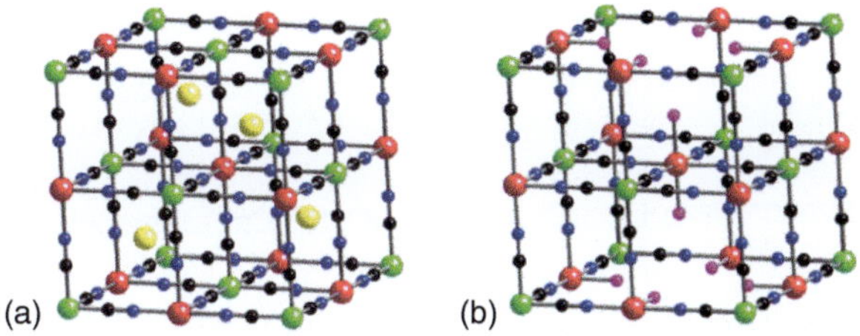

Figure 5.2 Unit cell of the soluble (a) and insoluble (b) PB compounds [24]. The iron(II) ions are shown in green, the iron(III) ions in red, the potassium ions in yellow, the carbon in black, the nitrogen in blue and the water oxygen in pink. Source: Grandjean et al. [24]. Copyright 2016, The Royal Society of Chemistry.

as the formula of soluble PB, whereas $Fe^{III}{}_4[Fe^{II}(CN)_6]_3$ is insoluble [20, 21]. These names do not refer to the real solubility in water but rather indicate the ease with which potassium ions peptize, and both forms are highly insoluble ($K_{ps} = 10^{-40}$) [22, 23]. For the soluble form, the high-spin ferric and low-spin ferrocyanide sites are each octahedrally surrounded by –NC and –CN units, respectively, with the K^+ counterion being in an interstitial site. While for the insoluble form, one-fourth of the $Fe^{II}(CN)_6$ units are missing, with their nitrogen sites occupied instead by water molecules coordinated to Fe^{III} sites and with as many as eight additional water molecules located interstitially, as shown in Figure 5.2 [24–26]. However, it is hard to ensure which of these two stoichiometries represent the PB film on the electrodes of ECDs, which hampers the demonstration of EC mechanisms [27].

5.4 Electrochromic Mechanism

PB film shows a broad absorption band from 500 to 900 nm and a strong absorption peak at 690 nm, resulting in a deep blue color with a large absorbance at the oxidized state [28]. Square openings and vacancies in the cubic framework give PB a porous structure, which is similar to conductive zeolites for insertion/extraction of ions [29, 30]. PB is also multi-EC and has four redox states. While PB itself is blue, it can be reduced to colorless Prussian white (PW) $K_2Fe^{II}[Fe^{II}(CN)_6]$, sometimes also called Everitt's salt (ES); it can also be oxidized to Berlin green (BG) $Fe^{III}{}_3[Fe^{III}(CN)_6]_2[Fe^{II}(CN)_6]$ or an even more highly oxidized state, Prussian yellow (PY) $Fe^{III}[Fe^{III}(CN)_6]$, as shown in Figure 5.3a [31, 33–35]. Complete Cyclic Voltammograms (CVs) for PB is shown in Figure 5.3b [32]. There were two coupled redox peaks, which were the anodic peaks at c. 0.36 V (A^{I}_{PB}) and c. 1.03 V (A^{II}_{PB}) and the cathodic peaks at c. 0.09 V (C^{II}_{PB}) and c. 0.82 V (C^{II}_{PB}). The first and second redox couples were related to the BG ↔ PB and PB ↔ PW (ES) reactions, respectively.

Since the color of the PB family changes with their oxidation state and shows good reversible cycling abilities, these products can be widely used in EC fields. The applied PB films need to possess the following properties: (i) low capacity per surface unit, (ii) significant color change between different states, (iii) fast response, and (iv) very high stability (at least tens or hundreds of thousand cycles). Theoretically, the color change between PB and PW undergoes a single-step insertion and extraction of K^+ ions during the electrochemical process. In comparison, the

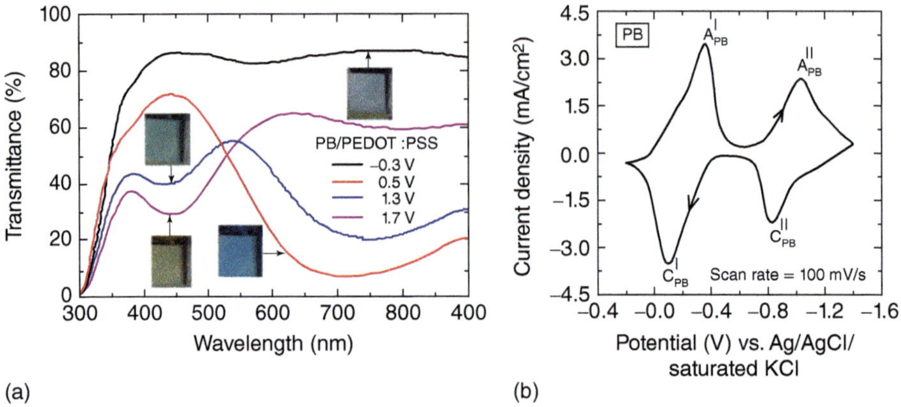

Figure 5.3 Transmittance spectra at different potentials with corresponding pictures [31] (a) and CVs from −0.2 to 1.4 V in 0.5 M KCl aqueous solution at a scan rate of 100 mV/s [32] (b) for PB. Source: (a) Liao et al. [31]. Copyright 2016 Elsevier; (b) Kuo et al. [32]. Copyright 2009, Wiley.

color change between BG and PW allows for a dual step insertion and extraction of K^+ ions, potentially leading to nearly double the capacity of K^+ ions [36, 37]. The EC procedure of the PB thin films can be generally summarized in the following equations.

The redox processes for soluble PB were described as [22]

$$KFe^{III}[Fe^{II}(CN)_6] + e^- + K^+ \leftrightarrow K_2Fe^{II}[Fe^{II}(CN)_6] \tag{5.1}$$

PB : blue/cyan thin film   PW : transparent thin film

$$3KFe^{III}[Fe^{II}(CN)_6] - 2e^- - 2K^+ \leftrightarrow KFe^{III}{}_3[Fe^{III}(CN)_6]_2[Fe^{II}(CN)_6] \tag{5.2}$$

PB : blue/cyan thin film   BG : green film

The redox processes for insoluble PB were described as [27, 34]

$$Fe^{III}{}_4[Fe^{II}(CN)_6]_3 + 4e^- + 4K^+ \leftrightarrow K_4Fe^{II}{}_4[Fe^{II}(CN)_6]_3 \tag{5.3}$$

PB : blue/cyan thin film   PW : transparent thin film

$$Fe^{III}{}_4[Fe^{II}(CN)_6]_3 - 3e^- + 3X^- \leftrightarrow Fe^{III}{}_4[Fe^{III}(CN)_6]_3(X)_3 \tag{5.4}$$

PB : blue/cyan thin film   PY : yellow/yellowish film

For electroneutrality, the reduction procedure of PB, both the soluble and the insoluble, requires insertion of K^+ (or another electrolyte cation of appropriate size) into the lattice, and this cation dependence has already been confirmed by experimental results. For soluble PB, part of the lattice K^+ departs from the lattice during the oxidation process, and the oxidative charge is two-thirds of the reductive charge. While for insoluble PB, an electrolyte anion enters the lattice during the oxidation process, and the oxidative charge is three-fourths of the reductive charge.

5.5 Synthesis

The synthetic technology of PB plays an important role in its practical application. The uniformity of thin films in large area is essential for high-tech applications. Nowadays, two different preparation methods, called direct and indirect, are used to synthesize either the soluble PB or the insoluble PB by precipitation from aqueous solution, which make the research on the synthetic technology of PB more complex. These two methods are outlined in Figure 5.4 [24].

For the direct method, soluble or insoluble PB is prepared by precipitation of the extremely insoluble product either by combining an Fe^{III} salt with a ferrocyanide salt, such as $FeCl_3 + K_4[Fe^{II}(CN)_6]$ [38], or by combining an Fe^{II} salt with a ferricyanide salt, such as $FeCl_2 + K_3[Fe^{III}(CN)_6]$ [39]. For the indirect method, PW $K_2Fe^{II}[Fe^{II}(CN)_6]$ is firstly prepared by combining an Fe^{II} salt with a ferrocyanide salt, such as $FeCl_2 + K_4[Fe^{II}(CN)_6]$ and then oxidized, often with H_2O_2, to form the final soluble or insoluble PB [40]. In either method, when an excess of the iron salt is used, insoluble PB will be prepared, whereas if a 1 : 1 molar ratio is used, soluble PB can be obtained [41, 42].

PB nanoparticles are prepared by a step growth mechanism, which is different from most nanoparticles that are fabricated by controlled condensation methods. Individual unit crystals are obtained from the complexation of two ions, which are individually and separately soluble in water. At room temperature, step growth occurs in three dimensions when solutions of the two ions are mixed together, and crystals nucleate and increase in size. Referring to the gradual growth mechanism of polymer, stoichiometric equalization between reactants is a necessary condition for large and uniform crystal growth of PB (although this deduction is not accurate due to the heterogeneous polymerization during PB synthesis). The crystal size will possibly be reduced when the reactant ratio is adjusted to large excess of a single reactant. A typical synthesis reaction for PB nanoparticles is shown in

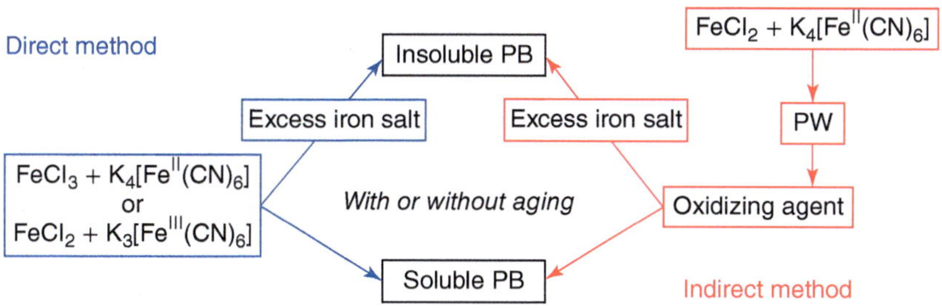

Figure 5.4 A schematic of the direct (blue) and indirect (red) synthetic methods of PB. Source: Grandjean et al. [24]. Copyright 2016, The Royal Society of Chemistry.

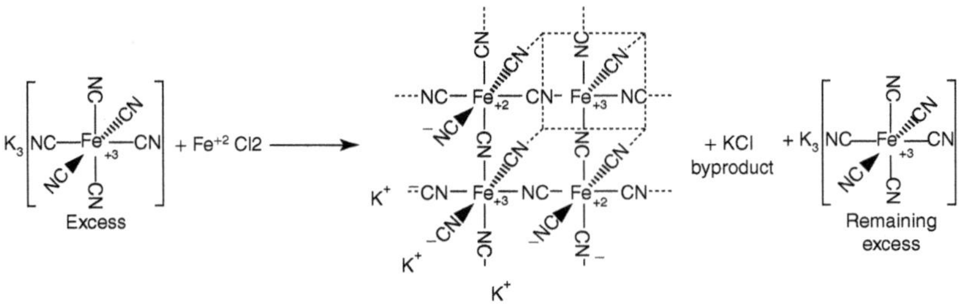

Figure 5.5 Reaction scheme for PB synthesis [43]. Source: DeLongchamp and Hammond [43]. Copyright 2004, Wiley.

Table 5.1 Most commonly used synthetic procedures for PB thin films.

Synthetic procedure	Description	References
Electrochemical cycling	Cyclic voltammetry in a $K_3Fe(CN)_6 + FeCl_3$ solution	[28, 44, 45]
	Cyclic voltammetry in a $K_3Fe(CN)_6$ solution	[46]
Electrochemical galvanostatic	Applied constant current for a fixed period in $K_3Fe(CN)_6 + FeCl_3$ solution	[47–51]
Electrochemical potentiostatic	Applied constant potential for a fixed period in $K_3Fe(CN)_6 + FeCl_3$ solution	[17, 52–54]
Hydrothermal	$K_3Fe(CN)_6 + C_6H_{12}O_6 + HCl$	[55, 56]
Spin coating	PB nanoparticles that were obtained from spontaneous reaction between Fe^{3+} and $Fe(CN)_6^{4-}$, dispersed in solvents to form a uniform suspension and then spin-coated to form thin film	[57, 58]
Spray coating	Soluble PB powder, $KFe^{III}[Fe^{II}(CN)_6]{\cdot}xH_2O$, dispersed in organic solvents to form a uniform colloidal suspension and then spray-coated to form thin film	[59]
Dip coating	Substrate was dipped in the mother liquor of PB, such as $Fe(OAc)_2$ and $Fe^{3+} + Fe(CN)_6^{4-}$, washed and dried	[60, 61]
Immersion	Substrates were immersed into a PB colloidal suspension, which was obtained from spontaneous reaction between Fe^{2+} and $Fe(CN)_6^{3-}$	[62]
Photochemical	With illumination (365 nm) for 5 h in a solution (pH 1.6) of 1 mM $Fe(CN)_6^{3-}$ + 0.2 M KCl	[63]

Figure 5.5 [43]. With this synthetic process, a dark blue aqueous suspension can be prepared, which does not scatter light and is infinitely stable without stirring or adding stabilizer. The long-term stability of the dispersion indicates that PB particles are primarily $KFe^{III}[Fe^{II}(CN)_6]$ and therefore ionized and stable. In view of the wide application potential, it has attracted more and more attention to fabricate well-performing PB thin films for device applications. Table 5.1 shows the most commonly used synthetic procedures for PB thin films on various electrode surfaces.

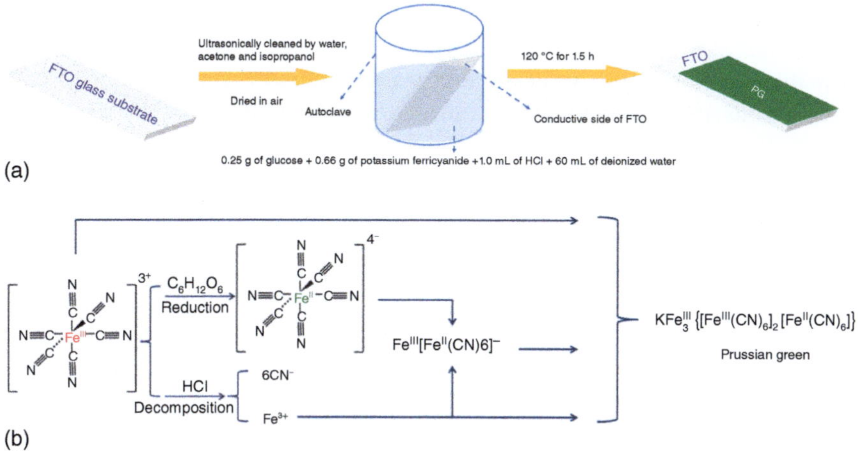

Figure 5.6 Schematic diagram for the hydrothermal process (a) and reaction mechanism (b) of PB films [55, 56]. Source: (a) Li et al. [56]. Copyright 2019, Wiley; (b) Qian et al. [55]. Copyright 2018, Wiley.

Due to easy operation and being less time consuming, electrodeposition and spin coating have always been used for preparing PB films in the past few decades, especially the former one. The first straightforward formation of PB films was introduced by Neff and coworkers based on electrochemical mechanism [17, 22, 64]. However, PB films prepared by electrodeposition or spin coating are easy to peel off from the substrates after several electrochemical cycles in aqueous electrolyte solution, which largely restrict the application of PB. Nowadays, hydrothermal method is an important method for preparing metal oxide EC thin film, because the prepared thin film shows strong adhesion to substrates, such as WO_3 films, NiO films, and ZnO films. As PB is not one of the common inorganic compounds but one of the coordination compounds, its film growth procedure is different from metal oxides. Thus, it is still a challenge to directly grow a PB film by hydrothermal method. A typical hydrothermal process for synthesizing PB films is shown in Figure 5.6.

The electrochemical deposition method and the hydrothermal method may not be easily used in large-scale production. Comparatively, spray coating is an efficient, suitable method to prepare large-area ECDs and can be used for patterning. It is mentioned earlier that PB nanoparticles, which can be uniformly dispersed in water, could be easily prepared by a simple method. Thus, the aqueous solution of water dispersible PB (wPB) or "ink" can be sprayed on electrodes to fabricate a complementary ECD, such as glasses coated with transparent conductive films. And the film thickness of PB can be preciously controlled by changing the spray coating conditions. Therefore, the spray-coating technology of wPB enables the industrialization of large-area ECDs, which can be applied for energy-saving windows and sunglasses.

For the aforementioned chemical and electrochemical methods to synthesize PB and its analogs, free metal ions and ferrocyanide/ferricyanide ions exist simultaneously in the solution, resulting in an excess of supporting ions, such as K^+, NH_4^+, Rb^+, Cs^+, and so on. Compacted PB film can also be prepared by a photochemical method from an acidic ferricyanide solution. Under illumination, ferricyanide ions are photochemically reduced to ferrocyanide ions and then coordinate with the free iron ions dissociated from ferricyanide in an acidic medium to form PB on the surface of electrode. This method can also be used to synthesize PB crystals in solution with ferricyanide and iron ions, which may allow the use of standard photolithography techniques to fabricate patterned films. Besides, PB films can also be prepared by an assembly-based processing technique, which is recently called the layer-by-layer (LBL) method. PB films are created alternately exposing an ionized substrate to dilute aqueous solutions/dispersions with opposite attraction affinities. Although the LBL PB films show finer thickness control and better uniformity than traditional electrochemical methods, fully inorganic PB film are brittle and polycrystalline with limited flexibility. Thus, flexible PB film composites of macromolecular complexes are fabricated by LBL assembling metal hexacyanoferrate nanoparticle/polymer nanocomposites to address the limitation of their compositional variation and application.

5.6 Electrochromic Devices (ECDs)

The ECD is typically composed of working electrode, counter electrode, and electrolyte. Much attention has been recently focused on the development of ECDs, which can be rapidly switched from a dark to a transparent state with large contrast. Although, PB ECDs can present four colors under different potentials, their performance still has a lot of room for improvement. An complementary ECD is composed of more than one coloring materials, with great advantage, such as fast coloration, high coloration efficiency, and low energy consumption. Hence, the complementary ECD shows great potential for numerous future applications due to its high contrast in absorption.

In 1993, De Paoli et al. fabricated a solid-state ECD based on PB and polyaniline (PANI) with a crude elastomer (Hydrin 200) containing $LiClO_4$ as an elastomeric electrolyte [65]. The optical memory of the device under open circuit was tested, as shown in Figure 5.7. Results showed that the colored form of the device exhibited no spectral variation after 10 hours under open circuit, while the bleached form exhibited a pronounced spectral variation in the first 10 minutes and almost become colored after six hours. It is indicated that the decreasing chromic ability of the device was mainly caused by the bad stability of the reduced form of PB and PANI.

In 1993, Jelle et al. reported an complementary ECD by combining PANI, PB, and tungsten oxide (WO_3) with poly(2-acrylamido-2-methylpropane-sulfonic acid) (PAMPS) as a solid organic polymer electrolyte binding the three EC materials together [66]. Transmission spectra in the 290–3300 nm wavelength region for the ECD at different applied potentials were recorded and showed good light modulation. Under a voltage between −1800 and +1600 mV, the transmission

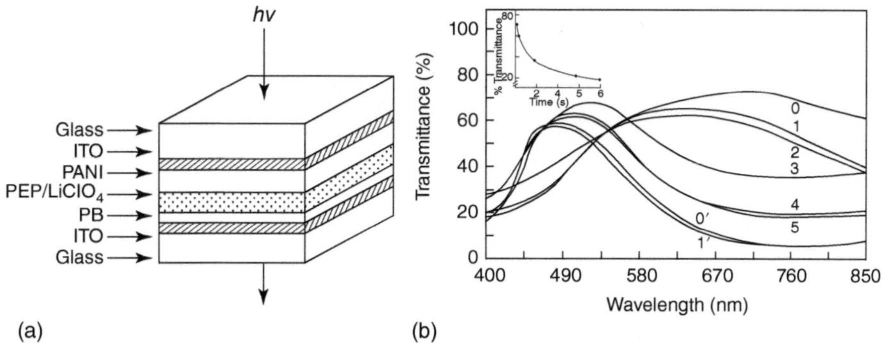

Figure 5.7 The structure (a) and transmittance spectra (b) of the PANI/PB-complementary ECD [65]. Colored state: 0′-initial, 1′–10 hours; bleached state: 0-initial, 1–10 minutes, 2–20 minutes, 3–2 hours, 4–6 hours, and 5–8 hours. The insert shows the transmittance variation at 650 nm of bleached state as a function of time. Source: Duek et al. [65]. Copyright 2004, Wiley.

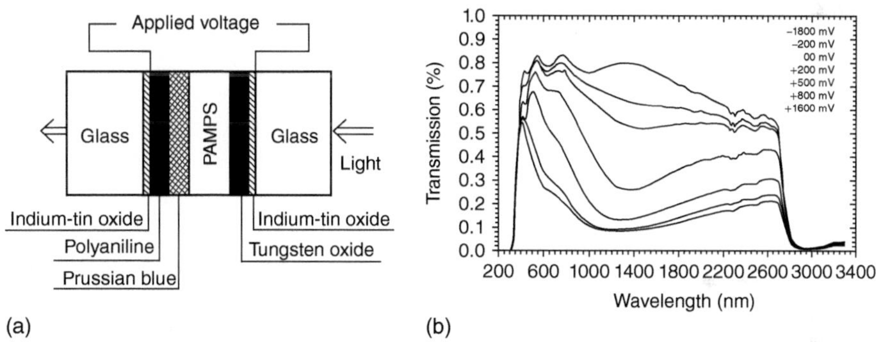

Figure 5.8 The structure (a) and transmittance spectra (b) of the PANI/PB/WO_3-complementary ECD [66]. Source: Jelle et al. [66]. Copyright 1993, Wiley.

changes from 0.76 to 0.11 at 1000 nm. The ECD is able to regulate 49% of the total solar energy, as shown in Figure 5.8.

In 1998, Su et al. synthesized all solid-state ECD of PB and electrodeposited WO_3 film with different gel electrolytes (PMMA, PEO, and PVC) [67–70]. They studied the influence of temperature and gelation time on the conductivity of the gel electrolytes and the EC performance of the ECDs. Results demonstrated that with these gel electrolytes, the ECDs featured excellent electrochromism and memory characteristics. Under a square-wave potential between −0.5 and 1 V, the coloring and bleaching duration of the devices were 60 and 30 seconds, respectively, as shown in Figure 5.9. The blue color of the oxidized form of PB and the reduced form of WO_3 film remained almost unchanged after maintaining the ECDs at open circuit state for more than one week, showing a transmittance change of only 1% per day.

In 2004, Hammond and coworker fabricated a multiple color ECD by LBL assembled PANI/PB thin films [71]. The film growth is induced by the intrinsic electrostatic attraction between the polycation PANI and a negatively ionized PB

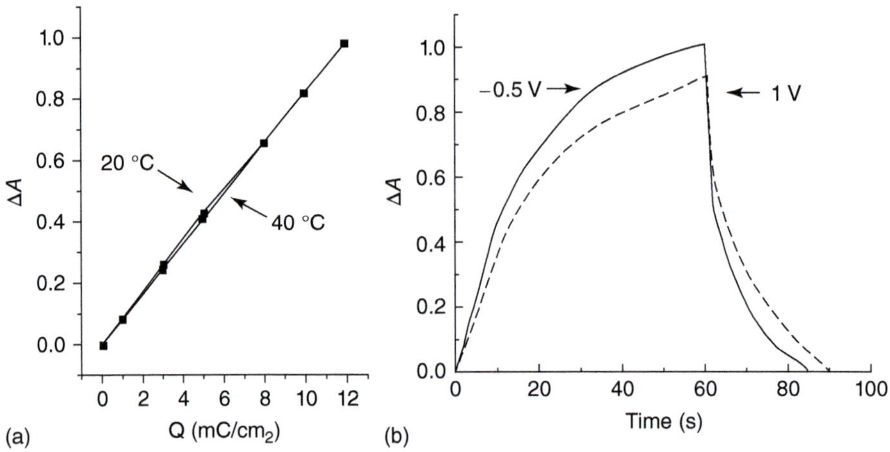

Figure 5.9 (a) The plot of optical density change at 650 nm vs the amount of charge injected into the ECD and (b) optical density change of the ECDs at 650 nm: solid line, 20 °C and broken line, 40 °C [68]. Source: Su et al. [68]. Copyright 1998, Wiley.

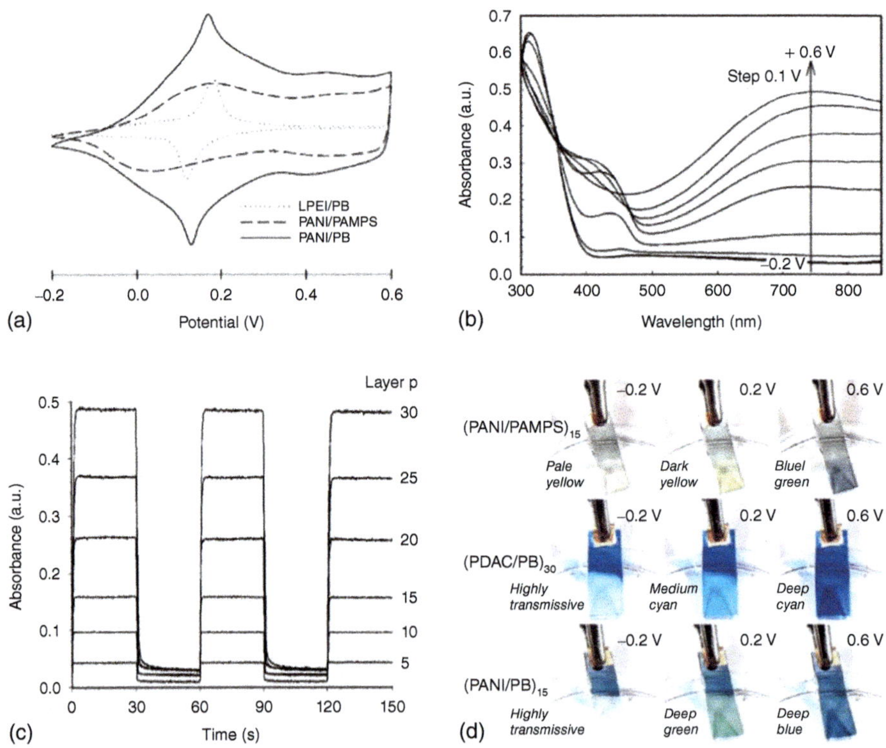

Figure 5.10 CV (a), absorbance (b), switching (c), and chromic (d) properties of the PANI/PB ECD [71]. Source: Source: Reproduced with permission of DeLongchamp and Hammond [71]. Copyright 2004, American Chemical Society.

nanoparticle dispersion, resulting in excellent smoothness and a classical linear increase in film thickness with assembly exposure steps. Electrochemical and spectrophotometric characterization confirms the distinct and noninteracting contributions from PANI and PB and reveals that both are fully electrochemically accessible even in thick, high contrast films (Figure 5.10a). The absorbance of the PANI/PB ECD increases as potential becomes more anodic from −0.2 to 0.6 V, with the biggest change at 750 nm (Figure 5.10b). Due to the presence of conducting PANI, electron transport is significantly enhanced, and the PANI/PB ECD features a fast color change response speed (Figure 5.10c). High contrast was also achieved due to the high density of EC material within the electrode films and the substantial extinction of each chromophore, resulting in an uncolored to green to blue transition (Figure 5.10d). These results validate an LBL assembly-based intermixing strategy for the design of multiple-hue EC electrode films.

In 2016, Viñuales and coworkers fabricated an complementary ECD with a transmittance change up to 71% at 600 nm by using PB on ITO/PET as the counter electrode and viologen-modified TiO_2 films on FTO/glass [72]. The TiO_2/PB ECD

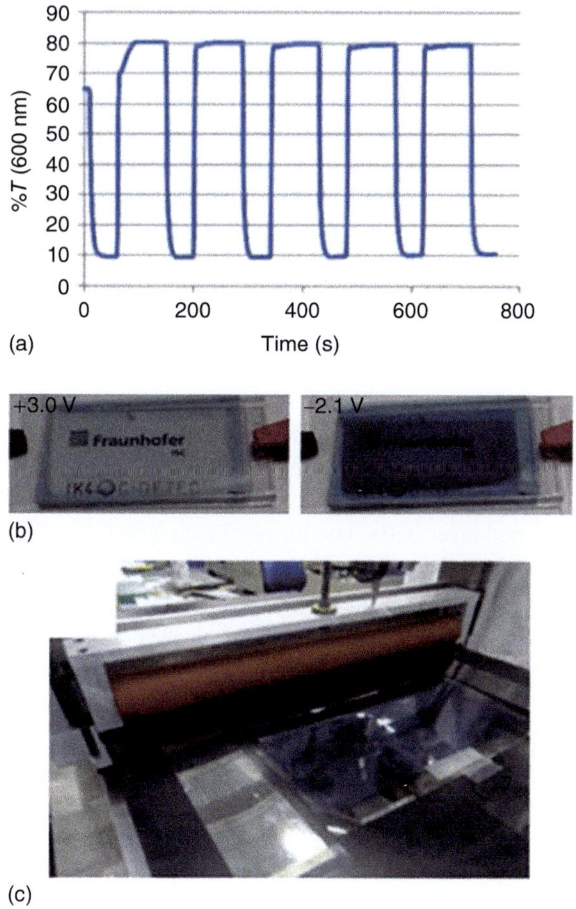

Figure 5.11 Switching response at 600 nm (a) and corresponding photographs in the bleached and colored states (b) of a TiO_2/PB ECD and fabrication of large-area flexible all-solid ECDs (c) [72]. Source: Reproduced with permission of Alesanco et al. [72]. Copyright 2016, Wiley.

switches from colorless at +3.0 V to deep blue at −2.4 V, with a transmittance of only 9% in the colored state (Figure 5.11b). And it is very competitive that the TiO_2/PB ECD exhibited comparatively short switching times, which are seven seconds for coloring process and three seconds for bleaching process, as illustrated in Figure 5.11a. Large-area flexible all-solid ECDs with the structure of PET/ITO/viologen-TiO_2/polymer electrolyte/PB/ITO/PET were also fabricated by laminating in a semiautomatic homemade laboratory coater controlled by air pressure, as shown in Figure 5.11c. The fabricated large-area (30 cm × 40 cm) all-solid ECD can reversibly switch between a colorless bleached state with a transmittance of 75.4% at 600 nm and a homogeneously deep blue colored state with a transmittance of 43.8% at 600 nm, resulting in a transmittance change close to 32%. The aforementioned results demonstrate the suitability of the reported technology for the production of flexible all-solid operational large-area ECDs.

5.7 Nanocomposites

PB is a comparatively stable EC material with reversible color change performance between blue and colorless at low potential. However, slow response time is always the vital drawback of PB ECDs, and other performances, like contrast, also influence the extent of their application. These problems can be improved by applying the concept of nanotechnology. But PB nanoparticles are hard to form a solid and even EC film by a mild preparation method, while organic polymer can be used to improve the mechanical property of PB film and fabricate large-area flexible ECDs. Thus PB nanocomposites have attracted more and more attention for better EC performance and electrochemical stability.

In 2004, DeLongchamp and coworker synthesized some nanocomposite thin films of PB and polycation poly(ethylene imine) (LPEI) by a LBL assembly method. LPEI was adopted with the purpose of improving the ionic conductivity and coloration speed of PB films [43]. Up to 50 layers of LPEI and PB nanoparticles were deposited using aqueous solutions on the indium tin oxide (ITO) glass with a LPEI/PB molar repeat ratio of 1.48. Assembly was very well behaved, with linear correlations in film thickness, faradaic charge capacity, and light absorbance with the number of adsorbed layer pairs. The EC performance of thick LPEI/PB nanocomposite films is better than that of PB films, exhibiting competitive switching speed and superior contrast. The absorbance spectra under different potentials and the corresponding colors of a 50-layer pair LPEI/PB film is shown in Figure 5.12a,b. With the potential increased from 0.6 to 1.5 V, the peak absorbance at 700 nm gradually decreased and the absorbance at 450 nm gradually increased, resulting in visible color changes. However, repeated switching between 0.6 and 1.5 V causes a steady decrease in overall absorbance with each cycle (Figure 5.12c). The electrochemical properties of the LPEI/PB nanocomposite films changed in a sheared cell with a continuous electrolyte flow, which is considered as a form of controlled release and might be considered for pharmaceutical delivery (Figure 5.12d). In conclusion, the PB/LPEI nanocomposite thin films possess an unusual functional combination of

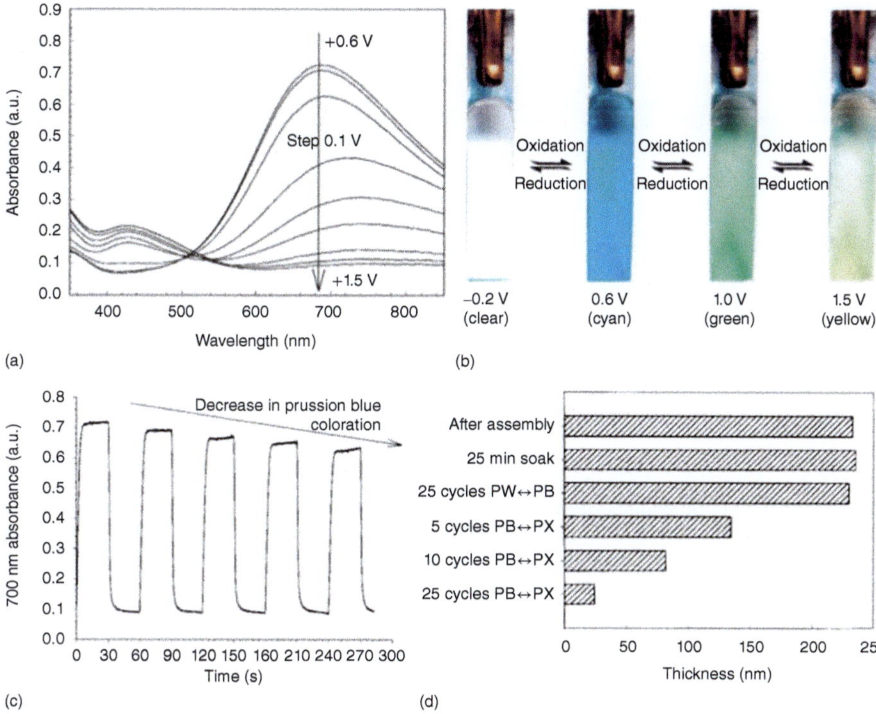

Figure 5.12 Absorbance spectra under different potentials (a), chromic properties (b), spectroscopic monitoring of potential switching in a quiescent cuvette cell (c), and film thickness after switching in a shear cell (d) of a 50-layer pair LPEI/PB film [43]. Source: Reproduced with permission of DeLongchamp and Hammond [43]. Copyright 2004, Wiley.

high-performance electrochromism for displays and controllable dissolution for drug delivery.

In 2007, Cheng et al. fabricated a nanocomposite PB film to synchronously improve the contrast and switching time of the regular PB film [73]. The nanocomposite PB film consists of ITO nanoparticles as a medium layer for PB to gain larger operative reaction surface area in electrolyte, as shown in Figure 5.13a. As shown in Figure 5.13b, the CV curve of the nanocomposite PB film has two pair of peaks, while the CV curve of the regular PB film shows only a symmetrical redox pair. Compared with the regular PB film, the corresponding peaks of PB ↔ PW in the nanocomposite PB film shift to a lower potential and possess higher current densities. That can be attributed to the ITO nanoparticles, which act as a porous conductive template for PB, and this structure can efficiently reduce the impedance of PB film and increase the efficiency of electron transmission. As seen from Figure 5.13c,d, when ±0.5 V was applied for 30 seconds for each step, the maximum transmittance contrast of the nanocomposite PB film was 78% at 750 nm, much higher than that of the regular PB film, 60% at 750 nm. Besides, the switching speed can be substantially improved by ITO nanoparticles (Figure 5.13e,f). The bleaching speed of the nanocomposite PB film is eight times faster than that of the regular PB

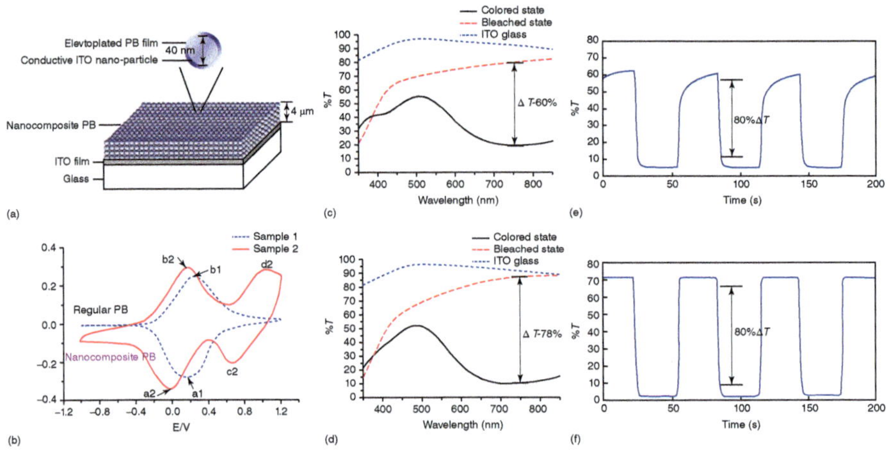

Figure 5.13 Structure (a), CV (b), transmittance spectra (d), and switching response at 750 nm (f) of the nanocomposite PB film and transmittance spectra (c) and switching response at 750 nm (e) of the regular PB film [73]. Source: Cheng et al. [73]. Copyright 2007, Wiley.

film, and though same coloring time, the nanocomposite PB film can accomplish 15% higher transparent contrast than that of the regular PB film. In conclusion, the nanocomposite PB film that provides porous structure to enlarge operative reaction area not only efficiently enhances transmittance contrast but also greatly improves switching time.

In 2011, Taya and coworkers fabricated a new ECD with fast switching, high contrast, and dark colored state based on viologen and PB/antimony tin oxide (ATO) nanocomposites [10]. The structure of the novel ECD is shown in Figure 5.14a. The anodic electrode was composed of compact PB/ATO nanocomposites, which were synthesized by using a galvanostatic electrodeposition process of PB based on a transparent nanostructured ATO nanoparticle film. The cathodic electrode was composed of viologen anchoring on titanium dioxide (TiO_2) nanoparticles, which were formed by modifying the nanocrystalline TiO_2 thin films on conducting glass by a monolayer of viologen with two anchoring groups, which can be much strongly adsorbed onto the surface of TiO_2 nanoparticles. A 2.5 cm × 2.5 cm ECD was fabricated by sandwiching a polymer gel electrolyte between the anodic and cathodic layers. As seen in Figure 5.14b–e, the ECD shows a high contrast of 64.8% at 600 nm, really dark colored state with a very low transmittance of 0.1% at 600 nm, high coloration efficiency of 912 cm^2/C at 600 nm, fast switching speeds (600 and 720 ms for coloring and bleaching, respectively), and good cycling stability. The good EC performance of the novel ECD can be attributed to ATO nanoparticles, which act as inter-conductive materials.

In 2012, Son and coworkers investigated the EC performance of the electrochemically deposited PB nanoparticles on transparent graphene film [74]. A transparent and high-quality graphene film mainly consisting of single-layered graphene was prepared by a CVD method, and then PB nanoparticles were electrodeposited on

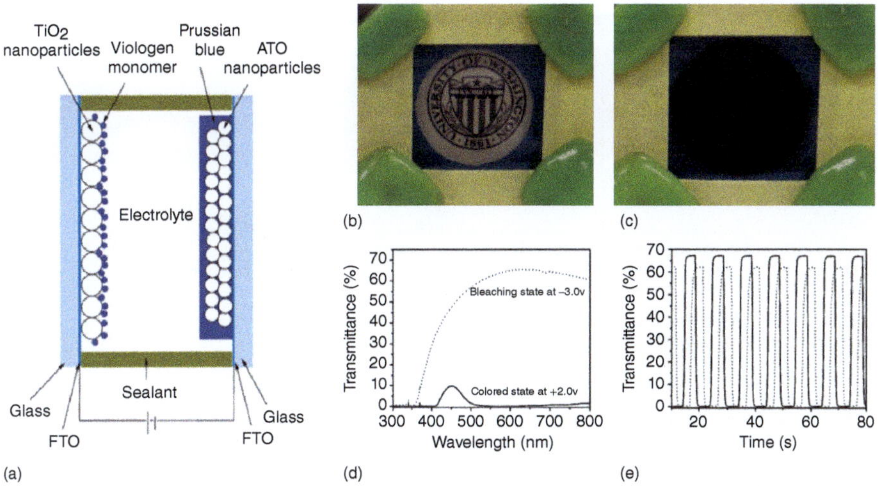

Figure 5.14 Structure (a), bleached state (b), colored state (c), transmittance spectra (d), and switching response at 600 nm (e) of the ECD based on viologen and PB/ATO nanocomposites [10]. Source: Reproduced with permission of Rong et al. [10]. Copyright 2011, Wiley.

the graphene film through cyclic voltammetry, changing the scan-cycle number from 10 to 100 to control the amount of loaded PB nanoparticles. The thickness of PB nanoparticles that were prepared by 40 scans is 420 nm (Figure 5.15d). The UV–visible absorbance and transmittance spectra showed that the maximum absorption peak of PB nanoparticle graphene was at 720 nm (Figure 5.15c). The absorbance at 720 nm and the size of the PB nanoparticles increased with increasing scan-cycle number (Figure 5.15a,b), while the response speed decreased with increasing scan-cycle number (Figure 5.15e). The response times of graphene-based EC systems were compared with those of the commercial ITO electrode. It is demonstrated that the response times of the PB materials on graphene are about half as fast as those on the ITO electrode, indicating that PB nanoparticles on transparent graphene film are a promising candidate system for smart (flexible) window devices.

In 2013, Zarbin and coworker prepared carbon nanotubes (CNTs)/PB films and tested their EC properties for ECDs [75]. The thicknesses (and transmittances) of the CNTs/PB films were controlled by changing the amount of both CNTs and PB. The CNTs/PB films were fabricated by firstly synthesizing CNTs films on ITO electrodes by a drop-casting technique and then depositing PB nanoparticles on the surface of the CNTs by a traditional electrochemical method. The CNTs/PB films exhibit a morphology of PB nanocubes combined with the spaghetti-like arrangement of CNTs films, which is significantly different from the ITO/PB film (without CNTs) synthesized by the same method, as shown in Figure 5.16a,b. The two redox pairs related to the PW ↔ PB and PB ↔ BG processes can be clearly seen from the CV curve in Figure 5.16c, and the absorbance of the CNTs/PB films increased with a more positive potential with the biggest absorbance change at

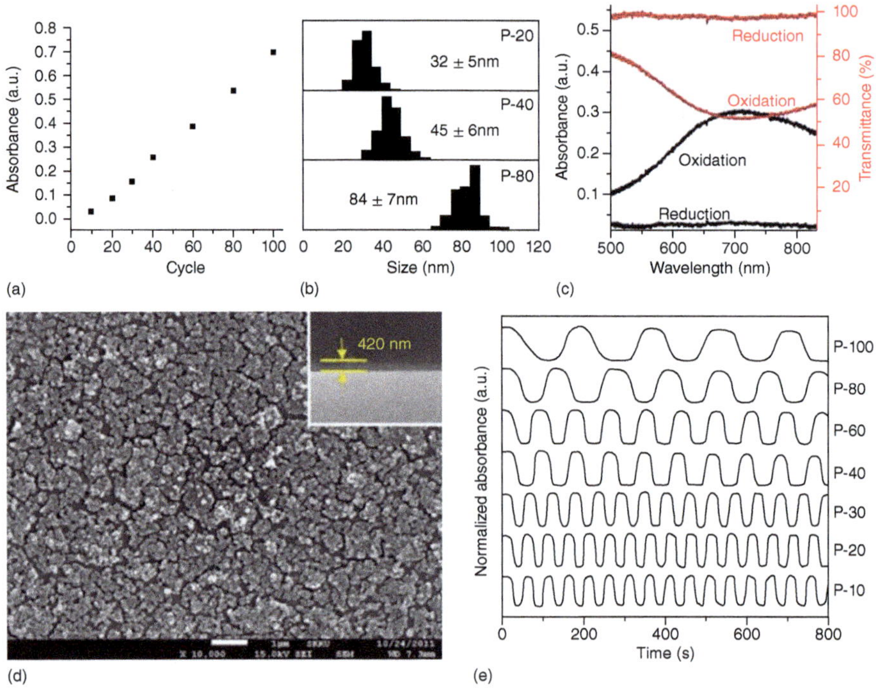

Figure 5.15 Absorbance at 720 nm (a), size distribution (b), and absorbance changes at 720 nm (e) of PB nanoparticles deposited on graphene with different scan-cycle numbers and electrochromic properties (c) and SEM images (d) of PB-40 nanoparticles [74]. Source: Reproduced with permission of Ko et al. [74]. Copyright 2012, The Royal Society of Chemistry.

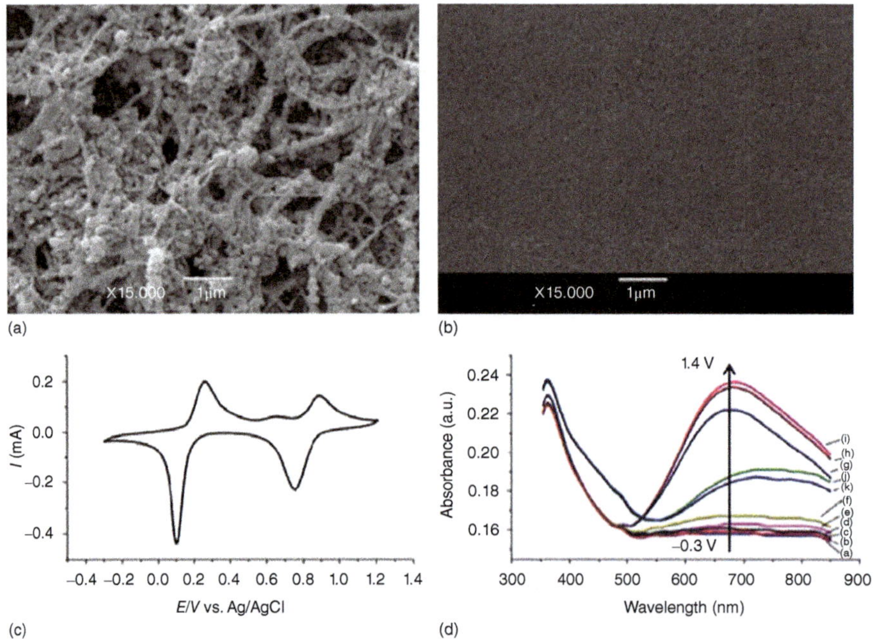

Figure 5.16 SEM images of the CNTs/PB film (a) and the ITO/PB film (b) and CV curve (c) and absorbance spectra (d) of the CNTs/PB film [75]. Source: Reproduced with permission of Nossol and Zarbin [75]. Copyright 2013, Wiley.

680 nm (Figure 5.16d). The CNT/PB films with different thicknesses presented coulombic efficiencies between 91% and 94%, coloration efficiency between 61.1 and 138.6 cm^2/C, response time between 1.1 and 3.2 seconds, and oxidation/reduction charge between 0.17 and 0.69 mC/cm, which enable them to be suitable for technological applications.

In 2016, Hu et al. synthesized a new nanocomposite of PB-PANI:polystyrene sulfonate (PB-PANI:PSS) by incorporating both PANI and PB EC materials [76]. The PB-PANI:PSS nanocomposite is synthesized by two-step polymerization, and the nanocomposite thin film is prepared by a wet-coating method on the conductive ITO glass. As seen in Figure 5.17a–c, the PB-PANI:PSS thin film shows three optical states: highly transparent in a reduced state (−0.5 V vs Ag/AgCl), green color at +0.2 V contributed by PANI, and blue-green color at +0.5 V contributed by both PANI and PB. The absorbance spectra in Figure 5.17d exhibit that the PB-PANI:PSS film possesses a broader absorbance range from 500 to 720 nm than that of pure PANI and PB, resulting in better EC properties. To evaluate this wide-range absorbance material, an averaged transmittance (T_{ave}) was established. When the operation voltage changed between −0.5 and +0.5 V, the ΔT_{ave} value of PB-PANI:PSS thin film was 52.3% with switching times of 8.1 and 13.3 seconds for bleaching and darkening, respectively (Figure 5.17e). And its coloration efficiency was calculated to be 76.6 cm^2/C at the averaged visible wavelength. This PB-PANI:PSS composite thin film could potentially be useful for multicolor EC applications because the ease of processing using water may facilitate the preparation of large-area thin films.

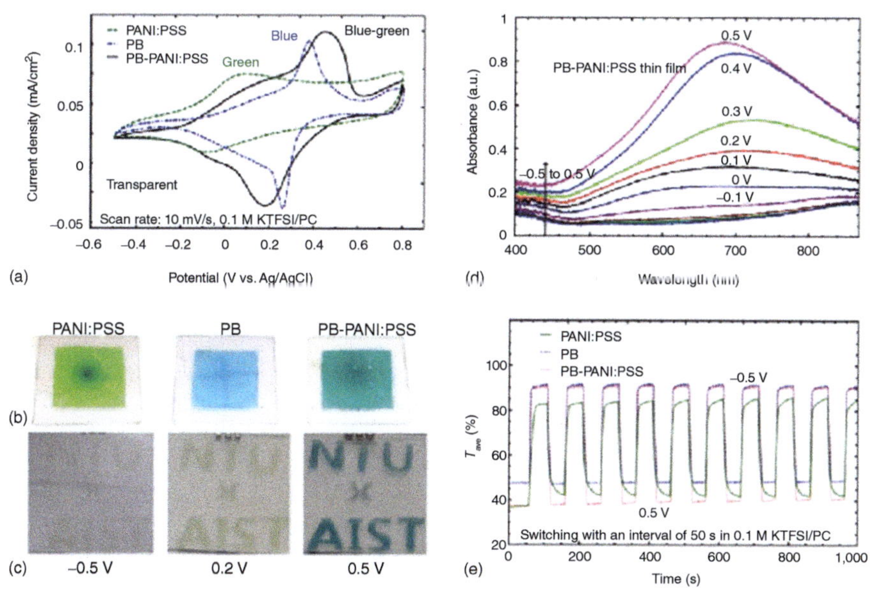

Figure 5.17 CV curves (a), photos (b), and switching response of T_{ave} (e) for PANI:PSS, PB, and PB-PANI:PSS thin films and patterned images (c) and cycling stability (d) of PB-PANI:PSS thin film [76]. Source: Reproduced with permission of Hu et al. [76]. Copyright 2016, The Royal Society of Chemistry.

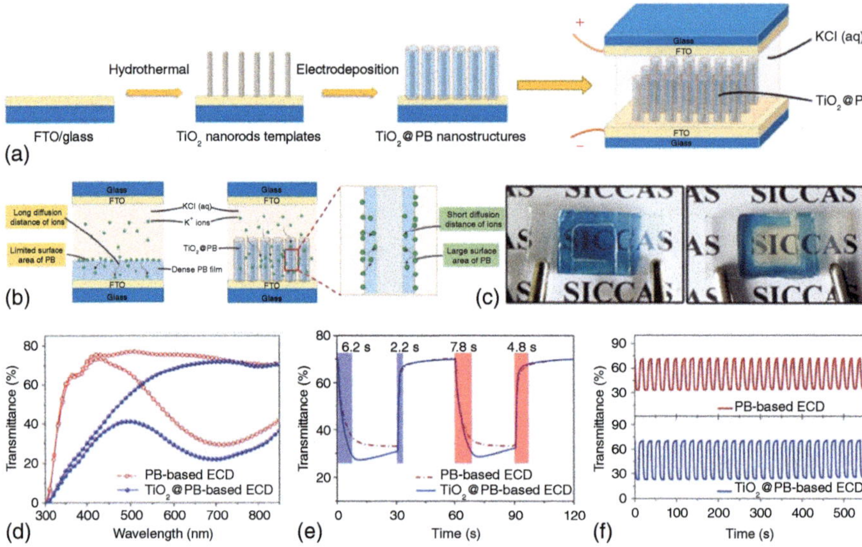

Figure 5.18 (a) Fabrication procedure of the ECD based on TiO_2@PB core–shell nanostructures; K^+ ions insertion/extraction (b), transmittance spectra (d), and transmittance response at 700 nm by applying ± 2.0 V for 30 seconds (e) and 10 seconds (f) in dense PB film and TiO_2@PB core–shell nanostructures; and (c) photographs of TiO_2@PB-based ECD in colored and bleached states [77]. Source: Reproduced with permission of Chen et al. [77]. Copyright 2017, Elsevier.

In 2017, Li and coworkers fabricated an ECD based on novel porous TiO_2@PB core–shell nanostructures as electrodes [77]. The core–shell nanostructures, in which TiO_2 nanorods serve as supporting templates of PB, were formed by coating PB on TiO_2 nanorod templates through two-step soft chemical methods, as shown in Figure 5.18a. Due to the distinct porous morphology of the nanostructures, the surface area of PB becomes much larger and the diffusion depth of ions is notably reduced, resulting in a higher ion storage capacity in EC process than that of the dense PB film, as shown in Figure 5.18b. The ECD based on the TiO_2@PB core–shell nanostructures exhibited a stronger optical contrast (48% at 700 nm, Figure 5.18d), faster response speed (6.2 seconds for coloring and 2.2 seconds for bleaching, Figure 5.18e), and higher coloration efficiency (131.5 cm^2/C at 700 nm, Figure 5.18f), which is significantly superior to the ECD based on the dense PB film. The corresponding colored and bleached states of the ECD based on the TiO2@PB-based ECD are presented in Figure 5.18c.

5.8 PB Analogs

As the research field of PB expanded and deepened, people gradually began to study the PB analogs (PBA), which can be described with the formula of $A_kM'[M''(CN)_6]_l{\cdot}nH_2O$, in which k and l are integers [4]. A can be chosen from Li, Na, K, Rb, and so on. M′ and M″ are transition metals, such as Fe, Mo, Cu, Ni,

Co, Zn, and so on, which can be one or two kinds of elements to replace the Fe element of PB. All the PBA have almost the same structure as PB, which is also a face-centered cubic structure. Nowadays, the most widely studied PBA with EC properties is potassium hexacyanoferrate, which can be described with the formula of $K_xM_y[Fe^{III}(CN)_6]$ (M is Ni, Co, Zn, Ru, etc.). Under positive voltage, different PBA show different color changes, which is really rich to attract many researchers. The color variation of PBA can be summarized in the following equations [4]:

$$(A^A(Fe^{III}[Fe^{II}(CN)_6]))_a + aB^+ + e^- \rightarrow A^A(B^IFe^{II}[Fe^{II}(CN)_6])_a \tag{5.5}$$

PB → blue/cyan thin film   PW → transparent thin film

$$A^A(Fe^{III}[Fe^{II}(CN)_6])_a \rightarrow A^A{}_{1/3}(Fe^{III}[Fe^{II}(CN)_6])_{a/3}(Fe^{III}[Fe^{III}(CN)_6])_{2a/3} + 2/3A^{a+} + 2a/3e^- \tag{5.6}$$

PB → blue/cyan film   BG → green film

$$A^A(Fe^{III}[Fe^{II}(CN)_6])_a \rightarrow Fe^{III}[Fe^{III}(CN)_6] + A^{a+} + ae^- \tag{5.7}$$

PB → blue/cyan thin film   PY → yellow/yellowish thin film

Cobalt(II) hexacyanoferrate(III, II) (CoHCF) is a typical PBA, which shows unique EC performance: its color is dependent not only on the oxidation potential but also on the nature of the counter cations sorbed from electrolyte during reduction. The redox process of cobalt hexacyanoferrate is proposed to describe as the general equation of $X_2Co^{II}[Fe^{II}(CN)_6] \leftrightarrow XCo^{II}[Fe^{III}(CN)_6]$, where X^+ stands for K^+ or Na^+. In 1995, Kulesza et al. synthesized cobalt(II) hexacyanoferrate(II) films based on a coagulation procedure and studied their redox behavior and chromic properties when exposed to aqueous electrolytes containing different cations [78]. The color of the "normal" system $Co^{II}{}_2Fe^{II}(CN)_6{\cdot}nH_2O$ is deep green, which is free of counter cations. But when the system is reduced, it is olive brown in the presence of hydrated Cs^+ (or K^+) ions and becomes a green color in the presence of H^+, Na^+, or Li^+ ions, whose absorbance spectra are shown in Figure 5.19a. In 1996, Zamponi and coworkers synthesized cobalt(II) hexacyanoferrate(II) films based on an electrodeposition method and studied their voltammetric behavior and spectroelectrochemical identity in potassium and sodium electrolytes [79]. Results show that $X_2Co^{II}[Fe^{II}(CN)_6]$ precipitates exhibit different colors upon incorporation of different X alkali metal cations. The oxidized film is purple brown in both electrolytes, but following reduction, the system turns olive brown in 1 M KCl and becomes green in 1 M NaCl as shown in Figure 5.19b. In 1998, Malik and coworkers deposited CoHCF films onto optically transparent gold-covered foil by a slow coagulation method, and the films show reversible EC behavior in potassium salt electrolyte [80]. The voltammetric and spectroelectrochemical results indicated that two CoHCF forms exist, presumably $K_2Co^{II}[Fe^{II}(CN)_6]$ and $KCo^{II}{}_{1.5}[Fe^{II}(CN)_6]$. They are differently colored and they

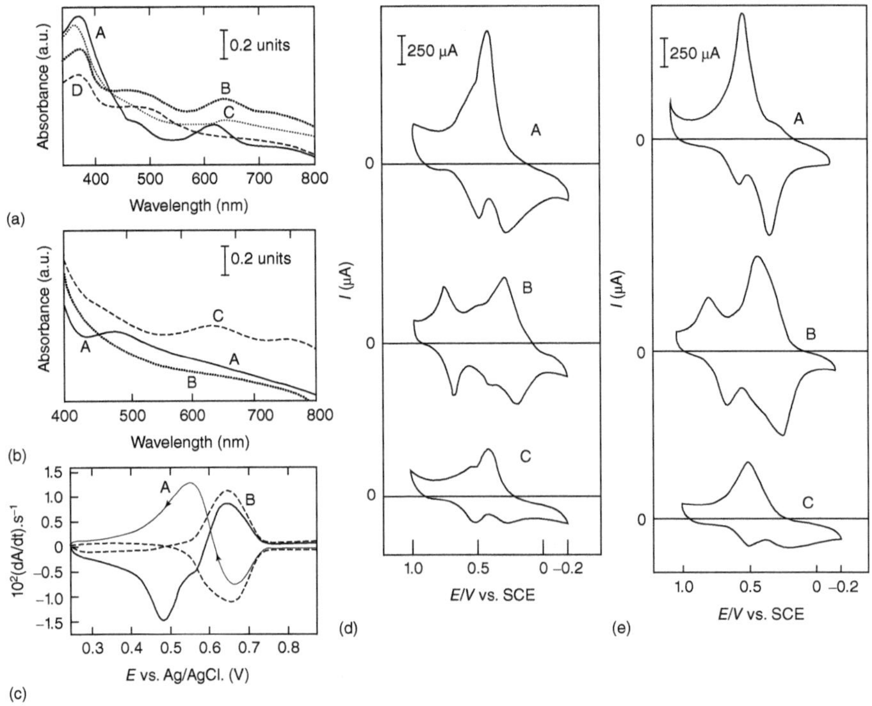

Figure 5.19 (a) Visible absorbance spectra of the reduced CoHCF films, which were deposited on SnO_2-coated glass and exposed to (A) 1 M HCl, (B) 1 M LiCl, (C) 1 M NaCl, and (D) 1 M CsCl [78]. Source: Kulesza et al. [78]. Copyright 1995, Wiley. (b) Visible absorbance spectra of the oxidized CoHCF film in 1 M KCl and 1 M NaCl (A) and the reduced state in 1 M KCl (B) and 1 M NaCl (C) [79]. Source: Kulesza et al. [79]. Copyright 1996, ECS – The Electrochemical Society. (c) Time-derivative cyclic voltabsorptograms of CoHCNFe film recorded at (A) 360 and (B) 410 nm [80]. Source: Kulesza et al. [80]. Copyright 1998, Wiley. CV response of CoHCF-modified electrode (d) and CoHCF + CTAB-modified electrode (e) in (A) 0.1 M KCl, (B) 0.1 M NaCl, and (C) 0.1 M KCl, with electrode being subjected to them in the given order [81]. Source: Vittal and Gomathi [81]. Copyright 2002, American Chemical Society.

undergo distinct, reversible, and EC reactions, which can be clearly evident from spectroelectrochemical measurements, particularly from voltabsorptometry, which involves monitoring of the time-derivative signal of absorbance as a function of the linearly scanned potential, as shown in Figure 5.19c. In 2002, Gomathi and coworker introduced a cationic surfactant, cetyltrimethylammonium bromide (CTAB), to modify electrodes with CoHCF film [81]. It is found that CTAB can improve the voltammetric currents and reversibility of redox reactions, resulting in beneficial influence on the electrochemical properties of the CoHCF-modified electrode. For the same pulse width employed, the diffusion coefficient value of CoHCF + CTAB film is always greater than that of CoHCF film both in KCl and in NaCl supporting electrolytes, as shown in Figure 5.19d,e.

PBA nanoparticles with uniform shape and size can be prepared by reverse microemulsions based on nanometer-sized water droplets as the reaction medium.

To obtain good EC properties, it is essential for PBA nanoparticles to form an organization of higher order in ECDs. PB forms stable colloidal solution during its formation reaction, even without any stabilizer. Thus, controlled organization of PB nano-colloids can be obtained from self-assembly using the Langmuir–Blodgett (LB) technique based on oppositely charged poly-electrolytes or lipids. However, similar methods cannot be used for the self-organization of PBA, due to the unstable colloids in aqueous solution, even at a very low concentration of reactants. In 2004, Ganguly and coworkers synthesized a sodium hexametaphosphate (HMP) stabilized nickel hexacyanoferrate (NiHCF) nanoparticles in aqueous solution and successfully extracted them into an organic phase using CTAB as the surfactant to form LB films [82]. The extraction procedure is shown in Figure 5.20a. As seen from the TEM image and DLS results in Figure 5.20b,c, the average size of these surfactant-encapsulated nanoparticles in the organic phase is about 22 nm, which does not change in repeated evaporation and re-extraction processes of the organic phase. The pressure-area isotherms of the organic phase in an LB trough (Figure 5.20d) indicates that stable monolayer of the surfactant-encapsulated NiHCF nanoparticles forms at the air–water interface with water as subphase. Multilayers of the surfactant-encapsulated NiHCF nanoparticles with reversible redox properties can be obtained by the LBL method, and their voltammetric response increases with the number of deposited layers (Figure 5.20e).

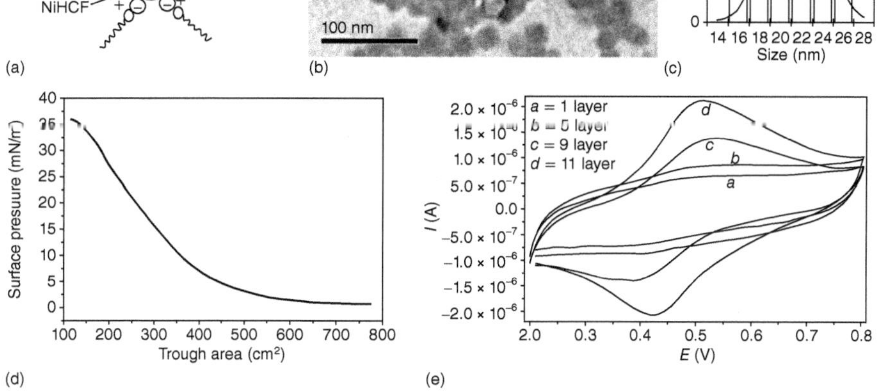

Figure 5.20 (a) Schematic representation of extraction of NiHCF nanoparticles from the aqueous phase to the organic phase, (b) TEM images of NiHCF nanoparticles from dried organic phase, particle-size distribution (c) and Pressure–area isothermf (d) of NiHCF nanoparticles extracted in the organic phase; and (e) cyclic voltammograms of 1, 5, 9, and 11 layers of NiHCF nanoparticles on an ITO-coated glass electrode [82]. Source: Reproduced with permission of Bagkar et al. [82]. Copyright 2004, The Royal Society of Chemistry.

Indium hexacyanoferrate (InHCF) is one of the poly-EC materials with light electrochromism of colorless-to-yellowish color change accompanied with the insertion and extraction of monovalent cations, such as K^+ and Na^+. With this feature, InHCF can potentially be considered as a quasi-transparent counter electrode for the application in ECDs. The redox reactions for the InHCF film in the presence of K^+ ions describe the general equation of $In[Fe^{III}(CN)_6]+K^++e^- \leftrightarrow KIn[Fe^{II}(CN)_6]$. In 2002, Ho and coworkers fabricated a complementary EC system based on PB and InHCF, which possessed a reversible, blue-to-yellowish electrochromism [83]. The overall EC reaction for the PB-InHCF ECD can be described by the equation $KFe[Fe(CN)_6] + KIn[Fe(CN)_6] \leftrightarrow K_2Fe[Fe(CN)_6] + In[Fe(CN)_6]$. The operating voltages for the fully bleached and fully colored states of the PB-InHCF ECD were determined to be 1.2 and 0 V (InHCF vs PB), respectively (Figure 5.21a), much wider than a single material. The major transmittance modulation occurs within a much narrower voltage window (0.9–0.5 V, seen in Figure 5.21b). Besides, the PB-InHCF ECD exhibits high electrochemical compatibility with the K-PAMPS electrolyte, without obvious coulometric capacity lose after cycling for 100 times, as shown in Figure 5.21c. In 2008, Chen et al. prepared a precoloring-free PANI-InHCF ECD with the structure of ITO/PANI/K-PAMPS/InHCF/ITO and investigated the influence of coloring voltage on its optical performance and cycling stability [84]. Results show that different coloring voltages resulted in different optical properties and cycling stabilities, as shown in Figure 5.21d,e. Both the bleached and colored states of the PANI-InHCF ECD could be attained reversibly by applying a voltage ranging from +1.5 to −1.5 V (InHCF vs PANI). With the bleaching voltage of 1.5 V, when the coloring voltage is more positive than −1.2 V (InHCF vs PANI), the

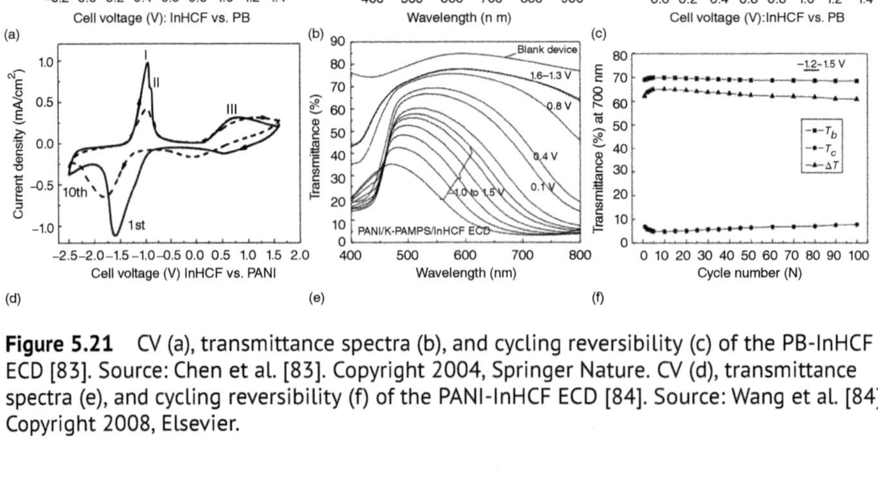

Figure 5.21 CV (a), transmittance spectra (b), and cycling reversibility (c) of the PB-InHCF ECD [83]. Source: Chen et al. [83]. Copyright 2004, Springer Nature. CV (d), transmittance spectra (e), and cycling reversibility (f) of the PANI-InHCF ECD [84]. Source: Wang et al. [84]. Copyright 2008, Elsevier.

PANI-InHCF ECD exhibits a transmittance change of about 60% at 700 nm and can be colored reversibly for at least 100 cycles (Figure 5.21f).

Zinc hexacyanoferrate (ZnHCF) is another PBA. With the formula of $A_yFe[Fe(CN)6]_x$, ZnHCF is constituted by cost-effective zinc cations and hexacyanoferrate anions, which also features a reversible redox behavior, facile ion insertion/extraction kinetics, transparent optical property, and electrocatalytic ability as InHCF does. Thus, ZnHCF can be used as a colorless counter electrode in ECD fields, which is more cost-effective than InHCF. However, a typical electrodeposited ZnHCF thin film suffers from poor cycling stability, low surface charge density, poor deposition yield, and uniformity problems, especially when it is deposited on an ITO glass substrate. Thus, many attempts have been made to improve the electrochemical properties of ZnHCF. In 2010, Chen and coworker dispersed ZnHCF nanoparticles by a transparent conducting polymer matrix to prepare a transparent counter electrode for a poly(3-methylthiophene) (PMeT) ECD [85]. The polymer matrix is commercial poly(3,4-ethylenedioxythiophene):poly(styrenesulfonate) (PEDOT:PSS) conducting ink. Thus, ZnHCF nanoparticles can be deposited on the ITO glass by low-cost wet-coating methods, such as spin coating and screen printing, which are also appropriate for flexible ECD fabrication. Results show that the ZnHCF/PEDOT:PSS electrode possesses considerably larger surface charge density and higher stability than the electrodeposited ZnHCF films. With the structure shown in Figure 5.22d, the ZnHCF/PEDOT:PSS-PMeT ECD could be reversibly switched between a red state under a potential more positive than 0.8 V and a gray state under a potential more negative than 0 V, accompanied with the maximum transmittance modulation of 45.3% at 750 nm and the response time for coloring less than one second (Figure 5.22b,c). Although the electrodeposited PMeT film alone could be cycled several hundred times only, the charge capacity of the ECD could retain 95% of its original value after 10 000 cycles of cyclic voltammetry aging test, as shown in Figure 5.22a.

Although incorporation with conducting polymers can enhance the cycling stability and film-forming ability of ZnHCF, there is much room to improve the cost and contrast of ECD. In 2013, Kawamoto and coworkers conducted a research to improve the redox reaction of ZnHCF by size-controlled nanotechnology [86]. They synthesized dense thin films of ZnHCF nanoparticles smaller than 100 nm by ultracentrifuge processing at 9000–20 000g, with no binding material. When the acceleration rate in the centrifuge process becomes higher, the particle size of ZnHCF becomes smaller, while the current density and transferred charge density of the ZnHCF film becomes greater. The ZnHCF film with the average particle size of 58.8 nm showed the largest current density and the maximum transferred charge density of 5.0 mC/cm^2, which were considerably stable after cycling for 10 000 times, as seen in Figure 5.22e,f. In 2014, Leung and coworkers prepared an ECD with high contrast and low-driving voltage based on a triphenylamine dendritic polymer (PG1) and ZnHCF [87]. They synthesized water-dispersible ZnHCF nanoparticles and fabricated a uniform thin film with high transparency and sufficient charge capacity by spin coating, without the assistance of polymers. When the ZnHCF film was used as counter electrode, the ECD exhibits a two-stage

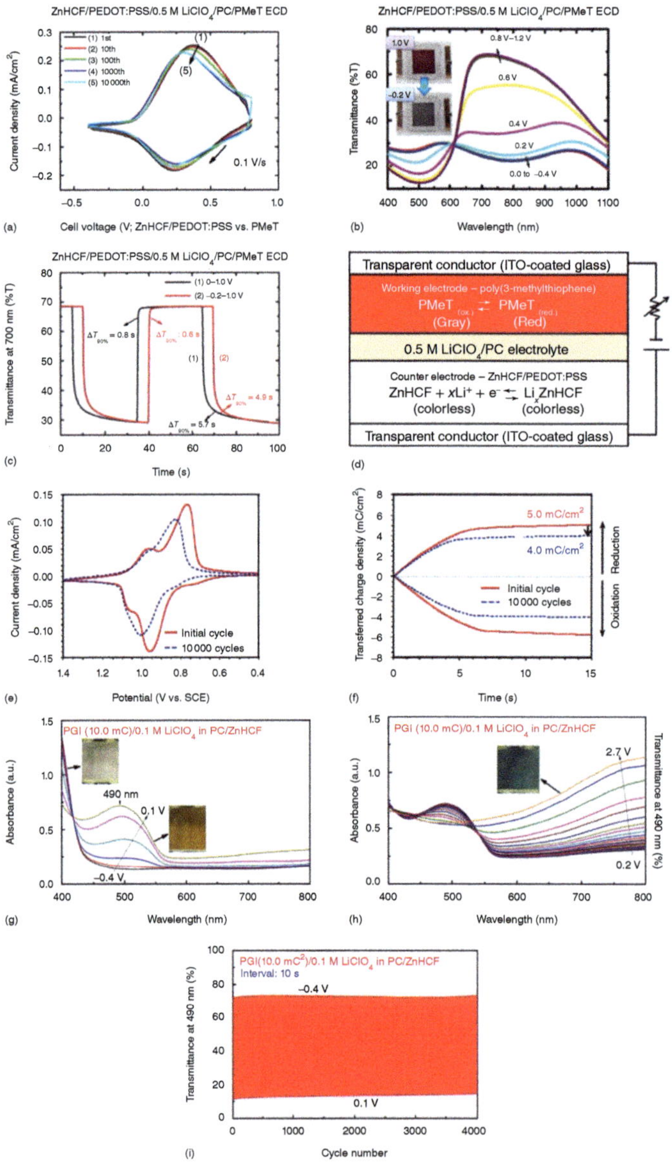

Figure 5.22 CV (a), transmittance spectra (b), transmittance response at 750 nm (c), and structure (d) of the ZnHCF/PEDOT:PSS-PMeT ECD [85]. Source: Reproduced with permission of Hong and Chen [85]. Copyright 2010, Elsevier. Cycle stability of CV (e) and transferred charge density (f) for the ZnHCF film with the average particle size of 58.8 nm [86]. Source: Lee et al. [86]. Copyright 2013, API Publishing. UV-vis spectrum in the first transition (g), second transition (h), and cycling stability for the first transition (i) of the PG1-ZnHCF ECD [87]. Source: Reproduced with permission of Kao et al. [87]. Copyright 2014, Elsevier.

electrochromism from transparent to red and to blue, accompanied with significant color changes, 60% at 490 nm from the bleached state to the first colored state and 63% at 740 nm from the bleached state to the second colored state (Figure 5.22g,h). It is worth noting that a very narrow potential window of 0.5 V is required for the first transition of the ECD, and no significant decay in performance was present after 4000 cycles of the first transition (Figure 5.22i).

In 2012, Chen and coworker demonstrated that the PEDOT:PSS-assisted spin coating is an effective method for preparing a uniform, thick enough PBA film on ITO for EC applications [88]. Highly transparent nano-ZnHCF/PEDOT:PSS films were synthesized and used as counter electrodes for ECDs. They found that the electrochemical performance of the composite film is largely influenced by the precipitation concentration for preparing the ZnHCF nanoparticles and the PEDOT:PSS addition amount for spin coating. As shown in Figure 5.23A,B, the charge capacity of the film spin coated with 5 mM nano-ZnHCF and 20 vol% PEDOT:PSS is much higher than the other samples, and the films spin coated

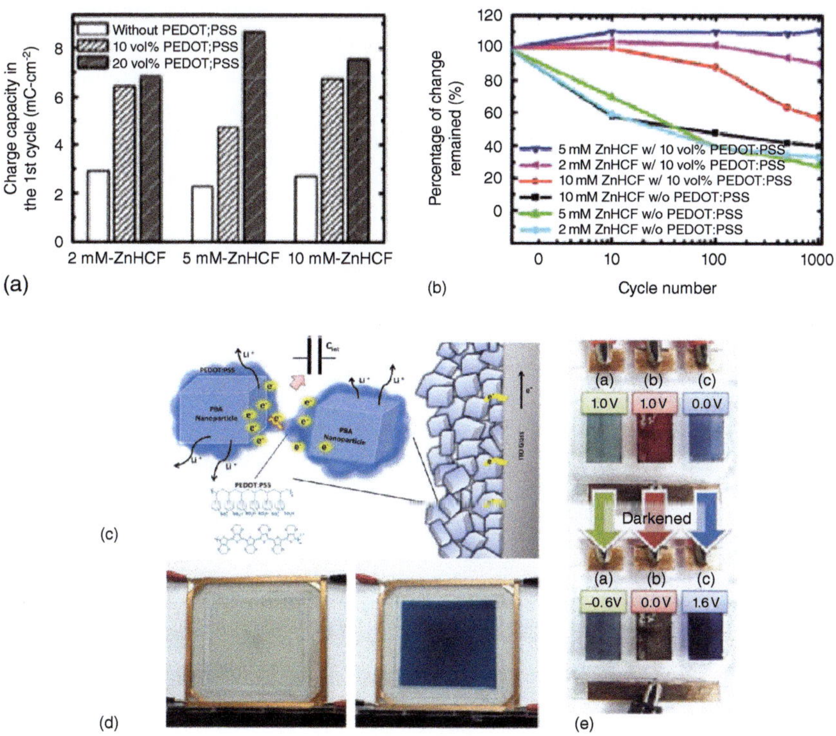

Figure 5.23 Anodic charge capacities in the first CV cycles (A) and percentages of charge capacities remained as a function of increasing CV cycle number (B) for each nano-ZnHCF film under different ZnHCF precipitation concentrations and PEDOT:PSS additive volumes and (C) schematic illustration of the proposed charge-transfer process in a PEDOT:PSS-assisted spin coating of nano-ZnHCF film: the WO_3 window prototype (D) and the RGB device prototype (E) based on the nano-ZnHCF/PEDOT:PSS counter electrode [88]. Source: Reproduced with permission of Hong and Chen [88]. Copyright 2012, Elsevier.

with PEDOT:PSS all possess higher charge capacity and cycling stability than those without PEDOT:PSS. It is also found that PEDOT:PSS not only serves as a transparent conductive binder for nano-ZnHCF but also acts as an electron storage for facile electron transfer between the ZnHCF nanoparticles (Figure 5.23C). The nano-ZnHCF/PEDOT:PSS can work as an effective transparent counter electrode for both EC windows and displays. As shown in Figure 5.23D,E, for practical applications, a WO_3 window prototype with a maximum transmittance modulation of 61.6% at 700 nm and a polymeric red, green, blue (RGB) device prototype with a low driving voltage of 1.6 V are successfully fabricated with the nano-ZnHCF/PEDOT:PSS counter electrode.

In 2016, Chen and coworkers synthesized a cathodically coloring material, molybdate hexacyanoferrate (MoOHCF), based on molybdate (MoO), and fabricated PB/MoOHCF electrochromic device (PMECD) with its electrodeposited film [89]. The EC redox reaction of MoOHCF can be described as the equation of $H_3[Fe^{III}(CN)_6]{\cdot}2MoO_3 + Na^+ + e^- \leftrightarrow H_3Na[Fe^{II}(CN)_6]{\cdot}2MoO_3$. As seen from Figure 5.24a–c, the MoOHCF, whose transmittance gradually increases with increasing potential, presents a brownish red color at its reduced state (0.51 V) and becomes yellowish at the oxidized state (1.17 V), resulting in a strong transmittance modulation of 48% at 480 nm, fast response time (2.1 seconds), and high cycling stability, which remains 95% of ΔT_{max} after 300 cycles. By using the MoOHCF film as counter electrode, a small-size PMECD smart window prototype was assembled based on an EC sol–gel PB/PEDOT:PSS thin film. As shown in Figure 5.24d–f, the PMECD exhibits a deep green (almost black) to yellowish (almost colorless) color transition, accompanied with a broad-band transmittance modulation of solar irradiation, which is about 40% optical transmittance

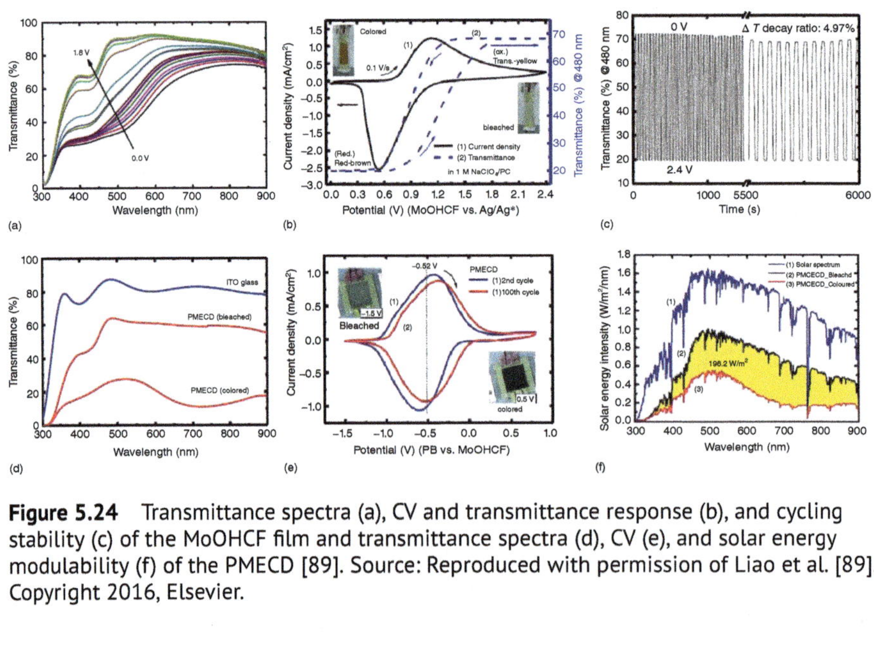

Figure 5.24 Transmittance spectra (a), CV and transmittance response (b), and cycling stability (c) of the MoOHCF film and transmittance spectra (d), CV (e), and solar energy modulability (f) of the PMECD [89]. Source: Reproduced with permission of Liao et al. [89]. Copyright 2016, Elsevier.

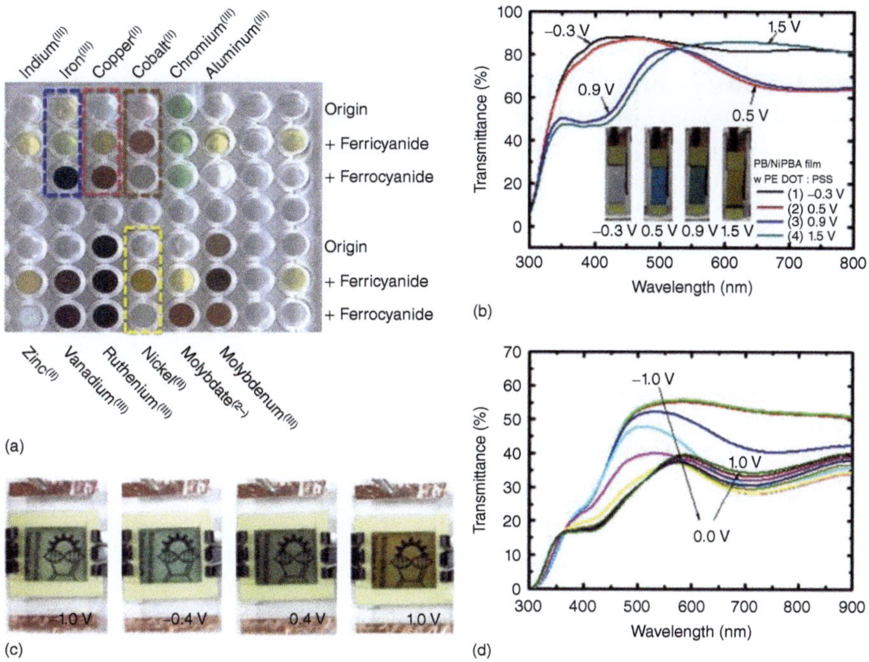

Figure 5.25 (a) A color palette of the PBA precipitates, (b) transmittance spectra and photos of the PB/Ni-PBA film at different potentials, color changes (c), and corresponding transmittance spectra (d) of the PB/NiPBA//Cu-PBA ECD at different potentials. Source: Reproduced with permission of Liao et al. [31]. Copyright 2016, Elsevier.

for almost the whole visible region, blocking out 192.2 W/m^2 solar irradiation energy.

In 2016, Chen and coworkers reported a color palette of PBA precipitates and devoted to apply it in innovating multicolor thin film and ECDs [31]. The color palette was constructed by simply mixing metal ions with ferricyanide or ferrocyanide, and the colors of PBA precipitates in oxidized state and reduced state are shown in Figure 5.25a. By spin coating of the aqueous inks of PBA nanoparticles with 10 vol% PEDOT:PSS as additive, homogeneous PB, Co-PBA, Ni-PBA, and Cu-PBA thin films with distinct reversible EC features were also synthesized. Based on the color palette and the EC properties of the PBA films, a novel composite EC film, which presents multicolor EC reactions from colorless, blue, and green to yellow with high visual contrasts (Figure 5.25b), is deduced and simply prepared by interchanged spin coating of PB and Ni-PBA nanoparticles. Besides, a full-PBA multicolor ECD prototype is assembled based on a PB/Ni-PBA working electrode and a Cu-PBA counter electrode to obtain a cooperative modulation of the visible light by PB (600–800 nm), Ni-PBA (350– nm), and Cu-PBA (500–600 nm). As the applying potential changes from −1.0 to 1.0 V (PB/Ni-PBA vs Cu-PBA), the ECD exhibits multicolor EC reactions from transparently yellowish, pale yellow, and blue to red, accompanied with a solar energy modulation of 125.01 W/m^2 when the voltage switches from −1.0 to 0.4 V, as shown in Figure 5.25c,d.

5.9 Multifunctional Applications

When the electrode surface in sensor is modified with PB, H_2O_2 can be detected at an applied potential around 0.0 V vs. Ag/AgCl with little electrochemical interferences, due to the particular characteristic of PB for catalyzing H_2O_2 reduction. Thus, PB has been widely used in H_2O_2 mediator-based probes, which is possible to have a practical application in clinical, environmental, and food analysis [90]. There are also many researchers working on the development of PB biosensors for the detection of other materials, such as glucose, lactate, cholesterol, galactose, choline, and so on. However, most of the PB sensors are based on amperometric detection, which cannot be seen by naked eyes. Nowadays, PB sensors with visual display ability have attracted more and more attention. Another recent trend is the assembly of self-powered PB sensors, ECDs, or multifunctional devices.

In 2013, Han and coworkers simply synthesized a polyethyleneimine (PEI)-functionalized graphene in an alkaline solution, as shown in Figure 5.26a [91]. The graphene sheets played an important role as connector to assemble the active amino groups of PEI and provided a suitable environment for further functionalization (Figure 5.26b). The PEI–graphene showed excellent dispensability in water, and a [PB/PEI–graphene]$_n$ multilayer film was constructed using a LBL complexation method. The electrocatalytic behavior of [PB/PEI–graphene]$_{10}$

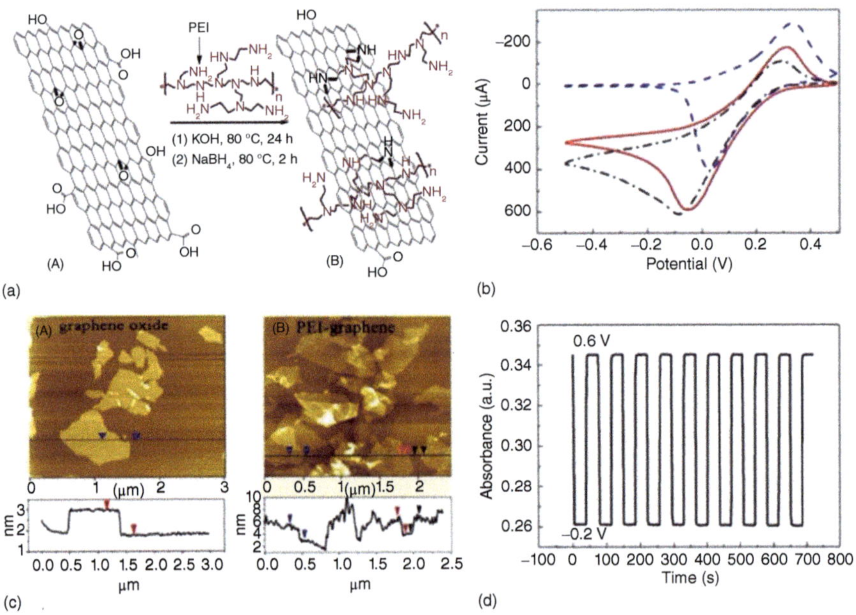

Figure 5.26 Synthesis procedure (a) and AFM (b) of PEI–graphene; (c) CV of the [PB/PEI–graphene]$_{10}$ film in N_2-saturated potassium hydrogen phthalate without (i)/with (ii) H_2O_2 and [PB/PEI]$_{10}$ film (iii) with H_2O_2; and dynamic absorbance response (d) of the [PB/PEI–graphene]$_{10}$ film [91]. Source: Reproduced with permission of Shan et al. [91]. Copyright 2013, Elsevier.

film toward H_2O_2 has also been investigated. As seen from Figure 5.26c, with the presence of H_2O_2 in the electrolyte of potassium hydrogen phthalate, the $[PB/PEI–graphene]_{10}$ film showed a more positive reductive potential than that of the $[PB/PEI]_{10}$ film, which could be attributed to the excellent electron transfer ability of graphene, and thus facilitated the electrocatalysis of H_2O_2 in the reaction system. In addition, the $[PB/PEI–graphene]_{10}$ film had preferable transparency and demonstrated excellent optical switching between PW (low absorbance, −0.2 V) and PB (high absorbance, 0.6 V), as shown in Figure 5.26d. It can be concluded that the $[PB/PEI–graphene]_n$ film showed excellent EC property and good electrocatalysis toward H_2O_2, which are promising for electrosensing applications.

Some reports are focused on solar-driven self-powered sensors. In 2014, Dong and coworkers fabricated a novel self-powered UV photodetector by combining Pt-modified TiO_2 nanotubes and PB-modified ITO, in which the existence of UV could be judged easily by the naked eye with the aid of PB for EC display (Figure 5.27a) [92]. More importantly, the device could also self-recover without UV light illumination. Its color changed from blue to colorless upon UV light illumination whereas the color reverted back to blue in the dark (Figure 5.27b,c). In 2017, Yu and coworkers fabricated a solar-driven photoelectrochemical (PEC) sensing platform for real time and sensitive detection of cellular H_2S by in situ generation of CdS quantum dots (QDs) on the surface of branched TiO_2 (B-TiO_2) nanorods combined with a PB EC display [93]. The sensing mechanism of this platform was based on detection photocurrent intensity generated by the CdS QDs decorate B-TiO_2 nanorod arrays and the colorimetric signal response of the PB, as shown in Figure 5.27d. Due to the CdS-B-TiO_2 heterostructures, a significant enhancement in the photocurrent was obtained, thus leading to high sensitivity of the H_2S level in buffer and cellular environments. PB captures the photoelectron generated from the photoelectrode and transforms to PW with a blue to colorless transition, resulting in a visual system for the quantum photoelectric effect. Thus, detection could also be achieved by measuring the color change of PB. A linear dependence between the absorbance and the logarithm of S^{2-} level was obtained in the range from 1.0 nM to 5 mM, as shown in Figure 5.27e.

Some reports are focused on bio-driven self-powered sensors. In 2014, Joensson -Niedziolka and coworkers have created the first self-powered EC sensor, which gives quantitative information about the electrolyte concentration [94]. The device was assembled by a nanocarbon-based anode for sensing of ascorbic acid (AA) in conjunction with a PB EC display for self-power, as shown in Figure 5.28a. The oxidation of AA on this anode occurs at a quite low overpotential, which enables the anode to be connected to a biocathode to form an AA/O_2 biofuel cell, which functions as a self-powered biosensor. When the AA is oxidized at the anode, the PB in the display is reduced and loses its blue color and becomes transparent. The display can easily be regenerated by connecting it to the biocathode, which returns the PB to its oxidized form. Besides, the rate of change from blue to colorless is dependent on the concentration of AA. So the concentration of AA in the sample can be determined by monitoring the rate of decolorization, as shown in Figure 5.28b. By comparing the color of the PB display after exposure to a sample of orange juice

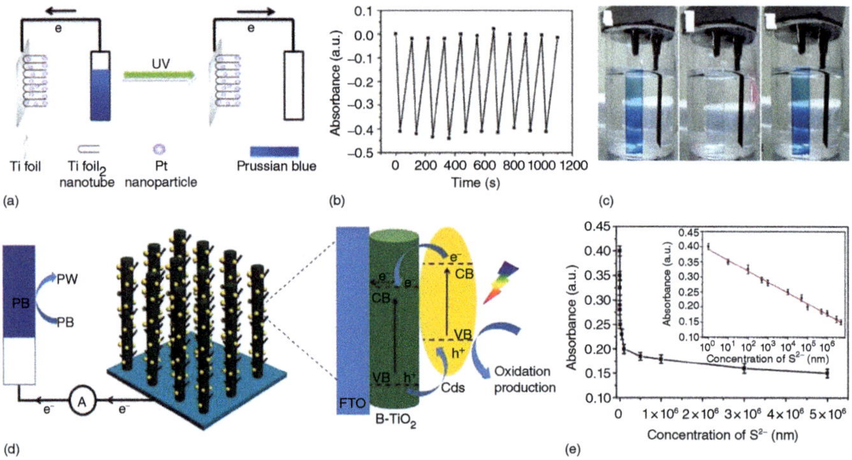

Figure 5.27 (a) General working principle on the self-powered visual UV photodetector, absorption response (b), and color changes (c) of PB at "on" and "off" states [92]. Source: Reproduced with permission of Han et al. [92]. Copyright 2014, The Royal Society of Chemistry. (d) General working principle of the visible PEC sensing platform based on solar driven. (e) The calibration curve of absorption spectra of PB after incubation with varied S^{2-} concentrations [93]. Source: Wang et al. [93]. Copyright 2017, Elsevier.

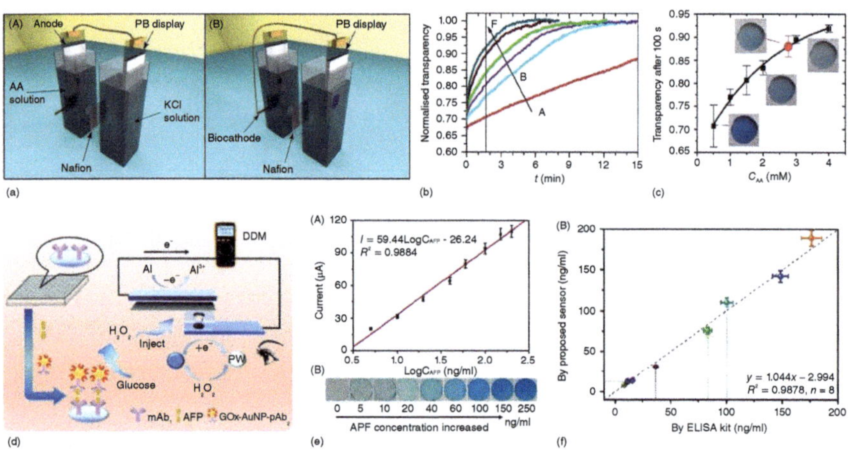

Figure 5.28 A schematic illustration of the self-powered biosensor (a): (A) the configuration used for AA detection and (B) the configuration for regeneration of the PB display. (b) The color change of the PB display vs. time with the anode immersing in 0.5, 1, 1.5, 2, 3, and 4 mM AA from (A) to (F). (c) Normalized transparency of the PB display obtained after 100 seconds plotted vs. AA concentration [94]. Source: Reproduced with permission of Zloczewska et al. [94]. Copyright 2014, Elsevier. Schematic illustration (d), calibration curve (e-A), and corresponding photographs (e-B) of the Al/PB-based self-powered ECD sensing system. (f) Comparison of the results for human serum specimens obtained from the Al/PB-based self-powered ECD sensing system and a human AFP ELISA kit [95]. Source: Reproduced with permission of Yu et al. [95]. Copyright 2018, The Royal Society of Chemistry.

for 100 seconds with the calibration images in Figure 5.28c, the authors determined the AA concentration in the juice to 2.77 ± 0.5 mM, which is in good agreement with the value on the label. In 2018, Tang et al. fabricated an enzyme sensing platform for the simple, rapid, and sensitive detection of disease-related biomarkers (α-fetoprotein [AFP] was used as a model) by coupling an Al/PB-based self-powered ECD with a digital multimeter readout, as shown in Figure 5.28d [95]. With the immunoreaction and the addition of the glucose substrate, the incorporated glucose oxidase catalyzes the decomposition of glucose to generate H_2O_2. Upon injection of the produced H_2O_2 into the single cell, colorless PW, which is reduced from PB in the Al/PB cell reaction, converts back to blue PB and amplify the energy output, which can be simply quantified by currents and color changes of PB (Figure 5.28e). The accuracy of this method was proved credible by analyzing eight clinical serum samples with a commercial human AFP ELISA kit as reference (Figure 5.28f).

Some reports are related to use PB ECD for fluorescence controlling. In 2017, Wang and coworkers proposed a highly efficient fluorescence switch system based on a closed bipolar electrode (C-BPE) system, as shown in Figure 5.29a [96]. In the fluorescence switch system, Au nanoclusters (Au NCs) are immobilized on one pole of the bipolar electrode (BPE) and acts as the fluorescent donor. Due to the spectral overlap between the absorbance spectrum of PB and the emission spectrum of Au NCs, fluorescence quenching ("off" state) was induced by the inner filter effect.

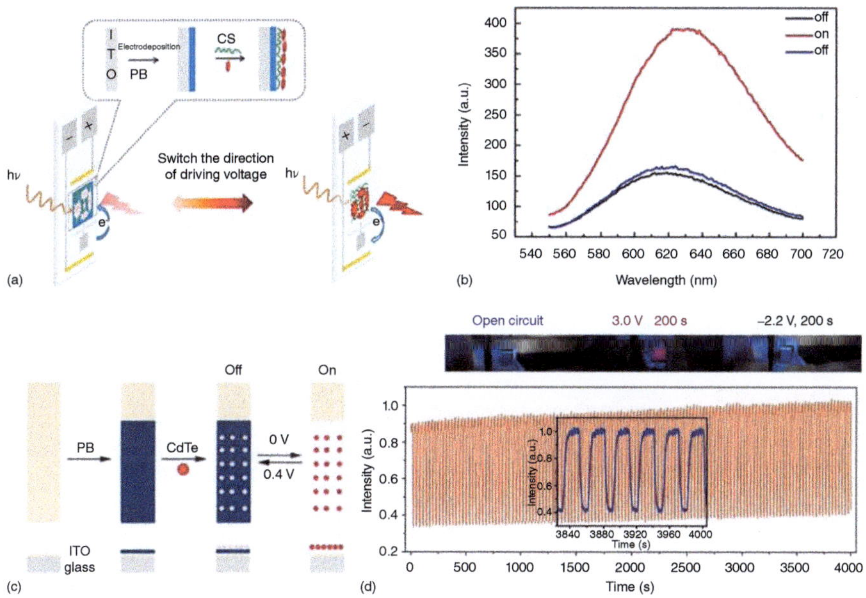

Figure 5.29 Working principle (a) and the fluorescence spectrum (b) for the fluorescence switch system [96]. Source: Reproduced with permission of Xing et al. [96]. Copyright 2017, American Chemical Society. Schematic diagram (c) and fluorescence intensity upon cycling between 0 and 0.4 V (d) of the ITO/PB/CdTe electrode, with the inset diagram representing the last five cycles [97]. Source: Ma et al. [97]. Copyright 2019, The Royal Society of Chemistry.

The fluorescence of the Au NCs can be easily recovered ("on" state) by switching the polarity of driving voltage because of the electrochemical reversible redox reaction between PB and PW. The on–off ratio of the integrated fluorescence switch was up to 2.7 (Figure 5.29b), and a good stability after 10 on–off cycles was obtained by the reasonable design of C-BPE and optimization of driving voltage. In 2019, Dong and coworkers fabricated an EC fluorescence switching ITO/PB/CdTe QDs system (Figure 5.29c) with excellent anti-fatigue performance [97]. Due to the ideal overlap between the absorbance spectrum of PB and the fluorescence spectrum of CdTe QDs, the fluorescence intensity can be regulated and controlled by changing the applied potential. The optimized ITO/PB/CdTe EC fluorescence switching system demonstrated high fluorescence contrast (up to 64%), short response time (10–20 seconds), and particularly superior anti-fatigue property. The fluorescence switching contrast remained almost unchanged after 133 consecutive cycles, as shown in Figure 5.29d.

The latest research is dealing with multifunctional PB devices for EC window, sensors, and energy storage. In 2017, Li and coworkers developed a high-performance electrochromic energy storage device (EESD), which realizes the multifunctional combination of electrochromism and energy storage by constructing tungsten trioxide monohydrate ($WO_3{\cdot}H_2O$) nanosheets and PW film as asymmetric electrodes, as shown in Figure 5.30a [98]. When a positive bias is applied on the PW electrode, Li^+ and electrons are extracted from the PW matrix to form PB. Meanwhile $WO_3{\cdot}H_2O$ is reduced to $LiWO_3{\cdot}H_2O$ by acquiring electrons, resulting in a deep blue color. Conversely, a bleaching procedure occurs when a negative bias is applied. As a complementary ECD, the EESD exhibits broad optical modulation, ultrafast response time, great coloration efficiency, and long cycle life. When the device is used as a supercapacitor, the reactions in the charging process should be PW → PB and $WO_3{\cdot}H_2O \rightarrow LiWO_3{\cdot}H_2O$, which is inversed in the discharging process. Thus, the proposed EESD is able to visually monitor the level of stored energy by color variation and combine with commercial solar cells to constitute

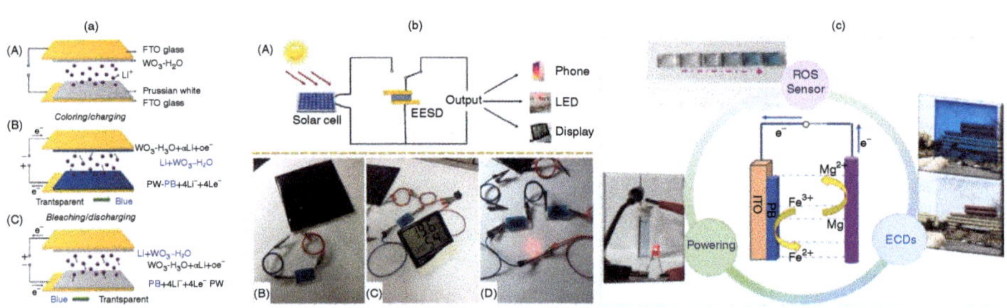

Figure 5.30 (a) Schematic diagrams of operating mechanisms of (A) the EESD in (B) coloring/charging and (C) bleaching/discharging processes. (b) Integrated the EESD with silicon-based solar cells: (A) the circuit diagram of the smart operating system, (B) solar cell is charging for EESD, (C) one EESD can independently drive an LCD screen, and (D) two EESDs in series can lighten a red LED [98]. Source: Reproduced with permission of Bi et al. [98]. Copyright 2017, The Royal Society of Chemistry. (c) Schematic illustration of the self-rechargeable battery-driven device for ECDs, ROS sensor, and energy storage [99]. Source: Reproduced with permission of Zhai et al. [99]. Copyright 2019, American Chemical Society.

an intelligent operating system, which would realize the adjustment of indoor sunlight and the improvement of physical comfort totally by the rational utilization of solar energy without additional electricity (Figure 5.30b). In 2019, Zhai et al. fabricated a self-powered ECD, which realizes the multifunctional combination of electrochromism, energy storage, and quantitative reactive oxygen species (ROS) sensing, as shown in Figure 5.30c [99]. The self-powered ECD, which is composed of PB and Mg as the cathode and anode, respectively, exhibits fast self-charging and high-power density output for continuous and stable energy supply. Benefitting from the particular electrochemical properties of PB, it can be used not only for structuring self-rechargeable batteries but also as an ECD for highly sensitive self-powered ROS sensing and visual analysis.

References

1 Hwang, J.Y., Myung, S.T., and Sun, Y.K. (2017). Sodium-ion batteries: present and future. *Chemical Society Reviews* 46 (12): 3529–3614.

2 Ponrouch, A., Frontera, C., Barde, F., and Palacin, M.R. (2016). Towards a calcium-based rechargeable battery. *Nature Materials* 15 (2): 169.

3 Tian, H.J., Gao, T., Li, X.G. et al. (2017). High power rechargeable magnesium/iodine battery chemistry. *Nature Communications* 8: 14083.

4 Paolella, A., Faure, C., Timoshevskii, V. et al. (2017). A review on hexacyanoferrate-based materials for energy storage and smart windows: challenges and perspectives. *Journal of Materials Chemistry A* 5 (36): 18919–18932.

5 Hermes, M., Lovric, M., Hartl, M. et al. (2001). On the electrochemically driven formation of bilayered systems of solid Prussian-blue-type metal hexacyanoferrates: a model for Prussian blue vertical bar cadmium hexacyanoferrate supported by finite difference simulations. *Journal of Electroanalytical Chemistry* 501 (1–2): 193–204.

6 Miecznikowski, K., Chojak, M., Steplowska, W. et al. (2004). Microelectrochemical electronic effects in two-layer structures of distinct Prussian blue type metal hexacyanoferrates. *Journal of Solid State Electrochemistry* 8 (10): 868–875.

7 Sitnikova, N.A., Komkova, M.A., Khomyakova, I.V. et al. (2014). Transition metal hexacyanoferrates in electrocatalysis of H_2O_2 reduction: an exclusive property of Prussian blue. *Analytical Chemistry* 86 (9): 4131–4134.

8 Xu, L., Zhang, G.Q., Chen, J. et al. (2013). Spontaneous redox synthesis of Prussian blue/graphene nanocomposite as a non-precious metal catalyst for efficient four-electron oxygen reduction in acidic medium. *Journal of Power Sources* 240: 101–108.

9 Fu, L., You, S.J., Zhang, G.Q. et al. (2011). PB/PANI-modified electrode used as a novel oxygen reduction cathode in microbial fuel cell. *Biosensors & Bioelectronics* 26 (5): 1975–1979.

10 Rong, Y., Kim, S., Su, F.Y. et al. (2011). New effective process to fabricate fast switching and high contrast electrochromic device based on viologen and Prussian blue/antimony tin oxide nano-composites with dark colored state. *Electrochimica Acta* 56 (17): 6230–6236.

11 Mortimer, R.J. and Varley, T.S. (2012). In situ spectroelectrochemistry and colour measurement of a complementary electrochromic device based on surface-confined Prussian blue and aqueous solution-phase methyl viologen. *Solar Energy Materials and Solar Cells* 99: 213–220.

12 Nelson, K.J. and Miller, J.S. (2008). Incorporation of substitutionally labile $[V^{III}(CN)_6]^{3-}$ into Prussian blue type magnetic materials. *Inorganic Chemistry* 47 (7): 2526–2533.

13 Byrd, H., Chapman, B.E., and Talley, C.L. (2013). Prussian blue coated electrode as a sensor for electroinactive cations in aqueous solutions. *Journal of Chemical Education* 90 (6): 775–777.

14 Chu, Z.Y., Shi, L., Liu, Y. et al. (2013). In-situ growth of micro-cubic Prussian blue–TiO_2 composite film as a highly sensitive H_2O_2 sensor by aerosol co-deposition approach. *Biosensors & Bioelectronics* 47: 329–334.

15 Kaneko, M., Hara, S., and Yamada, A. (1985). A photoresponsive graphite electrode coated with Prussian blue. *Journal of Electroanalytical Chemistry* 194 (1): 165–168.

16 Kaye, S.S. and Long, J.R. (2005). Hydrogen storage in the dehydrated Prussian blue analogues $M_3[Co(CN)_6]_2$ (M = Mn, Fe, Co, Ni, Cu, Zn). *Journal of the American Chemical Society* 127 (18): 6506–6507.

17 Neff, V.D. (1978). Electrochemical oxidation and reduction of thin-films of Prussian blue. *Journal of the Electrochemical Society* 125 (6): 886–887.

18 Pandey, P.C. and Pandey, A.K. (2013). Novel synthesis of Prussian blue nanoparticles and nanocomposite sol: electro-analytical application in hydrogen peroxide sensing. *Electrochimica Acta* 87: 1–8.

19 Torad, N.L., Hu, M., Imura, M. et al. (2012). Large Cs adsorption capability of nanostructured Prussian Blue particles with high accessible surface areas. *Journal of Materials Chemistry* 22 (35): 18261–18267.

20 Inoue, H. and Yanagisawa, S. (1974). Bonding nature and semiconductivity of Prussian blue and related compounds. *Journal of Inorganic and Nuclear Chemistry* 36 (6): 1409–1411.

21 Lundgren, C.A. and Murray, R.W. (1988). Observations on the composition of Prussian blue films and their electrochemistry. *Inorganic Chemistry* 27 (5): 933–939.

22 Ellis, D., Eckhoff, M., and Neff, V.D. (1981). Electrochromism in the mixed-valence hexacyanides. 1. Voltammetric and spectral studies of the oxidation and reduction of thin-films of Prussian blue. *The Journal of Physical Chemistry* 85 (9): 1225–1231.

23 Mortimer, R.J. and Rosseinsky, D.R. (1984). Iron hexacyanoferrate films: spectroelectrochemical distinction and electrodeposition sequence of soluble (K+-containing) and insoluble (K+-free) Prussian blue, and composition changes in polyelectrochromic switching. *Journal of the Chemical Society Dalton Transactions* (9): 2059–2061.

24 Grandjean, F., Samain, L., and Long, G.J. (2016). Characterization and utilization of Prussian blue and its pigments. *Dalton Transactions* 45 (45): 18018–18044.

25 Ludi, A. (1970). Structural chemistry of polynuclear transition metal cyanides. *Chimia* 24 (12): 445.

26 Zhou, P.H., Xue, D.S., Luo, H.Q., and Chen, X.G. (2002). Fabrication, structure, and magnetic properties of highly ordered Prussian blue nanowire arrays. *Nano Letters* 2 (8): 845–847.

27 Itaya, K., Uchida, I., and Neff, V.D. (1986). Electrochemistry of polynuclear transition-metal cyanides – Prussian blue and its analogs. *Accounts of Chemical Research* 19 (6): 162–168.

28 Yang, D.J., Hsu, C.Y., Lin, C.L. et al. (2012). Fast switching Prussian blue film by modification with cetyltrimethylammonium bromide. *Solar Energy Materials and Solar Cells* 99: 129–134.

29 Agnihotry, S.A., Singh, P., Joshi, A.G. et al. (2006). Electrodeposited Prussian blue films: annealing effect. *Electrochimica Acta* 51 (20): 4291–4301.

30 Guo, Y.Z., Guadalupe, A.R., Resto, O. et al. (1999). Chemically derived Prussian blue sol–gel composite thin films. *Chemistry of Materials* 11 (1): 135–140.

31 Liao, T.C., Chen, W.H., Liao, H.Y., and Chen, L.C. (2016). Multicolor electrochromic thin films and devices based on the Prussian blue family nanoparticles. *Solar Energy Materials and Solar Cells* 145: 26–34.

32 Kuo, T.H., Hsu, C.Y., Lee, K.M., and Ho, K.C. (2009). All-solid-state electrochromic device based on poly(butyl viologen), Prussian blue, and succinonitrile. *Solar Energy Materials and Solar Cells* 93 (10): 1755–1760.

33 Itaya, K., Shibayama, K., Akahoshi, H., and Toshima, S. (1982). Prussian-blue-modified electrodes – an application for a stable electrochromic display device. *Journal of Applied Physics* 53 (1): 804–805.

34 Itaya, K., Ataka, T., and Toshima, S. (1982). Spectroelectrochemistry and electrochemical preparation method of Prussian blue modified electrodes. *Journal of the American Chemical Society* 104 (18): 4767–4772.

35 Mortimer, R.J. and Rosseinsky, D.R. (1983). Electrochemical polychromicity in iron hexacyanoferrate films, and a new film form of ferric ferricyanide. *Journal of Electroanalytical Chemistry* 151 (1-2): 133–147.

36 Padigi, P., Thiebes, J., Swan, M. et al. (2015). Prussian green: a high rate capacity cathode for potassium ion batteries. *Electrochimica Acta* 166: 32–39.

37 Hegner, F.S., Galan-Mascaros, J.R., and Lopez, N. (2016). A database of the structural and electronic properties of Prussian blue, Prussian white, and Berlin green compounds through density functional theory. *Inorganic Chemistry* 55 (24): 12851–12862.

38 Chu, Z.Y., Liu, Y., Jin, W.Q. et al. (2009). Facile fabrication of a Prussian Blue film by direct aerosol deposition on a Pt electrode. *Chemical Communications* (24): 3566–3567.

39 Zamponi, S., Kijak, A.M., Sommer, A.J. et al. (2002). Electrochemistry of Prussian Blue in silica sol–gel electrolytes doped with polyamidoamine dendrimers. *Journal of Solid State Electrochemistry* 6 (8): 528–533.

40 Samain, L., Silversmit, G., Sanyova, J. et al. (2011). Fading of modern Prussian blue pigments in linseed oil medium. *Journal of Analytical Atomic Spectrometry* 26 (5): 930–941.

41 Jomma, E.Y., Bao, N., and Ding, S.N. (2016). Electrochemical sensors for hydroperoxides based on Prussian blue. *Current Analytical Chemistry* 12 (6): 512–522.

42 Dacarro, G., Grisoli, P., Borzenkov, M. et al. (2017). Self-assembled monolayers of Prussian blue nanoparticles with photothermal effect. *Supramolecular Chemistry* 29 (11): 823–833.

43 DeLongchamp, D.M. and Hammond, P.T. (2004). High-contrast electrochromism and controllable dissolution of assembled Prussian blue/polymer nanocomposites. *Advanced Functional Materials* 14 (3): 224–232.

44 Chi, Q.J. and Dong, S.J. (1995). Amperometric biosensors based on the immobilization of oxidases in a Prussian blue film by electrochemical codeposition. *Analytica Chimica Acta* 310 (3): 429–436.

45 Garjonyte, R. and Malinauskas, A. (1999). Operational stability of amperometric hydrogen peroxide sensors, based on ferrous and copper hexacyanoferrates. *Sensors and Actuators B: Chemical* 56 (1-2): 93–97.

46 Jaffari, S.A. and Turner, A.P.F. (1997). Novel hexacyanoferrate(III) modified graphite disc electrodes and their application in enzyme electrodes – Part I. *Biosensors & Bioelectronics* 12 (1): 1–9.

47 Tung, T.S. and Ho, K.C. (2006). Cycling and at-rest stabilities of a complementary electrochromic device containing poly(3,4-ethylenedioxythiophene) and Prussian blue. *Solar Energy Materials and Solar Cells* 90 (4): 521–537.

48 Jiao, Z.H., Wang, J.M., Ke, L. et al. (2012). Electrochromic properties of nanostructured tungsten trioxide (hydrate) films and their applications in a complementary electrochromic device. *Electrochimica Acta* 63: 153–160.

49 Chen, K.-C., Hsu, C.-Y., Hu, C.-W., and Ho, K.-C. (2011). A complementary electrochromic device based on Prussian blue and poly(ProDOT-Et_2) with high contrast and high coloration efficiency. *Solar Energy Materials and Solar Cells* 95 (8): 2238–2245.

50 Jiao, Z., Song, J.L., Sun, X.W. et al. (2012). A fast-switching light-writable and electric-erasable negative photoelectrochromic cell based on Prussian blue films. *Solar Energy Materials and Solar Cells* 98: 154–160.

51 Assis, L.M.N., Ponez, L., Januszko, A. et al. (2013). A green-yellow reflective electrochromic device. *Electrochimica Acta* 111: 299–304.

52 Pinheiro, C., Parola, A.J., Pina, F. et al. (2008). Electrocolorimetry of electrochromic materials on flexible ITO electrodes. *Solar Energy Materials and Solar Cells* 92 (8): 980–985.

53 Li, M.L., Wang, Z.F., Liang, J.Y. et al. (2018). A chemical/molecular 4-input/2-output keypad lock with easy resettability based on red-emission carbon dots-Prussian blue composite film electrodes. *Nanoscale* 10 (16): 7484–7493.

54 Fan, M.-S., Kao, S.-Y., Chang, T.-H. et al. (2016). A high contrast solid-state electrochromic device based on nano-structural Prussian blue and poly(butyl viologen) thin films. *Solar Energy Materials and Solar Cells* 145: 35–41.

55 Qian, J., Ma, D., Xu, Z. et al. (2018). Electrochromic properties of hydrothermally grown Prussian blue film and device. *Solar Energy Materials and Solar Cells* 177: 9–14.

56 Li, F., Ma, D., Qian, J. et al. (2019). One-step hydrothermal growth and electrochromic properties of highly stable Prussian green film and device. *Solar Energy Materials and Solar Cells* 192: 103–108.

57 Lu, H.-C., Kao, S.-Y., Chang, T.-H. et al. (2016). An electrochromic device based on Prussian blue, self-immobilized vinyl benzyl viologen, and ferrocene. *Solar Energy Materials and Solar Cells* 147: 75–84.

58 Hara, S., Tanaka, H., Kawamoto, T. et al. (2007). Electrochromic thin film of Prussian blue nanoparticles fabricated using wet process. *Japanese Journal of Applied Physics* 46 (36–40): L945–L947.

59 Yan, R.T., Liu, L.J., Zhao, H. et al. (2016). A TCO-free Prussian blue-based redox-flow electrochromic window. *Journal of Materials Chemistry C* 4 (38): 8997–9002.

60 Schott, M., Lorrmann, H., Szczerba, W. et al. (2014). State-of-the-art electrochromic materials based on metallo-supramolecular polymers. *Solar Energy Materials and Solar Cells* 126: 68–73.

61 Kumar, A., Peters, E.C., and Burghard, M. (2014). Self-assembled magnetic nanoparticles of Prussian blue on graphene. *RSC Advances* 4 (35): 18061–18064.

62 Liu, X., Zhou, A., Dou, Y. et al. (2015). Ultrafast switching of an electrochromic device based on layered double hydroxide/Prussian blue multilayered films. *Nanoscale* 7 (40): 17088–17095.

63 Hu, Y.L., Yuan, J.H., Chen, W. et al. (2005). Photochemical synthesis of Prussian blue film from an acidic ferricyanide solution and application. *Electrochemistry Communications* 7 (12): 1252–1256.

64 Rajan, K.P. and Neff, V.D. (1982). Electrochromism in the mixed-valence hexacyanides. 2. Kinetics of the reduction of ruthenium purple and Prussian blue. *The Journal of Physical Chemistry* 86 (22): 4361–4368.

65 Duek, E.A.R., De Paoli, M.A., and Mastragostino, M. (1993). A solid-state electrochromic device based on polyaniline, Prussian blue and an elastomeric electrolyte. *Advanced Materials* 5 (9): 650–652.

66 Jelle, B.P., Hagen, G., and Nodland, S. (1993). Transmission spectra of an electrochromic window consisting of polyaniline, Prussian blue and tungsten-oxide. *Electrochimica Acta* 38 (11): 1497–1500.

67 Su, L.Y., Hong, Q.Y., and Lu, Z.H. (1998). All solid-state electrochromic window of Prussian blue and electrodeposited WO_3 film with PMMA gel electrolyte. *Journal of Materials Chemistry* 8 (1): 85–88.

68 Su, L.Y., Wang, H., and Lu, Z.H. (1998). All-solid-state electrochromic window of Prussian blue and electrodeposited WO_3 film with poly(ethylene oxide) gel electrolyte. *Materials Chemistry and Physics* 56 (3): 266–270.

69 Su, L.Y., Wang, H., and Lu, Z.H. (1998). All solid-state electrochromic smart window of electrodeposited WO_3 and Prussian blue film with PVC gel electrolyte. *Supramolecular Science* 5 (5-6): 657–659.

70 Su, L.Y., Xiao, Z.D., and Lu, Z.H. (1998). All solid-state electrochromic window of electrodeposited WO_3 and Prussian blue film with PVC gel electrolyte. *Thin Solid Films* 320 (2): 285–289.

71 DeLongchamp, D.M. and Hammond, P.T. (2004). Multiple-color electrochromism from layer-by-layer-assembled polyaniline/Prussian Blue nanocomposite thin films. *Chemistry of Materials* 16 (23): 4799–4805.

72 Alesanco, Y., Palenzuela, J., Tena-Zaera, R. et al. (2016). Plastic electrochromic devices based on viologen-modified TiO_2 films prepared at low temperature. *Solar Energy Materials and Solar Cells* 157: 624–635.

73 Cheng, K.-C., Chen, F.-R., and Kai, J.-J. (2007). Electrochromic property of nano-composite Prussian Blue based thin film. *Electrochimica Acta* 52 (9): 3330–3335.

74 Ko, J.H., Yeo, S., Park, J.H. et al. (2012). Graphene-based electrochromic systems: the case of Prussian blue nanoparticles on transparent graphene film. *Chemical Communications* 48 (32): 3884–3886.

75 Nossol, E. and Zarbin, A.J.G. (2013). Electrochromic properties of carbon nanotubes/Prussian blue nanocomposite films. *Solar Energy Materials and Solar Cells* 109: 40–46.

76 Hu, C.-W., Kawamoto, T., Tanaka, H. et al. (2016). Water processable Prussian blue-polyaniline: polystyrene sulfonate nanocomposite (PB-PANI:PSS) for multi-color electrochromic applications. *Journal of Materials Chemistry C* 4 (43): 10293–10300.

77 Chen, Y., Bi, Z., Li, X. et al. (2017). High-coloration efficiency electrochromic device based on novel porous TiO_2@Prussian blue core–shell nanostructures. *Electrochimica Acta* 224: 534–540.

78 Kulesza, P.J., Malik, M.A., Zamponi, S. et al. (1995). Electrolyte-cation-dependent coloring, electrochromism and thermochromism of cobalt(II) hexacyanoferrate(III, II) films. *Journal of Electroanalytical Chemistry* 397 (1-2): 287–292.

79 Kulesza, P.J., Malik, M.A., Miecznikowski, K. et al. (1996). Countercation-sensitive electrochromism of cobalt hexacyanoferrate films. *Journal of the Electrochemical Society* 143 (1): L10–L12.

80 Kulesza, P.J., Zamponi, S., Malik, M.A. et al. (1998). Spectroelectrochemical characterization of cobalt hexacyanoferrate films in potassium salt electrolyte. *Electrochimica Acta* 43 (8): 919–923.

81 Vittal, R. and Gomathi, H. (2002). Beneficial effects of cetyltrimethylammonium bromide in the modification of electrodes with cobalt hexacyanoferrate surface films. *Journal of Physical Chemistry B* 106 (39): 10135–10143.

82 Bagkar, N., Ganguly, R., Choudhury, S. et al. (2004). Synthesis of surfactant encapsulated nickel hexacyanoferrate nanoparticles and deposition of their Langmuir–Blodgett film. *Journal of Materials Chemistry* 14 (9): 1430–1436.

83 Chen, L.C., Huang, Y.H., and Ho, K.C. (2002). A complementary electrochromic system based on Prussian blue and indium hexacyanoferrate. *Journal of Solid State Electrochemistry* 7 (1): 6–10.

84 Wang, J.-Y., Yu, C.-M., Hwang, S.-C. et al. (2008). Influence of coloring voltage on the optical performance and cycling stability of a polyaniline–indium hexacyanoferrate electrochromic system. *Solar Energy Materials and Solar Cells* 92 (2): 112–119.

85 Hong, S.-F. and Chen, L.-C. (2010). A red-to-gray poly(3-methylthiophene) electrochromic device using a zinc hexacyanoferrate/PEDOT:PSS composite counter electrode. *Electrochimica Acta* 55 (12): 3966–3973.

86 Lee, K.M., Tanaka, H., Kim, K.H. et al. (2013). Improvement of redox reactions by miniaturizing nanoparticles of zinc Prussian blue analog. *Applied Physics Letters* 102 (14): 141901.

87 Kao, S.-Y., Lin, Y.-S., Chin, K. et al. (2014). High contrast and low-driving voltage electrochromic device containing triphenylamine dendritic polymer and zinc hexacyanoferrate. *Solar Energy Materials and Solar Cells* 125: 261–267.

88 Hong, S.-F. and Chen, L.-C. (2012). Nano-Prussian blue analogue/PEDOT:PSS composites for electrochromic windows. *Solar Energy Materials and Solar Cells* 104: 64–74.

89 Liao, H.-Y., Liao, T.-C., Chen, W.-H. et al. (2016). Molybdate hexacyanoferrate (MoOHCF) thin film: a brownish red Prussian blue analog for electrochromic window application. *Solar Energy Materials and Solar Cells* 145: 8–15.

90 Ricci, F. and Palleschi, G. (2005). Sensor and biosensor preparation, optimisation and applications of Prussian Blue modified electrodes. *Biosensors & Bioelectronics* 21 (3): 389–407.

91 Shan, C., Wang, L., Han, D. et al. (2013). Polyethyleneimine-functionalized graphene and its layer-by-layer assembly with Prussian blue. *Thin Solid Films* 534: 572–576.

92 Han, L., Bai, L., and Dong, S. (2014). Self-powered visual ultraviolet photodetector with Prussian blue electrochromic display. *Chemical Communications* 50 (7): 802–804.

93 Wang, Y., Ge, S., Zhang, L. et al. (2017). Visible photoelectrochemical sensing platform by in situ generated CdS quantum dots decorated branched-TiO_2 nanorods equipped with Prussian blue electrochromic display. *Biosensors & Bioelectronics* 89: 859–865.

94 Zloczewska, A., Celebanska, A., Szot, K. et al. (2014). Self-powered biosensor for ascorbic acid with a Prussian blue electrochromic display. *Biosensors & Bioelectronics* 54: 455–461.

95 Yu, Z., Cai, G., Ren, R., and Tang, D. (2018). A new enzyme immunoassay for alpha-fetoprotein in a separate setup coupling an aluminium/Prussian blue-based self-powered electrochromic display with a digital multimeter readout. *Analyst* 143 (13): 2992–2996.

96 Xing, H., Zhang, X., Zhai, Q. et al. (2017). Bipolar electrode based reversible fluorescence switch using Prussian blue/Au nanoclusters nanocomposite film. *Analytical Chemistry* 89 (7): 3867–3872.

97 Ma, Q., Zhang, H., Chen, J. et al. (2019). Reversible regulation of CdTe quantum dots fluorescence intensity based on Prussian blue with high anti-fatigue performance. *Chemical Communications* 55 (5): 644–647.

98 Bi, Z., Li, X., Chen, Y. et al. (2017). Large-scale multifunctional electrochromic-energy storage device based on tungsten trioxide monohydrate nanosheets and Prussian white. *ACS Applied Materials & Interfaces* 9 (35): 29872–29880.

99 Zhai, Y., Li, Y., Zhang, H. et al. (2019). Self-rechargeable-battery-driven device for simultaneous electrochromic windows, ROS biosensing, and energy storage. *ACS Applied Materials & Interfaces* 11 (31): 28072–28077.

6

Electrochromic Conjugated Polymers (ECPs)

6.1 Introduction

Along with the exploitation of special optical and electrochemical properties of conducting polymers around 1980s after the initial discovery [1, 2], electrochromic (EC) characteristics of a series of conjugated polymers were revealed [3] and gave rise to the rapid emerge of this new class of electrochromisms, which led a strong wave in this field and have been broadly investigated to date. The alternated single and double bonds on the extended conjugated structure allow for easy charge delocalization and endow these conjugated polymers unique electron conductivity, which makes redox process in these species possible. The electrochemical oxidation or reduction allow the transformation of optical absorption properties and induce diverse color changes.

As type III materials [4], EC conducting polymers show a number of advantages compared to other EC systems. Thanks to the feature being operated in thin film state, EC polymers can retain certain color state without further electric signal input once switched to show better "optical memory" effect, which causes the low energy-consumption operation together with the merit of relatively low driven voltage. Besides, easy modifications on polymer structures can yield broad color gamut with abundant tints and also bring design space for solution processability for large-scale, flexible, low-cost and environmental friendly manufactures, which is important for future practical applications.

To date, conjugated polymer EC system has been well developed, from better understandings on mechanisms to completed color palette with soluble or electro-deposited polymers, and even full-color display samples or roll-to-roll fabricated flexible devices. In this chapter, we will introduce basic information and development of this area, especially the approaches to full-color display including material design and processing, which is one of the important application orientations.

6.1.1 Common Categories and Operation Mechanism

Generally, several types of electrochromic conjugated polymers have been mainly investigated: polymers based on derivates of thiophene, pyrrole, furan, carbazole,

Organic Electronics for Electrochromic Materials and Devices, First Edition. Hong Meng.

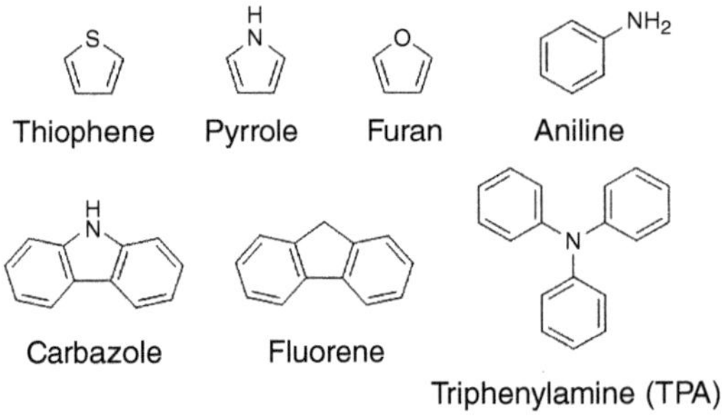

Figure 6.1 Structure of common aromatic molecules utilized in electrochromic polymers.

fluorene, aniline, and triphenylamine (TPA) (Figure 6.1). Each of them has gone through long-time development, whereas several classes have become the frequently referred elements in current research of ECPs, which will be mentioned in the following parts of this Sections 6.2–6.4.

Most of these conjugated polymers are p-type electroactive compounds owing to the electron-donor conjugated structures, which follow a p-type doping process to result in optical changes when electrochemically oxidized. Upon oxidation, the extraction of electrons induces the formation and migrates positive charges on polymer chains and the insertion of counter anions from the electrolyte, resulting in localized structural distortions of polymer backbones and making the geometry change from a benzenoid-like to a quinoid-like structure. The charge carriers created by localized distortion-induced defects were referred as terms such as polarons and bipolarons [5]. A positive polaron is a radical cation and a bipolaron can be related to the dication. Regarding this, as examples, the illustrations of doping process for polythiophene, polypyrrole (PPy), and polyaniline are shown in Figure 6.2.

Accordingly, electronic band structures also experience a series of evolutions during these processes, alternating the situation of electron transition as well as optical characterizations, which are illustrated in Figure 6.3. The electronic bandgaps (E_g) comprise the electronic transition from the top of the valence band to the bottom of the conduction band (Figure 6.3a), and the value can be estimated from the onset wavelength of π–π^* transition absorption (Figure 6.3b). Upon electrochemical oxidation, once the polaron state emerges, two localized electronic states with levels positioned symmetrically about midgap are created, which produce two new growing absorption bands around 1.5 and 1.0 eV at the cost of the highest energy absorption intensity (see E_a, E_b in Figure 6.3a, ii). While the applied potential keeps stepping towards higher value to form higher concentration of bipolarons, only one sub-gap transition is allowed (E_c in Figure 6.3a, iii), which can be corresponded to the enhanced low-energy absorption bands. Actually, the doping process yielding different charge carriers relies on the oxidation level and might be a combination of polarons and bipolarons. Besides, the energy for π–π^* transition is also increased during doping, which results in the corresponding absorption shifted to higher energy. At a high potential (1.0 V, for example), polymer containing a high density of bipolarons would be resulted, of which the intermediate energy level can be represented by bands due to the large number of possible states (Figure 6.3a, iv). Furthermore, if possible, at high doping levels the intra-gap bipolaron bands will become sufficiently

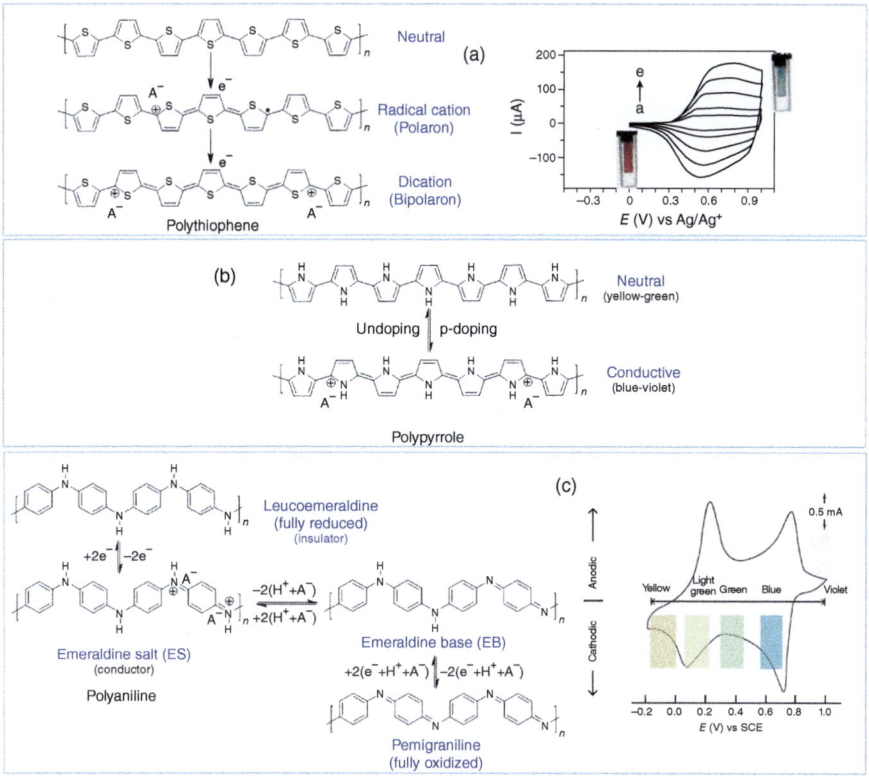

Figure 6.2 Diagrams of doping process for polythiophene, polypyrrole, and polyaniline attached with their corresponding color changes as well as CV curves. (a) Picture of neutral and oxidized polythiophene films electrochemically deposited in boron fluoride ethylether (BFEE) and its CV plots tested in TBAP (0.1 M)/CAN solution under different scan rates. Source: Reproduced with permission of Ref. [6] from Copyright © 2003, WILEY-VCH. (b) Shapes for indicating the color change of polypyrrole film during redox as well as CV curves of PPy-PF_6 film electrodeposited onto Pt disk electrode tested in $TBAPF_6$/PC and ionic liquid electrolyte [BMIM][PF_6]. (c) CV scan and film coloration of PANI at different potentials. Source: Adapted with permission of Ref. [7] from Copyright © 2015, John Wiley and Sons.

broad that they intersect with the valence and conduction bands as illustrated in Figure 6.3a, v. Eventually, if over-high potentials are applied, polymers will undergo an irreversible degradation due to over oxidation [5, 8–10].

More information can be revealed by the cyclic voltammogram (CV) curve containing the two-electron redox electrochemical proceeds. The onset oxidation potential ($E_{\text{onset,ox}}$) as well as redox peak potentials ($E_{p,\text{ox}}$ and $E_{p,\text{red}}$) can be located. However, the latter are not really discernible in most cases owing to the broad outline, which probably consisted of several immerged peaks. One research on P3HT disclosed a more specific electrochemical process (Figure 6.4) [5]. Curves for four different oxidation processes (marked as ox 1–4) with maximum at 496, 667, 898, and 1265 mV, respectively, were plotted by Gaussian functions to result a fitting plot (pink) and compare that with experimental curve (black). In terms of the physical

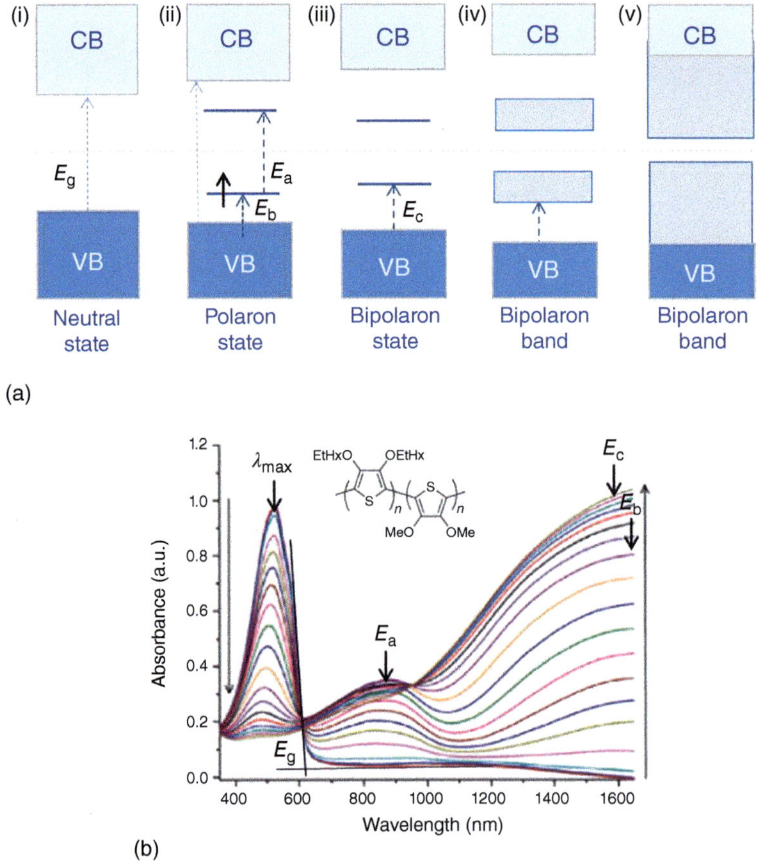

Figure 6.3 (a) Electronic band evolution for nondegenerate ground-state conjugated polymers based on the comprehension of references [5, 8, 10]. (b) Optical absorption spectra transformation of thiophene-based polymer during oxidation. Source: Reproduced with permission from [12], Copyright © 2010, American Chemical Society.

meaning of the four oxidation processes, ox 4 was considered as a capacitive current and/or irreversibility resulted from over-oxidation of the polymer. For other three processes, the number of charges per thiophene unit during each stage and the whole oxidation were found as a resort. The ox 1 showed 0.05 charges per unit, indicating a rare and spread distribution. It is known that the situation of one positive charge is localized over four to five thiophene units (0.2–0.25) and be denoted as a polaron, so ox 1 and 2 were related to the formation of polarons. Furthermore, ox 3 with a 40% doping level indicated the appearance of bipolarons with the density of two charges over five units.

6.1.2 Synthetic Methods

For thiophene derivates, even if electrochemical polymerization is convenient to carry out, chemical methods for polymerization are quite important for fabricating

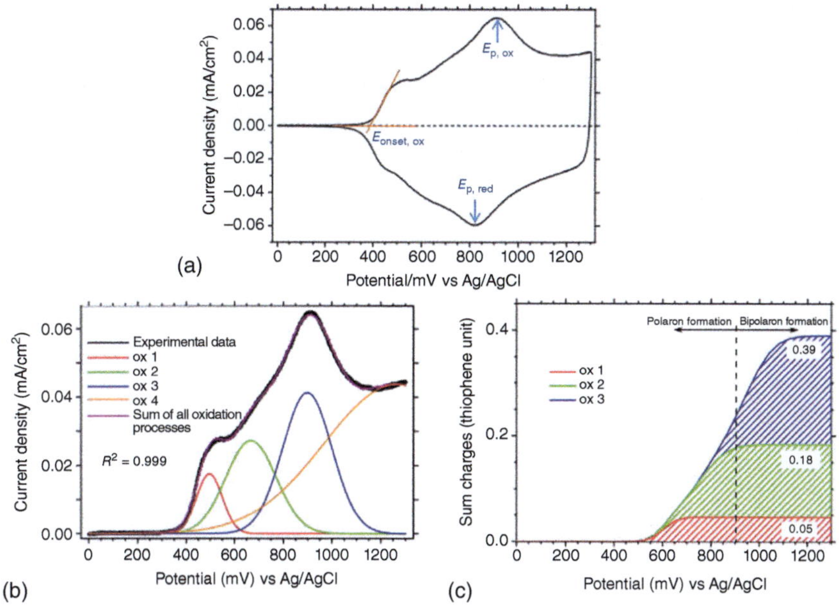

Figure 6.4 Investigations on the electrochemical process of P3HT. Source: Reproduced with permission of Ref. [5] from Copyright © 2016 Wiley-VCH Verlag GmbH & Co. KGaA, Weinheim.

complex copolymers and catering to the vastly required solution-processable polymers. Generally, there are mainly four synthetic tools for coupling monomers together and forming polymers with desirable molecular weight.

Firstly, $FeCl_3$ oxidative polymerization was frequently used to form homo- or irregular copolymers with an easy and convenient operation. By stirring solution of Iron(III) chloride and relative monomers at room temperature under ambient atmosphere for hours, doped polymer can be obtained after precipitation and washing in methanol. Then the doped polymer is reduced in stirring chloroform solution where hydrazine monohydrate is dropwise added in, and the final product is yielded after repeated precipitation and dissolution. This method is practical for ratio optimizing experiments, but the uncertain molecular structure is one of the drawbacks. Examples can be seen in Figure 6.5.

Suzuki and Stille coupling are efficient to synthesize copolymers based on different monomers. Corresponding instances are shown in Figure 6.6. One of the monomers should be brominated, while another one should be bis-substituted by boric acid/boric acid ester for Suzuki polycondensation or organic tin regents for Stille coupling. For the former, reactants, Pd catalyst, and phosphorus ligands in toluene along with base aqueous solution are mixed together under the help of phase-transfer catalyst in N_2 or Ar atmosphere. The coupling reaction in Stille style requires the palladic catalysis in refluxed toluene under protection atmosphere. However, the reaction for the toxic stannane-substituted precursor monomer brings troubles during the reaction and post-treatments. These two classes of coupling

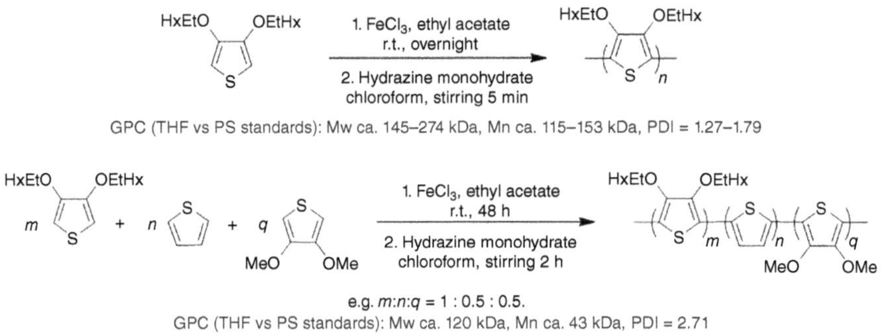

Figure 6.5 Synthesis of thiophene-based homopolymer and copolymer via oxidative polymerization as examples selected from references [11, 12].

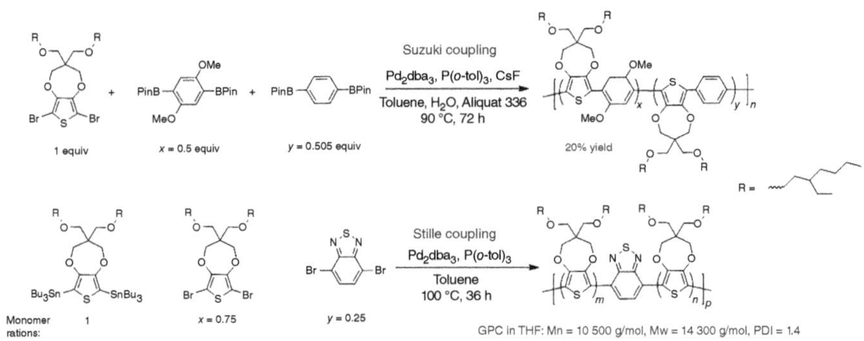

Figure 6.6 Examples for copolymerizing ECPs via Suzuki (upper) or Stille (down) coupling selected from references [13, 14].

method are efficient for obtaining relative certain repeating segment structure and satisfied molecular weight but might make the synthetic routes more complicated.

In recent years, another frequently admired approach has been drawing attentions of chemists in EC conjugated polymer areas. It is direct (hetero)arylation polymerization (abbreviated as DHAP or DArP), namely, C–H direct coupling. As an environmentally benign method, besides the tin-free merit, the absence of phosphine avoids its incorporation into the polymer backbone, and a low residual metal content gives the simplified purification process, which is favorable for reducing costs. Some reviews towards DHAP can be referred [15, 16]. Demands for qualified polymers for optoelectronic research can be satisfied well, and merits of DHAP can be revealed by comparison with conventional polymerization methods. For instance, the utilization of DHAP on copolymers of thiophene and arylenes were developed to obtain decent yield and molecular weight without introducing any phosphine ligands [17] (Figure 6.7a). Lower metallic element residual contents in polymers resulted via DHAP were proved [18] (Figure 6.7b), superior photovoltaic performances were achieved in DHAP copolymers as compared with traditional Stille coupling [19] (Figure 6.7c), and shorter synthetic route was planned [20] (Figure 6.7d). Not only for polymers, small molecules can also be formed from

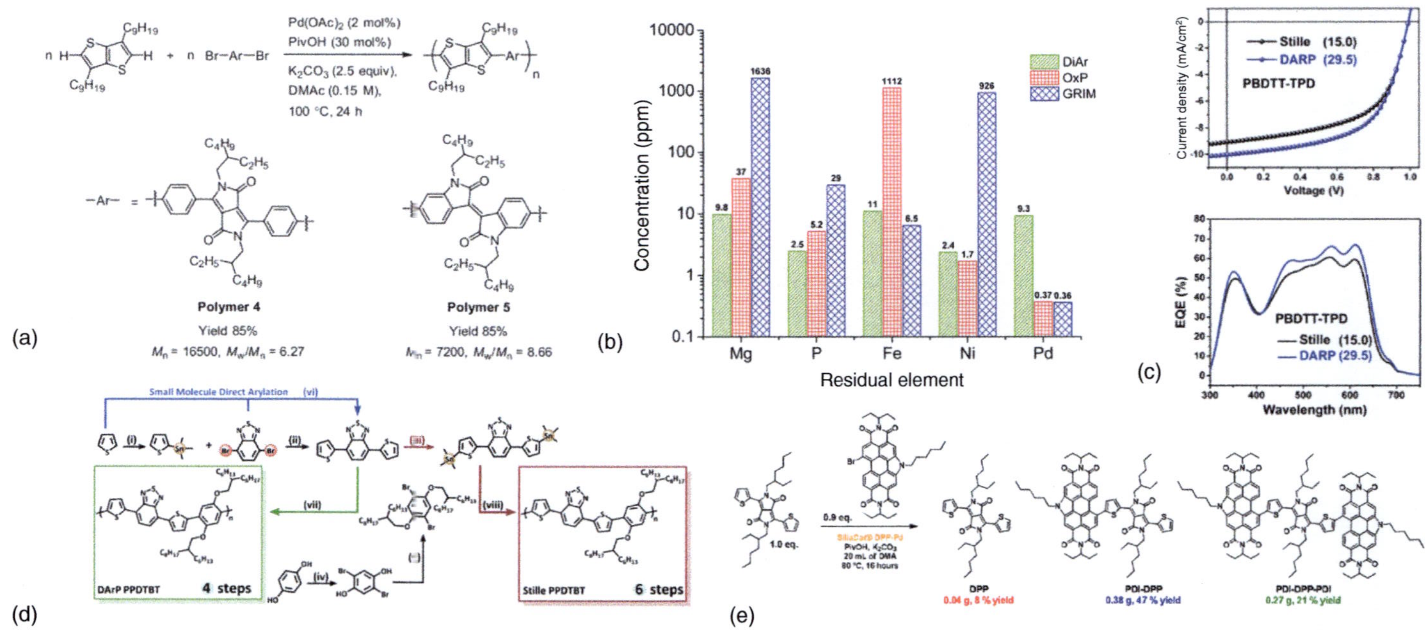

Figure 6.7 Diagrams for research works employing DHAP. Source: Adapted with permission from (a) Ref. [17], Copyright © The Chemical Society of Japan(CSJ); (b) Ref. [18], Copyright © 2013, American Chemical Society; (c) Ref. [19], Copyright © 2016, American Chemical Society; (d) Ref. [20], from Copyright © 2016 Wiley Periodicals, Inc.; (e) Ref. [21], Copyright © the Centre National de la Recherche Scientifique (CNRS) and The Royal Society of Chemistry.

mono-brominized precursor via DHAP methods [21] (Figure 6.7e). Regarding to synthetic methodologies, Barry C. Thompson and collaborators have attributed a lot, including the flow synthesis setup as commercially available reactor [22], the influence of solvents for reaction [23], and the copper catalyzed synthesis method [24], etc. Interestingly, a kind of robust DHAP reaction in toluene/water biphasic conditions was developed for user-friendly utilizing [25]. To date, DHAP has been broadly used in synthesis of ECPs, especially when dealing with donor–acceptor (D–A) copolymers [26–28].

6.2 Thiophene-Based Conjugated Electrochromic Polymers

6.2.1 Introduction

Among several species of conjugated polymers, thiophene-based polymers have been widely investigated from early researches about its basic properties and doping mechanisms to latest developed derivatives towards practical applications. Following the heated studied polymers based on thiophene unit and its 3- or 4-substituted derivatives [3, 6, 29–34], booming polymers containing alkylenedioxy-substituted analogs has brought notable results and took a great part of current EC polymer domain.

The most familiar building block is 3,4-(ethylenedioxy)thiophene (EDOT), the homopolymer of which can transform from deep blue to transmissive light blue upon oxidation and successfully commercialized owing to its moderate bandgap, low oxidation potential, and high transparency in visible region. Derivates from EDOT, 3,4-(propylenedioxy)thiophene (ProDOT) have become a new "star" brick making corresponding polymers achieve better processability, more clear transmissive oxidation state, higher optical contrast, and faster response [35, 36]. Also, in this family, ordinary dialkoxy-thiophene (AlkDOT) derivates also play an important role in color-control projects [12]. Moreover, with sophisticated design, 3,4-phenylenedioxythiophene (PheDOT) with the replacement of the ethylenedioxy bridge by a phenylenedioxy one developed by Jean Roncali, Igor F. Perepichka, and coauthors showed special electronic and optical properties, which is interesting for design of organic semiconductors (Figure 6.8) [37–42].

Besides, fused thiophenes were also introduced as building blocks of ECPs, thanks to their excellent performance as active materials in devices, such as OTFT, OPV, and thermoelectronics. The rigid electron-rich structure endows these building blocks low oxidation potentials and excellent charge mobility, which can benefit the performance promotion of conducting polymers. The optical, electronic, and electrochemical properties of thienothiophene (TT)-based polymers were exploited in 1980s [43–46], and subsequent studies on substituted TT revealed more relationships between molecular structure, conformation, and property. It is worth noting that the homo-polymerized 3,6-dimethoxy-thieno[3,2-*b*]thiophene (DMTT) was claimed to be a potential alternative to PEDOT, which attributed

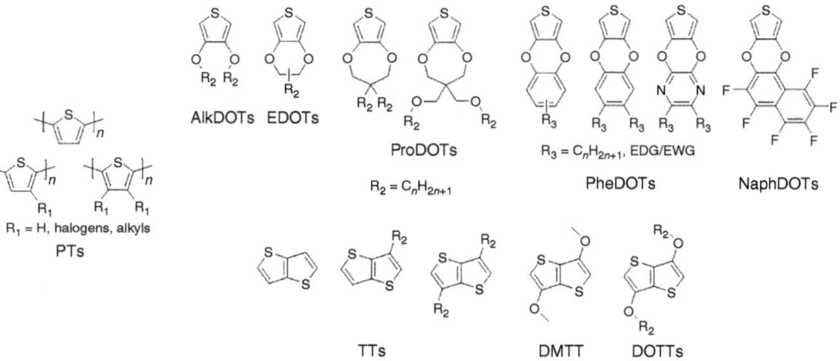

Figure 6.8 Commonly utilized building blocks for thiophene-based electrochromic conjugated polymers.

to the self-rigidification of the conjugated structure resulted from the confirmed non-covalent S–O intramolecular interactions [47]. The great potential of DMTT as a building block for improving the performance of electrochromisms have been revealed by Meng's group [48–51]. After modification of long side chains for better solubility, dialkyoxy-thieno[3,2-*b*]thiophenes (DOTTs) were well introduced to EC polymers and showed advantages like enhanced solution processability, low onset oxidation potential, and high coloration efficiency [28, 52–54].

During the process of exploring thiophene-based polymers, a vast variety of colored thin films were yielded under the direction of optical-control strategies so that the color palette has been gradually fulfilled. Especially for primary and secondary colors, efforts towards purer tint, higher contrast, and better performance have been made well.

Besides dealing with coloration, recent progresses of thiophene-based ECPs towards functionalizing via introducing side chains with functional groups will also be included. For example, polymer films were developed, which can be processed from environmentally benign solvents, and redoxed in aqueous-based devices, as well as cross-linkable for multistep solution–process device fabrications.

In this section, the aforementioned will be selectively introduced to give a picture of the development and corresponding strategies of thiophene-based ECPs.

6.2.2 Color-Tuning Strategies for Thiophene-Based Polymers

Electrochromic conjugated polymers constituted mainly by thiophene derivatives normally display a cathodic coloration (i.e. colored under neutral state and bleached under oxidation state) or multicolor transformations. Aiming at utilizing ECPs in practical applications not only for smart windows and digital tags but also hopefully for reflective-type full-color display devices, fulfilling the color palette with solution-processable ECPs has been pursued in this decade. Based on color mixture principles, polymers with additive and subtractive primary colors (RGB and CMYK) were preferentially considered and synthesized, towards which a set of structural design strategies have been concluded in some reviews [55–57]. This section will briefly introduce the thought about structure–color relationships and the mainstream optical-control methodologies, which have produced materials with absorption spectrum throughout the visible light region. Besides, specifically color-tuning examples towards representative colors will be showed in Section 6.2.3.

To understand the correlations between molecular structure and observed color, we can consider following this: the molecular structure of polymer will influence the highest occupied molecular orbital (HOMO)/LUMO energy and the energy gap between them (bandgap), which change the energy of absorption light and shift the corresponding spectrum, finally resulting in the color variation. The relationship between spectrum and coloration of polymers can be easily understand following the complementary color theories. However, the relevance between structure and spectrum is rather complicated and important for chemists to understand. Therefore, efforts including a variety of experiments and computational explanations have been made to figure out the effect of steric repulsion control, building block cooperation, substituent group, etc. on electronic band structures and optical properties. Behind these, backbone and side-chain engineering are utilized as main approaches to subtle property control.

6.2.2.1 Steric Effects

Tuning the steric hindrances of polymer skeleton has become the most frequently used strategy for optical control, especially after the establishment of a series of common thiophene-based building blocks like XDOTs. Substituent onto backbone rings, like methyl or halogen group, bulky groups and straight or branched side chains or a backbone brick itself with bulky structure, can cause the inter-ring steric interaction and twist the polymer chain, which will shorten the effective conjugation length and thus enhance the energy bandgap, inducing the shift of absorption band towards a higher energy region and vice versa. Here we raise some representative examples.

Differing levels of inter-ring strain can be induced in polymers based on dioctylthiophene in alternation with dioxythiophenes (DOTs) with various side chains. The cooperation with EDOT shows the slightest torsion and lowest oxidation potential

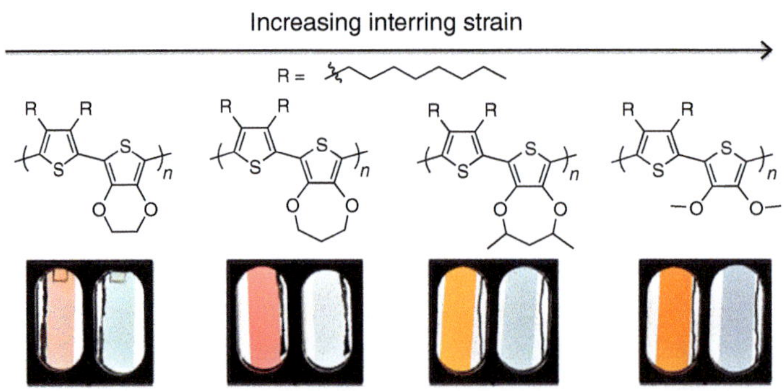

Figure 6.9 Effect of interring strain. Source: Adapted with permission from [27]. Copyright © 2018, American Chemical Society.

while the introduction of dimethoxythiophene (DMOT) raises the inter-ring strain and induces a high oxidation potential as well as large bandgap. These slightly structural changes allowed for varying inter-ring steric strain between these polymers and led to distinct changes in color (Figure 6.9) [27].

Another exploration on controlling steric interactions was constructing alternating polymers with different arrangements utilizing XDOTs bricks. The torsion degree between neighboring XDOT units as depicted in Figure 6.10 indicates that connecting two 3,4-di(2-ethylhexyloxy)thiophene (AcDOT) together will induce relatively strong conformational twisting, whereas the strain will be relaxed through the incorporation with ProDOT and EDOT, which can lead to a better planarity and extended conjugation. Based on this, regular copolymers were prepared in different manners of repeat unit combination. The yield color of resulted polymer films showed a vast span from a high energy band as orange to a long wavelength region absorption as purple and blue [58]. It is worth noting that normally an orange film would contain benzene analogs for raising energy gap and a blue film requiring a low bandgap would be fabricated by an electron D–A structure. However, the aforementioned steric strain-relaxation strategy realized such effect with excellent adjustability only by all-donor thiophene-based structure, which is beneficial for maintaining a low oxidation potential and eliminate the residual visible absorption when fully oxidized (a common negative effect of D–A polymers).

Long and bulky side chains have been hung onto backbone of ECPs to increase the solution processability. Meanwhile, strong steric repulsion can also be introduced between adjacent bulky groups especially for branched chains. This was demonstrated in the case of polymers consisting of AcDOT and DMOT as shown in Figure 6.11. For homopolymerized AcDOT, the branched alkyl groups can induce steric distortion of the neutral polymer backbone, shortening the conjugation length and subsequently shifting the absorbance to a high energy band (480 nm). When randomly incorporated with short side-chain modified DMOT, the steric repulsion was relaxed, so the bandgap was slightly narrowed (2.00 eV) and the maximum absorption was significantly shifted to 525 nm [12, 56].

6.2.2.2 Substituent and Electronic Effects

Beside the long chain strategy for solubility, introducing substituent with different electronic effects is also a main topic of side-chain engineering. Once directly incorporated onto the aromatic unit in the main chain, the electron-donating group (EDG) or electron-withdrawing group (EWG) will perturb the molecular orbitals through either inductive or mesomeric effects. Generally, EDG raises the HOMO energy level and decreases the oxidation potential, whereas the EWG could reduce the LUMO energy so that a decreased bandgap might be resulted [59]. For example alkoxyl group is a class of EDG, which is very frequently employed in thiophene-based polymers. The corporation of oxy-substituent can not only slightly narrow the bandgap and lower the oxidation potential but also make it more possible to reach the stabilized high p-doping levels with the closed-shell bipolaron structure, thanks to their electron-donating properties, which play a critical role in realization of the transparent oxidation states [60]. On the other hand, typical

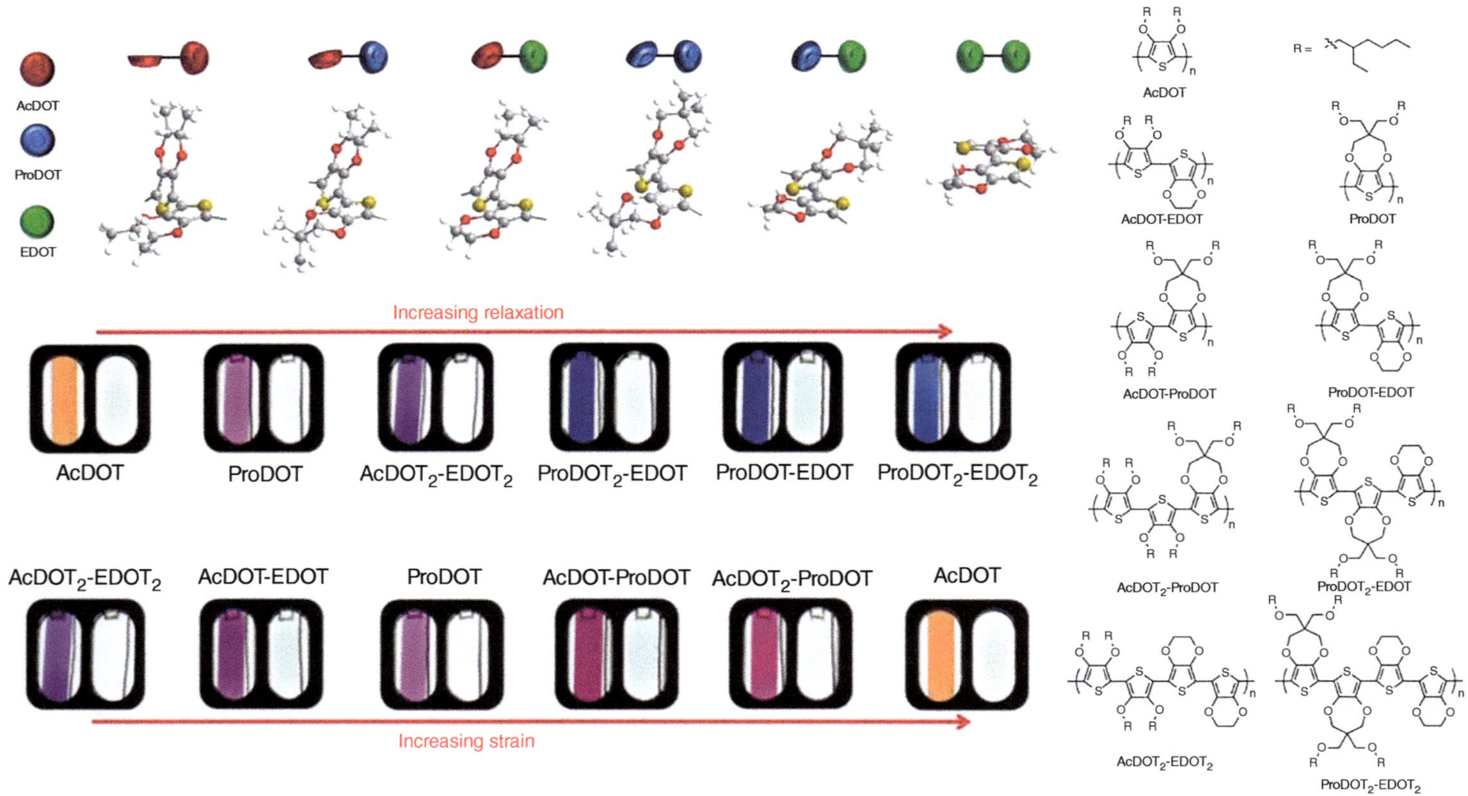

Figure 6.10 Steric strain-increasing or relaxing strategy. Source: Reproduced from Ref. [58] with permission from Copyright © The Royal Society of Chemistry.

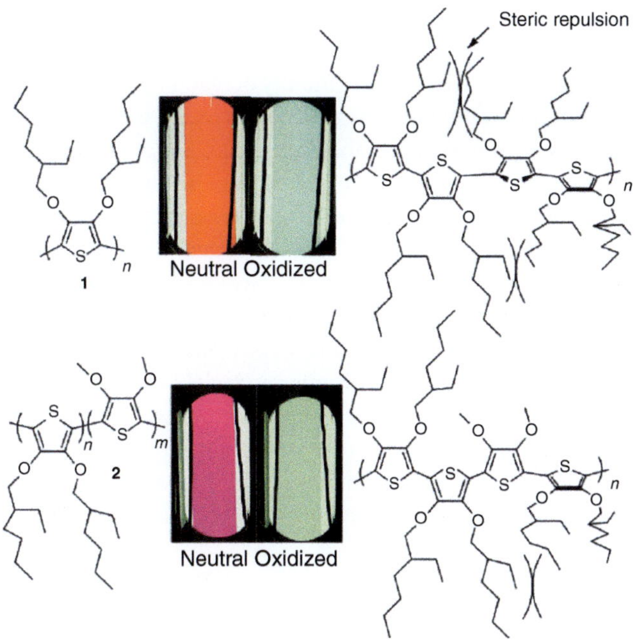

Figure 6.11 Effect of bulky side chains. Source: Adapted with permission from [56]. Copyright © 2011, American Chemical Society.

EWGs like cyano, nitro, halogens, trifluoromethyl, and pyridyl can also make subtle control on energy level when added onto the backbone [37, 38, 51].

6.2.3 Typical Colored Polymers

In this section we will present some examples of polymers containing thiophene derivates with either colored-to-transmissive transformations or multicolored stepwise change. As thiophene-based polymers with all-donor structure are cathodically coloring materials, attention has been paid to the color of their neutral films, normally about the completing of color palette with primary colors of both RGB and cyan, magenta, yellow (CMY) system and color purity or saturation. The strategies of color control for certain color series are herein briefly concluded. On the other hand, practical applications require a highly transparent and colorless film in the bleached state. Besides, it is surely important to behave good performances such as low drive voltage, rapid switching, high coloration efficiency, and long-term working stability in the air ambient. Towards this, approachable data of the listed examples will be collected for a referential comparison as well as an insight of the current progress extent.

Moreover, some electrochromophores displaying multiple color states during redox are included. Even considered as having lower potential for applications than colored-to-transmissive counterparts, multicolored ECPs still arouse curiosity for research, and some representative results like three-in-one RGB transformation and green-to-soil switching are worthy of consideration.

6.2.3.1 Yellow and Orange

It is challenging to realize the desirable cathodically colored-to-transmissive polymer with yellow or orange neutral tint owing to the difficulty of reaching high energy absorption and then mostly shift it into the near-infrared (NIR) region upon doping. For most of yellow or orange objects, a maximum absorbance at wavelength smaller than 500 nm is required. Aiming at this target, incorporation of benzene derivates has become a main method to widen the energy gap and blue shift the maximum absorption. In recent years, progresses in this filed have emerged, of which some works are shown in Figure 6.12.

Starting with the homopolymerized AcDOT named as **ECP-Orange**, the strong steric repulsion between the branched side chains twists the polymer backbone and just shifts the absorption into the high-energy region [12]. The first yellow-to-transmissive polymer **(ECP-Yellow)** was yielded according to the design conception of the corporation of *p*-phenylene moieties, which has intrinsically high aromaticity and subtle steric repulsion from ortho-C–H units to allow a strong blue-shifted absorption while remaining capable for strong bipolaronic absorption extending beyond 1600 nm [61].

Following this, an extended exploration was carried out on ProDOT-based copolymers in corporation with electron-rich arylenes like fluorine, carbazole, and pyrene for further optimization and deeper understanding of structure–property relationships for yellow ECPs. It was found that increasing the electron richness of the arylene led to a decrease of the oxidation potential while maintaining a high bandgap due to steric interactions. Among the resulted products, three are selected and presented here owing to their vibrant yellow or orange neutral tint and highly transmissive even colorless oxidized film. **ProDOT-Cbz** shows a relative ideal situation with an oxidation potential lower than **ProDOT-Ph** and nearly colorless oxidized film. Besides, addition of electron-donating oxygen atoms on phenylene rings or the introduction of additional ProDOT units can decrease oxidation potentials to a level lower than 300 mV (vs Ag/Ag^+). However, the decrease in oxidation potential comes at the cost of red-shifting or broadening of the overall spectrum and yielding orange hues [13]. However, there still remained challenges of the balancing between achieving large changes in the absorption profile while maintaining sufficiently low oxidation potentials. To address this problem, in 2016, electron-rich methoxy units were introduced onto the phenylene to raise the HOMO level and decrease the oxidation potential, and meanwhile, AcDOT or ProDOT systems were kept for either a greater torsional strain and increasing bandgap or a raised HOMO that lowers the bandgap. Moreover, the pendant methoxy group replacing the hydro atoms on phenylene can also play a role of preventing unwanted chemical reactivity during oxidation like cross-linking and preserve better redox stability [62].

Coming there, the strategies for governing the bandgap and redox properties in the set of DOT copolymers with high energy absorption have been generally elucidated, and satisfied products presenting vivid yellow to highly transmissive have been attained. But even though, regarding to real-world applications, there still was a demand for greater long-term stability of this series. Accordingly, polymers based on repeated units of dialkylthiophene (DAT) in alternation with DOTs with various

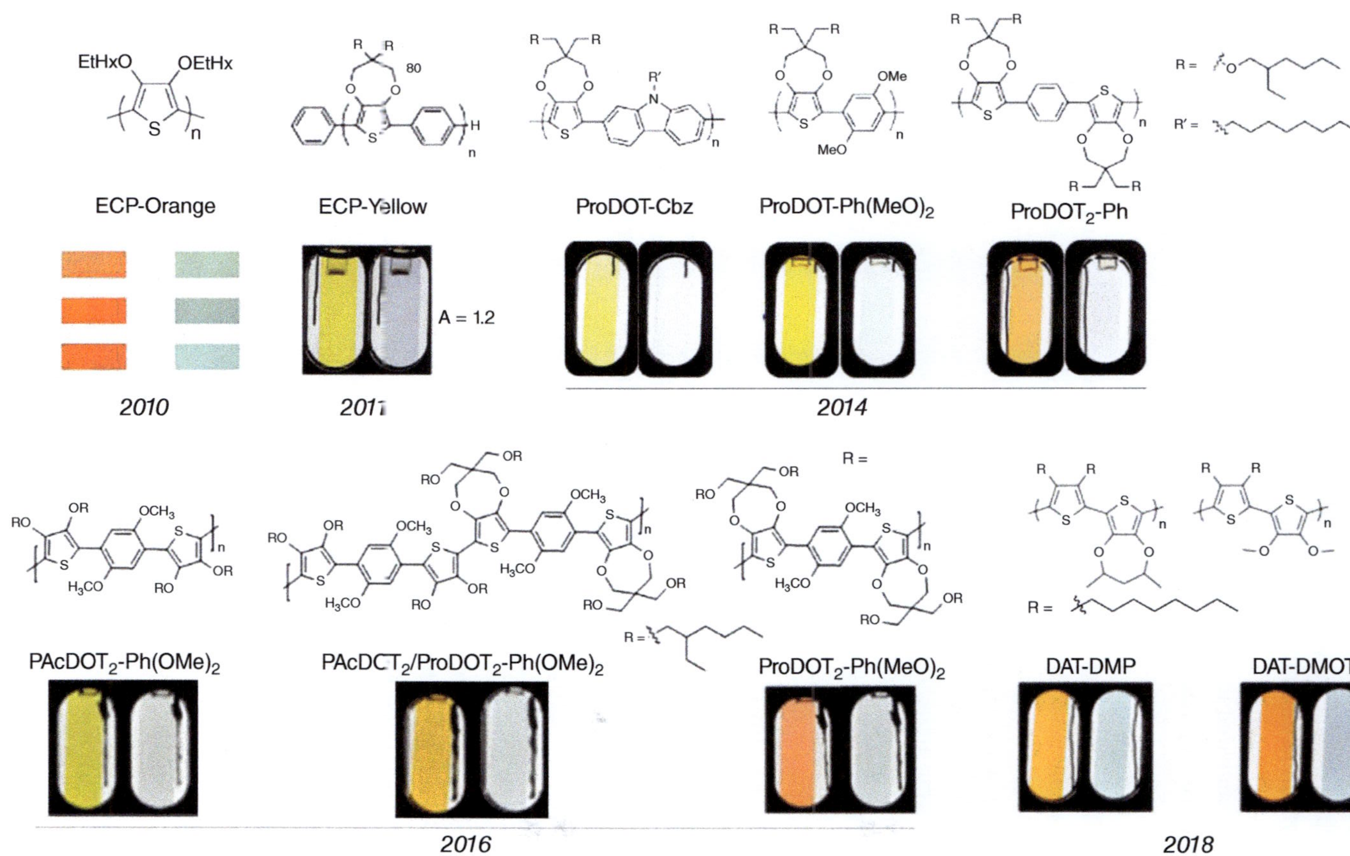

Figure 6.12 Progress of yellow and orange ECPs. Source: Adapted with permission from [12]. Copyright © 2010, American Chemical Society, [61]. Copyright © The Royal Society of Chemistry, [13]. Copyright © 2014, American Chemical Society, [62]. Copyright © 2016, American Chemical Society, [27]. Copyright © 2018, American Chemical Society.

side chains were synthesized to overcome the issues with the switching stability in wide-gap phenylene-containing ECPs [27]. DAT monomer 3,4-dioctylthiophene provides interring strain with alkyl chains directly appended to the backbone to reach wide optical gaps while blocking possible reactive sites for the radical cations. The octyl groups allow for the use of DOT monomers without side chains and ensure the output of soluble processable polymers. Increased redox switching stability was revealed compared to previous wide-gap ECPs with minimal drops in optical contrast over 1000 cycles.

6.2.3.2 Red

During the long-term developing of thiophene-based EC polymers, it is not rare to find a material neutrally colored in red or similar tints. For instance homopolymers of thiophene derivates show deep red in neutral film but transforms to blue with a relative low transmittance to be a multicolored material [3, 6, 29]. This kind of residue blue tint is also a common problem for this series of ECPs, which requires a large bandgap approximately higher than 2.0 eV. Besides, a true-red color is actually not easy to achieve since the maximum absorption should be settled in the cyan region, which covers a narrow wavelength range according to the complementary color theory. This might be the reason that the resulted film often contains some impure tint inclining into either orange or magenta hues.

Regarding a pure red tint, fused ring blocks were incorporated in alternation with EDOTs, for example, ladder-like brick IDT was employed and deep red neutral film was obtained, but the color changed to black when oxidized **(PT-Red-2)** [63]. Fortunately, in another case, the introduction of phenanthrocarbazole induced a red-to-transmissive polymer **(PT-Red-3)** successfully with desirable performances, but the film was formed by electrochemical polymerization, which is not compatible with solution processing [64]. Aiming at soluble true-red ECPs with ideal transparent state, one approach is turning to the combination of DOT analogs again and making use of the steric strain or relaxation strategy. The **ECP-Red** developed by Reynolds' group in 2010 is an example using dimethoxythiophene monomers as spacer to release the repulsion between branched side chains of diethylhexyloxythiophene monomers **(PT-Red-4)** [12], which was further improved by Chunye Xu and coworkers **(PT-Red-5)** [11]. Slightly different, another polymer consisting DAT and unsubstituted ProDOT was obtained during the pursuing for yellow materials as mentioned in Section 6.2.3.1, where the DAT plays the role of enhancing bandgap and ProDOT responses for low oxidation potential and low energy absorption with its good planarity so that a balance was built to produce a red-to-transmissive film **(PT-Red-6)** [27]. Moreover, through the cooperation of a rigid thieno[3,2-*b*]thiophene unit that provides desirable carrier conductivity as well as low energy absorption and a strongly aromatic dimethoxybenzene monomer for the increase of bandgap, neutral red polymer was also acquired, which exhibited rapid switching to the pale-blue oxidation state **(PT-Red-7)** [53]. The structure of aforementioned polymers together with photos for their color transformations are put in Figure 6.13.

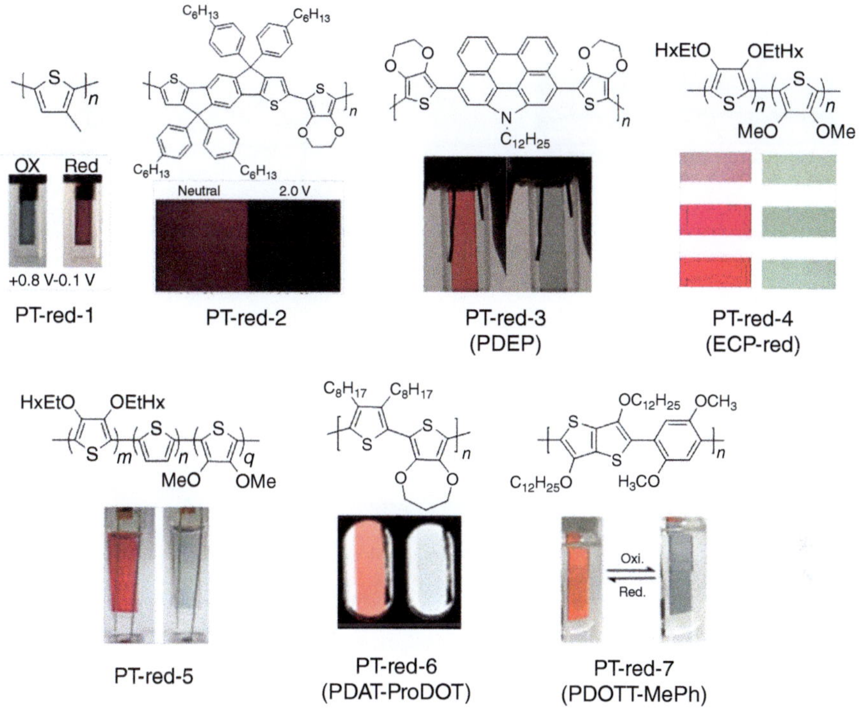

Figure 6.13 Typical example of red ECPs. Source: Adapted with permission from Ref. [3], Copyright © 2003 Wiley-VCH Verlag GmbH & Co. KGaA, Weinheim; Ref. [63], Copyright © 2015, Springer Nature; Ref. [64], Copyright © The Royal Society of Chemistry 2016; Ref. [12], Copyright © 2010, American Chemical Society; Ref. [11], the Centre National de la Recherche Scientifique (CNRS) and The Royal Society of Chemistry; Ref. [27], Copyright © 2018, American Chemical Society; Ref. [53], Copyright © 2018, American Chemical Society.

6.2.3.3 Magenta and Purple

As an important element of subtractive primaries (cyan, magenta, yellow, and black), magenta ECPs have been early placed on the agenda for realizing full-color printing and displays. It seems not difficult to discover magenta products in the family of thiophene-based polymer, since similar hues can be achieved by homo-polymerized thiophene derivates with simple synthetic design.

One representative is **PProDOTs**, towards which four polymers with different side chains were developed in 2004, exhibiting optical bandgap vary from 1.80 to 1.96 eV and absorption band around 600 nm to give magenta or purple tints [36]. Among them **PProDOT(CH_2OEtHx)$_2$** with branched alkoxy-side group, which was referred as **ECP-Magenta**, was magenta in as-sprayed film, and the neutral color can shift into purple upon redox switching (Figure 6.14a), showing high contrast up to 80% and excellent coloration efficiency of 1235 cm^2/C. Using this material, large-scale roll coating process and self-powered OPV/electrochromic device (ECD) module [65] (Figure 6.15a) as well as integrated solar-powered EC window [66] were developed, dual-active subtractive color mixing ECD was

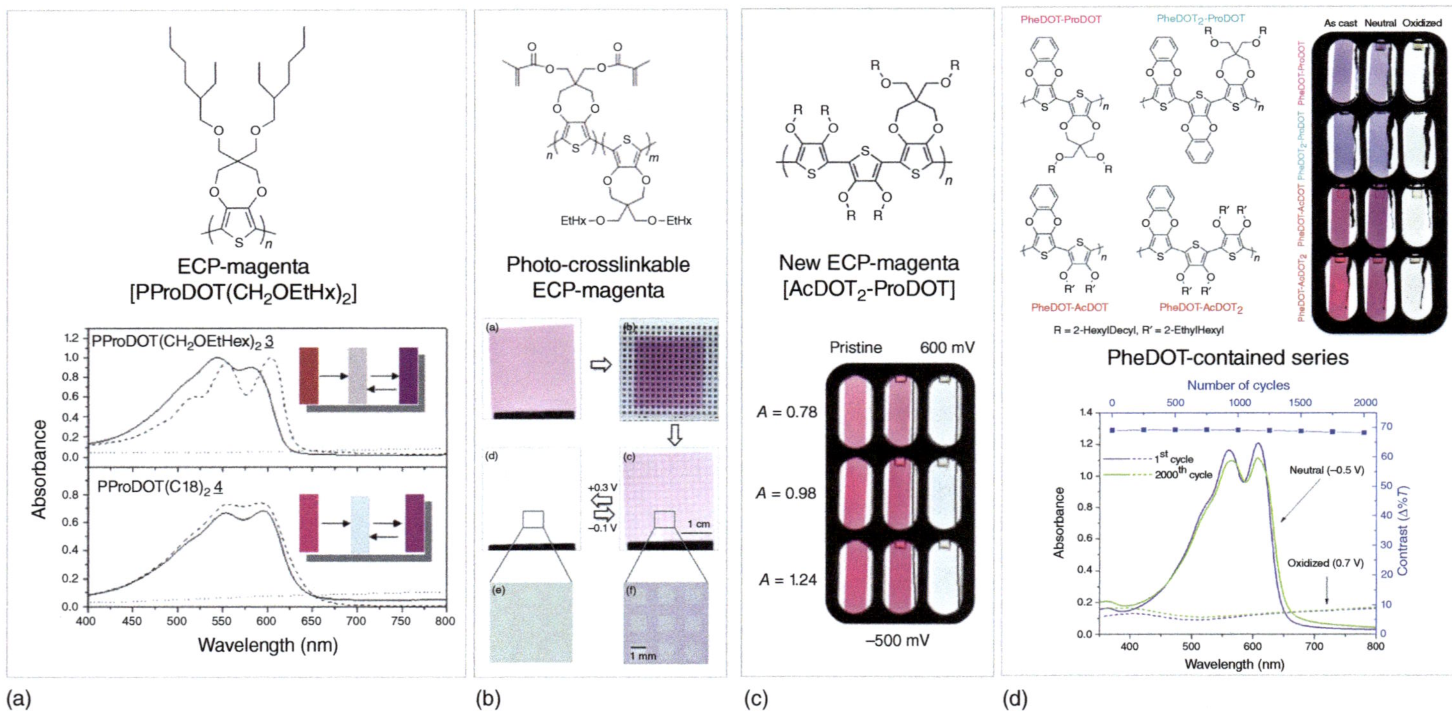

Figure 6.14 Molecular structure, diagram, and photography of electrochromic properties for representative magenta (main) and purple ECPs. Source: Adapted with permission from (a) Ref. [36], Copyright © 2004, American Chemical Society; (b) Ref. [69], Copyright © 2013 WILEY-VCH Verlag GmbH & Co. KGaA; (c) Ref. [58], Copyright © The Royal Society of Chemistry 2015; and (d) Ref. [71], Copyright © The Royal Society of Chemistry 2018.

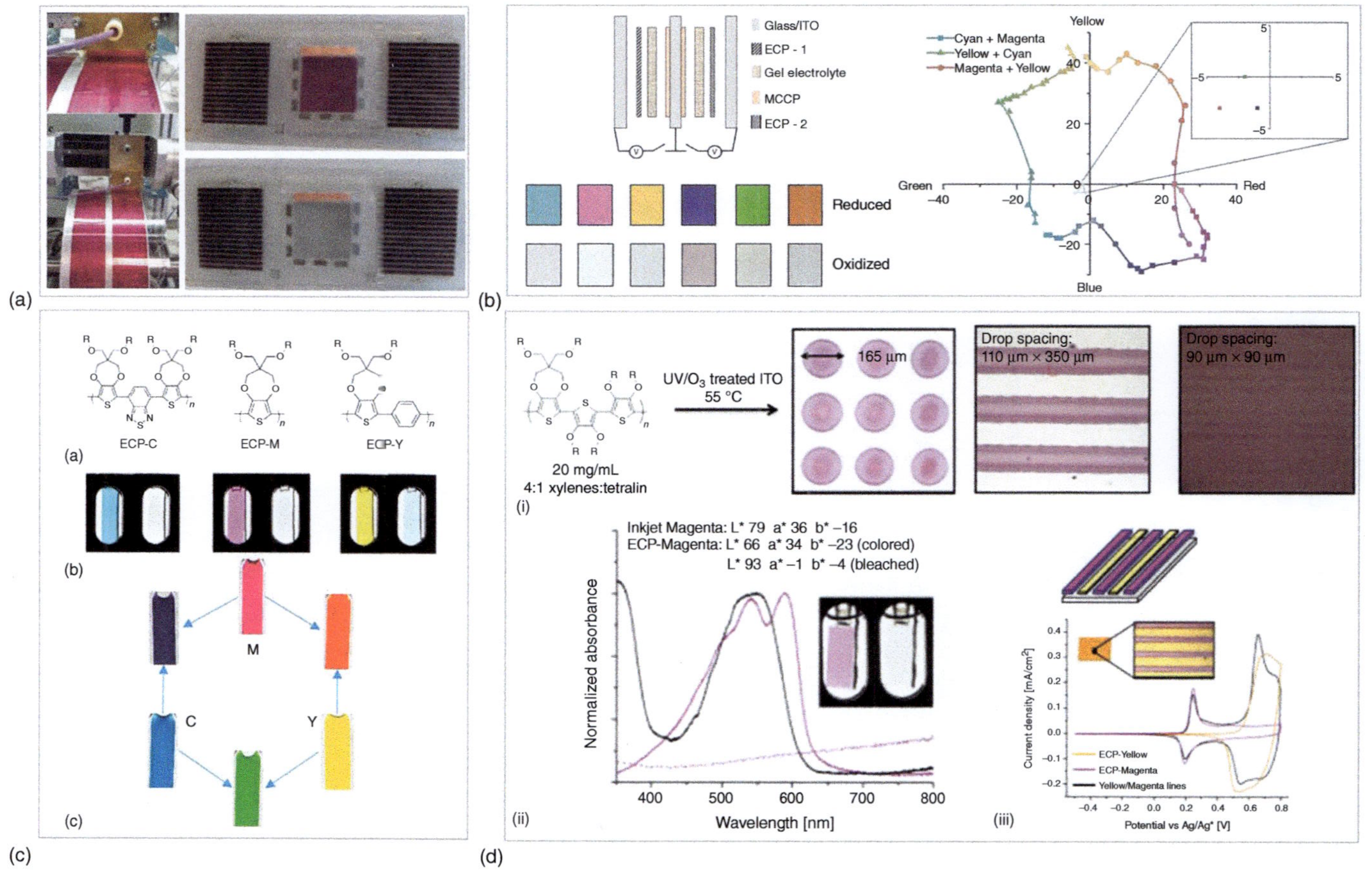

Figure 6.15 Device or processing developed based on magenta ECPs. Source: Adapted with permission from (a) Ref. [65], Copyright © 2012 Wiley Periodicals, Inc.; (b) Ref. [67], Copyright © 2014, American Chemical Society; (c) Ref. [68], Copyright © 2015, American Chemical Society; and (d) Ref. [70], Copyright © 2016 WILEY-VCH Verlag GmbH & Co. KGaA, Weinheim.

fabricated to generate a wide color gamut [67] (Figure 6.15b), and solution co-processed ECP mixtures were also demonstrated [68] (Figure 6.15c). Based on **ECP-Magenta**, a polymer modified with photo-cross-linkable methacrylate side groups for direct photopatterning process was further developed [69] (Figure 6.14b).

Ten years later, a **new ECP-Magenta** was born in 2015 with the structure containing AcDOT and ProDOT constituents **($AcDOT_2$–ProDOT)** [58] (Figure 6.14c). It was selected from the set of poly DOTs as a fruit of subtle steric tuning for color control. In addition to the desirable EC performances, more importantly, it achieved a magenta neutral state with color value of (L^*, a^*, b^* = 56, 59, −16) just close to the magenta value (52, 50, −15) defined by Munsell, and the absorption spectral can remain nearly unchanged after broken-in. This made great advantage, which allows more accurate subtractive CMY color mixing. Accordingly, an attempt towards practical inkjet printing was based on this material [70]. From the solution of **$AcDOT_2$–ProDOT**, uniform droplets, lines, and homogeneous film can be formed via inkjet printing process (Figure 6.15d, i), and the spectrum plot of printed film is similar to the commercial magenta ink (Figure 6.15d, ii). Device containing alternating lines of ECP-Magenta and ECP-Yellow can yield a composite color with an overlapped CV curve (Figure 6.15d, iii).

Furthermore, recently, an infrequently noticed monomer 3,4-phenylenedioxythiophenes **(PheDOTs)** was applied to soluble thiophene-based ECP family [71]. The planarity and steric interactions of **PheDOTs** is similar to **EDOTs**. However, it shows a significantly lower electron density on the thiophene ring owing to the π-electrons delocalizing into the phenylene ring, which results in increasing stabilities compared to conventional DOTs. Magenta and purple polymers were synthesized via DHAP, which can be switched into a highly transparent colorless oxidized state, and the corresponding cycling stability was proved excellent (Figure 6.14d).

6.2.3.4 Black

Though colorful electrochromisms have been gradually enriched, black EC materials are urgently in demand as a special class to impel practical applications such as displays, smart sunglasses, and energy-saving windows. These require the device reaching significant optical contrasts and high-degree achromatic black appearance, which normally needs a broaden and strong absorption in visible region that can transform into transparent when doped, posing challenges to researchers in this field. Although efforts have been made from physical standpoint, however, copolymers with neutral “black” nature can present better performance once designed. For example, blending mixtures consisting of cyan and magenta polymers showed contrasts up to 50%, but corresponding copolymers composing D–A segments reached a higher contrast and more black neutral state as well as faster response [72]. The first black-to-transmissive copolymer **PT-Black1** was reported by Reynolds and coworkers in 2008 approached by D–A strategy, which developed copolymer backbone through “chemical mixing” of donor segments with optical absorption complementing the deficiency of the two-band absorption from the D–A segments [73]. Following, backbone engineering was utilized to fine-tune the spectra of XDOTs-based D–A polymers (**PT-Black2 and PT-Black3**), broadening the absorption spectra into more flat curves in visible band and resulting darker neutral film and better bleached state [14, 26]. Other than XDOTs series, thienothiophene-based polymers also showed potential for black electrochromisms using complementary strategy to offset the absorption valley and reach a satisfactory "black" neutral state (**PT-Black4)** [54]. Interestingly, established on the well-developed basement of color-to-transmissive polymers with CMY together with orange and green tints, polymer blends were developed again, and achromatic hues were successfully yielded (**PT-Black5**) [74]. Aforementioned progresses are displayed in Figure 6.16, revealing promising potential of thiophene-based black copolymers as crucial materials for future applications.

6.2.3.5 Multicolored

There are several manners to reach multicolored transformation. For polythiophene family, two-state coloration switching from yellow, red [6], or green [28] to a bluish state is common as long as the spectra are not significantly shifted to remain some absorption in visible region, such as poly(3-methylthiophene). And during the doping process, if the transition induced by special polarons or bipolarons can form a distinct profile, then some intermediate color states different from both neutral and fully doped state will appear, but this is difficult to predict and control precisely, towards which **ProDOT-Fl** could be an example [13]. Besides, fabricating D–A backbone is an efficient approach to yield a great diversity of multicolored electrochromisms. Levent Toppare's group attributed amount of attempts in this field, which can be referred in their relative reviews [75, 76]. Not only can multiple colored states occur in the D–A type polymer in the process of p-type doping but also bipolar property might happen, which means the polymer can also be switched to more color states via n-type doping, raising **PTBTTh** and **PTBTPh** for an example [77]. Apart from these, decorating electrochromes on the ECP backbones can also play a role in multiple coloring, for example, an azobenzene, a coumarin, or a fluorescein was attached to the terthienyl chain followed by electrochemical polymerization [78], showing interesting color changes including red, blue, green, brown, orange, and dark gray states. Materials mentioned in this paragraph are shown in Figure 6.17.

In terms of potential for practical applications, besides the novel three-in-one polymer displaying RGB color during redox [79], green-to-soil-colored ECPs also drew attentions for camouflage applications. With a D–A structure bearing quinoxaline as the acceptor, green-brown switchable EC materials were synthesized, along with which a fabric chameleonic ECD was constructed as a typo for real utilizing as adaptive camouflage in the conditions of forest and sands [80] (Figure 6.18a). Similarly, D–A polymers constituted by thiophene and thieno[3,4-*b*]pyrazine alternating with carbazole unit also showed green-to-brown transformations upon oxidation, and based on one of the polymers, multicolored device with the help of P3HT was fabricated to show the adaptive camouflage ability [81] (Figure 6.18b). Moreover, a dithienyl-pyrrole-based polymer with soil-like brown, vegetation-like green, and sea-like blue color changes exhibited the possibility for multi-scenario applications (Figure 6.18c) [82].

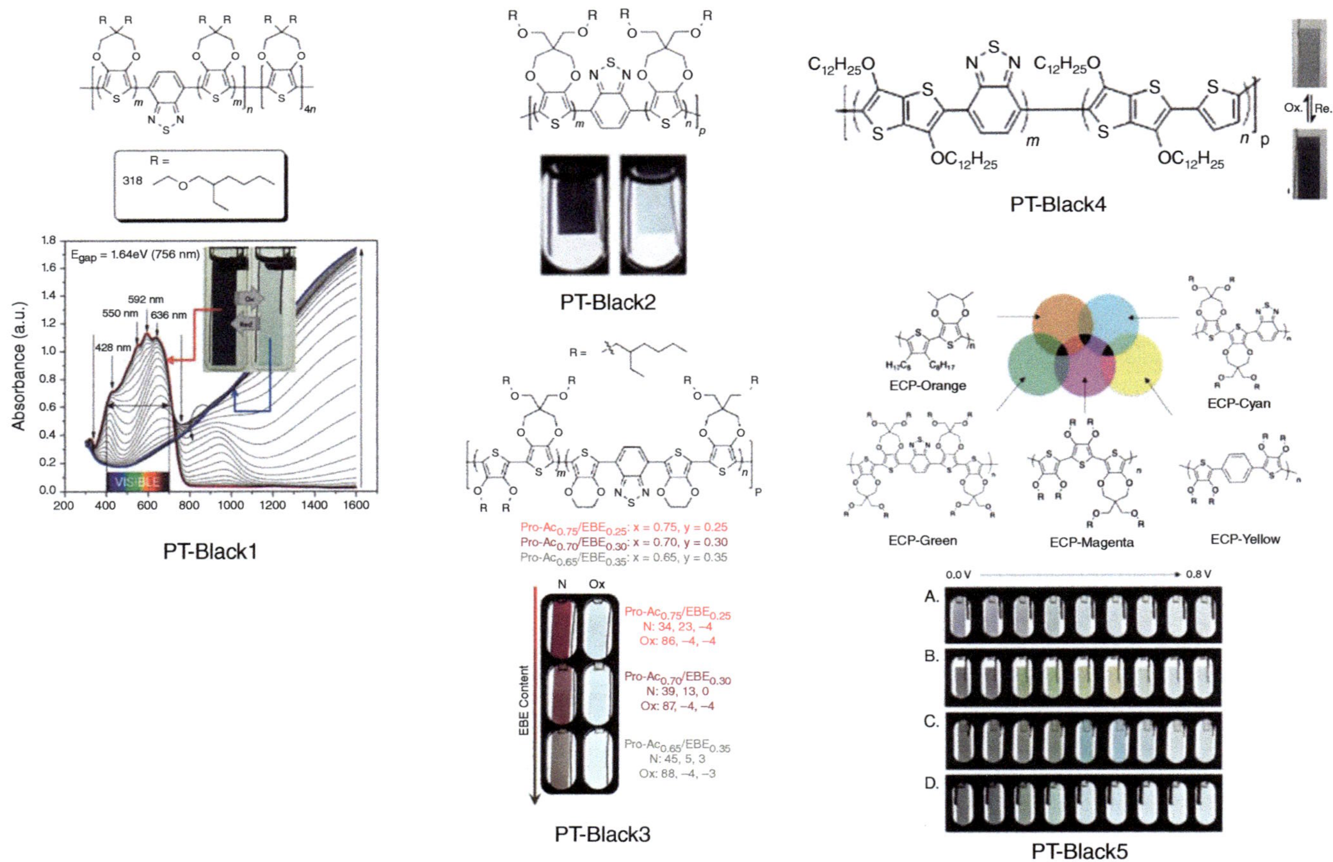

Figure 6.16 Molecular structure and displayed color changes of thiophene-based black electrochromisms. Source: Reproduced with permission from Ref. [73], Copyright © 2008, Springer Nature; Ref. [14], Copyright © 2010 WILEY-VCH Verlag GmbH & Co. KGaA, Weinheim; Ref. [26], Copyright © 2016, American Chemical Society; Ref. [54], Copyright © The Royal Society of Chemistry; Ref. [74], Copyright © 2018 WILEY-VCH Verlag GmbH & Co. KGaA, Weinheim.

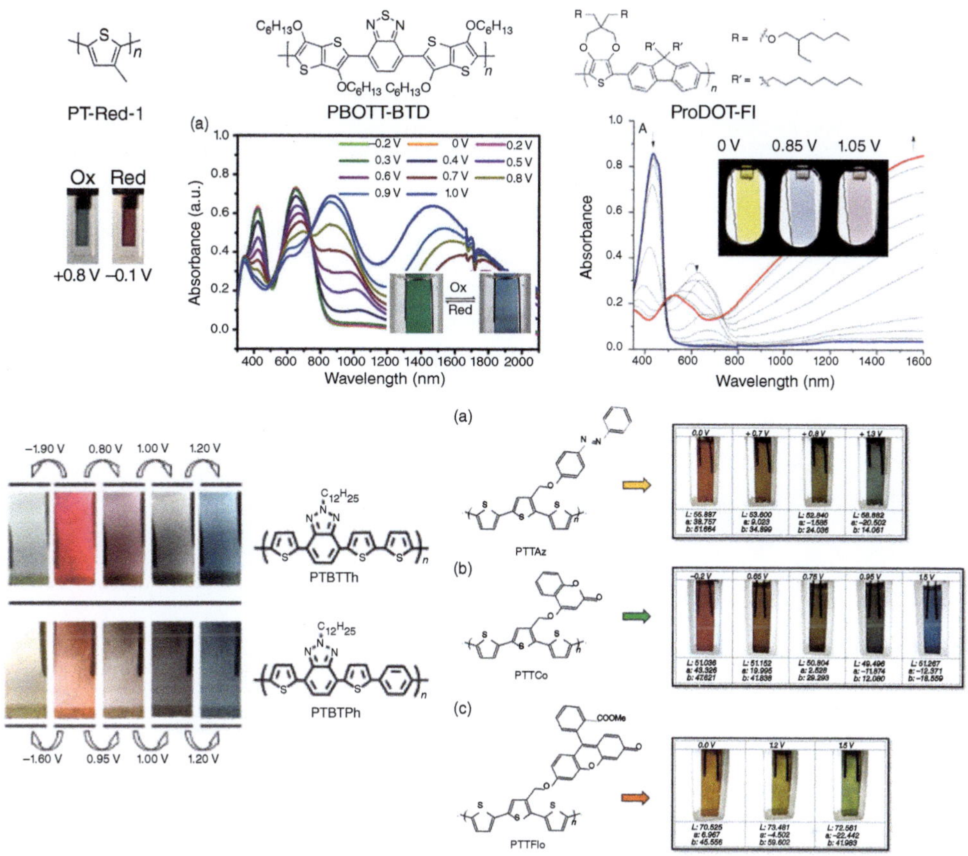

Figure 6.17 Multicolored electrochromisms. Source: Reproduced with permission from Ref. [6], Copyright © 2003 WILEY-VCH Verlag GmbH & Co. KGaA, Weinheim; Ref. [28], Copyright © 2016, American Chemical Society; [13], Copyright © 2014, American Chemical Society; Ref. [77], Copyright © The Royal Society of Chemistry 2011; Ref. [78], Copyright © 2013 Elsevier B.V.

6.2.3.6 Anodically Coloring Polymers

High-bandgap polymers normally show their neutral absorption in ultraviolet (UV) region while retaining little absorbance in visible region, which result in the colorless or highly transmissive appearance in neutral state and become colored when oxidized, that is, anodically coloring systems. Comparing to cathodically coloring species, anodical one can more easily reach beneficial colorless transparent state and thus higher optical contrast, which is difficult for cathodically polymers owing to the inevitable absorption residue in visible region when oxidized.

A major set of anodically coloring polymers is triarylamine-based polymers, which can remain high bandgap while being easily oxidized at the nitrogen center and form stable radical cation to show vibrant distinct coloring states. However, research on triarylamine-based conjugated-type polymers is limited, so examples like polyester or polyamine are raised here. One review in 2012 has involved a number of works

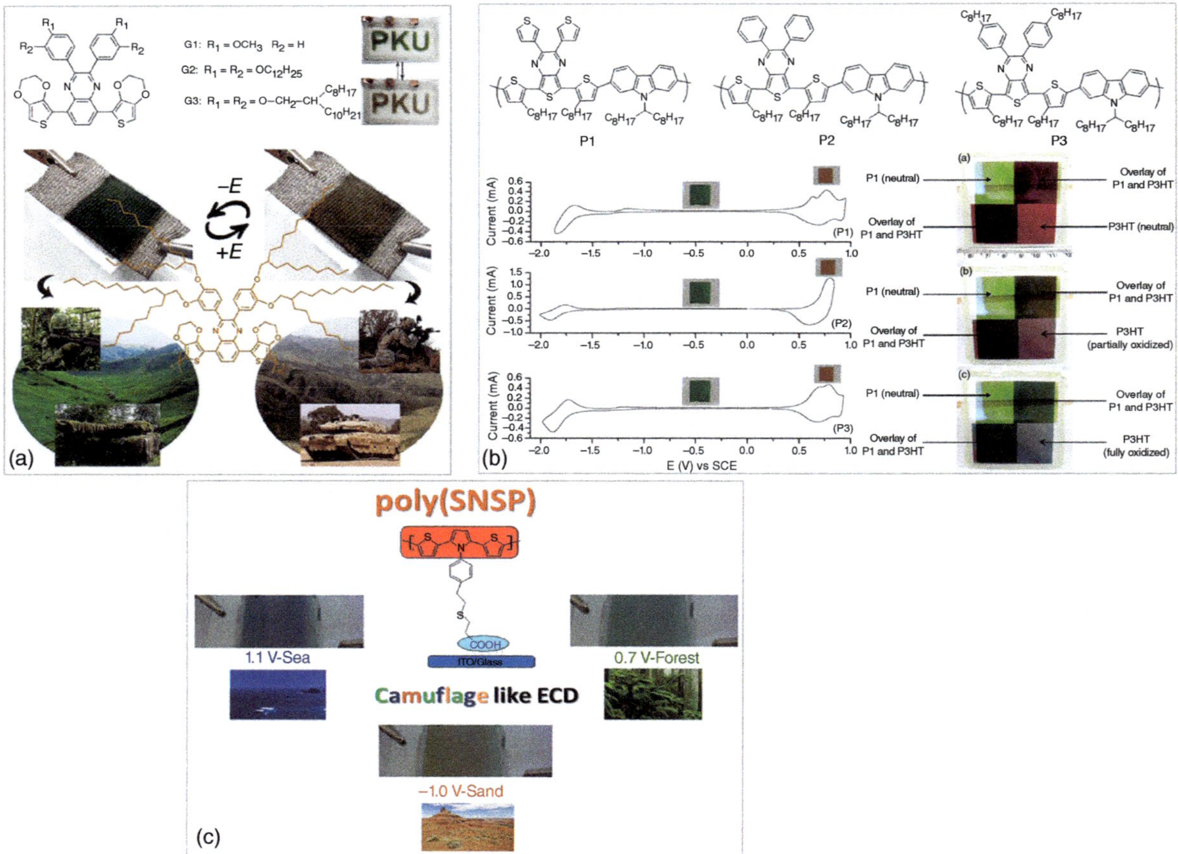

Figure 6.18 Green to soil-like colored materials and relative devices. Source: Reproduced with permission from (a) Ref. [80], Copyright © The Royal Society of Chemistry; (b) Ref. [81], Copyright © 2009, American Chemical Society; (c) Ref. [82], Copyright © 2018 Elsevier B.V.

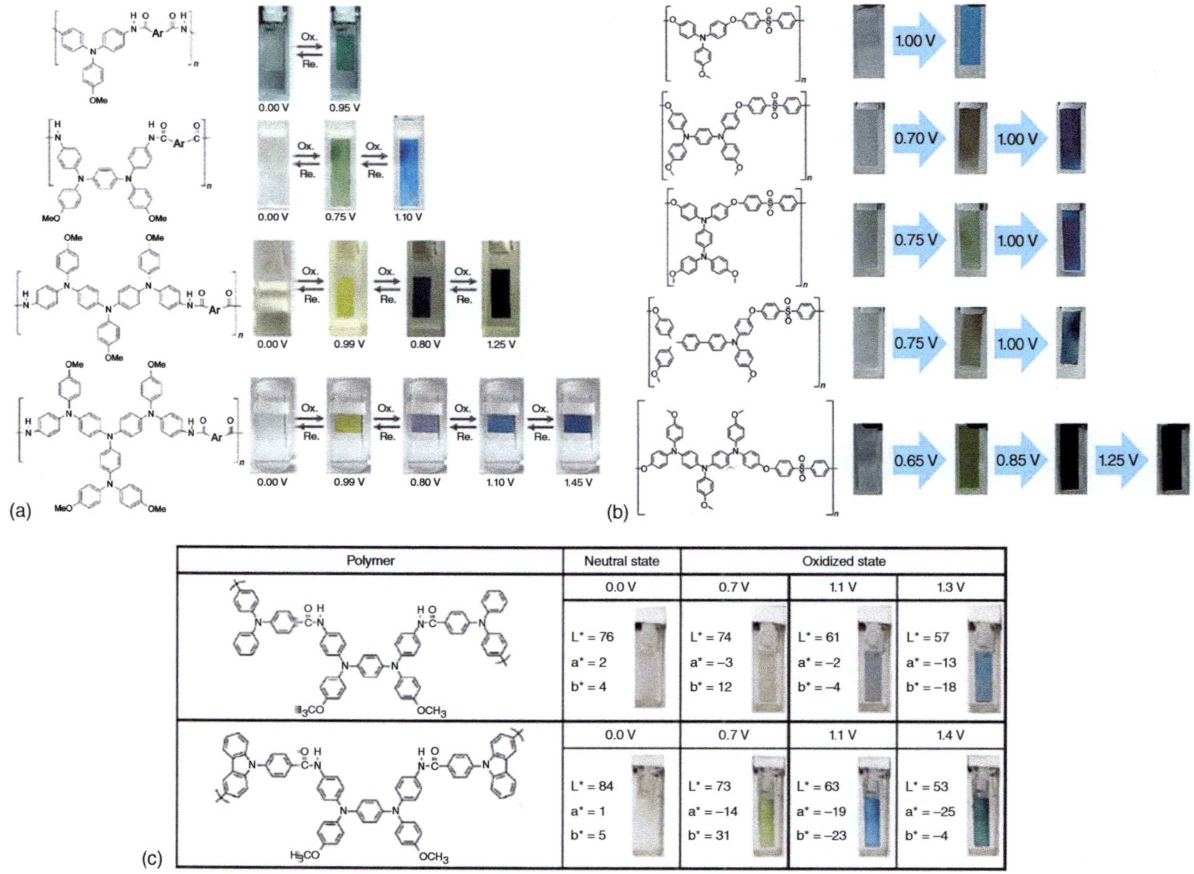

Figure 6.19 Triarylamine-based anodic electrochromic polymers. Source: (a) Adapted with permission from Ref. [83], Copyright © The Royal Society of Chemistry 2012; (b) Adapted with permission from Ref. [84], Copyright © The Royal Society of Chemistry 2019, and (c) Adapted with permission from Ref. [85], Copyright © 2018 The Electrochemical Society.

in this area like polyamines displaying multicolor changes shown in Figure 6.19a [83]. Similarly, polyethers containing triarylamine units were reported as shown in Figure 6.19b [84]. Also, polyamines based on bis-triarylamine blocks were constructed to show multiple color states (Figure 6.19c) [85].

Besides, polymers containing carbazoles and fluorenes can also be fabricated as anodic coloring electrochromisms, which can switch from colorless or transmissive yellow neutral state to blue, green, or darker tints. Even though, the number of stories about anodic conjugated EC polymers is still limited. For more examples for non-conjugated polymer please refer to Chapter 7.

6.2.4 Water- or "Green Solvents"-Soluble ECPs

During the past several decades, organic-solvent-soluble conducting polymers have been extensively developed. Previously, solubility in some common organic solvents was regarded as the singular path available to processability of conducting polymers. Since the first demonstration of two polymers, sodium poly(3-thiophene-β-ethansulfonate) (P3-ETSNa) and sodium poly(3-thiophene-δ-butanesulfonate) (P3-BTSNa) by F. Wudl and coworkers in 1987, which were soluble in water both in the doped and undoped states [86], many scientists paid a great deal of attention to develop water-soluble (WS) polymers. In general, these polymers are functionalized with ionic pendant groups such as sulfonate ($-SO_3^-$), carboxylate ($-CO_2^-$), phosphonate ($-PO_3^-$), and ammonium ($-NR_3^+$) to allow the organic backbone to be soluble in water. The significant advantages of WS conducting polymers are to minimize environmental impact and processing costs, hence promoting the practical commercial viability (Figure 6.20).

The first water spray-processable π-conjugated polymer electrochrome was reported by J. R. Reynolds and coworkers in 2010. As shown in Figure 6.21, they utilized a homopolymer of 3,4-propylendioxythiophene (ProDOT) bearing alkyl

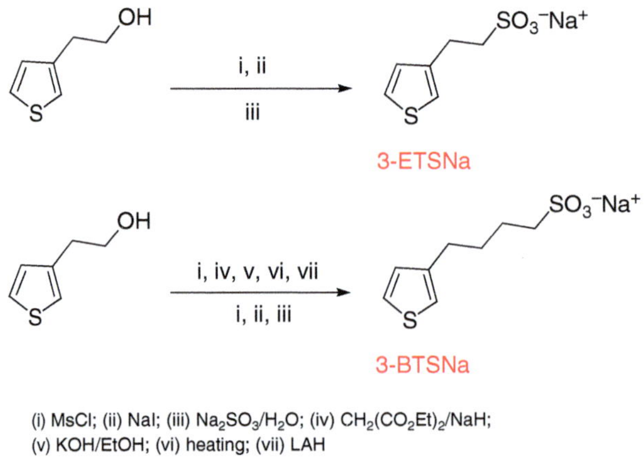

Figure 6.20 Chemical structure of the monomers 3-ETSNa and 3-BTSNa. Source: Reproduced with permission of Ref. [86] from Copyright © 1987, American Chemical Society.

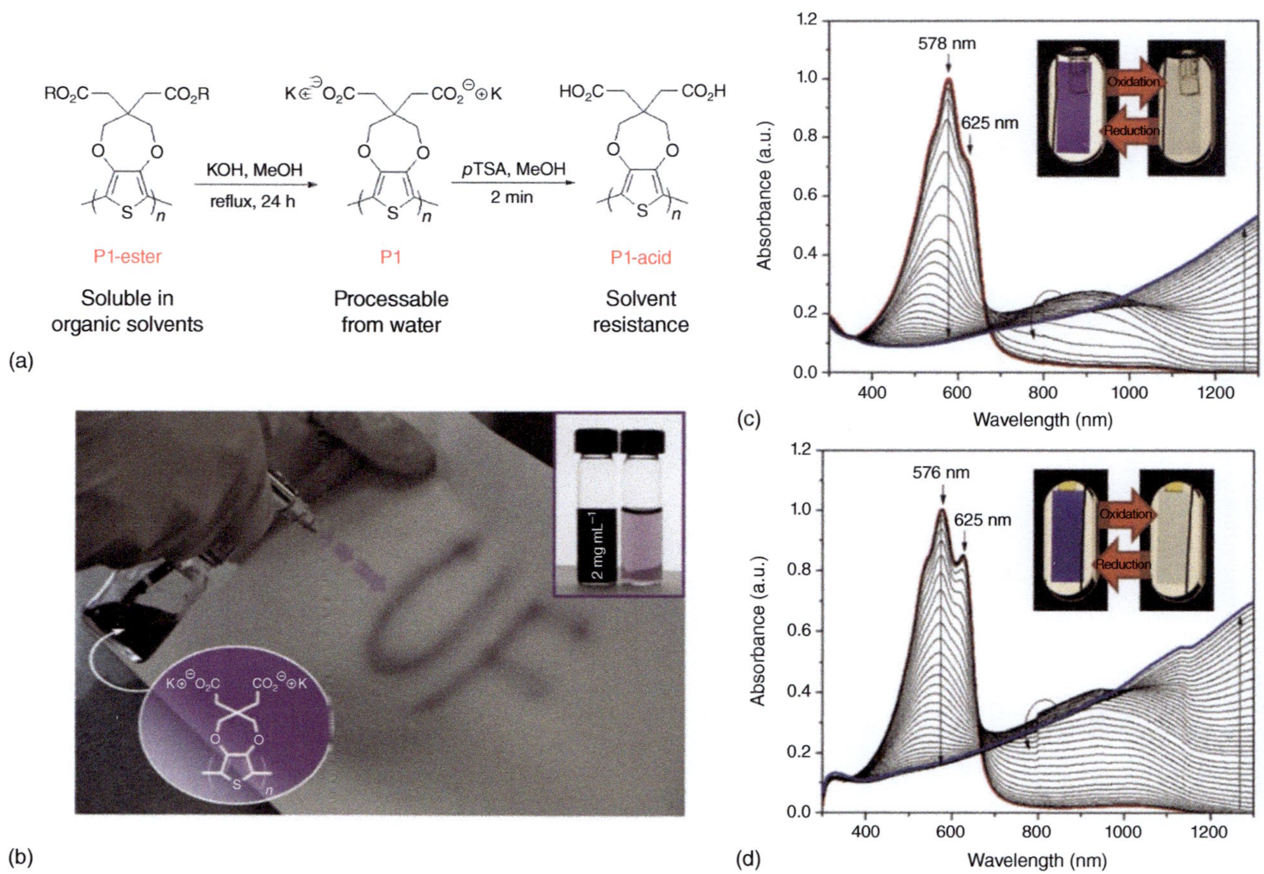

Figure 6.21 (a) Chemical structure of the polymers. (b) Concentrated water solutions of PProDOT-salt (P1) are made and simply processed using a conventional air-brush. Spectro-electrochemistry of (c) PProDOT-ester (P1-ester). The film was spray-cast onto ITO-coated glass from a toluene solution (2 mg/ml). (d) PProDOT-acid (P1-acid). The film was spray-cast onto ITO-coated glass from a water solution of PProDOT-salt (2 mg/ml), and the deposited polyionic salt was neutralized by immersion in a MeOH solution of *p*TSA (c. 1 mg/ml). Source: Reproduced with permission of Ref. [87] from Copyright © 2010 WILEY-VCH Verlag GmbH & Co. KGaA, Weinheim.

ester side chains and cleaved these solubilizing groups to obtain ionic π-conjugated polymer salts. These polymers were soluble in organic solvents due to the ester groups, while they achieved good water solubility in their carboxylate salt form. These feature allowed them to be spray processable from water. Upon subsequent neutralization of the spray-cast films, the EC properties of the defunctionalized PProDOT were fully recovered. Both films were redox cycled until a stable and reproducible electrochemical switch was reached. The defunctionalization process did not significantly alter the neutral-state purple-blue of the electrochrome, and no appreciable difference in transmissivity was observed between the doped state of the PProDOT-ester (P1-ester) and that of the PProDOT-acid (P1-acid) [87].

In order to develop WS D–A copolymers where an unfunctionalized acceptor repeat unit is incorporated for fine-tuning of the absorption spectra, J. R. Reynolds and coworkers incorporated the ProDOT with four carboxylate ester groups and 2,1,3-benzothiadiazole (BTD) units to yield two EC polymers: ECP-Blue-A (P2-ester) and ECP-Blue-R (P3-ester). Subsequent saponification of the ester groups afforded polymer carboxylate salts (P2 and P3) with water solubility greater than 4 mg/ml. Thus, the saponified polymer salts of P2 and P3 could be processed from aqueous solutions into thin films by spray-casting. Upon the subsequent neutralization, the resulting polymer acid films were both solvent resistant (SR) and could be electrochemically switched between the colored state and a transmissive state in a KNO_3/water electrolyte solution. P2-acid and P3-acid showed the optical contrast of 38% at 655 nm and 39% at 555 nm (Figures 6.22–6.24) [88].

To further explore the scope and versatility of the aforementioned method for a broader range of application requiring redox activity, a 3,4-propylenedioxythiophene

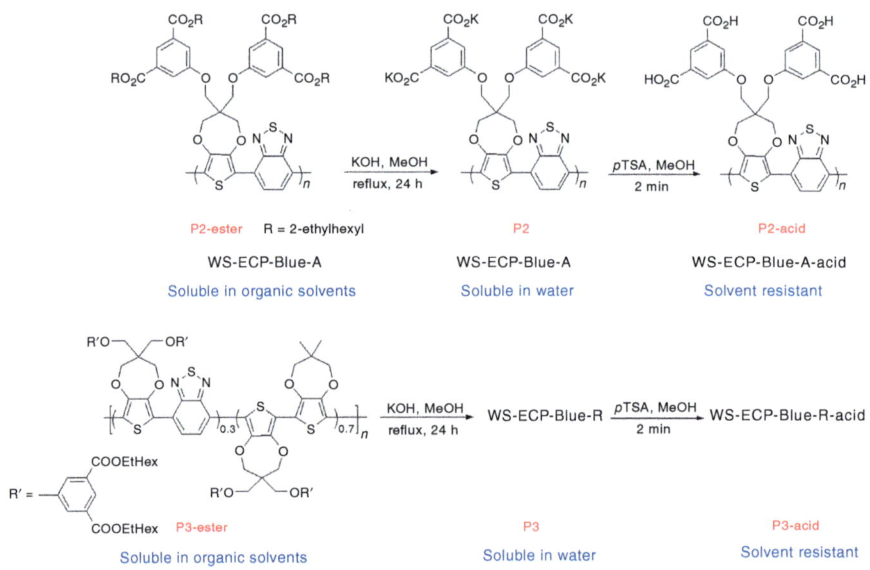

Figure 6.22 Chemical structure of the polymers P2 and P3 and the corresponding polymer salts. Source: Reproduced with permission of Ref. [88] from Copyright © 2012 American Chemical Society.

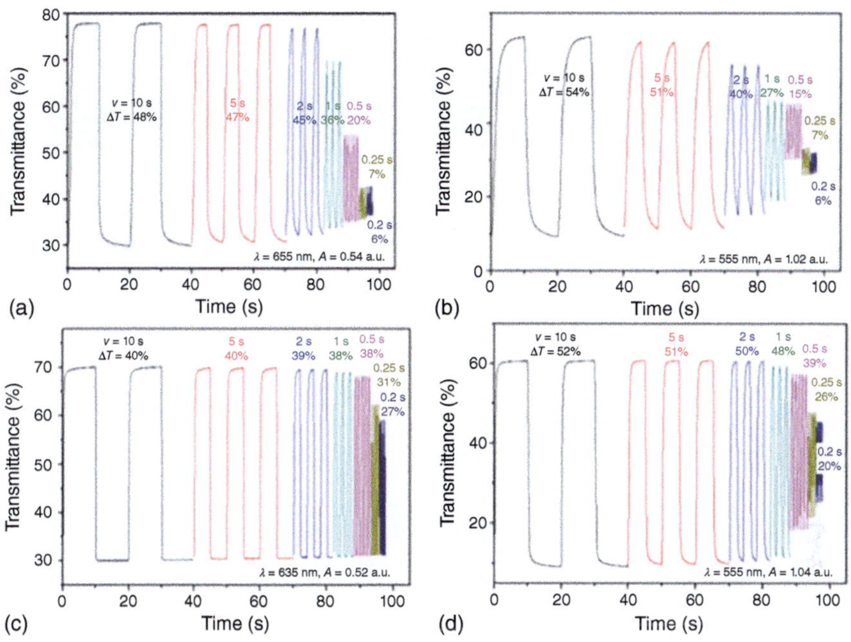

Figure 6.23 Square-wave potential-step chronoabsorptometry of (a) ECP-Blue-A (P2-ester) (monitored at 655 nm, switching potential range: −0.31 to +0.74 V vs Fc/Fc$^+$); (b) ECP-Blue-R (P3-ester) (monitored at 555 nm, switching potential range: −0.49 to +0.56 V vs Fc/Fc$^+$) in 0.2 M LiBTI/PC electrolyte solution; (c) WS-ECP-Blue-A-acid (P2-acid) (monitored at 635 nm, switching potential range: −0.1 to +0.95 V vs Ag/AgCl); and (d) WS-ECP-Blue-R-acid (P3-acid) (monitored at 555 nm, switching potential range: −0.2 to +0.95 V vs Ag/AgCl) in 0.2 M KNO_3/water electrolyte solution [88]. Reproduced with permission of Ref. [88] from Copyright © 2012 American Chemical Society.

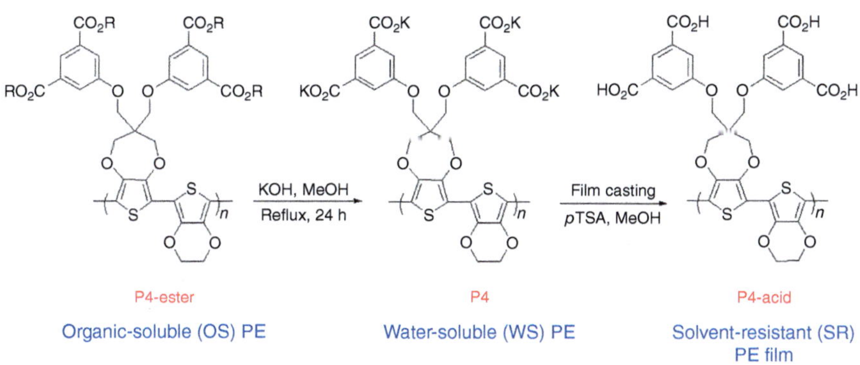

Figure 6.24 Chemical structure of the polymer P4 and the corresponding polymer salts. Source: Reproduced with permission of Ref. [89] from Copyright © 2017 American Chemical Society.

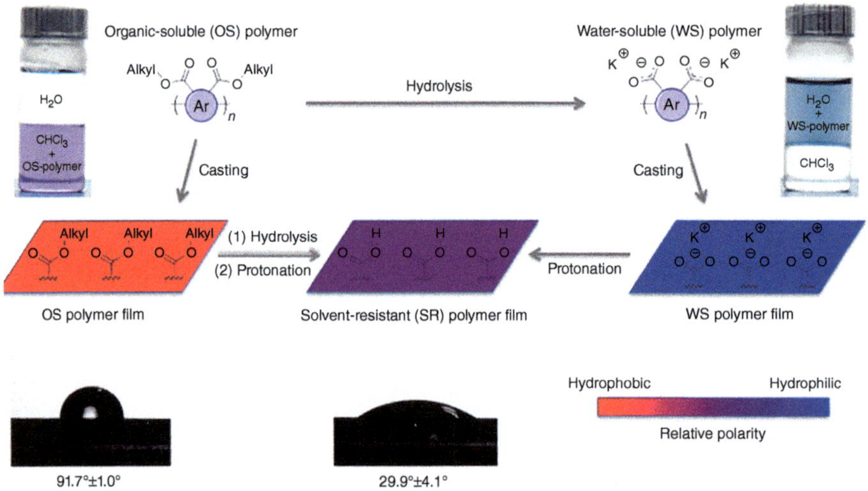

Figure 6.25 General overview of the conversion of an organic soluble polymer (P4-ester) to the water-soluble polymer salt form (P4). Source: Reproduced with permission of Ref. [89] from Copyright © 2017 American Chemical Society.

ProDOT monomer bearing multiple alkyl ester-based side chains with the carbonyl side of the ester closest to the propylene bridge was polymerized with a heteroaryl dihydrogen species (EDOT) using DHAP conditions to yield an organic soluble (OS) polymer (P4). The ester side chains were hydrolyzed to the corresponding conjugated polyelectrolytes, and the polymeric carboxylate salt was dissoluble in water (up to 10 mg/ml). This WS polymer (P4) could be cast using various methods to form films, which could be further protonated via a dilute acid wash. The resulting film, now bearing carboxylic acid side chains, was insoluble in both organic and aqueous media and was termed SR. Figure 6.25 demonstrated the general overview of the conversion process [89].

As shown in Figure 6.26, this material (P4-acid) was effective as a high-contrast EC material with a vibrant blue neutral state (λ_{max} = 613 nm) and a highly transmissive oxidized state. Rapid EC switching (~0.2 seconds) was achieved. Besides, supercapacitors incorporating this SR copolymer (P4-acid) exhibited symmetric charge/discharge behavior at currents of up to 20 A/g (1 seconds discharge) and were able to maintain >75% of the initial capacitance over 175 000 cycles using 0.5 M NaCl/water as the device electrolyte. The combination of aqueous solubility, excellent EC properties, high capacitance, and stability in aqueous media suggests that this material may be useful in multiple electrochemical applications such as ECDs, supercapacitors, and biocompatible devices [89].

C. Xu and coworkers also developed a highly regiosymmetric and water-processable ProDOT derivative (Figure 6.27) switching between blue and transmissive regime. Water processability of partially hydrolyzed PProDOT salt (8 mg/ml) offered green chemistry for industrial scale device applications. Blue to transmissive switching EC polymer (P5-acid) showed a high contrast (56% at 580 nm) and relatively fast switching speed (1.8 seconds) [90].

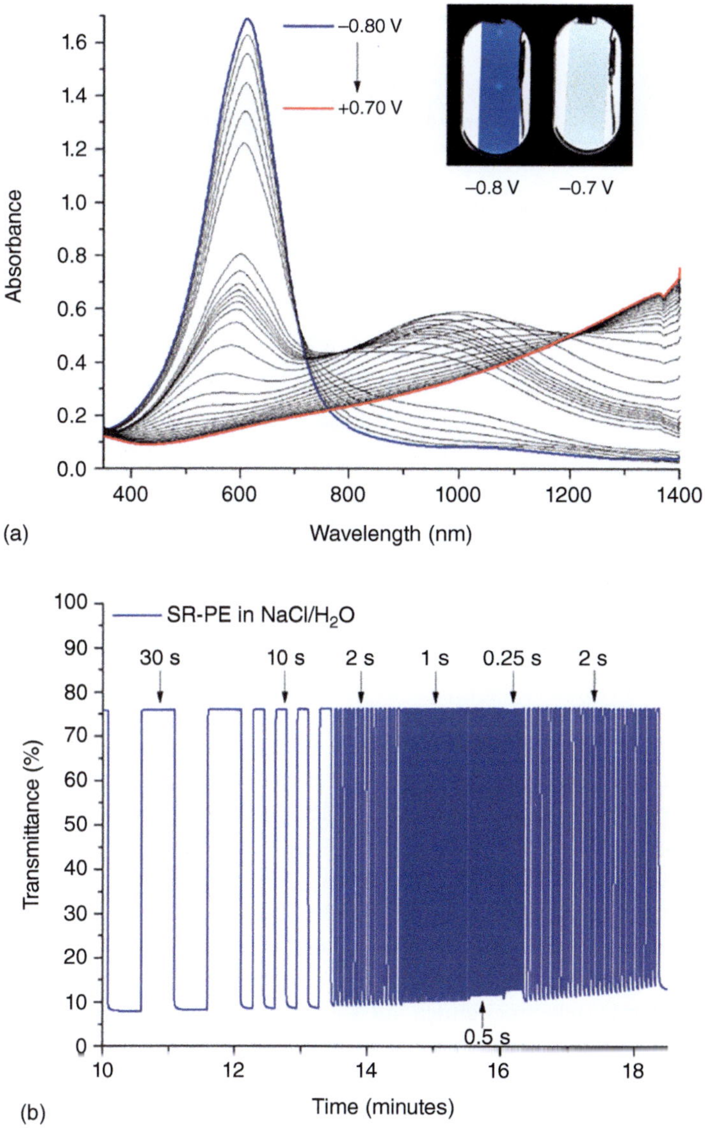

Figure 6.26 (a) Spectra recorded in 50 mV increments from −0.80 to +0.70 V vs Ag/AgCl and photos taken at −0.80 and + 0.70 V. (b) Transmittance at λ_{max} as a function of switching time of a spray-cast film of SR-PE (P4-acid) in 0.5 M NaCl/water switching from the charge neutral state (−0.80 V vs Ag/AgCl) to the oxidized state (0.70 V vs Ag/AgCl). Source: Reproduced with permission of Ref. [89] from Copyright © 2017 American Chemical Society.

Other efforts have been made to develop new methods to increase the water solubility of EC polymers, like introducing the nanostructure. K.-C. Ho and coworkers reported a new method to obtain a water-processable Prussian blue-polyaniline: polystyrene sulfonate (PB-PAIN:PSS) nanocomposite. This composite film exhibited multicolor electrochromism with three optical states. The film became highly

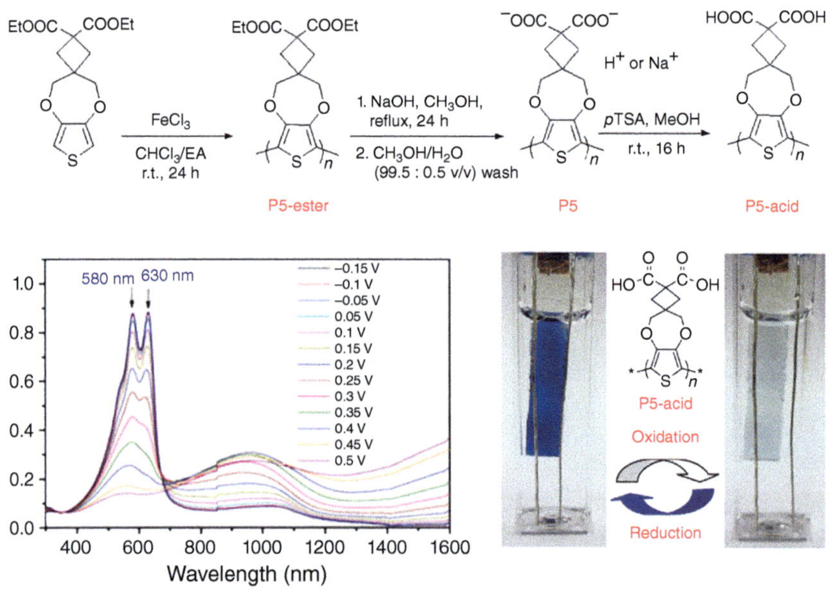

Figure 6.27 Electrochromic properties of water-processable ProDOT derivative (P5-acid). Source: Reproduced with permission of Ref. [90] from Copyright © 2015 American Chemical Society.

transparent in the neutral state (−0.5 V vs Ag/AgCl), with a green color at +0.2 V contributed by PANI and a blue-green color at +0.5 V [91]. X. Xing et al. developed water-dispersed electrochromic polymer nanoparticles (WDENs) prepared with miniemulsion process for EC polymer, poly[2,3-bis-(3-octyloxyphenyl)quinoxaline-5,8-diyl-alt-thiophene-2,5-diyl]. The nanoparticles (NPs) electrode showed much faster switching speed than the compacted electrode (e.g. 2.10 seconds vs 24.15 seconds for coloring and 8.65 seconds vs 25.95 seconds for bleaching at 0.4 V and 1.30 seconds vs 9.20 seconds for coloring and 1.7 seconds vs 2.90 seconds for bleaching at 1.0 V). The microelectrochemical measurement indicated much more efficient counter-ion diffusion between the electrolyte and the NP films than the compacted films, which resulted in much faster EC switching [92].

Design EC active polymers soluble in benign organic solvent (ethyl acetate, etc.), water, or alcohols have aroused wide concern in recent years owing to the urgent demands for environmental-friendly and low-cost large-scale fabrications. Besides, the ECP films compatible with aqueous electrolytes are also required. Regarding this, several paths have been exploited during the past decades. The first one is charged polymers, obtained by polymerizing the ECP in the presence of a WS polyelectrolyte host, such as PEDOT:PSS, but this method remains some limitation. Other routes are related with functionally modified conjugated polymers. One example is introducing ionic pendant, like $-SO_3^-$, $-NR_3^+$, etc., similar to which salinization can also increase the water solubility. However, these WS polymers couldn't be immersed and cycled in the aqueous electrolytes. To achieve water solubility and water switchability simultaneously, a strategy is to synthesize polymers (intrinsically soluble in organic solvents) decorated by side groups, which can be directly transformed into WS analogs via some treatments. This allows the easy synthesis and post-treatments in organic solvents and water processability for film formation. The aforementioned statements have been introduced in detail in book chapters published in 2015 and 2019 [7, 93].

Recently, four-ester functional side chains were hung on the backbone of poly ProDOT synthesis via DHAP to achieve organic solvent solubility and aqueous compatibility in the same time. The resulted polymer ProDOT-TetEster was spray-coated as dissolved in environmentally sustainable solvent 2-methyltetrahydrofuran, and the hydrophilic surface of film was proved by a measured contact angle of 90°, indicating a decent aqueous compatibility. The redox activity was desirable in 0.1 M aqueous NaCl, while the CV curves of polymers with ethylhexyl ether (ProDOT-OEtHx) and butyloctyl ester (ProDOT-BOE) chains were unnoticeable when compared in the same condition (Figure 6.28). In all, the addition couple of ester groups make the polymer attainable from benign synthesis, processable from benign solvent, and compatible with aqueous electrolyte [94].

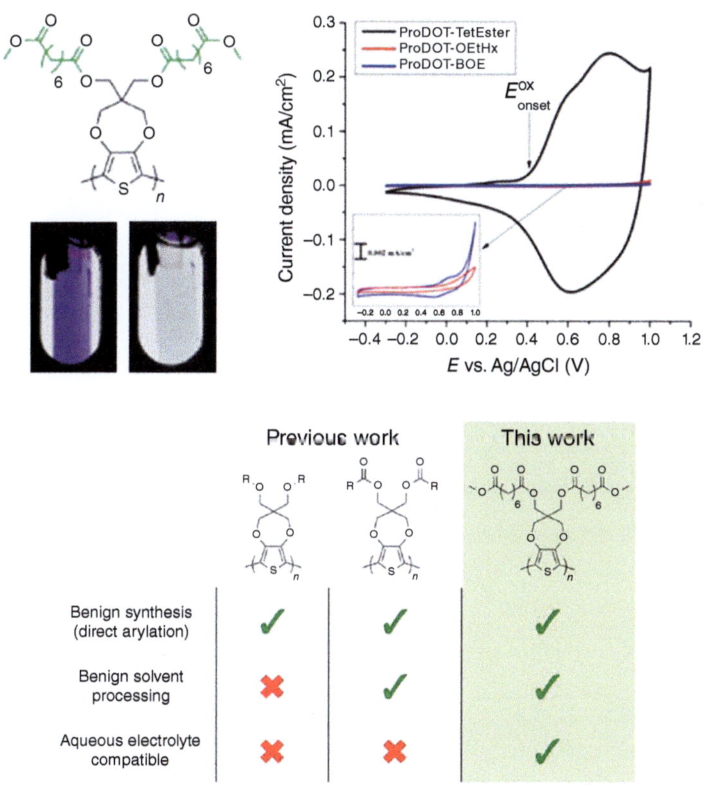

Figure 6.28 Solubility tuning of ProDOT with four-ester functional side chains. Source: Reproduced with permission of Ref. [94] from Copyright © 2018 American Chemical Society.

6.3 Polypyrroles-Based Conjugated Electrochromic Polymers

6.3.1 Introduction

Among conducting polymers, polypyrrole (PPy) is by far the most extensively studied polymer because of the multiple advantages such as low oxidation potential, water solubility, low cost of its monomer (pyrrole), and so on [95]. In 1968, PPy was prepared by oxidation of pyrrole in sulfuric acid as a black powder and shown to be a conducting polymer [96]. After that, workers at IBM found that PPy could be obtained by electrochemical polymerization. Due to the low oxidation potential of pyrrole, electropolymerization can be carried out in aqueous [97, 98] or organic electrolytes [99]. As a result of the initial oxidation, the radical cation of monomer is formed and reacts with other monomers present in solutions to form oligomeric products and then polymer (Figure 6.29). The mechanism has been described by Diaz and coworkers [100], which was later supported by the theoretical studies based on the correlation between the reactivity and the unpaired electron density of the radical cations of Waltman et al. [101, 102].

After the electropolymerization, PPy can be obtained. Among the conducting polymers, PPy was the first one to be investigated for its optoelectronic properties, whose electronic structure could be reversibly changed with doping. Energy diagram of PPy

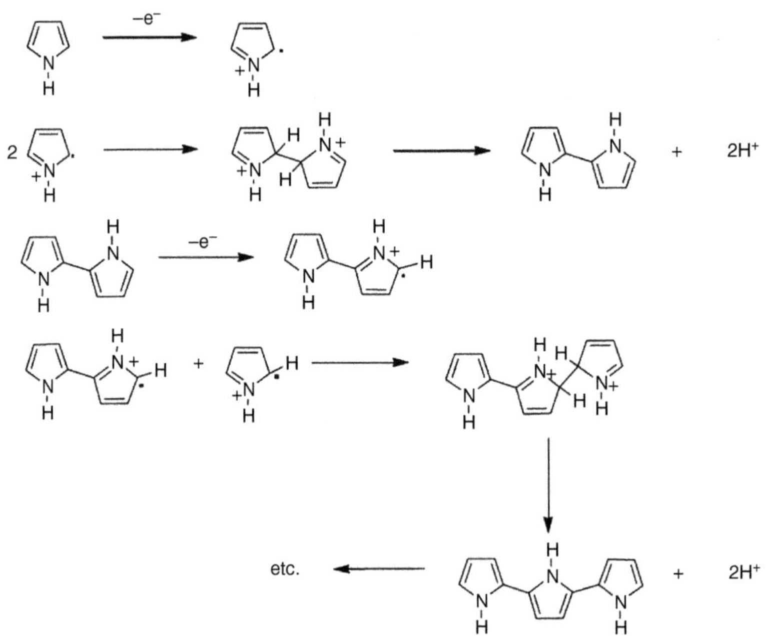

Figure 6.29 Proposed mechanism for electropolymerization of pyrrole by Diaz and coworkers [100] Source: Reproduced with permission of Ref. [100] from Copyright © 1983 Published by Elsevier B.V.

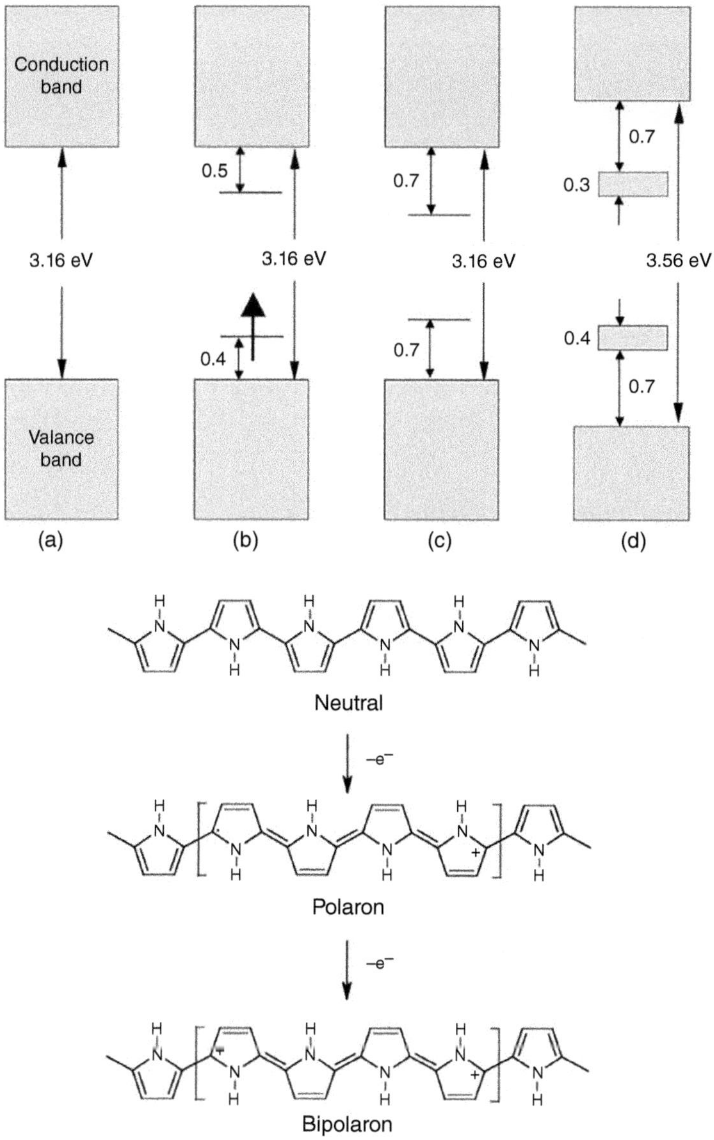

Figure 6.30 Electronic energy diagrams and structures for (a) neutral PPy, (b) polaron, (c) bipolaron, and (d) fully doped PPy. Source: Reproduced with permission of Ref. [104] from Copyright © The Royal Society of Chemistry 2014.

in neutral form is represented in Figure 6.30a. Upon oxidation of neutral PPy, the local relaxation of the benzoid structure towards quinoid-like structure occurs, creating radical cations (polarons) [9]. The formation of polarons induces two new energy levels, which are symmetrically positioned within the bandgap. Thus, two new electronic transitions at longer wavelengths emerge. Further oxidation results in the formation of bipolarons, which are the charge carriers of the coupled cations (dication).

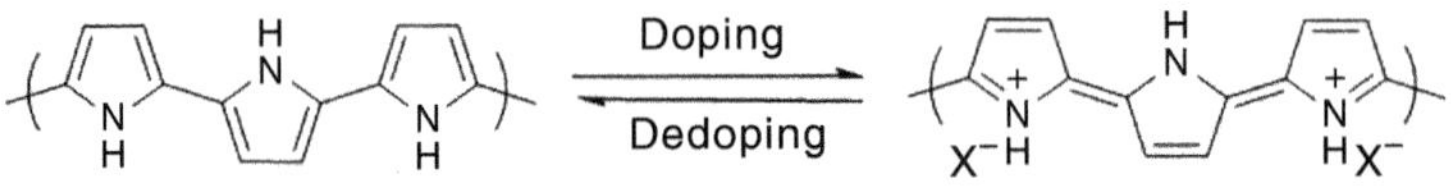

Figure 6.31 Electrochromism of polypyrroles.

With an empty lower energy state, bipolarons are signified by the broad, low energy absorptions due to the transitions from the top of the valence band (Figure 6.30c). As the polymer is oxidized further, the bipolaronic energy state overlaps and forms intermediate band structures as given in Figure 6.30d. Thus, the optical state (and the color) of the polymer is altered by the formation of new electronic states (p-type doping), which results in the development of the lower-energy absorptions [103].

6.3.2 Electrochromic Properties of Polypyrroles (PPy)

The optical bandgap of PPy is reported to be ~2.7 eV, which can be switched from neutral yellow to blue-violet in the oxidized form. However, only the extremely thin films of PPy could be completely dedoped with a low contrast ratio which limits the application of PPy on electrochromism. Moreover, during the switching, structural defects arising from β couplings would result in conjugation breaks and thus a low degree of stability (Figure 6.31) [105].

6.3.3 Tuning of Electrochromic Properties of Polypyrrole (PPy)

It is well known that the poor solubility and processability of polypyrroles prevent their industrial application. Various substituents to improve their solubility have been carried out. Moreover, since the coloration in the conducting polymers mainly depends on the bandgap and their band structure, enhanced EC properties would be achieved through the tailor-made functionalization of repeat units for polypyrrole. The technology development of polypyrrole derivatives is shown in Figure 6.32.

6.3.3.1 Structural Modification

N-Substituted Polypyrroles

A series of poly(*N*-alkyl pyrrole)s with various substituents from methyl to butyl to phenyl were developed, all exhibiting yellow to brown-black electrochromism upon oxidation [106]. Differed from poly(*N*-methylpyrrole), poly(*N*, *N*′-dimethyl-2,2′-bipyrrole) polymerized from the dimer *N*,*N*′-dimethyl-2,2′-bipyrrole possessed 30 times higher conductivity, exhibiting pale yellow with a blue-shifted onset of absorbance at the neutral state and gray in fully oxidized form [107, 108]. Moreover, when the benzylideneamino groups were introduced to PPy, they showed neutral red to blue in the oxidized form (Figure 6.33) [109].

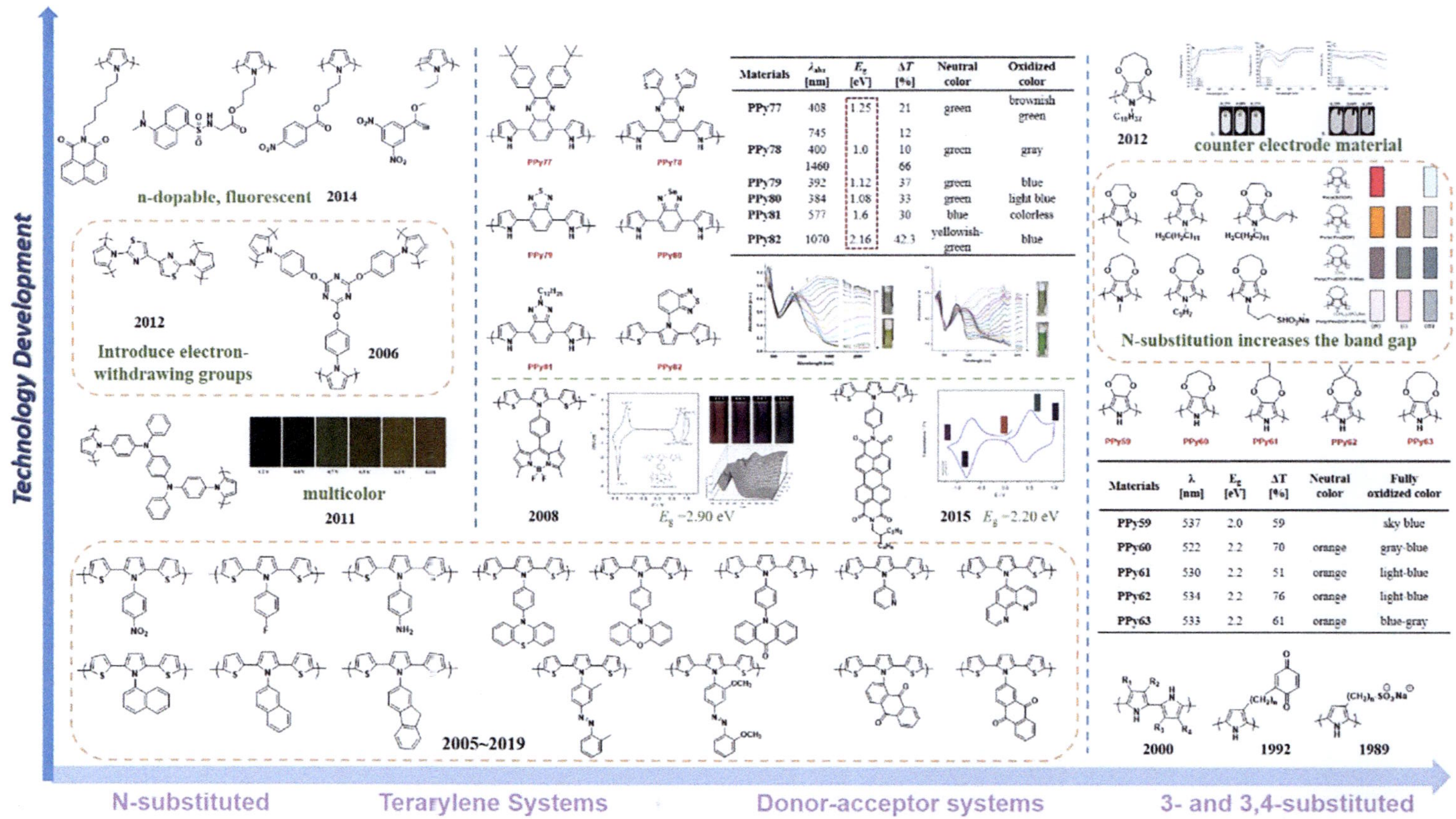

Figure 6.32 Technology development of pol/pyrrole derivatives.

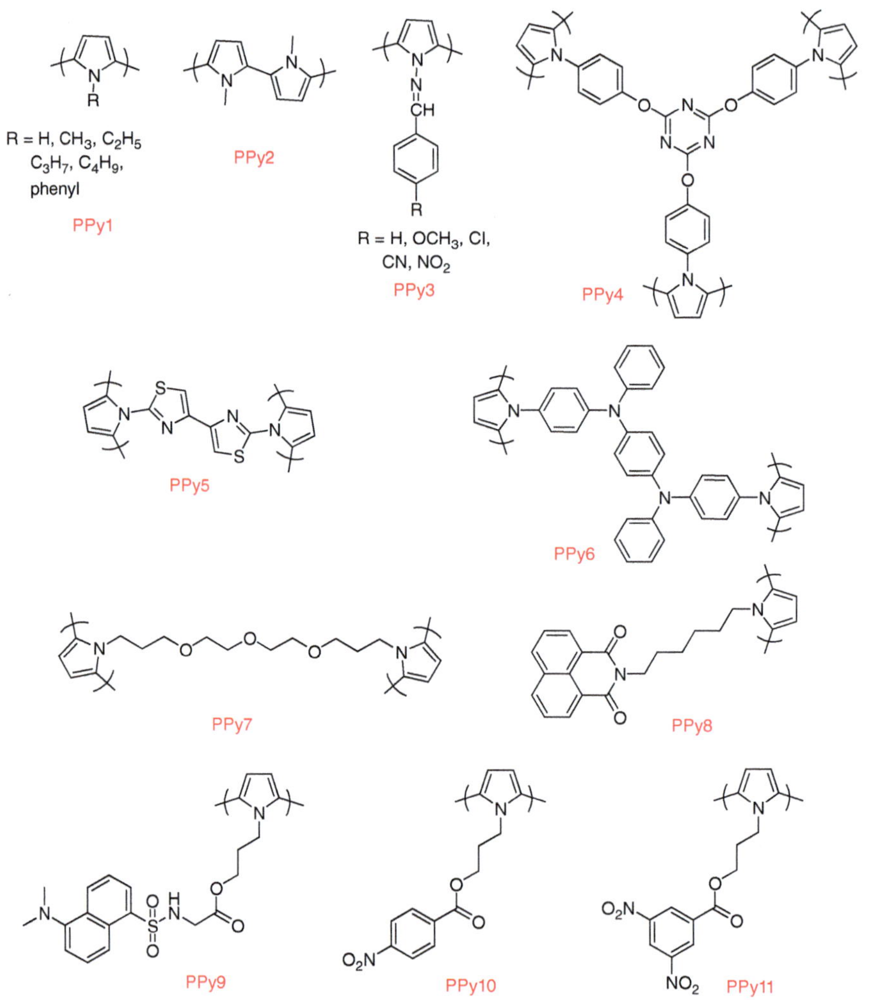

Figure 6.33 Chemical structure of N-substituted polypyrrole derivatives.

Another polypyrrole derivative, electropolymerized from a star-shaped monomer, displayed red and turquoise colors in the fully oxidized and reduced state. The optical contrast were measured at 700 nm and found to be 20% with a switching time of 1.1 seconds [110]. Polypyrrole derivatives containing bithiazole moieties were electropolymerized both in the absence and presence of 3,4-ethylenedioxythiophene. The copolymers displayed multichromism within a wide span of the visible spectrum. Short switching times of 0.6 seconds and high optical contrast of 54% were both achieved by bithiazole-based copolymers [111]. C. Zhang and coworkers reported TPA-based polypyrrole derivatives (PPy6) with six various colors under different potentials, exhibiting high optical contrast of 41% at 852 nm and 52% at 617 nm and fast switching time of 1.3 seconds at 410 nm, 1.4 seconds at 852 nm, and 0.6 seconds at 617 nm [112]. An etheric member of N-linked polybispyrroles

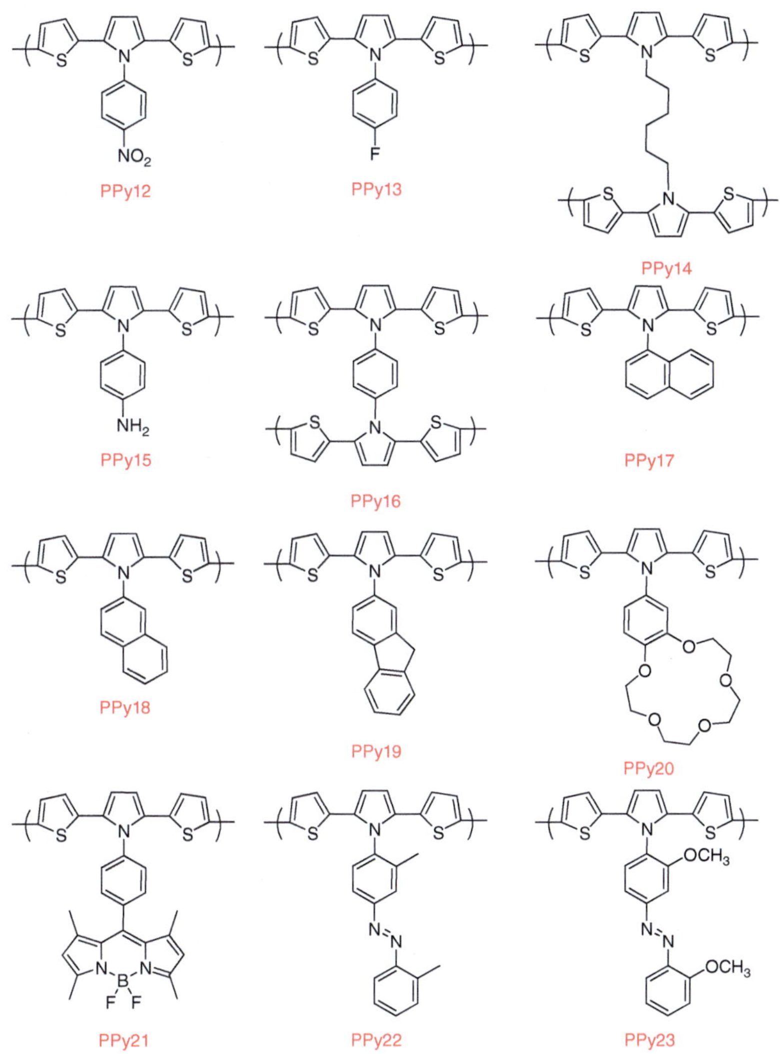

Figure 6.33 (*Continued*)

(PPy7) was electro-oxidized in water to form the polymer film [113]. They showed transparent yellow in the neutral state, light pink in the intermediate state, and blue in the oxidized state (Figure 6.34).

Apparently, most of the polypyrrole derivative introduced earlier was p-dopable. To achieve n-dopable, soluble, florescent, and EC polypyrrole derivatives, P. Camurlu et al. incorporated 1,8-naphthalimide moiety to pyrrole via alkyl spacer group [114]. This polymer, PPy8, exhibited maxima emission at 471 and 543 nm. Besides, PPy8 displayed 33.37% of the optical contrast at 704 nm with a switching time of 1.63 seconds. Another fluorescent polypyrrole derivative PPy9 bearing a dansyl substituent was synthesized by the electropolymerization with the addition of 20% boron trifluoride diethyl etherate to the electrolyte [115]. PPy9 with green

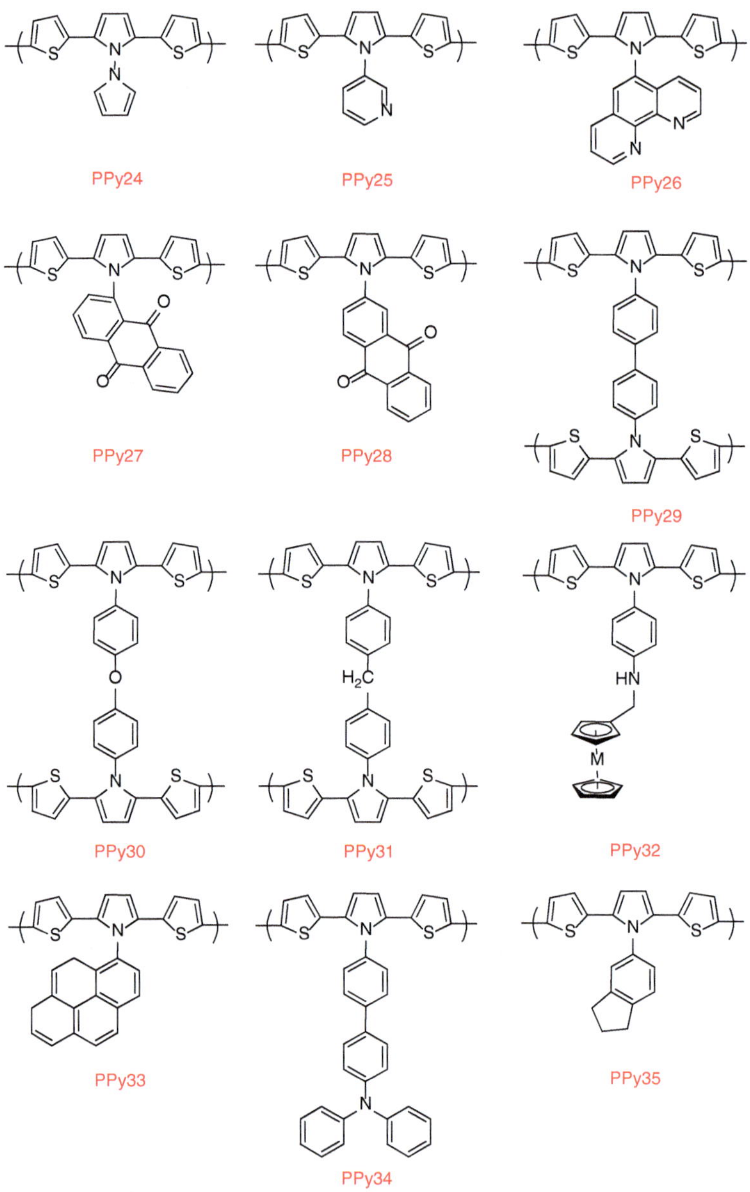

Figure 6.33 *(Continued)*

emission exhibited greenish-yellow in the neutral state and grayish-green in the oxidized state. Furthermore, two fluorescent nitrobenzoyl-based polypyrrole derivatives, PPy10 and PPy11, were reported [116]. Their polymer films both exhibited EC behavior with color changes from pale green (reduced state) to dark gray (oxidized state) and also a significant absorbance in the infrared region (oxidized state) (Figure 6.35 Table 6.1).

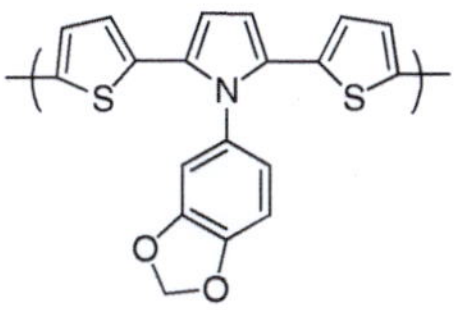

PPy36

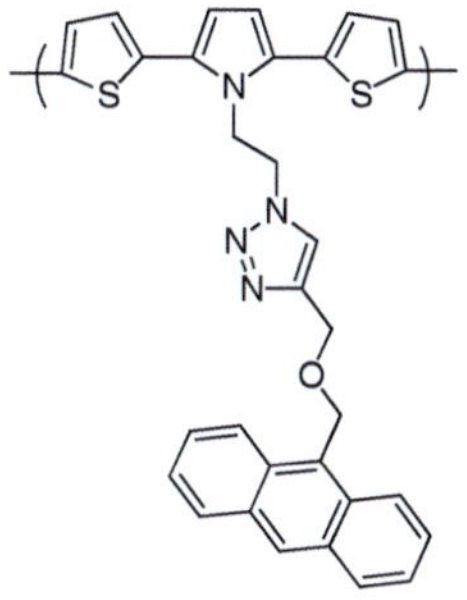

PPy37

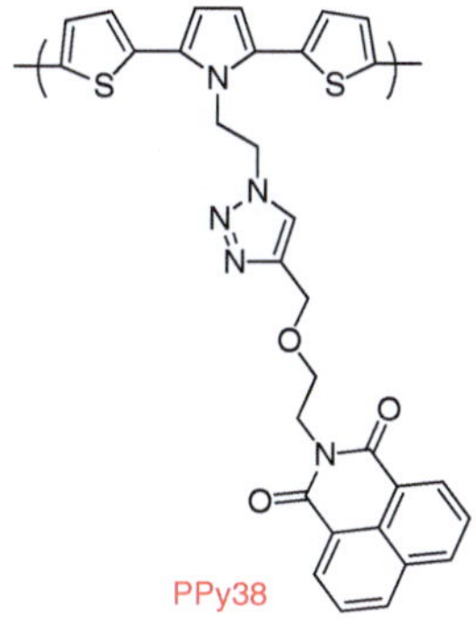

PPy38

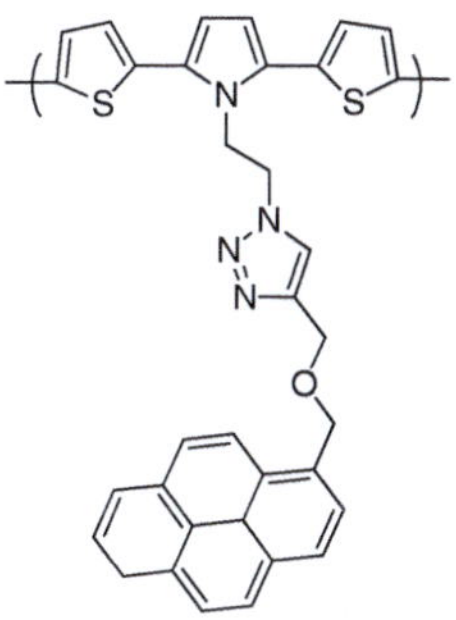

PPy39

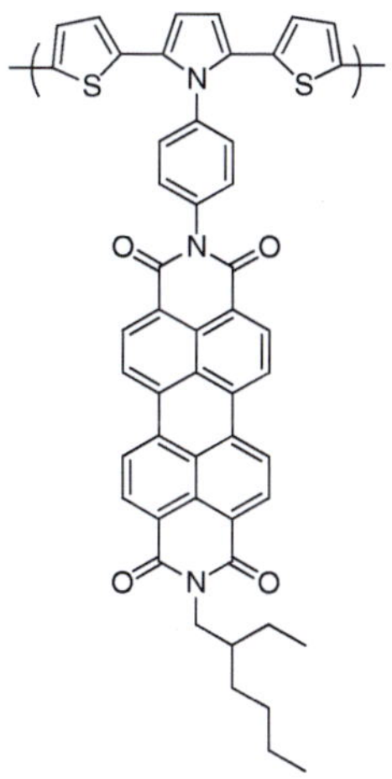

PPy40

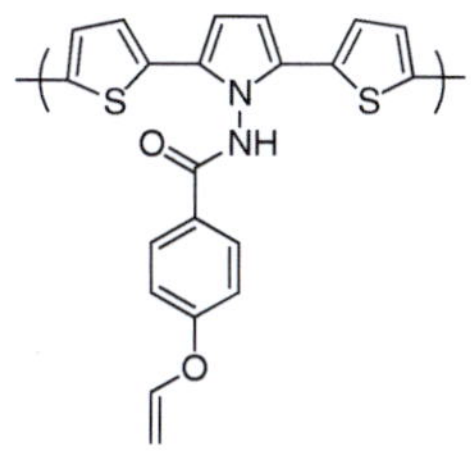

PPy41

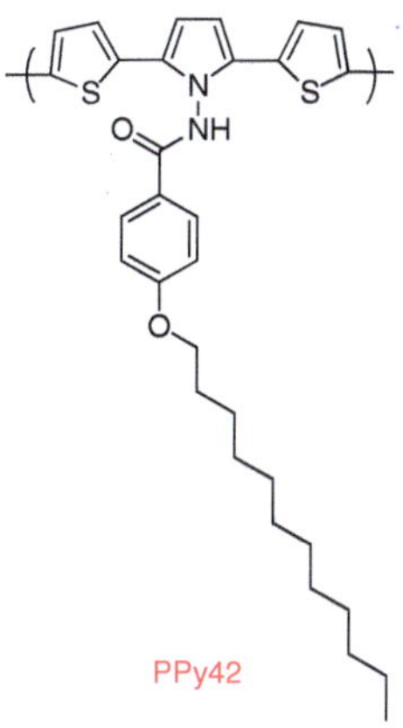

PPy42

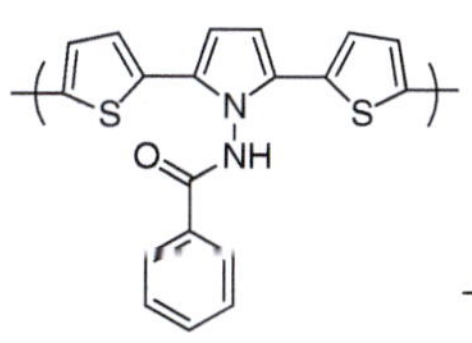

PPy43

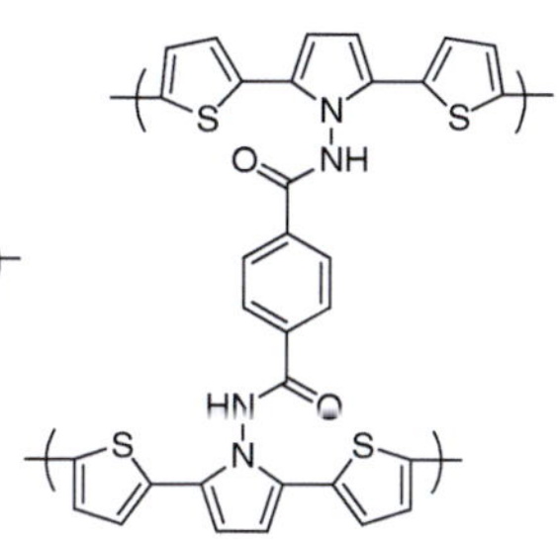

PPy44

Figure 6.33 *(Continued)*

Since Ferraris and Skiles obtained the 2,5-bis(2-thienyl)-1*H*-pyrrole polymer (PSNS) by the direct polymerization of thiophene and pyrrole derivative in 1987, many researchers have reported a series of different polymers based on 2,5-bis(2-thienyl)-1*H*-pyrroles (SNS). The oxidation potentials of poly(2,5-dithienyl-pyrrole) derivatives are about 0.7 V vs Ag/AgCl [117, 118], revealing that the incorporation of a pyrrole ring between two thiophene units in SNS monomer could decrease the oxidation potentials of polymer films because of the 1.5

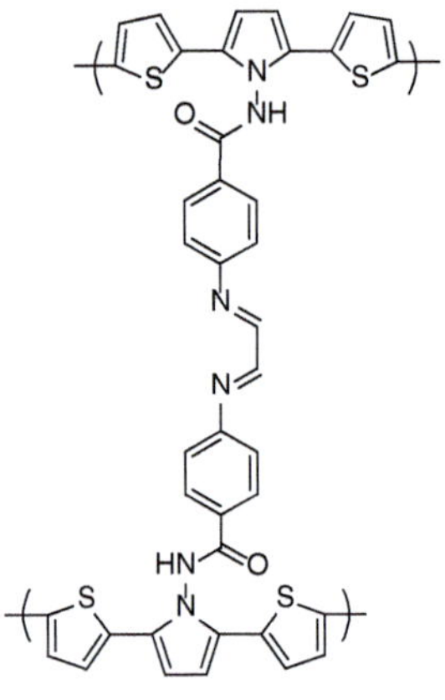

PPy45

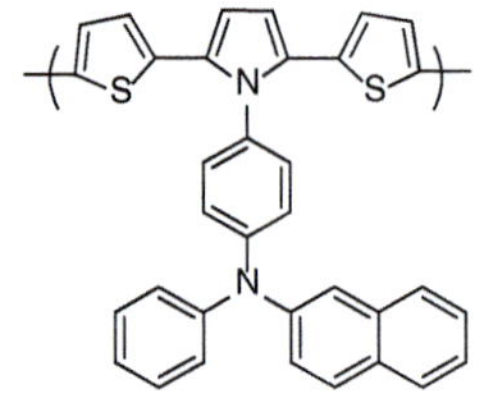

PPy46

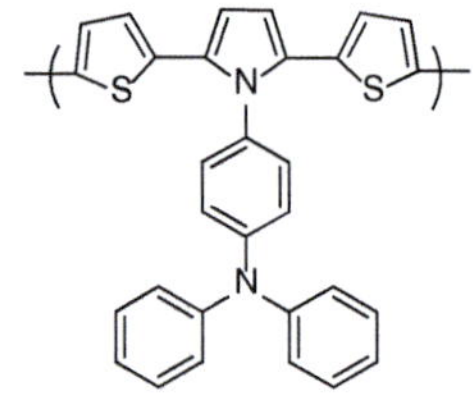

PPy47

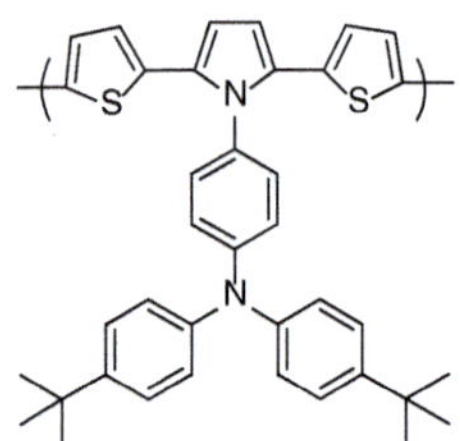

PPy48

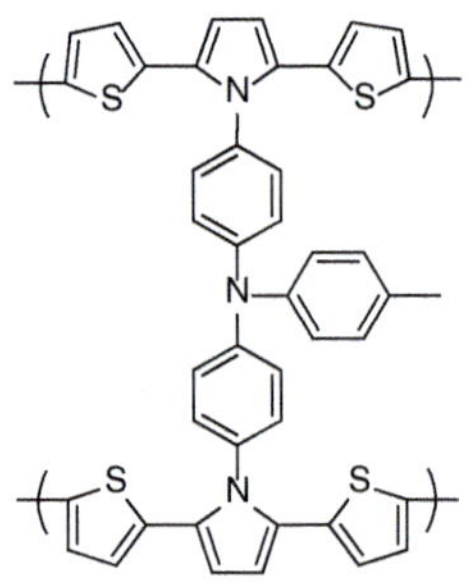

PPy49

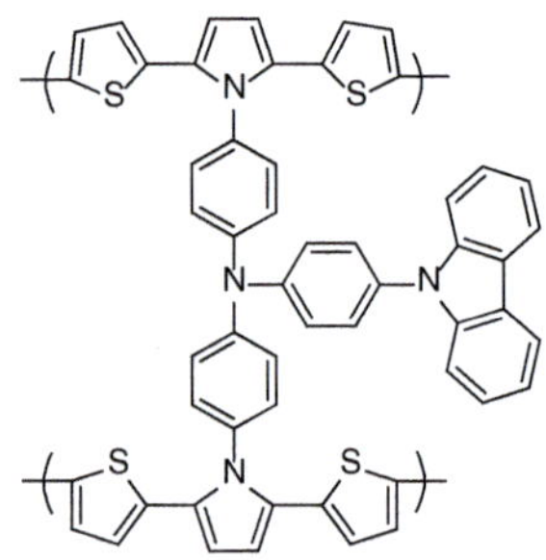

PPy50

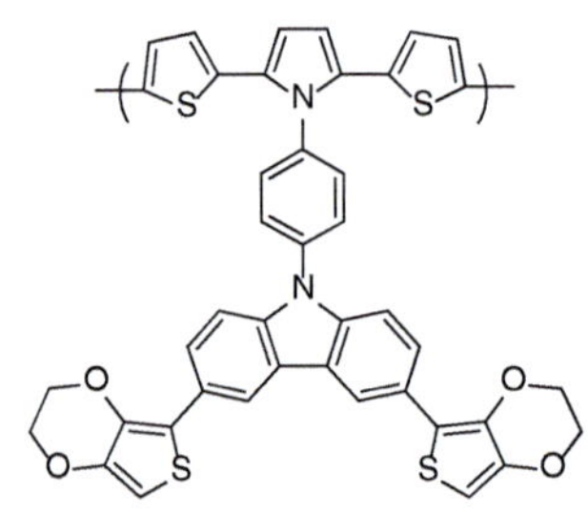

PPy51

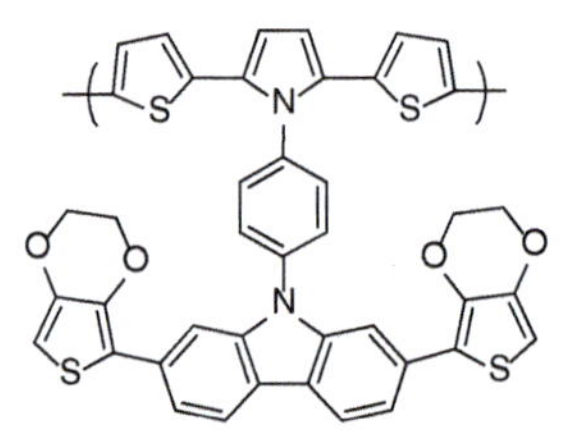

PPy52

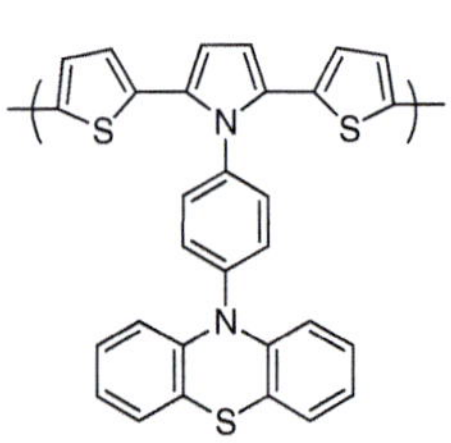

PPy53

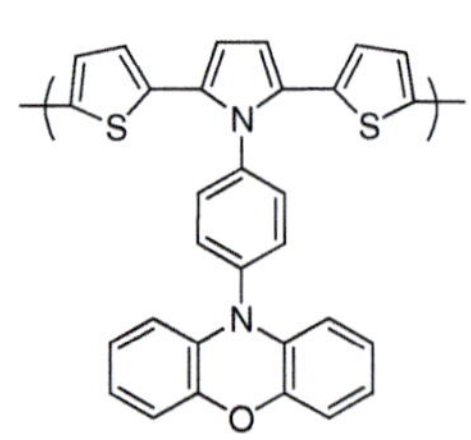

PPy54

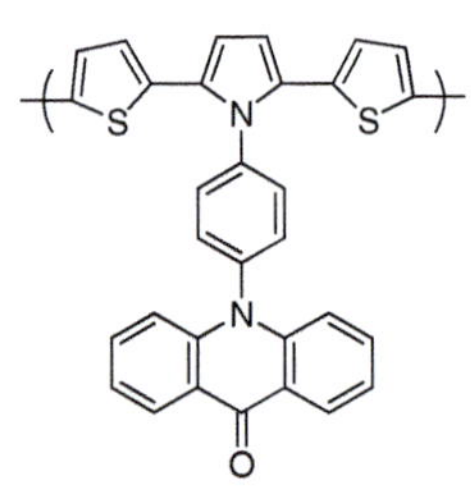

PPy55

Figure 6.33 (*Continued*)

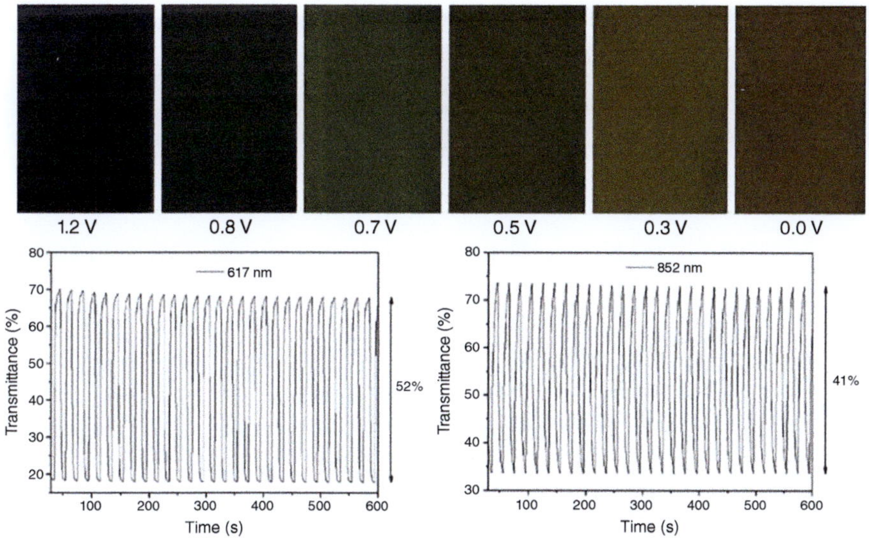

Figure 6.34 Color changes and the optical contrast of polymer PPy6 at different wavelengths. Source: Reproduced with permission Ref. [112] from Copyright © 2011 Elsevier Ltd.

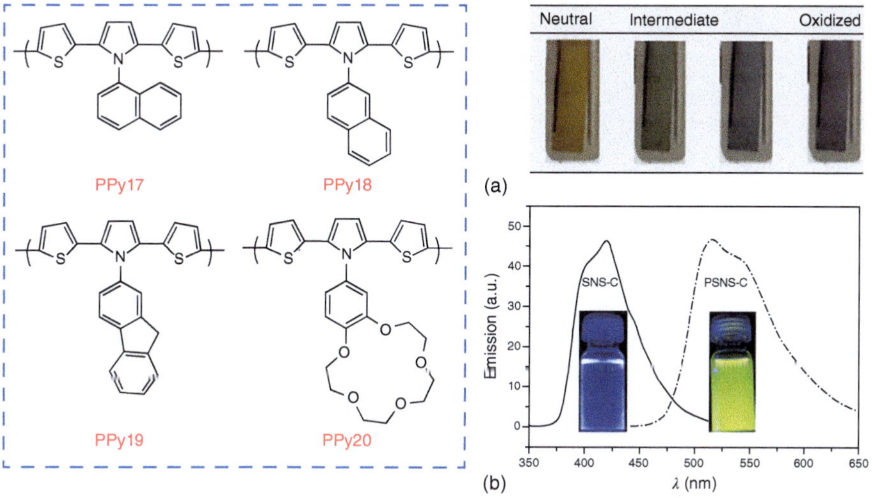

Figure 6.35 (a) Color changes of polymers PPy17, PPy18, PPy19, and PPy20. (b) Emission spectra of PPy20 excited at 415 nm in acetone. Source: Reproduced with permission of Ref. [124] from Copyright © 2008 Elsevier Ltd.

and 0.9 V vs Ag/AgCl for polythiophene and polypyrrole, respectively. In order to develop various high-performance EC polymers, L. Toppare and coworkers reported large numbers of polymers based on SNS, which has been proven a convenient strategy to minimize the steric influence of the central substituent through the addition of the thiophene spacers. A soluble polymer

Table 6.1 Electrochromic properties of N-substituted polypyrrole derivatives.

Materials	λ_{abs} (nm)	E_g (eV)	ΔT (%)	Neutral color	Fully oxidized color
PPy1	—	—	—	Yellow	Brown–black
PPy2	—	—	—	Pale yellow	Gray
PPy3	—	—	—	Red	Blue
PPy4	700	2.97	20	Red	Turquoise
PPy5	—	2.59	—	Yellow	Gray
PPy6	617		52	Brown	Blue
	852		41		
PPy7	315	2.71	16.77	Transparent yellow	Blue
	484		17.53		
PPy8	704	2.99	33.37		
PPy9		2.65	< 20	Greenish-yellow	Grayish-green
PPy10		2.78		Pale green	Gray
PPy11		2.75		Pale green	Gray
PPy12	400	2.15	13		
	860		23		
PPy13	398	1.94	2.3		
PPy14	335	2.16	28	Yellow	Blue
PPy15	376	2.12	20.7	Yellow	Blue
PPy16	410	2.03	8		
	1230		41		
PPy17	423	2.33	18.2	Yellow	Violet
PPy18	400	2.40	10.5	Yellow	Violet
PPy19	445	2.18	21.7	Yellow	Violet
PPy20	444	2.14	17.8	Yellow	Violet
PPy21	351	2.90	12.1	Purple	Gray green
	525		17.7		
	670		28.6		
PPy22	435	2.31		Yellow-green	Dark green
PPy23	430	2.34		Mustard color	Green
PPy24	444	2.21	18	Orange	Violet
	661		33		
PPy25	430	2.1	42	Brownish-yellow	Dark blue
PPy26	451	1.85	31	Greenish-yellow	Light blue
PPy27	780	2.20	18	Gray	Dark blue-purple
PPy28	780	2.13	16.5	Gray	Dark blue-purple
PPy29	430	1.95	5	Orange	Blue-grayish
	750		26		

Table 6.1 (Continued)

Materials	λ_{abs} (nm)	E_g (eV)	ΔT (%)	Neutral color	Fully oxidized color
PPy30	430	2.15	10	Yellow-greenish	Blue
	750		32		
PPy31	430	2.08	7	Orange	Violet-grayish
	750		22		
PPy32	440	2.02	13.7	Yellow	Blue
PPy33	450	2.27	1.2	Yellowish orange	Blue
PPy34	810	2.17	53.18	Yellow	Blue
	1310		69.82		
	1550		70.51		
PPy35	566	2.37	8.8	Yellowish-green	Blue
	952		53.4		
PPy36	640	2.25	25.0	Yellow	Blue
	962		64.8		
PPy37	370	2.57	7.09	Green	Blue
	900		29.28		
PPy38	338	3.15	1.73	Light yellow	Blue
PPy39	860	3.36	20.28	Light green	Blue
PPy40	900	2.20	45	Orange	Blue
PPy41	430	1.99	23	Reddish orange	Blue
	830		42		
PPy42	900	1.92	60		
PPy43	430	2.1	< 30	Yellow	Blue
PPy44	430	2.2	40	Greenish-yellow	Blue
	850		87		
PPy45	435	2.02	54	Yellowish	Blue
PPy46		2.6	23	Yellow	Violet
PPy47		2.61	24	Yellow	Violet
PPy48		2.56	13	Yellow	Violet
PPy49		2.27	19	Yellow	Violet
PPy50		2.21	17	Yellow	Violet
PPy51	617	2.32	33	Light yellow	Light purple
PPy52	620	2.21	27	Orange	Blue
PPy53	1100	2.28	13.5	Yellow–green	Purple red
PPy54	560	2.36	8.34	Greenish	Purple
PPy55	603	2.20	13.31	Yellow-green	Purple

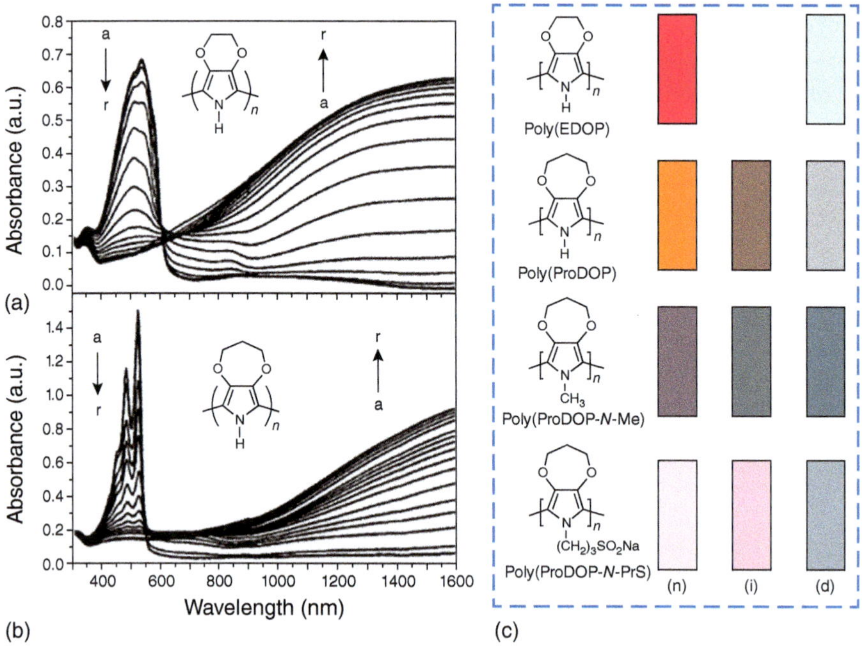

Figure 6.36 (A) Spectroelectrochemistry of PEDOP in 0.1 M $LiClO_4$/PC at applied potentials of (a) −1.33, (b) −1.23, (c) −1.13, (d) −1.03, (e) −0.98, (f) −0.93, (g) −0.88, (h) −0.83, (i) −0.78, (j) −0.73, (k) −0.68, (l) −0.63, (m) −0.53, (n) −0.43, (o) −0.33, (p) −0.23, (q) −0.13, and (r) +0.07 V vs Fc/Fc^+. (B) Spectroelectrochemistry of PProDOP in 0.1 M $LiClO_4$/PC at applied potentials of (a) −1.13, (b) −1.03, (c) −0.93, (d) −0.88, (e) −0.83, (f) −0.73, (g) −0.62, (h) −0.58, (i) −0.53, (j) −0.48, (k) −0.43, (l) −0.38, (m) −0.33, (n) −0.23, (o) −0.13, (p) −0.03, (q) +0.07, and (r) +0.17 V vs Fc/Fc^+. Source: Reproduced with permission of Ref. [151] from Copyright © 2000, American Chemical Society. (C) Colors of various polypyrrole derivatives. Source: Reproduced with permission of Ref. [150] from Copyright © 2006 WILEY-VCH Verlag GmbH & Co. KGaA, Weinheim.

PPy12 was electropolymerized from 1,4-nitrophenyl-2,5-di(2-thienyl)-1*H*-pyrrole ($SNSNO_2$), which revealed transmittance changes of 13% and 23% for 400 and 860 nm [119]. Then they continued to develop a new soluble polymer PPy13 but which had poor optical contrast of 2.3% at 398 nm [120]. Another polymer PPy14 was synthesized by electrochemical oxidative polymerization of 1,6-bis(2,5-di[thiophen-2-yl]-1*H*-pyrrol-1-yl)hexane. It showed yellow in the fully reduced (0.0 V) and switched to a blue color in oxidized states (+1.0 V) [121]. A novel monomer, 4-(2,5-di[thiophen-2-yl]-1*H*-pyrrol-1-yl)benzenamine, was electrochemically polymerized in the presence of $LiClO_4$ and $NaClO_4$ (1 : 1) as the supporting electrolyte in acetonitrile. This polymer PPy15 had an electronic bandgap of 2.12 eV with a yellow color in the fully reduced form and a blue color in the fully oxidized form [122]. PPy16 exhibited a switching time of 0.7 seconds with the optical contrast of 8% at 410 nm and 41% at 1230 nm (Figure 6.36) [123].

Furthermore, A. Cihaner et al. developed a series of conducting polymers based on SNS, PPy17, PPy18, PPy19, and PPy20. They all exhibited yellow in the neutral state, green and blue in the intermediate state, and violet in the oxidized state.

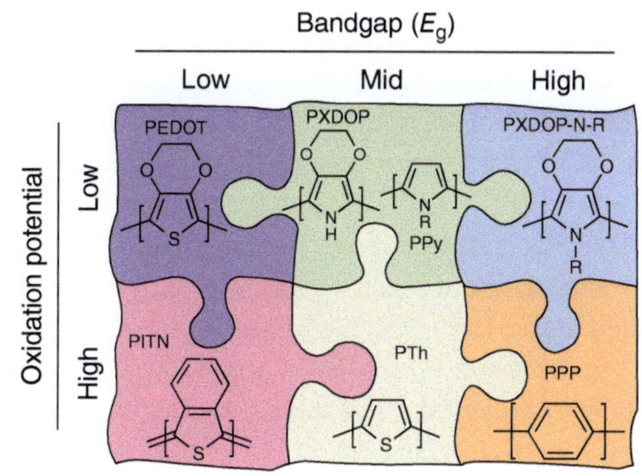

Figure 6.37 Completing the puzzle. Shown in blue, the N-substituted PXDOP family of polymers exhibit high bandgaps coupled with low oxidation potentials. Source: Reproduced with permission of Ref. [150] from Copyright © 2006 WILEY-VCH Verlag GmbH & Co. KGaA, Weinheim.

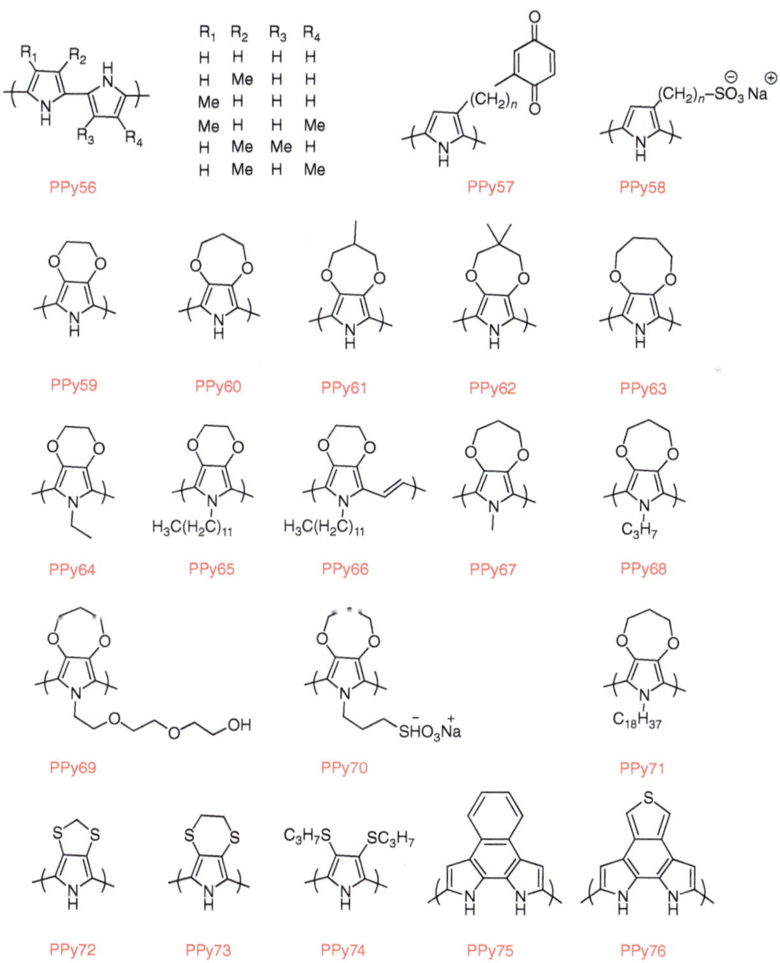

Figure 6.38 Chemical structure of 3,4-substituted polypyrrole derivatives.

These polymers are soluble in organic solvents and all are fluorescent (Figure 6.37) [117, 124, 125]. 4,4-difluoro-4-bora-3*a*,4*a*-diaza-*s*-indacene dye was incorporated to SNS at N-position, and the corresponding polymer PPy21 had a bandgap pf 2.9 eV. The transmittance changes between both states (neutral and oxidized) were found as 12.1% for 351 nm and 17.7% for 525 nm in the visible region as well as 46.2% for 1050 nm in the NIR region. Response time was found to be 1.0 seconds [126]. Moreover, azo dyes were incorporated into an PSNS polymer by A. Cihaner et al. to obtain two dithienylpyrrole-based polymers, PPy22 and PPy23 [118]. Both of their monomers exhibited photoisomerism properties under irradiation at 360 nm. Upon oxidation, the color changes from yellowish green to dark green and mustard color to green for PPy22 and PPy23, respectively. The polymer PPy24 synthesized via electropolymerization of 1-(1*H*-pyrrol1-yl)-2,5-di(thiophen-2-yl)-1*H*-pyrrole demonstrated EC behavior, e.g. orange in the neutral state and violet in the oxidized state. It also possessed a yellowish orange light-emitter property (Figure 6.38) [127].

Y. Shim and coworkers reported two SNS-based polymers, PPy25 and PPy26, which incorporated 3-pyridinyl and 1,10-phenanthroline moieties in the central pyrrole ring. PPy25 displayed brownish-yellow color in the neutral state and dark blue in the fully oxidized state, whereas PPy26 was greenish-yellow color in the neutral state and light blue in the fully oxidized state. Both exhibited a fast optical switching time within 1.0 seconds (Figure 6.39 and Table 6.2) [128]. X. Fu and coworkers reported two new polymers PPy27 and PPy28 bearing anthraquinone units, and they both showed similar EC behavior, yellow in the reduced state, gray in the neutral state, and blue in the oxidized state with the contrast of 18% and 16.5% at 780 nm [129]. They then developed three double dithienylpyrroles derivatives, PPy29, PPy30, and PPy31. Except their multicolor EC behaviors, they all exhibited different intensity emission bands at different wavelengths. Compared to PPy29 and PPy31, PPy30 was less stable to thermal effects due to the unshared electron pairs of the O atom (Figure 6.40) [130]. To enhance the stability against humidity and air, P. Camurlu et al. reported a new ferrocene-substituted EC polymer, PPy32 with a bandgap of 2.02 eV. It displayed yellow to blue coloration with poor contrast of 13.7% upon doping [131]. S. Tirkes and A. Cihaner and coworkers reported a new polymer bearing pyrene appendages at N-position of pyrrole. PPy33 changed from yellowish orange to greenish-yellow, to green/blue, and finally to blue with a response time within 1.0 seconds [132]. Another new polymer PPy34 containing SNS and TPA was developed by C. Li and coworkers. Under the applied potential from 0 to 1.4 V,

Figure 6.39 UV/vis–NIR transmittance spectra for the neutral (a), intermediate (b), and oxidized (c) states of four PPy71 films of increasing thickness, spray-cast onto ITO glass. The polymer films were switched in a cuvette with a platinum wire counter electrode and a Ag/Ag^{+} reference electrode, at potential of −0.25, 0, and 0.4 V, respectively. Photographs show the film of 70 nm (d) and 343 nm (e) thickness at the voltages indicated. Source: Reproduced with permission of Ref. [155] from Copyright © The Royal Society of Chemistry 2012.

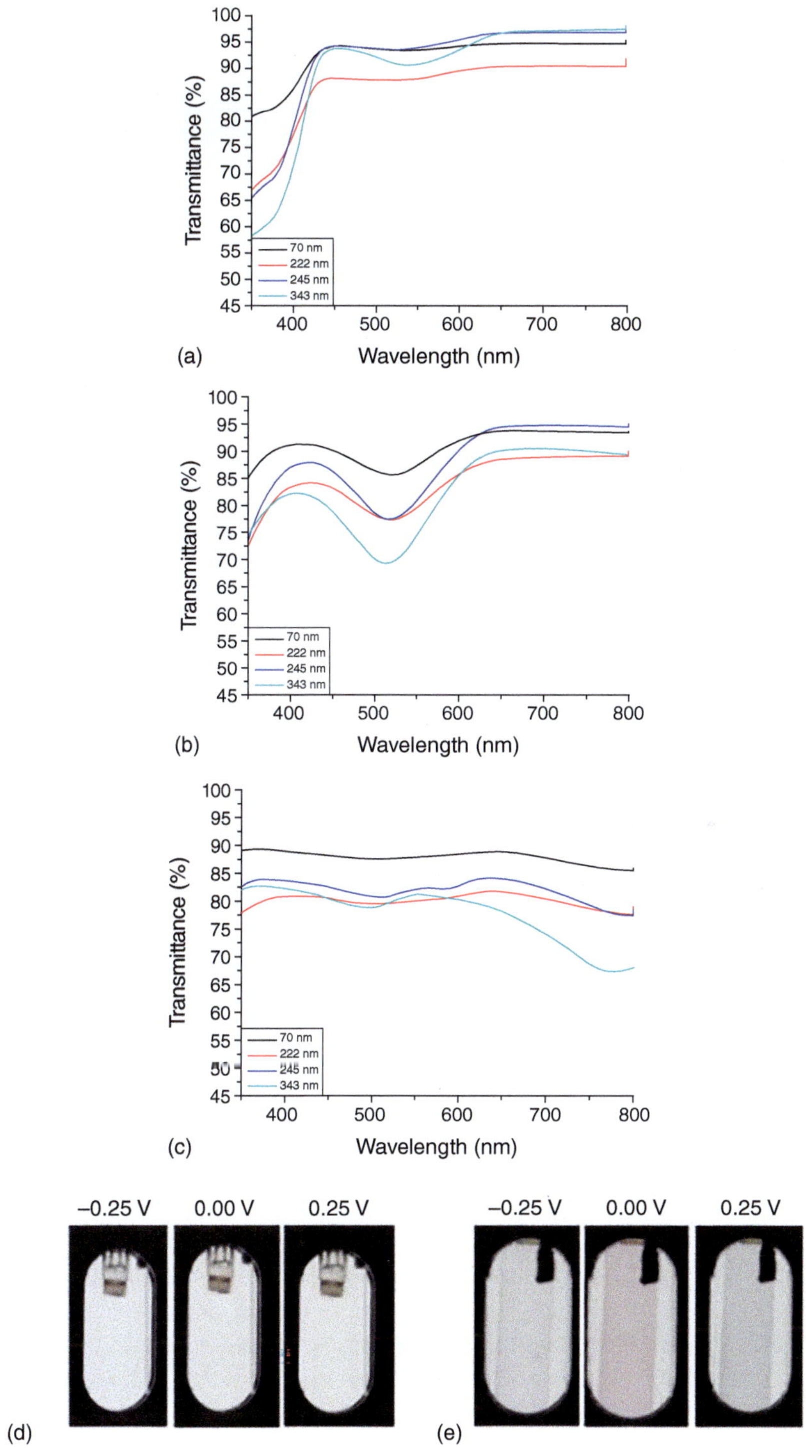
Transmittance (%)
Wavelength (nm)
70 nm
222 nm
245 nm
343 nm
(a)
(b)
(c)
−0.25 V
0.00 V
0.25 V
(d)
(e)

Table 6.2 Electrochromic properties of 3,4-substituted polypyrrole derivatives.

Materials	λ_{abs} (nm)	E_g (eV)	ΔT (%)	Reduced color	Neutral color	Fully oxidized color
PPy59	537	2.0	59		Red	Sky blue
PPy60	522	2.2	70	Orange	Brown	Gray-blue
PPy61	530	2.2	51	Orange	Red-brown	Light-blue
PPy62	534	2.2	76	Orange	Red-brown	Light-blue
PPy63	533	2.2	61	Orange		Blue-gray
PPy64		2.1		Green	Violet	Gray
PPy65	312	3.26				
PPy66	659	1.57			Dark blue-black	Yellow-greenish
PPy67	476	3.0		Deep purple	Dark green	Blue
PPy68	306	3.4		Colorless	Pink	Gray-blue
PPy69	306	3.4		Colorless	Pink	Blue
PPy70	340	2.9		Colorless	Pink	Gray-blue
PPy71	350	2.96		Colorless	Pink	Slight gray
PPy72	254	1.8				
PPy73	232	2.1				
PPy74	240	2.3				
PPy75					Yellow	Black
PPy76					Yellow-mauve	Black
PPy77	408	1.25	21		Green	Brownish green
	745		12			
PPy78	400	1.0	10		Green	Gray
	1460		66			
PPy79	392	1.12	37		Green	Blue
PPy80	384	1.08	33		Green	Light blue
PPy81	577	1.6	30		Blue	Colorless
PPy82	1070	2.16	42.3		Yellowish-green	Blue

PPy34 exhibited reversible multicolor in the visible region (yellow, light green, magenta, and blue) as well as excellent EC properties in the NIR region with high contrast ratio ($\Delta T = 70.5\%$ in 1550 nm, $\Delta T = 67.9\%$ in 1310 nm). A very short response time was also achieved [133]. T. Wu and coworkers introduced indan and 1,3-benzodioxole groups to the central pyrrole ring of polymer backbone, which were denominated as PPy35 and PPy36. PPy35 showed yellowish-green in the neutral state and could be oxidized to blue color (Figure 6.41). Two oxygen atoms in the 1,3-benzodioxole group of PPy36 increased the electron-donating ability of the SNS unit, and PPy36 showed yellow and blue colors in neutral and highly oxidized states [134]. P. Camurlu and coworker developed a new

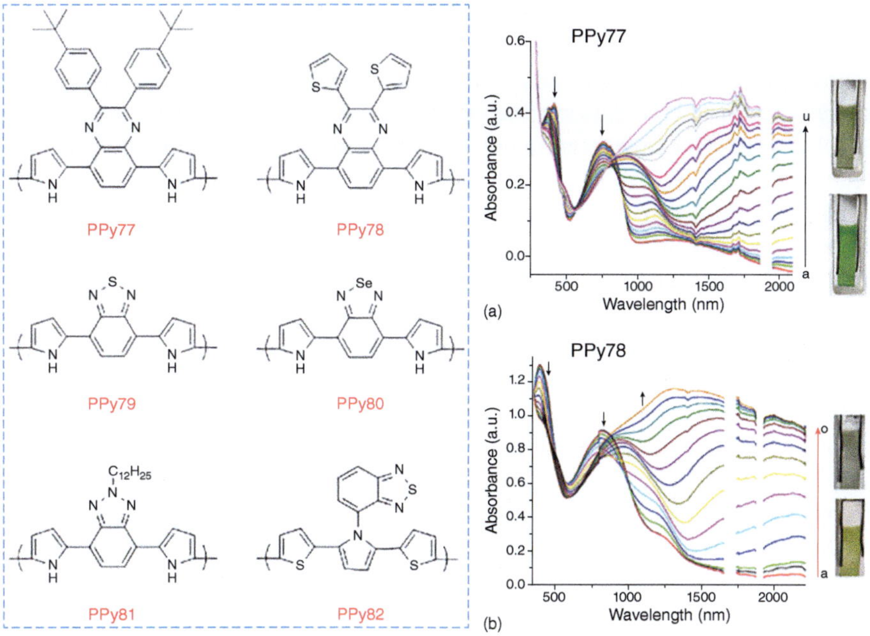

Figure 6.40 Chemical structure of D–A type polypyrrole derivatives and the electrochromic properties of (a) PPy77. Source: Reproduced with permission of Ref. [159] from Copyright © 2009 Elsevier B.V. (b) PPy78. Reproduced with permission of Ref. [158] from Copyright © 2009 Elsevier Ltd.

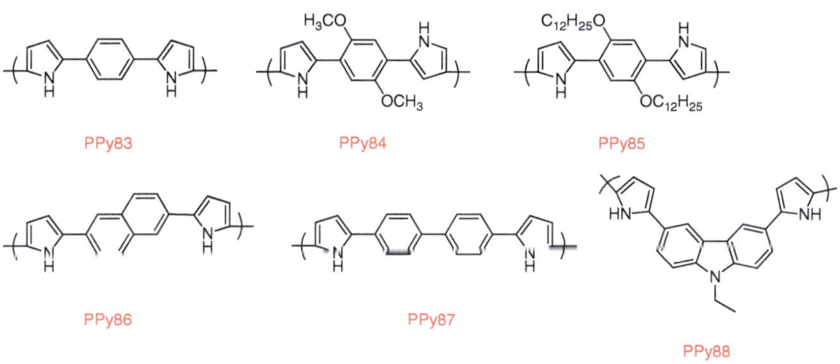

Figure 6.41 Chemical structure of polypyrrole-based terarylene systems.

SNS-based polymer PPy37 containing anthracene unit, which displayed green to blue coloration in 1.40 seconds with a coloration efficiency of 78 cm^2/C [135]. He also disclosed the construction of both p- and n-dopable SNS-based polymer, PPy38 with 1,8-naphthalimide moiety. PPy38 had a bandgap of 3.15 eV and displayed yellow-blue color variation upon doping [136]. Besides, another SNS derivative PPy39 was developed to investigate the effect of pendant pyrene on the optical and electronic properties. PPy39 had a bandgap of 3.36 eV and displayed light green in the neutral state, which could be switched to blue color in the oxidized state in less than 2.48 seconds [137]. S. Koyuncu and coworkers introduced perylenediimide (PDI) acceptor as pendant side chain to the N-position of SNS monomer, which was then deposited onto ITO-glass surface via electrochemical polymerization process. This polymer PPy40 exhibited ambipolar multi-EC behavior including purple and violet–red–khaki–blue colors in both anodic and cathodic regime only between −1.2 and 1.0 V. Fast response of 0.5 seconds, high contrast of 45% at 900 nm, and coloration efficiency of 254 cm^2/C were achieved. Due to the presence of phenylene spacer between SNS-donor and PDI-acceptor moiety, the electronic interaction at the excited state leads to efficient fluorescence quenching [138].

To improve the polymers' optical properties, M. Ak and H. Soyleyici and coworkers utilized hydrazine instead of amine to obtain a new SNS-based polymer PPy41 with the lowest bandgap of 1.99 eV compared with the other SNS derivatives. The color of PPy41 revealed three different hues: blue (1.0 eV), greenish (0.4 V), and reddish orange (−0.2 eV). The ΔT and switching time were 23% and 2.5 seconds for 430 nm and 42% and 2.5 seconds for 875 nm, respectively [139]. To further enhance the solubility of conductive polymers, M. Ak and coworkers reported a novel polymer in which amid-substituted 2,5-dithienylpyrrole moiety incorporated with alkoxy spacer group. This polymer PPy42 revealed optical contrast of 60% at 900 nm with 1.5 seconds of the switching time [140]. A similar amide-substituted SNS-based polymer PPy43 was designed, showing the contrast of <30% at 430 nm and high coloration efficiency of 277.77 cm^2/C at 900 nm [141]. Besides, terephthalamide was utilized to synthesize double-sided, symmetric amide substituted polymer PPy44 containing SNS moiety. Thus, PPy44 has a superstructure that provides three dimensional electrical conductivity. This polymer was greenish-yellow and became blue color when the potential increased to 1.7 V. The transmittance changes were calculated as 40% at 430 nm and 87% at 850 nm [142]. Moreover, T. Soganci also synthesized another double-sided symmetric amide-substituted SNS-based polymer PPy45. It turned from yellowish to blue as the applied potential increased with a high contrast of 54% at 435 nm [143]. Thus, similar EC phenomenon could be observed for amide-substituted SNS-based polymers.

Given that triarylamine compounds have attracted much interest as EC materials because of their high mobilities and low ionization potentials, H. Niu and coworkers reported a series of polymers, PPy46, PPy47, PPy48, PPy49, and PPy50, consisting both SNS and different triarylamine units. They all had an approximate optical contrast and became from yellow to dusty blue to violet. But the switching speed of PPy46, PPy47 and PPy48 are faster than that of PPy49 and PPy50, which may be attributed to their more smaller steric hindrance [144].

To realize multi-EC properties, B. Hu and L. Jin and coworkers integrated SNS units into carbazole-EDOT derivative to obtain the asymmetric structure compounds, which were electropolymerized to form two polymers, PPy51 and PPy52. PPy51 showed light yellow and PPy52 showed orange at 0 V. As the potentials increase, both PPy51 and PPy52 films showed an absorption wave around 800 nm (0.4–0.9 V), which can be attributed to oxidation of carbazole-phenyl-pyrrole units of polymer, and the color of PPy51 changed to light green (0.8 V), while yellow-green

(0.6 V) and green (0.9 V) can be seen for PPy52. With the potentials further increasing, the absorption wave in NIR region increased obviously, PPy51 and PPy52 films revealing light purple and blue color, respectively [145]. They also reported three novel SNS-based polymers, PPy53, PPy54, and PPy55 including phenothiazine, phenoxazine and 9(10*H*)-acridone, respectively. All the polymer films could show multicolor change, especially for PPy53, which displayed yellow-green (0 V), green (0.8 V), and purple red color (1.2 V) [146].

6.3.3.2 3- and 3,4-Substituted Polypyrroles

Substitution through 3- and 4-positions of pyrroles plays an important role on the suppression of the undesirable α-β couplings during polymerization, which would result in conjugation breaks and a low degree of stability of polymer films. In order to ensure the polymerization through 2- and 5-positions, it is significant to modify at the 3- and 4-positions of pyrroles to gain a film with fewer defects.

Only a few reports on poly(3-alkylpyrrole)s have been published, such as poly(methyl substituted bipyrrole)s (PPy56) [147], poly(3-hydroquinonylpyrrole) (PPy57) [148], and poly(33 alkylsulfonate pyrrole)s (PPy58) [149]. These polymers could achieve effective bandgap control. Methyl-substituted bipyrroles could display a transmissive-yellow to gray-blue color transition upon oxidation. Substitution at the 3-position has a less pronounced effect on the oxidation potential and conductivity of the polymers [148].

Since the demonstration of the poly(3,4-alkylenedioxythiophene) (PXDOT) family, breakthrough has been made to develop a variety of PXDOT derivatives with low bandgap and high stability. PEDOT showed transparent to dark blue upon oxidation and had a bandgap of 1.6 eV, significantly lower than polythiophene. The decline in the bandgap of the polymer was attributed to the tendency of the oxygen atoms to donate electron density, which results in an increase in HOMO of the p-system and leads to a decline in the bandgap. Encouraged by these researches, a series of easily oxidized PPy derivatives with the attachment of alkylenedioxy bridge to the 3,4-positions of pyrroles was born. For instance, Reynold and coworkers reported poly(3,4-ethylenedioxypyrrole) (PEDOP, PPy59) with a contrast of 59% between neutral red and oxidized sky blue [105, 150]. It possessed a bandgap of 2.0 eV. To investigate the effect of different ring size from six-membered to seven- and eight-membered ring in pyrrole derivatives, poly(3,4-propylenedioxypyrrole) (PProDOP, PPy60), MePProDOP (PPy61) and Me_2PProDop (PPy62) were developed. They all had a bandgap of 2.2 eV and were switched from neutral orange to red-brown and finally to a light blue [150]. It can be seen that a striking number of color states could be achieved by making only minimal changes to the polymer structure. PEDOP (PPy59) is a red neutral polymer, whereas PProDOP (PPy60) is orange. Poly(3,4-butenyldioxypyrrole) (PBuDOP, PPy 63) with eight-membered ring in pyrrole structure can also exhibit the multicolored character [151].

Furthermore, *N*-alkyl-substituted PEDOP derivatives PPy64 containing the 3,4-dioxysubstitution structural fragment and electron-rich pyrrole ring was synthesized by A. Kraft et al. This polymer PPy64 had a low bandgap of 2.1 eV, which can be switched from green (reduced state) to violet (intermediate state) to gray (oxidized state) [152]. Kim et al. introduced a vinyl group into the polymer backbone to realize polymer poly(1-dodecyl-3,4-ethylenedioxy-2,5-pyrrolylene vinylene) (PPy66) with a lower bandgap of 1.57 eV, compared to 3.26 eV of PPy65. PPy66 presented dark blue-black in the neutral state and became transparent light yellow-greenish when doped [153].

N-substitution of PProDOP (PPy60) with several alkyl chains has been proven a method to increase the bandgap up to 3.0 eV, resulted from the presence of side chains inducing a twist of the conjugated backbone and the decrease of π-conjugation. Because of the high bandgap, these polymers are colorless in their neutral state. As the polymers were gradually oxidized, absorption processes attributed to polarons and bipolarons emerged into the visible region of the spectrum, giving the polymer color. For instance, PPy67 and PPy68 revealed both higher bandgap compared to PPy60 due to the distortion of the polymer backbone, giving rise to a decline in the effective conjugation [154]. Besides, PPy69 and PPy70 with high bandgap displayed multicolored electrochromism switched from neutral colorless state to blue state [154]. Therefore, N-alkylated PXDOT derivatives possess high bandgaps and high conductivity, making them strong candidates as anodically coloring materials in ECDs. Another polymer PPy71 with high solubility exhibited colorless in the neutral state. With the increasing potential, the polymer's transparency decreased and it showed faint pink [155].

As a direct analog of the PEDOP, the synthesis and properties of poly(3,4-ethylene-dithiopyrrole) (PPy72) and its analogs (PPy73 and PPy74) were reported. They have a bandgap of 1.8, 2.1, and 2.3 eV, respectively, which were lower than PPy and comparable to that of PPy59. From the in-situ conductivity measurements, these polymers can be switched between a neutral nonconductive and an oxidized conductive state [156]. T. Swager and coworkers reported a new β-linked pyrrole monomer, which could lead to the incorporation of polycyclic aromatic residues into a polymer backbone through electrochemical polymerization. This polymer PPy75 was yellow in the neutral state and oxidized black owing to its significant absorption in the visible region. Another analog PPy76 showed yellowish-mauve and also gave a black film when it was doped [157].

6.3.3.3 Donor–Acceptor Approach

Pyrrole, also as a conventional donor unit, has been widely utilized to construct D–A-type polymers. Different acceptor units, such as quinoxaline [158, 159], benzothiadiazole [160], and benzoselenadiazole [160], have been introduced into the pyrrole-based backbone. Several such kinds of polymers have been reported and all possess narrow bandgaps with two distinct p–p* transition absorption bands at ~400 and ~750 nm in the neutral state. Therefore, these polymers exhibited green color and could be switched to a bluish state upon oxidation. For example, L. Toppare and coworkers reported a neutral green polymer, poly(2,3-bis(4-*tert*-butylphenyl)-5,8-di(1*H*-pyrrol-2yl) quinoxaline) (PPy77), with a narrow bandgap of 1.25 eV, which revealed two distinct absorption bands at 408 and 745 nm. PPy77 showed the optical contrast of 21% at 408 nm and 12% at 745 nm [159]. Furthermore, they developed polymer (PPy78)-bearing pyrrole and quinoxaline units to realize low oxidation potential of +0.70 V. This polymer has green color in the neutral state and gray color at the oxidized state [158]. Subsequently, polymers PPy79 and PPy80 were investigated. Though they had different electron-deficient component, both achieved low bandgap (1.12 and 1.08 eV) and high optical contrasts (37% at 392 nm and 33% at 384 nm), displaying neutral green color to blue color at the oxidized state [160]. 2-Dodecyl benzotriazole as acceptor unit was also introduced in the main chain of polypyrrole by L. Toppare and coworkers This polymer PPy81 was found to be both p- and n-dopable. It revealed 30% of the optical contrast at 577 nm between the reduced blue color and highly transmissive color in oxidized state with a short switching time of 0.5 seconds, making this polymer a nice candidate in EC display technologies [161]. In the discussion earlier, SNS derivatives attract wide attention because of their high-performance EC properties. Thus, benzothiadiazole unit was incorporated to the central pyrrole ring of SNS to construct a new D–A-type polymer PPy82. PPy82 revealed yellow in neutral states (0 V), green in intermediate states (1.2 V), and blue in highly oxidized states (1.4 V) [162]. Incorporation of EWGs to pyrrole-based polymer could lower the bandgap and tune the color.

6.3.3.4 Terarylene Systems

It has been proven that terarylene monomers that combine two heterocycles with a central aromatic unit could not only reduce the polymerization potential but also provide an opportunity to tune the optical and redox properties. Besides the series of 2,5-bis(2-thienyl)-1*H*-pyrroles (SNS)-based polymers and D–A type polymers, which both possess terarylene structure discussed earlier, arylene derivatized poly(bis[pyrrol-2-yl] arylene)s are another new collection of polypyrrole-based EC materials.

A family of poly(bis[pyrrol-2-yl] arylene)s that contain phenyl (PPy83), 2,5-dialkoxy substituted phenyls (PPy84 and PPy85), naphthyl (PPy86), and biphenyl (PPy87) were developed [163]. Electrochemical studies revealed that PPy83 and PPy84 exhibited bandgaps of 2.4 and 2.3 eV, respectively, which are smaller compared to those for polypyrrole (2.7 eV). Lowering of the bandgap was attributed to the increased structural regularity induced by the presence of arylene units. In another studies, carbazole-containing poly(bis[pyrrol-2-yl] arylene)s have a higher bandgap of 2.5 eV due to the incorporation of carbazole into the main chain limiting the effective conjugation length. This polymer exhibited similar EC properties to PPy [164]. Significantly lower polymerization potential could be achieved compared to that of directly polymerizing carbazole when introduced with the pyrrole units symmetrically into two sides.

6.4 Polycarbazole-Based Conjugated Electrochromic Polymers

6.4.1 Introduction

Another interesting multicolor system is the carbazole-based polymers. Carbazole is a tricyclic fused aromatic hydrocarbon containing two benzene rings fused to a pyrrole ring. Since the first demonstration of electrochemical polymerization of carbazole by Ambrose and Nelson in 1968 [165], the mechanism of electrochemical oxidative polymerization has been widely investigated, generally considered through the coupling of 3, 6, and 9 positions in Figure 6.42 [166]. At a potential of ca. +1.2 V vs SCE, carbazole can be oxidized to form the cation radical in a one-electron process. But due to the instability of cation radical, it tends to couple with another cation radical or parent molecule and form the more stable dicarbazyl, accompanied by the loss of two protons, one per carbazole unit. As a result of coupling, two different products are obtained in Figure 6.42, 9,9′-dicarbazyl (minor product) and 3,3′-dicarbazyl (main product). For 9,9′-dicarbazyl, it is difficult to oxidize because the oxidation process displays two peaks at high potential, approximately 1.45 and 1.8 V vs SCE. This oxidation is irreversible in Figure 6.42b and the products are still unknown [165, 167]. Besides, carbazole oxidation has been proven to be pH sensitive, and a low proton concentration is needed for the formation of 9,9′-dicarbazyl.

For 3,3′-dicarbazyl, it undergoes two successive, reversible, and one-electron oxidation steps to the cation radical and the quinoid-like dication. These processes could be observed in the CV with two peaks at approximately 0.8 and 1.1 V vs SCE. The chronoamperometric investigation of carbazole oxidation indicates that usually 2.5–2.8 electrons are involved in the whole process. After the electrochemical oxidation and then coupling reactions, oligomers (3,3′-dimers) will be coated on the electrode surface. However, there are two main disadvantages for the formation of polymer films. On the one hand, the low molecular weight oligomers with comparatively high solubility are difficult to form films on the electrode surface. With the same preparation process in MeCN, films obtained by Bargon et al. were very poor and easily cracked. It has been proven that oxidation of carbazole in protic acid medium can achieve films with high stability [168]. On the other hand, the high applied oxidation potential of direct electropolymerization of carbazole compared to its corresponding polymers would cause some degradation of polymer, resulting in the slow polymerization process. Thus, in order to deal with this problem, various carbazole derivatives have been developed to reduce the oxidation potential and promote the electrochemical polymerization to achieve high-quality films through the substitution at 3-, 6-, or 9-positions [169].

6.4.2 Electrochromic Properties of Polycarbazoles (PCARB)

In 1988, Berthold Schreck and coworkers first observed the phenomenon of electrochromism in polycarbazoles (PCARBs), which was electrochemically deposited in aqueous $HClO_4$. It has been revealed that reduced PCARB films show pale yellowish with low conductivity (>10^{-9} S/cm), whereas the oxidized films are green with enhanced conductivity (>10^{-3} S/cm) [168].

6.4.3 Electrochromic Properties of Polycarbazoles Derivatives

Carbazole is a material with a wide bandgap of 5.38 eV, which leads to the high oxidation potential. To decrease its bandgap, carbazole could be easily functionalized

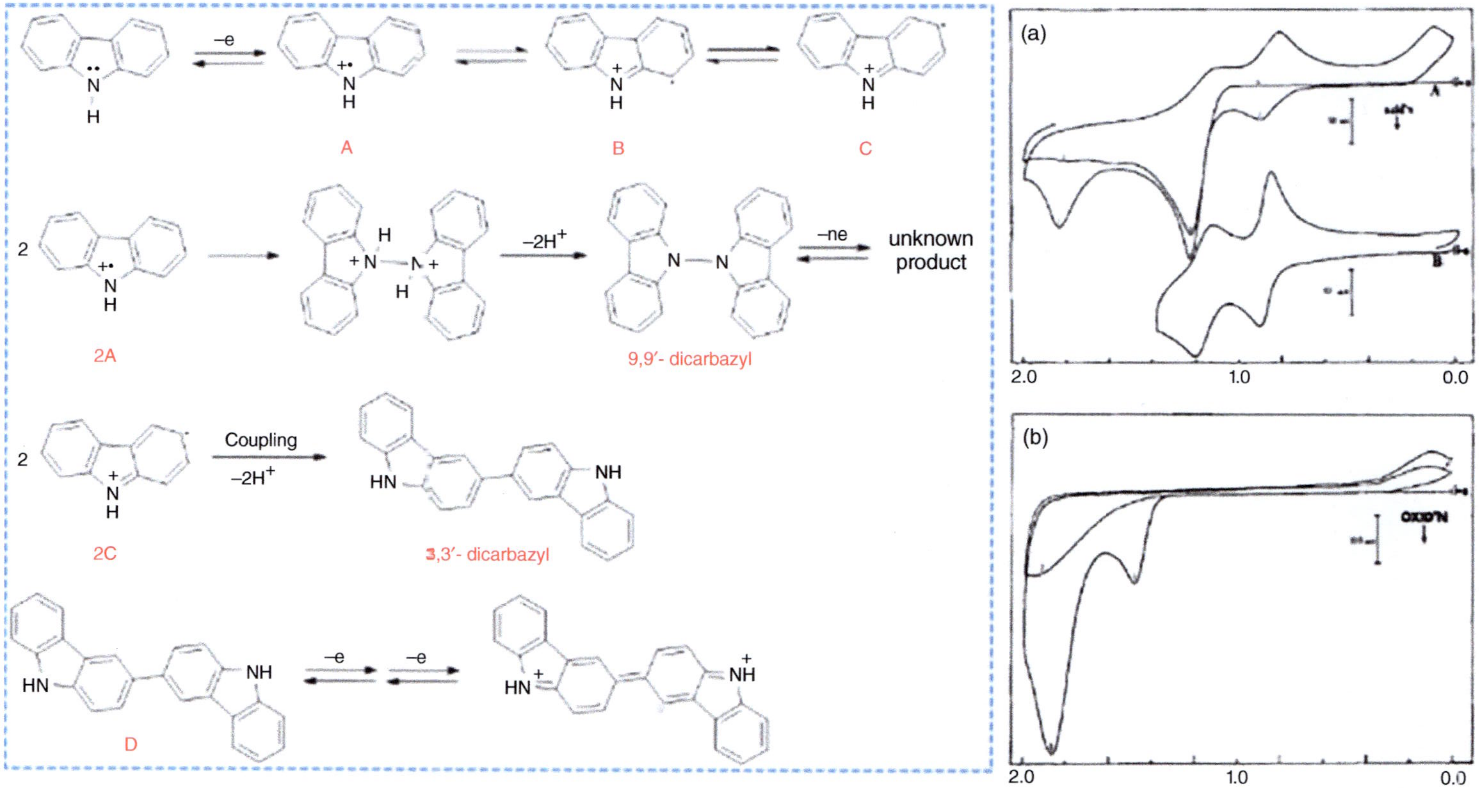

Figure 6.42 Mechanisms of electrochemical oxidation of carbazole. Cyclic voltammograms of (a) (A) 2.0×10^{-3} M carbazole in 0.1 M TEAP/MeCN and (B) saturated solution of 3,3′-dicarbazyl in 0.1 M TEAP/MeCN. (b) 1.0×10^{-3} M 9,9′-dicarbazyl in 0.1 M TEAP/MeCN. Abscissa is volts vs SCE. Source: Reproduced with permission of Ref. [165] from Copyright © 1968, ECS - The Electrochemical Society.

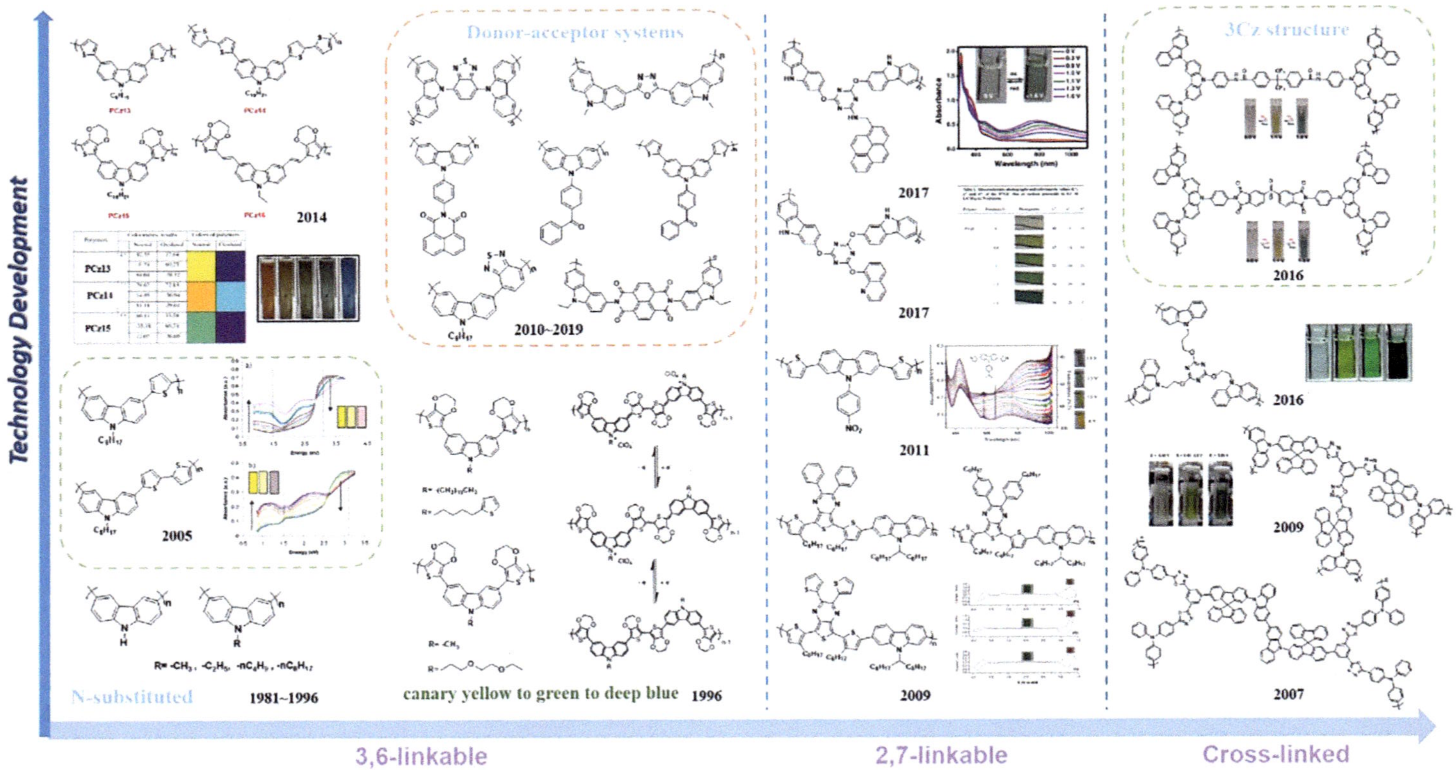

Figure 6.43 Technology development of polycarbazole derivatives.

at the 3- and 6-positions or 2- and 7-positions, and a wide variety of alkyl and aryl chains could be introduced to the nitrogen atom. It could be covalently linked into polymeric systems, both in the main chain as building blocks or in a side chain as pendant groups. The technology development of PCARB derivatives is shown in Figure 6.43.

6.4.3.1 Linear Polycarbazole Derivatives

Plenty of traditional carbazole-based polymers with multicolor changes have been designed during the past three decades. Similar to the PPy, the N-position can be easily substituted with a wide variety of functional groups. EC properties of poly-*N*-vinylcarbazoles were observed by Monvernay et al. [170] in the 1980s and studied by Chevrot and coworkers [171, 172] in the 1990s. They were colorless in the neutral state and green in the oxidized state. Besides, alkyl groups with different chain length (methyl, ethyl, *n*-butyl, and *n*-octyl) have been introduced to the nitrogen atom [173, 174], and the modified monomers were then electrodeposited at the 3- and 6-positions onto a conducting substrate. They displayed two stable color changes during the redox processes, colorless to green ($0.70\,V < E < 0.90\,V$ vs SCE) and green to blue ($1.00\,V < E < 1.3\,V$ vs SCE). But better solubility of oligomers in the electrochemical medium influenced the formation of oligomers of different conjugation and chain length. The EC properties of these N-substituted PCARB derivatives are similar to that of pure PCARBs. Alkyl groups with different chain length had little effect on their electrochromism.

It is well known that the bandgap could be decreased by the increase of conjugation on the polymer backbone. Reynolds and coworkers modified N-substituted carbazoles with two EDOT at its 3- and 6-positions. Two monomers, 3,6-bis[2-(3,4-ethylenedioxy)thienyl]-*N*-methylcarbazole (BEDOT-NMCz) and 3,6-bis[2-(3, 4-ethylenedioxy)thienyl]-*N*-ethoxyethoxyethylcarbazole (BEDOT-NGCz), were synthesized. Low oxidation potential of monomers was observed and both had good film-forming properties through electropolymerization. Their polymers PCz6 and PCz7 both exhibited two distinct redox processes with first redox potential at 0.2 V and the second at 0.5 V vs Ag/Ag^+. They gradually changed to green upon mildly oxidizing potential from transparent canary yellow in the reduced state and then turning deep blue at higher potentials. There was no significant difference of EC properties for these two polymers with the alkyl and glyme substituted at the nitrogen atom [175]. Subsequently, to investigate the function of different alkyl groups substituted at the nitrogen atom, a series of similar poly(EDOT-N-substituted carbazoles) have been investigated in Figure 6.44. Likewise, the fully reduced form of these polymers were canary yellow. Upon mild oxidation, the films became green and in the fully oxidized form, blue [164].

Therefore, polymers containing 3,6-linked carbazole moieties have been proven as an interesting multicolor system. Reynolds and coworker [176] described the oxidation process of 3,6-linked carbazole-containing polymers in Figure 6.44, which took poly(carbazole-*co*-phenylene) as example. He revealed that a 3,6-linked carbazole moiety in the main chain of polymers served as a conjugation break (Figure 6.45). An electron is removed from the conjugated backbone upon the initial

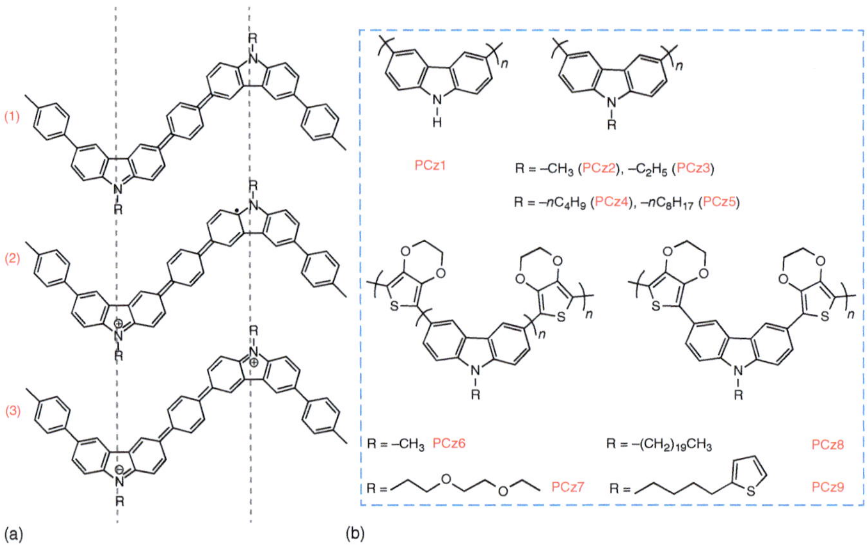

Figure 6.44 (a) A segment of poly(carbazole-co-phenylene) depicting (1) neutral state, (2) radical cation state, and (3) dication state. Source: Reproduced with permission of Ref. [176] from Copyright © 2005 American Chemical Society. (b) Chemical structure of polymers.

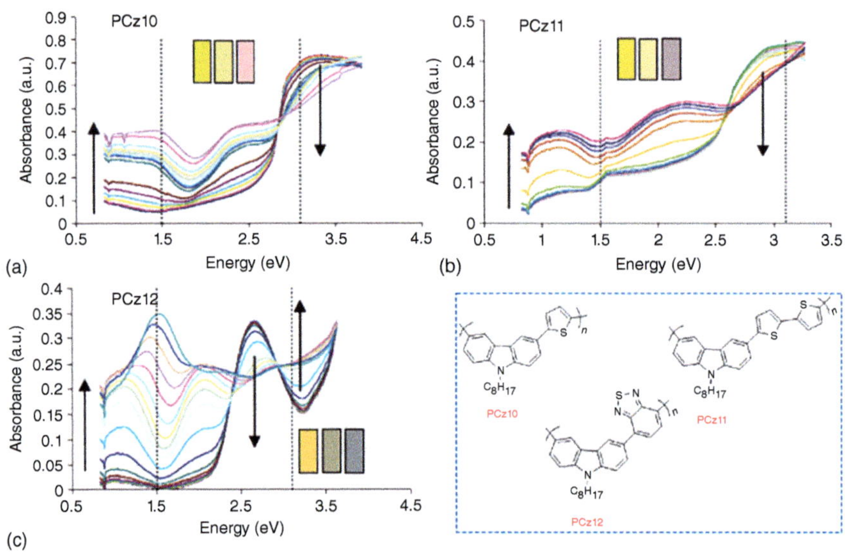

Figure 6.45 Spectroelectrochemical spectra and colors of (a) PCz6 (0.4–1.35 V vs Ag/Ag^{+}), (b) PCz7 (0.3–1.15 V vs Ag/Ag^{+}), and (c) PCz8 (0.65–1.5 V vs Ag/Ag^{+}). Potential steps are 50 mV. Source: Reproduced with permission of Ref. [176] from Copyright © 2005 American Chemical Society.

oxidation and a radical cation will be formed. Due to the break of conjugation, the radical cations are separated from each other and can't combine. Upon further oxidation, another electron is removed, forming dications. Thus, 3,6-linked PCARB derivatives possessed two distinct oxidation states leading to a three-color effect, with separate colors present for the neutral, radical cation, and dication states. For example, three new 3,6-linked carbazole-containing polymers with various comonomers used along with the carbazole unit were designed. After the chemical syntheses of these polymers, they were spray-cast onto ITO-glass slides, giving uniform films, which exhibited different EC properties. These thiophene-containing polymers, PCz10 and PCz11, showed very similar color in their neutral states, as both were bright yellow. When converted to the radical cation state, PCz10 took on a faint green color and PCz11 showed an orange hue. Further oxidized to higher potential, PCz10 presented a faint pink color and PCz11 became a rather neutral gray, whereas for PCz12 with D–A structure, it showed orange in the neutral state, green–gray as a radical cation, and a neutral gray in its most oxidized state. Thus these polymers showed various color in their different oxidation states including yellow, green, pink, and various grays [176]. The method of introducing various groups, p-type or n-type units, to the 3- and 6-positions of carbazole can expand the palette of available colors of their polymers (Figure 6.46).

3,6-Linked PCARB derivatives (PCz13, PCz14 and PCz15) were further investigated to explore the effect of different thiophene derivatives on the polymer backbone by P. Data et al. [177]. PCz13 was proved to be the best for electrochromism, presenting yellow in the undoped state and turning to purple when it was doped with ΔT = 44% of the contrast and 225.10 cm^2/C of the coloration efficiency. PCz15 with low bandgap of 1.68 eV displayed ΔT = 44% but lower coloration efficiency of 130.27 cm^2/C. PCz15 turned from neutral green to oxidized purple. Yonghong Hu and coworkers [178] attempted to insert the vinylene between 3,6-linked *N*-ethyl-carbazole as the inner unit and EDOT as the end unit. This vinylene-linked EDOT-carbazole monomer could be electrochemically oxidized at a very low onset potential of 0.41 V vs SCE (Figure 6.47). The polymer PCz16 showed multicolored electrochromism: brick-red in the reduced state, sky-blue in the oxidized state, and three intermediate colors, brown at 0.2 V, olive green at 0.4 V, and blue-green at 0.6 V vs SCE. Improved optical contrast (50% at 759 nm, 27% at 470 nm) and higher coloration efficiency (224.1 cm^2/C at 759 nm and 383.7 cm^2/C at 470 nm) were achieved compared to PCz15, indicating that the embedding vinylene could effectively tune the bandgap and EC properties (Figure 6.48).

Since the D–A method developed by E. E. Havinga and coworkers [179] has been proven as the most efficient way to tune the optical bandgap, conjugated polymers with carbazole as electron donor and various acceptor units have been an important focus in the field of EC research. For example, 1,8-naphalimide acceptor unit was linked to the nitrogen atom of carbazole, and the monomer was electropolymerized onto ITO-coated glass. The polymer PCz17 has the optical bandgap of 2.44 eV, showing yellowish green color at the neutral state and turned into dark green during the oxidation process. During the reduction process, anion radicals were formed on the carbonyl groups of the 1,8-naphthalimide moiety,

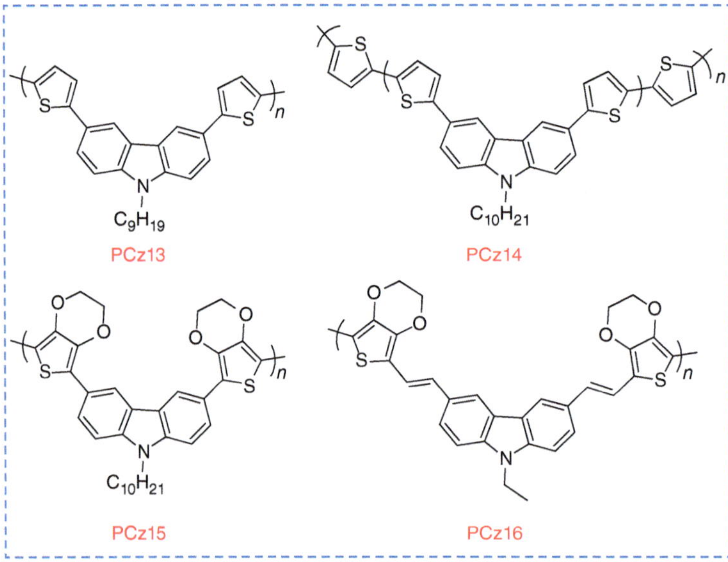

Polymers	Colorimetric results			Colors of polymers	
		Neutral	Oxidized	Neutral	Oxidized
PCz13	*L**	92.55	37.84		
	*a**	−1.79	60.27		
	*b**	84.04	−78.12		
PCz14	*L**	79.67	72.13		
	*a**	14.89	−36.94		
	*b**	81.18	−29.61		
PCz15	*L**	66.11	33.58		
	*a**	−35.18	66.71		
	*b**	12.07	−70.05		

(a)

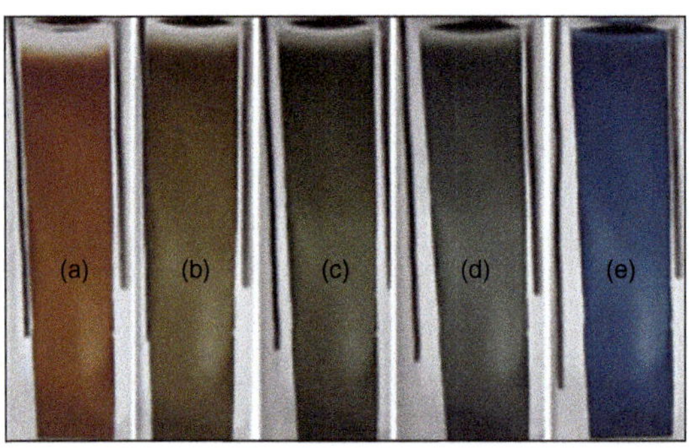

(b)

Figure 6.46 (a) Colorimetric data for polymers PCz13, PCz14 and PCz15. Source: Reproduced with permission of Ref. [177] from Copyright © 2013 Elsevier Ltd. (b) The electrochromic behavior of PCz16 film on ITO-coated glass in 0.1 M $LiClO_4$/ACN at applied potentials (*V*): (a) −0.4, (b) 0.2, (c) 0.4, (d) 0.6, and (e) 0.9 V. Source: Reproduced with permission of Ref. [178] from Copyright © 2014 Elsevier Ltd.

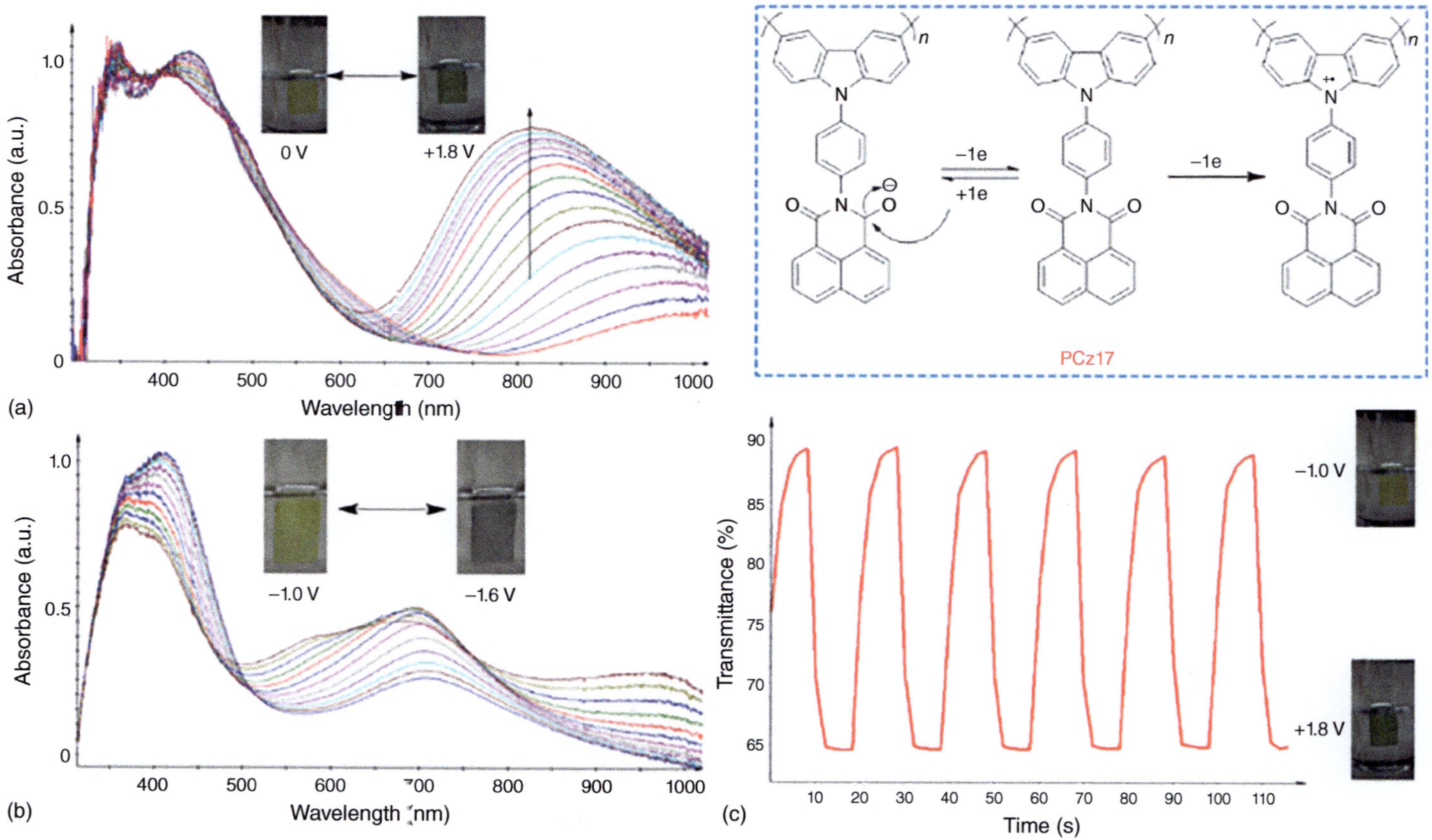

Figure 6.47 Electronic absorption spectra and the colors of polymer PCz17 on ITO-coated glass in 0.1 M $TBAPF_6$/MeCN from (a) 0 to +1.8 V and (b) −1.0 to −1.6 V. (c) Electrochromic switching optical absorbance monitored at 825 nm for PCz17 film between −1.0 and +1.8 V. Source: Reproduced with permission of Ref. [180] from Copyright © 2010 Elsevier Ltd.

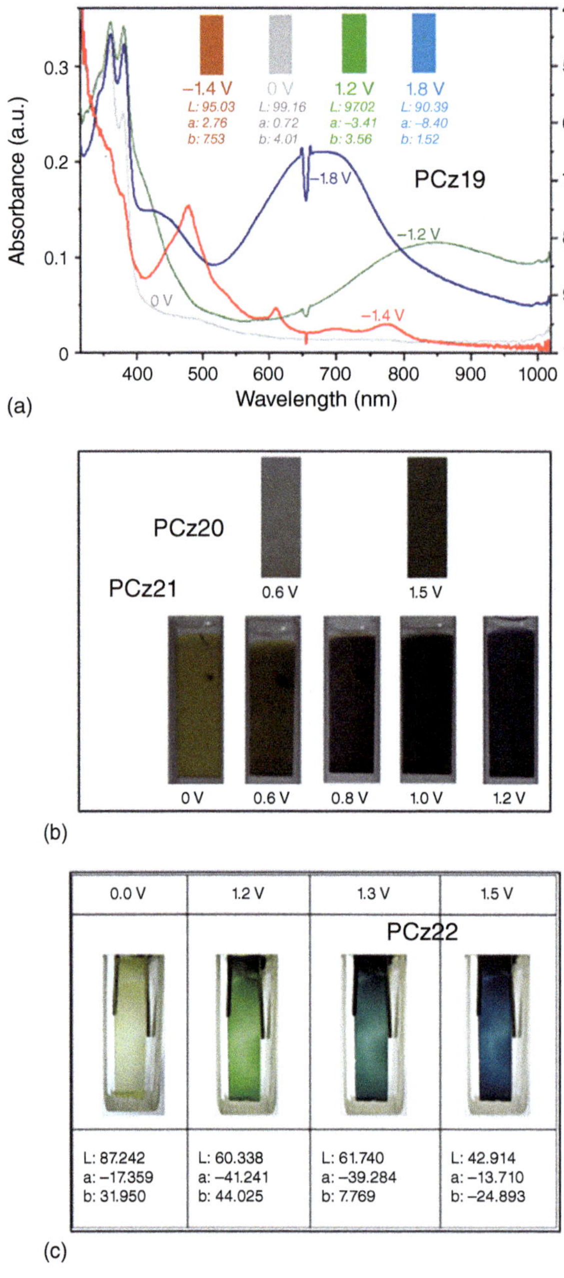

Figure 6.48 (a) Electronic absorption, transmittance behaviors, and colors of polymer PCz19 on ITO-coated glass at various applied potentials. Source: Reproduced with permission of Ref. from Copyright © 2012 Elsevier Ltd. (b) Multichromic behaviors of PCz20 and PCz21 films [183]. Reproduced with permission of Ref. [183] from Copyright © 2012 Elsevier B.V. (c) PCz22 film at various applied potentials [184]. Reproduced with permission of Ref. [184] from Copyright © 2015 Published by Elsevier B.V.

turning the polymers into blue [180]. The optical contrast was $\Delta T = 25\%$ monitored at 825 nm. E. Ozdemir and coworkers reported a new D–A carbazole derivative containing *N*-2-(2-ethylhexyl)-1,8-napthalimide as acceptor pendant, which was then electrochemically polymerized to obtain polymer PCz18. But the absorption bands did not change too much with the applied potential and there was no EC effect. The copolymer achieved by electrochemical copolymerization of this monomer and EDOT gave purple color, turning into turquoise upon oxidation [181].

Moreover, a strong EWG, naphthalene bisimide, has been also utilized to functionalize carbazoles to achieve a new D–A system by F. B. Koyuncu and coworkers [182], and then it was polymerized electrochemically onto ITO-coated glass surface to give the stable films. The polymer PCz19 film exhibited reversible ambipolar redox behavior of both cathodic and anodic regime. Thus, the transparent film in the neutral state converted to green and turquoise resulting from the association of polaronic and bipolaronic species from carbazolylium cation radicals. During the reduction process, the transparent color turned into reddish brown. Besides, another n-type material, phenyl-methanone, has been also introduced into the N-position of carbazole to decrease its bandgap [183]. Two polymers, PCz20 and PCz21, were synthesized through electropolymerization from their corresponding monomers in TBAP/CAN solution. However, poor optical contrast (25% and 41%) and switching time (three and two seconds) were found for both polymers. Low bandgap of 2.3 eV for PCz21 could be realized. Y. A. Udum et al. [184] synthesized a D–A-type polymer, PCz22 bearing oxadiazole as the A moiety. The polymer films had no sharp absorption peak in the visible range in the neutral state with a bandgap of 2.2 eV, giving a yellow color. Applying a potential of 1.2 V, the polymer films displayed two absorption bands at 480 and 720 nm, giving a green color, and can be further switched to a blue color due to increased absorption at 600 nm.

Interestingly, S. Hsiao et al. reported two series of electroactive polyamides (PCz23 and PCz24) with pendant carbazole groups. They all showed well-defined and reversible redox couples upon electrochemical oxidation, together with a strong color change from a colorless neutral form to blue or green oxidized forms. PCz24 series polymers exhibited better redox stability and EC properties as compared to the corresponding analogs PCz23 without *tert*-butyl substituents on the active sites of carbazole unit [185].

Compared to polymers based on 3,6-linked carbazole, 2,7-linked PCARBs generally have lower bandgap due to the longer effective conjugation length, which allows polymeric materials that are absorptive in the visible region [186]. However, the synthesis of poly(2,7-carbazole)s was not straightforward due to the unreactive 2 and 7 positions of carbazole. Several approaches have been developed to overcome the barrier to obtain various poly(2,7-carbazole) derivatives.

Leclerc and coworkers [81] reported a series of new poly(2,7-carbazole) derivatives, PCz25, PCz26, and PCz27 in Figure 6.49. They all reflected green light in their neutral state and showed transmissive brown when oxidized, which were compatible for adaptive camouflage for military assets. In another study, E. Ozdemir and coworkers developed a new poly-2,7-di-2-thienyl-9*H*-carbazole derivative PCz28 bearing nitro ($-NO_2$) subunit. Because of electron-withdrawing effect of $-NO_2$

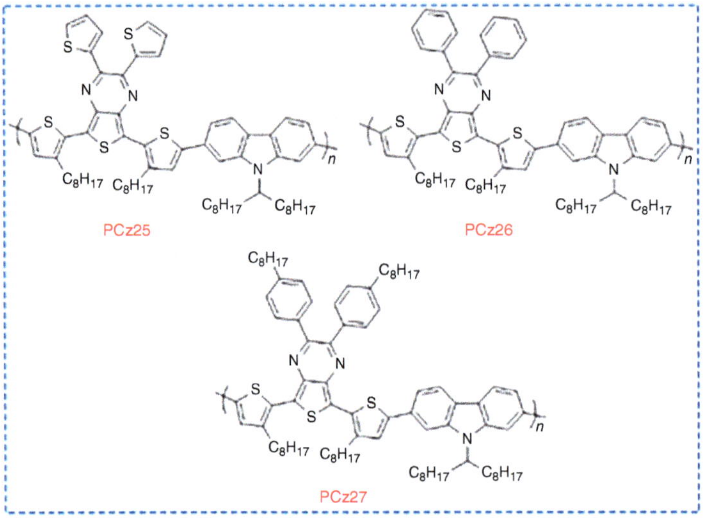

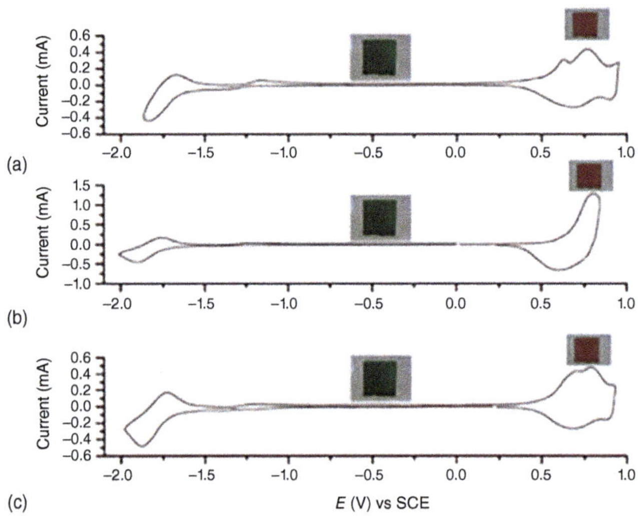

Figure 6.49 Cyclic voltammograms and color changes of (a) PCz25, (b) PCz26, and (c) PCz27. Source: Reproduced with permission of Ref. [81] from Copyright © 2009 American Chemical Society.

moiety, the oxidation potential of the monomer is higher than that of its analogs without $-NO_2$ moiety. PCz28 could exhibit a multi-EC behavior, presenting orange in neutral state and converting to green, turquoise, and dark blue with the increasing positive potentials [187]. Moreover, triazine moiety with electron-withdrawing ability was also introduced into poly(2,7-carbazole) derivatives. And thus two polymers PCz29 with 8-hydroxyquinoline and PCz30 with 1-pyrenemethylamine were obtained. PCz29 revealed color changes between transparent and dark green in the reduced and oxidized states, respectively. Switching time and optical contrast were measured as 2.1 seconds and 58% at 680 nm [188]. Differently, PCz30 changed from a transparent reduced state at 0 V to a blue oxidized state at +1.6 V. Switching time and optical contrast were calculated as 3.5 seconds and 21% [189].

6.4.3.2 Cross-Linked Polycarbazoles Derivatives

It is well known that carbazole moiety can realize efficient cross-linkage through electropolymerization. Various cross-linked PCARBs derivatives have been widely designed and investigated for the EC application. Similar to the linear PCARB derivatives, D–A interactions are the effective method to tune the EC properties. F. Fungo and coworkers [190] developed a novel D–A bipolar polymer PCz31 with 9.9′-spirobifuorene as the core, 1,3,4-oxadiazoles as the A moieties, and TPA and carbazole as the D groups. Electropolymerization of its monomer was carried out through the TPA moieties at low oxidation potentials (<0.70 V) and cross-linking by the 3- and 6-positions of carbazole groups at higher potentials (> 0.70 V). This polymer film exhibited canary yellow at 0 V, organic at 0.6 V, and green at 0.9 V with the high contrast of 71.9%. F. Fungo and coworkers [191] continued to develop a new cross-linked EC polymer PCz32 from a bipolar starburst monomer. It showed reversible electrochemical processes accompanied with stable color changes from transparent to green to light blue. Because of electrochemically homogeneity, PCz32 exhibited good conductivity and strong adherence to the electrode.

In another study, J. Zhao and coworkers [192] modified BTD with two carbazoles at its 2- and 5-positions, giving the polymer PCz33 via electrochemical polymerization of the monomer. Compared to the linear polymer PCz12 containing benzothiadiazole and N-substituted carbazole in the main chain, PCz33 exhibited an absorption band at 470 nm with a bandgap of 2.26 eV due to the π–π^* transition, giving a neutral yellow color. As the potential was increased, the absorption bands at 420 and 800 nm increased gradually, leading to a green color. Upon further oxidation, films turned turquoise blue.

Besides D–A type polymers, other cross-linked polymers with multiple color have been investigated. For example, two branched electroactive monomers with 2,5-di(2-thienyl)-1*H*-pyrrole substituted to the 3- and 6-position of carbazole were directly coated onto the ITO surface by using electrochemical polymerization process [193]. The polymer PCz35 with longer side alkyl chains exhibited superior EC properties compared to that of PCz34. PCz35 showed high contrast ratio of 68% at 850 nm, switching time of 0.8 seconds, and high coloration efficiency of 352 cm^2/C. More importantly, this polymer possessed high stability, retaining the properties by 97.1% after 5000 cycles owing to the cross-linked structure. Thus, longer alkyl chain at the N-position of carbazoles may greatly influence their properties.

In another study, a new phosphazene/carbazole hybride dendron was synthesized and then directly coated onto ITO-glass surface to provide a dendritic polymer network, PCz36. PCz36 presented colorless in the neutral state and converted to green after applied positive potentials with 51% of the optical contrast. Neutral colorless ECD was constructed in the sandwich configuration of ITO/PCz36/gel electrolyte/PEDOT/ITO, revealing a short response time of 0.5 seconds and high coloration efficiency of 981 cm^2/C [194]. M. Ak and coworkers reported another novel polymer PCz37 through electrochemical polymerization from a triazine-based,

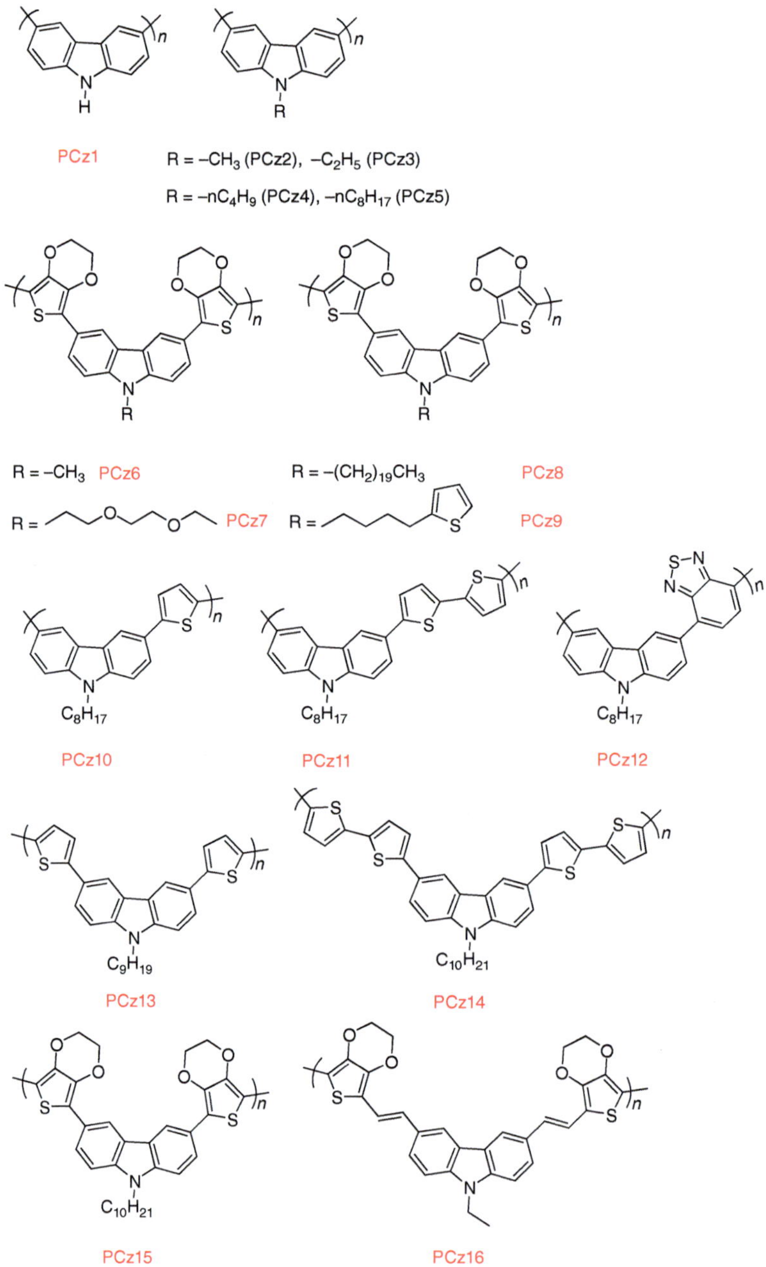

Figure 6.50 Chemical structure of linear polycarbazole derivatives.

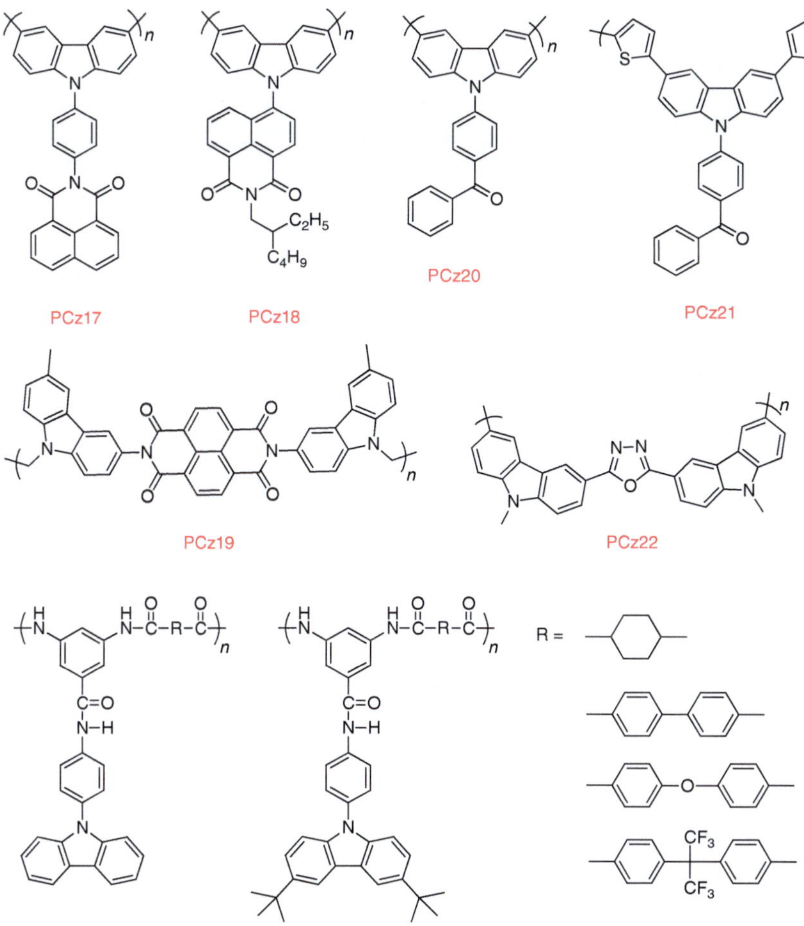

Figure 6.50 (*Continued*)

star-shape monomer, which contains 2,4,6-trichloro-1,3,5-triazine as the core and 2-(9*H*-carbazol-9yl)ethanol as the arms. PCz37 had turquoise color in the oxidized state and was transparent in the neutral state. Optical contrast and switching time were calculated as 70.67% and 2 seconds, respectively (Figures 6.50–6.53) [195].

B. Carbas et al. introduced tetraphenylethylene moiety into carbazole-based polymer, yielding the polymer PCz38, which displayed transparent to military green color in 1.8 seconds with an optical contrast of 23% at 770 nm [196]. F. Fungo and coworkers developed a series of peripherally carbazole functionalized dendrimeric starburst monomers, which were then electropolymerized onto ITO-glass surface to generate polymer films PCz39, PCz40, and PCz41. Seen in Figure 6.54, PCz39 showed transparent with one absorption peak at 346 nm in the neutral state. As the potential increased, it converted to pale yellow and then green. With regard to PCz40, which contained electroactive central core TPA moiety, similar behavior

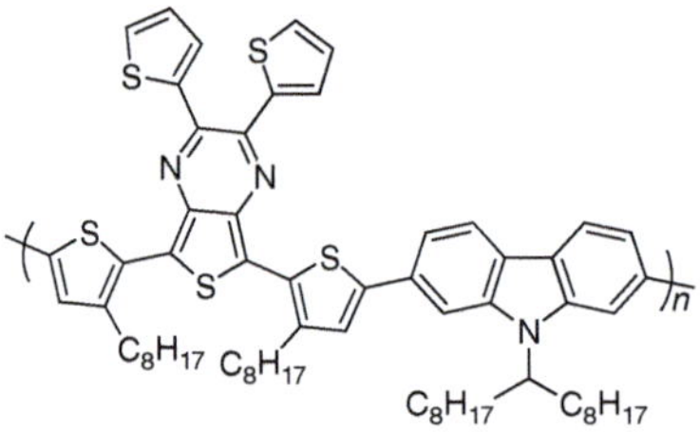

PCz25

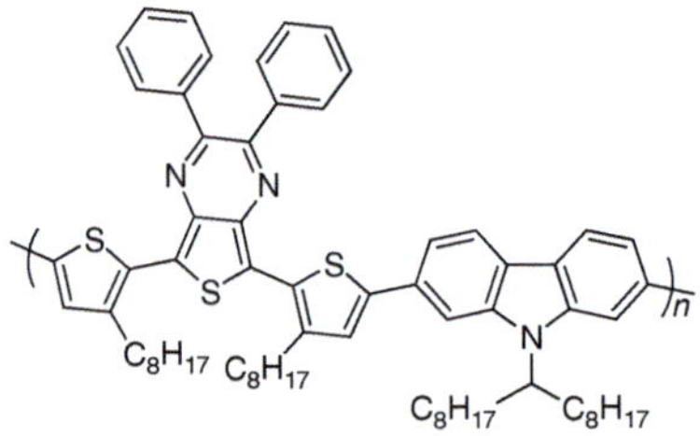

PCz26

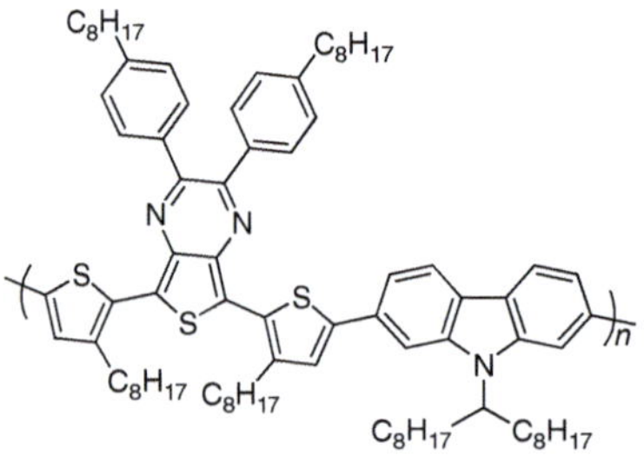

PCz27

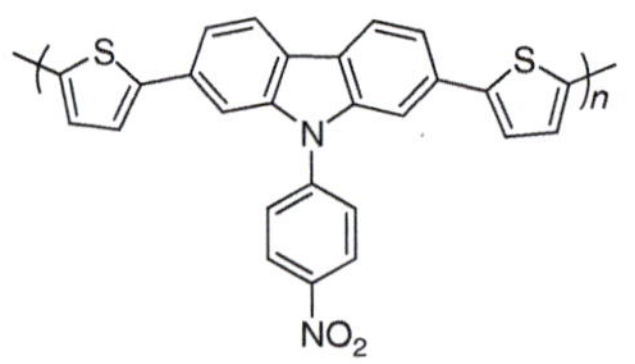

PCz28

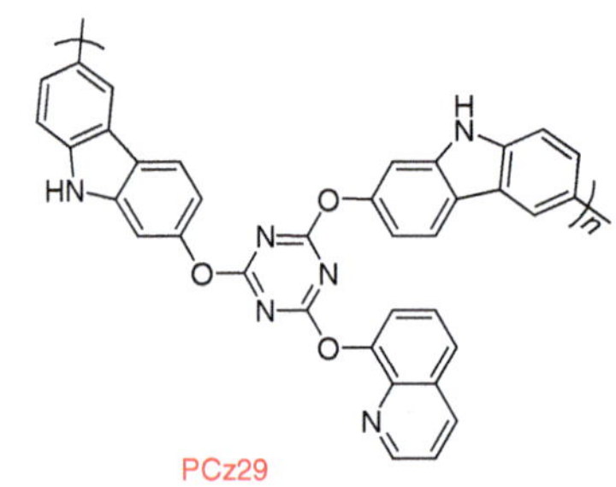

PCz29

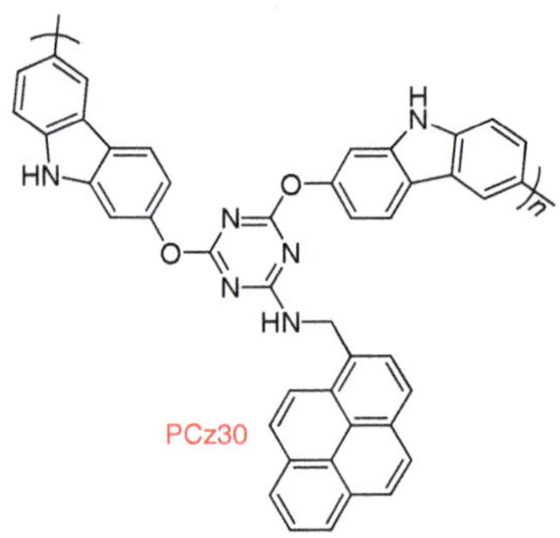

PCz30

Figure 6.50 *(Continued)*

was obtained, but the observed changes in coloration were less intense than that of PCz39. This study also indicated that PCz41 could not remain on the electrode surface because of the high solubility of its dimers, preventing the possibility of studying the spectroelectrochemical properties [197]. Besides, C. Kuo and Lee reported poly(1,3,5-tris[*N*-carbazolyl]benzene) (PCz42), and three copolymers based on 1,3,5-tris(*N*-carbazolyl)benzene (tnCz) and 2,2′-bithiophene (bTp) were electrochemically synthesized based on different molar ratio. The polymer P(tnCz1-bTp2) film with a tnCz/bTp feed molar ratio of 1/2 showed four colors, light orange at 0 V, yellowish-orange at 0.7 V, yellowish-green at 0.8 V, and blue at 1.1 V from the neutral state and oxidized states. UV–visible absorption spectra of PCz42 and P(tnCz1-bTp2) on ITO electrodes as function of applied potentials are shown in Figure 6.55 [198].

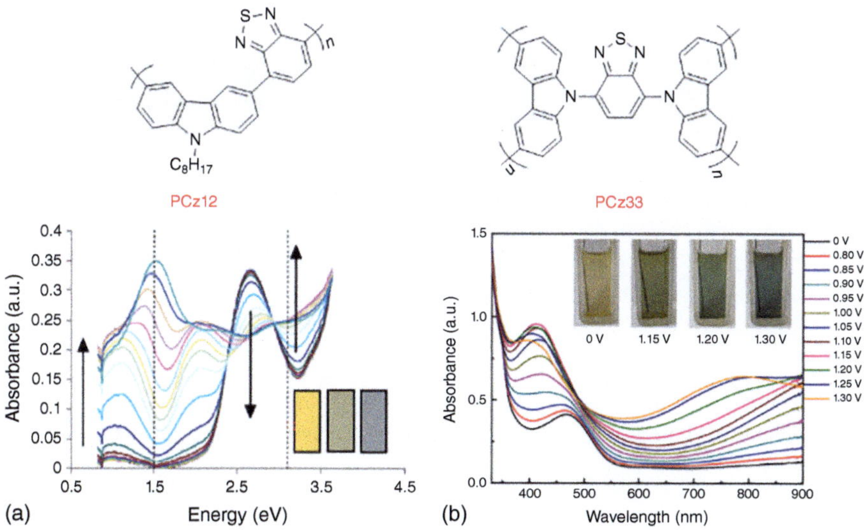

Figure 6.51 Spectroelectrochemical spectra and colors of (a) PCz12 and (b) PCz33 at different potentials. Source: Reproduced with permission of Ref. [176] from Copyright © 2005 American Chemical Society. Reproduced with permission of Ref. [192] from Copyright © 2012 Elsevier B.V.

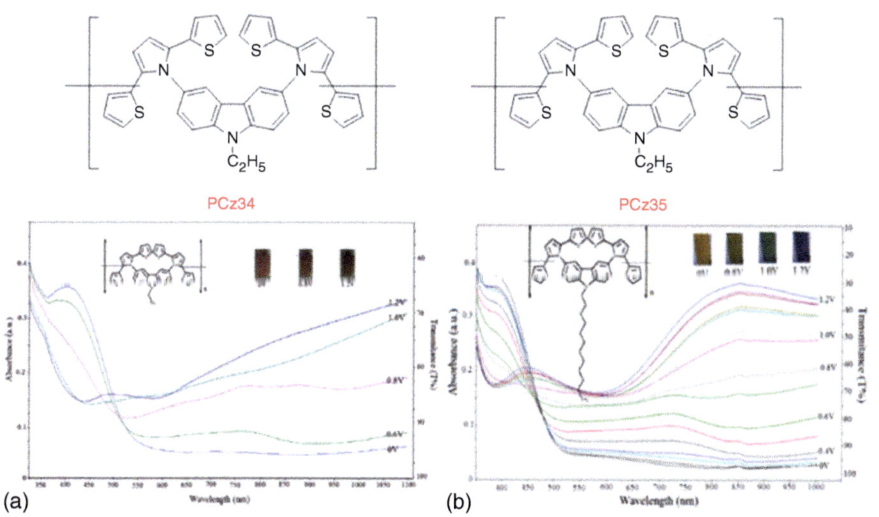

Figure 6.52 Spectroelectrochemical spectra and color changes of (a) PCz34 and (b) PCz35 onto the ITO-coated glass. Source: Reproduced with permission of Ref. [193] from Copyright © 2011 American Chemical Society.

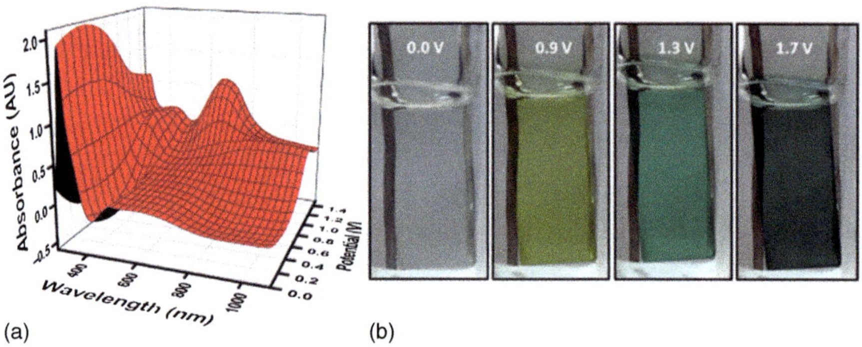

Figure 6.53 (a) Spectroelectrochemical spectra of PCz37 between −0.3 and + 1.5 V in $LiClO_4$/ACN. (b) Electrochromic properties of PCz37 between −0.3 and + 1.7 V in $LiClO_4$/CAN. Source: Reproduced with permission of Ref. [195] from Copyright © the Owner Societies 2016.

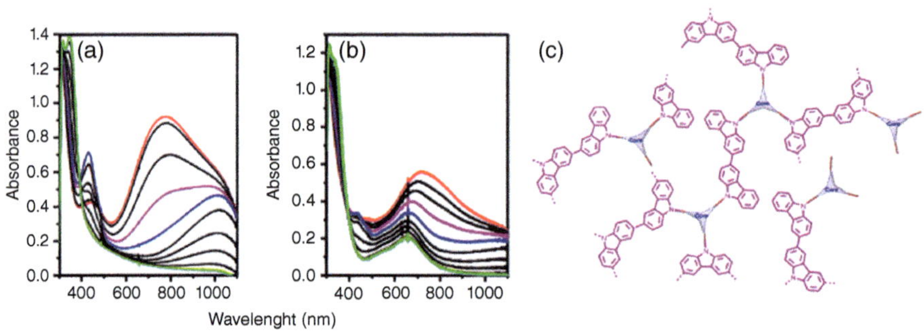

Figure 6.54 UV–visible absorption spectra of (a) PCz39 and (b) PCz40 films on an ITO electrode as function of applied potentials. (c) Electrochemical polymerization of end-capped CBZ dendrimers. Source: Reproduced with permission of Ref. [197] from Copyright © 2016 Elsevier Ltd.

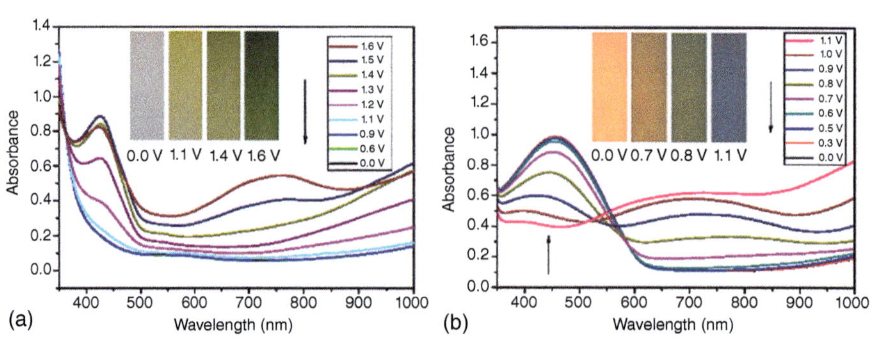

Figure 6.55 UV–visible absorption spectra of (a) PCz42 and (b) P(tnCz1–bTp2) films on an ITO electrode as function of applied potentials [198].

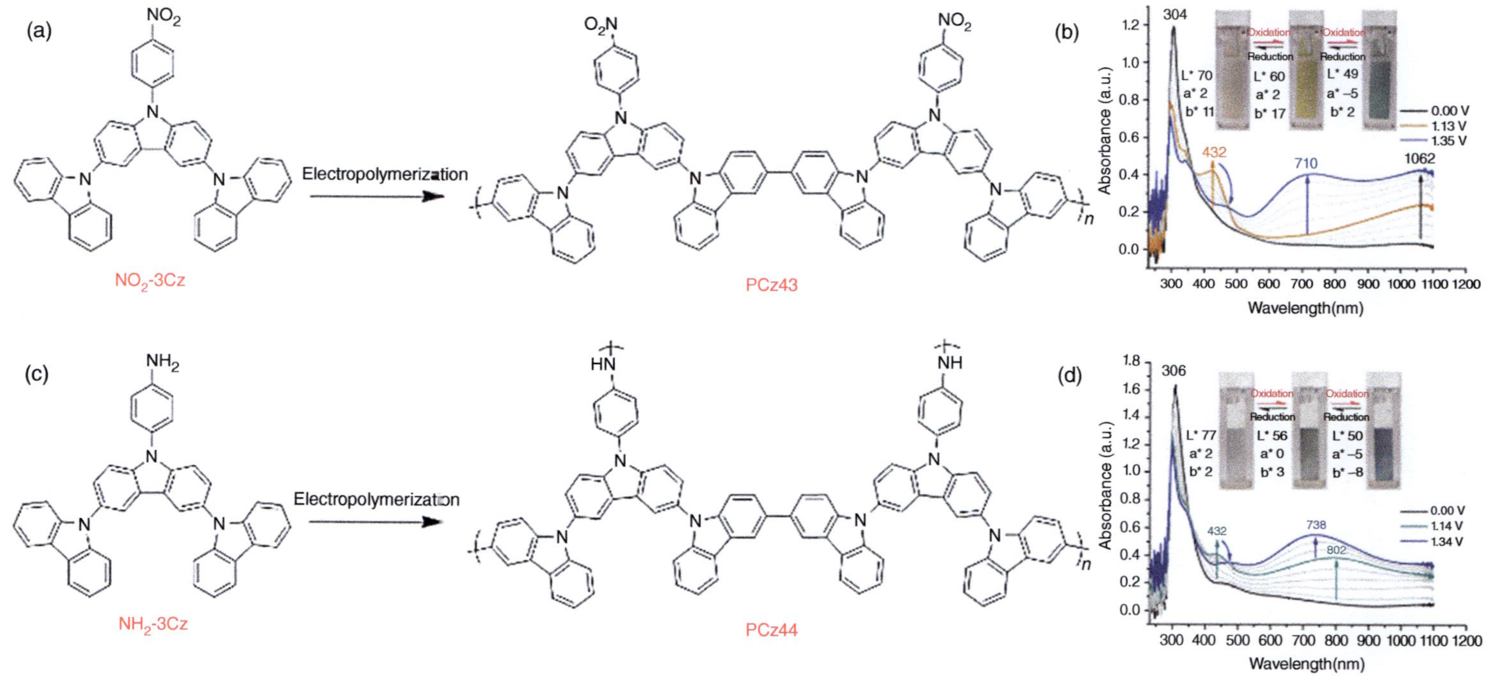

Figure 6.56 Electropolymerization of (a) NO_2−3Cz and (c) NH_2−3Cz. Spectroelectrograms and color changes of (b) PCz43 and (d) PCz44 thin films on ITO-coated glass in 0.1 M Bu_4NClO_4/MeCN at various applied voltages. Source: Reproduced with permission of Ref. [169] from Copyright © The Royal Society of Chemistry 2016.

Furthermore, the unit 3,6-bis(*N*-carbazolyl) carbazole with tri-carbazole structure has been a focus in the field of polymer research. It has been extensively utilized in the polymer structure for the application of supercapacitors. S. H. Hsiao and Lin synthesized two compounds, 3,6-di(carbazol-9-yl)-*N*-(4-nitrophenyl)carbazole (NO_2-3Cz) and 3,6-di(carbazol-9-yl)-*N*-(4-aminophenyl)carbazole (NH_2-3Cz), and then they were electropolymerized on the electrode surface in an electrolyte solution via the oxidative coupling reactions to obtain the polymer films. He proposed the anodic oxidation pathways of polyaniline-like and biscarbazole moieties. Both of these two polymers exhibited reversible electrochemical oxidation processes, with significant EC behavior. The color of PCz43 films changed from the pale yellow neutral state to yellow-green as a radical cation and then to blue when fully oxidized. With regard to PCz44 films, the color changed from colorless to pale green and finally to blue upon oxidation. Lower onset oxidation potential and higher EC stability of PCz44 than that of PCz43 were due to the presence of polyaniline-like structure (Figures 6.56–6.58) [169].

Similar to the tri-carbazole structure, triphenylamine-di-(*N*-carbazole) moiety, which combines two kinds of electron-rich groups, triarylamine and carbazole, is easily oxidized to form stable polarons associated with noticeable changes of colors. Polymer PCz45 derived from the monomer dithiafulvenyl-triphenylamine-di-(*N*-carbazole) was designed by L. Deng and coworkers. PCz45 exhibited light yellow at the neutral state and changed its color to green to blue gray upon oxidation. But poor switching time (t_c = 4 seconds, t_b = 8 seconds) and coloration efficiency (23.77 cm^2/C) were obtained [199]. In order to compare these two similar electroactive moieties both with two end-capped carbazoles, S. H. Hsiam et al. reported two polymers containing 3,6-bis(*N*-carbazolyl) carbazole and triphenylamine-di(*N*-carbazole) moieties. The polymer PCz47 exhibited a lower first oxidation potential than PCz46, resulted from the center TPA unit of PCz47 being much easier to oxidize than the *N*-(4-methoxyphenyl)carbazole unit of PCz46. PCz46 exhibited significant color changes, from colorless to yellow and to

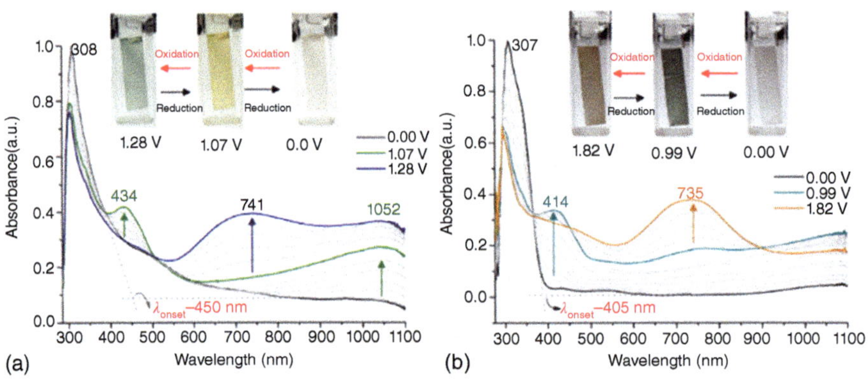

Figure 6.57 Spectroelectrochemical spectra and color changes of (a) PCz46 and (b) PCz47 onto the ITO-coated glass in 0.1 M Bu_4NClO_4/MeCN at various applied potentials (vs Ag/AgCl). Source: Reproduced with permission of Ref. [200] from Copyright © 2015, Elsevier B.V.

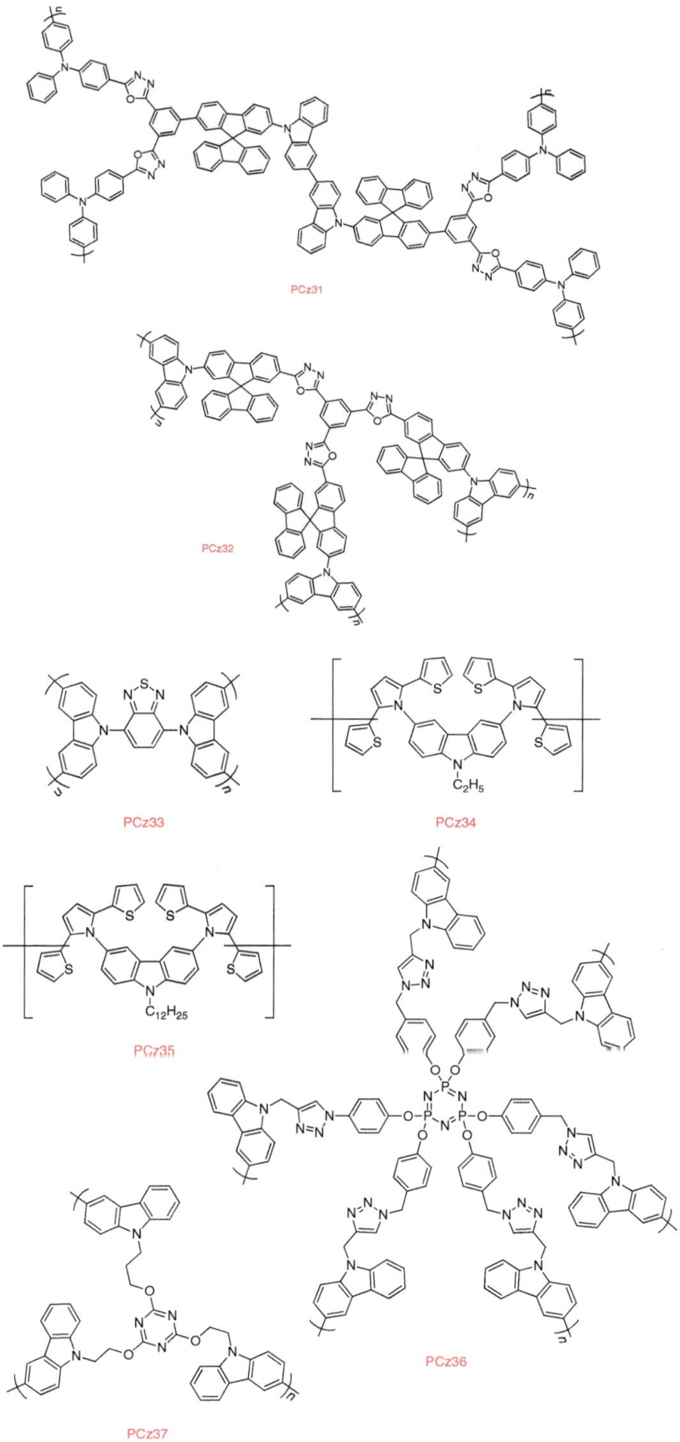

Figure 6.58 Chemical structure of cross-linked polycarbazole derivatives.

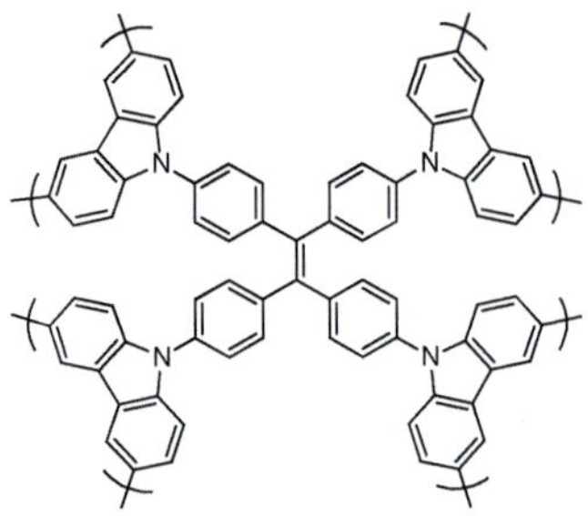

PCz38

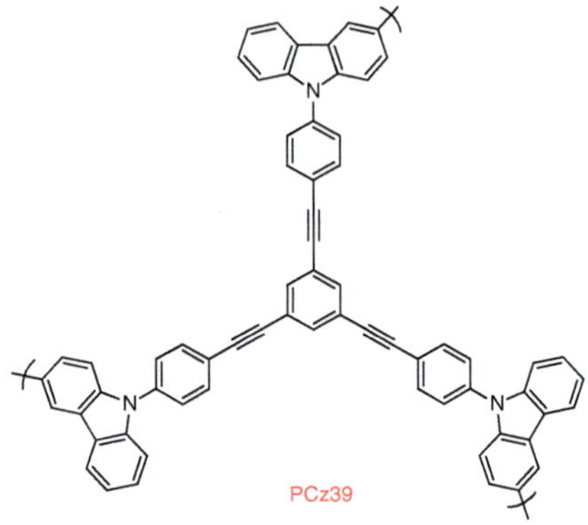

PCz39

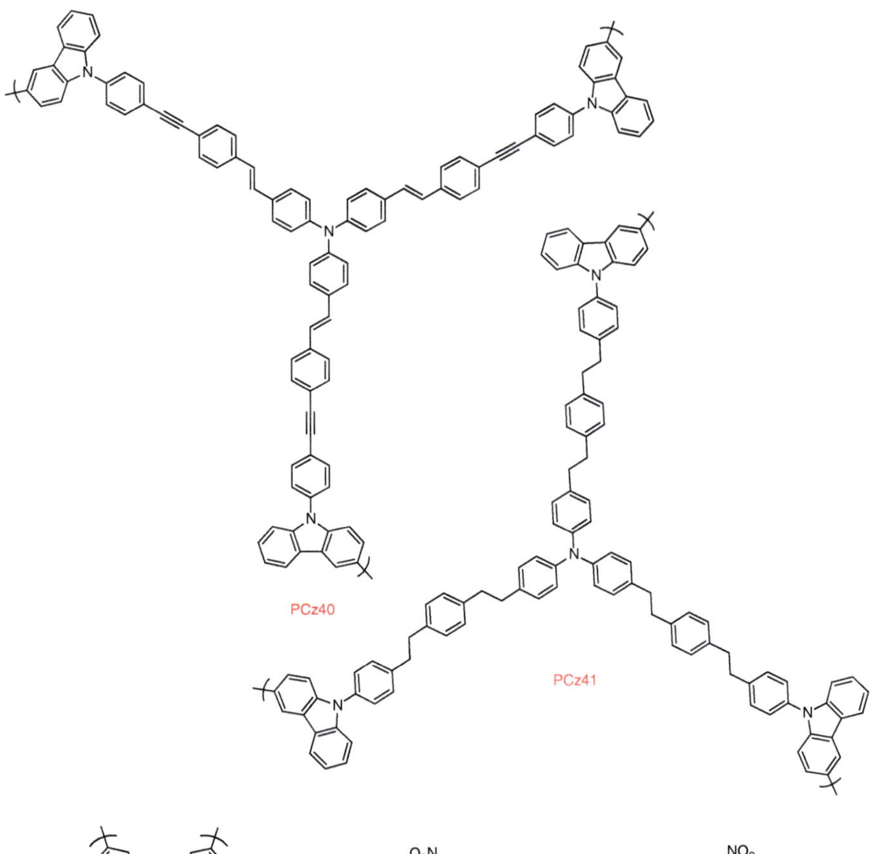

PCz40

PCz41

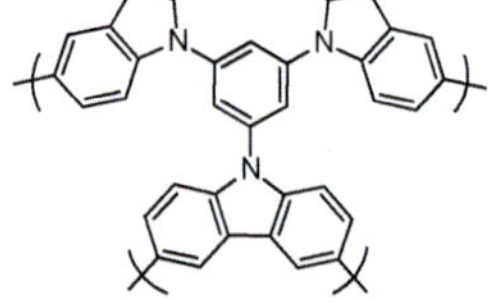

PCz42

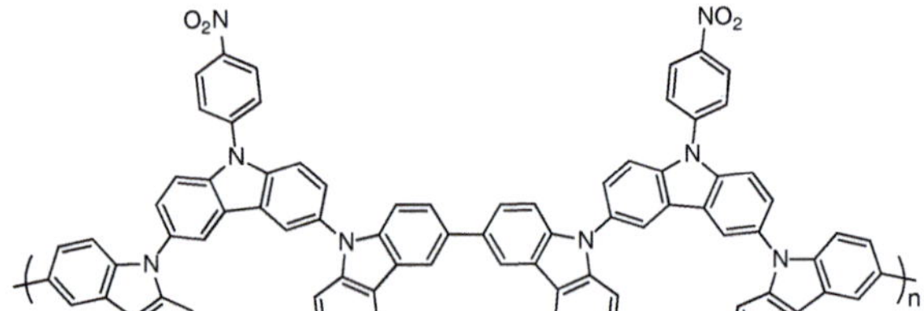

PCz43

Figure 6.58 (*Continued*)

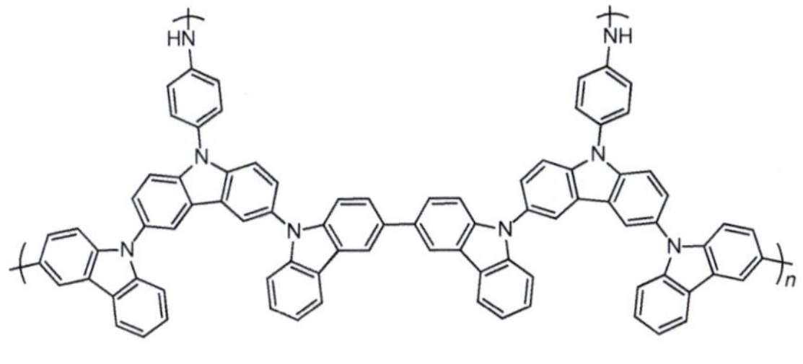

PCz44

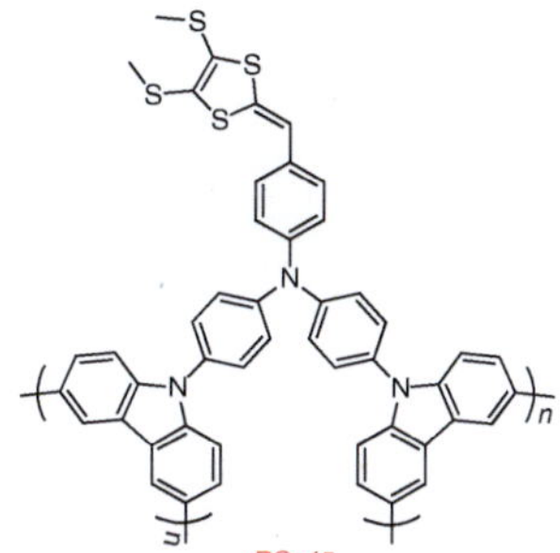

PCz45

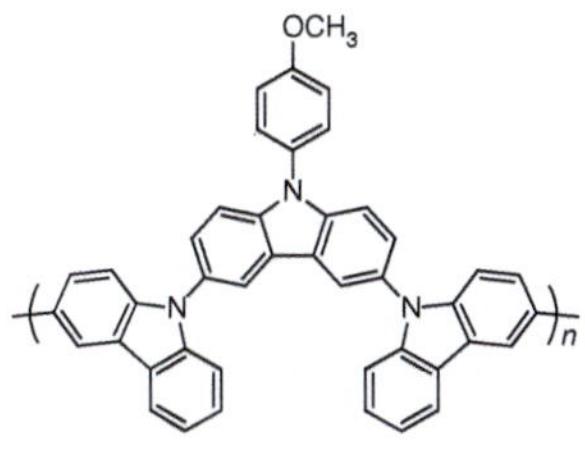

PCz46

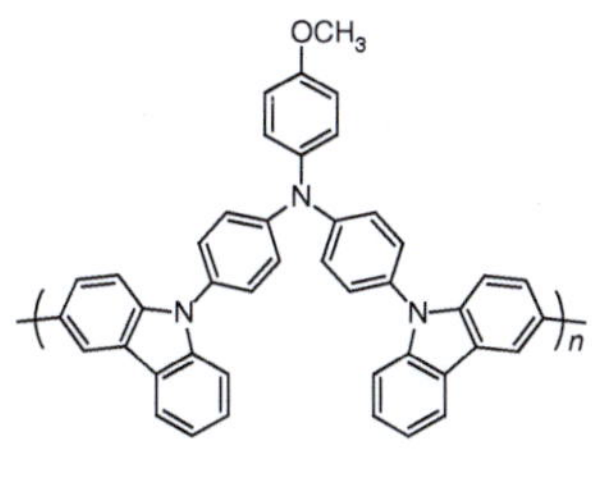

PCz47

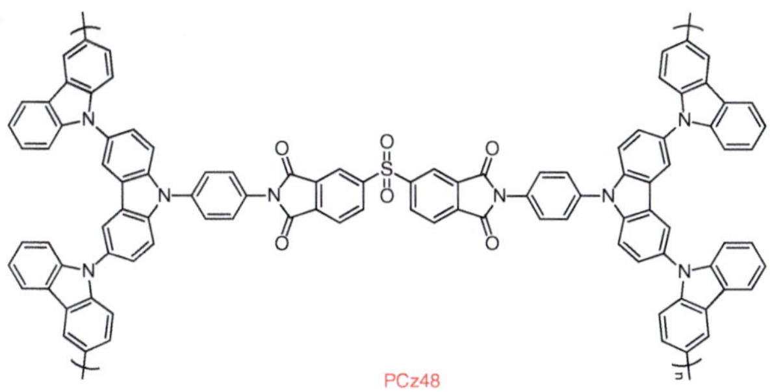

PCz48

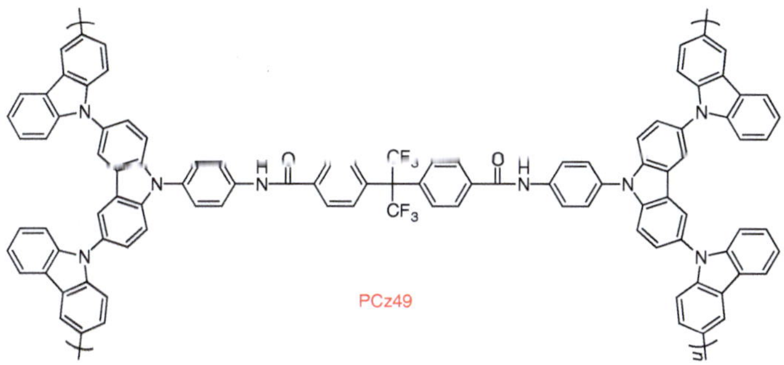

PCz49

Figure 6.58 *(Continued)*

green, while PCz47 turned from colorless to dark green and then to brown with the increasing potential. Both had high redox activity [200].

S. Hsiao and Lin reported two cross-linked polymer films PCz48 and PCz49, which were electropolymerized from diamide- and diimide-cored *N*-phenylcarbazole dendrons on the electrode surface in an electrolyte solution. They both showed two reversible oxidation redox couples accompanied by a color change from the colorless neutral state to the yellow-green and blue oxidized forms in the range of 1.11–1.34 V [201].

References

1 Garnier, F., Tourillon, G., Gazard, M., and Dubois, J.C. (1982). *Journal of Electroanalytical Chemistry* 135: 173.

2 Hideki Shirakawa, E.J.L., MacDiarmid, A.G., Chiang, C.K., and Heeger, A.J. (1977). *Journal of the Chemical Society, Chemical Communications* 578.

3 Garnier, F., Tourillon, G., Garzard, M., and Dubois, J.C. (1983). *Journal of Electroanalytical Chemistry* 148: 299.

4 Mortimer, R.J., Dyer, A.L., and Reynolds, J.R. (2006). *Displays* 27: 2.

5 Enengl, C., Enengl, S., Pluczyk, S. et al. (2016). *ChemPhysChem* 17: 3836.

6 Alkan, S., Cutler, C.A., and Reynolds, J.R. (2003). *Advanced Functional Materials* 13: 331.

7 R. J. Mortimer, D. R. Rosseinsky, P. M. S. Monk, Wiley-VCH , Weinheim, Germany, 2015.

8 Harbeke, G., Baeriswyl, D., Kiess, H., and Kobel, W. (1986). *Physica Scripta* T13: 302.

9 Bredas, J.L. and Street, G.B. (1985). *Accounts of Chemical Research* 18: 309.

10 Child, A.D., Sankaran, B., Larmat, F., and Reynolds, J.R. (1995). *Macromolecules* 28: 6571.

11 Chen, X., Xu, Z., Mi, S. et al. (2015). *New Journal of Chemistry* 39: 5389.

12 Dyer, A.L., Craig, M.R., Babiarz, J.E. et al. (2010). *Macromolecules* 43: 4460.

13 Kerszulis, J.A., Amb, C.M., Dyer, A.L., and Reynolds, J.R. (2014). *Macromolecules* 47: 5462.

14 Shi, P., Amb, C.M., Knott, E.P. et al. (2010). *Advanced Materials* 22: 4949.

15 Pouliot, J.R., Grenier, F., Blaskovits, J.T. et al. (2016). *Chemical Reviews* 116: 14225.

16 Mercier, L.G. and Leclerc, M. (2013). *Accounts of Chemical Research* 46: 1597.

17 Kumada, T., Nohara, Y., Kuwabara, J., and Kanbara, T. (2015). *Bull. Chem. Soc. Jpn.* 88: 1530.

18 Estrada, L.A., Deininger, J.J., Kamenov, G.D., and Reynolds, J.R. (2013). *ACS Macro Letters* 2: 869.

19 Dudnik, A.S., Aldrich, T.J., Eastham, N.D. et al. (2016). *Journal of the American Chemical Society* 138: 15699.

20 Livi, F., Gobalasingham, N.S., Thompson, B.C., and Bundgaard, E. (2016). *Journal of Polymer Science, Part A: Polymer Chemistry* 54: 2907.

21 Breukelaar, W.B., McAfee, S.M., and Welch, G.C. (2018). *New Journal of Chemistry* 42: 1617.

22 Gobalasingham, N.S., Carle, J.E., Krebs, F.C. et al. (2017). *Macromolecular Rapid Communications* 38.

23 Pankow, R.M., Ye, L., Gobalasingham, N.S. et al. (2018). *Polymer Chemistry* 9: 3885.

24 Pankow, R.M., Ye, L., and Thompson, B.C. (2018). *Polymer Chemistry* 9: 4120.

25 Grenier, F., Goudreau, K., and Leclerc, M. (2017). *Journal of the American Chemical Society* 139: 2816.

26 Kerszulis, J.A., Bulloch, R.H., Teran, N.B. et al. (2016). *Macromolecules* 49: 6350.

27 Christiansen, D.T. and Reynolds, J.R. (2018). *Macromolecules* 51: 9250.

28 Li, W., Guo, Y., Shi, J. et al. (2016). *Macromolecules* 49: 7211.

29 Aradilla, D., Casanovas, J., Estrany, F. et al. (2012). *Polymer Chemistry* 3: 436.

30 Giglioti, M., Trivinho-Strixino, F., Matsushima, J.T. et al. (2004). *Solar Energy Materials & Solar Cells* 82: 413.

31 Catia Arbizzani, A.B., Mastragostino, M., Zanelli, A. et al. (1995). *Advanced Materials* 7: 571.

32 Yoshino, K., Love, P., Onoda, M., and Sugimoto, R.-I. (1988). *Japanese Journal of Applied Physics* 27: L2388.

33 Hayes, W., Kaneto, K., Wong, K., and Yoshino, K. (1985). *Journal of Physics C: Solid State Physics* 18: L555.

34 Hattori, W.H.T., Wong, K., Kaneto, K., and Yoshino, K. (1984). *Journal of Physics C: Solid State Physics* 17: L803.

35 Welsh, D.M., Kumar, A., Meijer, E.W., and Reynolds, J.R. (1999). *Advanced Materials* 11: 1379.

36 Reeves, B.D., Grenier, C.R.G., Argun, A.A. et al. (2004). *Macromolecules* 37: 7559.

37 Krompiec, M.P., Baxter, S.N., Klimareva, E.L. et al. (2018). *Journal of Materials Chemistry C* 6: 3743.

38 Darmanin, T., Godeau, G., Guittard, F. et al. (2018). *ChemNanoMat* 4: 1.

39 Poverenov, E., Sheynin, Y., Zamoshchik, N. et al. (2012). *Journal of Materials Chemistry* 22: 14645.

40 Grenier, C.R., Pisula, W., Joncheray, T.J. et al. (2007). *Angewandte Chemie, International Edition* 46: 714.

41 Perepichka, I.F., Roquet, S., Leriche, P. et al. (2006). *Chemistry* 12: 2960.

42 Roquet, S., Leriche, P., Perepichka, I. et al. (2004). *Journal of Materials Chemistry* 14: 1396.

43 Taliani, C., Zamboni, R., Danieli, R. et al. (1989). *Physica Scripta* 40: 781.

44 Danieli, R., Taliani, C., Zamboni, R. et al. (1986). *Physica Scripta* 13: 325.

45 Corradini, A., Marinangeli, A.M., and Mastragostino, M. (1986). *Abstracts of Papers of the American Chemical Society* 192: 45.

46 Brédas, J.L., Elsenbaumer, R.L., Chance, R.R., and Silbey, R. (1983). *The Journal of Chemical Physics* 78: 5656.

47 Turbiez, M., Frere, P., Leriche, P. et al. (2005). *Chemical Communications* 2005: 1161.

48 Zhu, M., Li, W., Xu, P. et al. (2017). *Science China. Chemistry* 60: 63.

49 Xu, P., Murtaza, I., Shi, J. et al. (2016). *Polymer Chemistry* 7: 5351.

50 Shi, J., Zhu, X., Xu, P. et al. (2016). *Macromolecular Rapid Communications* 37: 1344.

51 Shao, S., Shi, J., Murtaza, I. et al. (2017). *Polymer Chemistry* 8: 769.

52 Hergue, N., Frere, P., and Roncali, J. (2011). *Organic & Biomolecular Chemistry* 9: 588.

53 Yin, Y., Li, W., Zeng, X. et al. (2018). *Macromolecules* 51: 7853.

54 Li, W., Ning, J., Yin, Y. et al. (2018). *Polymer Chemistry* 9: 5608.

55 Beaujuge, P.M. and Reynolds, J.R. (2010). *Chemical Reviews* 110: 268.

56 Amb, C.M., Dyer, A.L., and Reynolds, J.R. (2011). *Chemistry of Materials* 23: 397.

57 Dyer, A.L., Thompson, E.J., and Reynolds, J.R. (2011). *ACS Applied Materials & Interfaces* 3: 1787.

58 Kerszulis, J.A., Johnson, K.E., Kuepfert, M. et al. (2015). *Journal of Materials Chemistry C* 3: 3211.

59 Cheng, Y.-J., Yang, S.-H., and Hsu, C.-S. (2009). *Chemical Reviews* 109: 5868.

60 Lin, C., Endo, T., Takase, M. et al. (2011). *Journal of the American Chemical Society* 133: 11339.

61 Amb, C.M., Kerszulis, J.A., Thompson, E.J. et al. (2011). *Polymer Chemistry* 2: 812.

62 Cao, K., Eric Shen, D., Sterholm, A.M.Ö. et al. (2016). *Macromolecules* 49: 8498.

63 Cho, C.M., Ye, Q., Neo, W.T. et al. (2015). *Journal of Materials Science* 50: 5856.

64 Atakan, G. and Gunbas, G. (2016). *RSC Advances* 6: 25620.

65 Jensen, J., Dam, H.F., Reynolds, J.R. et al. (2012). *Journal of Polymer Science Part B: Polymer Physics* 50: 536.

66 Dyer, A.L., Bulloch, R.H., Zhou, Y. et al. (2014). *Advanced Materials* 26: 4895.

67 Bulloch, R.H., Kerszulis, J.A., Dyer, A.L., and Reynolds, J.R. (2014). *ACS Applied Materials & Interfaces* 6: 6623.

68 Bulloch, R.H., Kerszulis, J.A., Dyer, A.L., and Reynolds, J.R. (2015). *ACS Applied Materials & Interfaces* 7: 1406.

69 Jensen, J., Dyer, A.L., Shen, D.E. et al. (2013). *Advanced Functional Materials* 23: 3728.

70 Oesterholm, A.M., Shen, D.E., Gottfried, D.S., and Reynolds, J.R. (2016). *Advanced Materials Technologies* 1: 1600063.

71 Ponder, J.F., Schmatz, B., Hernandez, J.L., and Reynolds, J.R. (2018). *Journal of Materials Chemistry C* 6: 1064.

72 Vasilyeva, S.V., Beaujuge, P.M., Wang, S. et al. (2011). *ACS Applied Materials & Interfaces* 3: 1022.

73 Beaujuge, P.M., Ellinger, S., and Reynolds, J.R. (2008). *Nature Materials* 7: 795.

74 Savagian, L.R., Österholm, A.M., Shen, D.E. et al. (2018). *Advanced Optical Materials* 6: 1800594.

75 Balan, A., Baran, D., and Toppare, L. (2011). *Polymer Chemistry* 2: 1029.

76 Gunbas, G. and Toppare, L. (2012). *Chemical Communications* 48: 1083.

77 Hızalan, G., Balan, A., Baran, D., and Toppare, L. (2011). *Journal of Materials Chemistry* 21: 1804.

78 Yigit, D., Udum, Y.A., Gullu, M., and Toppare, L. (2014). *Journal of Electroanalytical Chemistry* 712: 215.
79 Balan, A., Baran, D., Gunbas, G. et al. (2009). *Chemical Communications (Cambridge)* https://doi.org/10.1039/b912482a6768.
80 Yu, H., Shao, S., Yan, L. et al. (2016). *Journal of Materials Chemistry C* 4: 2269.
81 Beaupre, S., Breton, A.-C., Dumas, J., and Leclerc, M. (2009). *Chemistry of Materials* 21: 1504.
82 Koyuncu, S. and Koyuncu, F.B. (2018). *Reactive and Functional Polymers* 131: 174.
83 Yen, H.-J. and Liou, G.-S. (2012). *Polymer Chemistry* 3: 255.
84 Wu, J.-T., Fan, Y.-Z., and Liou, G.-S. (2019). *Polymer Chemistry* 10: 345.
85 Hsiao, S.-H. and Lu, H.-Y. (2018). *Journal of the Electrochemical Society* 165: H638.
86 Patil, A.O., Ikenoue, Y., Wudl, F., and Heeger, A.J. (1987). *Journal of the American Chemical Society* 109: 1858.
87 Beaujuge, P.M., Amb, C.M., and Reynolds, J.R. (2010). *Advanced Materials* 22: 5383.
88 Shi, P., Amb, C.M., Dyer, A.L., and Reynolds, J.R. (2012). *ACS Applied Materials & Interfaces* 4: 6512.
89 Ponder, J.F., Österholm, A.M., and Reynolds, J.R. (2017). *Chemistry of Materials* 29: 4385.
90 Chen, X., Liu, H., Xu, Z. et al. (2015). *ACS Applied Materials & Interfaces* 7: 11387.
91 Hu, C.-W., Kawamoto, T., Tanaka, H. et al. (2016). *Journal of Materials Chemistry C* 4: 10293.
92 Xing, X., Zeng, Q., Vagin, M. et al. (2018). *Nano Energy* 47: 123.
93 Xu, J.W., Chua, M.H., and Shah, K.W. (2019). *Electrochromic Smart Materials: Fabrication and Applications*. The Royal Society of Chemistry.
94 Collier, G.S., Pelse, I., and Reynolds, J.R. (2018). *ACS Macro Letters* 7: 1208.
95 Wang, L.-X., Li, X.-G., and Yang, Y.-L. (2001). *Reactive and Functional Polymers* 47: 125.
96 Patil, A.O., Heeger, A.J., and Wudl, F. (1988). *Chemical Reviews* 88: 183.
97 Kuwabata, S., Yoneyama, H., and Tamura, H. (1984). *Bulletin of the Chemical Society of Japan* 57: 2247.
98 Beck, F., Michaelis, R., Schloten, F., and Zinger, B. (1994). *Electrochimica Acta* 39: 229.
99 Bazzaoui, M., Bazzaoui, E.A., Martins, L., and Martins, J.I. (2002). *Synthetic Metals* 130: 73.
100 Genies, E.M., Bidan, G., and Diaz, A.F. (1983). *Journal of Electroanalytical Chemistry and Interfacial Electrochemistry* 149: 101.
101 Waltman, R.J. and Bargon, J. (1984). *Tetrahedron* 40: 3963.
102 Waltman, R.J. and Bargon, J. (1986). *Canadian Journal of Chemistry* 64: 76.
103 Skotheim, T.A. and Reynolds, J. (2006). *Conjugated Polymers: Theory, Synthesis, properties, and Characterization*. CRC press.
104 Camurlu, P. (2014). *RSC Advances* 4: 55832.

105 Gaupp, C.L., Zong, K., Schottland, P. et al. (2000). *Macromolecules* 33: 1132.
106 Diaz, A.F., Castillo, J., Kanazawa, K.K. et al. (1982). *Journal of Electroanalytical Chemistry and Interfacial Electrochemistry* 133: 233.
107 Gazotti, W.A., Paoli, M.A.D., Casalbore-Miceli, G. et al. (1999). *Journal of Applied Electrochemistry* 29: 757.
108 De Paoli, M.A., Casalbore-Miceli, G., Girotto, E.M., and Gazotti, W.A. (1999). *Electrochimica Acta* 44: 2983.
109 Murakami, Y. and Yamamoto, T. (1999). *Polymer Journal* 31: 476.
110 Ak, M., Ak, M.S., and Toppare, L. (2006). *Macromolecular Chemistry and Physics* 207: 1351.
111 Camurlu, P. and Gültekin, C. (2012). *Smart Materials and Structures* 21: 025019.
112 Ouyang, M., Wang, G., and Zhang, C. (2011). *Electrochimica Acta* 56: 4645.
113 Mert, O., Demir, A.S., and Cihaner, A. (2013). *RSC Advances* 3: 2035.
114 Camurlu, P., Eren, E., and Gültekin, C. (2012). *Journal of Polymer Science, Part A: Polymer Chemistry* 50: 4847.
115 Almeida, A.K.A., Dias, J.M.M., Silva, A.J.C. et al. (2014). *Electrochimica Acta* 122: 50.
116 Coelho, E.C.S., Nascimento, V.B., Ribeiro, A.S., and Navarro, M. (2014). *Electrochimica Acta* 123: 441.
117 Cihaner, A. and Algı, F. (2008). *Electrochimica Acta* 53: 2574.
118 Cihaner, A. and Algı, F. (2009). *Electrochimica Acta* 54: 1702.
119 Varis, S., Ak, M., Tanyeli, C. et al. (2006). *European Polymer Journal* 42: 2352.
120 Arslan, A., Türkarslan, Ö., Tanyeli, C. et al. (2007). *Materials Chemistry and Physics* 104: 410.
121 Ak, M., Şahmetlioğlu, E., and Toppare, L. (2008). *Journal of Electroanalytical Chemistry* 621: 55.
122 Yildiz, E., Camurlu, P., Tanyeli, C. et al. (2008). *Journal of Electroanalytical Chemistry* 612: 247.
123 Kiralp, S., Camurlu, P., Gunbas, G. et al. (2009). *Journal of Applied Polymer Science* 112: 1082.
124 Cihaner, A. and Algı, F. (2008). *Electrochimica Acta* 54: 665.
125 Cihaner, A. and Algı, F. (2008). *Journal of Electroanalytical Chemistry* 614: 101.
126 Cihaner, A. and Algı, F. (2008). *Electrochimica Acta* 54: 786.
127 Cihaner, A., Mert, O., and Demir, A.S. (2009). *Electrochimica Acta* 54: 1333.
128 Hwang, J., Ik Son, J., and Shim, Y.-B. (2010). *Solar Energy Materials & Solar Cells* 94: 1286.
129 Wang, G., Fu, X., Huang, J. et al. (2010). *Electrochimica Acta* 55: 6933.
130 Wang, G., Fu, X., Huang, J. et al. (2011). *Electrochimica Acta* 56: 6352.
131 Camurlu, P., Bicil, Z., Gültekin, C., and Karagoren, N. (2012). *Electrochimica Acta* 63: 245.
132 Tirkeş, S., Mersini, J., Öztaş, Z. et al. (2013). *Electrochimica Acta* 90: 295.
133 Lai, J.-C., Lu, X.-R., Qu, B.-T. et al. (2014). *Organic Electronics* 15: 3735.
134 Wu, T.-Y., Liao, J.-W., and Chen, C.-Y. (2014). *Electrochimica Acta* 150: 245.
135 Guven, N. and Camurlu, P. (2015). *Polymer* 73: 122.

136 Camurlu, P. and Karagoren, N. (2013). *Journal of the Electrochemical Society* 160: H560.

137 Guven, N., Camurlu, P., and Yucel, B. (2015). *Polymer International* 64: 758.

138 Sefer, E., Bilgili, H., Gultekin, B. et al. (2015). *Dyes and Pigments* 113: 121.

139 Soganci, T., Ak, M., Giziroglu, E., and Söyleyici, H.C. (2016). *RSC Advances* 6: 1744.

140 Soganci, T., Soyleyici, H.C., Giziroglu, E., and Ak, M. (2016). *Journal of the Electrochemical Society* 163: H1096.

141 Soganci, T., Soyleyici, S., Soyleyici, H.C., and Ak, M. (2017). *Journal of the Electrochemical Society* 164: H11.

142 Soganci, T., Soyleyici, S., Soyleyici, H.C., and Ak, M. (2017). *Polymer* 118: 40.

143 Soganci, T. (2019). *Journal of the Electrochemical Society* 166: H12.

144 Cai, S., Wen, H., Wang, S. et al. (2017). *Electrochimica Acta* 228: 332.

145 Hu, B., Li, C., Liu, Z. et al. (2019). *Electrochimica Acta* 305: 1.

146 Hu, B., Li, C.-Y., Chu, J.-W. et al. (2019). *Journal of the Electrochemical Society* 166: H1.

147 Benincori, T., Brenna, E., Sannicolò, F. et al. (2000). *Chemistry of Materials* 12: 1480.

148 Kon, A.B., Foos, J., and Rose, T.L. (1992). *Chemistry of Materials* 4: 416.

149 Havinga, E., Ten Hoeve, W., Meijer, E., and Wynberg, H. (1989). *Chemistry of Materials* 1: 650.

150 Walczak, R.M. and Reynolds, J.R. (2006). *Advanced Materials* 18: 1121.

151 Schottland, P., Zong, K., Gaupp, C.L. et al. (2000). *Macromolecules* 33: 7051.

152 Kraft, A., Rottmann, M., Gilsing, H.-D., and Faltz, H. (2007). *Electrochimica Acta* 52: 5856.

153 Kim, I.T., Lee, J.Y., and Lee, S.W. (2004). *Chemistry Letters* 33: 46.

154 Sönmez, G., Schwendeman, I., Schottland, P. et al. (2003). *Macromolecules* 36: 639.

155 Knott, E.P., Craig, M.R., Liu, D.Y. et al. (2012). *Journal of Materials Chemistry* 22: 4953.

156 Li, H., Lambert, C., and Stahl, R. (2006). *Macromolecules* 39: 2049.

157 Nadeau, J.M. and Swager, T.M. (2004). *Tetrahedron* 60: 7141.

158 Taskin, A.T., Balan, A., Epik, B. et al. (2009). *Electrochimica Acta* 54: 5449.

159 Celebi, S., Balan, A., Epik, B. et al. (2009). *Organic Electronics* 10: 631.

160 Baran, D., Oktem, G., Celebi, S., and Toppare, L. (2011). *Macromolecular Chemistry and Physics* 212: 799.

161 Baran, D., Balan, A., Esteban, B.M. et al. (2010). *Macromolecular Chemistry and Physics* 211: 2602.

162 Wu, T.-Y. and Li, J.-L. (2016). *RSC Advances* 6: 15988.

163 Tsuie, B., Reddinger, J.L., Sotzing, G.A. et al. (1999). *Journal of Materials Chemistry* 9: 2189.

164 Sotzing, G.A., Reddinger, J.L., Katritzky, A.R. et al. (1997). *Chemistry of Materials* 9: 1578.

165 Ambrose, J.F. and Nelson, R.F. (1968). *Journal of the Electrochemical Society* 115: 1159.

166 Desbene-Monvernay, A., Lacaze, P.-C., and Dubois, J.-E. (1981). *Journal of Electroanalytical Chemistry and Interfacial Electrochemistry* 129: 229.

167 Karon, K. and Lapkowski, M. (2015). *Journal of Solid State Electrochemistry* 19: 2601.

168 Mengoli, G., Musiani, M.M., Schreck, B., and Zecchin, S. (1988). *Journal of Electroanalytical Chemistry and Interfacial Electrochemistry* 246: 73.

169 Hsiao, S.-H. and Lin, S.-W. (2016). *Polymer Chemistry* 7: 198.

170 Desbene-Monvernay, A., Lacaze, P.C., Dubois, J.E., and Desbene, P.L. (1983). *Journal of Electroanalytical Chemistry and Interfacial Electrochemistry* 152: 87.

171 Faïd, K., Adès, D., Siove, A., and Chevrot, C. (1994). *Synthetic Metals* 63: 89.

172 Faid, K., Siove, A., Ades, D., and Chevrot, C. (1993). *Synthetic Metals* 55: 1656.

173 Siove, A., Ades, D., N'Gbilo, E., and Chevrot, C. (1990). *Synthetic Metals* 38: 331.

174 Chevrot, C., Ngbilo, E., Kham, K., and Sadki, S. (1996). *Synthetic Metals* 81: 201.

175 Reddinger, J.L., Sotzing, G.A., and Reynolds, J.R. (1996). *Chemical Communications* https://doi.org/10.1039/CC99600017771777.

176 Witker, D. and Reynolds, J.R. (2005). *Macromolecules* 38: 7636.

177 Data, P., Zassowski, P., Lapkowski, M. et al. (2014). *Electrochimica Acta* 122: 118.

178 Wang, K., Zhang, T., Hu, Y. et al. (2014). *Electrochimica Acta* 130: 46.

179 van Mullekom, H.A.M., Vekemans, J.A.J.M., Havinga, E.E., and Meijer, E.W. (2001). *Materials Science & Engineering R: Reports* 32: 1.

180 Koyuncu, F.B., Koyuncu, S., and Ozdemir, E. (2010). *Electrochimica Acta* 55: 4935.

181 Koyuncu, F.B., Koyuncu, S., and Ozdemir, E. (2011). *Synthetic Metals* 161: 1005.

182 Baycan Koyuncu, F. (2012). *Electrochimica Acta* 68: 184.

183 Hu, B., Zhang, Y., Lv, X. et al. (2013). *Journal of Electroanalytical Chemistry* 689: 291.

184 Udum, Y.A., Hızlıateş, C.G., Ergün, Y., and Toppare, L. (2015). *Thin Solid Films* 595: 61.

185 Hsiao, S.-H., Wang, H.-M., Lin, J.-W. et al. (2013). *Materials Chemistry and Physics* 141: 665.

186 Morin, J.-F., Leclerc, M., Adès, D., and Siove, A. (2005). *Macromolecular Rapid Communications* 26: 761.

187 Baycan Koyuncu, F., Koyuncu, S., and Ozdemir, E. (2011). *Organic Electronics* 12: 1701.

188 Karatas, E., Guzel, M., and Ak, M. (2017). *Journal of the Electrochemical Society* 164: H463.

189 Guzel, M., Karatas, E., and Ak, M. (2017). *Journal of the Electrochemical Society* 164: H49.

190 Natera, J., Otero, L., Sereno, L. et al. (2007). *Macromolecules* 40: 4456.

191 Natera, J., Otero, L., D'Eramo, F. et al. (2009). *Macromolecules* 42: 626.

192 Xu, C., Zhao, J., Wang, M. et al. (2013). *Thin Solid Films* 527: 232.

193 Baycan Koyuncu, F., Sefer, E., Koyuncu, S., and Ozdemir, E. (2011). *Macromolecules* 44: 8407.

194 Sezgin, M., Ozay, O., Koyuncu, S. et al. (2015). *Chemical Engineering Journal* 274: 282.

195 Guzel, M., Soganci, T., Ayranci, R., and Ak, M. (2016). *Physical Chemistry Chemical Physics* 18: 21659.

196 Carbas, B.B., Odabas, S., Türksoy, F., and Tanyeli, C. (2016). *Electrochimica Acta* 193: 72.

197 Mangione, M.I., Spanevello, R.A., Minudri, D. et al. (2016). *Electrochimica Acta* 207: 143.

198 Kuo, C.W. and Lee, P.Y. (2017). *Polymers (Basel)* 9.

199 Zhou, P., Wan, Z., Liu, Y. et al. (2016). *Electrochimica Acta* 190: 1015.

200 Hsiao, S.-H. and Hsueh, J.-C. (2015). *Journal of Electroanalytical Chemistry* 758: 100.

201 Hsiao, S.-H. and Lin, S.-W. (2016). *Journal of Materials Chemistry C* 4: 1271.

7

TA-Based Electrochromic Polyimides and Polyamides

7.1 Introduction

7.1.1 Aromatic Polyimides and Polyamides

Since the initiation of the investigation on polymers incorporating aromatic moieties into the polymer backbone in the 1950s [1], these polymers have drawn considerable attention of the research community and industrial community due to their outstanding mechanical, chemical, and thermal stability. Aromatic polyimides (PIs) and polyamides (PAs) are generally regarded as the most successful types of aromatic polymers. Especially, the commercialization of several kinds of polyimides (Kapton and Upilex R) and aromatic polyamides (Nomex and Kevlar) [2] for their high glass transition temperature, high thermal stability, mechanical resistance, and chemical resistance in the past few years has encouraged the efforts to further develop high-performance polymers and investigate their novel physical chemical properties (Scheme 7.1). The correlation between electrochemical properties and chemical structures of different polyimides was first described in 1990. Furthermore, PMDA-ODA (Kapton) turned out to exhibit reversible redox reactions and electrochromic (EC) properties with color transitions from yellow (neutral state) to green (anion radical state) and to violet (dianion state) [3]. However, the advancement of PIs/PAs-based EC materials was limited by poor processability, which was caused by their high rigidity of the backbone and strong intermolecular interactions.

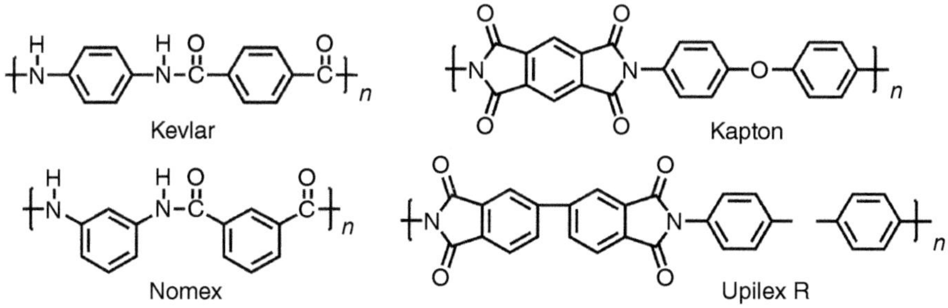

Scheme 7.1 Chemical structures of some typical commercialized PIs and PAs.

Organic Electronics for Electrochromic Materials and Devices, First Edition. Hong Meng.

7.1.2 Triarylamine-Based Aromatic Polymers

Endowed with high thermal stability, acceptable processability, reasonable hole mobility, and other versatile properties, the triarylamine (TA) groups are promising building blocks applied in organic photovoltaic [4], organic light-emitting diodes [5–7], organic lasers [8–10], and industrial dyes. Nowadays, TA derivatives are also regarded as qualified candidates for EC devices. In general, reversible electrochemical behaviors are observed in TA derivatives, which indicate suitable electrochemical stability for EC application. In addition, attributing to the electron-rich nitrogen center of TA groups, it's feasible to form a cation state by a mono-electron oxidation process for TA derivatives. Basically, arylamines are colorless unless they have charge transfer (CT) interactions with the electronic acceptor species. Therefore, an obvious color change from the colorless/pale yellow neutral state to colored oxidized state can be observed accompanied with the oxidation process. Thus, the synthesis and EC behavior of various TA-based polymers have been reported in the literature in the past few years [11].

Among these TA-based polymers, TA-based PIs/PAs are two important classes of materials due to their outstanding and desirable characteristics. The incorporation of bulky and packing-disruptive TA groups not only improved the solubility but also kept the excellent thermal stability. The combination of TA and PI was first achieved in 1991 with 4,4′-diaminotriphenylamine (TPA) and various tetracarboxylic dianhydrides in order to improve the solubility of aromatic PIs [12–14]. The resulting soluble polyimides PI 1 (Scheme 7.2) exhibited high glass-transition temperatures and reasonable solubility in many polar solvents. On the other hand, the initiation of investigation on TA-based polyamide PA 1 was in 1990 when new aromatic polyamides were prepared from 4,4′-diaminotriphenylamine and aromatic dicarboxylic acids [15].

As the advancing in synthesis of aromatic polymers [16] and diamine monomers [17, 18], many kinds of novel TA-based PIs/PAs with desirable EC properties and moderate solubility were successfully synthetized and characterized. The first aromatic polyimides integrating interesting EC properties containing TPA groups (Figure 7.1) were disclosed in 2005 from a newly synthesized diamine, *N*,*N*-bis(4-aminophenyl)-*N*′,*N*′-diphenyl-1,4-phenylenediamine, and various tetracarboxylic dianhydrides by polycondensation reactions [19]. The solution-cast PI films exhibited two well-defined oxidation redox couples (at 0.78 and 1.14 V vs Ag/AgCl) with distinct color changes from pale yellowish to green and then blue during oxidative scanning. Since then, Liou, Hsiao and other groups developed numerous TPA-based EC PIs/PAs. Most of the PIs/PAs were solution processible and thermally stable with excellent adhesion with indium tin oxide-coated glass electrode and good electrochemical stability. Now the TA-based PIs/PAs are considered as great anodic EC materials due to proper oxidation potentials, electrochemical stability, and thin film formability.

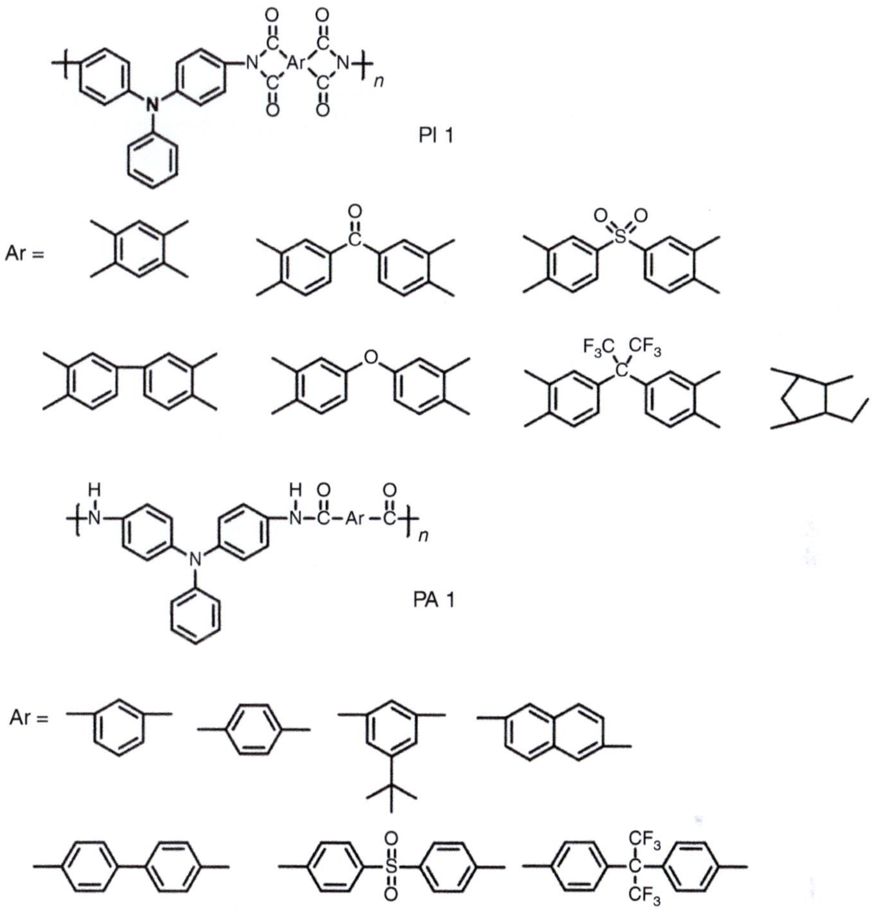

Scheme 7.2 Chemical structures of the first TA-based polyimides and TA-based polyamides.

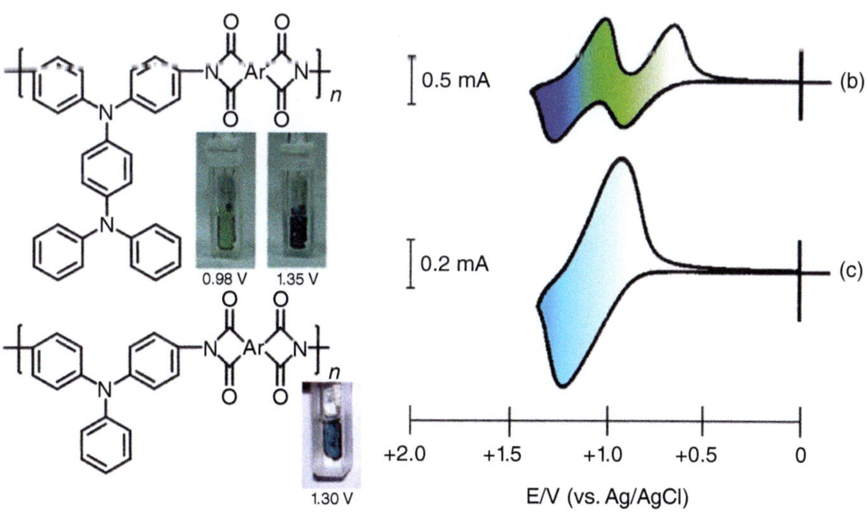

Figure 7.1 Cyclic voltammetric diagrams of the first EC PI and its EC behavior. Source: Adapted with permission from Ref. Cheng et al. [19]. Copyright (2005) American Chemical Society.

7.1.3 Electrochemical and Electrochromic Behaviors of MV Triarylamine Systems

Normally, TPA-based PIs/PAs exhibited EC characteristics when scanning potentials positively. The electron-rich nitrogen atoms play an important role in the EC behavior of TPA-based polymers. Thus, among the numerous redox centers, TPA system is of great interest in investigation on hole-transfer processes between redox centers. In order to illustrate the role of TPA groups in EC materials, the intramolecular electron transfer (ET) interaction had been studied extensively in the TPA-contained mixed-valence (MV) systems [20]. In 1968, Robin and Day classified MV compounds containing two (or more) redox centers into three types: (class I) no coupling between the different redox states, which are completely localized, (class II) moderate coupling between the MV centers, and (class III) complete delocalization and strong electronic coupling over the two redox centers [21].

The MV class I/II/III transition and electrochemistry of the synthesized model compounds were investigated for the bridged TPA system with various N–N distances and intramolecular ET capability according to the work by Liou and coworkers [22] (Scheme 7.3). The tetraphenyl-*p*-phenylenediamine (TPPA) cation radical has been reported as a symmetrical delocalized class III structure with a strong electronic coupling, whereas tetraphenylbenzidine (TPB) cation radical was demonstrated as a class II structure [23] with a weak electronic coupling; furthermore, the electroactive ether-linked bis(TPA) monomer can be characterized as MV class I due to the completely isolated two redox centers. The absorption spectrum of the three kinds of monomers at various applied bias (Figure 7.2) indicates that MV class II and class III systems exhibit intervalence CT absorption bands in the near-infrared (NIR) region, while there is no obvious intervalence CT-state absorption in class I system. The corresponding polymer of the MV class I system also shows independent electrochemical behavior of the two separated redox centers where two electrons are removed simultaneously during the redox process. The investigation on the electrochemistry and MV I/II/III transition of the corresponding materials with various N–N distances can guide the molecular design for multicolor/multifunctional EC materials.

7.2 Development of TA-Based Electrochromic Polyimides and Polyamides

Towards the practical applications of EC materials, the performance requirements of the EC materials such as low driving voltage, rapid respond time, high color contrast ratio, and long cycle life should be fulfilled. On the other hand, TA-based polyimides and amides with extensive color transitions and multifunctional applications are much in demand in the research community and industrial community. Basically, side group engineering and backbone modulation are two accessible approaches to fine tune the functions and colors of TPA-based PIs/PAs due to their capability of being chemically modified easily (Figure 7.3).

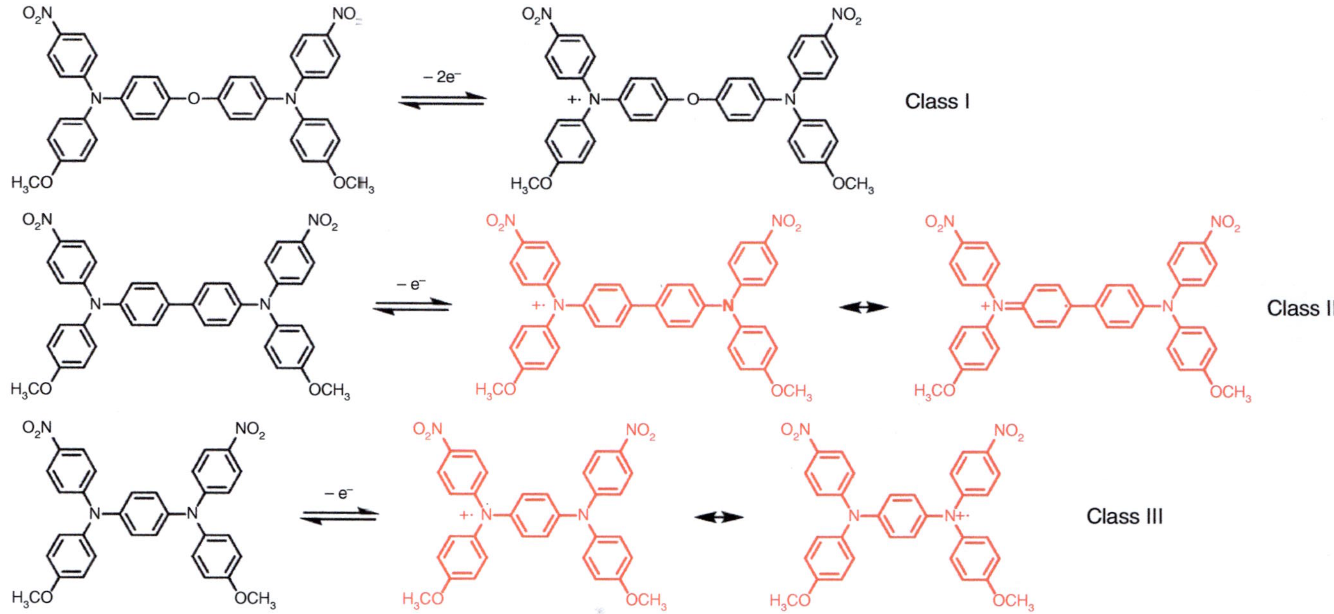

Scheme 7.3 The oxidation pathways of MV class I–III monomers.

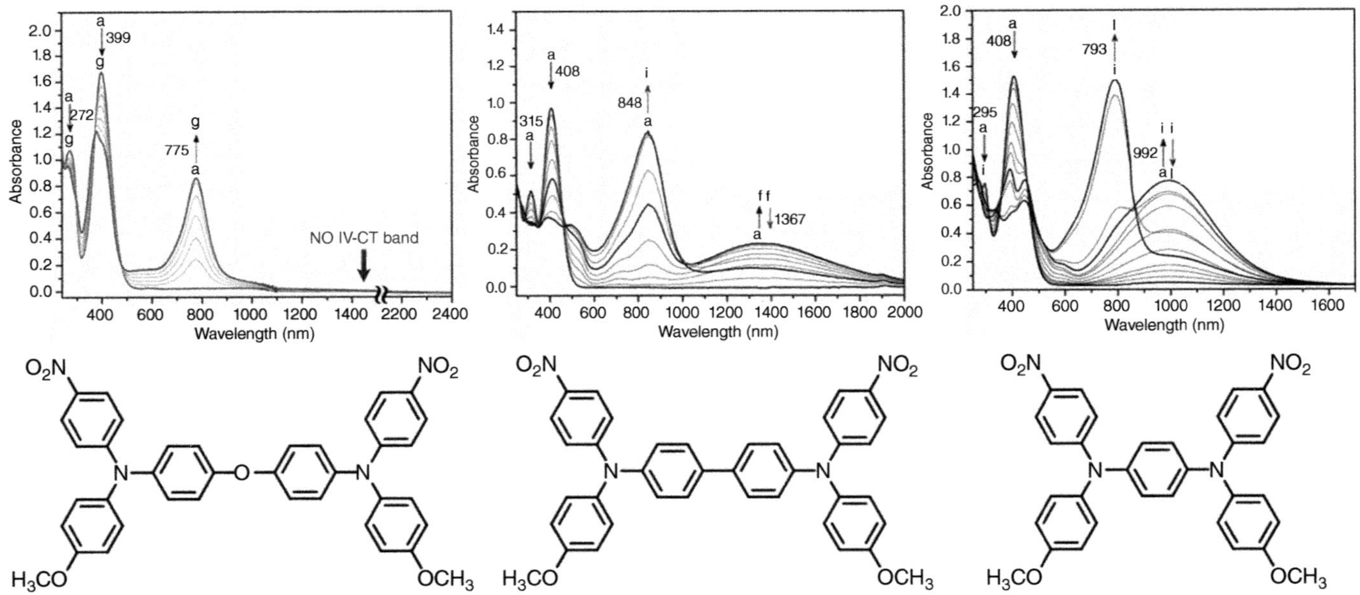

Figure 7.2 Absorption spectral change of MV class I–III compounds at various applied potential. Source: Adapted with permission from Ref. Yen et al. [22]. © 2011 John Wiley & Sons.

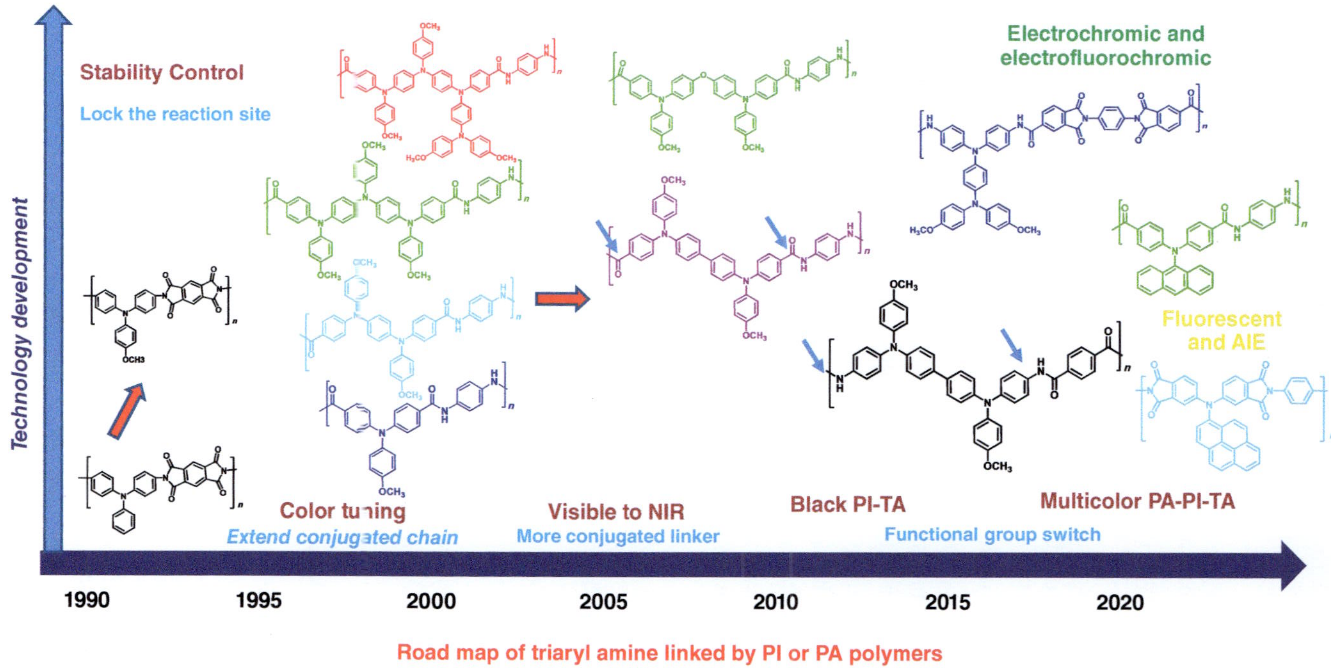

Figure 7.3 Development of TA-based electrochromic PIs/PAs.

7.2.1 Side Group Engineering

7.2.1.1 Introduction of Protecting Groups

The electrogenerated cation radical of TPA is not stable and will form TPB through a tail to tail coupling (Scheme 7.4) [24, 25]. In order to improve the electrochemical stability of TPA-based PIs/PAs, typical protecting groups such as methoxyl group and methyl group are incorporated on the active sites of TPA groups. When the electron-donating substituents are introduced to the para-position of TPA, the coupling reactions were prevented by affording stable cationic radicals [26, 27] (Scheme 7.5 PA 2a, PA 2d). Furthermore, the oxidation potentials of the electroactive polymers were lowered through the introduction of the electron-donating groups. Besides, meta-substituted and ortho-substituted TPA also exhibited improved electrochemical and EC stability (Scheme 7.5 PA 2b, PA 2c). The 3,5-dimethyl substituents might interfere sterically with the coupling reaction between the TPA radical cations upon anodic oxidation and thus lead to an increased redox stability of the polyamides [28]. Additionally, it is interesting to emphasize that the additional substitution of ortho methoxyl group on the TPA unit gives the resulting polyamides another stable oxidized state; thus, two differently colored redox states of the polymers can be accessed through electrochemical oxidation (Figure 7.4) [29]. However, Me_3TPA-polyamides (Scheme 7.5 PA 2e) showed higher oxidation potential and lower electrochemical stability than MeTPA-polymers (Scheme 7.5 PA 2d), which could be attributed to the higher steric hindrance of ortho substituents

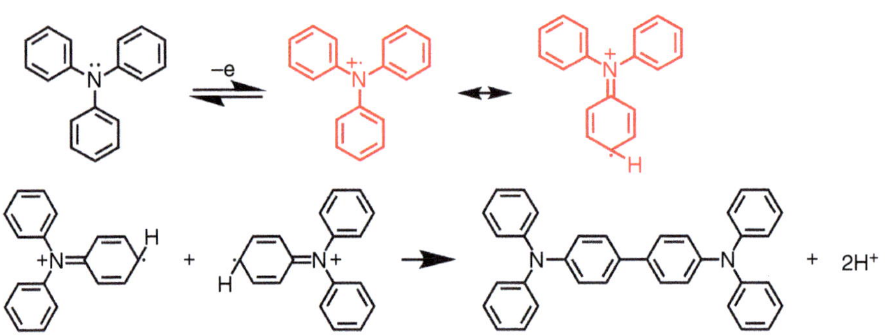

Scheme 7.4 The oxidation process of TPA and the reaction scheme of unstable TPA cation radicals.

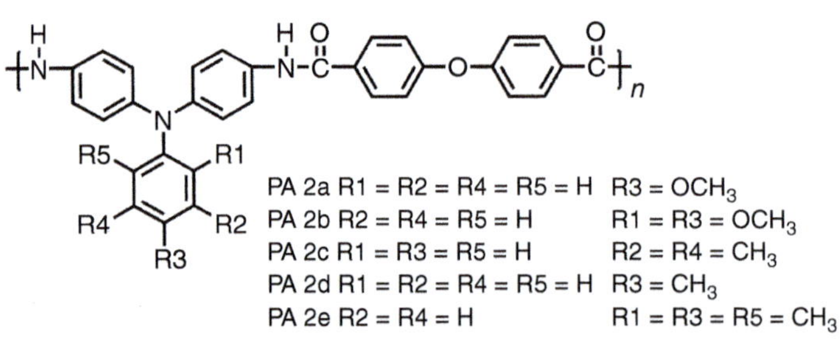

Scheme 7.5 TPA-based polyamides with protecting groups.

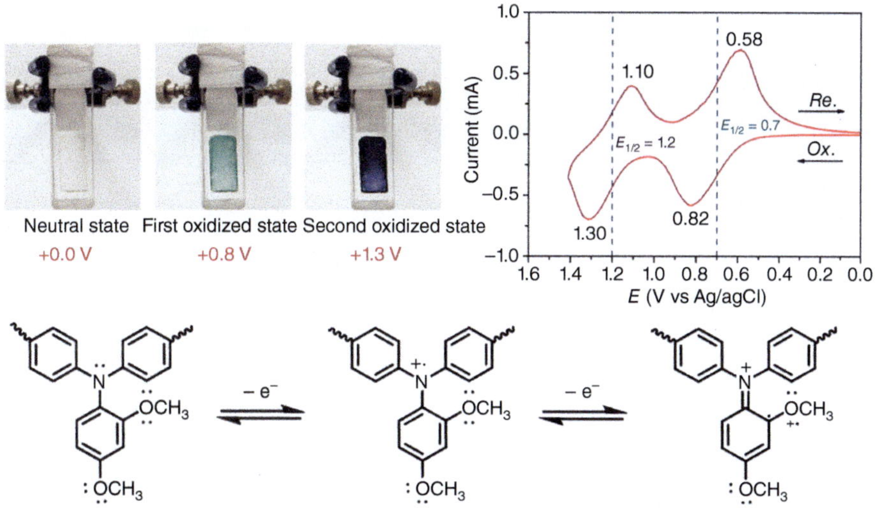

Figure 7.4 The cyclic voltammetric diagram and color changes in electrochemical cell at indicated electrode voltages of the PA 2b thin film. Source: Adapted with permission from Ref. Hsiao et al. [29]. © 2010 John Wiley & Sons.

in Me_3TPA moieties where resonance stabilization of cation radical became much more difficult for the Me_3-substituted phenyl ring [30].

7.2.1.2 Introduction of Electroactive Groups to Achieve Color Tuning of EC Material

EC materials with multiple interesting color changes are crucial for both future practical applications and investigation on the electrochemical/EC behaviors of EC materials. Thanks to the intrinsic electron-rich characteristic of TPA-based monomer, which makes chemical modification at active sites of TPA groups feasible and effective, we can tune the electrochemical and EC properties by introducing specific side groups. In general, when side groups with increased electroactive sites are introducedxs to the TA-containing monomers, increasing coloring stages could be characterized in the corresponding polymers (Scheme 7.6) [26, 31–33].

Novel solution-processable PIs containing TPA and pendant viologen moieties exhibited interesting color transitions [34]. The obtained multicolored EC PI films revealed ambipolar electrochemical behavior with coloration changes from a transmissive neutral state to the cyan/magenta/yellow redox states, implying great potential for application in smart windows and displays (Figure 7.5).

Increasingly, attention of the optical changes has been focused extending from the NIR (800–2000 nm) through the microwave regions of the spectrum, which could be exploitable for environmental control (heat gain or loss) in buildings. Incorporation of pendant phenothiazine group is an approach for tuning the coloration transition, and the prepared polyamides have great potential for applications in both the NIR and visible regions [35] (Figure 7.6 PA 4a). The intermolecular/intramolecular ET between bridged phenothiazine centers in the MV cation radicals state leads

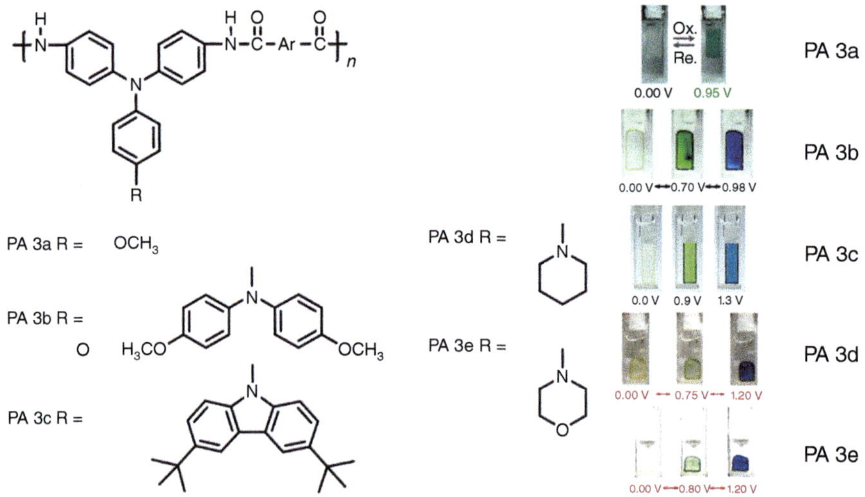

Scheme 7.6 Introduction of increased electroactive groups towards multiple color EC polyamides. Source: (**PA 3a**) Adapted with permission from Ref. [27]. Copyright (2007) Royal Society of Chemistry; (**PA 3b**) Adapted with permission from Ref. [19]. Copyright (2005) American Chemical Society; (**PA 3c**) Adapted with permission from Ref. [31]. Copyright (2013) Royal Society of Chemistry; (**PA 3d**) Adapted with permission from Ref. [32]. Copyright (2016) Elsevier; (**PA 3e**) Adapted with permission from Ref. [33]. Copyright (2015) Wiley.

to the absorption bands in the NIR region. The polyamides PA 3c (Scheme 7.6) also reveal valuable EC characteristics such as enhanced absorption in the NIR region upon oxidation, green and blue EC behaviors, fast switching time, and high electrochemical/electrochromic reversibility [31]. The introduction of the (3,6-ditert-butylcarbazol-9-yl)triphenylamine segment into the aramid backbone not only affords high T_g and good thermal stability but also leads to good solubility and electrochemical stability of the aramids. In addition, aromatic polyimides with pendant benzimidazole group (Figure 7.6 PI 3) are potential candidates as NIR EC materials due to a new absorption peak emerged at 847 nm in the UV–Vis–NIR spectrum in oxidized state, good thermal property and appropriate energy levels of HOMO and LUMO [36].

7.2.1.3 Introduction of Side Groups to Achieve Electrofluorochromic Materials

Electrofluorochromism (EFC) deals with the electrically driven reversible optical changes in the fluorescence. Reversible switching of the optical states by electrochemical or photochemical conversion of UV–Vis or photoluminescence spectra is a desirable field of investigation in optoelectronic devices such as displays, sensors, or optical memories.

The highly efficient luminophore groups were incorporated in order to improve the durability, thermal stability, and fluorescence of PAs/PIs. Combined with aliphatic backbone modulation, the introduction of fluorescent side groups can result in several EC PIs/PAs with electrochemical fluorescence switching function (Scheme 7.7) [37–42]. The electrofluorescent film of PA 7c revealed a high contrast

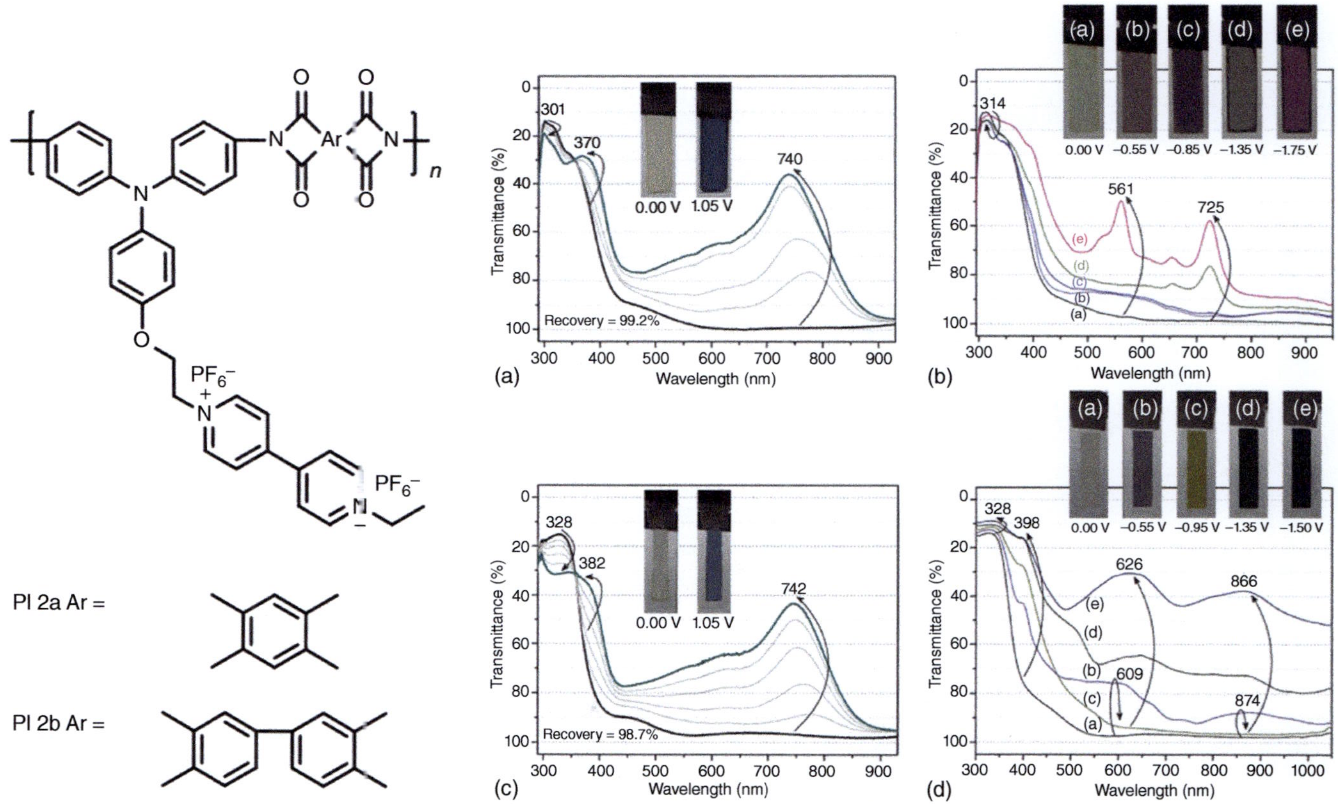

Figure 7.5 Electrochromic behavior of multiple color PI 2a and PI 2b thin films. Source: Adapted with permission from Ref. Yen et al. [34]. © 2017 John Wiley & Sons.

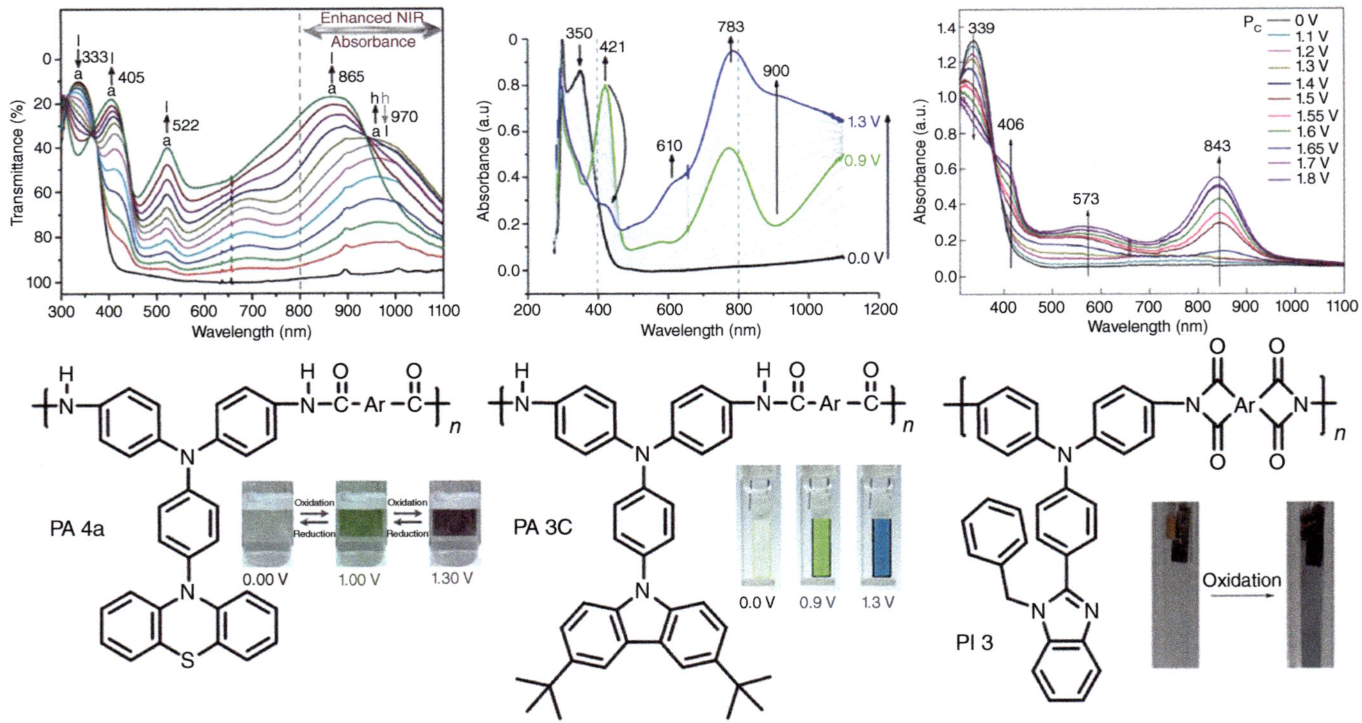

Figure 7.6 Electrochromic behaviors of the NIR EC PIs/PAs with pendant groups. Source: **(PA 4a)** Adapted with permission from Ref. [35]. Copyright (2010) Royal Society of Chemistry; **(PA 3c)** Adapted with permission from Ref. [31]. Copyright (2013) Royal Society of Chemistry; **(PI 3)** Adapted with permission from Ref. [36]. Copyright (2016) Springer Nature.

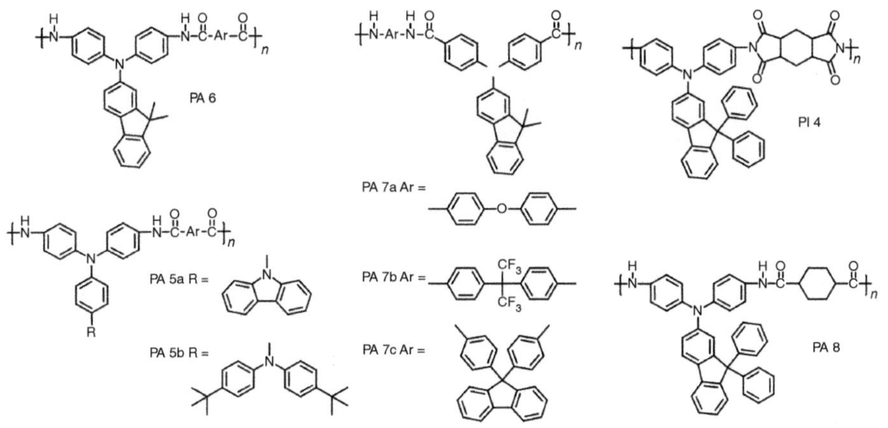

Scheme 7.7 Chemical structures of EFC TA-based PIs/PAs by introduction of luminophores.

ratio of 221.4 due to the excited transition from HOMO to LUMO delocalized over the DPA-based benzoyl moiety.

However, conventional EFC devices suffered limited fluorescent/non-fluorescent on/off ratio due to the low photoluminescent efficiency of the EFC materials. Recently, aggregation-enhanced emission (AEE), which are attributed to the restriction of intramolecular rotation in the condensed phase and almost non-fluorescent in solution, paved a new avenue for the design and synthesis of efficient solid-state emitters [43, 44]. Incorporated with the AEE concept, polyimide CN-PI and polyamide CN-PA, which have the AEE feature and are highly emissive in the solid state with a PL quantum yield of up to 65%, were reported (Figure 7.7). The electrochemical fluorescence switching of CN-PI from a fluorescent neutral state to a non-fluorescent monocation radical state at a low working potential, giving rise to the highest contrast ratio of 151.9.

7.2.1.4 Introduction of Other Functional Side Groups to Achieve Multiple Functions EC Material

Novel electron-donating triphenylamine (TPA)-containing electroactive functional polyimides with isolated donor–acceptor (PI 5) and non-isolated donor–acceptor (PI 6) structure electron-withdrawing pendant anthraquinone moiety exhibited static random access memory (SRAM) and dynamic random access memory (DRAM) behaviors, respectively [45] (Scheme 7.8). Meanwhile, the polymer films exhibited EC characteristics with a color change from neutral colorless or pale yellowish to blue and green at applied potentials, ranging from 0.00 to 1.40 V.

Very recently, Wang and his coworkers prepared a series of novel multiple stimuli-responsive smart polymers bearing TA and naphthalimide moieties [46] (Scheme 7.9). These novel polyimides (PI 7) are able to respond to four different stimuli (light, pH, explosives, and voltage) and perform four functions (light detectors, electrochromism, fluorescence detectors, and memory devices). Furthermore, the smart materials are suitable for complicated environment.

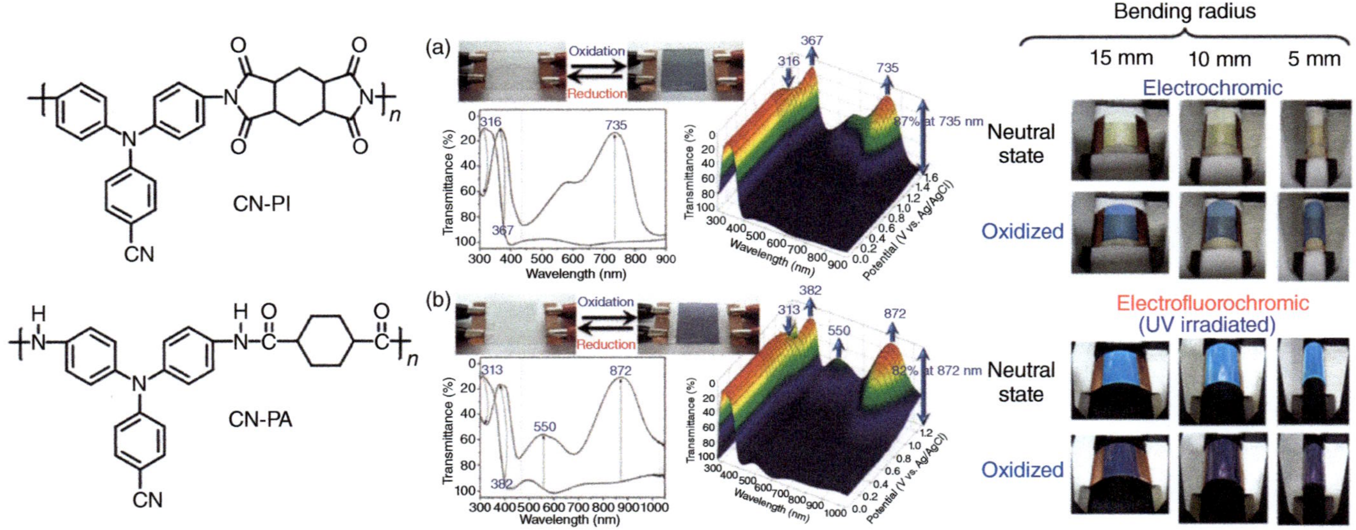

Figure 7.7 EC properties of CN-PA/CN-PI and EFC behavior of CN-PI. Source: Adapted with permission from Ref. [43]. Copyright (2013) Royal Society of Chemistry.

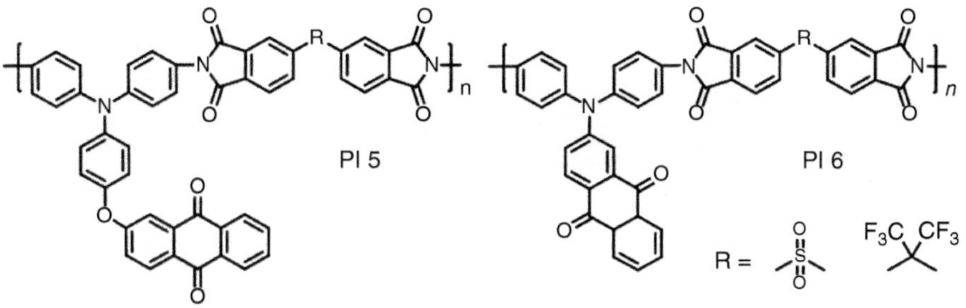

Scheme 7.8 Chemical structures of polyimides with dual functions of memory and EC behavior.

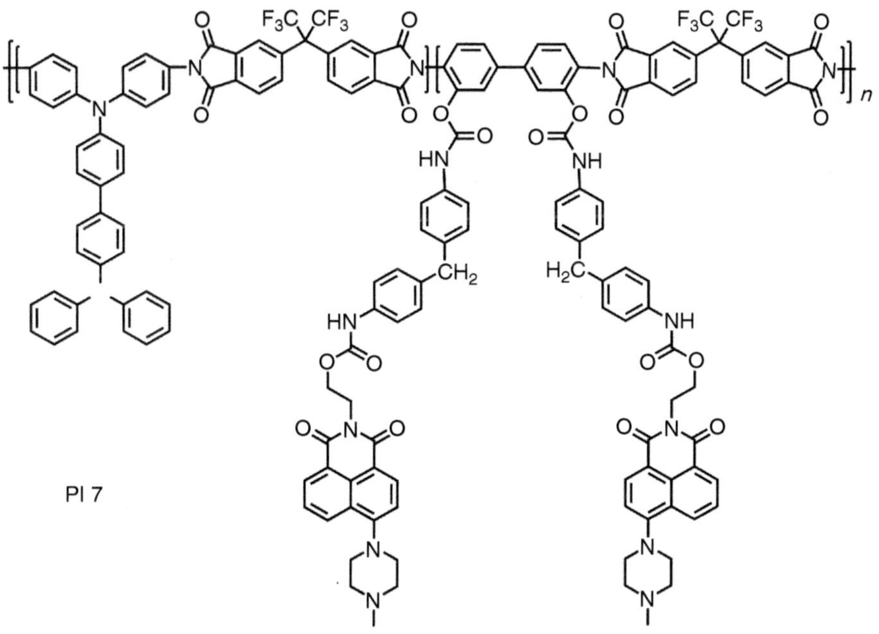

Scheme 7.9 Novel multifunctional polyimides with multiple stimuli-responsive behaviors (pH, voltage, light, and explosives).

7.2.2 Backbone Modulation

7.2.2.1 Extending the Polymer Backbone by Introducing More Electroactive Groups

In order to be useful for applications, EC materials must exhibit long-term stability, multiple colors with the same material, and large changes in transmittance (large ΔT %) between their bleached and colored states. Extending the polymer backbone by introducing more packing-disruptive and electroactive TPA groups along the polymer chain is a useful strategy to obtain better solubility in polar solvents, thin film formability, and multiple color transitions in TPA-based PIs/PAs.

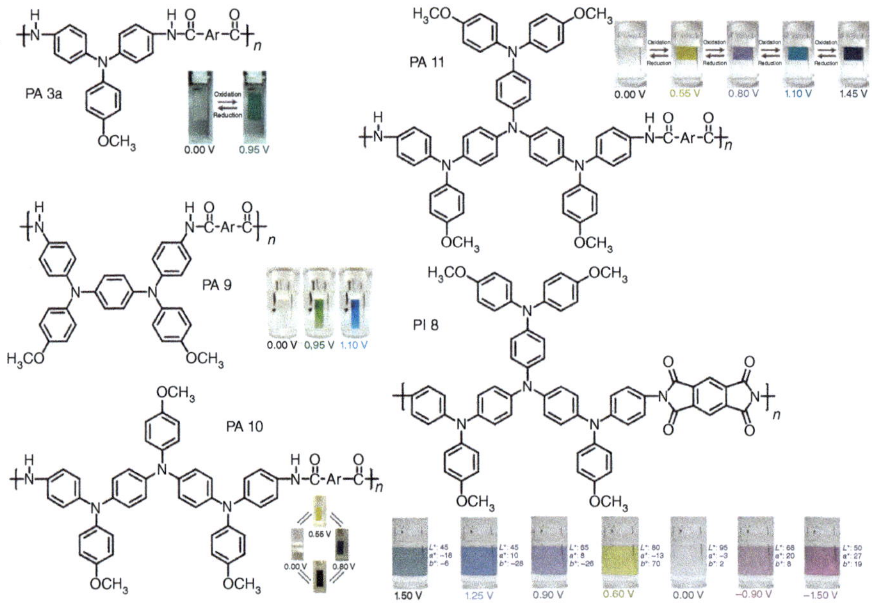

Scheme 7.10 Typical multiple color PIs/PAs with increased electroactive TPA groups along the polymer backbone. Source: (**PA 3a**) Adapted with permission from Ref. [27]. Copyright (2007) Royal Society of Chemistry; (**PA 9**) Adapted with permission from Ref. [47]. Copyright (2009) American Chemical Society; (**PA 10**) Adapted with permission from Ref. [48]. Copyright (2013) American Chemical Society; (**PA 11**) Adapted with permission from Ref. [49]. Copyright (2016) American Chemical Society; (**PI 8**) Adapted with permission from Ref. [50]. Copyright (2015) Royal Society of Chemistry.

In the past few years, Liou and Hsiao and coworkers have developed many kinds of multiple color PIs/PAs by extending the TA-containing backbone (Scheme 7.10) [47–52]. The CV (Figure 7.8a) indicates that the derived polymers containing more TPA units show multiple redox stages and consequently exhibit multiple color transitions. Moreover, the polymer films showed both reversible electrochemical oxidation and electrochromism with high contrast ratio in the NIR region. The broad absorption in NIR region was the characteristic result due to IV–CT excitation between states in which the positive charge is centered at different nitrogen atoms (Figure 7.8b). It is worthy of note that the polyimide film of PI 8, which contains multiple TPA units and pyromellitimide unit along the backbone, exhibited ambipolar electrochemical behavior and multiple colors at different applied potentials due to its four oxidation and two reduction states [49].

Since the concept "black-to-transmissive" was first termed by Reynolds and coworkers in 2008 [53], black EC materials are in urgent demand for practical applications such as displays, windshields, smart sunglasses, and windows. However, the development of black electrochromism still remains a great challenge due to the complexity of absorbing spectrum modulation to cover the whole visible region homogeneously [54]. By the copolymerization of two TA-based polyamides, the absorbing bands and reflected coloration could be merged and offer the potential

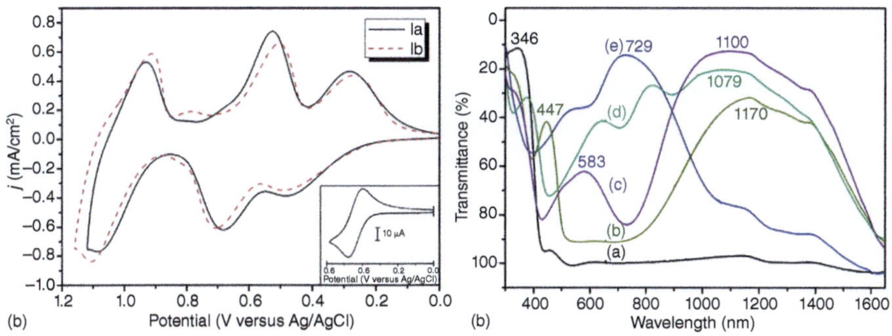

Figure 7.8 (a) Cyclic voltammograms of polyamide PA 11 film onto an indium tin oxide (ITO)- coated glass substrate in CH_3CN containing 0.1 M TBAP at scan rate of 0.05 V/s. (b) Absorption spectral change of PA 11 in CH_3CN solution containing 0.1 M TBAP at various applied potentials between 0.0 and 1.45 (V vs Ag/AgCl). Source: Adapted with permission from Ref. [49]. Copyright (2016) American Chemical Society.

for electrochromic copolymers (ECPs) possessing either highly saturate or darker colors [55]. In addition, the novel EC devices consisted of electroactive PAs with TPPA and TPB units in the backbone and heptyl viologen (HV) in the supporting electrolyte through an absorption-complementary approach demonstrated in a high L^* change (ΔL^*) of 81% and a significant transmittance change (ΔT) of 60% in the visible region (Figure 7.9) [56].

7.2.2.2 Introduction of Amide Linkage into Polyimide Backbone

The poor processability of aromatic polyimides is the critical problem for their optoelectronic applications. Introducing amide linkage into polyimide backbone is therefore an effective approach to overcome this drawback. One of the more facile approaches to prepare poly(amide-imide)s (PAIs) with high molecular weights is the direct polycondensation between imide-bearing dicarboxylic acids with aromatic diamines following the Yamazaki–Higashi phosphorylation technique. Researchers synthesized several aromatic PAIs, which offer a proper compromise between excellent processability and thermal stability (Scheme 7.11) [57–61]. In addition, these resulted films showed reversible electrochemical oxidation redox couples with good EC stability and high contrast of optical transmittance change.

7.2.2.3 Introduction of Ether Linkage into PIs/PAs Backbone

The introduction of flexible ether linkage into the polymer backbone is another popular strategy to increase solubility and lower melting/softening temperature. It has been generally recognized that aryl–ether linkage imparts properties such as better solubility and melt-processing characteristics. TA-based solution-processable poly(ether-imide)s (PEIs) and poly(ether-amide)s (PEAs) can be systematically differentiated by various monomers. Several types of TPA-based solution-processable PEIs were derived from TPA-containing bis(ether-anhydride) (Scheme 7.12 PEI 1) or bis(ether-amine) (Scheme 7.12 PEI 2) monomers. Meanwhile, PEAs could be prepared from bis(p-aminophenoxy)TA (Scheme 7.12 PEA 1)

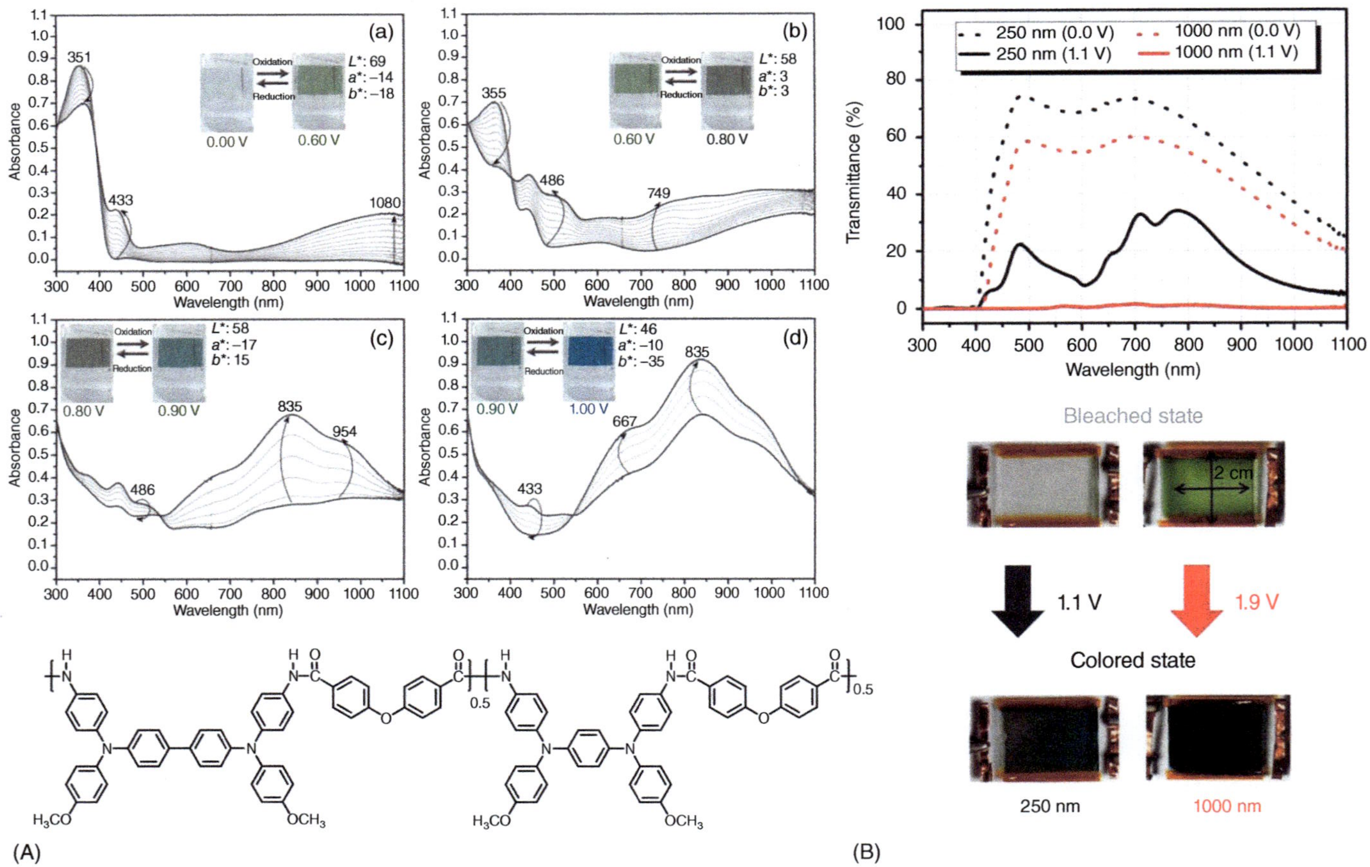

Figure 7.9 TA-based black-to-transmissive EC copolymer (A) and EC devices consisted of three electroactive species (B). Source: Left chart: Adapted with permission from Ref. [55]. Copyright (2011) Royal Society of Chemistry; Right chart: Ref. [56], https://www.nature.com/articles/am201757, licensed under CC BY 4.0.

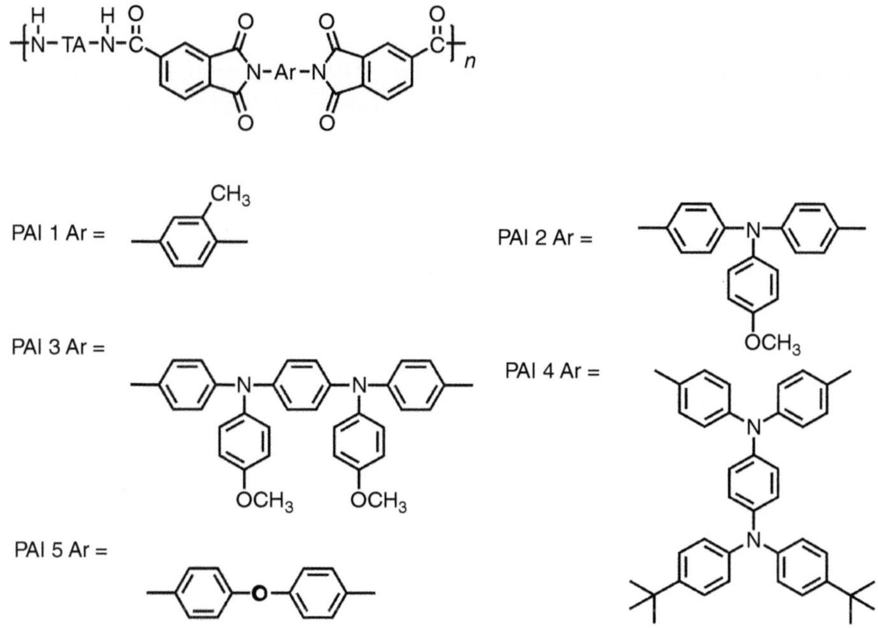

Scheme 7.11 TA-based EC poly(amide-imide)s.

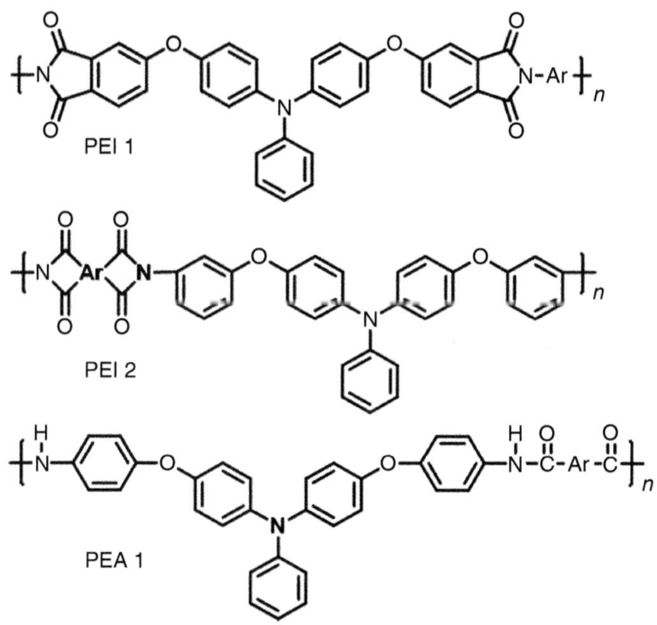

Scheme 7.12 TA-based EC poly(ether-imide)s and poly(ether-amide)s.

via the phosphorylation technique. The resulting PEIs/PEAs exhibited good solubility in many organic solvents. The good solubility of these PEIs and PEAs can be attributed to the presence of three-dimensional TPA units, together with the flexible ether linkage along the polymer backbone. Besides the basic characterizations, the PEI and PEA thin films revealed excellent electrochemical and EC stability.

7.2.2.4 Introduction of Alicyclic Structures into PIs/PAs Backbone

Most of the aromatic PIs revealed too low photoluminescence quantum yield (PLQY) to be used as fluorescent materials. The most critical issue is CT transition between electron-donating diamine and electron-accepting dianhydride moieties of PIs after UV–vis irradiation, resulting in red-shifted and suppressed PL intensity. Based on the aforementioned theory, PL efficiency of PIs could be effectively improved by introducing alicyclic diamines or dianhydrides into the PIs to reduce CT interactions and enhance LE transitions.

Zhou and coworkers prepared polyamide containing bis(diphenylamino)-fluorene moieties with cycloaliphatic and aromatic dicarboxylic acids (Scheme 7.13) [62]. Compared with the polyamide derived from aromatic dicarboxylic acids, the

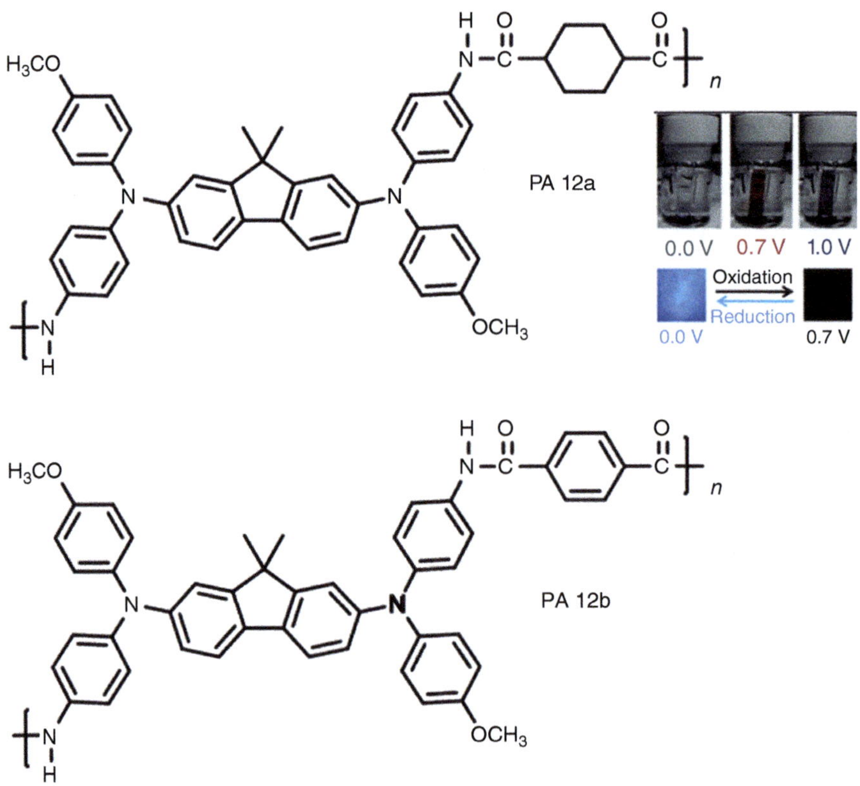

Scheme 7.13 TA-based EFL polyamides with cycloaliphatic (PA 12a) and aromatic (PA 12b) dicarboxylic acids. Source: Adapted with permission from Ref. [62]. Copyright (2016) Royal Society of Chemistry.

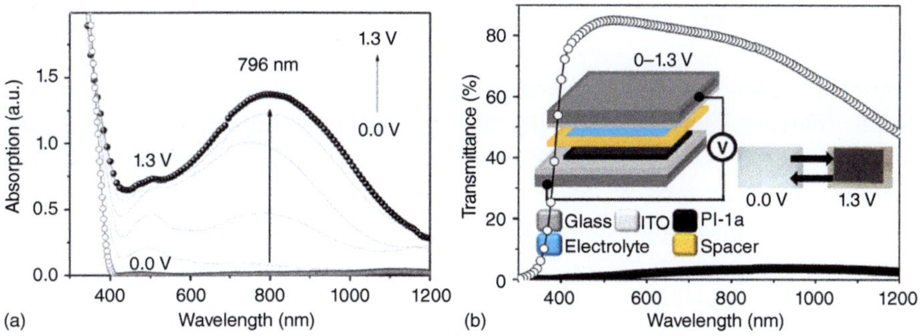

Figure 7.10 (a) Electrochromic behaviors of EC polyimides PI 8a and (b) device transmittance of ITO/PI 8a/electrolyte/ITO. Source: Ref. Zhang et al. [63], https://www.nature.com/articles/s41467-019-09054-8, licensed under CC BY 4.0.

polyamide derived from cycloaliphatic dicarboxylic acid exhibited strong fluorescence with quantum yield up to 50.2%. Furthermore, the fluorescence could be reversibly modulated by electrochemical redox with superior stability and a high contrast ratio of 152.

The introduction of alicyclic structures can also significantly reduce the charge transfer complex (CTC) formation and thus enhance the visible transparency in neutral state. Therefore, it is advantageous to introduce alicyclic structures for higher contrast ratio of the EC PIs/PAs. Recently, Liaw and coworkers prepared colorless-to-colorful EC PIs [63], exemplified by PI 8a (colorless to black) and PI 8b (colorless to blue), with specific molecular designs (Scheme 7.14). The colorless-to-colorful EC PIs with ultrahigh contrast could be especially practical in EC displays and optical areas (Figure 7.10).

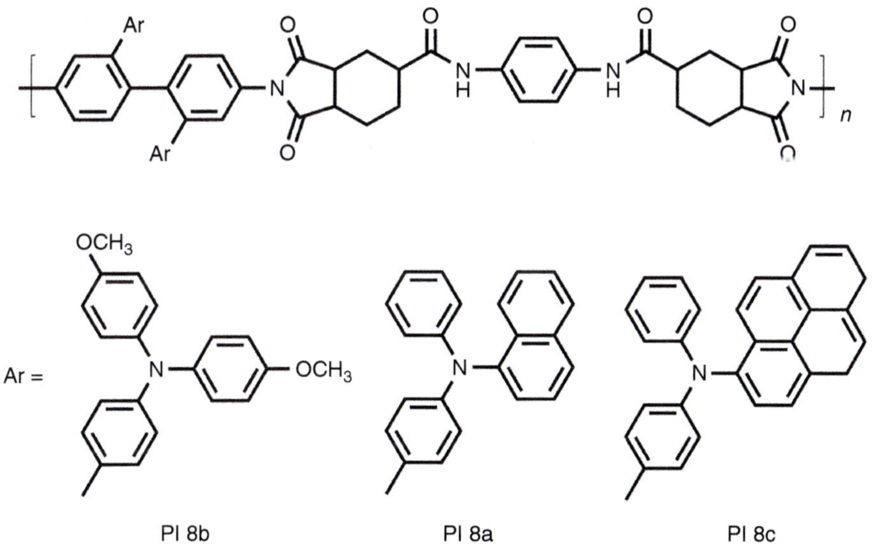

Scheme 7.14 TA-based high-contrast-ratio EC PIs with alicyclic nonlinear diimides.

7.3 Conclusions

In this chapter, we mainly summarized the recent development of TA-based EC materials. TA-based derivatives are promising materials in organic optoelectronics attributing their versatile properties. With emphasis on the design strategies to achieve high-performance, multicolored, and multifunctional TA-based PIs/PAs, many kinds of typical polymers were mentioned in this context. Side group engineering and backbone modulation are two accessible methods to unlock the potential in future applications of TPA-based PIs/PAs. Based on the excellent and interesting results of these EC materials, we believe that the optimization of their devices could further improve the performances and fully exploit the potential of these EC devices in various applications such as self-darkening mirrors, smart windows, EC displays, data storage devices, sunglasses, wearable devices, multifunctional detectors, and EC fabrics. High-performance TA-based EC PIs/PAs with multiple functions and multiple color transitions will carve themselves a useful niche in the coming smart and environment-friendly society of human being.

References

1 Hill, R. and Walker, E. (1948). Polymer constitution and fiber properties. *Journal of Polymer Science* 3 (5): 609–630.

2 Yang, H. (1993). *Kevlar Aramid Fiber*. Wiley.

3 Viehbeck, A., Goldberg, M., and Kovac, C. (1990). Electrochemical properties of polyimides and related imide compounds. *Journal of The Electrochemical Society* 137 (5): 1460–1466.

4 Hu, Z., Fu, W., Yan, L. et al. (2016). Effects of heteroatom substitution in spiro-bifluorene hole transport materials. *Chemical Science* 7 (8): 5007–5012.

5 Fang, Z., Chellappan, V., Webster, R.D. et al. (2012). Bridged-triarylamine starburst oligomers as hole transporting materials for electroluminescent devices. *Journal of Materials Chemistry* 22 (30): 15397–15404.

6 Chou, M.-Y., M-k, L., Su, Y.O. et al. (2004). Electropolymerization of starburst triarylamines and their application to electrochromism and electroluminescence. *Chemistry of Materials* 16 (4): 654–661.

7 Ego, C., Grimsdale, A.C., Uckert, F. et al. (2002). Triphenylamine-Substituted polyfluorene – a stable blue-emitter with improved charge injection for light-emitting diodes. *Advanced Materials* 14 (11): 809–811.

8 Chénais, S. and Forget, S. (2012). Recent advances in solid-state organic lasers. *Polymer International* 61 (3): 390–406.

9 Chen, Q.-D., Fang, H.-H., Xu, B. et al. (2009). Two-photon induced amplified spontaneous emission from needlelike triphenylamine-containing derivative crystals with low threshold. *Applied Physics Letters* 94 (20): 201113.

10 Holzer, W., Penzkofer, A., and Hörhold, H.-H. (2000). Travelling-wave lasing of TPD solutions and neat films. *Synthetic Metals* 113 (3): 281–287.

11 Yen, H.-J. and Liou, G.-S. (2018). Recent advances in triphenylamine-based electrochromic derivatives and polymers. *Polymer Chemistry* 9 (22): 3001–3018.

12 Okamoto, K.-i., Tanaka, K., Kita, H. et al. (1992). Gas permeability and permselectivity of polyimides prepared from 4, 4′-diaminotriphenylamine. *Polymer Journal* 24 (5): 451.

13 Oishi, Y., Ishida, M., Kakimoto, M.A. et al. (1992). Preparation and properties of novel soluble aromatic polyimides from 4, 4′-diaminotriphenylamine and aromatic tetracarboxylic dianhydrides. *Journal of Polymer Science Part A: Polymer Chemistry* 30 (6): 1027–1035.

14 Vasilenko, N., Akhmet'eva, Y.I., Sviridov, Y.B. et al. (1991). Soluble polyimides based on 4, 4′-diaminotriphenylamine. Synthesis, molecular mass characteristics and solution properties. *Polymer Science USSR* 33 (7): 1439–1450.

15 Oishi, Y., Takado, H., Yoneyama, M. et al. (1990). Preparation and properties of new aromatic polyamides from 4,4′-diaminotriphenylamine and aromatic dicarboxylic acids. *Journal of Polymer Science Part A: Polymer Chemistry* 28 (7): 1763–1769.

16 Imai, Y. (1996). Recent advances in synthesis of high-temperature aromatic polymers. *Reactive and Functional Polymers* 30 (1–3): 3–15.

17 Goodson, F.E., Hauck, S.I., and Hartwig, J.F. (1999). Palladium-catalyzed synthesis of pure, regiodefined polymeric triarylamines. *Journal of the American Chemical Society* 121 (33): 7527–7539.

18 Marcoux, J.-F., Wagaw, S., and Buchwald, S.L. (1997). Palladium-catalyzed amination of aryl bromides: use of phosphinoether ligands for the efficient coupling of acyclic secondary amines. *The Journal of Organic Chemistry* 62 (6): 1568–1569.

19 Cheng, S.-H., Hsiao, S.-H., Su, T.-H., and Liou, G.-S. (2005). Novel aromatic poly (amine-imide) s bearing a pendent triphenylamine group: synthesis, thermal, photophysical, electrochemical, and electrochromic characteristics. *Macromolecules* 38 (2): 307–316.

20 Lambert, C. and Nöll, G. (1999). The class II/III transition in triarylamine redox systems. *Journal of the American Chemical Society* 121 (37): 8434–8442.

21 Robin, M.B. and Day, P. (1968). Mixed valence chemistry-a survey and classification. In: *Advances in Inorganic Chemistry and Radiochemistry*, vol. 10 (eds. H.J. Emeléus and A.G. Sharpe), 247–422. Elsevier.

22 Yen, H.-J., Guo, S.-M., Liou, G.-S. et al. (2011). Mixed-valence class I transition and electrochemistry of bis(triphenylamine)-based aramids containing isolated ether-linkage. *Journal of Polymer Science Part A: Polymer Chemistry* 49 (17): 3805–3816.

23 Szeghalmi, A.V., Erdmann, M., Engel, V. et al. (2004). How delocalized is *N,N,N′,N′*-tetraphenylphenylenediamine radical cation? An experimental and theoretical study on the electronic and molecular structure. *Journal of the American Chemical Society* 126 (25): 7834–7845.

24 Nelson, R. and Adams, R. (1968). Anodic oxidation pathways of substituted triphenylamines. II. Quantitative studies of benzidine formation. *Journal of the American Chemical Society* 90 (15): 3925–3930.

25 Seo, E.T., Nelson, R.F., Fritsch, J.M. et al. (1966). Anodic oxidation pathways of aromatic amines. Electrochemical and electron paramagnetic resonance studies. *Journal of the American Chemical Society* 88 (15): 3498–3503.

26 Liou, G.-S. and Chang, C.-W. (2008). Highly stable anodic electrochromic aromatic polyamides containing *N*,*N*,*N*′,*N*′-tetraphenyl-*p*-phenylenediamine moieties: synthesis, electrochemical, and electrochromic properties. *Macromolecules* 41 (5): 1667–1674.

27 Chang, C.-W., Liou, G.-S., and Hsiao, S.-H. (2007). Highly stable anodic green electrochromic aromatic polyamides: synthesis and electrochromic properties. *Journal of Materials Chemistry* 17 (10): 1007–1015.

28 Hsiao, S.-H. and Wu, C.-N. (2016). Electrochemical and electrochromic studies of redox-active aromatic polyamides with 3,5-dimethyltriphenylamine units. *Journal of Electroanalytical Chemistry* 776: 139–147.

29 Hsiao, S.-H., Liou, G.-S., Kung, Y.-C., and Hsiung, T.-J. (2010). Synthesis and properties of new aromatic polyamides with redox-active 2,4-dimethoxytriphenylamine moieties. *Journal of Polymer Science Part A: Polymer Chemistry* 48 (15): 3392–3401.

30 Yen, H.-J., Guo, S.-M., and Liou, G.-S. (2010). Synthesis and unexpected electrochemical behavior of the triphenylamine-based aramids with ortho- and para-trimethyl-protective substituents. *Journal of Polymer Science Part A: Polymer Chemistry* 48 (23): 5271–5281.

31 Hsiao, S.-H., Wang, H.-M., and Liao, S.-H. (2014). Redox-stable and visible/near-infrared electrochromic aramids with main-chain triphenylamine and pendent 3,6-di-*tert*-butylcarbazole units. *Polymer Chemistry* 5 (7): 2473–2483.

32 Hsiao, S.-H., Hsiao, Y.-H., Kung, Y.-R. et al. (2016). Triphenylamine-based redox-active aramids with 1-piperidinyl substituent as an auxiliary donor: enhanced electrochemical stability and electrochromic performance. *Reactive and Functional Polymers* 108: 54–62.

33 Hsiao, S.-H., Hsiao, Y.-H., and Kung, Y.-R. (2016). Highly redox-stable and electrochromic aramids with morpholinyl-substituted triphenylamine units. *Journal of Polymer Science Part A: Polymer Chemistry* 54 (9): 1289–1298.

34 Yen, H.J., Tsai, C.L., Chen, S.H., and Liou, G.S. (2017). Electrochromism and nonvolatile memory device derived from triphenylamine-based polyimides with pendant viologen units. *Macromolecular Rapid Communications* 38 (9): 1600715.

35 Yen, H.-J. and Liou, G.-S. (2010). Enhanced near-infrared electrochromism in triphenylamine-based aramids bearing phenothiazine redox centers. *Journal of Materials Chemistry* 20 (44): 9886–9894.

36 W-a, C., J-w, C., H-j, N. et al. (2016). Synthesis and electrochromic properties of polyimides with pendent benzimidazole and triphenylamine units. *Chinese Journal of Polymer Science* 34 (9): 1091–1102.

37 Sun, N., Zhou, Z., Chao, D. et al. (2017). Novel aromatic polyamides containing 2-diphenylamino-(9,9-dimethylamine) units as multicolored electrochromic and high-contrast electrofluorescent materials. *Journal of Polymer Science Part A: Polymer Chemistry* 55 (2): 213–222.

38 Meng, S., Sun, N., Su, K. et al. (2017). Optically transparent polyamides bearing phenoxyl, diphenylamine and fluorene units with high-contrast of electrochromic and electrofluorescent behaviors. *Polymer* 116: 89–98.

39 Sun, N., Meng, S., Zhou, Z. et al. (2016). High-contrast electrochromic and electrofluorescent dual-switching materials based on 2-diphenylamine-(9,9-diphenylfluorene)-functionalized semi-aromatic polymers. *RSC Advances* 6 (70): 66288–66296.

40 Sun, N., Feng, F., Wang, D. et al. (2015). Novel polyamides with fluorene-based triphenylamine: electrofluorescence and electrochromic properties. *RSC Advances* 5 (107): 88181–88190.

41 Liou, G.-S., Chen, H.-W., and Yen, H.-J. (2006). Synthesis and photoluminescent and electrochromic properties of aromatic poly(amine amide)s bearing pendent *N*-carbazolylphenyl moieties. *Journal of Polymer Science Part A: Polymer Chemistry* 44 (13): 4108–4121.

42 Hsiao, S.-H., Liou, G.-S., and Wang, H.-M. (2009). Highly stable electrochromic polyamides based on *N*,*N*-bis(4-aminophenyl)-*N*′,*N*′-bis(4-*tert*-butylphenyl)-1,4-phenylenediamine. *Journal of Polymer Science Part A: Polymer Chemistry* 47 (9): 2330–2343.

43 Yen, H.J. and Liou, G.S. (2013). Flexible electrofluorochromic devices with the highest contrast ratio based on aggregation-enhanced emission (AEE)-active cyanotriphenylamine-based polymers. *Chem Commun (Camb).* 49 (84): 9797–9799.

44 Yen, H.J., Chen, C.J., and Liou, G.S. (2013). Novel high-efficiency PL polyimide nanofiber containing aggregation-induced emission (AIE)-active cyanotriphenylamine luminogen. *Chemical Communications (Cambridge)* 49 (6): 630–632.

45 Hu, Y.-C., Chen, C.-J., Yen, H.-J. et al. (2012). Novel triphenylamine-containing ambipolar polyimides with pendant anthraquinone moiety for polymeric memory device, electrochromic and gas separation applications. *Journal of Materials Chemistry* 22 (38): 20394–20402.

46 Zhang, X., Lu, Q., Yang, C. et al. (2019). Multi-stimuli responsive novel polyimide smart materials bearing triarylamine and naphthalimide groups. *European Polymer Journal* 112: 291–300.

47 Yen, H.-J. and Liou, G.-S. (2009). Solution-processable novel near-infrared electrochromic aromatic polyamides based on electroactive tetraphenyl-*p*-phenylenediamine moieties. *Chemistry of Materials* 21 (17): 4062–4070.

48 Liou, G.-S. and Lin, H.-Y. (2008). Synthesis and electrochemical properties of novel aromatic poly (amine–amide) s with anodically highly stable yellow and blue electrochromic behaviors. *Macromolecules* 42 (1): 125–134.

49 Yen, H.-J., Lin, H.-Y., and Liou, G.-S. (2011). Novel starburst triarylamine-containing electroactive aramids with highly stable electrochromism in near-infrared and visible light regions. *Chemistry of Materials* 23 (7): 1874–1882.

50 Chen, C.-J., Yen, H.-J., Hu, Y.-C., and Liou, G.-S. (2013). Novel programmable functional polyimides: preparation, mechanism of CT induced memory, and ambipolar electrochromic behavior. *Journal of Materials Chemistry C* 1 (45): 7623–7634.

51 Yen, H.-J., Chen, C.-J., and Liou, G.-S. (2013). Flexible multi-colored electrochromic and volatile polymer memory devices derived from starburst triarylamine-based electroactive polyimide. *Advanced Functional Materials* 23 (42): 5307–5316.

52 Huang, L.-T., Yen, H.-J., and Liou, G.-S. (2011). Substituent effect on electrochemical and electrochromic behaviors of ambipolar aromatic polyimides based on aniline derivatives. *Macromolecules* 44 (24): 9595–9610.

53 Beaujuge, P., Ellinger, S., and Reynolds, J. (2008). The donor–acceptor approach allows a black-to-transmissive switching polymeric electrochrome. *Nature Materials* 7 (10): 795.

54 Li, W., Ning, J., Yin, Y. et al. (2018). Thieno[3,2-*b*]thiophene-based conjugated copolymers for solution-processable neutral black electrochromism. *Polymer Chemistry* 9 (47): 5608–5616.

55 Yen, H.-J., Lin, K.-Y., and Liou, G.-S. (2011). Transmissive to black electrochromic aramids with high near-infrared and multicolor electrochromism based on electroactive tetraphenylbenzidine units. *Journal of Materials Chemistry* 21 (17): 6230–6237.

56 Liu, H.-S., Pan, B.-C., Huang, D.-C. et al. (2017). Highly transparent to truly black electrochromic devices based on an ambipolar system of polyamides and viologen. *NPG Asia Materials* 9 (6): e388.

57 Hsiao, S.-H., Teng, C.-Y., and Kung, Y.-R. (2015). Synthesis and characterization of novel electrochromic poly(amide-imide)s with *N*,*N*′-di(4-methoxyphenyl)-*N*,*N*′-diphenyl-*p*-phenylenediamine units. *RSC Advances* 5 (113): 93591–93606.

58 Wang, H.-M. and Hsiao, S.-H. (2010). Multicolor electrochromic poly(amide-imide)s with *N*,*N*-diphenyl-*N*′,*N*′-di-4-tert-butylphenyl-1,4-phenylenediamine moieties. *Polymer Chemistry* 1 (7): 1013–1023.

59 Hsiao, S.-H., Liou, G.-S., Kung, Y.-C., and Lee, Y.-J. (2010). Synthesis and characterization of electrochromic poly(amide–imide)s based on the diimide-diacid from 4,4′-diamino-4″-methoxytriphenylamine and trimellitic anhydride. *European Polymer Journal* 46 (6): 1355–1366.

60 Cheng, S.-H., Hsiao, S.-H., Su, T.-H., and Liou, G.-S. (2005). Novel electrochromic aromatic poly(amine–amide–imide)s with pendent triphenylamine structures. *Polymer* 46 (16): 5939–5948.

61 Zhang, H., Niu, H., Ji, Y. et al. (2013). Preparation and characterization, stable bismaleimide-triarylamine polymers with reversible electrochromic properties. *Spectrochimica Acta. Part A, Molecular and Biomolecular Spectroscopy* 111: 204–210.

62 Sun, N., Meng, S., Chao, D. et al. (2016). Highly stable electrochromic and electrofluorescent dual-switching polyamide containing bis(diphenylamino)-fluorene moieties. *Polymer Chemistry* 7 (39): 6055–6063.

63 Zhang, Q., Tsai, C.Y., Li, L.J., and Liaw, D.J. (2019). Colorless-to-colorful switching electrochromic polyimides with very high contrast ratio. *Nature Communications* 10 (1): 1239.

8

Metallo-Supermolecular Polymers

8.1 Introduction

A unique class of hybrid electrochromic (EC) material – metallo-supramolecular polymers – has attracted attention and shown benefits as both inorganic and organic EC materials. The synthesis of various functional organic ligands with a single metal ion or hetero-metal ions for application to electronic papers and smart windows has been reported. Reports also show that different metal centers can directly affect polymer characteristics, including viscosity, magnetic moment, fluorescence, and even electrochromism.

Metallo-supramolecular polymers have several electronic transitions based on the metal complex moieties. Firstly, the electronic transitions originate from the metal ion center, mainly including the d–d* transitions in transition metal complexes (such as Fe, Ru, Co, Zn, and Cu) and the d–f* transitions in lanthanide metal complexes (such as Eu) and plus the π–π^* transition from the π-conjugated organic ligand. Therefore, metallo-supramolecular polymers are expected to have various electronic and optical properties based on these transitions. The most typical electronic transition of a metallo-supramolecular polymer is metal-to-ligand charge transfer (MLCT), which is from d orbitals of the transition metal to π^* orbitals of the ligand, as shown in Figure 8.1. In contrast, the ligand-to-metal charge transfer (LMCT) is electronic transitions from π orbitals of the ligand to d* orbitals of the transition metal. Meanwhile, the color of metallo-supramolecular polymers is decided by the bandgap of MLCT, which is controllable by selecting the metal ion species, because different metal ions show different highest occupied molecular orbital (HOMO) potentials. The most distinguished difference in the polymer structure between the conventional organic polymers and metallo-supramolecular polymers is that the chain length of metallo-supramolecular polymers is not fixed in solution unlike the conventional polymers, because the complexation is an equilibrium reaction. In addition, by utilizing the intramolecular metal–ligand and/or metal–metal interactions through the polymer chain, some unique electronic and optical properties, which are difficult to be realized using organic polymers, are expectable in metallo-supramolecular polymers.

Organic Electronics for Electrochromic Materials and Devices, First Edition. Hong Meng.

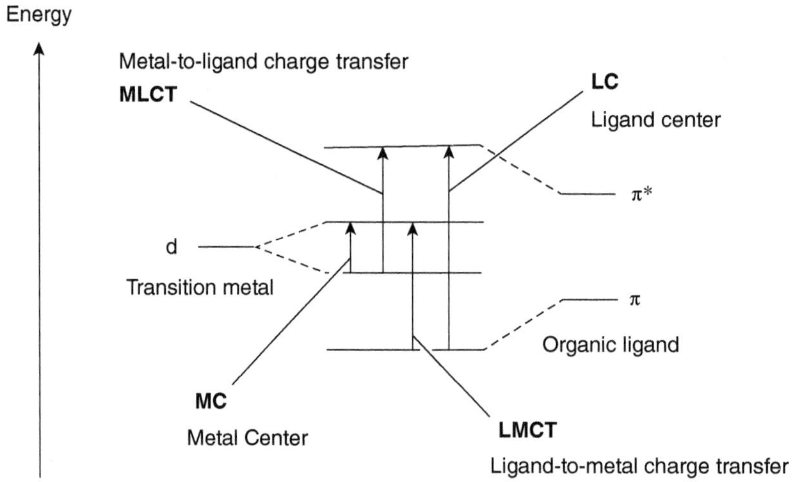

Figure 8.1 Electronic transitions in transition metal complexes. Source: Higuchi [1]. Reproduced with permission of Springer Nature.

8.2 Single Metallic System

8.2.1 Fe(II)- and Ru(II)-Based Metallo-Supramolecular Polymers

The MLCT in the visible region occurs from HOMO of Fe(II) (or Ru(II)) to the lowest unoccupied molecular orbital (LUMO) of the ligand in the complex moieties of the polymers. When the Fe(II) (or Ru(II)) ions are oxidized to Fe(III) (or Ru(III)), an electron is removed from the metal ion, and the charge transfer to the ligand does not happen anymore due to the increase of bandgap by the lowering of the HOMO potential of the metal ions (Figure 8.2a). The MLCT absorption of the Fe(II) and Ru(II) complex moieties appeared around 580 and 520 nm, exhibiting a purple and orange color, respectively. The relevant one-electron reversible redox process from Fe(II)/Ru(II) to Fe(III)/Ru(III) was observed on the cycle voltammetry (CV) curve (Figure 8.2b,c).

The Fe(II)- or Ru(II)-based metallo-supramolecular polymers were usually synthesized by using bis(terpyridine)s as a typical ligand (Figure 8.3a) [2]. To further investigate the influence of electrochromic behaviors by the ligand and metal ion, different substituents (H, OCH_3, Br) at the pyridine periphery of the ligands, different spacer lengths, and different metal ions (Fe and Co) were designed, and corresponding metallo-supramolecular polymers poly-Fe(L1–L5) and poly-Co(L1–L5) were synthesized (Figure 8.3b) [5]. The following conclusions can be drawn: (i) the colors of polymers are readily changed either by the choice of the metal ion or by the design of the ligands, and (ii) metal ion or the design of the ligands both the switching rates and stability of metallo-supramolecular coordination polyelectrolytes (MEPEs) is affected by the electronic nature of the substituents in the pyridine periphery of the ligands. Electron-donating groups give rise to faster switching rates, while electron-withdrawing ones lower the rates

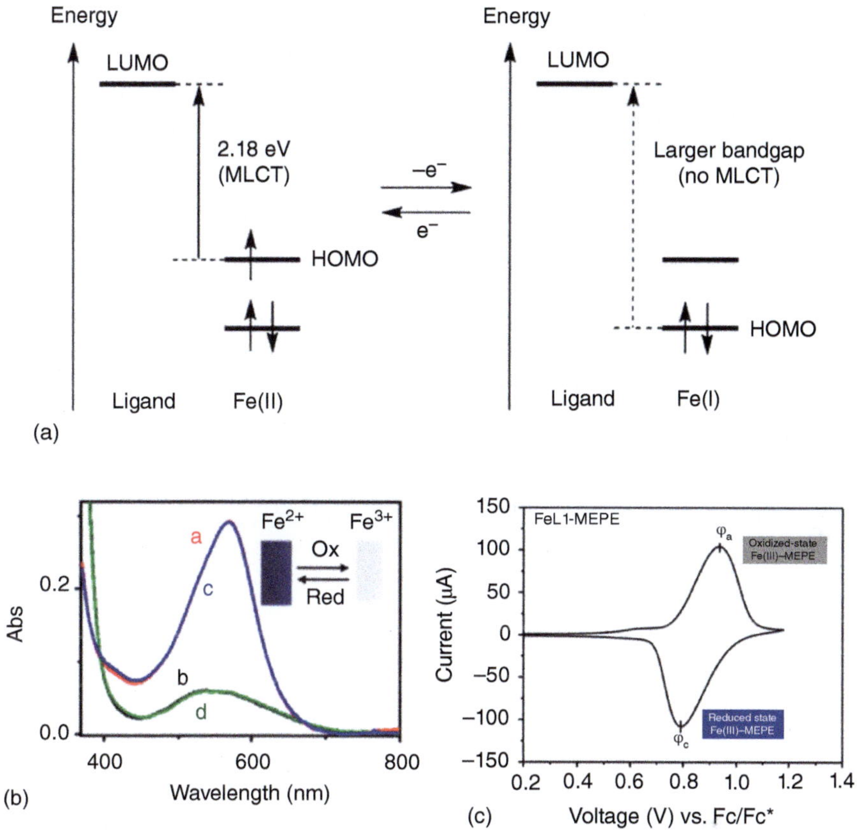

Figure 8.2 (a) A possible mechanism for the electrochromic change in Fe-based metallo-supramolecular. Source: Higuchi [2]. Reproduced with permission of Royal Society of Chemistry. (b) The typical MLCT related with metal ion Fe happened around 580 nm. Source: Shankar et al. [3]. Reproduced with permission of American Chemical Society. (c) A typical CV curve of one-electron Fe(II)/Fe(III) redox process. Source: Pai et al. [4]. Reproduced with permission of American Chemical Society.

and decrease the stability; (iii) the optical memory is affected mainly by steric factors near the metal center. Bulky groups lead to an enhanced optical memory by protecting the metal center from being reduced; (iv) larger conjugated spacers connecting the terpyridine moieties result in a higher molar absorbance of the MLCT transition. Another example reported by Dirk G. Kurth and coworkers shows that the change of ligand from 1,4-bis[2,6-bis(2-pyridyl)pyrimidin-4-yl]benzene to 5,5′-bis[2,6-bis(2-pyridyl)pyrimidin-4-yl]-2,2′-bithiophene can affect the electrochromic properties (Figure 8.3c) [4]. FeL2–MEPE in its oxidized state exhibits a broad absorption band covering the near-infrared (NIR) region (c. 1500 nm) due to the LMCT transition originating due to charge delocalization in the bithiophene spacer. It also shows a high absolute fluorescence quantum yield ($\Phi_f = 82\%$).

By exchanging the counter anion to a fluorescent dye anion (sulforhodamine B [SRB]), Hideyuki Maki and coworkers reported a Fe(II)-based

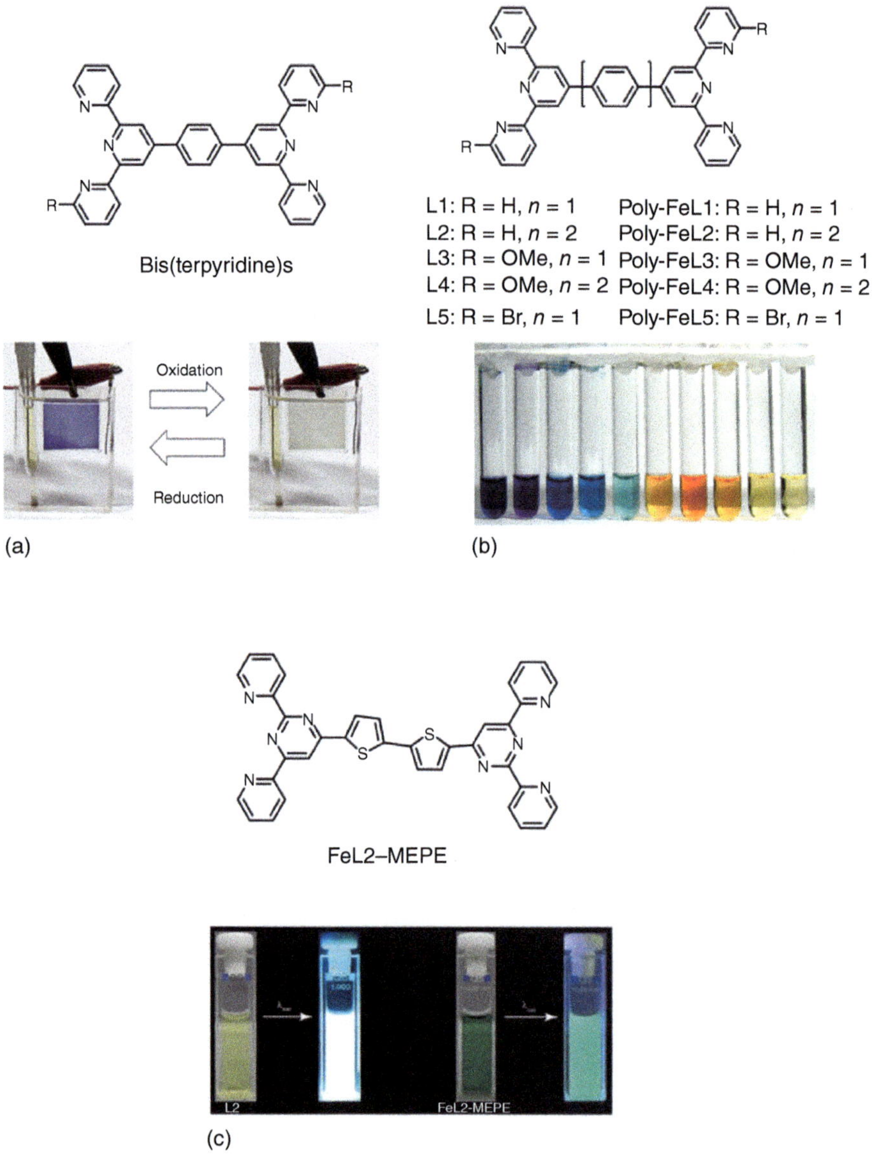

Figure 8.3 (a) The molecular structure of ligand bis(terpyridine)s and the photographic image shows electrochromic phenomenon of polyFeL1. Source: Higuchi [2]. Reproduced with permission of Royal Society of Chemistry. (b) The molecular structure and the photographic image shows the solution colors of poly-Fe(L1–L5) and poly-Co(L1–L5). Source: Han et al. [5]. Reproduced with permission of John Wiley and Sons. (c) The molecular structure and the photographic image shows the colors of L2, FeL2–MEPE. Source: Pai et al. [4]. Reproduced with permission of American Chemical Society.

metallo-supramolecular polymer (polyFe) electrochemically switchable photoluminescence (Figure 8.4) [6]. A solid-state device using the obtained polymer (polyFe–SRB) shows electrochromic properties based on the electrochemical redox of Fe(II) ions by applying at ±2.8 V. The device shows electrochemical switching of photoluminescence at 584 nm by applying ±2.8 V. This switching was caused by the appearance and disappearance of the MLCT absorption of polyFe–SRB. It is considered that the MLCT absorbs the excitation light at 560 nm or the emission from SRB.

Those metallo-supramolecular polymers mentioned earlier are one-dimensional (1D) linear structures. Furthermore, the introduction of a three-dimensional (3D) structure to the metallo-supramolecular polymer backbone is anticipated to affect the surface morphology, solubility, intrafacial and/or interfacial electron transfer speed, maximum absorption wavelength, and thermal properties. 3D structures were synthesized by the stepwise complexation of a Fe(II) salt with different ratios of a linear bis(terpyridine) ligand and a branched tris(terpyridine) ligand (Figure 8.5) [7]. By changing the molar ratio of L1 and L2, significantly different surface morphologies and EC properties were observed. A $FeL1_{85\%}L2_{15\%}$ film with a highly porous surface (pore size: approximately 30–50 nm in diameter) showed the best EC properties among the 3D and 1D polymers (response time for the coloring, 0.19 seconds; response time for the bleaching, 0.36 seconds; ΔT, 50.7%; η, 383.4 cm^2/C), probably because of the smooth anion transfer inside the porous polymer film.

8.2.2 Co^{II}-Based Metallo-Supramolecular Polymers

Co^{II} ion has six coordination sites with an octahedral structure, like Fe^{II} and Ru^{II} ions. Therefore the commonly used ligand bis(terpyridine)s was introduced to coordinate with Co^{II} ion. Dirk G. Kurth and coworkers described the preparation of two novel Co^{II}–bis-terpyridine metallo-supramolecular polymers prepared from a pyridine ring-substituted and unsubstituted ligand (Figure 8.6A) [8]. These polymers showed significant color change and high reversibility. More importantly, they studied the effects of structural factors of the ligands and found that introduction of electron-donating groups at the pyridine ring affects remarkably the charge transfer and electrochemical properties. Co ion is known to have unique properties with water molecules. Co^{II} is likely to form a hexaaqua complex ($Co(H_2O)_6^{2+}$) in aqueous solution. Moreover, mixed-valence Co^{II}/Co^{I} supramolecular complexes have been found to be stable in an aqueous environment [12]. Masayoshi Higuchi and coworkers demonstrated a Co^{II}–bisterpyridine metallo-supramolecular polymer film (polyCo, Figure 8.6B) showing electrochromic switches in aqueous electrolyte solutions [9]. With the reduction of the metal center from Co^{II} to Co^{I}, the polymer thin film exhibits a stable deep black color ($L^* = 32$, $a^* = 0$, $b^* = -8$), which switches in an electrolyte solution of water rather than an organic solvent. Increasing the pH of the aqueous solution results in the thin film displaying further bleaching in the oxidized state. As a consequence, the polyCo thin film exhibits black-to-transmissive electrochromism over a wide attenuation range of 300–1700 nm. The polyCo film

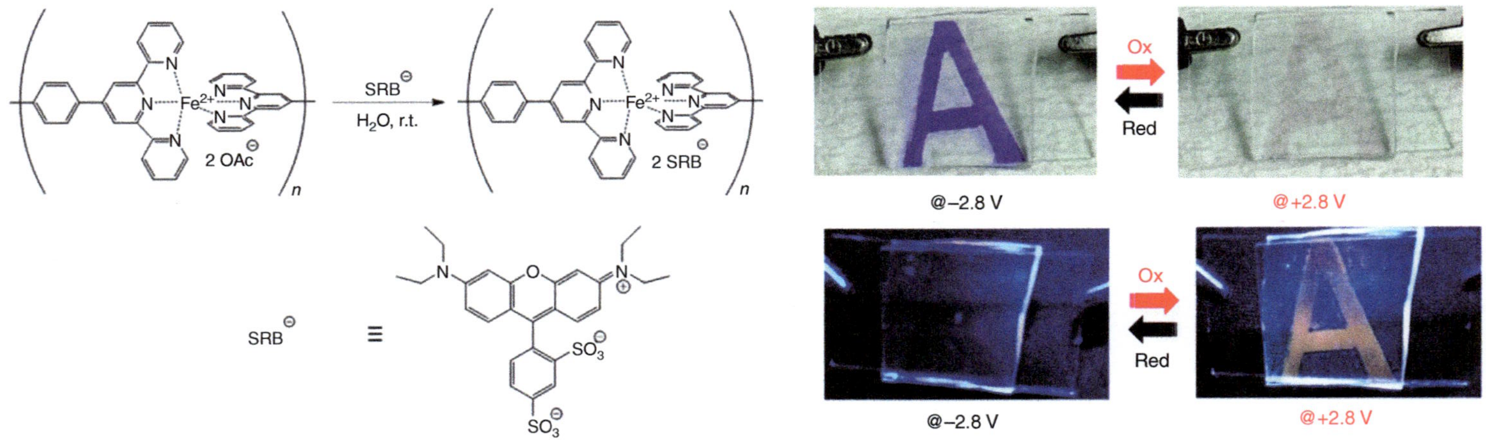

Figure 8.4 Preparation of the Fe(II)-based metallo-supramolecular polymer with SRB anions (polyFe–SRB) and corresponding images of the electrochromic and photoluminescence switching under 365 nm UV light irradiation. Source: Suzuki et al. [6]. Reproduced with permission of Royal Society of Chemistry.

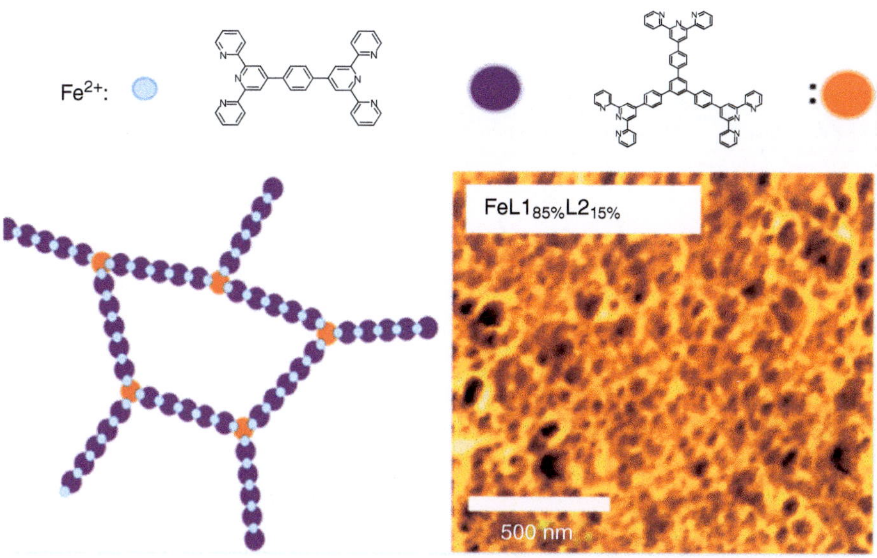

Figure 8.5 Illustration of the 3D structure metallo-supramolecular polymer. Source: Hu et al. [7]. Reproduced with permission of American Chemical Society.

switched in pH 13 aqueous electrolyte can thus be achieved with a transmittance change of 74.3% (81.9–7.6%) at 550 nm. However, the response times become longer in pH 13 solution (t_b = 23.6 seconds; t_d = 31.5 seconds) in comparison to those in a pH 7 solution (t_b = 3.3 seconds; t_d = 6.0 seconds) because of the weakened electron transfer ability associated with higher concentrations of hydroxide ions.

A series of bimetallic $Co^{II}Co^{II}$ complexes (R = H, Me, Ph) were synthesized, incorporating two co-centers within a pyridazine-based building block (Figure 8.6C) [10]. Those metallo-supramolecular polymers showed multi-electron transformations of five redox states, which ascribed to $Co^{II}Co^{II}/Co^{II}Co^{I}$, $Co^{II}Co^{I}/Co^{I}Co^{I}$, $Co^{II}Co^{II}/Co^{II}Co^{III}$, and $Co^{II}Co^{III}/Co^{III}Co^{III}$ couples. This reversible multi-electron transformation makes the polymers a potential application on electrochemical energy storage. Moreover, an electrochemically active conjugated Co^{III} polymer (CP) and a non-conjugated Co^{III} polymer (NCP) were used to fabricate a unique memristive device by molecular engineering and demonstrated that the leakage current tuning in the device is 100 times more efficient than that in a standard device (Figure 8.6D) [11]. The redox switching of an organic-metallic hybrid polymer generated bistable states with a large on/off ratio, which supports random flip-flops for several hours.

8.2.3 Zn^{II}-Based Metallo-Supramolecular Polymers

Zn^{II} ions also have an octahedral geometry with a coordination number of six, like Fe^{II}, Ru^{II}, and Co^{II} ions. Besides, the Zn^{II}-based polymers show outstanding photoluminescent (PL) and electroluminescent (EL) properties [13, 14]. Masayoshi Higuchi and coworkers reported a series of Zn^{II}-based metallo-supramolecular

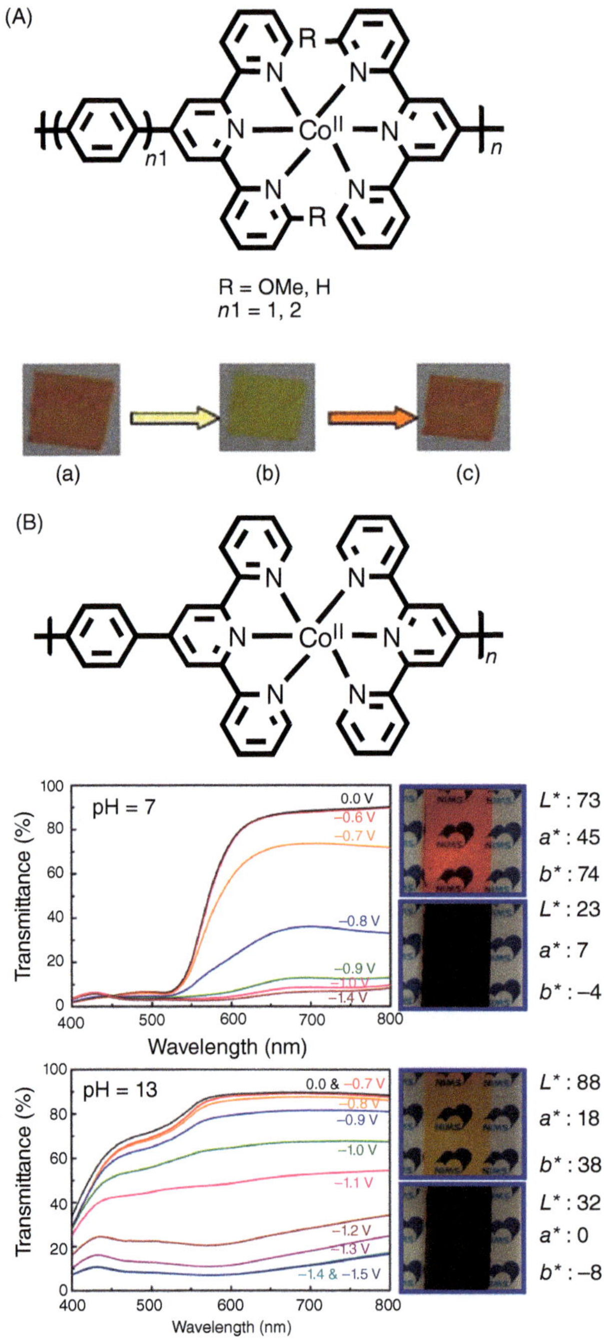

Figure 8.6 (A) Molecular structure of Co^{II}–bis-terpyridine metallo-supramolecular polymers and corresponding film's color change. Source: Han et al. [8]. Reproduced with permission of Elsevier. (B) Molecular structure of polyCo and black electrochromic behavior. Source: Hsu et al. [9]. Reproduced with permission of American Chemical Society. (C) Molecular structure of bimetallic Co^{II}Co^{II} complexes and CV curve. Source: Szymczak et al. [10]. Reproduced with permission of Royal Society of Chemistry. (D) Molecular structure of Co^{III} polymers. Source: Modified from Bandyopadhyay et al. [11].

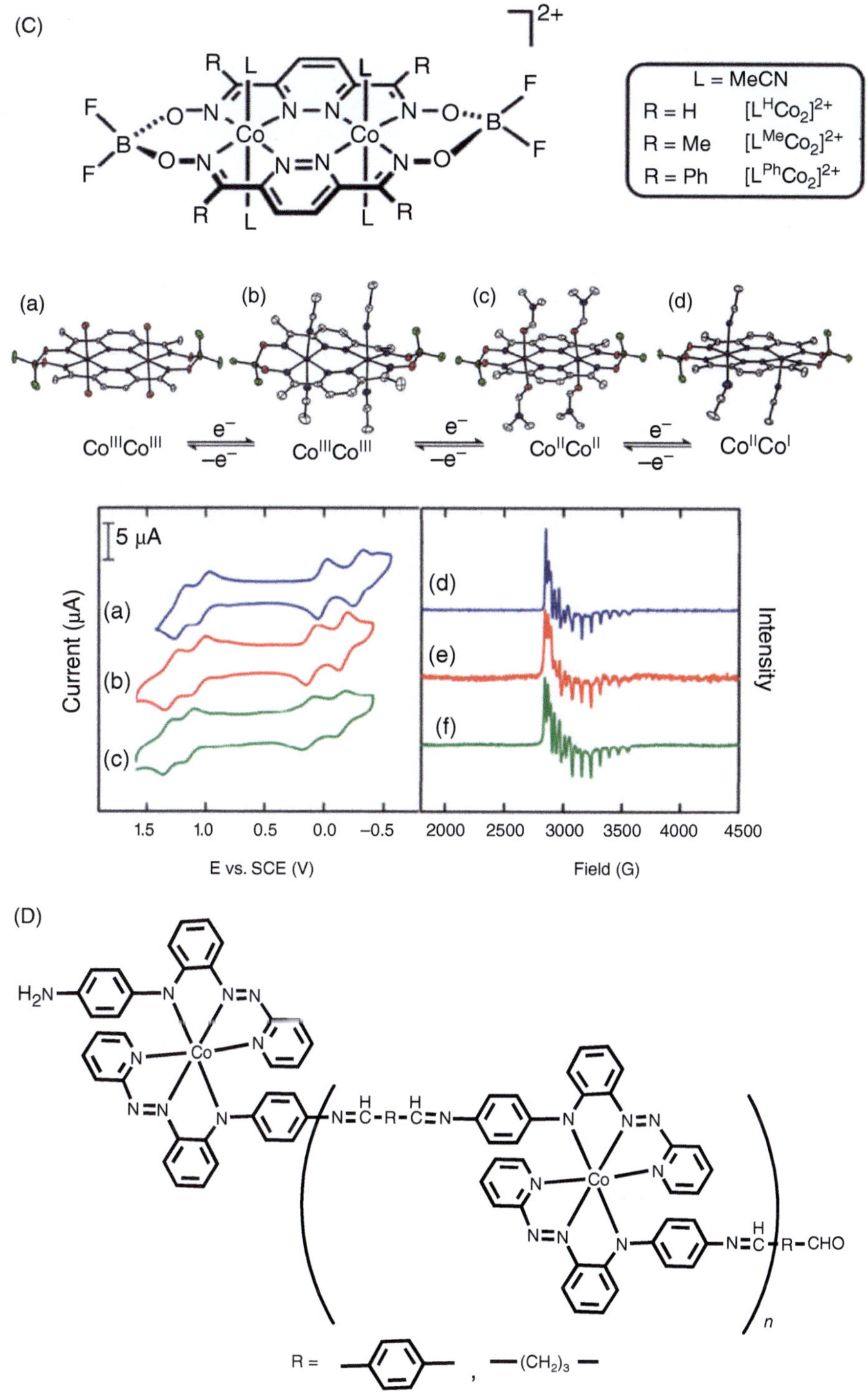

Figure 8.6 (*Continued*)

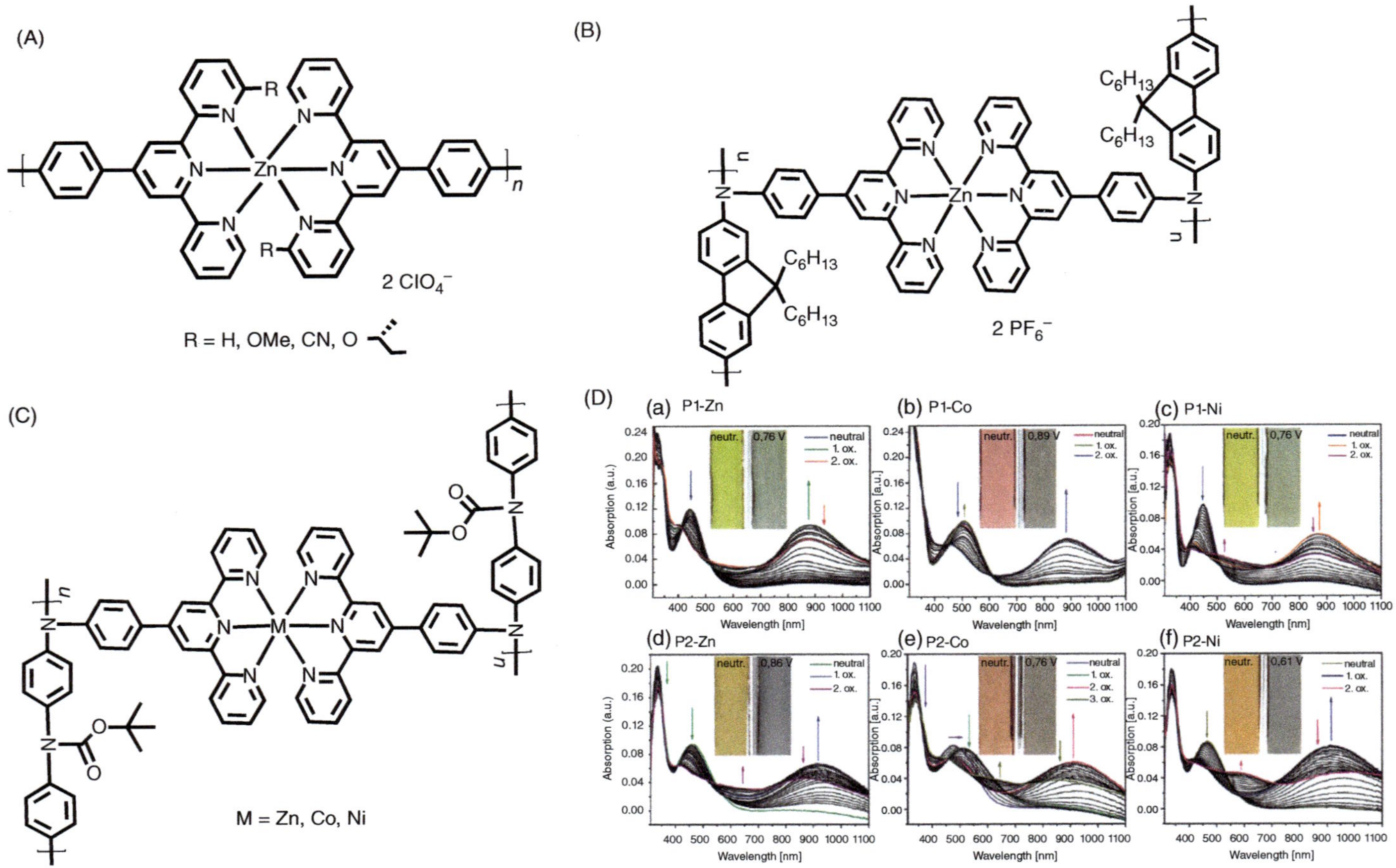

Figure 8.7 (A) Molecular structure of Zn^{II}-based metallo-supramolecular polymers [15]. (B) Molecular structure of Zn^{II}–polyiminofluorene–terpyridine [16]. (C) Molecular structure of the *tert*-butyloxycarbonyl (boc)-modified Zn, Co, and Ni polymers and corresponding spectroelectrochemistry [17]. Source: Reproduced with permission of American Chemical Society.

polymers, synthesized by 1 : 1 complexation of $Zn(ClO_4)_2$ and bis(terpyridine)s bearing electron-donating or -withdrawing groups at the 6-position of the pyridine moiety (Figure 8.7A) [15]. The polymer films showed different luminescence colors from blue to cyan and green based on the large Stokes shift, which was caused by the strong interchain π–π stacking. It is anticipated that the ligand modification at the 6-position of the terpyridine moiety greatly enhanced the stacking, due to the dipole moment change between the coordination site and the spacer in the ligand. Bernd Tieke and coworkers. coordinated Zn^{II} ion with polyiminofluorene–terpyridine ligand and prepared the ultrathin films with electrochromic properties (Figure 8.7B) [16]. The multicolor change from yellow to orange to blue happened due to the oxidation of nitrogen atoms in the backbone. Importantly, the terpyridine (tpy) group is at the side chain position, which enables immobilization of the polymer chains via complex formation with the metal ions. Thus, a highly porous and rigid network structure is formed comparable with metal-organic frameworks (MOFs), which allow for rapid ion transfer and fast switching times. Later they introduced the *tert*-butyloxycarbonyl (boc) protecting group to modify the polyiminofluorene–terpyridine ligand and coordinated with Zn and Co and Ni ions (Figure 8.7C) [17]. The existence of boc group could prevent *N*-nitroso formation and improve the solubility of the polymer in common organic solvents; meanwhile the solution exhibits a strong fluorescence. This result demonstrates that the change of metal ion and aromatic units in the backbone influences the color, oxidation potential, and electrochromic behavior of metallo-supramolecular polymer.

8.2.4 Cu-Based Metallo-Supramolecular Polymers

(a) Cu^{I}-based metallo-supramolecular polymers
In 1996, Matthias Rehahn and coworker firstly used 2- or 9-position linking a bis(1,10-phenanthroline) ligand with a rigid spacer ((4,4″-bis(9-aryl)-2-*o*-phenanthrolinyl)-2′,5′-dihexyl-*p*-terphyl) to construct a homogeneous, high-molecular-mass Cu^{I}-based metallo-supramolecular polymer (Figure 8.8a) [18]. Then in 2014, Kuo-Chuan Ho and coworkers chose Cu^{I} as the metal ion center to coordinate with bis-terpyridine. A cathodically coloring metallo-supramolecular electrochromic material (MEPE–Cu(I)) was synthesized and studied (Figure 8.8b) [19]. The electrochromic mechanism involves the redox reaction of Cu^{I}/Cu^{II} pair (light purple-blue to transparent) and MLCT. A fast response time of less than one second, a coloration efficiency value of 260 cm^2/C and transmittance change (ΔT) of 20% at 580 nm was recorded. An anodic EC material polyaniline–carbon nanotube (PANI-CNT) was chosen to act as the ionic storage layer to fabricate ECD with MEPE–Cu(I) and achieved faster switching time and long-term stability.
(b) Cu^{II}-Based Metallo-Supramolecular Polymers
A main advantage for introducing Cu^{II} ions into metallo-supramolecular polymers is achieving green electrochromism because Cu^{II} complexes are known to be green. Masayoshi Higuchi and coworkers used 1,10-phenanthroline ligand

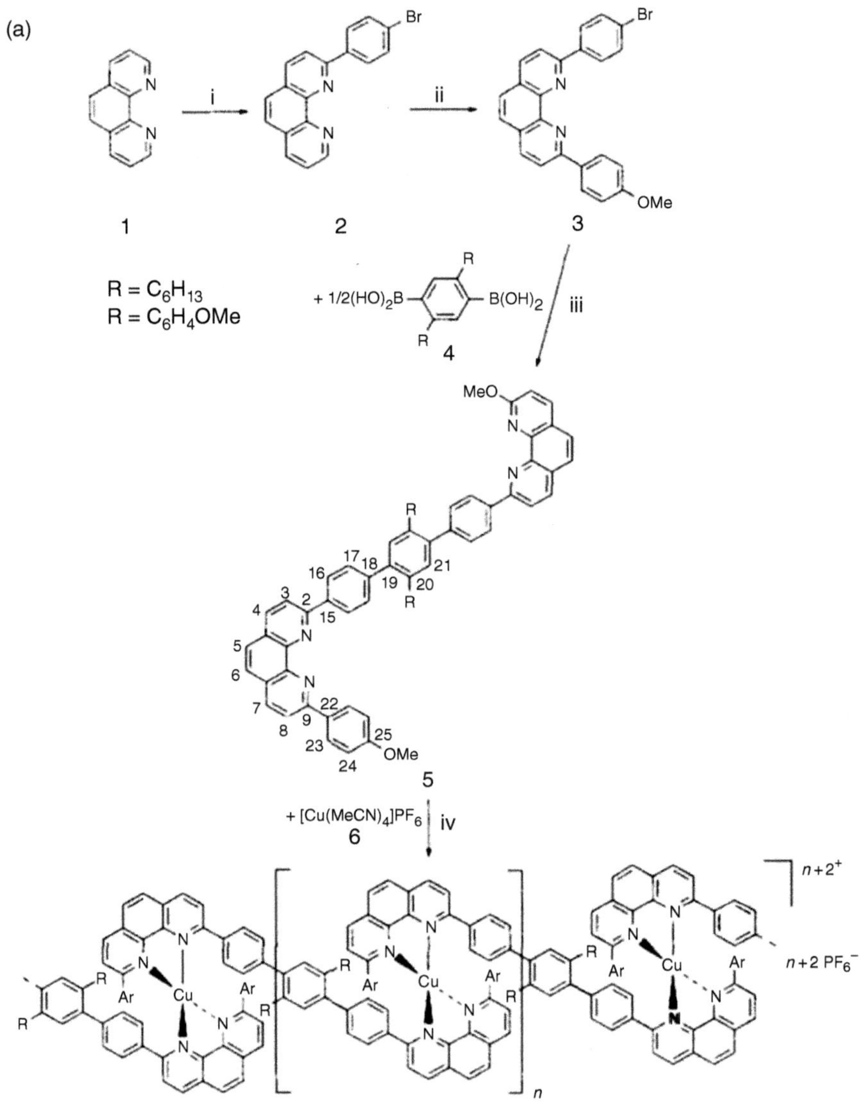

Figure 8.8 (a) Synthetic route of Cu^{I}–*o*-phenanthroline metallo-supramolecular polymer. Source: Velten and Rehahn [18]. Reproduced with permission of Royal Society of Chemistry. (b) Synthetic route of MEPE–Cu(I) and corresponding CV curve. Source: Chen et al. [19]. Reproduced with permission of Elsevier. (c) Synthetic route of CuL1 and CuL2, corresponding visible reflectance spectrum, CIE 1931 chromaticity diagram, CV, and color changes. Source: Hossain et al. [20]. Reproduced with permission of John Wiley and Sons.

(b)

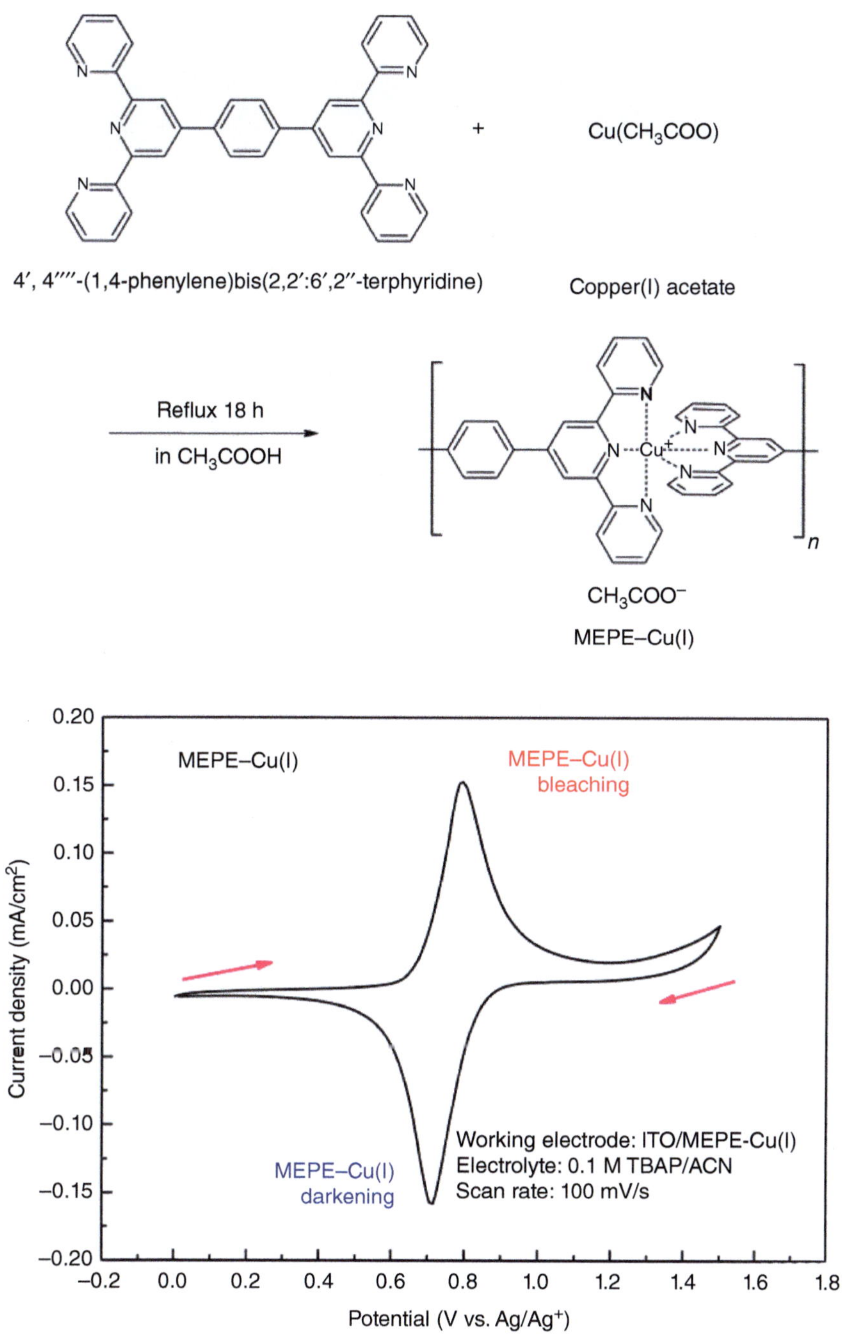

Figure 8.8 *(Continued)*

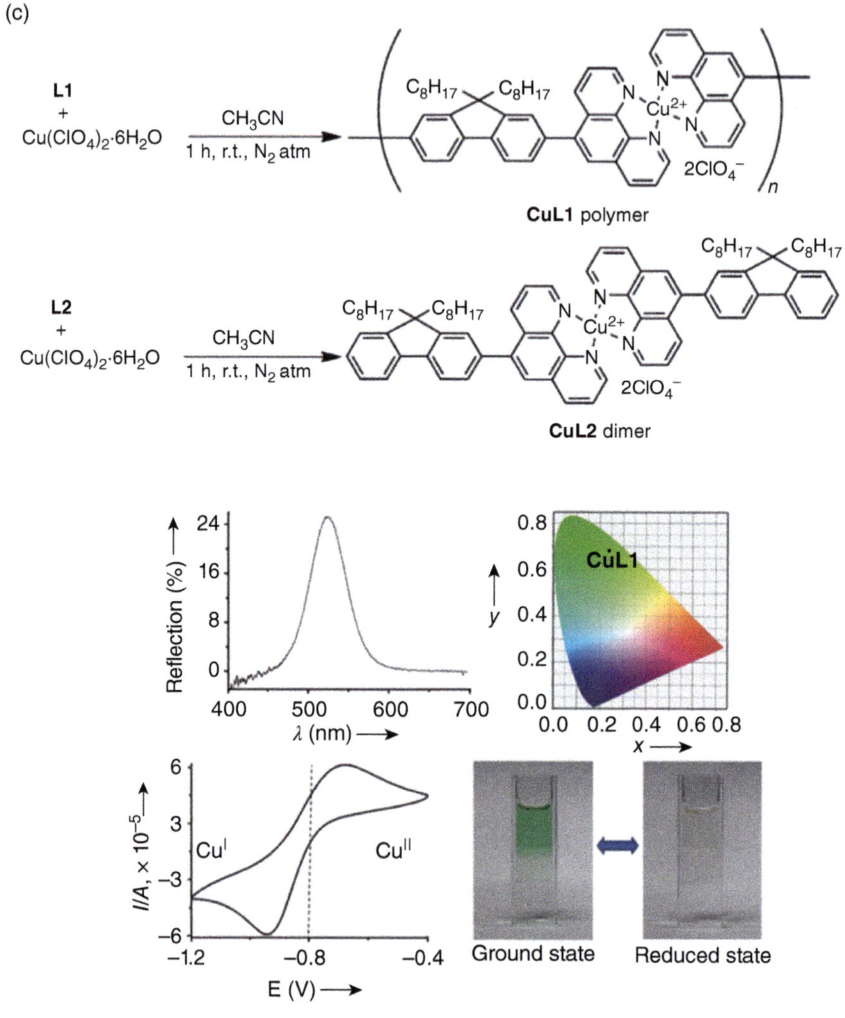

Figure 8.8 (*Continued*)

with fluorene to synthesize a green Cu^{II}-based metallo-supramolecular polymer (CuL1 and CuL2) (Figure 8.8c) [20]. The CuL1 polymer showed a broad MLCT band between 600 and 900 nm, which is centered at 721 nm with a International Commission on Illumination (CIE) 1931 value ($x = 0.236$, $y = 0.675$) and which lies in the green region. The cathodic response in the case of the polymer was due to the reduction of Cu^{II} to Cu^{I} (green to colorless), making it attractive as a green-colored electrochromic material (ECM).

8.2.5 Eu^{III}-Based Metallo-Supramolecular Polymers

Lanthanide metal complexes have unique fluorescent properties. Therefore by introducing Europium (Eu) metal into the metallo-supramolecular polymers, dual

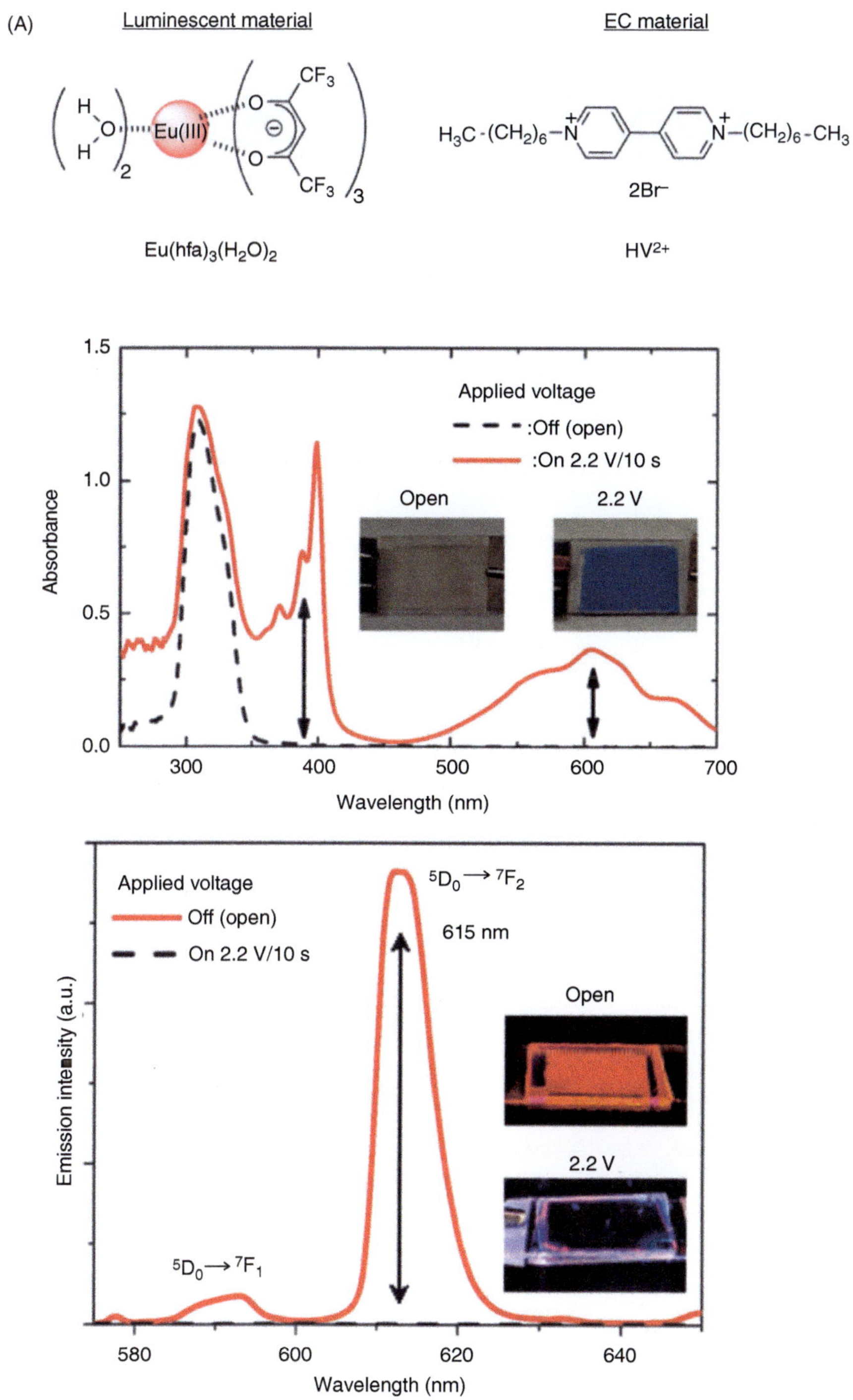

Figure 8.9 (A) Molecular structure of $Eu(hfa)_3(H_2O)_2$ and HV^{2+} and corresponding absorption and emission spectra. Source: Nakamura et al. [21]. Reproduced with permission of Royal Society of Chemistry. (B) Molecular structure of polyEu, UV–vis spectra of ligand complexed with $Eu(NO_2)_3$, and acid–base vapor-induced luminescence switching of a polyEu film. Source: Sato and Higuchi [22]. Reproduced with permission of Royal Society of Chemistry.

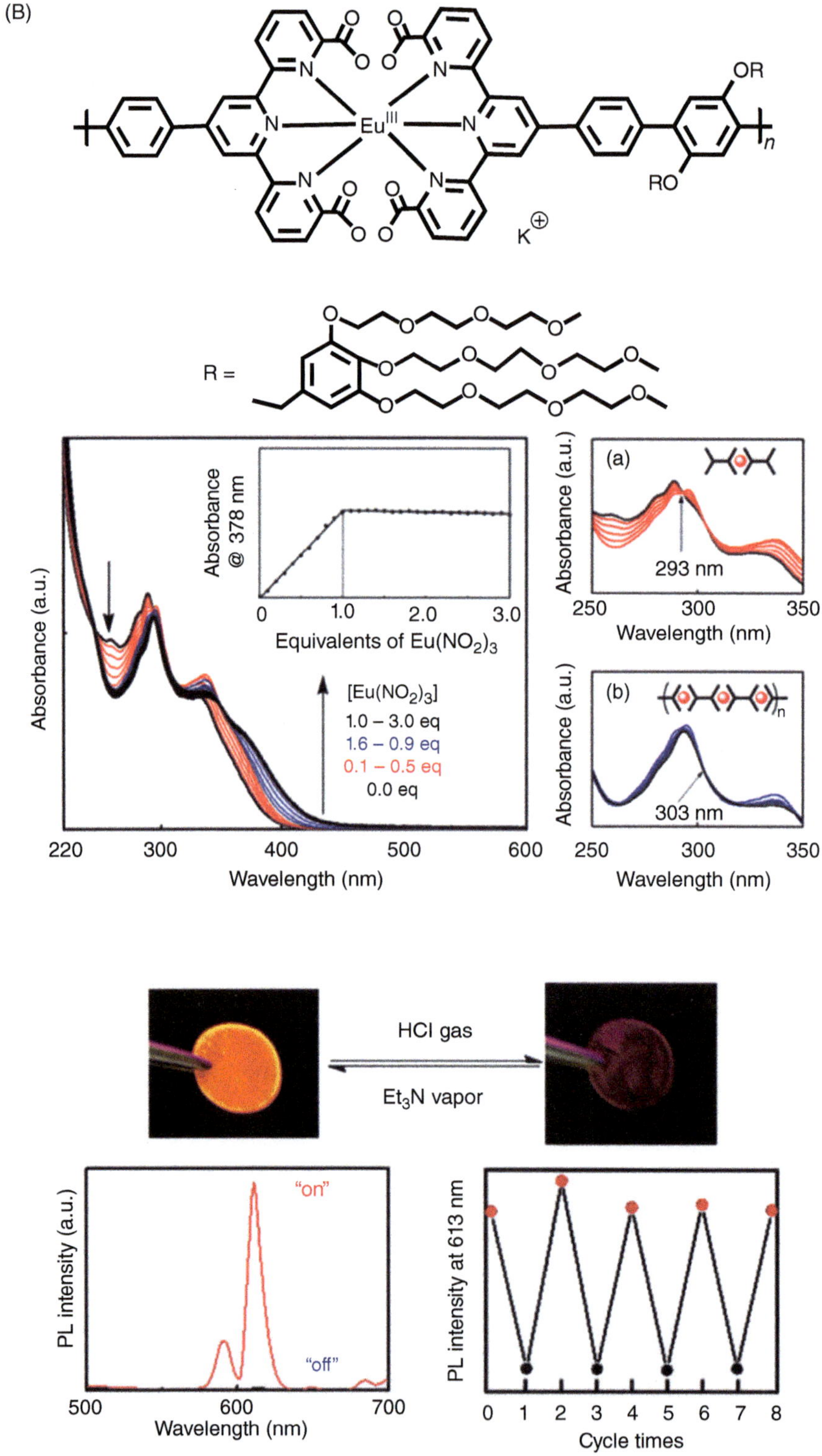

Figure 8.9 (*Continued*)

functions including electrochromism and photoluminescence could be achieved. In 2011, Norihisa Kobayash and coworkers used a luminescent Eu^{III} complex $Eu(hfa)_3(H_2O)_2$ and an electrochromic molecule diheptyl viologen HV^{2+} as a composite material (Figure 8.9A) [21] to realize both luminescence (red emission) and coloration control (colorless to blue). They demonstrated that the photoluminescence of the Eu(III) complex was controlled by the electrochemical coloration of HV^{2+} via intermolecular energy transfer mechanisms. In 2012, Masayoshi Higuchi and coworker reported an Eu-based metallo-supramolecular polymer (polyEu) by coordination polymerization of the Eu ion and a bis(dicarboxylic acid-containing terpyridine) ligand (Figure 8.9B) [22]. This metallo-supramolecular polymer showed reversible photoluminescence switching by exposure to acid–base vapors. Upon selective excitation at 365 nm, a free-standing film of polyEu emitted a red luminescence corresponding to the Eu metal. When this film was exposed to a HCl gas-enriched environment for five seconds, the red luminescence disappeared from the naked eye. After subsequent exposure of the non-luminescent film to triethylamine (Et_3N) for five seconds, the red luminescence was recovered as seen by the naked eye.

8.3 Hetero-Metallic System

In general, metallo-supramolecular polymers involve only one metal ion species. However, in order to get more unique properties, two metal ion species are introduced to the polymer chain regularly. The neighboring metal to metal interactions are anticipated to appear in the polymer. The formation of heteronuclear metal complexes using the different coordination environments of typical ligands has been widely investigated [16, 23–29]. For example, Hui Wang and coworkers reported a heteronuclear Fe^{II}–Cu^{I} complex by hydrothermal synthesis of Cu, Fe salts, and bidentate diimine ligands [16]. Masayoshi Higuchi and coworker alternatingly introduced Fe^{II} and Eu^{III} ions to a metallo-supramolecular polymers backbone with the help of an unsymmetrical ligand bearing both terpyridine and carboxylate-substituted terpyridine moieties [29]. Interestingly, the polymer showed unique and reversible "on–off" switching of the $Eu^{(III)}$ luminescence by electrochemical redox of the Fe ions in the solid-state.

Then many researches focused on alternate introduction of two transition metal ions to the polymer backbone, mainly because the transition metal ions have similar coordination structures such as those with octahedron or tetrahedron geometries and shapes. In 2007, Masayoshi Higuchi and Dirk G. Kurth reported the first concept of electrochromic heterometallo-supramolecular polymers, which consist of Fe^{II} and Co^{II} ions and which produce three distinct colors: red-purple, blue, and colorless upon voltage charging (Figure 8.10A) [30]. Next a heterometallo-supramolecular polymer with alternating Cu^{I} and Fe^{II} ions (polyCuFe) was synthesized by simple 0.5 : 0.5 : 1 complexation of Cu^{I} and Fe^{II} salts with an unsymmetrical ditopic ligand by Masayoshi Higuchi and coworkers [31]. A solid-state device with a polyCuFe film and a gel electrolyte layer showed multicolor electrochromic behavior (from

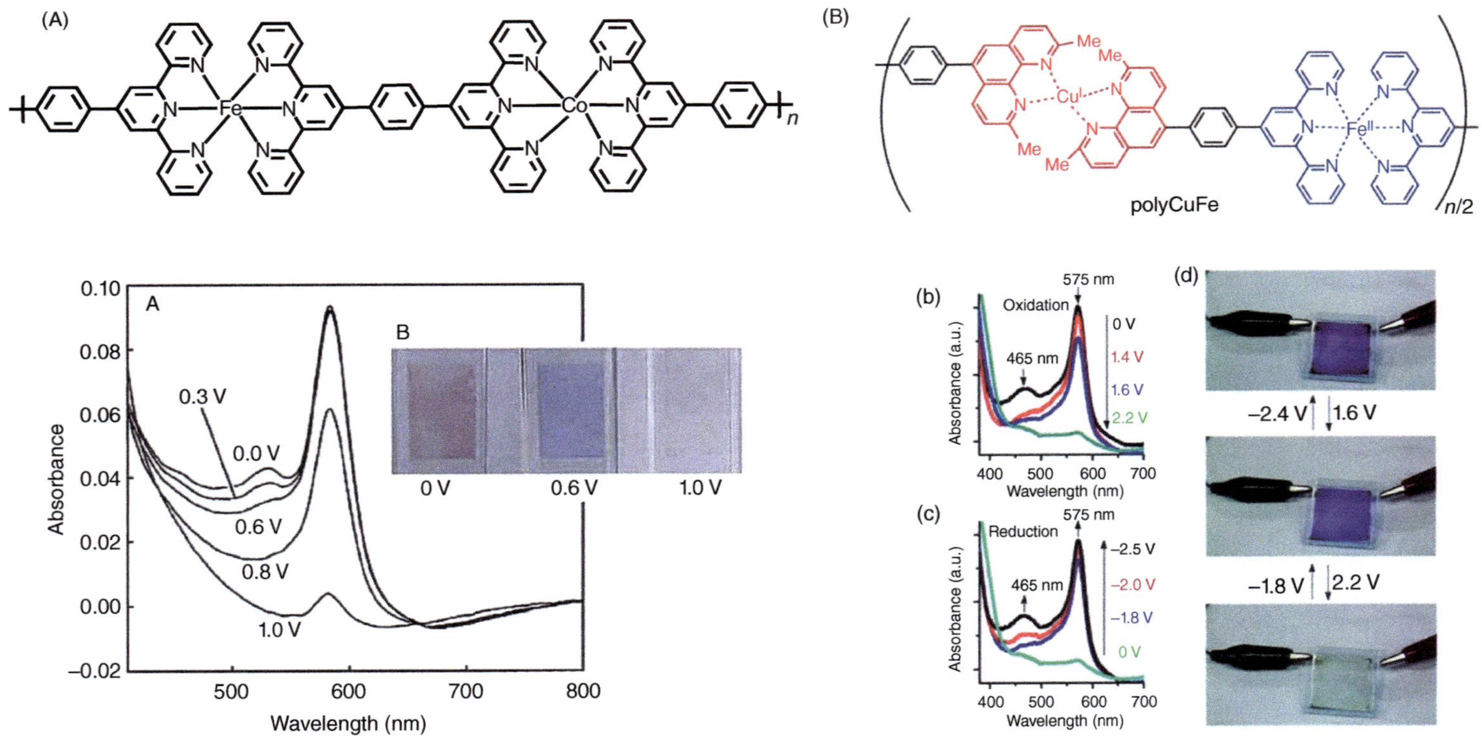

Figure 8.10 (A) Molecular structure of heterometallo-supramolecular polymers Co(II)/Fe(II)−MEPE and corresponding spectral and color changes. Source: Higuchi and Kurth [30]. Reproduced with permission of John Wiley and Sons. (B) Molecular structure of polyCuFe and corresponding spectral and color changes of electrochromic device. Source: Hossain et al. [31]Reproduced with permission of John Wiley and Sons. (C) Molecular structure of Fe^{II}/Ru^{II}-based heterometallo-supramolecular polymers and corresponding spectral and color changes. Source: Hu et al. [32]. Reproduced with permission of Royal Society of Chemistry. (D) Molecular structure of *trans*-polyPtFe and *cis*-polyPtFe and color changes of electrochromic device. Source: Chakraborty et al. [33]. Published by Royal Society of Chemistry. (E) Molecular structure of the heterometallo-supramolecular polymer with Eu^{III} and Fe^{II} ions (polyEuFe). Source: Modified from Sato and Higuchi .[29].

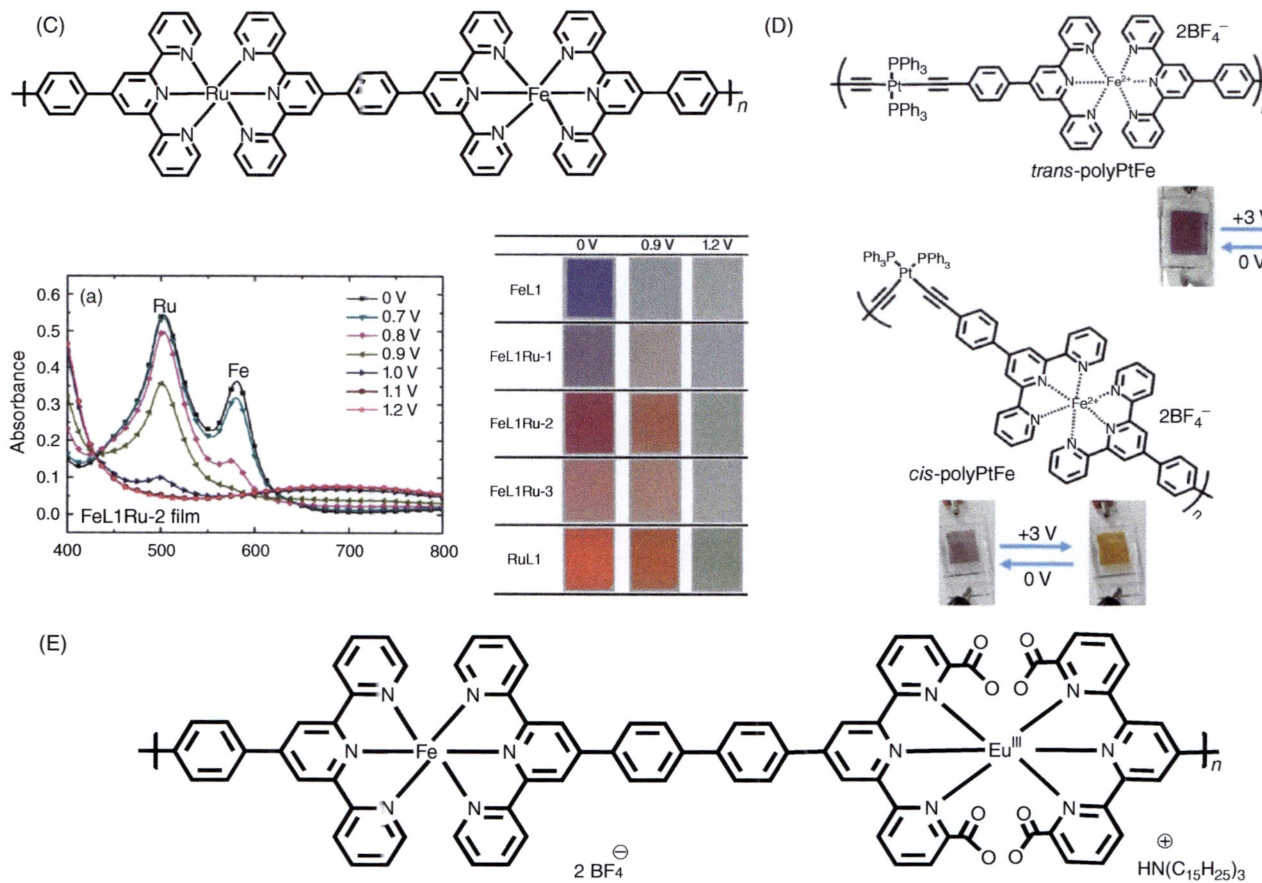

Figure 8.10 *(Continued)*

purple to blue to colorless) due to the stepwise decrease of the MLCT absorptions in the Cu^{I} and Fe^{II} complex moieties with the increase of the voltage applied to the device (Figure 8.10B). Considering the Fe^{II} and Ru^{II} have a similar octahedral coordination structure with six coordination sites, a series of Fe^{II}/Ru^{II}-based heterometallo-supramolecular polymers (polyFe$_{0.75}$Ru$_{0.25}$L1, polyFe$_{0.5}$Ru$_{0.5}$L1, and polyFe$_{0.25}$Ru$_{0.75}$L1) were synthesized by changing the molar ratio of Fe(II) and Ru(II) to 0.75 : 0.25, 0.5 : 0.5, and 0.25 : 0.75 (Figure 8.10C) [32]. An electronic interaction between the two metal ions through the p-conjugated spacer unit of the bis(terpyridine) ligand was observed in the CV diagram. The polymers showed multicolor changes (purple to orange to green), rapid switching time (within two seconds), and high coloration efficiency (242.1 cm^2/C at 508 nm), and the superior durability to the color changes (at least 10 000 cycles) was confirmed. Later, they continued to introduce Pt^{II} and Fe^{II} ions into heterometallo-supramolecular polymers. One important point is they prepared the *trans*- and *cis*-conformational heterometallo-supramolecular polymers (*cis*-polyPtFe and *trans*-polyPtFe) in a precise way, using the 1 : 1 complexation of Fe(II) ions with *cis*- or *trans*-conformational organo-Pt(II) ligands (Figure 8.10D) [33]. The conformational difference between cis- and trans-greatly changed the morphology, crystallinity, ionic conductivity, electrochromic properties, and redox-triggered fluorescence of the polymers. In order to introduce dual function such as luminescence and coloration control into one heterometallo-supramolecular polymer, a Eu^{III}- and Fe^{II}-based heterometallo-supramolecular polymer (polyEuFe) was precisely prepared via stepwise complexation of the metal ions to a new unsymmetrical ligand with both a dicarboxylate-substituted terpyridine and an unsubstituted terpyridine (Figure 8.10E) [29]. PolyEuFe showed reversible "on–off" switching of the Eu^{III} luminescence by the electrochemical redox of the Fe^{II} ions in the solid-state.

8.4 The Fabrication Method of Metallopolymer Film

8.4.1 Layer-by-Layer Self-Assembly and Dip-Coating Methods

Metallopolymers can be simply produced on transparent conducting substrates with layer-by-layer (LbL) self-assembly and dip coating. While LbL self-assembly is useful in producing homogeneous thin films, other coating methods such as dip coating are more convenient for thicker films. Dirk G. Kurth and coworkers presented the metal ion-induced self-assembly of iron(II)-acetate with the rigid ditopic ligand Fe–MEPE thin films, which show excellent electrochromic properties [34]. Further, they carried out a detailed study of LbL deposition and dip coating [35]. Compared with the LbL-assembled Fe–MEPE thin films, the dip-coated Fe–MEPE layers exhibit faster electron transport through the electrochromic layer, and thus the color change is faster. The reason is that the ion exchange and electron transport in LbL–Fe–MEPE/PSS films are slowed down by the sodium polystyrenesulfonate (PSS) layers, which represent negatively charged barriers. Thus, the results show that Fe–MEPE is a very attractive material for electrochromic devices, which can also be built up with flexible substrates such as the tin-doped indium oxide (ITO)-coated polyethylene terephthalate (PET) foil.

Wendel A. Alves and coworkers [36] firstly reported the electrochromic behavior of iron complexes derived from tetra-2-pyridyl-1,4-pyrazine (TPPZ). They used LbL self-assembly method to form the very stable supramolecular complex poly[Fe(TPPZ)FeCN] on the ITO substrate, which probably has a structure similar to Prussian blue (PB). The modified electrodes exhibited excellent electrochromic behavior with both intense and persistent coloration as well as a chromatic contrast of approximately 70%. In addition, this system achieved high electrochromic efficiency (over 70 cm^2/C at 630 nm) and a response time, which could be measured in milliseconds. The electrode was cycled more than 103 times, indicating excellent stability.

Further, thin-film chemistry and surface engineering are useful strategies to form a functional monolayer on the substrate surface. Milko E. van der Boom and coworkers used *p*-chloromethylphenyl-trichlorosilane to treat the substrate and formed the siloxane-based coupling layers (Figure 8.11a) [37]. Subsequently, the colorless chlorobenzyl-functionalized substrates were immersed in a solution of the ruthenium complex. The new ruthenium-based monolayers strongly adhere to the glass and silicon substrates and are insoluble in common organic solvents. And the low-voltage operation and the stability of the monolayer make it a suitable candidate for nonvolatile memory devices. Then in 2015, they used a dip-coating process to generate molecular assemblies. MA1–4 were obtained by alternating the deposition of $PdCl_2$ and iron polypyridyl complexes 1–4 (Figure 8.11b) from solution on a pyridine-terminated monolayer [3]. The combination of low-voltage operation (~1.0 V), fast response time of ~0.30 seconds, high contrast ratios (41%), and high values of coloration efficiency (955–1488 cm^2/C) makes it an excellent EC assembly. Next, they demonstrated a versatile approach for the formation of electrochromic nanoscale assemblies on transparent conductive oxides on both rigid and flexible substrates. By varying the central metal ion of the polypyridyl complexes (Os, Ru, and Fe) and their ligands and by mixing these complexes, coatings with a wide range of colors can be achieved. These coatings cover a large area of RGB (red, green, and blue) color space (Figure 8.11c) [38]. Later Antonino Gulino and coworkers followed this siloxane surface engineering method to design hetero-bimetallic complex-based monolayer comprising two redox-active data-saving bits (Ru^{2+} and Os^{2+}) and conductive imidazole aromatic spine (Figure 8.11d) [39]. Notably, rapid response time was observed for this charge-based bit storage/release (i.e. 0.49 seconds/0.38 seconds for $Os^{2+/3+}$, 0.59 seconds/0.26 seconds for Os^{3+} and $Ru^{2+/3+}$). This ternary-state readout provides a vision to integrate the next generation of "smart electro-optical devices" viable for multistate memory.

8.4.2 Electropolymerized Conducting Metallopolymers

To construct metallo-supermolecular polymers film, electropolymerization is also considered as an effective method. Typically, some functional groups such as vinyl, thiophene, and triphenylamine are designed as the terminal active sites for

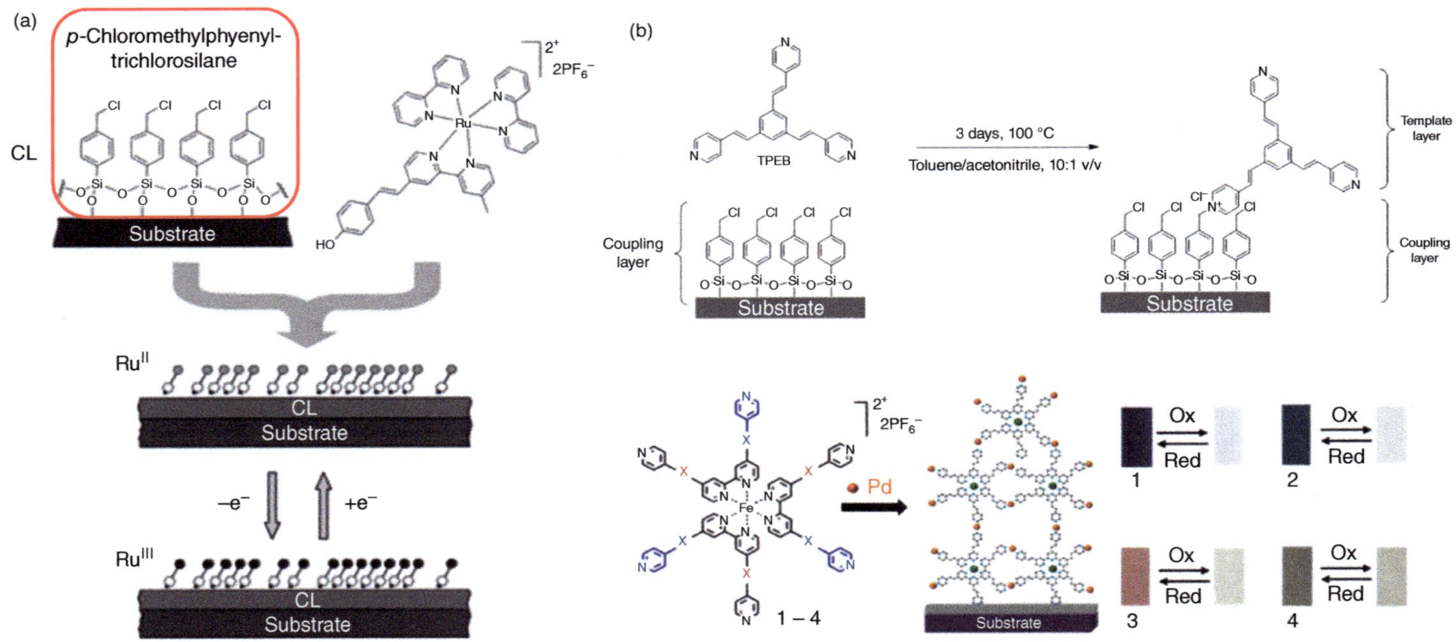

Figure 8.11 (a) Formation of ruthenium-based monolayers on ITO-coated glass. Source: Shukla et al. [37]. Reproduced with permission of John Wiley and Sons. (b) Formation of the iron-based molecular assemblies MA1–4. Source: Shankar et al. [3]. Reproduced with permission of American Chemical Society. (c) Schematics of spin coating of TCOs on glass or PET with electrochromic assemblies by depositing alternating layers of $PdCl_2$ and iron, ruthenium, or osmium polypyridyl complexes. Source: Elool Dov et al. [38]. Reproduced with permission of American Chemical Society. (d) Representation of electro-optical tuning of the 1-ITO monolayer under precise potential command. Source: Kumar et al. [39]. Reproduced with permission of Royal Society of Chemistry.

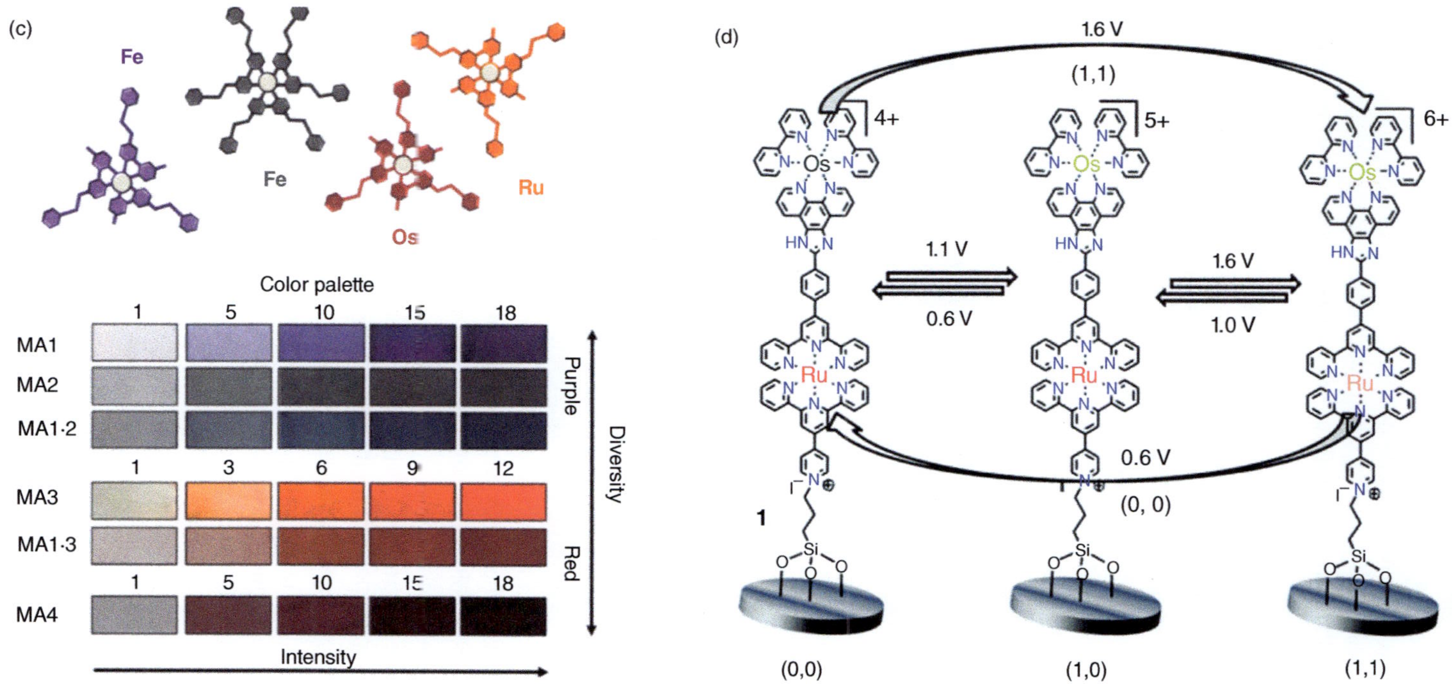

Figure 8.11 (*Continued*)

electrochemical polymerization (Figure 8.12), providing a feasible and controllable way to process large areas of polymers.

In 2008, Robert J. Forster and coworkers reported a metallopolymer containing Ru–O^- coordinated quinoid ligands, using vinyl as electropolymerize group **(1)** [40]. The thin film exhibited reversible three-color change from wine red (reduced) to red-orange (mixed composition) and then to light green (oxidized) in the visible region and an irreversible change in the NIR region. Later, a series of Ru^{II} complexes with N–N bidentate ligands functionalized with vinyl (**2**–**5**) [41–44] have been electropolymerized by Yuwu Zhong and coworkers. These metallopolymer films displayed intense intervalence charge transfer (IVCT) absorptions around 1300, 2050, 1165, and 1070 nm, which presented potential NIR electrochromism application in many civilian and military aspects.

Thiophene is another widely used functional group for electropolymerization. In 2009, Martin T. Lemaire and coworker reported the synthesis and electrical transport properties of a hybrid spin crossover conducting poly(terthiophene) metallopolymer (**6**) [45]. Afterward they prepared a conducting and electrochromic metallopolymer film with a color change from orange to black under applied potential (0.6 V vs Ag), using bidentate π-extended derivatives of the ligand *N*-phenyl-2-pyridinalimine (ppi) (**7**) [46]. Richard A. Jones and coworkers exhibited three Fe^{II} bis(terpyridine)-based complexes with thiophene, bithiophene, and 3,4-ethylenedioxythiophene side chains (**8**) [47]. The electropolymerization provides convenient and controllable film fabrication, and the film exhibited great contrast with a change of transmittance percentage ($\Delta T\%$) of 40% and a high coloration efficiency of 3823 cm^2/C with a switching time of one second. Masaaki Abe et al. developed a Ru^{II} porphycene complex by oxidative electropolymerization of thiophene group (**9**) [48]. The polymeric film exhibits remarkably stable and reproducible electrochromism with high coloration efficiency (CE) values as well as a 1D multiporphycene array through axial coordination. M. Kasım Şener et al. firstly introduced an electropolymerizable EDOT to unsymmetrical metallophthalocyanine ZnPc2 (**10**) [49]. The copolymer showed a greenish-yellow color in the reduced state and a dark blue color in the oxidized state as a film. An optical contrast ratio of 44% at 471 nm was achieved with a fast switching time (2.5 seconds).

Triphenylamine can provide a pair of lone electron on N atom, which can readily form mono radical cation and dual radical cation to lead the corresponding electropolymerization. Dongfang Qiu and coworkers incorporated an electron-donating diphenylamine (DPA) moiety at the 4′-position of the electron-withdrawing terpyridine (TPY) to form a donor–acceptor (D-A) system, which is expected to induce the appearance of an intense n–π^* transition in UV region and much increased MLCT transition of its corresponding Fe(II) complex in visible region (**11**) [50]. Later, Xutang Tao and coworkers synthesized a series of terpyridine–triphenylamine complexes (**12**) [51], and the corresponding polymer films have been electropolymerized to investigate electrochromic properties with attention paid to the metal ion effect and ligand conjugation effect upon the electrochromic colors, memory ability, and the durability of electrochromic films. Yuwu Zhong and coworkers envisaged that when a bisruthenium complex with two terminal diphenyl

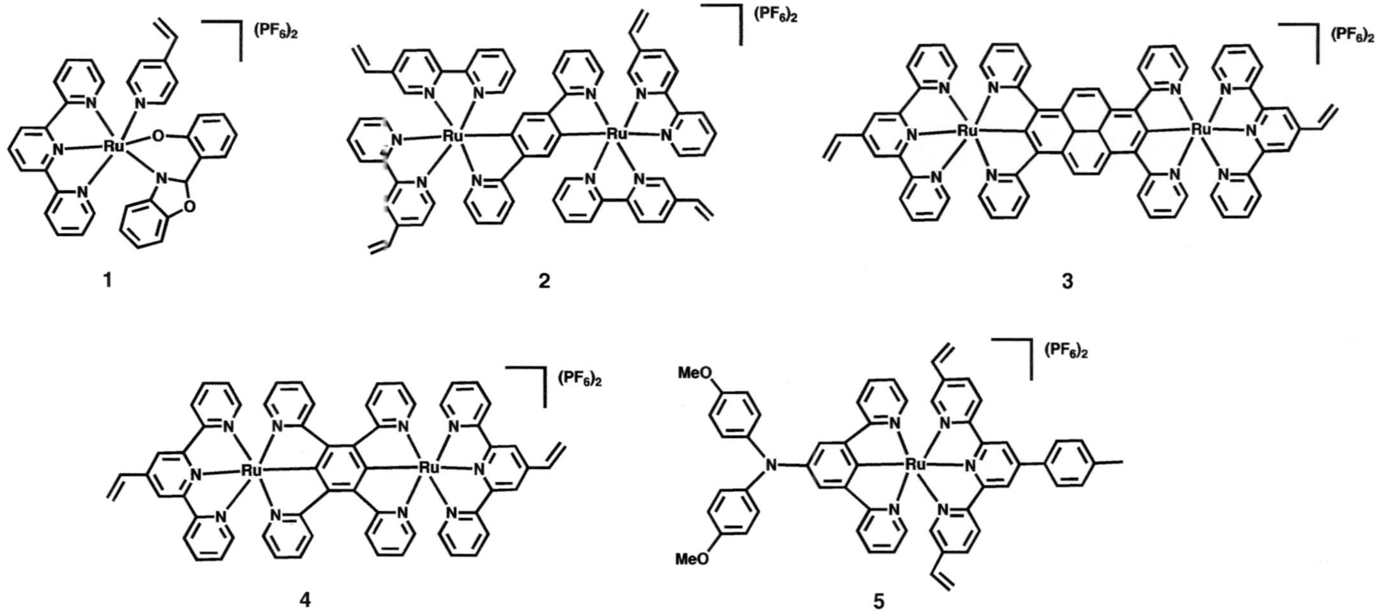

Figure 8.12 Molecular structure of electropc lymerizable monomer.

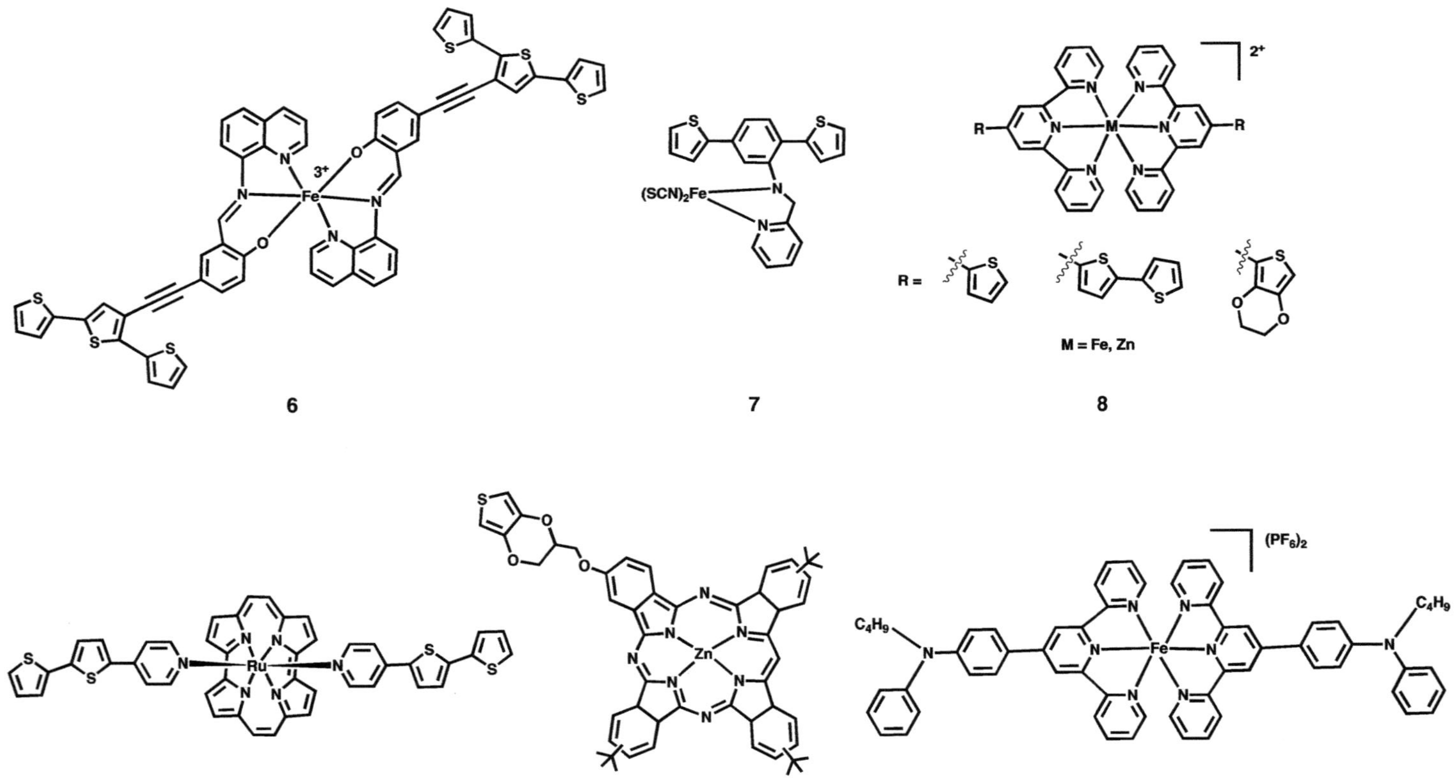

6 7 8

9 10 11

Figure 8.12 (Continued)

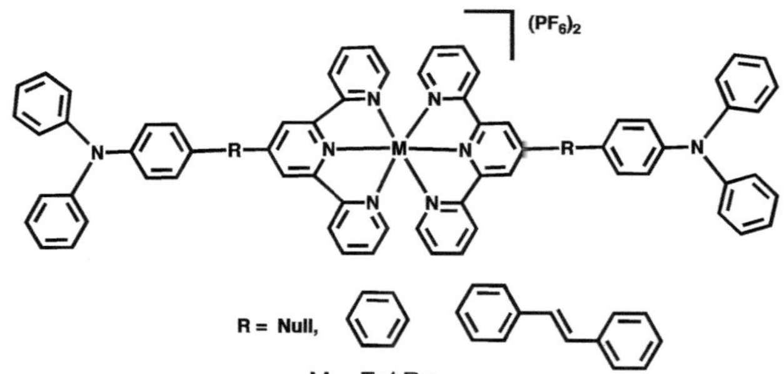

M = Fe′ Ru
12

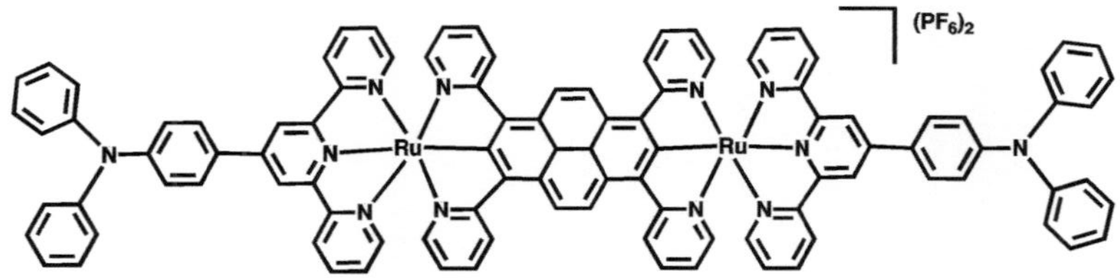

13

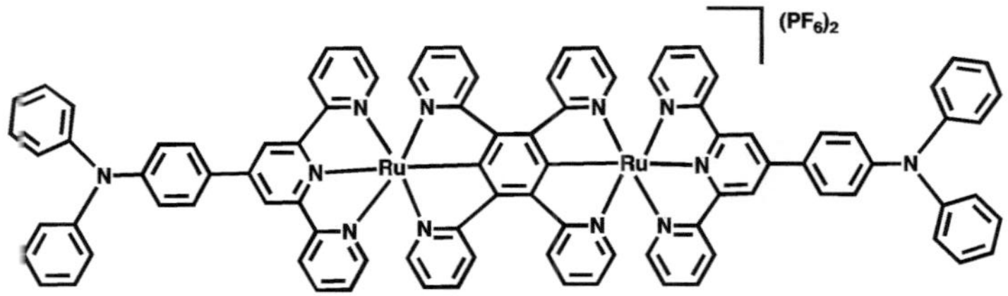

14

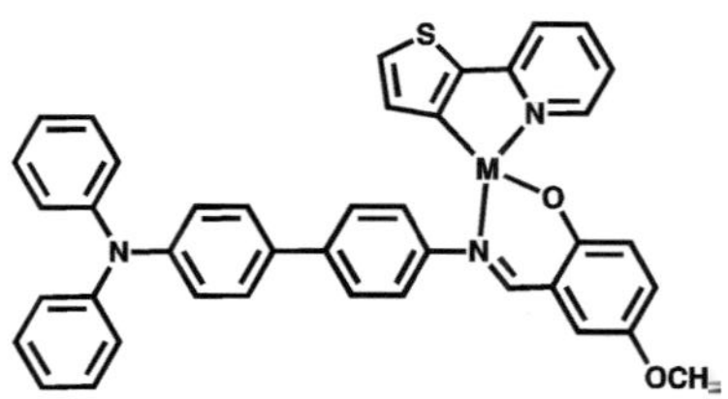

M = Pd, Pt
15

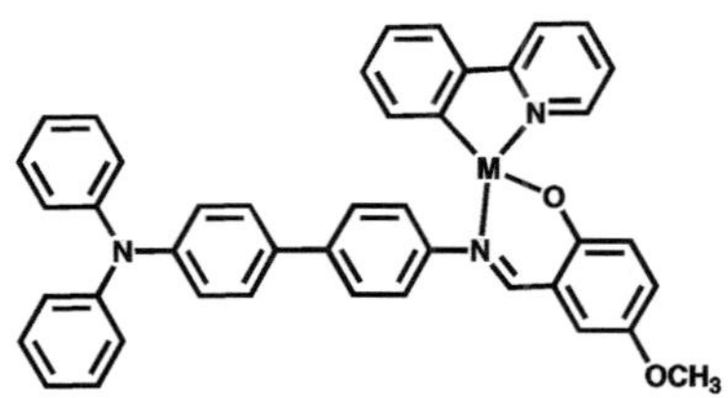

M = Pd, Pt
16

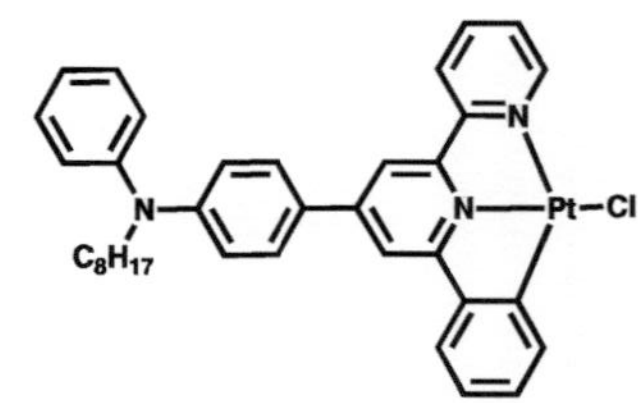

17

Figure 8.12 (*Continued*)

amino groups was subjected to head-to-tail electropolymerization (**13**,**14**) [52], the polymeric film could be deposited on electrode surfaces accompanied by the formation of biphenyl-bridged bis-amine units. As a result, those metallopolymers displayed more than three redox processes as well as multicolor electrochromism, which are very useful for some practical applications such as color displays, optical communication, and e-papers. Nicolas Godbert and coworkers reported a series of cyclometalated complexes with metal ion Pd^{II} or Pt^{II} (**15**,**16**) [53, 54]. They used a triphenylamine functionalized Schiff base to conduct the oxidative electropolymerization. The obtained metallopolymeric thin films present stable reversible redox behavior and typical cauliflower-like textures with a higher

Table 8.1 Summary of the electrochromic behavior of the mentioned electropolymerizable metallopolymers.

Polymer	λ_{test} (nm)	ΔT (%)	$t_{switching}$ (s)	CE (cm^2/C)	Color	References
1	760	—	—	—	Red to green to orange	[40]
2	1300	41	15–20 s	200	—	[41]
3	2050	35	Ox: 6 s Red: 2 s	220	Green to yellow to orange	[42]
4	1165	40	Ox: 6 s Red: 5 s	250	Blue to red to green	[43]
5	1070	52	5–10 s	—	Purple to brown to blue	[44]
6	—	—	—	—	—	[45]
7	—	—	—	—	Orange to black	[46]
8	596	40	1 s	3823	Blue to transparent	[47]
9	609	—	—	178	Blue to green	[48]
10	471	44	2.5 s	—	Yellow to blue	[49]
11	579	20.3	Coloration: 1.1 s Bleaching: 2.2 s	164	Purple to pale yellow	[50]
12	519, 498, and 493	12, 20, and 40	11, 9, and 3 s	—	Orange to yellow Purple to blue	[51]
13	2150	50	15–20 s	550	Brown to orange	[52]
14	1185	35	—	—	Blue to green	[52]
15	864	65	Coloration: 20 s Bleaching: 10.5 s		Yellow to bleached	[53]
16	834	28	—	—	Bleached to dark	[54]
17	664	70.5	Coloration: 5.8 s Bleaching: 5.2 s	441–489	Orange to black	[55]

tendency of the Pt complexes to promote the 3D growth. Meanwhile the polymers show NIR absorption in the oxidized state, high contrast ratios (up to 65%), and low response times. A novel class of Pt^{II} metal coordination complex with arylamine-functionalized cyclometalating ligands were synthesized by Dongfang Qiu et al. (**17**) [55]. The polymers showed efficient electropolymerization behavior to form an electrochromic film with times of 5.8 seconds for the coloration step and 5.2 seconds for the bleaching step, optical contrast of 70.5%, and high coloration efficiency of 441.4 cm^2/C (Figure 8.12 and Table 8.1).

The electrochromic performance of the above-mentioned materials have been summarized in Table 8.1.

8.5 Conclusion

In this chapter, we summarized the recent development of metallo-supramolecular polymers according to the types of metal ion center. The mostly investigated Fe(II)- and Ru(II)-based metallo-supramolecular polymers, other transition metal complexes (such as Co, Zn, Cu) and even the lanthanide metal complex such as Eu have been explored as metallo-supramolecular polymers. By changing the metal ion types and combing hetero-metallic systems, the metallo-supramolecular polymers make it easy to tune the electrochromic colors including RGB, CMY (cyan, magenta, and yellow) and black, as well as multicolor. More importantly, a unique role of metallo-supramolecular polymers is the application as NIR electrochromic materials, due to the intense charge transfer transitions of transition metals in the NIR region, which makes the metallo-supramolecular polymer becoming attractive research filed among many kinds of electrochromic materials. Furthermore, the commonly used fabrication method of metallopolymer film including LbL self-assembly, dip coating, and electropolymerization method have been reviewed. Among them, electropolymerization of monomers was found to be a good in situ method to control the growth of metallopolymers on an electrode surface, but its application is limited by the small areas of electrodes. On the other hand, self-assembly and dip coating are more suitable for large-area fabrication. All in all, the metallo-supramolecular polymers have undergone rapid development these years. The improvement of electrochromic properties and a combination of other functions such as photoluminescence have been achieved through the researcher's efforts.

References

1 Higuchi, M. (2019). Introduction. In: *Metallo-Supramolecular Polymers*, NIMS Monographs (ed. N. Ohashi). Tokyo: Springer.

2 Higuchi, M. (2014). *Journal of Materials Chemistry C* 2: 9331.

3 Shankar, S., Lahav, M., and van der Boom, M.E. (2015). *Journal of the American Chemical Society* 137: 4050.

4 Pai, S., Moos, M., Schreck, M.H. et al. (2017). *Inorganic Chemistry* 56: 1418.
5 Han, F.S., Higuchi, M., and Kurth, D.G. (2007). *Advanced Materials* 19: 3928.
6 Suzuki, T., Sato, T., Zhang, J. et al. (2016). *Journal of Materials Chemistry C* 4: 1594.
7 Hu, C.W., Sato, T., Zhang, J. et al. (2014). *ACS Applied Materials & Interfaces* 6: 9118.
8 Han, F.S., Higuchi, M., Akasaka, Y. et al. (2008). *Thin Solid Films* 516: 2469.
9 Hsu, C.-Y., Zhang, J., Sato, T. et al. (2015). *ACS Applied Materials & Interfaces* 7: 18266.
10 Szymczak, N.K., Berben, L.A., and Peters, J.C. (2009). *Chemical Communications* https://doi.org/10.1039/B913946J6729.
11 Bandyopadhyay, A., Sahu, S., and Higuchi, M. (2011). *Journal of the American Chemical Society* 133: 1168.
12 Schull, T.L., Henley, L., Deschamps, J.R. et al. (2007). *Organometallics* 26: 2272.
13 Yu, S.C., Kwok, C.C., Chan, W.K., and Che, C.M. (2003). *Advanced Materials* 15: 1643.
14 Chen, Y.-Y., Tao, Y.-T., and Lin, H.-C. (2006). *Macromolecules* 39: 8559.
15 Sato, T., Pandey, R.K., and Higuchi, M. (2013). *Dalton Transactions* 42: 16036.
16 Maier, A., Rabindranath, A.R., and Tieke, B. (2009). Fast-switching electrochromic films of zinc polyiminofluorene-terpyridine prepared upon coordinative supramolecular assembly. *Advanced Materials* 21: 959–963. doi:https://doi.org/10.1002/adma.200802490.
17 Maier, A. and Tieke, B. (2012). *The Journal of Physical Chemistry B* 116: 925.
18 Velten, U. and Rehahn, M. (1996). *Chemical Communications* https://doi.org/10.1039/CC99600026392639.
19 Chen, W.-H., Chang, T.-H., Hu, C.-W. et al. (2014). *Solar Energy Materials and Solar Cells* 126: 219.
20 Hossain, M.D., Sato, T., and Higuchi, M. (2013). *Chemistry - An Asian Journal* 8: 76.
21 Nakamura, K., Kanazawa, K., and Kobayashi, N. (2011). *Chemical Communication (Cambridge)* 47: 10064.
22 Sato, T. and Higuchi, M. (2012). *Chemical Communication (Cambridge)* 48: 4947.
23 Dietrich-Buchecker, C., Colasson, B., Fujita, M. et al. (2003). *Journal of the American Chemical Society* 125: 5717.
24 Imbert, D., Cantuel, M., Bünzli, J.-C.G. et al. (2003). *Journal of the American Chemical Society* 125: 15698.
25 Piguet, C., Hopfgartner, G., Bocquet, B. et al. (1994). *Journal of the American Chemical Society* 116: 9092.
26 Thomas, J.A. (2011). *Dalton Transactions* 40: 12005.
27 Fenton, H., Tidmarsh, I.S., and Ward, M.D. (2010). *Dalton Transactions* 39: 3805.
28 Argent, S.P., Adams, H., Harding, L.P. et al. (2005). *New Journal of Chemistry*: 29.
29 Sato, T. and Higuchi, M. (2013). *Chemical Communication (Cambridge)* 49: 5256.
30 Higuchi, M. and Kurth, D.G. (2007). *The Chemical Record* 7: 203.
31 Hossain, M.D., Zhang, J., Pandey, R.K. et al. (2014). *European Journal of Inorganic Chemistry* 2014: 3763.

32 Hu, C.-W., Sato, T., Zhang, J. et al. (2013). *Journal of Materials Chemistry C* 1: 3408.

33 Chakraborty, C., Pandey, R.K., Rana, U. et al. (2016). *Journal of Materials Chemistry C* 4: 9428.

34 Schott, M., Lorrmann, H., Szczerba, W. et al. (2014). *Solar Energy Materials and Solar Cells* 126: 68.

35 Schott, M., Szczerba, W., and Kurth, D.G. (2014). *Langmuir* 30: 10721.

36 da Silva, C.A., Vidotti, M., Fiorito, P.A. et al. (2012). *Langmuir* 28: 3332.

37 Shukla, A.D., Das, A., and van der Boom, M.E. (eds.) (2005). *Angewandte Chemie, International Edition* 44: 3237.

38 Elool Dov, N., Shankar, S., Cohen, D. et al. (2017). *Journal of the American Chemical Society* 139: 11471.

39 Kumar, A., Chhatwal, M., Mondal, P.C. et al. (2014). *Chemical Communication (Cambridge)* 50: 3783.

40 Zeng, Q., McNally, A., Keyes, T.E., and Forster, R.J. (2008). *Electrochimica Acta* 53: 7033.

41 Nie, H.J. and Zhong, Y.W. (2014). *Inorganic Chemistry* 53: 11316.

42 Yao, C.J., Yao, J., and Zhong, Y.W. (2012). *Inorganic Chemistry* 51: 6259.

43 Yao, C.J., Zhong, Y.W., Nie, H.J. et al. (2011). *Journal of the American Chemical Society* 133: 20720.

44 Cui, B.-B., Yao, C.-J., Yao, J., and Zhong, Y.-W. (2014). *Chemical Science* 5: 932.

45 Djukic, B. and Lemaire, M.T. (2009). *Inorganic Chemistry* 48: 10489.

46 Djukic, B., Seda, T., Gorelsky, S.I. et al. (2011). *Inorganic Chemistry* 50: 7334.

47 Liang, Y., Strohecker, D., Lynch, V. et al. (2016). *ACS Applied Materials & Interfaces* 8: 34568.

48 Abe, M., Futagawa, H., Ono, T. et al. (2015). *Inorganic Chemistry* 54: 11061.

49 Göktuğ, Ö., Soganci, T., Ak, M., and Şener, M.K. (2017). *New Journal of Chemistry* 41: 14080.

50 Bao, X., Zhao, Q., Wang, H. et al. (2013). *Inorganic Chemistry Communications* 38: 88.

51 Fan, C., Ye, C., Wang, X. et al. (2015). *Macromolecules* 48: 6465.

52 Yao, C.J., Zhong, Y.W., and Yao, J. (2013). *Inorganic Chemistry* 52: 10000.

53 Ionescu, A., Aiello, I., La Deda, M. et al. (2016). *ACS Applied Materials & Interfaces* 8: 12272.

54 Ionescu, A., Lento, R., Mastropietro, T.F. et al. (2015). *ACS Applied Materials & Interfaces* 7: 4019.

55 Qiu, D., Bao, X., Zhao, Q. et al. (2013). *Journal of Materials Chemistry C* 1: 695.

9

Metal-Organic Framework (MOF)- and Covalent Organic Framework (COF)-Based Electrochromism (EC)

9.1 Introduction

Metal-organic framework (MOF) and covalent organic framework (COF) are highly crystalline materials and normally contain regular nanopores formed by their molecular structure. MOFs have drawn growing research interest since the 1990s [1–7], and COFs also have been a hot topic since it was first synthesized in 2005 [8–11]. Compared with the porous amorphous materials, MOFs and COFs can have regular nano channels, which can act as the highway for the transport of guest molecules, therefore making them superior in many applications, such as supercapacitor, catalyst, gas adsorption, and separation. However, their applications in EC devices are underdeveloped, though they have great potential. The first EC MOF was reported in 2013 [12], and there was no report about EC COF until late 2019 [13, 14]. Up to now, reports in this area are still very limited. In this chapter, the development of EC MOFs and EC COFs are introduced and summarized, and the roadmap was shown in (Figure 9.1). We hope that it will arouse more extensive attention and promote the development in this research area.

9.2 Current Studies in EC MOFs

MOFs are a kind of crystalline coordination polymer with microporous structures, which is formed by self-assembly of metal nodes and organic linkers through coordination bonds. Normally, MOF material contains periodic regular nanopores and shows large specific surface areas. Due to the diversity of metal nodes and organic linkers, the structure and property of MOFs are highly designable and easily tunable. Therefore MOFs can meet the requirements of various applications [16, 17]. MOFs have been an active research topic in recent 20 years, and they have shown great prospects in a number of applications, including gas storage [18], chemical separation [19], catalysis [20], drug delivery [21], and sensing [22]. Very recently, some optoelectronic-active or/and -conductive MOFs have been reported and drawn much attention, because of their potential application in advanced optoelectronic devices [23–27].

Organic Electronics for Electrochromic Materials and Devices, First Edition. Hong Meng.

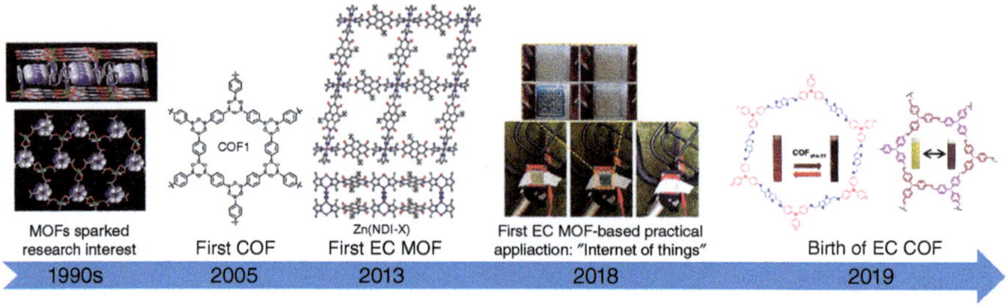

Figure 9.1 The roadmap of EC MOFs and EC COFs. Source: Adapted with permission from Ref. Yaghi et al. [4]. Copyright 1995, Springer Nature; Ref. [12]. Copyright 2013, Wiley; Ref. [15]. Copyright 2018, American Chemical Society; Ref. Hao et al. [13]. Copyright 2019, American Chemical Society; Ref. [14]. Copyright 2019, Elsevier.

Some essential features of MOFs give them advantages in EC, including designable precise molecular structure, simple self-assembly synthesis, and porous structure that facilitate the entry and exit of guest molecules, such as electrolyte ions. However, the reports about the EC application of MOFs are very rare. The reasons may be that (i) MOFs are insoluble materials. The solvothermal methods generally produce MOFs as bulk crystals or microcrystalline powders. It is difficult to process MOFs into homogeneous films, which are desired in devices. (ii) For most of the MOFs that have been reported, their molecular skeletons are non-conductive, and the metal nodes and organic linkers are not redox active.

The research on the EC application of MOFs is still in its infancy. The very limited reports in this area will be introduced in detail in this section.

9.2.1 The Organic Linkers in EC MOFs

9.2.1.1 NDI-Based Organic Linkers

Currently, redox-active naphthalene diimide (NDI) derivatives are the most used organic linkers in EC MOFs [12, 15, 28, 29]. NDI derivatives are classic n-type semiconductors. Their intramolecular charge transfer properties vary with substituents, resulting in different redox properties and colors [30].

In 2013, Prof. M. Dincă's group in Massachusetts Institute of Technology (MIT) reported the first research about the EC application of MOFs [12]. NDI derivatives and Zn^{2+} were applied as the organic linkers and metal nodes, respectively. Three MOF materials (Zn(NDI-H), Zn(NDI-SEt), and Zn(NDI-NHEt)) with different substituents on NDI linkers were synthesized, and their EC properties were explored.

To fabricate homogeneous films of the MOFs, the fluorine-doped tin oxide (FTO)-coated glass substrates were placed in the synthesis reaction mixture. A layer of 1 μM thick homogeneous MOF film was grown on the substrate surface in situ, in four hours. The film was adsorbed on the substrate so tight that it retains even after ultrasonic treatment in *N,N*-Dimethylformamide (DMF). The simulated structures of the three MOFs matched well with powder X-ray diffraction (PXRD) and are shown in Figure 9.2.

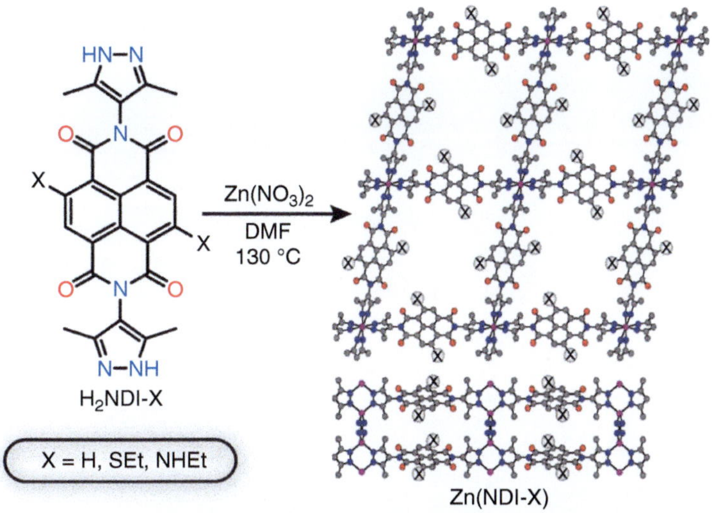

Figure 9.2 The synthesis of the simulated structure of Zn(NDI-H), Zn(NDI-SEt), and Zn(NDI-NHEt). Source: Adapted with permission from Ref. Wade et al. [12]. Copyright 2013, Wiley.

In cyclic voltammetry (CV) measurements, the film of Zn(NDI-H) and Zn(NDI-SEt) shows two quasi-reversible reduction peaks around −0.5 and −1 V, corresponding to $[NDI]^{0/\cdot-}$ and $[NDI]^{\cdot-/2-}$ redox couples, respectively. However, Zn(NDI-NHEt) shows only one moderately reversible reduction peak around −0.9 V. Zn(NDI-H) is stable in the CV cycles. However, the peak current of Zn(NDI-SEt) gradually increases in 50 cycles. Probably, the substituent groups on Zn(NDI-SEt) impede the electrolyte ions from getting access into the channels of MOFs to some extent, at the beginning of several cycles. On the other hand, with the cycle number increasing, the film of Zn(NDI-NHEt) gradually delaminated, resulting in a decrease in peak current. PXRD tests of the three MOFs confirm their crystal structure remained after the CV test, suggesting their good stability. The color changes and the corresponding voltages are shown in Figure 9.3. The coloration efficiencies (CEs) of these three MOFs are higher than 100 cm^2/C at 470–500 nm (blue region), and it is a little lower (60–90 cm^2/C) if calculated from 600 to 700 nm. The CE values are comparable with EC organic polymers.

This research reported by Dincă's group confirms the feasibility of MOF materials as EC materials for the first time, and a facile method for in situ growth of MOF films on substrates is introduced. This work has great significance to the following research of EC MOFs.

In 2016, Prof M. Dincă's group reported another two NDI-based MOFs with different metal nodes, Mg and Ni, named as Mg-NDISA and Ni-NDISA, respectively [28]. Their synthesis and simulated structure are shown in Figure 9.4a. These two MOFs exhibited one-dimensional (1D) channels with a diameter of about 3.3 nm. These large channels can benefit the migration of the electrolyte ions in the EC application.

The MOF films were grown on FTO-coated glass substrates through the same solvothermal method introduced earlier [12]. With a faster nuclear rate, Ni-NDISA

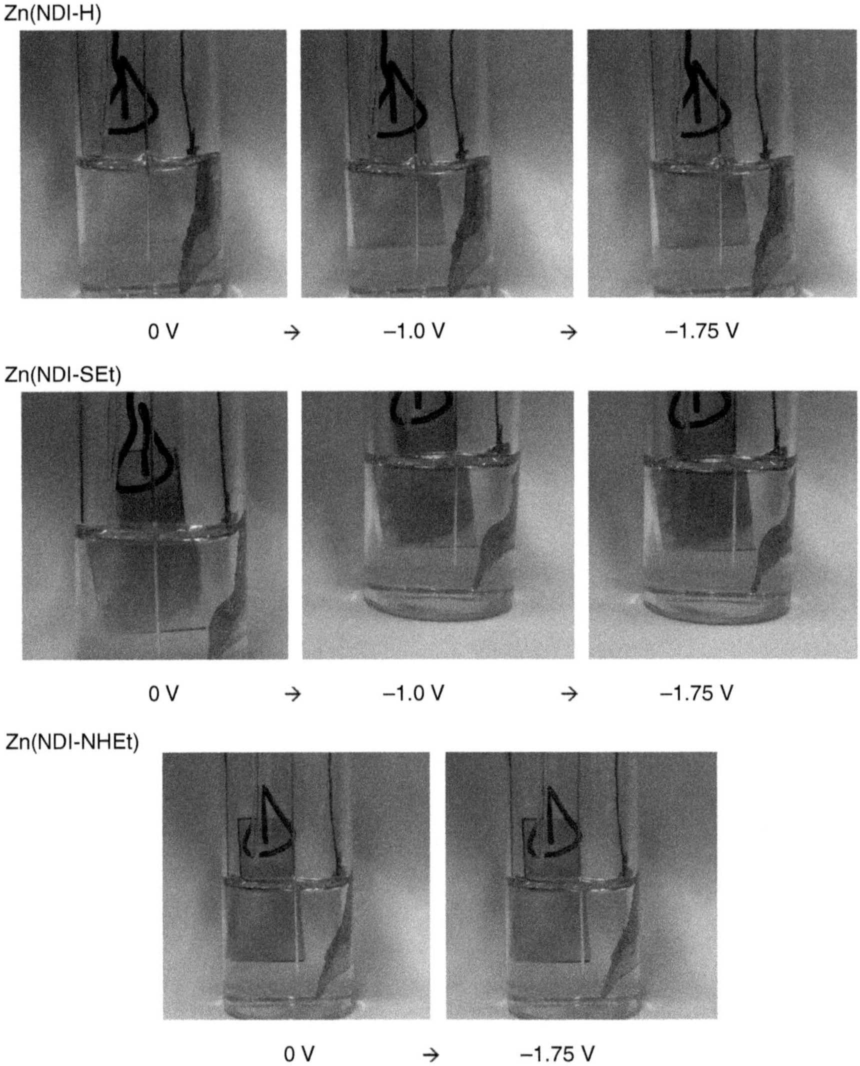

Figure 9.3 The color changes and the corresponding voltages of Zn(NDI-H), Zn(NDI-SEt), and Zn(NDI-NHEt). Source: Adapted with permission from Ref. [12]. Copyright 2013, Wiley.

processes a denser and homogeneous film. On the other hand, the film of Mg-NDISA exhibits a fiber-like morphology, which induces significant scattering in the UV–vis absorption spectra. Due to the quality of the optical spectra, only Ni-NDISA was investigated in EC applications.

Ni-NDISA shows two-step reversible reduction corresponding to the formation of $[NDI]^{\cdot -}$ and $[NDI]^{2-}$. Coincidently, $[NDI]^{\cdot -}$ and $[NDI]^{2-}$ exhibit almost complementary absorption over the visible spectrum, as shown in Figure 9.4b,c. When the quantity of $[NDI]^{\cdot -}$ is approximately equal to $[NDI]^{2-}$, the film turns almost black (as shown in Figure 9.4d, at the potential of −2 V), which is desirable for EC applications.

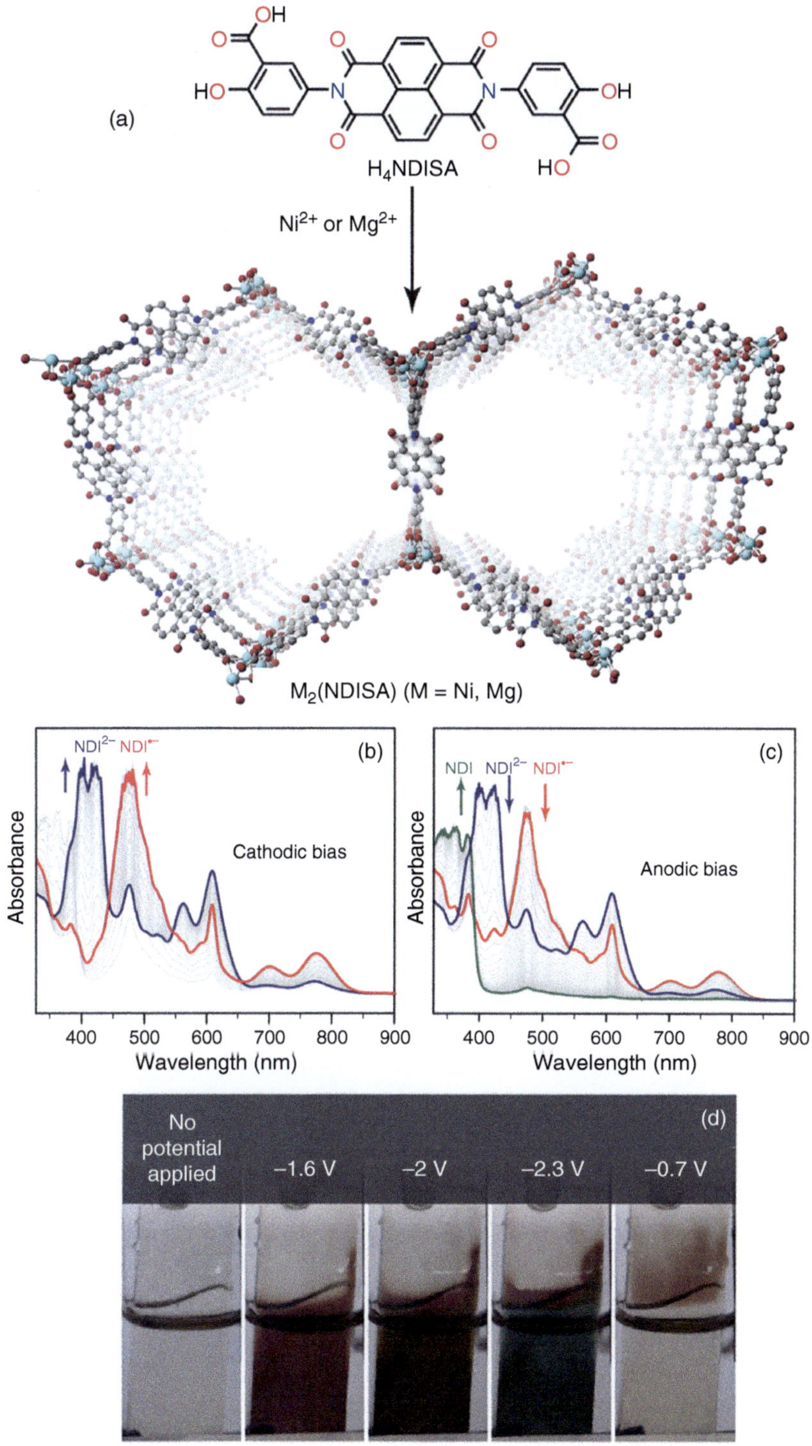

Figure 9.4 (a) Synthesis and simulated structure of M-NDISA. Spectroelectrochemical spectra of Ni-NDISA: (b) reducing from −0.5 to −2.0 V, (c) oxidizing from −2 to −0.5 V, and (d) reversible color switching and the corresponding potentials. Source: Adapted with permission from Ref. AlKaabi et al. [28]. Copyright 2016, Elsevier.

Ni-NDISA exhibits a high Brunauer–Emmett–Teller (BET) surface area of about 1722 m^2/g, suggesting its mesoporosity, which can benefit the transport of the electrolyte ions. Therefore, Ni-NDISA shows fast color switching in the EC cycles: 7 seconds for $[NDI]^0/[NDI]^{\cdot-}$ transition and 23 seconds for $[NDI]^{\cdot-}/[NDI]^{2-}$, which is comparable with common polymer materials. Ni-NDISA shows a CE value higher than 100 cm^2/C in the blue region similar to their previously reported NDI-based EC MOFs [12].

9.2.1.2 Other Organic Linkers

Besides NDI derivatives, some other organic units were also reported as the organic linkers in EC MOFs.

In 2013, Prof Omar K. Farha, Joseph T. Hupp, and coworkers in Northwestern University reported an EC MOF material, named NU-901, constructed of pyrene-based organic linkers (H_4TBAPy [1,3,6,8-tetrakis(*p*-benzoic acid)pyrene]) and Zr_6O_{16} cluster-based nodes, as shown in Figure 9.5 [31]. NU-901 was grown on FTO-coated glass substrates through a similar solvothermal method reported by Prof M. Dincă's group [12]. However differently, the substrates were pretreated in an H_4TBAPy solution before the MOF thin film growth to ensure good contact between the FTO substrate and the thin film composed of NU-901 nanocrystals. NU-901 films were homogeneous, continuous, and crackless with no gaps between the MOF films and FTO. The molecular structure of NU-901 is shown in Figure 9.5a and confirmed by X-ray diffraction (XRD) and simulation. NU-901 exhibits diamond-shaped channels, and the pore size is 29 Å in length and 12 Å in width, which is large enough for the transport of common electrolyte ions.

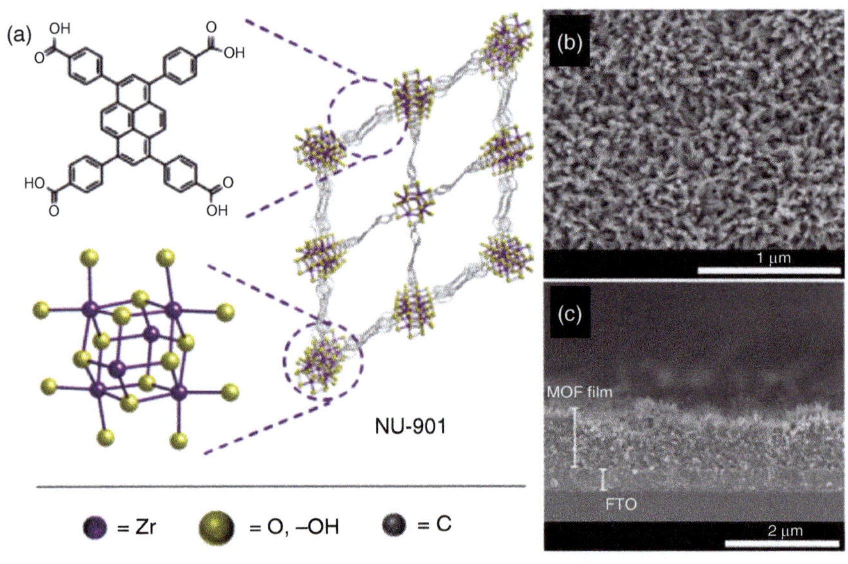

Figure 9.5 (a) Simulated molecular structure of NU-901 and the scanning electron microscope (SEM) image of the film of NU-901: (b) plane view and (c) cross-section. Source: Adapted with permission from Ref. Kung et al. [31]. Copyright 2013, American Chemical Society.

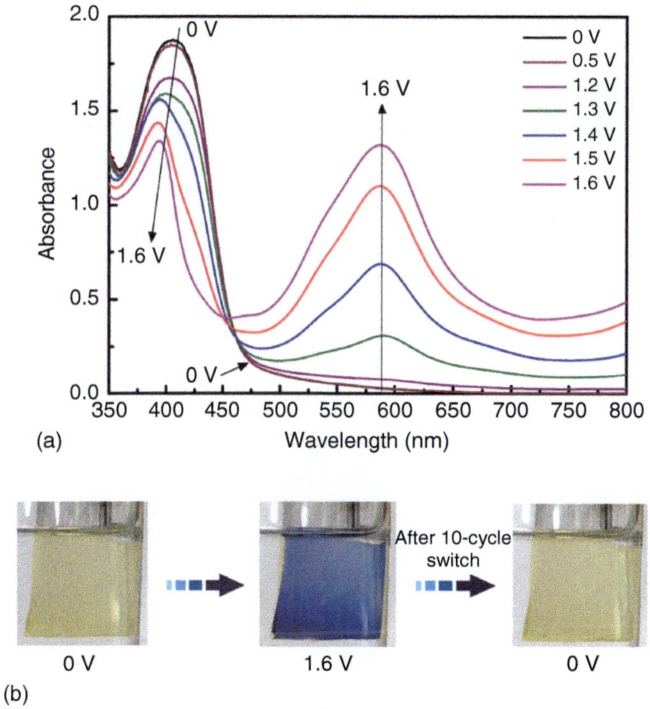

Figure 9.6 (a) Absorption spectrum change and (b) color change of NU-901 films in the CV cycles. Source: Reproduced with permission from Ref. Kung et al. [31]. Adapted with permission from Ref. [31]. Copyright 2013, American Chemical Society.

NU-901 films exhibit one reversible oxidization peak in the CV measurement. With the potential increase from 0 to 1.6 V, the peak at 405 nm decreases and the peak at 587 increases gradually, in the absorption spectra, concomitantly the color changes from yellow (neutral state) to deep blue (oxidation state), as shown in Figure 9.6. XRD pattern of the films after 10 cycles reveals that the crystal structure is retained. The color change arises from the formation of the pyrene radicals under oxidation, which is confirmed by the electron paramagnetic resonance (EPR) spectrum. The MOF framework can effectively prevent the dimerization of the pyrene radicals, which the H_4TBAPy ligand suffers in the EC cycles. The limiting transmittance change ($\Delta T\%$) of NU-901 film is 62% and decreases to 38% after 60 EC switch cycles. The decreasing $\Delta T\%$ is due to the degradation of the pyrene radicals by water from the environment. The porous MOF framework structure endows NU-901 films fast switching times, 5 seconds for bleaching and, 12 seconds for coloration. The CE value is as high as 204 cm^2/C at 587 nm.

Triarylamine (TPA) is a classic building block in organic electronics. Its three-dimensional (3D) molecular configuration, electron-donating property, and ease for modification make it extensively used in constructing kinds of functional materials [32]. TPA has easily manipulated redox state, and it can be facilely oxidized into a TPA radical. Thus it has great potential in EC applications [33, 34].

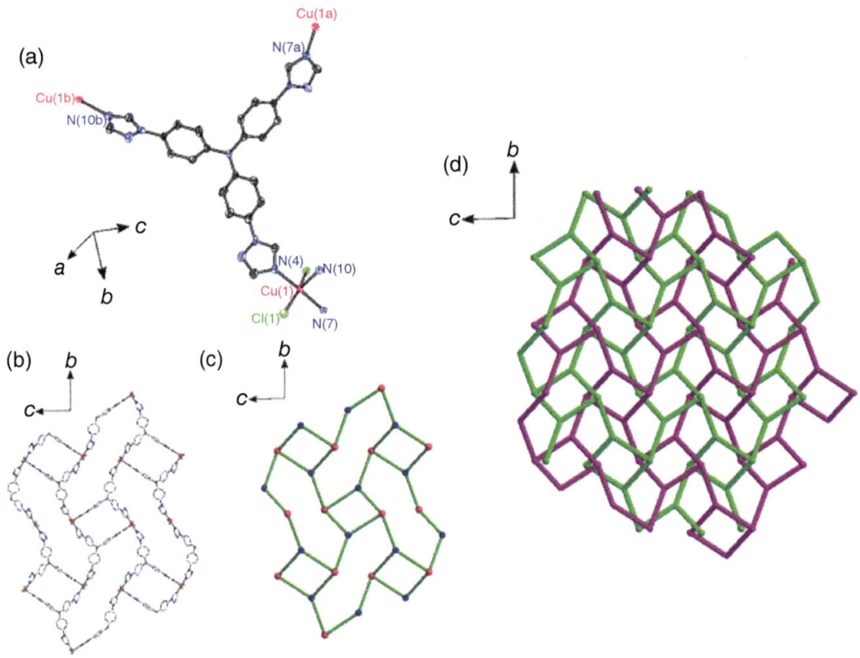

Figure 9.7 The framework structure of $[CuCl_2(TTPA)\cdot1.5DMF]_n$: (a) one TTPA organic linker coordinate with three Cu^{2+} metal nodes, (b) 2D framework structure, and (c) simplified symbolic 2D framework structure. (D) Extended twofold interpenetrating networks. Source: Adapted with permission from Ref. Ngue et al. [35]. Copyright 2019, Royal Society of Chemistry.

In 2019, Kuang-Lieh Lu, Ching-Wen Chiu, and their coworkers reported two EC MOFs, through applying a TPA derivative, tris(4-(1*H*-1,2,4-triazol-1-yl)phenyl)amine (TTPA), as organic linkers, and Mn^{2+} or Cu^{2+} as metal nodes, named as $[MnCl_2(TTPA)(DMF)]_n$ and $[CuCl_2(TTPA)\cdot1.5DMF]_n$, respectively [35]. These two MOFs shows very similar molecular framework structure. They both have two-dimensional (2D) framework structures, which possess twofold interpenetrating networks. The framework structure of $[CuCl_2(TTPA)\cdot1.5DMF]_n$ is shown in Figure 9.7. PLATON calculations found the void accessible volumes were 11.2% and 24.3% for $[MnCl_2(TTPA)(DMF)]_n$ and $[CuCl_2(TTPA)\cdot1.5DMF]_n$, respectively.

In the solid-state in situ spectroelectrochemical measurements, $[MnCl_2(TTPA)(DMF)]_n$ undergoes an irreversible color change from white to blue and further to dark brown, arising from the formation of TPA radical cations with the potential changing from 0 to 1.9 V, as shown in Figure 9.8a,b. When the potential is higher than 1.5 V, the framework begins to degrade. For $[CuCl_2(TTPA)\cdot1.5DMF]_n$, its color reversibility changes from yellow to dark greenish blue over the potential from 0 to 1.5 V, as shown in Figure 9.8c,d. The color change can run six cycles before its degradation. The poor EC performances of these two MOFs are probably caused by the highly interpenetrated framework structures, which cannot separate the TPA organic linkers effectively and impede the electrolyte ions diffused into the MOFs.

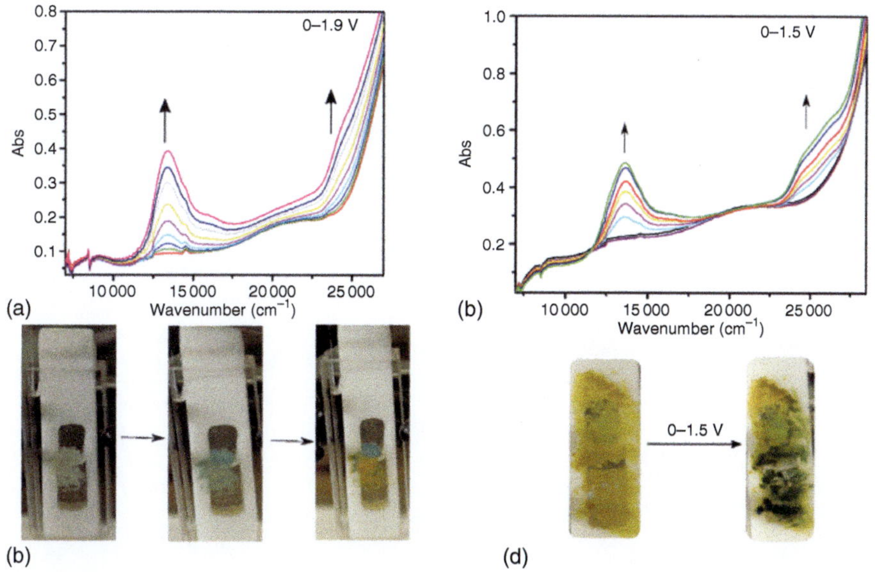

Figure 9.8 Solid-state in situ spectroelectrochemical spectra and color changes: (a,b) for $[MnCl_2(TTPA)(DMF)]_n$ and (c,d) for $[CuCl_2(TTPA)\cdot1.5DMF]_n$, respectively. Source: Adapted with permission from Ref. Ngue et al. [35]. Copyright 2019, Royal Society of Chemistry.

Only a few classes of organic molecules have been used as the linkers in EC MOFs, so far. However, we believe that any organic ligand with redox activity probability can be used in the EC MOFs after appropriate modification. There is a lot of room in this area.

9.2.2 The Transport of Electrolyte Ions in EC MOFs

The inherent regular homogeneous nanoporous structure of MOFs makes them ideal materials for fast ion transport, which is desired in EC applications. However, researches about how the ions transport in the MOFs are quite limited, though the EC MOFs exhibit fast color switching in EC (Figure 9.9).

In 2018, Prof. Hongzhi Wang and coworkers did a deep investigation about the ion transport in EC MOFs [15]. They designed and synthesized two NDI-based EC MOFs, named Ni-NDI-bis-isophthalate (Ni-BINDI) and Ni-NDI-bis-Salicylate (Ni-CHNDI), with different channel sizes: 1.0 and 3.3 nm, respectively. Their structures were shown in Figure 9.9A and B. Their films were fabricated through in situ growth in the reaction mixture. The films show acicular (nanorods) morphology, as shown in Figure 9.9C and D, which is suitable for EC application.

The ion transport process in these two MOFs was investigated in-depth, as shown in Figure 9.10. Generally, with the larger channel size, Ni-CHNDI allows the ions to move faster in it than Ni-BINDI. It is worth noting that, among the four kinds of cations (TBA^+ [tetra-*n*-butylammonium cation], Na^+, Li^+, and Al^{3+}), Na^+ shows the fastest diffusion coefficient. This is because the size of TBA^+ is the largest, so it suffers from the strongest steric resistance. For Li^+ and Al^{3+}, though their sizes

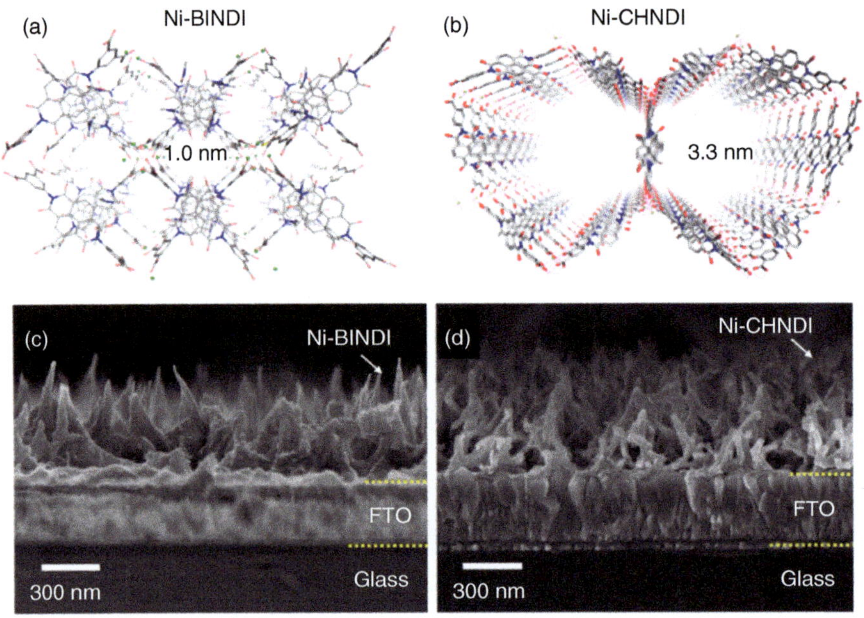

Figure 9.9 Molecular structure of Ni-BINDI (a) and Ni-CHNDI (b) and the morphology of their films (c) for Ni-BINDI and (d) for Ni-CHNDI. Source: Adapted with permission from Ref. Li et al. [15]. Copyright 2018, American Chemical Society.

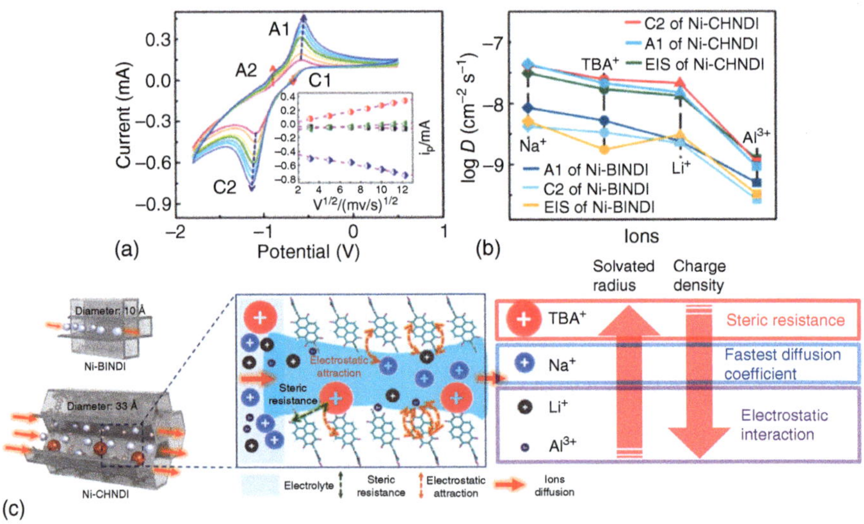

Figure 9.10 (a) CV curves of Ni-CHNDI at different scan rates (The peak current varies with the scanning rate.). (b) Diffusion coefficient of Na^+, TBA^+, Li^+, and Al^{3+} in Ni-BINDI and Ni-CHNDI. (c) Ion transport process in the MOFs. Source: Li et al. [15]. Copyright 2018, American Chemical Society.

Table 9.1 The apparent ion diffusion coefficients ($D_{app,CV}$) (cm^2/s) of Na^+, TBA^+, Li^+, and Al^{3+} in propylene carbonate in Ni-BINDI and Ni-CHNDI.

	$NaClO_4$	$TBAClO_4$	$LiClO_4$	$Al(ClO_4)_3$
Ni-BINDI	7.54×10^{-9}	6.12×10^{-9}	2.67×10^{-9}	1.86×10^{-10}
Ni-CHNDI	3.21×10^{-8}	1.03×10^{-8}	9.67×10^{-9}	1.21×10^{-9}

are smaller, their change densities are higher, and that makes them suffer from the stronger electrostatic interaction, induced from the stronger interaction between the cations and the reduced NDI in the MOF channels. Under the effects of both steric resistance and electrostatic interaction, Na^+ exhibits the fastest diffusion coefficient. The apparent ion diffusion coefficients of the four cations are summarized in (Table 9.1).

The EC performance of Ni-CHNDI and Ni-BINDI is shown in Figure 9.11. Both of them show that multicolor change varies between transparent, red, and dark blue, consistent with different redox states of NDI. For Ni-CHNDI, the color switching

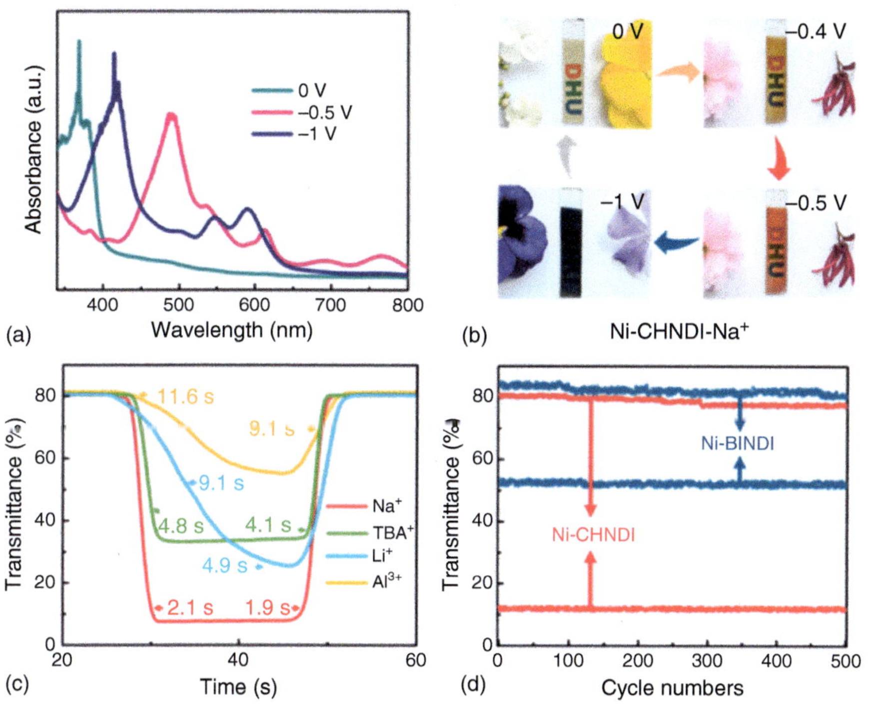

Figure 9.11 EC performance of Ni-BINDI and Ni-CHNDI. (a) UV–vis absorbance spectra of Ni-CHNDI films in $NaClO_4$/propylene carbonate system at different potentials. (b) Color changing of Ni-CHNDI at different potentials. (c) Transmittance changes of Ni-CHNDI in different electrolyte systems. (d) Cycling stability of Ni-BINDI and Ni-CHNDI devices. Source: Adapted with permission from Ref. [15]. Copyright 2018, American Chemical Society.

time in the Na^+ system is very fast, only about two seconds, with the transmittance change of 73% at 720 nm. In addition, Ni-CHNDI and Ni-BINDI have excellent stability. The performance of their EC devices shows little change after 500 cycles, and the devices can work under UV irradiations or high temperature for a long time.

Thanks to the excellent EC performance and stability, a "quick response code" smart display device was fabricated, and this product can be used as antitheft devices on shared bikes.

9.2.3 Special EC MOFs

9.2.3.1 Photochromic and Electrochromic Multi-Responsive MOF

In 2016, Prof. Lei Han in Ningbo University and his coworkers reported a novel photochromic and electrochromic MOF material, named NBU-3, a construct of tetrazole-based NDI derivative NDI-bis- tetrazole (NDI-ATZ) and Zn^{2+} [29]. NBU-3 processes a 2D planar structure (shown in Figure 9.12), which is confirmed by PXRD.

The films of NBU-3 were fabricated through a similar solvothermal method reported by Prof. M. Dincă's group. The FTO-coated glass substrates were submerged into the reaction solution and then a homogeneous 640 nm thick NBU-3 film was grown on the substrates in situ, after four hours of reaction at 80 °C.

NBU-3 undergoes a photo-induced color change (Figure 9.13). Its color turns from yellow to green under light and returns to yellow in the dark. PXRD revealed its crystal structure retains during the color change process. The color change

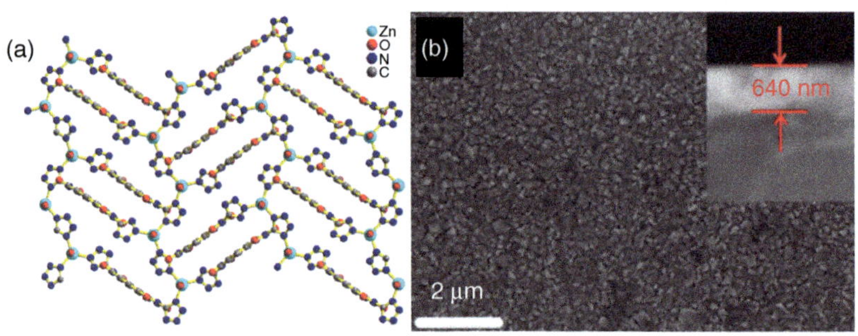

Figure 9.12 (a) Simulated molecular structure of NBU-3 and (b) the SEM image of NBU-3 film. Source: Adapted with permission from Ref. Xie et al. [29]. Copyright 2016, American Chemical Society.

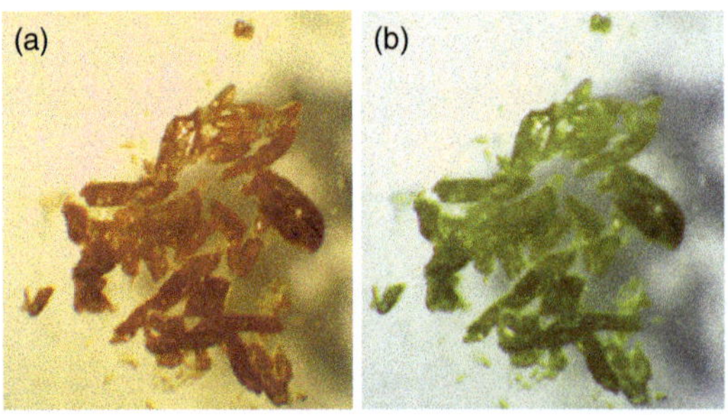

Figure 9.13 The photo-induced color change of NBU-3. Source: Adapted with permission from Ref. [29]. Copyright 2016, American Chemical Society.

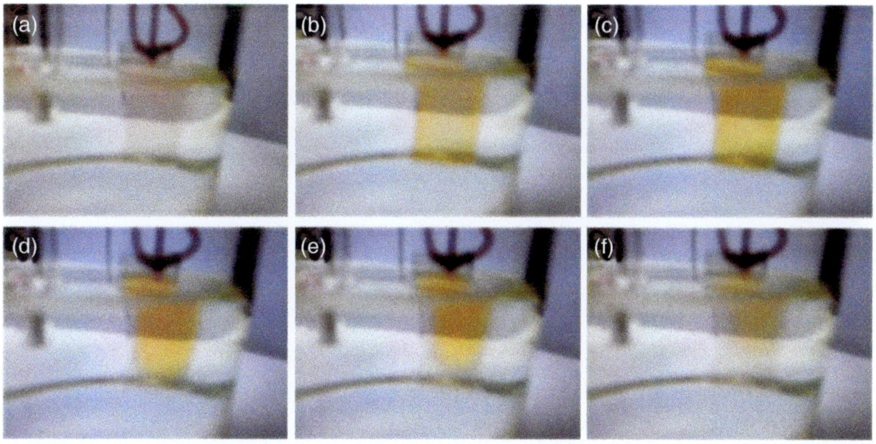

Figure 9.14 The reversible color change of NBU-3 in CV test. Source: Adapted with permission from Ref. [29]. Copyright 2016, American Chemical Society.

arises from the photo-induced radical generation of NDI moieties, and it has been confirmed by electron spin resonance (ESR) spectra. The NBU-3 film undergoes reversible color change from colorless to yellow at $E_{1/2} = -1.18\,V$ in the CV test (Figure 9.14). This color change is probably induced by the $[NDI]^{\cdot -/2-}$ redox couple. The color change can only run for six cycles, and thin-film loss happens after that.

NBU-3 is the first MOF that undergoes both photochromic and electrochromic behaviors.

9.2.3.2 MOF-Based Double-Sided EC Device and Other Color-Switching Mechanisms

In 2017, Aline Rougier and coworkers reported the first MOF-based double-sided EC device [36]. Two MOF materials, named HKUST-1 and ZnMOF-74, were employed in this device. HKUST-1 was synthesized from copper acetate and benzene-1,3,5-tricarboxylic acid, and ZnMOF-74 was from zinc acetate and 2,5-dihydroxyterephthalic (DHTP) acid in the heated tubular flow reactor via flow chemistry. Homogeneous 1.5 μm thick films of these two MOFs were individually deposited on indium tin oxide (ITO) substrate through doctor blading from the aqueous dispersion of the MOFs powders. Then the double-sided device was set up by applying the two MOF-coated ITO substrates as electrodes and lithium bis-trifluoromethanesulfonimide/1-ethyl-3-methylimidazolium-bis (trifluoromethyl-sulfnyl)imide) (LiTFSI-EMITFSI)-based polymethyl methacrylate (PMMA) as electrolyte membrane. The device structure can be presented as ITO/HKUST-1/LiTFSI-EMITFSI in PMMA/Li_xZnMOF-74/ITO and shown in Figure 9.15. The colors of the two sides can be simultaneously changed when applying a voltage because the oxidation of one side corresponds to the reduction of the other. When the voltage reversed, the color of the MOFs also reversed. The CV curves of this device show little change in 100 cycles of oxidation and reduction, suggesting the good stability of these two MOFs. Due to the highly

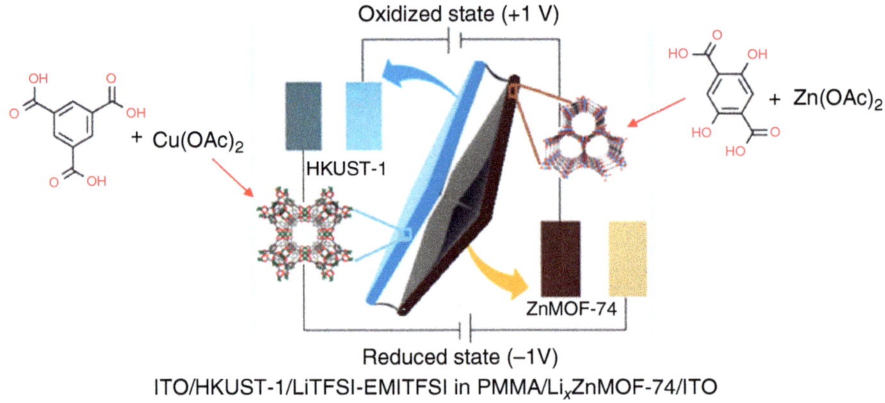

Figure 9.15 Structure of the double-sided device and the color change of the MOFs on the two sides. Source: Adapted with permission from Ref. Mjejri et al. [36]. Copyright 2017, American Chemical Society.

porous architecture of these MOF materials, this double-sided device show a very fast color-switching time of about eight seconds.

It's worth noting that the color-changing mechanisms of these two MOFs are different from that of other EC MOFs, which arises from the redox of the organic linkers. For HKUST-1, the color change between light blue and bright blue is due to the oxidation state changing of copper between Cu(II) and Cu(I). For ZnMOF-74, the color change between brown and yellow may arise from the charge transfer between the organic linker and the Zn metal nodes or due to the interaction of MOF with the electrolyte. These diverse color-changing mechanisms greatly enlarged EC MOF candidates and revealed new strategies to design EC MOFs.

9.2.3.3 EC Base on "Guest@MOF" Composite System

In 2019, Prof Jin-Chong Tan's group reported a "guest@MOF" composite system for EC application [37]. In this composite system, a redox-active molecule, DHTP acid, was inserted into the channels of Zn-MOF-74. The film of this composite was fabricated through several steps of dip coating, as shown in Figure 9.16B. The ITO was firstly coated with a thin layer of zinc to enhance the film adhesion. The structure of this composite DHTP@Zn-MOF-74 was confirmed by XRD. The thermo gravimetric analysis (TGA) results suggested 3 wt% of the DHTP guest molecules were loaded in the MOF channel (Figure 9.16).

The film of DHTP@Zn-MOF-74 composite exhibits a reversible color change between colorless transparent to magenta under the potential of 0 to −2.5 V. This EC behavior is different from previously reported Zn-MOF-74 (yellow $\rightleftharpoons$ brown, 2 to −1 V). The color change of DHTP@Zn-MOF-74 arises from the redox of the guest molecule DHTP, as shown in Figure 9.17. The authors suggested the first 15 cycles as activation cycles, due to the pore channels that are not fully accessible at the beginning of the CV test. With the cycle number increasing, the accessibility of the channels improved. This device can run 450 cycles before the current density falls to zero.

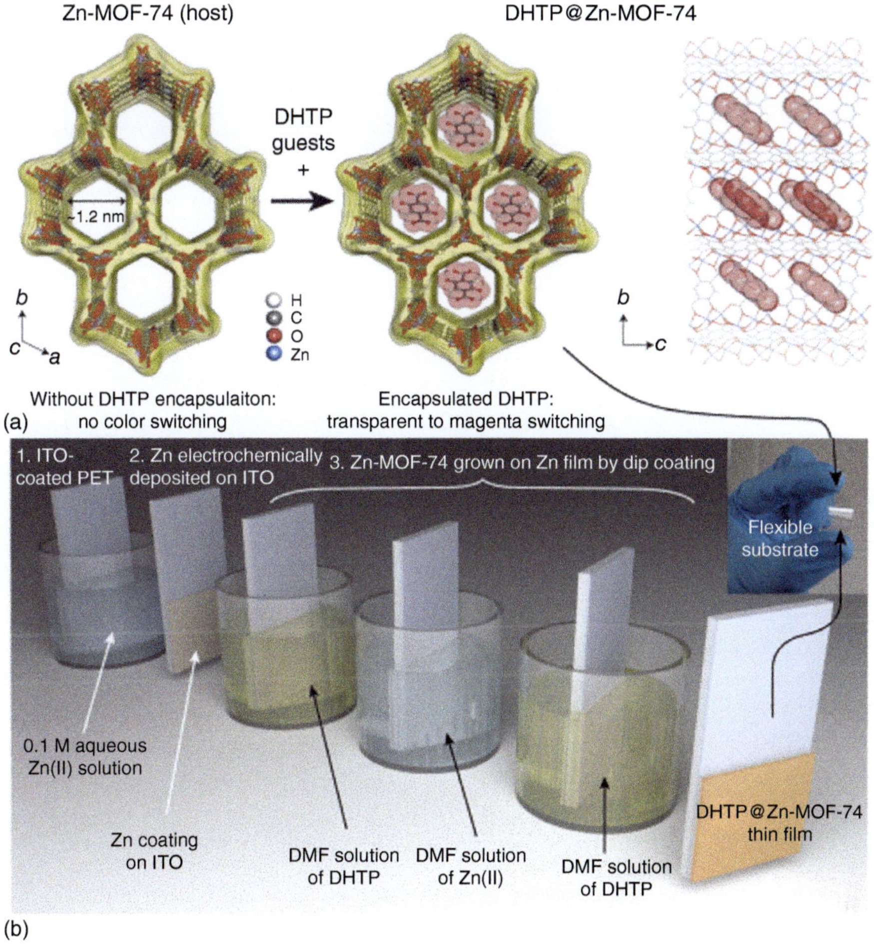

Figure 9.16 (a) Structure of the DHTP@Zn-MOF-74 composite. (b) The film fabrication method. Source: Ref. Chaudhari et al. [37], https://aip.scitation.org/doi/10.1063/1.5108948. Licensed under CC BY 4.0.

This guest@MOF composite gives great inspiration to design EC MOFs, especially to involve the large amount of redox-inactive MOFs in this field.

9.3 Current Studies in EC COFs

COFs are a kind of crystalline polymers with microporous structures, which is similar to MOFs. The difference is that the monomers of COFs are connected by covalent bonds. The covalent bonds are more stable than coordinate bonds; therefore better stability of COFs can be expected. However, COFs are less crystalline than MOFs, and single crystals of COFs are rarely reported [38, 39]. COFs have shown great potential in various applications, such as catalysts, energy storage, gas storage, and adsorption [40, 41]. However, COFs are less developed than MOFs in general.

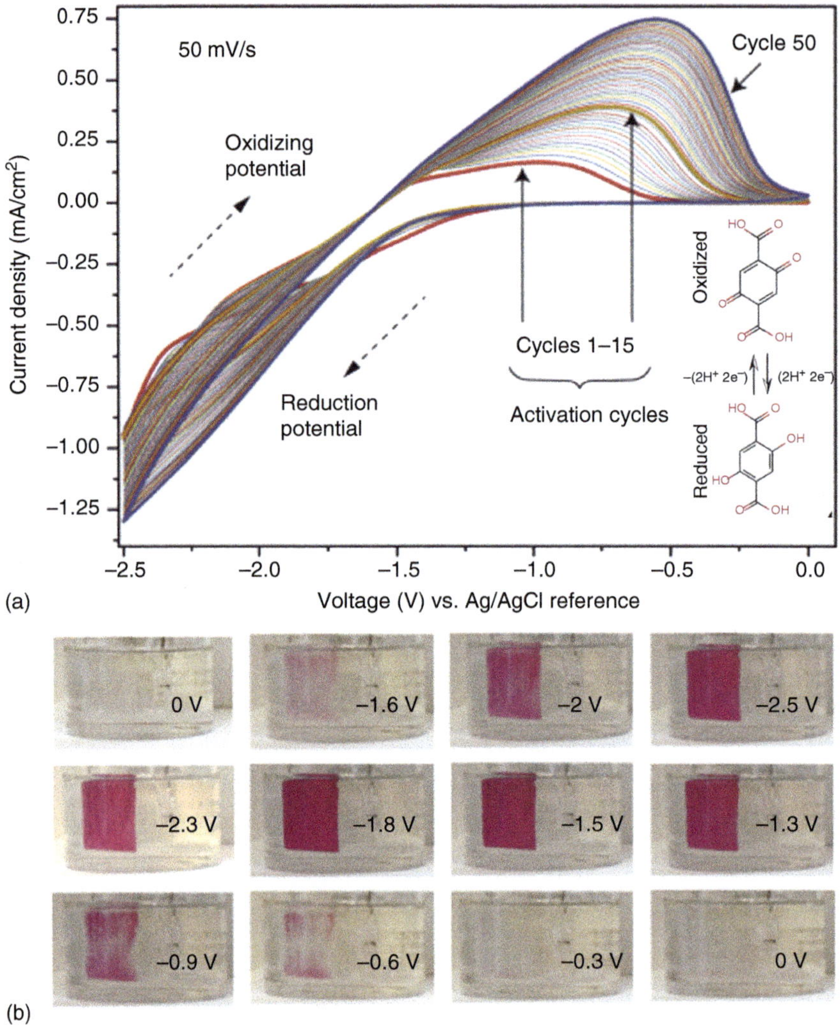

Figure 9.17 EC behavior of DHTP@Zn-MOF-74. (a) CV cycles and (b) color change under different potential. Source: Ref. Chaudhari et al. [37], https://aip.scitation.org/doi/10.1063/1.5108948. Licensed under CC BY 4.0.

Though patents of EC COFs have been reported since 2018 [42], research papers have just emerged very recently. The only reports about EC COFs were introduced in this section.

In 2018, the concept of EC COF was proposed by Zachary and coworkers in the patent [42]. They designed series of COF materials and device structures for EC application. However, the performance of the EC COFs was not introduced (Figure 9.18).

In 2019, the EC application performance of COF material was revealed by Dong Wang and coworkers [13]. Taking advantage of the electrochromic property of TPA, $COF_{3PA\text{-}TT}$ was selected for their study, and the molecular structure was shown in

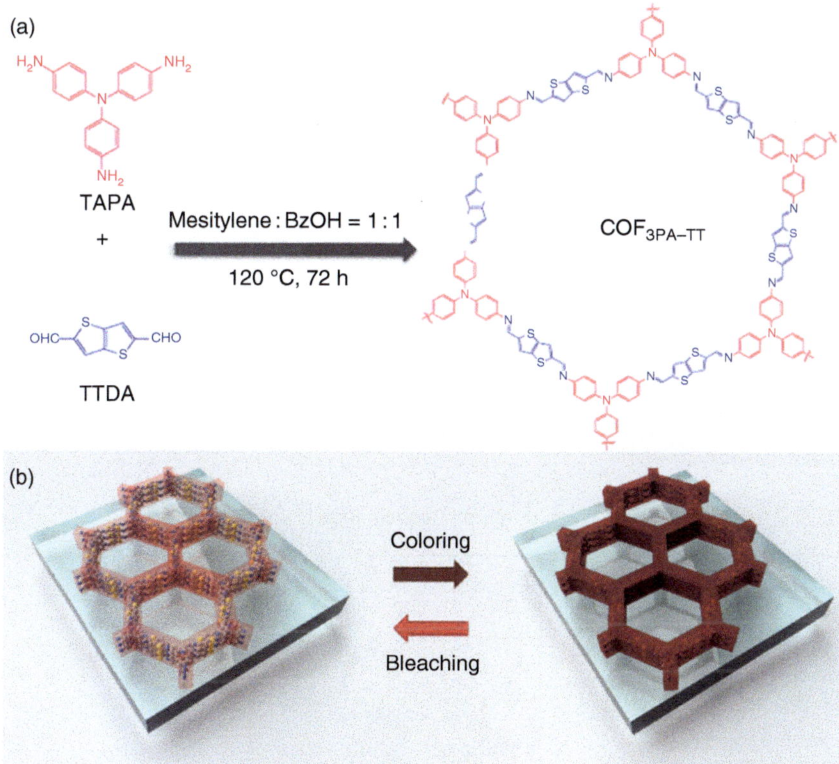

Figure 9.18 (a) Molecular structures of $COF_{3PA\text{-}TT}$ and the synthesis method. (b) Scheme of electrochromic $COF_{3PA\text{-}TT}$ film on ITO substrate. Source: Adapted with permission from Ref. Hao et al. [13]. Copyright 2019, American Chemical Society.

Figure 9.18A. The 120 nm thick $COF_{3PA\text{-}TT}$ film was directly grown on ITO through solvothermal method, with the pore structures perpendicular to the substrate, as schematically shown in Figure 9.18B. The crystallinity structure of the film was confirmed by grazing incidence wide-angle X-ray scattering (GIWAXS) and PXRD (Figure 9.19).

The electrochromic performance of the COF film was shown in Figure 9.19. Firstly, it was first investigated through CV and spectroelectrochemical techniques. In the CV curves in the current at 1.30 and 0.82 V changes linearly with the scanning rate, as shown in Figure 9.19a,b, suggesting that the redox reaction is not limited by the counterions' diffusion or the contact between the $COF_{3PA\text{-}TT\ film}$ and ITO electrode. The color of $COF_{3PA\text{-}TT}$ film can reversibly change between deep red and dark brown with the applied potential ranging from 0 to 1.4 V, as shown in Figure 9.19c. In the spectroelectrochemical spectra, when the potential rises from 0.35 to 1.05 V, the absorption peaks at 350 and 482 nm decrease, and the absorption between 400 and 585 nm as well as near-infrared (NIR) region increases gradually, as shown in Figure 9.19d and e. When the potential further rises to 1.85 V, the absorption in the NIR region decrease, as shown in Figure 9.19f. The

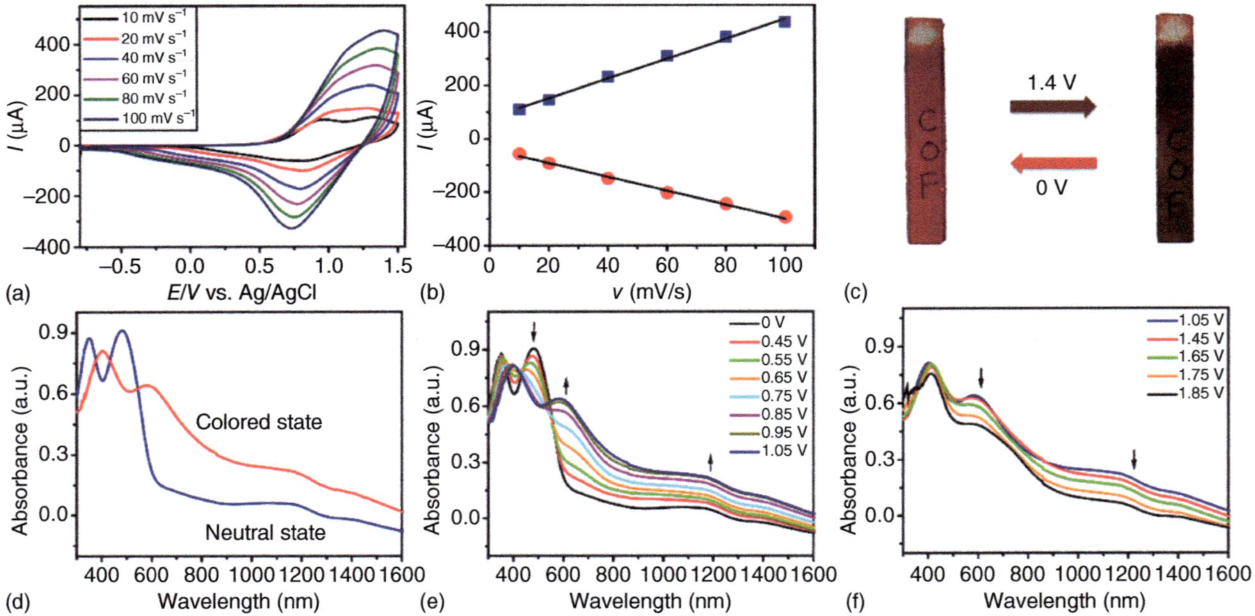

Figure 9.19 (a) CV curves of the $COF_{3PA\text{-}TT}$ EC electrodes measured at various scan rates. (b) Dependence of anodic and cathodic currents (at 1.30 and 0.82 V) on the scan rates. (c) Pictures of $COF_{3PA\text{-}TT}$ EC electrodes at 0 and 1.4 V. (d) Absorption spectra of $COF_{3PA\text{-}TT}$ EC electrodes at 0 and 1.4 V. (e,f) Absorption spectral changes of $COF_{3PA\text{-}TT}$ EC electrodes in 0.1 M $LiClO_4$/PC (propylene carbonate) recorded during oxidative spectroelectrochemical measurement. The applied potentials are referenced vs Ag/AgCl. Source: Adapted with permission from Ref. Hao et al. [13]. Copyright 2019, American Chemical Society.

authors assume that the absorption change between 400 and 600 nm attributed to the charge transfer from the TPA units to the thienothiophene–diimine units. The increased absorption in the NIR region attributes to the intervalence charge transfer (IVCT) effect between TPA cation radicals and TPA units, and the formation of TPA dications weakens the IVCT effect, which contributes to the decreased absorption in the NIR region under high potential.

To better evaluate the EC performance of COF material, they made a comparison with the amorphous 3PA-TT film. They found that the amorphous film showed very poor stability in the EC process, which was almost irreversible. The author attributes the better EC performance of $COF_{3PA\text{-}TT}$ film to the good π–π stacking structure and the additional channels for the exchange of electrons and ions provided by the oriented pore structure.

In the EC devices, the $COF_{3PA\text{-}TT}$ film shows higher $\Delta T\%$ (21% at 610 nm and 41% at 1300 nm), better stability, and a higher CE values (152 and 102 cm^2/C at 1300 and 610 nm) but slower response time (18/13 seconds and 17.5/10.5 seconds for coloring/bleaching at 1300 and 610 nm, 90% change in $\Delta T\%$) than its EC electrodes (15% and 16%, 56 and 104 cm^2/C, 20/2.5 seconds and 18.5/7 seconds for coloring/bleaching at 1300 and 610 nm). The reason for the performance difference is still not clear and needs more investigation.

Almost the same time as Dong Wang's report, Zhen Li's group also reported a study on EC COF [14]. They also chose a TPA-based COF:TATF. The TATF COF film was also grown on ITO glass through the solvothermal method. But here's the difference; the ITO surface was pretreated by 1,6-hexanediamine. As a result, the surface was amino-functionalized and then further modified with tris(4-formylphenyl)amine, as shown in Figure 9.20. They found that the pretreatment surface could greatly increase the adherence force of the TATF COF film on ITO glass. In comparison, the TATF COF film on untreated ITO could be easily removed by ultrasonic treatment.

The TATF COF film grown in an autoclave exhibits morphology consisting of dense nanofibers. The selected area electron diffraction (SAED) pattern and XRD analysis revealed the low crystallinity of the nanofiber film. However, the loose packing structure may be good for ion transport in the electrochromic switching process.

The TATF COF film shows the reversible color change between bright yellow and brown, as shown in Figure 9.21a. TATF COF shows several redox peaks in the CV curves, as shown in Figure 9.21b, and a possible pathway of the redox reaction of the TPA network as shown in Figure 9.21c. Calculated at 530 nm, TATF COF shows a fast switching time of 4.5 seconds/4.9 seconds for coloring/bleaching tested as 90% absorbance change, long memory time ($t_{1/2}$ up to 258 minutes), high CE of 115 cm^2/C, and good stability in long time cycling, as shown in Figure 9.21d.

These first two EC COFs have shown great performance in some aspects, such as switching speed and CE. Considering the highly designable molecular structure and topological structure as well as the ordered pore channels, COFs can be a new platform for the research and development of high-performance electrochromic materials.

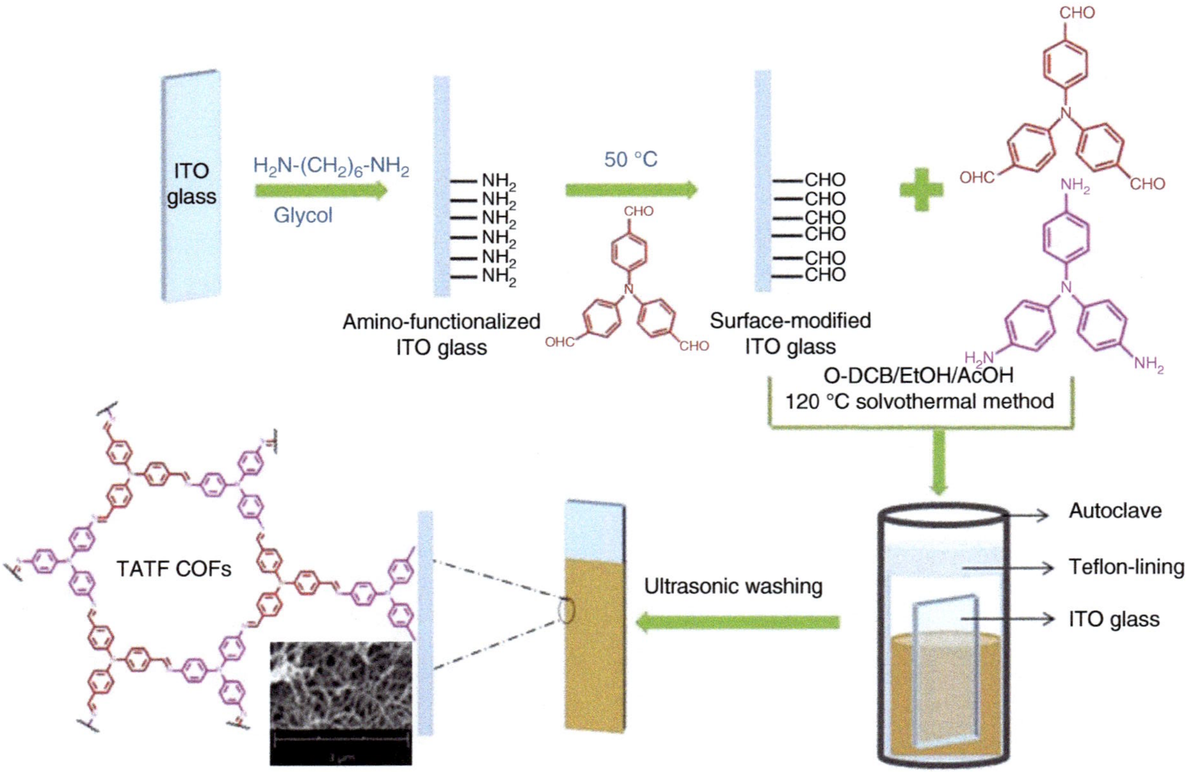

Figure 9.20 Growth TATF COFs nanofibers on ITO glass substrate. Source: Adapted with permission from Ref. Xiong et al. [14]. Copyright 2019, Elsevier.

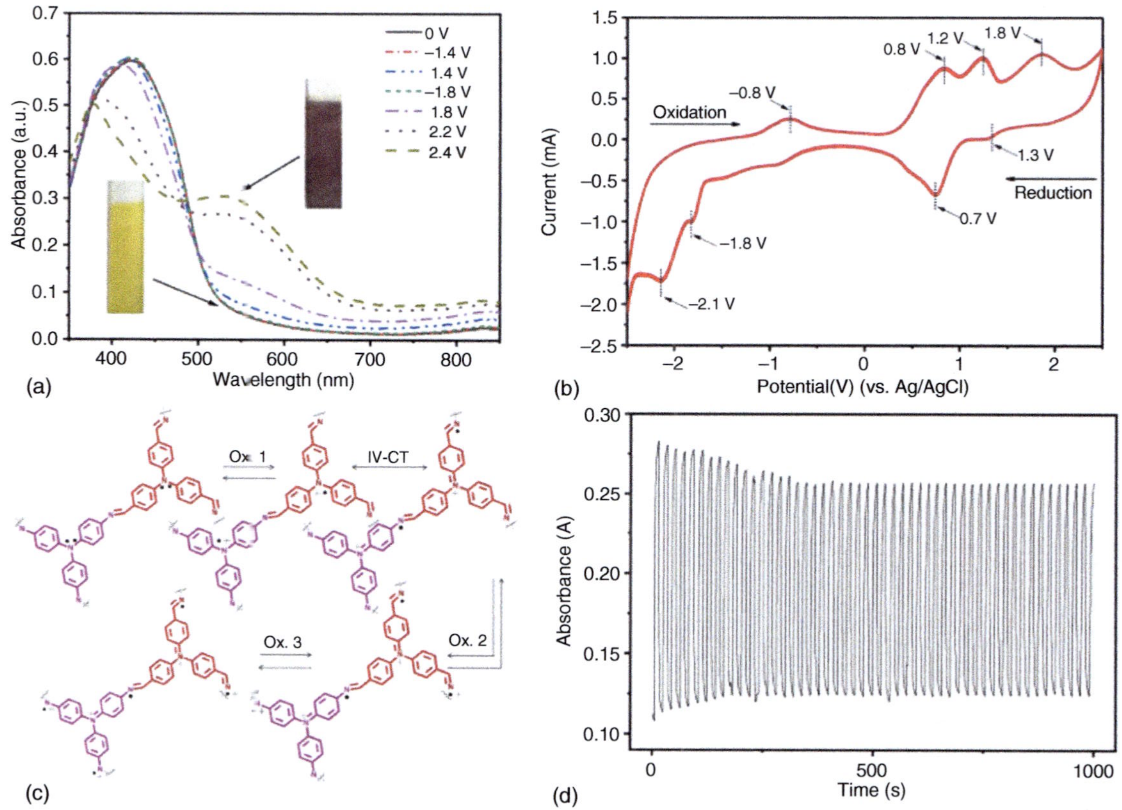

Figure 9.21 (a) UV–vis spectra and color changing of TATF COF films. (b) CV curve. (c) Redox process of TATF COF during the CV process. (d) Cycling performances for TATF COF EC device. Wavelength: 530 nm; potential: −2.5 to 2.5 V; interval: 10 seconds. Source: Adapted with permission from Ref. Xiong et al. [14]. Copyright 2019, Elsevier.

9.4 Conclusion and Prospect

Currently, the significance and value of MOFs/COFs in the EC field have not been fully revealed. They haven't shown obvious advantages over other typical EC materials (WO_3, polymers, etc.). However, its inherent structural features make them hold great potential in this field, especially in fast-switching EC devices.

Several issues need to be further investigated in EC MOFs/COFs:

(1) *Stability*: The stability of many reported EC MOFs/COF is not good enough. Though some of them have shown excellent stability not only in hundreds of CV cycles but also under long time UV irradiation and high temperature, the strategy to design stable EC MOFs/COFs is still not clear. The instability of EC MOFs may arise from the weak coordination bonds in MOFs. Upon this consideration, EC COFs may be a better choice. However, more researches in EC COFs are needed to confirm it.
(2) *Ion transport*: Fast ion transport is the biggest advantage of EC MOFs/COFs. The reported EC MOFs/COFs have shown fast switching time of about a few seconds. The ion transport process and interactions between ions and MOFs/COFs need to be further systematically investigated in-depth so that the switching time probably can be further improved.
(3) *Film fabrication*: MOFs/COFs are insoluble materials. Fabricating homogeneous films is the biggest challenge for EC MOFs/COFs. Most of the reported MOFs/COFs films are fabricated through in situ growth on the substrates in the reaction mixture, and the films often suffer from the scattering of light, which decreases their transparency. More practical methods need to be developed.
(4) *More samples*: The reported EC MOFs/COFs are quite limited (seven papers for EC MOFs and two papers for EC COFs), by now. More samples are needed to reveal more possibilities.

References

1 Yaghi, O.M. and Li, G. (1995). Mutually interpenetrating sheets and channels in the extended structure of [Cu(4,4′-bpy)Cl]. *Angewandte Chemie International Edition in English* 34 (2): 207–209.
2 Yaghi, O.M. and Li, H. (1995). Hydrothermal synthesis of a metal-organic framework containing large rectangular channels. *Journal of the American Chemical Society* 117 (41): 10401–10402.
3 Hoskins, B.F. and Robson, R. (1990). Design and construction of a new class of scaffolding-like materials comprising infinite polymeric frameworks of 3D-linked molecular rods. A reappraisal of the zinc cyanide and cadmium cyanide structures and the synthesis and structure of the diamond-related frameworks $[N(CH_3)_4][Cu^IZn^{II}(CN)_4]$ and CuI[4,4′,4″,4‴-tetracyanotetraphenylmethane]$BF_4{\cdot}xC_6H_5NO_2$. *Journal of the American Chemical Society* 112 (4): 1546–1554.

4 Yaghi, O.M., Li, G., and Li, H. (1995). Selective binding and removal of guests in a microporous metal-organic framework. *Nature* 378 (6558): 703–706.

5 Eddaoudi, M., Kim, J., Rosi, N. et al. (2002). Systematic design of pore size and functionality in isoreticular MOFs and their application in methane storage. *Science* 295 (5554): 469–472.

6 Li, H., Eddaoudi, M., O'Keeffe, M., and Yaghi, O.M. (1999). Design and synthesis of an exceptionally stable and highly porous metal-organic framework. *Nature* 402 (6759): 276–279.

7 Czaja, A.U., Trukhan, N., and Müller, U. (2009). Industrial applications of metal–organic frameworks. *Chemical Society Reviews* 38 (5): 1284–1293.

8 Côté, A.P., Benin, A.I., Ockwig, N.W. et al. (2005). Porous, crystalline, covalent organic frameworks. *Science* 310 (5751): 1166–1170.

9 Ding, S.-Y. and Wang, W. (2013). Covalent organic frameworks (COFs): from design to applications. *Chemical Society Reviews* 42 (2): 548–568.

10 Feng, X., Ding, X., and Jiang, D. (2012). Covalent organic frameworks. *Chemical Society Reviews* 41 (18): 6010–6022.

11 Kandambeth, S., Dey, K., and Banerjee, R. (2019). Covalent organic frameworks: chemistry beyond the structure. *Journal of the American Chemical Society* 141 (5): 1807–1822.

12 Wade, C.R., Li, M., and Dincă, M. (2013). Facile deposition of multicolored electrochromic metal-organic framework thin films. *Angewandte Chemie International Edition* 125 (50): 13619–13623.

13 Hao, Q., Li, Z.-J., Lu, C. et al. (2019). Oriented two-dimensional covalent organic framework films for near-infrared electrochromic application. *Journal of the American Chemical Society* 141 (50): 19831–19838.

14 Xiong, S., Wang, Y., Wang, X. et al. (2020). Schiff base type conjugated organic framework nanofibers: solvothermal synthesis and electrochromic properties. *Solar Energy Materials and Solar Cells* 209: 110438.

15 Li, R., Li, K., Wang, G. et al. (2018). Ion-transport design for high-performance Na^+-based electrochromics. *ACS Nano* 12 (4): 3759–3768.

16 Cui, Y., Li, B., He, H. et al. (2016). Metal-organic frameworks as platforms for functional materials. *Accounts of Chemical Research* 49 (3): 483–493.

17 Furukawa, H., Cordova, K.E., O'Keeffe, M., and Yaghi, O.M. (2013). The chemistry and applications of metal-organic frameworks. *Science* 341 (6149): 1230444.

18 Lin, Y., Kong, C., Zhang, Q., and Chen, L. (2017). Metal-organic frameworks for carbon dioxide capture and methane storage. *Advanced Energy Materials* 7 (4): 1601296.

19 Adil, K., Belmabkhout, Y., Pillai, R.S. et al. (2017). Gas/vapour separation using ultra-microporous metal-organic frameworks: insights into the structure/separation relationship. *Chemical Society Reviews* 46 (11): 3402–3430.

20 Chughtai, A.H., Ahmad, N., Younus, H.A. et al. (2015). Metal-organic frameworks: versatile heterogeneous catalysts for efficient catalytic organic transformations. *Chemical Society Reviews* 44 (19): 6804–6849.

21 Wu, M.-X. and Yang, Y.-W. (2017). Metal-organic framework (MOF)-based drug/cargo delivery and cancer therapy. *Advanced Materials* 29 (23): 1606134.

22 Lustig, W.P., Mukherjee, S., Rudd, N.D. et al. (2017). Metal-organic frameworks: functional luminescent and photonic materials for sensing applications. *Chemical Society Reviews* 46 (11): 3242–3285.

23 Stassen, I., Burtch, N., Talin, A. et al. (2017). An updated roadmap for the integration of metal-organic frameworks with electronic devices and chemical sensors. *Chemical Society Reviews* 46 (11): 3185–3241.

24 Zhao, W., Peng, J., Wang, W. et al. (2018). Ultrathin two-dimensional metal-organic framework nanosheets for functional electronic devices. *Coordination Chemistry Reviews* 377: 44–63.

25 Chueh, C.-C., Chen, C.-I., Su, Y.-A. et al. (2019). Harnessing MOF materials in photovoltaic devices: recent advances, challenges, and perspectives. *Journal of Materials Chemistry A* 7 (29): 17079–17095.

26 Lustig, W.P. and Li, J. (2018). Luminescent metal-organic frameworks and coordination polymers as alternative phosphors for energy efficient lighting devices. *Coordination Chemistry Reviews* 373: 116–147.

27 Usman, M., Mendiratta, S., and Lu, K.-L. (2017). Semiconductor metal-organic frameworks: future low-bandgap materials. *Advanced Materials* 29 (6): 1605071.

28 AlKaabi, K., Wade, C.R., and Dincă, M. (2016). Transparent-to-dark electrochromic behavior in naphthalene-diimide-based mesoporous MOF-74 analogs. *Chem* 1 (2): 264–272.

29 Xie, Y.-X., Zhao, W.-N., Li, G.-C. et al. (2016). A naphthalenediimide-based metal-organic framework and thin film exhibiting photochromic and electrochromic properties. *Inorganic Chemistry* 55 (2): 549–551.

30 Maniam, S., Higginbotham, H.F., Bell, T.D.M., and Langford, S.J. (2019). Harnessing brightness in naphthalene diimides. *Chemistry – A European Journal* 25 (29): 7044–7057.

31 Kung, C.-W., Wang, T.C., Mondloch, J.E. et al. (2013). Metal-organic framework thin films composed of free-standing acicular nanorods exhibiting reversible electrochromism. *Chemistry of Materials* 25 (24): 5012–5017.

32 Wang, J., Liu, K., Ma, L., and Zhan, X. (2016). Triarylamine: versatile platform for organic, dye-sensitized, and perovskite solar cells. *Chemical Reviews* 116 (23): 14675–14725.

33 Ma, X., Niu, H., Wen, H. et al. (2015). Synthesis, electrochromic, halochromic and electro-optical properties of polyazomethines with a carbazole core and triarylamine units serving as functional groups. *Journal of Materials Chemistry C* 3 (14): 3482–3493.

34 Cai, S., Wen, H., Wang, S. et al. (2017). Electrochromic polymers electrochemically polymerized from 2,5-dithienylpyrrole (DTP) with different triarylamine units: synthesis, characterization and optoelectrochemical properties. *Electrochimica Acta* 228: 332–342.

35 Ngue, C.-M., Liu, Y.-H., Wen, Y.-S. et al. (2019). Spectroelectrochemical studies of the redox active tris[4-(triazol-1-yl)phenyl]amine linker and redox state manipulation of Mn(ii)/Cu(ii) coordination frameworks. *Dalton Transactions* 48 (27): 10122–10128.

36 Mjejri, I., Doherty, C.M., Rubio-Martinez, M. et al. (2017). Double-sided electrochromic device based on metal-organic frameworks. *ACS Applied Materials & Interfaces* 9 (46): 39930–39934.

37 Chaudhari, A.K., Souza, B.E., and Tan, J.-C. (2019). Electrochromic thin films of Zn-based MOF-74 nanocrystals facilely grown on flexible conducting substrates at room temperature. *APL Materials* 7 (8): 081101.

38 Ma, T., Kapustin, E.A., Yin, S.X. et al. (2018). Single-crystal X-ray diffraction structures of covalent organic frameworks. *Science* 361 (6397): 48–52.

39 Evans, A.M., Parent, L.R., Flanders, N.C. et al. (2018). Seeded growth of single-crystal two-dimensional covalent organic frameworks. *Science* 361 (6397): 52–57.

40 Cao, S., Li, B., Zhu, R., and Pang, H. (2019). Design and synthesis of covalent organic frameworks towards energy and environment fields. *Chemical Engineering Journal* 355: 602–623.

41 Wang, H., Zeng, Z., Xu, P. et al. (2019). Recent progress in covalent organic framework thin films: fabrications, applications and perspectives. *Chemical Society Reviews* 48 (2): 488–516.

42 Erno, Zachary B.; Franz, Sue F.; Kloeppner, Leroy J.; Looman, Steven D. (2018). Electrochromic organic frameworks. WO2018204802 (A1), filed 04 May 2018, issued 08 November 2018.

10

Nanostructure-Based Electrochromism

10.1 Introduction

With the pioneering research on mesoporous materials by Mobil researchers in 1992, this area of scientific investigation has extended into numerous branches [1]. Many of them contribute significantly to the areas like catalysis, energy, environment, and biomedical research. Thousands of publications came out in large varieties of national and international journals. The most generalized definition of a porous material is a continuous and solid network material filled through voids. In the case of nanoporous materials, the pore or voids are of the order of ≈1–100 nm. The voids might be filled with a liquid or even a solid [2].

Actually the porous structure plays an important role for electrochromics as well. The switching speed of electrochromism is controlled by the efficiency of ion transport into/out of the EC film, which strongly depends on its microstructures and morphology. Therefore, to achieve fast and sufficient intercalation of ions, a high surface ratio of EC films is required. Numerous reports have demonstrated that porous structure can promote the infiltration of electrolyte and reduce the diffuse path and consequently speed up the reaction kinetics. Besides the electrochromic nanocomposites suffering periodic volume change caused by ion extraction and insertion during the electrochromic switching process, strong interfacial interactions are helpful for maintaining the structural stability of the nanocomposites and avoiding structural damage caused by the volume change. So the large interfacial interactions led by porous structure play an important role in improving the structure strength, mass transport, electron conduction, and electrochromic performances of nanocomposites. Therefore many scientists work on electrochromic nanocomposites combination of the advantages of various electrochromic materials.

In the area of electrochromism, porous materials can be classified as purely inorganic, hybrid organic–inorganic, and all-organic porous polymers. This chapter mainly highlights the organic electrochromic materials-related nanostructure application and the performance. Therefore, we do not discuss the inorganic electrochromic materials, such as 11-VI and 111-V semiconductor quantum dots CdSe, ZnSe [3], crystalized transition oxides like WO_3 [4, 5], and V_2O_5 [6], which are widely studied by scientists.

Organic Electronics for Electrochromic Materials and Devices, First Edition. Hong Meng.

10.2 Current Studies of Nanostructure in Electrochromism

Generally, porous structure application in organic electrochromics are classified into two types: the nanostructure support by non-electrochromic active materials and by electrochromic active materials. Furthermore, different templates and their function would be discussed. And the processing of nanostructured electrochromic materials will be shown in this part.

10.2.1 Non-Electrochromic Active Materials as a Template for ECs

10.2.1.1 Photonic Crystals as Templates for ECs

Photonic crystals (PCs) materials (Figure 10.1 as one example) with a periodic modulation in refractive index can be sources of exceptionally bright and brilliant reflected colors arising from coherent Bragg optical diffraction. One-, two-, and three-dimensional PCs could be available by means of colloidal self-assembly or nano-imprinting to achieve tunable structural color. Tuning of colloidal PC optical properties has been affected by the infilling of metals, insulators, semiconductors, and polymers of all types [7].

Due to the feasible self-assembly methods that produce three-dimensionally ordered arrays of monodisperse spheres, so-called colloidal crystals or synthetic

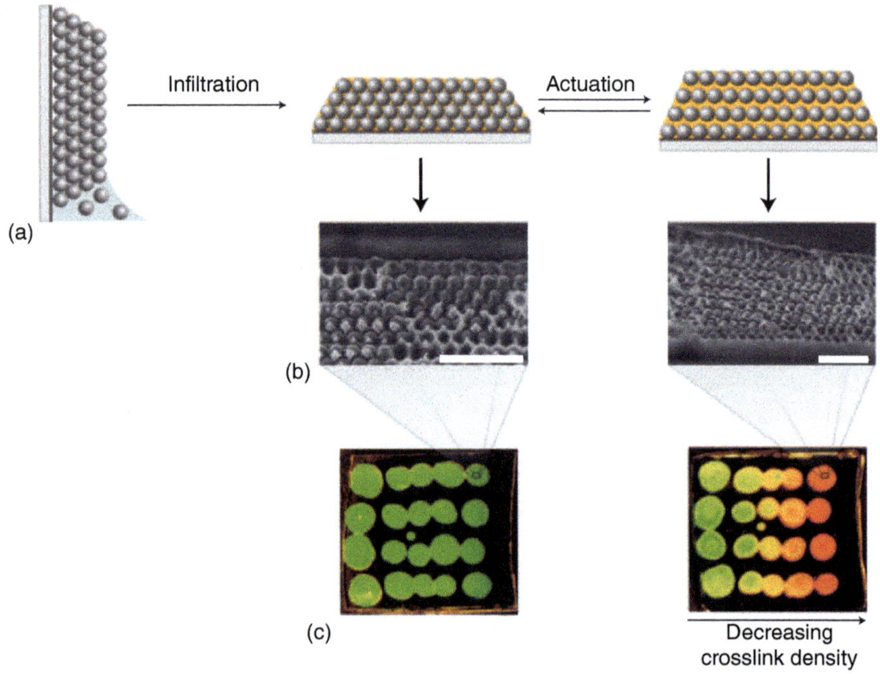

Figure 10.1 One example of fabrication and optical characteristics of thin-film electroactive PCs [7]. Source: Adapted with permission from Arsenault et al. [7]. Copyright 2007, Springer Nature.

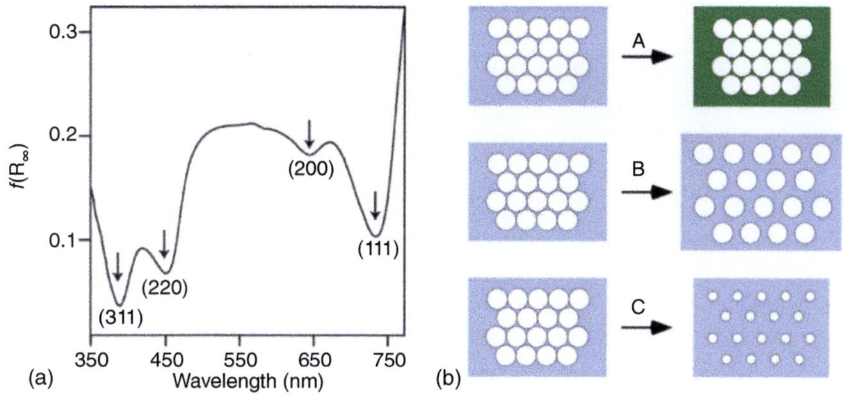

Figure 10.2 (a) UV–visible reflectance spectrum of a polycrystalline sample of mercaptopropyl silica infiltrated with 2-propanol. The *Y*-axis $f(R_\infty)$ is calculated from the Kubelka–Munk equation, which is a index for the strength of reflectance. (b) Mechanisms of tuning colors in colloidal crystals: (A) change of refractive index, (B) change of lattice constant, and (C) change of filling factor [8]. Source: Adapted with permission from Aguirre et al. [8]. © 2006, John Wiley & Sons.

opals have been studied intensively. One of the advantages of colloidal PCs is their versatility for changing the characteristic reflection peaks in their optical spectra (the "stop bands"). The optical reflection peaks are analogous to diffraction peaks in X-ray powder patterns but fall into the submicrometer range instead of the angstrom range (Figure 10.2a). In these materials, the stop band positions depend largely on three factors: (i) the refractive index contrast between two periodic media (solid spheres or air spheres and the surrounding phase), (ii) the lattice constant (the spacing between spheres), and (iii) the filling factor (volume of the spheres compared to the volume of the surrounding phase) (Figure 10.2b). The concept of tunable color in PCs is based on changing the stop band characteristics reversibly through variation of any of these parameters by applying external stimuli [8].

Therefore, many scientists focus on the PCs fabrication, nanostructure engineering, and color tuning properties of the PCs [9–11]. Large-scale fabrication of colloidal PCs by coating or printing has also been demonstrated based on the use of the latex particles with additional assembly force [12]. This fabrication of functional colloidal PCs will greatly extend their applications in various fields.

Therefore, it is predicted that the combinations of structurally colored PCs and electrochemically colored conducting polymers would produce new chromic colors. It is worth noting that although the optical characteristics could be tuning through structural engineering, however, real-time adjustability is really poor, which could be compensated by the optical switching property through a low-voltage loading on electro-responsive polymers. The color manipulation upon electrical stimulus was not based on the modification of the crystal periodicity but on the electrochromism of the conductive polymers, while the participation of structural color makes the optical character of the system more versatile.

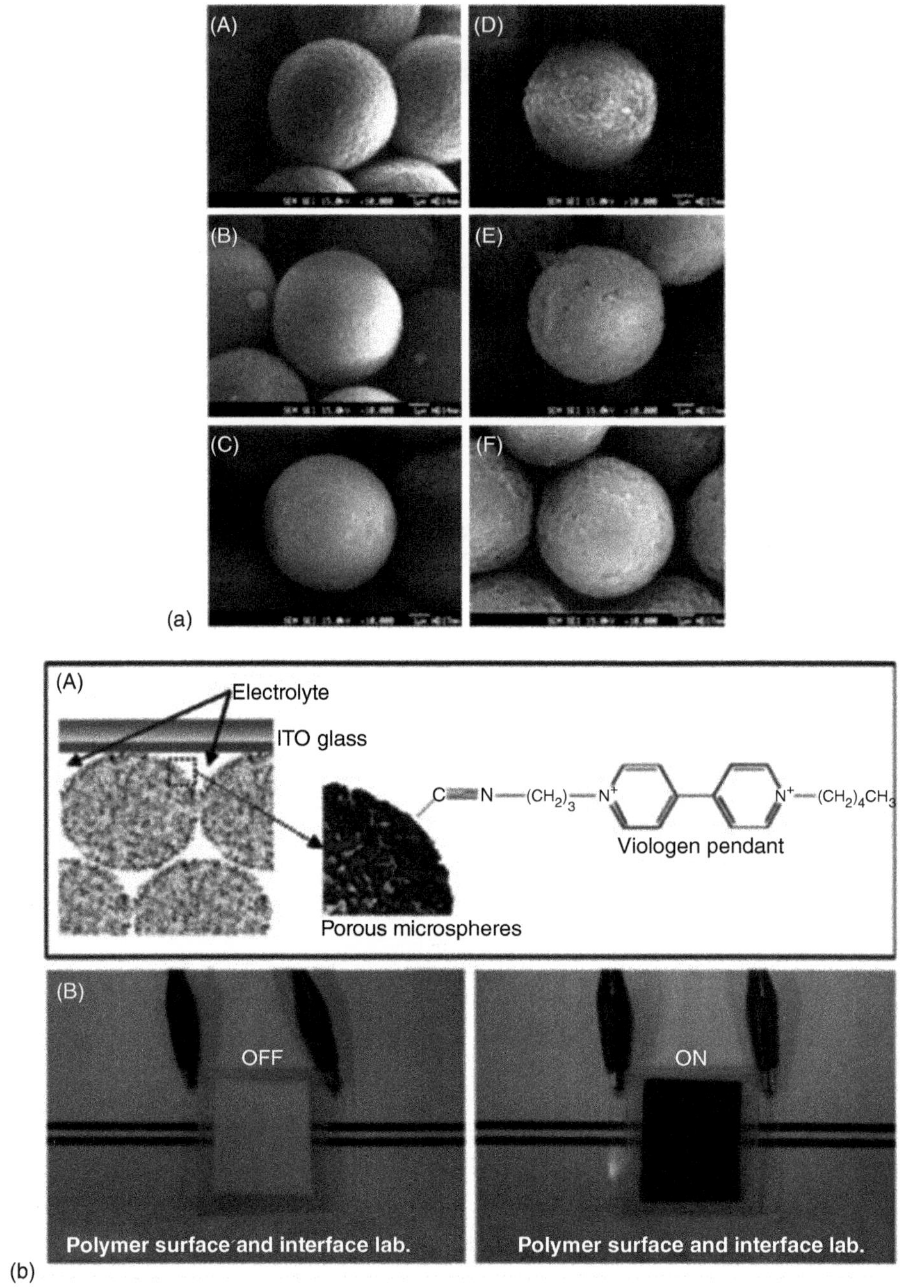

Figure 10.3 (a) SEM photographs of porous polymeric microspheres: 50% (A), 100% (B), and 150% (C) of the toluene contents used in seeded polymerization and viologen-modified porous polymeric microspheres POR05 (D), POR10 (E), and POR15 (F). (b) Schematic representation of the substrate particle in the manufactured device (A). Images of the device (B) in the OFF state (left) and in the ON state [13]. Source: (a) Adapted with permission from Ryu et al. [13]. Copyright 2006, Elsevier, (b) Ryu et al. [13]. © 2006, John Wiley & Sons.

In 2006, Jee-Hyun Ryu et al. applied viologen to modify porous polymeric microspheres, which is prepared by a seeded polymerization and refluxing process (Figure 10.3). The switching could achieve a millisecond response time. The porosity of the substrate particles is controlled by adjusting the toluene content. This could prove through the employment of nanostructure, and the electrochromic switching could largely be improved [13].

Based on this work, in 2008 Jee-Hyun Ryu et al. achieved multicolor electrochromic through viologens side group engineering [14]. The viologens containing different substituents were employed for multicolored ECD devices. Viologen molecules with cyanophenyl, salicylic, and alkyl groups as substituents were synthesized and reacted with methyl chloride groups on the surface of microspheres (Figure 10.4A,B). Comparing spectrophotometric and visual appearance results, it was confirmed that each ECD cell exhibited red, green, and blue colors within the potential when cyanophenyl, salicylic, and alkyl substituents were used on the viologen molecules, respectively (Figure 10.4C,D).

As we know, the main purpose of nanostructure application is either taking advantage of the structural color of PC nanoparticles or the enhancement of the switching of electrochromism. In this work, the authors did not mention the detailed characteristics in both aspects.

In 2013, with a similar structure but different materials, Dengteng Ge et al. fabricated the polyaniline (PANI) with ordered PS template through template-assisted electropolymerization. Both the ion diffusion performance and the correlation with optical switching properties are found to be greatly enhanced due to the modification of the chemical structure and ordered porous microstructure of PANI. Compared with the pristine PANI films, the ion diffusion rate of PANI/PS composites is greatly improved due to the change of their chain structure and crystalline structure originally from the confined growth under ordered templates. It also shows excellent ion diffusion reversibility and ion storage capacity because of the interconnected macropores with high volume fraction of electrolyte (as the scheme shows in Figure 10.5) [15].

In 2016, Tomoya Kuno et al. reported that the electro-responsive materials composed of ordered arrays of polyaniline@poly(methyl methacrylate) (PANI@PMMA) core–shell nanoparticles have been successfully prepared. The core–shell nanoparticles were synthesized by the deposition of PANI shells on the surfaces of the poly(methyl methacrylate) (PMMA) cores by the oxidative polymerization of anilinium chloride. Ordered arrays were then fabricated by using the fluidic cell method, because the ordered arrays and the PANI shells generate structural and electrochromic colors, respectively. These core–shell colloidal crystals exhibited colors resulting from the combined effects. The crystal colors depended greatly on the size of PANI@PMMA particles and could also be varied by the application of a voltage. The electrochromic colors of these arrays were found to be quite different from those exhibited by pure PANI films prepared by electrochemical oxidation [16].

It is worth noting in this work that the authors successfully achieved structural color turnability and electrochromic color tunability in one cell, which proved that this is a powerful tool for the preparation of new colored materials (Figure 10.6).

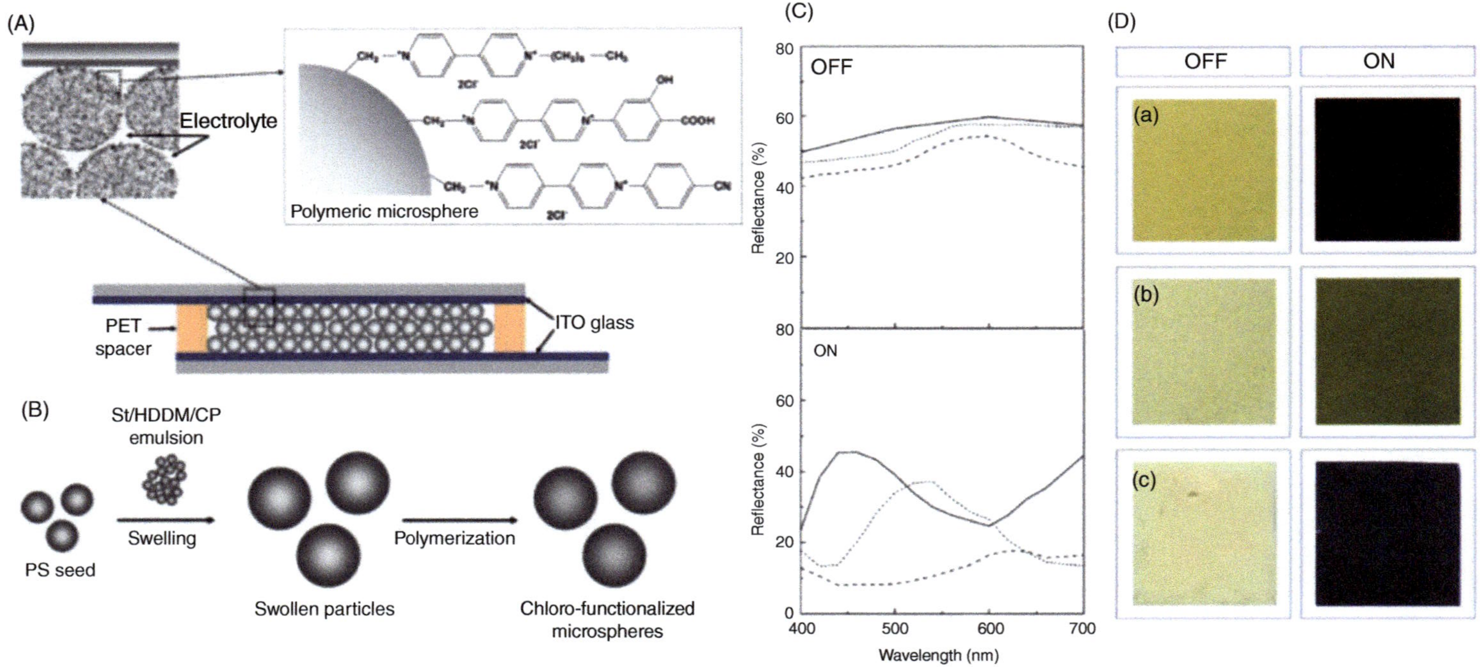

Figure 10.4 (A) Cross-sectional view of an R-ECD cell. (B) Schematic representation of seeded polymerization. (C) Spectrophotometric analysis of an R-ECD depending on the potential and (D) photo images of an R-ECD cell without and with potential [14]. Source: Ryu et al. [14]. © 2008, John Wiley & Sons.

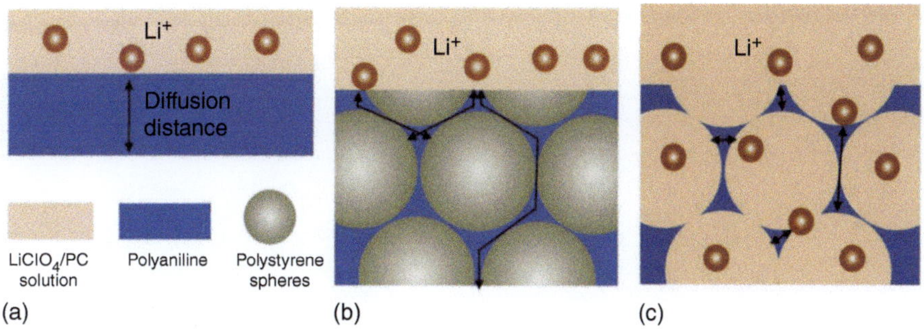

Figure 10.5 Schematic of ion diffusion paths in (a) PANI films and (b,c) PANI/PS. The brown spheres represent the Li+ and the double-headed arrow means the diffusion path of Li^{+15}. Source: Ge et al. [15]. © 2013, Elsevier.

Scientists Dan Zhou et al. prepared polyaniline/manganese dioxide (PANI/MnO_2) hybrid electrochromic films via one-pot potentiostatically anodic electrodeposition from an aqueous solution containing aniline and manganese sulfate. In the one-pot deposition process, electropolymerization of aniline and electrodeposition of MnO_2 occur simultaneously, while some of the resultant MnO_2 can act as an oxidizer to further initiate chemical polymerization of aniline, facilitating the rapid growth of the hybrid films. This unique deposition mechanism gives rise to significant differences between the hybrid and neat PANI films in morphology, structures, and electrochemical and electrochromic properties. Compared with neat PANI films, the hybrid film with optimal MnO_2 content shows much higher optical contrast, coloration efficiency, and cycling stability (as shown in Figure 10.7). The excellent performance of the hybrid film can be attributed to its unique porous morphology with interconnected small nanoparticles and donor–acceptor interactions between PANI and MnO_2 [17].

Besides, the nanostructure supported by MnO_2 also has electrochromic characteristics, giving a better carrier transport property than the previously introduced dielectric nanospheres (NSs), such as PS or PMMA.

In short summary, it could tell from the aforementioned introduction that many research works mainly focus on the very conventional PANI as the electrochromic active materials, which largely constrain the universal application of the processing methods. Besides, only few studies really employed the structural color characteristics of the PCs into the combined system, which definitely keeps large space for future research.

10.2.1.2 Plasmonic Structures as Templates for ECs

Plasmonic colors are structural colors that emerge from resonant interactions between light and metallic nanostructures. The engineering of plasmonic colors is a rapidly developing research field.

The phenomena is realized in both dielectric and metallic structures, ranging from all-dielectric PCs to metallic gratings. Comparing with dielectric nanostructures, the metallic plasmonic structure could easily be tuned through not only the size

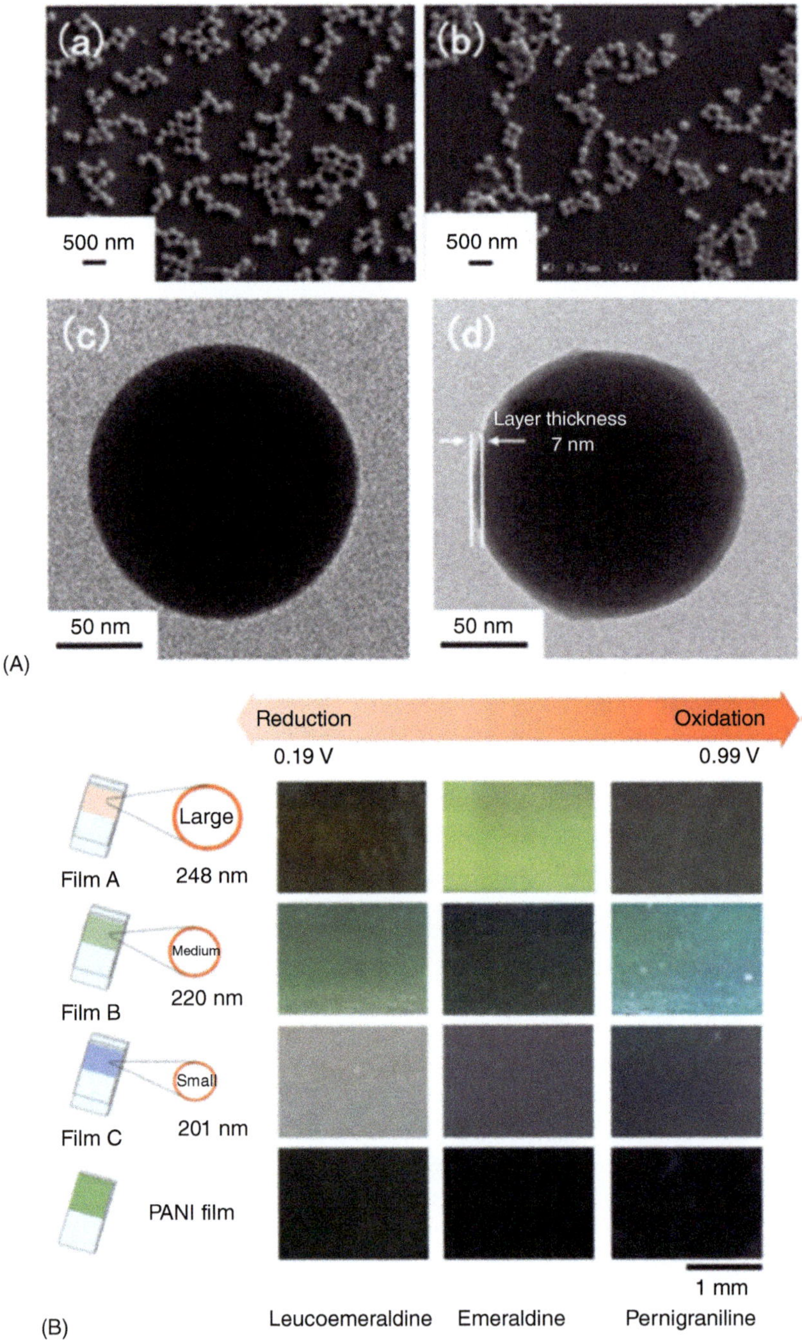

Figure 10.6 (A) SEM and TEM images of PMMA nanoparticles (left) and PANI@PMMA core–shell nanoparticles (right). (B) Electrochromic properties of the A, B, and C colloidal crystal films and the PANI@PMMA core–shell nanoparticles film [16]. Source: (A) Adapted with permission from Kuno et al. [16]. Copyright 2016, Wiley, (B) Kuno et al. [16]. © 2016, John Wiley & Sons.

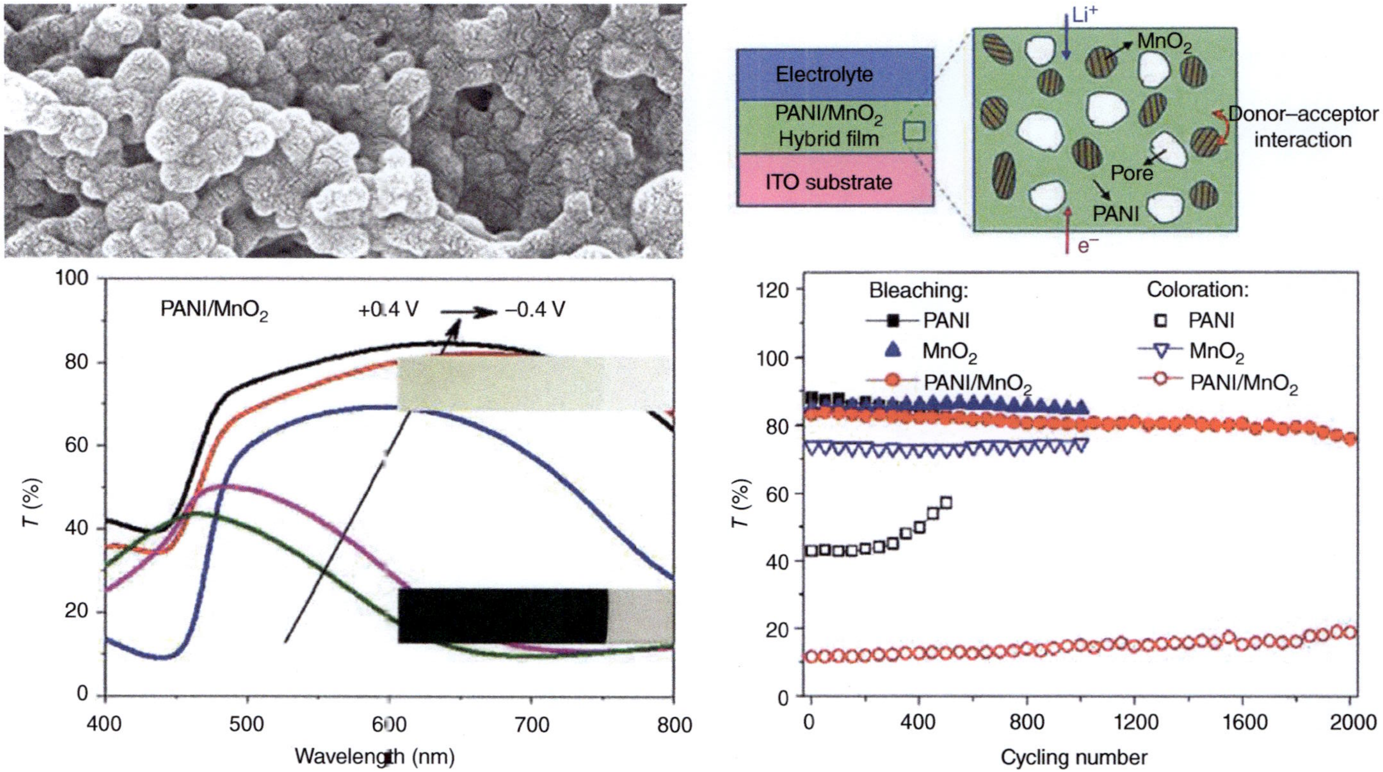

Figure 10.7 Polyaniline/manganese dioxide hybrid films are prepared via one-pot potentiostatically anodic electrodeposition and exhibit enhanced electrochromic properties, especially excellent cycling stability [17]. Source: Adapted with permission from Zhou et al. [17]. Copyright 2017, Royal Society of Chemistry.

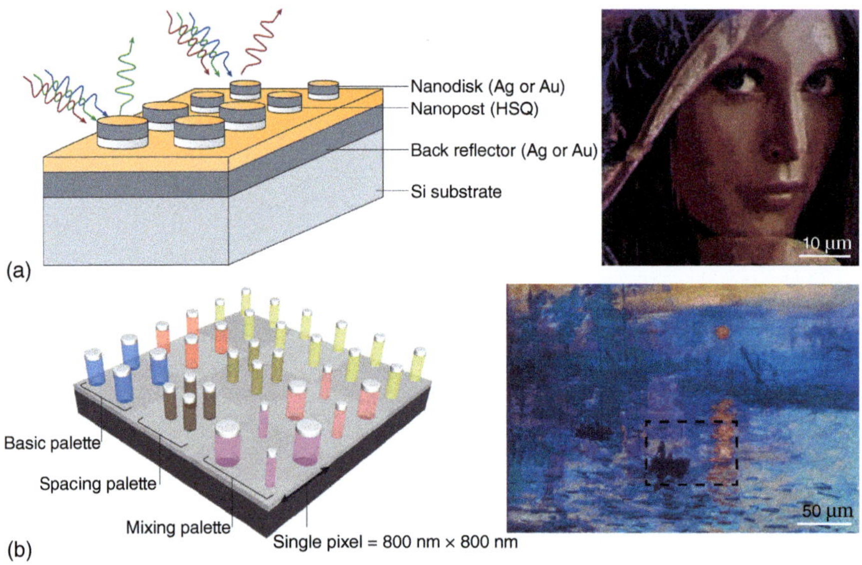

Figure 10.8 (a) Various pixel designs with the corresponding printed images patterned using electron-beam lithography [22]. (b) Aluminum-based plasmonic colors achieved by aluminum nanostructures [23]. Source: Adapted with permission from Tan et al. [23]. Copyright 2014, American Chemical Society.

of the structure but also the optical parameter of the working environment, which significantly broadens the control range of the optical characteristics [18, 19].

Besides, the enhanced optical transmission through subwavelength nanoapertures in metal films [20] has paved the way for the design and realization of silver-, gold-, and aluminum-based color filters. Because of their high plasma frequency and low losses [21], the high cost of these noble metals hinders the broad integration into practical applications. Therefore, aluminum is the most abundant metal in the Earth's crust, and it finds a wide range of practical applications.

As shown in Figure 10.8a [22], individual plasmonic pixels made of a metal disk (gold or silver) are raised above a holey metal back reflector by hydrogen silsesquioxane (HSQ) nanoposts. Each pixel consists of between one and four disks, which fit within a 250 nm square area corresponding to a resolution of around 100 000 dots/in. The resulting print of a photorealistic test image that goes under the name of Lena is shown. Figure 10.8b [23] shows that the combining aluminum nano-disks with different sizes and varying the spacing between them broadens the color gamut, making it possible to create an accurate micrometer-sized miniaturized reproduction.

One excellent application of the plasmonic structural combining electrochromics is reported by Ting Xu et al. [24]. By using plasmonic nanoslit arrays, they demonstrated high contrast, fast monochromatic, and full-color electrochromic switching using two different electrochromic polymers, electrodeposited PANI and poly(2,2-dimethyl-3,4-propylenedioxythiophene) (PolyProDOT-Me_2). The extremely thin coatings on metal structures with well-controlled thicknesses supports the scatter-free propagation of SPPs, leading to maximum interaction with

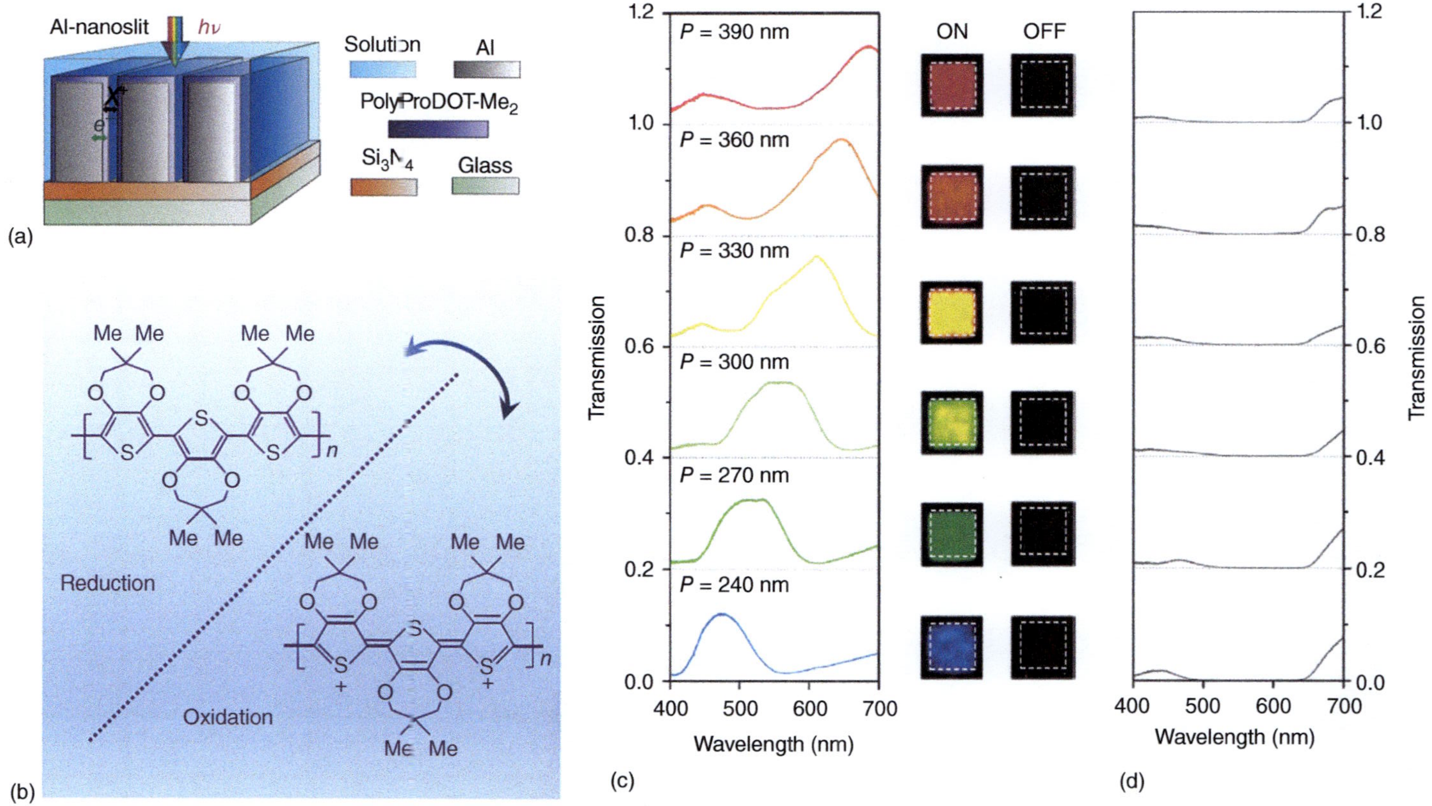

Figure 10.9 Full-color plasmonic electrochromic electrodes. (a) Schematic diagram of a plasmonic electrochromic electrode incorporating Al-nanoslit array with different sizes. (b) Chemical structures of PolyProDOT-Me_2 in the oxidized and reduced form. (c,d) Optical transmission spectra of PolyProDOT-Me_2-coated Al-nanoslit structures with respective values of slit period, along with corresponding optical micrographs of device [24]. Source: Xu et al. [24]. Licenced under CC BY 4.0.

the electrochromic films (Figure 10.9a,b). As a result, the plasmonic electrochromic switchable configurations retain the advantages of both fast switching speed and versatile color performance (Figure 10.9c,d).

This work could well prove that the plasmonic nanostructures hold a great potential for structural color technologies, and several of their key performance indicators exceed those of traditional color technologies.

Besides the nanostructure fabricated through the widely used e-beam lithography on substrates [25] or nano-imprinting [26], solution processing is successfully applied as the nanostructure supporting for electrochromics. In 2017, Wenzheng Lu et al. have realized the coating of a nanoscale PANI shell on differently shaped Au nanocrystals, including nanospheres (NSs), nanorods (NRs), and nanobipyramids (NBPs), and investigated their active electrochemical plasmonic switching

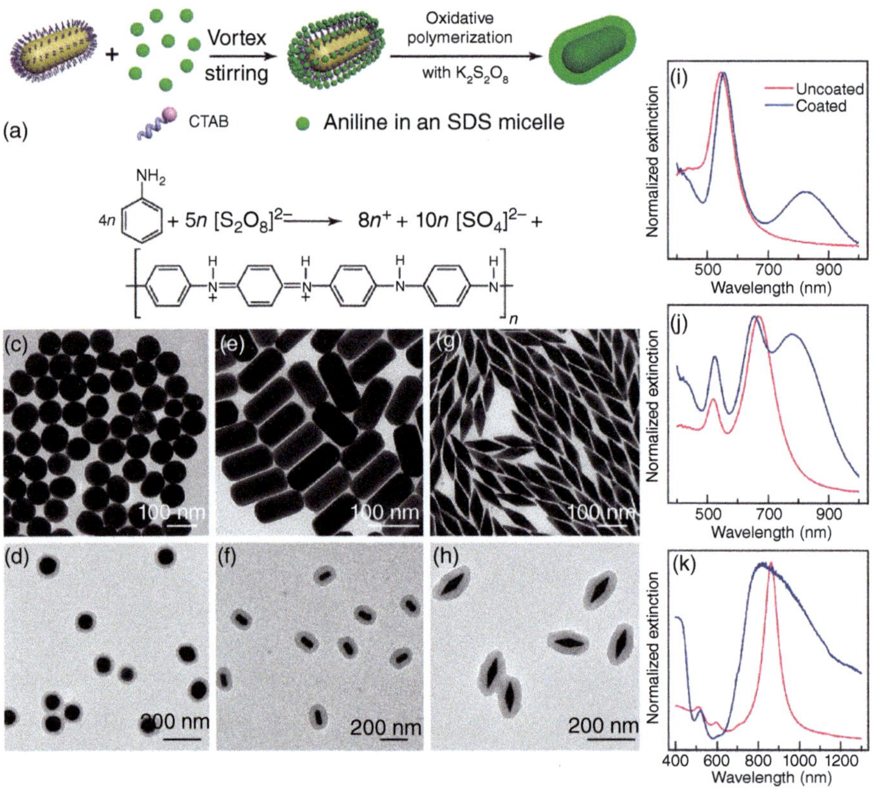

Figure 10.10 (Au nanocrystal)@PANI nanostructures. (a) Schematic illustrating the surfactant-assisted oxidative polymerization process for coating PANI on the Au nanocrystals. A gold NR is employed for illustration purposes. (b) Equation of the oxidative polymerization of aniline with persulfate as the oxidant. (c–h) TEM images of the uncoated (top) and PANI-coated (bottom) Au NSs, NRs, and NBPs, respectively. (i–k) Extinction spectra of the uncoated and PANI-coated Au NSs, NRs, and NBPs, respectively. Please note the different wavelength range for the Au NBPs from those for the Au NSs and NRs [26]. Source: Adapted with permission from Franklin et al. [26]. Copyright 2015, Royal Society of Chemistry.

performances. Plasmon peak shifts as large as 100 nm were obtained on the (Au NR)@PANI nanostructures. Such remarkable plasmon peak shifts have not been achieved before with merely PANI as an electrochemically controlling medium (as shown in Figure 10.10).

In short summary, plasmonic nanostructures hold great potential for structural color technologies, and several of their key performance indicators exceed those of traditional color technologies. Most remarkable are the opportunity for intriguingly high color printing resolution (exceeding 10^5 DPI) at high refresh rates (>50 Hz) and optical contrast [27] and the prospect for plasmonic color laser printing at the post-processing stage. Especially when it combines together with electrochromic materials, the system would give more interesting and versatile optical properties.

10.2.2 Nanostructured Electrochromic Materials in ECs

Since the last century, in the 1970s, inorganic transient metal oxides have been found that could apply as electrochromic materials. This research area developed very fast, and some materials such as WO_3 have already been commercialized as the smart windows in buildings and rearview mirrors in cars. The processing of the inorganic films has been widely investigated especially with solution-processed nanostructure for WO_3 [28, 29], NiO [30], and V_2O_5 [31]. However, the nanostructure formed by organic electrochromic active materials without using any templates is much more difficult and seldom reported, which is mostly attributed to the intrinsic property of organic materials. Comparing with inorganic materials, organic materials could not stand high temperatures during material synthesis or film processing. For keeping the electrochromic active characteristic, commonly the processing temperature should be lower than 100 °C, which largely confines the technic application in the nanostructure engineer of organic electrochromic materials.

In this chapter, the previous approaches introducing nanostructures have some disadvantages, such as templates involved, complicated process, and too high cost to commercialization. Moreover, electropolymerized electrochromic electrodes of poly(3,4-ethylenedioxythiophene) and PANI are not suitable for scaling-up production. A facile, low cost, and all-solution processing of chemical synthesized materials to introduce nanostructures into large area ECP electrode is urgently demanded. While inspecting those difficulties, in the recent years some excellent research works exhibit low-temperature and solution-processable way to achieve nanostructures of electrochromic materials.

In 2010, Sung Jong Yoo et al. [32] prepared PANI/PSS nanoparticles by chemical oxidation polymerization in aqueous solution (as shown in Figure 10.11). The potential of the PANI/PSS-NPs to be used as an anode electrode for electrochromic devices and the diffusion of Li^+ were investigated. A uniform electrochromic layer of PANI/PSS-NPs with a size of c. 28 nm could be obtained by a solution process, specifically spin coating. The PANI/PSS-NPs film has a high Li^+ diffusion coefficient (~7.7×10^{-9} cm^2/s) and low charge transfer resistance (~99 Ω), which result in it having a fast electrochromic response time (coloring time <1.7 seconds, bleaching time <2.4 seconds) and high coloration efficiency (>108 cm^2/C).

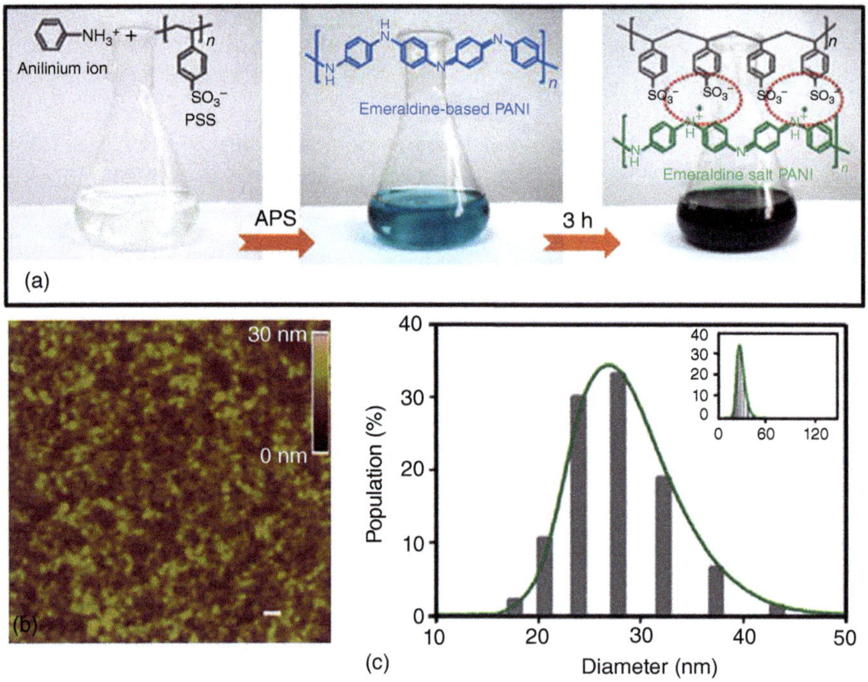

Figure 10.11 (a) Schematic diagram and photograph of preparation of PANI/PSS-NPs. (b) AFM image of PANI/PSS-NPs (c. 28 nm in diameter). (c) DLS distribution of PANI/PSS-NPs [32]. Source: Adapted with permission from Yoo et al. [32]. Copyright 2010, Elsevier.

In 2007, J. Jang reported the water-processed polyaniline-poly(4-styrene sulfonate) nanoparticles for the application of inkjet-printed chemical sensor applications. These chemical sensors show unprecedented sensitivity and rapid response time compared to the conventional PANI-based chemical sensors. This means it is very promising of its application on the electrochemistry device electrochromics [33].

In 2015, Raphael Dehmel et al. [34] reported that self-assembly of block copolymers into the gyroid morphology is replicated into 3D nanostructured conjugated polymers (as shown in Figure 10.12). Voided styrenic gyroidal networks are used as scaffolds for the electrodeposition of two poly(3,4-ethylenedioxythiophene) derivatives and poly(pyrrole). The careful choice of solvents and electrolytes allows the replication of the initial self-assembled morphology into self-supporting gyroidal conjugated polymer networks. The nanostructured films are employed to fabricate electrochromic devices, exhibiting excellent color contrast upon switching, with fast switching speeds. The fabrication of the electrically switchable zone plate demonstrated the reliability and scalability of this 3D nanopatterning approach without a template.

In 2018, Xing et al. [35] successfully prepared water-dispersed electrochromic polymer nanoparticles (WDENs), taking two commercialized ECPs as examples

Figure 10.12 Preparation of DG-structured conjugated polymer electrodes. (a) Schematic illustration of the gyroidal styrenic template, electropolymerization, and subsequent template dissolution yielding a freestanding conjugated polymer replica (R_1 = H, F and R_2 = H, CH_2OH). Cross-sectional SEM images of (b) freestanding, mesoporous PEDOTMeOH, and (c) PPy networks [34]. Source: (a) Dehmel et al. [34]. Licenced under CC BY 4.0, (b, c) Adapted with permission from Dehmel et al. [34]. Copyright 2015, Wiley.

(as shown in Figure 10.13), poly [2,3-bis-(3-octyloxyphenyl) quinoxaline-5,8-diyl-alt-thiophene-2,5-diyl] (TQ1) and region-regular poly(3-hexylthiophene-2,5-diyl) (P3HT). The nanoparticle electrode shows much faster switching speed than the compacted electrode (e.g. 2.10 vs 24.15 seconds for coloring, 8.65 vs 25.95 seconds for bleaching @0.4 V, 1.30 vs 9.20 seconds coloring, and 1.7 vs 2.90 seconds for bleaching @1.0 V). Moreover, the potentiality of WDENs for universal ECPs is demonstrated. The microelectrochemical measurement indicates much more efficient counter-ion diffusion between the electrolyte and the NP films than the compacted films, which results in much faster electrochromic switching. Besides the facile and eco-friendly processing of the WDENs, all solution and low cost fabrication of ECP NP films suggest their broad applications in commercial

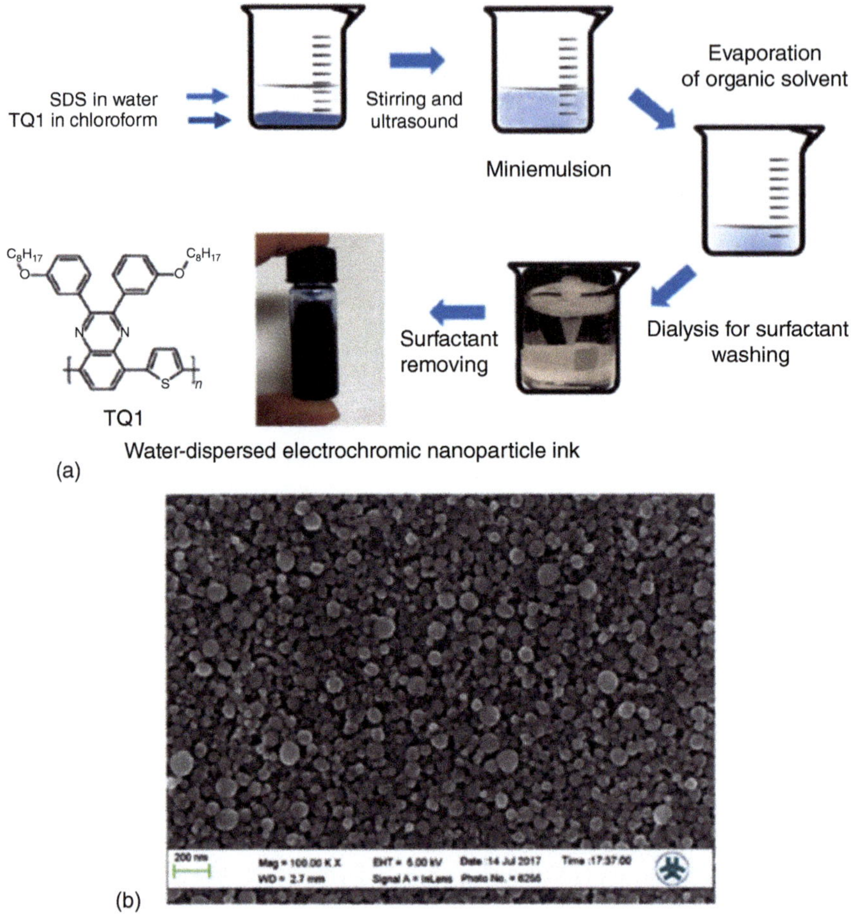

Figure 10.13 (a) The procedure of WDENs preparation and (b) SEM image of TQ1 NP film prepared from WDENs (size of particles is c. 50 nm in diameter) [35]. Source: (a) Xing et al. [35]. © 2018, Elsevier, (b) Adapted with permission from Xing et al. [35]. Copyright 2019, Elsevier.

production of polymer electrochromic display and great potential for other polymer electrochemical electronics.

Very recently, Sihang Zhang et al. [36] reported the solution-processable three-dimensional honeycomb-like PEDOT (H-PEDOT) nanostructure networks (as shown in Figure 10.14). The morphology, structure, and electrochromic performance of the H-PEDOT nanostructures were characterized. The H-PEDOT nanostructured film was assembled into patterned all-solid-state electrochromic devices. These patterned electrochromic devices show obvious color variation and fast switching speed, including larger optical contrast, faster response time (0.8 seconds for bleaching and 0.9 seconds for coloring), and higher coloration efficiency. The patterned electrochromic displays exhibit promising applications in energy-efficient displays, e-books, and e-cards.

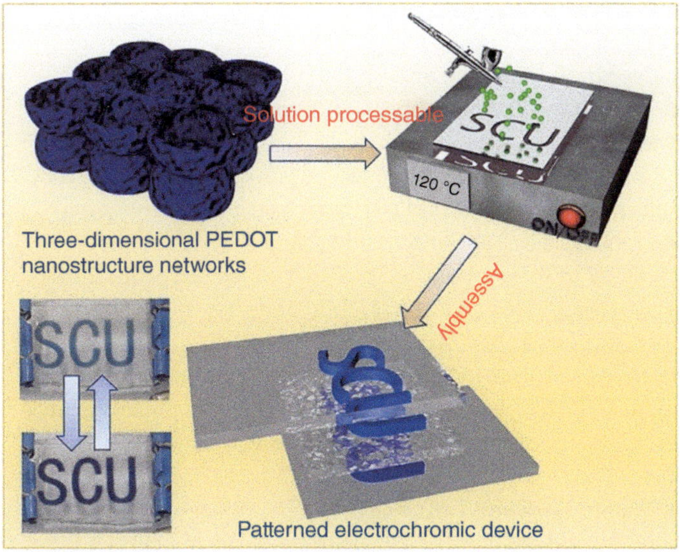

Figure 10.14 The scheme of the nanostructure PEDOT and the film processing [36]. Source: Adapted with permission from Zhang et al. [36]. Copyright 2020, Elsevier.

10.3 Conclusion and Prospect

Introducing the dielectric nanospheres or nanostructure could enhance the ionic diffusion efficiency, however, leading to lower electron/hole transport mobility. The plasmonic nanostructure could give a versatile optical performance, a faster electrochromic switching without a negative effect on the electron/hole transport mobility. The apparent idea of metal–template interaction will help to generate porous network in these materials, and conventional nonporous materials using surfactants molecules will open multiple branches in the research area of nanoporous solids. However, the e-beam lithography or nano-imprinting processing methods are expensive.

The most promising way is employing different types of a porous structure without any templates in polymers. So in the future, the in-depth study on the organic electrochromic materials processing technic is essential for the switching and stability performance of electrochromism, certainly, which could also be applied in catalysis, sensing, adsorption, optoelectronics, and light harvesting and may open a good opportunity for further research on many related research areas.

References

1 Kresge, C.T., Leonowicz, M.E., Roth, W.J. et al. (1992). Ordered mesoporous molecular sieves synthesized by a liquid-crystal template mechanism. *Nature* 359 (6397): 710–712.

2 Pal, N. and Bhaumik, A. (2013). Soft templating strategies for the synthesis of mesoporous materials: inorganic, organic-inorganic hybrid and purely organic solids. *Advances in Colloid and Interface Science* 189–190: 21–41.
3 Wang, C., Shim, M., and Guyot-Sionnest, P. (2001). Electrochromic nanocrystal quantum dots. *Science* 291 (5512): 2390.
4 Zhang, J., Tu, J.P., Cai, G.F. et al. (2013). Enhanced electrochromic performance of highly ordered, macroporous WO_3 arrays electrodeposited using polystyrene colloidal crystals as template. *Electrochimica Acta* 99: 1–8.
5 Badilescu, S. (2003). Study of sol–gel prepared nanostructured WO_3 thin films and composites for electrochromic applications. *Solid State Ionics* 158 (1-2): 187–197.
6 Tong, Z., Zhang, X., Lv, H. et al. (2015). From amorphous macroporous film to 3D crystalline nanorod architecture: a new approach to obtain high-performance V_2O_5 electrochromism. *Advanced Materials Interfaces* 2 (12): 1500230.
7 Arsenault, A.C., Puzzo, D.P., Manners, I., and Ozin, G.A. (2007). Photonic-crystal full-colour displays. *Nature Photonics* 1 (8): 468–472.
8 Aguirre, C.I., Reguera, E., and Stein, A. (2010). Tunable colors in opals and inverse opal photonic crystals. *Advanced Functional Materials* 20 (16): 2565–2578.
9 Lehoux, A., Ramos, L., Beaunier, P. et al. (2012). Tuning the porosity of bimetallic nanostructures by a soft templating approach. *Advanced Functional Materials* 22 (23): 4900–4908.
10 Zhang, Y., Wang, J., Huang, Y. et al. (2011). Fabrication of functional colloidal photonic crystals based on well-designed latex particles. *Journal of Materials Chemistry* 21 (37): 14113–14126.
11 Chen, X., Wang, L., Wen, Y. et al. (2008). Fabrication of closed-cell polyimide inverse opal photonic crystals with excellent mechanical properties and thermal stability. *Journal of Materials Chemistry* 18 (19): 2262–2267.
12 Zhang, K., Xu, L.L., Jiang, J.G. et al. (2013). Facile large-scale synthesis of monodisperse mesoporous silica nanospheres with tunable pore structure. *Journal of the American Chemical Society* 135 (7): 2427–2430.
13 Ryu, J.H., Lee, J.H., Han, S.J., and Suh, K.D. (2006). Novel electrochromic displays using monodisperse viologen-modified porous polymeric microspheres. *Macromolecular Rapid Communications* 27 (14): 1156–1161.
14 Ryu, J.-H., Lee, Y.-H., and Suh, K.-D. (2008). Preparation of a multicolored reflective electrochromic display based on monodisperse polymeric microspheres with *N*-substituted viologen pendants. *Journal of Applied Polymer Science* 107 (1): 102–108.
15 Ge, D., Yang, L., Tong, Z. et al. (2013). Ion diffusion and optical switching performance of 3D ordered nanostructured polyaniline films for advanced electrochemical/electrochromic devices. *Electrochimica Acta* 104: 191–197.
16 Kuno, T., Matsumura, Y., Nakabayashi, K., and Atobe, M. (2016). Electroresponsive structurally colored materials: a combination of structural and electrochromic effects. *Angewandte Chemie, International Edition* 55 (7): 2503–2506.
17 Zhou, D., Che, B., and Lu, X. (2017). Rapid one-pot electrodeposition of polyaniline/manganese dioxide hybrids: a facile approach to stable high-performance

anodic electrochromic materials. *Journal of Materials Chemistry C* 5 (7): 1758–1766.

18 Shrestha, V.R., Lee, S.-S., Kim, E.-S., and Choi, D.-Y. (2015). Polarization-tuned dynamic color filters incorporating a dielectric-loaded aluminum nanowire array. *Science Reports* 5: 12450.

19 Duempelmann, L., Casari, D., Luu-Dinh, A. et al. (2015). Color rendering plasmonic aluminum substrates with angular symmetry breaking. *ACS Nano* 9 (12): 12383–12391.

20 Ebbesen, T.W., Lezec, H.J., Ghaemi, H. et al. (1998). Extraordinary optical transmission through sub-wavelength hole arrays. *Nature* 391 (6668): 667–669.

21 Zeng, B., Gao, Y., and Bartoli, F.J. (2013). Ultrathin nanostructured metals for highly transmissive plasmonic subtractive color filters. *Science Reports* 3: 2840.

22 Kumar, K., Duan, H., Hegde, R.S. et al. (2012). Printing colour at the optical diffraction limit. *Nature Nanotechnology* 7 (9): 557.

23 Tan, S.J., Zhang, L., Zhu, D. et al. (2014). Plasmonic color palettes for photorealistic printing with aluminum nanostructures. *Nano Letters* 14 (7): 4023–4029.

24 Xu, T., Walter, E.C., Agrawal, A. et al. (2016). High-contrast and fast electrochromic switching enabled by plasmonics. *Nature Communications* 7 (1): 10479.

25 Jacobs, K.E. and Ferreira, P.M. (2015). Painting and direct writing of silver nanostructures on phosphate glass with electron beam irradiation. *Advanced Functional Materials* 25 (33): 5261–5268.

26 Franklin, D., Chen, Y., Vazquez-Guardado, A. et al. (2015). Polarization-independent actively tunable colour generation on imprinted plasmonic surfaces. *Nature Communications* 6 (1): 7337.

27 Peng, J., Jeong, H.-H., Lin, Q. et al. (2019). Scalable electrochromic nanopixels using plasmonics. *Science Advances* 5 (5): eaaw2205.

28 Afify, H., Hassan, S., Obaida, M. et al. (2019). Preparation, characterization, and optical spectroscopic studies of nanocrystalline tungsten oxide WO_3. *Optics and Laser Technology* 111: 604–611.

29 Bourdin, M., Mjejri, I., Rougier, A. et al. (2020). Nano-particles (NPs) of WO_3-type compounds by polyol route with enhanced electrochromic properties. *Journal of Alloys and Compounds* 823: 153690.

30 Xie, Z., Liu, Q., Lu, B. et al. (2019). Large coloration efficiency and fast response NiO electrochromic thin film electrode based on NiO nanocrystals. *Materials Today Communications* 21: 100635.

31 Chang, C.-C., Chi, P.-W., Chandan, P., and Lin, C.-K. (2019). Electrochemistry and rapid electrochromism control of MoO_3/V_2O_5 hybrid nanobilayers. *Materials (Basel)* 12 (15): 2475.

32 Yoo, S.J., Cho, J., Lim, J.W. et al. (2010). High contrast ratio and fast switching polymeric electrochromic films based on water-dispersible polyaniline-poly(4-styrenesulfonate) nanoparticles. *Electrochemistry Communications* 12 (1): 164–167.

33 Jang, J., Ha, J., and Cho, J. (2007). Fabrication of water-dispersible polyaniline-poly(4-styrenesulfonate) nanoparticles for inkjet-printed chemical-sensor applications. *Advanced Materials* 19 (13): 1772–1775.

34 Dehmel, R., Nicolas, A., Scherer, M.R., and Steiner, U. (2015). 3D nanostructured conjugated polymers for optical applications. *Advanced Functional Materials* 25 (44): 6900–6905.

35 Xing, X., Zeng, Q., Vagin, M. et al. (2018). Fast switching polymeric electrochromics with facile processed water dispersed nanoparticles. *Nano Energy* 47: 123–129.

36 Zhang, S., Zhao, Y., Du, Z. et al. (2020). Solution-processable three-dimensional honeycomb-like poly(3,4-ethylenedioxythiophene) nanostructure networks with very fast response speed for patterned electrochromic devices. *Solar Energy Materials & Solar Cells* 207: 110354.

11

Organic Electroluminochromic Materials

11.1 Introduction

Electroluminochromism is a phenomenon where materials exhibit reversible changes in photoluminescence whether emission intensity or wavelength under external electric stimulus [1], which has been surprisingly lagged behind for several decades compared with considerable research in electrochromism [2, 3]. While switching luminescence involves the modulation of triplet phosphorescent and singlet fluorescent emission, most electroluminochromic (ELC) materials exhibit changes in fluorescent emissions. Therefore, the term electrofluorochromic (EFC) is equivalent to ELC in most cases. The controllable luminescent switching via a facile electrochemical way makes ELC materials largely applicable for multifunctional and smart displays [4]. More importantly, such in situ and simultaneous detection of physical and electrochemical signals of redox materials makes it possible to have a deeper understanding of chemical reactions, electron transfer, or other processes, rendering ELC materials promising for sensing applications. The first paper concerned with ELC dates back to 1993 reported by Lehn and coworkers [5], where $Ru(bpy)_3^{2+}$ was chosen as fluorophore and quinone/hydroquinone was selected as redox couple. However, until 2004, Levillain, Sauvage, and coworkers designed a spectroelectrochemical cell and reported the EFC spectroelectrochemistry of perylene dipentylimide [6]. Audebert, Kim, and coworkers reported the first EFC window based on a tetrazine compound in 2006 [7]. Since then, many advances of ELC materials and devices have been achieved in the past decade, including small molecules and polymers (Figure 11.1).

Over the past 10 years, plenty of compounds exhibiting ELC properties have been published. The mechanisms corresponding to ELC performance differ from each class of materials, which will be discussed in detail subsequently. Recently, several reviews concerned with ELC materials and devices have appeared [1, 4, 8], providing a broad coverage of different aspects for researchers to design and optimize ELC materials and devices. In this chapter, conventional mechanisms of electroluminochromism and ELC performance parameters will be summarized. Meanwhile, four types of ELC materials will be discussed, including small organic molecules, emissive polymeric films, transition metal complexes, and nanocomposite films. The strategies to achieve high ELC performance and the optimization

Organic Electronics for Electrochromic Materials and Devices, First Edition. Hong Meng.

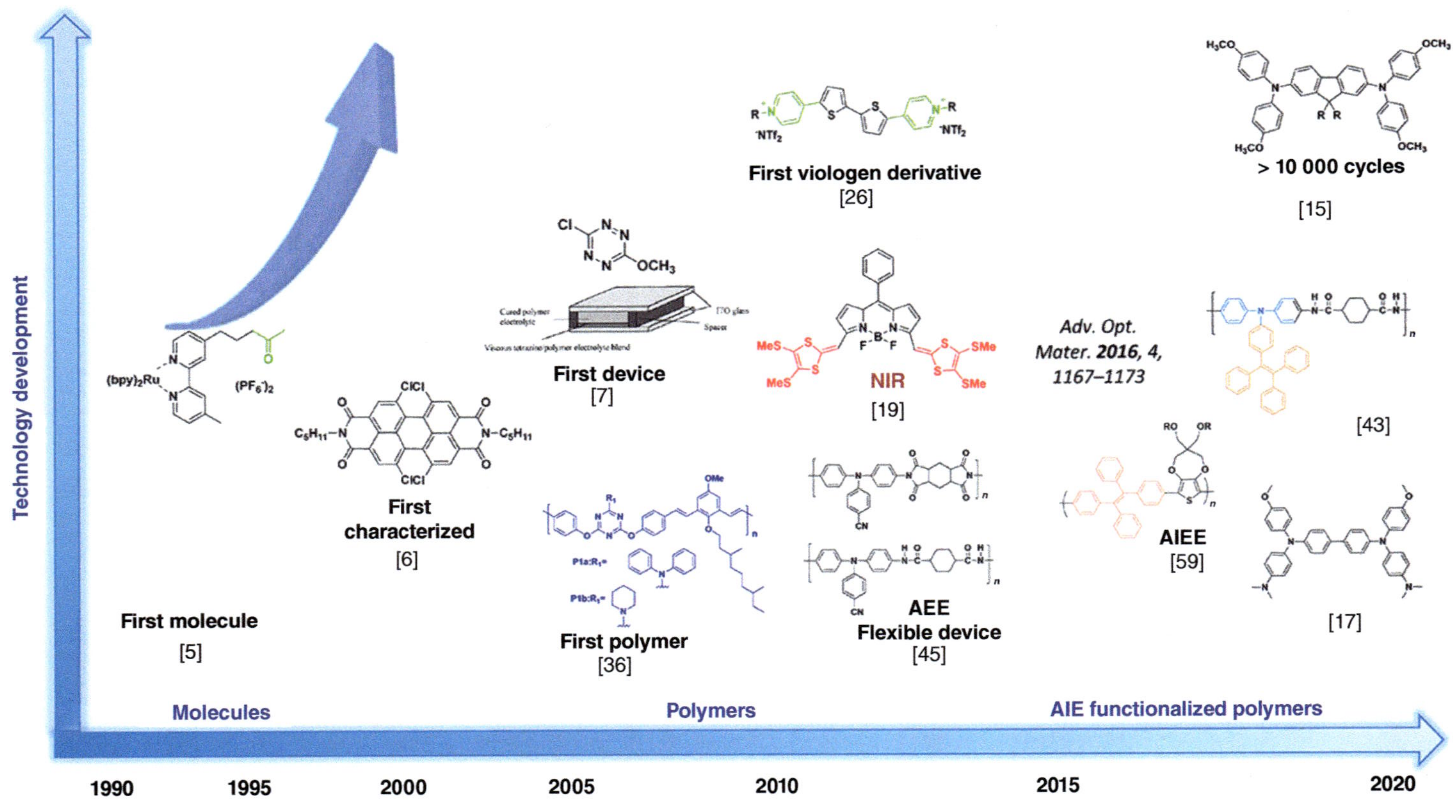

Figure 11.1 Road map of electroluminochromic materials.

of device fabrications of certain materials will be briefly discussed through some examples.

11.2 Conventional Mechanisms of Electroluminochromism

Among different ELC materials, there are three main mechanisms (Figure 11.2) that can be broadly distinguished: (i) intrinsic mechanism, (ii) electron transfer (ET) mechanism, and (iii) energy transfer (EnT) mechanism. Each mechanism corresponds to certain kinds of materials [1].

11.2.1 Intrinsic Mechanism

The first one is a very straightforward way to modulate the fluorescence of materials, which involves the intrinsically electroactive luminophores. During the process, switching fluorescence can be achieved by reversibly changing between different redox states, thus, considerable stabilities of ion radicals or dianionic (dicationic) species prerequisite for ELC performance of this kind of materials. Commonly, radicals should be non-fluorescent, due to their "open-shell" structure and only a few radicals have weak fluorescence. However, stronger emission has been observed in dianionic or dicationic species in some cases [9]. The typical examples for this

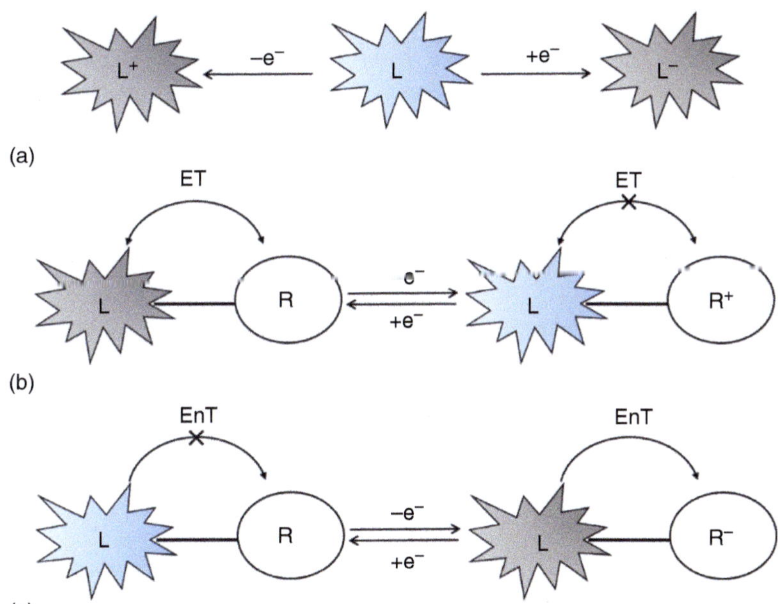

Figure 11.2 Working mechanisms of electroluminochromic materials. (a) Intrinsic mechanism; (b) Electron Transfer Mechanism (ET), and (c) Energy Transfer Mechanism (EnT).

mechanism are intrinsically switchable luminophores, like triarylamines [10] and tetrazines [7], and so forth.

11.2.2 Electron Transfer (ET) Mechanism

Apart from intrinsically switchable ELC materials, there is another type of ELC materials, redox group-controlled material, which are made of a fluorophore, linked to one or two redox moieties through a conjugated or a non-conjugated spacer, like dyads and triads. The switching fluorescent properties can be controlled by allowing or forbidding photo-induced electron or proton transfer between a redox-active moiety and a luminophore. Other materials for this type of mechanism involve polymers composed of multiple redox-active moieties and luminophores, and an ELC system consists of redox part and pH-sensitive luminescent materials [11]. According to the oxidizing or reducing character of redox moiety, the electron or proton transfer is allowed or not, and therefore the fluorescence is switched off or on.

11.2.3 Energy Transfer (EnT) Mechanism

In addition to the ET process, a mechanism for the redox group-controlled ELC materials is the energy transfer mechanism. There are two foundations for this class of ELC materials. Firstly, the partner redox moiety of the luminophore has to be chosen carefully, because the redox character of redox moiety and excited states of luminophore must be consistent. Secondly, an electrochromic reaction is responsible to change the overlap between absorption of the redox part and the emission of the luminophore for the switching light emission [1]. According to the existence of overlap or not, the light of luminophore can be switched off or on. Compared with ET mechanism, the EnT can happen in systems that have a much wider separation between redox moieties and luminophores, for example, several ELC composite films reported are based on the intermolecular EnT mechanism.

11.3 Electroluminochromic Performance Parameters

When dealing with an ELC material, the first characteristic that one should be concerned with is its emission intensity and its intensity changes; sometimes, the peak wavelength also changes. Thereafter, other aspects of the intensity variation should be considered. Among them are the degree of intensity change, how fast the intensity changes, and so on. Within the ELC research field, several terminologies have been adopted to refer to these key ELC performance parameters.

11.3.1 Emission Contrast

In general, the definitions of the terms are also followed as consistently as possible so as to standardize the experimental procedures and to allow easy comparison of performances of ELC materials across different research groups. However, since the ELC field has been rapidly developed just from 2006, the definition of intensity change has not been unified. There exist several definitions among the ELC community.

(1) Intensity ratio
The intensity ratio of an ELC material measures the degree of intensity change between its on and off states [12], and the contrast is defined by the intensity ratio of on/off state at a certain wavelength, which is the mostly adopted method in the ELC community.

$$\text{Emission Contrast} = \frac{I_{\text{on}}}{I_{\text{off}}}$$

(2) Intensity difference
The second method is defined as the difference between the fluorescence intensity in the on and off state, as the formula shows. The advantage of this method is easy to compare among different materials since all results are within 0–100% [13].

$$\text{Emission Contrast} = \frac{I_{\text{on}} - I_{\text{off}}}{I_{\text{on}}} = \Delta\text{FL}\%$$

(3) Ratio of spectrum area
The third method uses the ratio of the full emission spectral areas in the off and on states to demonstrate the emission contrast [13]. Even though it can reflect the light change of the whole spectrum and be calculated from the fluorescence spectrum, the ratio of the spectrum area is uncommonly used.

$$\text{Emission Contrast} = \frac{A_{\text{on}}}{A_{\text{off}}}$$

11.3.2 Switching Time

The switching time measures how fast the ELC material switches between the on/off states in response to applied bias. Similar to electrochromic materials, there are two distinct processes occurring in the emission change (transition from turn-on to turn-off states). The turn-off time t_{off} corresponds to the speed of the quenched process, whereby the ELC material changes from a state with a higher emission intensity to a state with a lower emission intensity. On the other hand, the turn-on time t_{on} corresponds to the speed of the lighted process, whereby the ELC material changes from a state with a lower emission intensity to a state with a higher emission intensity. Predominantly, the switching time is defined as the time taken for the emission contrast to switching to 95%/90% of its full switch [14].

11.3.3 Long-Term Stability/Cycle Life

The stability of an ELC material or device evaluates its ability to sustain the ELC performance under repeated redox cycling. One common approach for stability measurement is to record the number of redox cycles that the ELC material or device can sustain before remarkable degradation in emission contrast occurs [15]. This gives an approximation of its cycle life. As the redox cycleability of the ELC material or device is significantly affected or highly influenced by experimental conditions, such experimental parameters have to be reported clearly.

11.4 Classical Materials

In this chapter, the classical reported ELC materials will be discussed and summarized in detail with focus on the following categories: (i) organic molecules, (ii) transition metal complexes, (iii) organic polymers, and (iiii) nanocomposite ELC films. Various examples in each category and subcategory with ELC performance parameters will be presented.

11.4.1 Small Molecules

11.4.1.1 Small Molecular Dyads

Most of ELC dyads or triads are small molecules. The redox moiety in these molecules provides the electroactivity and the luminescence switches through ET or EnT process between the excited state of fluorophore and the redox moiety as described earlier. The redox moiety constituting dyads can be not only non-emissive redox moieties, like ferrocence (Fc) [16, 17], quinone [11, 18], tetrathiafulvalene (TTF) [19], and so on but also emissive redox moieties, like naphthalimide (NI) [20]. The fluorophore constituting dyads can choose from a variety of moieties, such as tetrazine [20, 21], boradiazaindacene [19], pyrene [16], and so on. Several small molecular ELC dyads consisting of tetrazine and TTF will be discussed in detail. Figure 11.3 shows typical structures of small molecular ELC dyads.

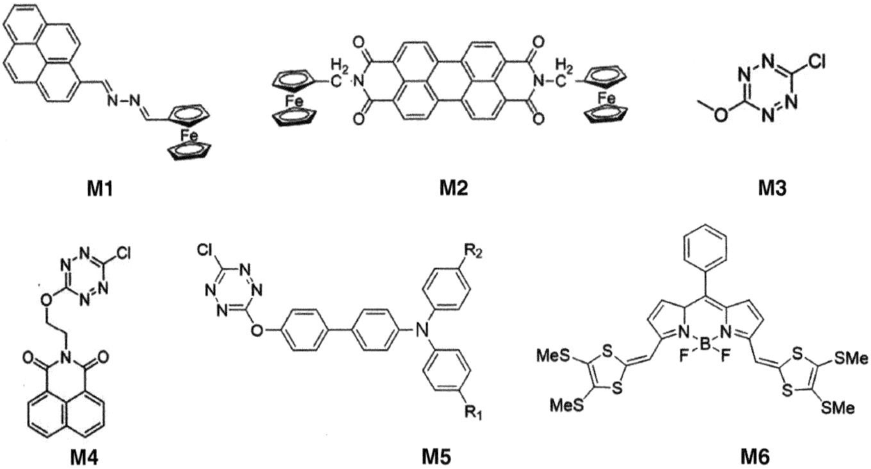

Figure 11.3 Typical structures of small molecular dyads.

Fc is a redox-active building block, which can be linked with some luminescent moieties to form ELC dyads or triads, like M1 [16] and M2 [17]. Such dyads and triads are likely to show intramolecular charge transfer (ICT) or ET process in their neutral state, which will be forbidden at Fc oxidation state, leading to switching luminescence. Martínez et al. used Fc as redox moiety and pyrene as fluorophore to obtain EFC dyads M1, which can be reversibly turned on/off through changing redox states of Fc. Similarly, the fluorescence intensity of M2 consisting of Fc and perylene diimide was reversibly switched by electrochemical oxidation and reduction process in sequence via suppressing and releasing the polyethylene terephthalate (PET) process.

Tetrazine compounds are one classical EFC small organic molecule, which were developed for nearly two decades. Tetrazine compounds are intrinsically ELC materials since they can achieve reversibly switching emission change because they can be reduced to non-emissive radical anions from natural emissive state, making tetrazines promising for ELC devices. The first ELC device based on M3 sandwiched between two indium tin oxide (ITO) glass electrodes has shown over 120 cycles of fluorescence switching at around 560 nm between +2 and −2 V without distinct degradation [21]. Apart from substituted tetrazines, ELC dyads consist of one tetrazine unit, which is well known. Kim, Audebert, and coworkers developed dyad M4 [20] as a dyad where chlorotetrazine (TZ) was linked to NI fluorophore via a non-conjugated ethoxy-linkage (Figure 11.4a–b). Three different states can be achieved in dyad M4: a neutral state, a TZ reduced state, and both a TZ and NI reduced state. The NI group can enhance the fluorescent efficiency by quasi-complete energy transfer from NI to TZ, so natural state of M4 is yellow emission. What's more, multiple emission color states can be obtained by blending (2-hydroxyethyl)-*N*-naphthalimide (HNI) (blue emission at neutral state) with M4 deriving from different reduction voltages of TZ and NI: white emission at neutral state, blue emission at −0.8 to −1.2 V, and dark above −1.4 V. After this, Audebert and coworkers reported a series of triphenylamine (TPA)–tetrazine dyads M5 subsequently [22], in which TPA moieties with different substitutes were attached to chlorotetrazine. A series of M5 [22] compounds were non-fluorescent at the neutral state owing to ET process from TPA moiety to tetrazine moiety. When TPA units were oxidized to radical cationic state, the PET process would be forbidden, leading to fluorescence turn-on of M5 compounds.

Tetrathiafulvalene–boradiazaindacene (TTF-BODIPY) compound M6 [19] was reported to access three-state "on–off–on" performance [19] (Figure 11.4c). The TTF-BODIPY undergoes two one-electron oxidations subsequently, leading to radical cationic and dicationic states accompanied by quenching of the fluorescence with λ_{max} at 803 nm and the occurring of fluorescence with λ_{max} at 1185 nm, respectively. Both of the two "on" states are in the near-infrared (NIR) region. The photophysical properties of NIR emissive aza-boron-dipyrromethene (aza-BDP) dyes incorporating nitrofluorene and alkoxy decorations were intensively investigated. Lim et al. studied NIR Electrofluorochromic Properties of Aza-Boron-dipyrromethene Dyes [23]. The NIR electrofluorochromic switching was visually observed through a visible light cut-off filter, featuring high fluorescence contrast.

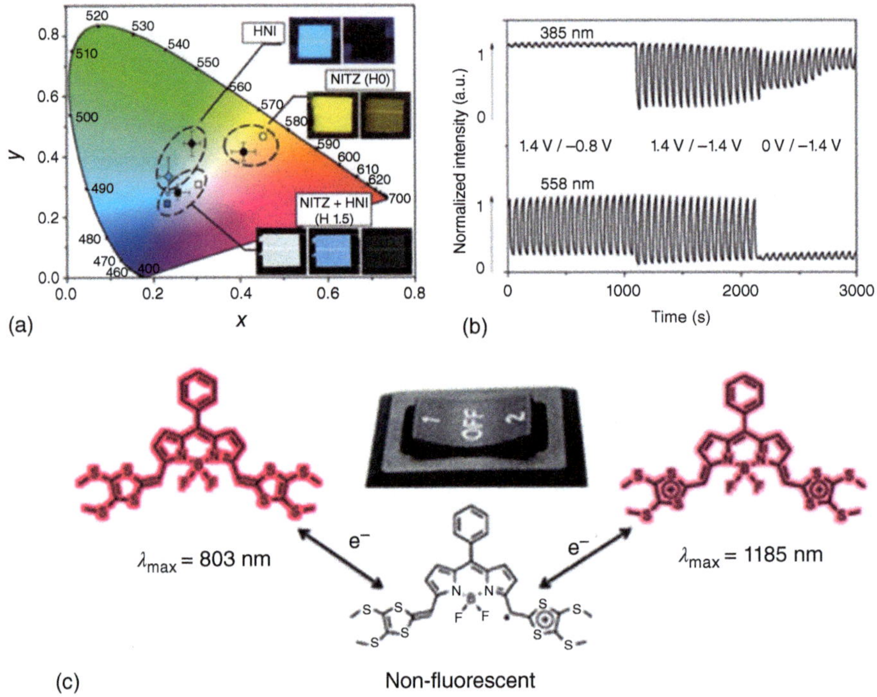

Figure 11.4 (a,b) Electroluminochromic performances of tetrazine dyads [20]. Source: Reproduced with permission from ref. Seo et al. [20]. © 2012 John Wiley & Sons. (c) ELC performances of tetrathiafulvalene dyads [19]. Source: Reproduced with permission from ref. Bill et al. [19]. Copyright 2014, The Royal Society of Chemistry.

11.4.1.2 Redox-Active Moiety and Luminophores System

Similar to dyads, the redox-active moiety and luminophores system is also composed of redox moiety and luminophores separated without the linked bridge, which is capable to realize switching of luminescence under applied external voltages. Zhang et al. reported another ELC system comprising redox-active *p*-benzoquinone (BQ) and pH-responsive fluorescent fluorescein. In this system, no emission was detected at neutral state while yellow fluorescence (λ_{max} at 535 nm) occurred at an applied negative potential of −1.5 V, and the devices showed excellent stability without degradation during 1500 cycles [11]. Following that, Zhang et al. designed an RGB pH-sensitive EFC system consisting a redox-active BQ moiety and 2-(2-(4-hydroxystyryl)-6-methyl-4*H*-pyran-4-ylidene)malononitrile, rhodol, and 7-hydroxycoumarin to produce red, green, and blue fluorescence, respectively. Figure 11.5 shows the ELC mechanism of this EFC system through indirectly tuning three fluorophores under applying electrical bias by using the reversible redox property of acid/base-mediated BQ and the pH-sensitivity of fluorophores (Figure 11.5a). The device based on this ELC system owns fast and reversible modulation of RGB-color fluorescence switching between on or off states, exhibiting great potential for encrypted information storage and multifunctional displays [18].

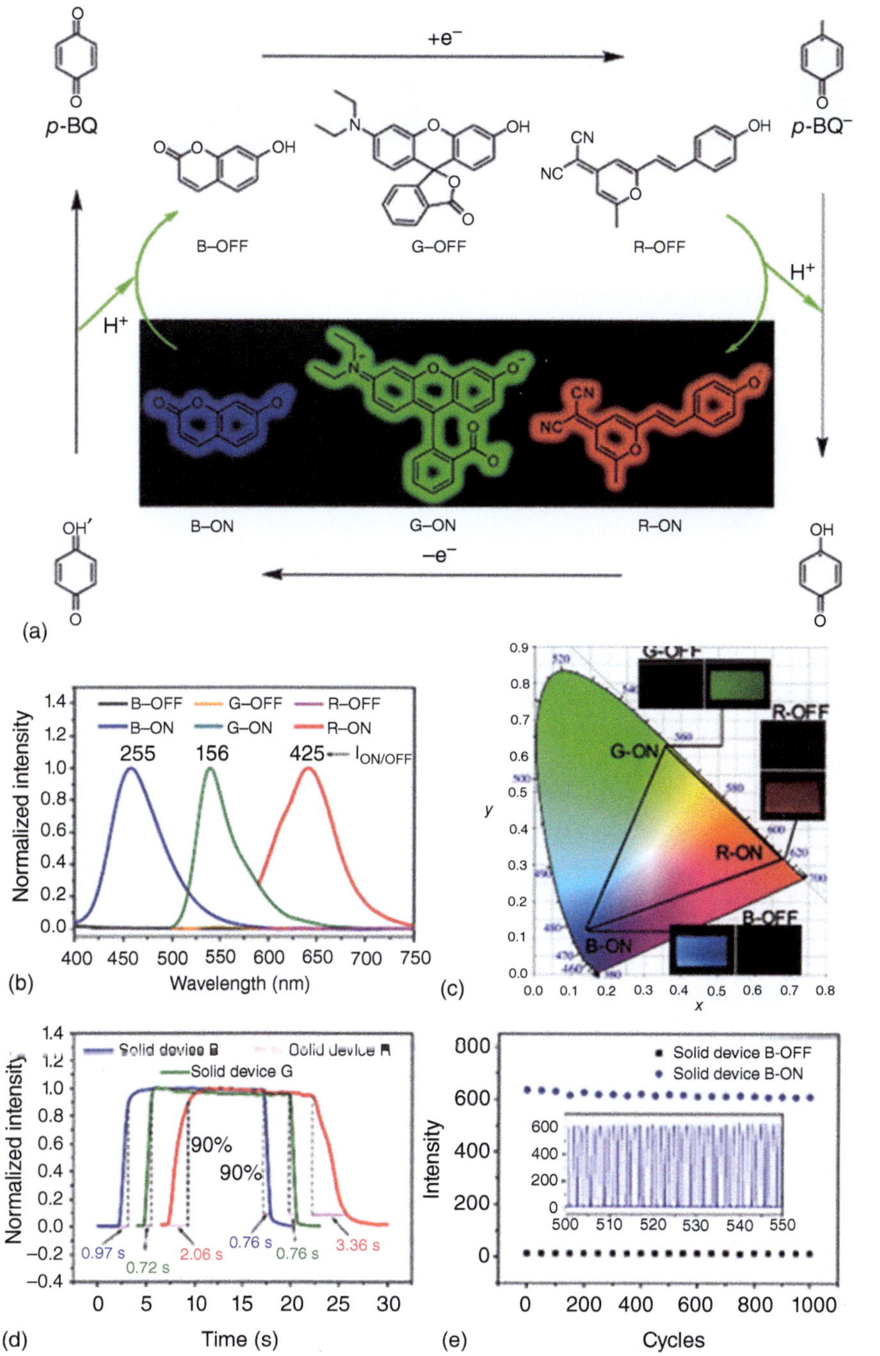

Figure 11.5 Redox-active moiety and PH-sensitive luminophores ELC system mechanism and performance [18]. Source: Reproduced with permission from ref. Wang et al. [18]. Copyright 2017, The Royal Society of Chemistry.

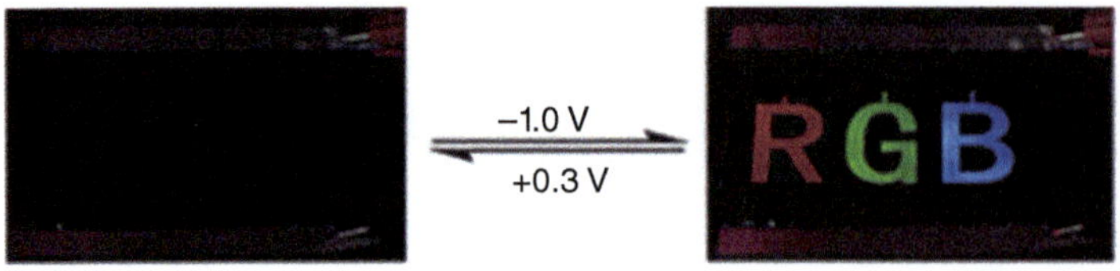

(f)

Figure 11.5 *(Continued)*

11.4.1.3 Electroactive Luminophores

Unlike dyads where switching luminescence is deriving from involved electron or proton transfer and energy transfer process, some fluorescent molecules are electroactive, and they are capable to modulate fluorescence properties between neutral and ionic states under external electric bias, which involve functionalized conventional fluorophores [12, 13], viologen derivatives [9, 13, 14], triarylamine-based compounds [15, 17], and so on. Figure 11.6 summarizes the typical structures of electroactive molecular luminophores.

Triphenylamine is a promising building block for ELC materials due to its excellent electrochemical properties and stability of radical cations under oxidation [11]. Alain-Rizzo, Miomandre, and coworkers synthesized a series of TPA derivatives with different substituents, investigating the effects on ELC performance. At neutral state, the fluorescence of six compounds are in the blue to green region and quenched effectively at oxidation state due to the formation of triphenylaminium radical cation [10].

After that, the same group reported TPA-based small ELC molecules M8 with functional *N*,*N*-dimethylaniline moiety on two phenyl groups [24]. It was found that M8 exhibited four stable redox states and three protonated states due to containing three nitrogen atoms, leading to color and emission change from one state to another state. Neutral M8 gave blue fluorescence with λ_{max} at 424 nm. When undergoing the first oxidation, significant fluorescence quenching cations formed. When undergoing the second oxidation, quenching was declined due to the appearance of non-radical dication accompanied by the presence of a new emission fluorescence with λ_{max} at 513 nm. When undergoing the third oxidation, radical trication formed leading to quenching of fluorescence with λ_{max} at 513 nm. Liu and coworkers reported ELC molecule M9 with indolo[3,2-*b*]carbazole core linked to two redox-active di(*p*-methoyphenyl)amine groups [25] (Figure 11.7a). M9 can be oxidized from natural state to 1^+ and 1^{2+} exhibiting color change from light yellow to red and blue. Meanwhile, its blue fluorescence could be reversibly turn off/on upon electrochemical oxidation and reduction, respectively. More recently, Giuseppina A. Corrente et al. reported a fluorene-based ELC molecule with two diarylamine as redox center and fluorescence unit as fluorescent center and electron coupling bridge, which allows direct ELC switching performance and exhibits extraordinary reversibility and cyclability of more than 10 000 cycles due to the stabilization of radical ions by electron coupling between two diarylamine redox centers [15] (Figure 11.7b).

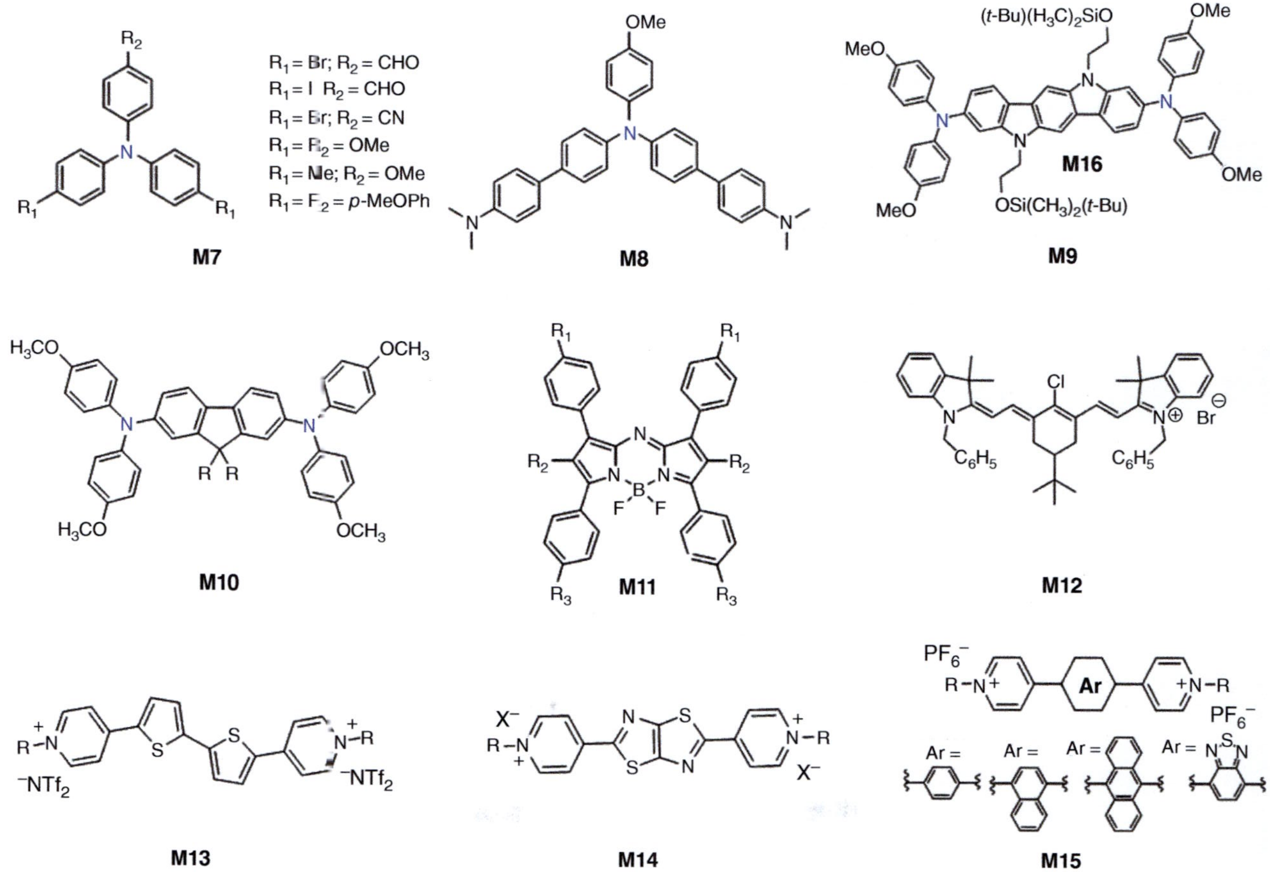

Figure 11.6 Typical structures of electroactive molecular luminophores.

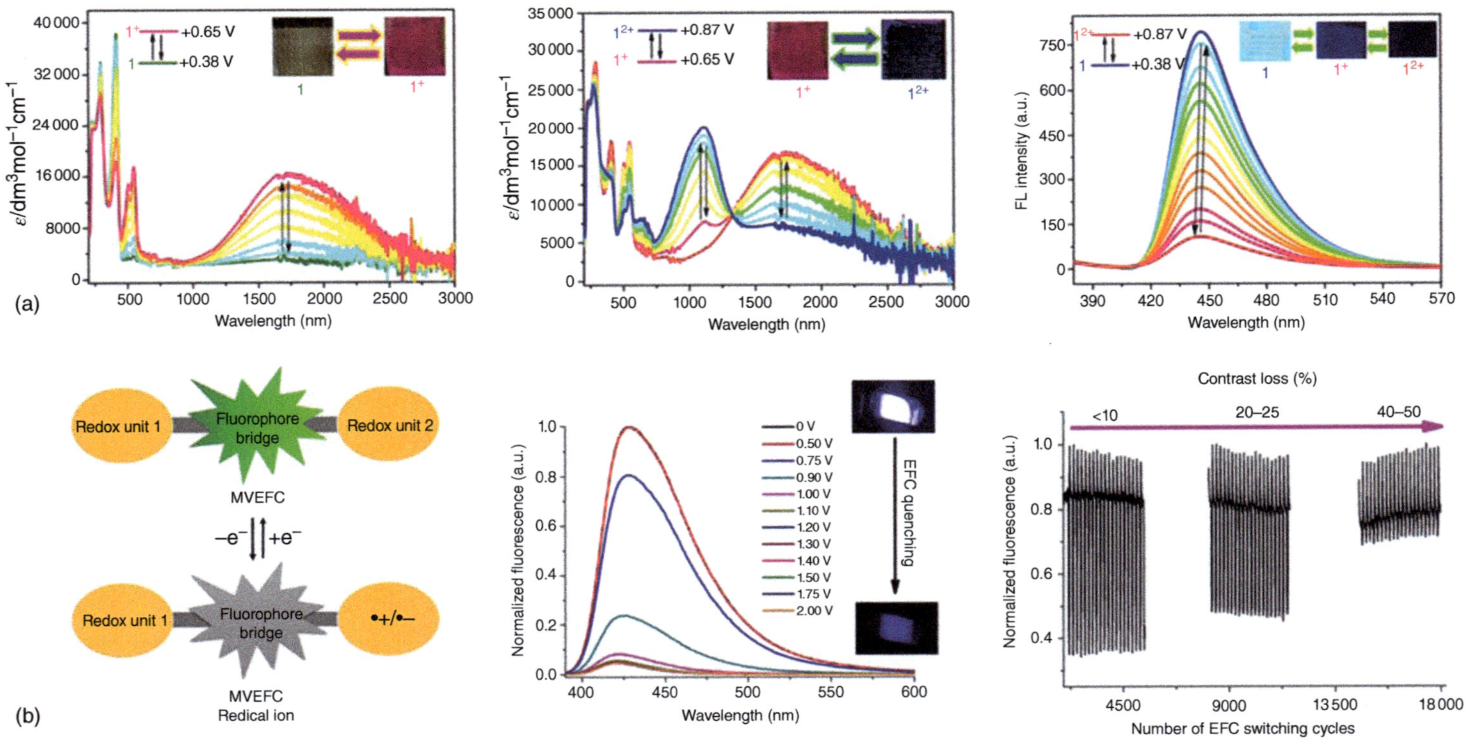

Figure 11.7 (a) EC and EFC performance of M9 [25]. Source: Reproduced with permission from ref. Zhang et al. [25]. Copyright 2017, Elsevier B.V. (b) EFC performance of M10 [15]. Source: Reproduced with permission from ref. Corrente et al. [15]. Copyright 2019, American Chemical Society.

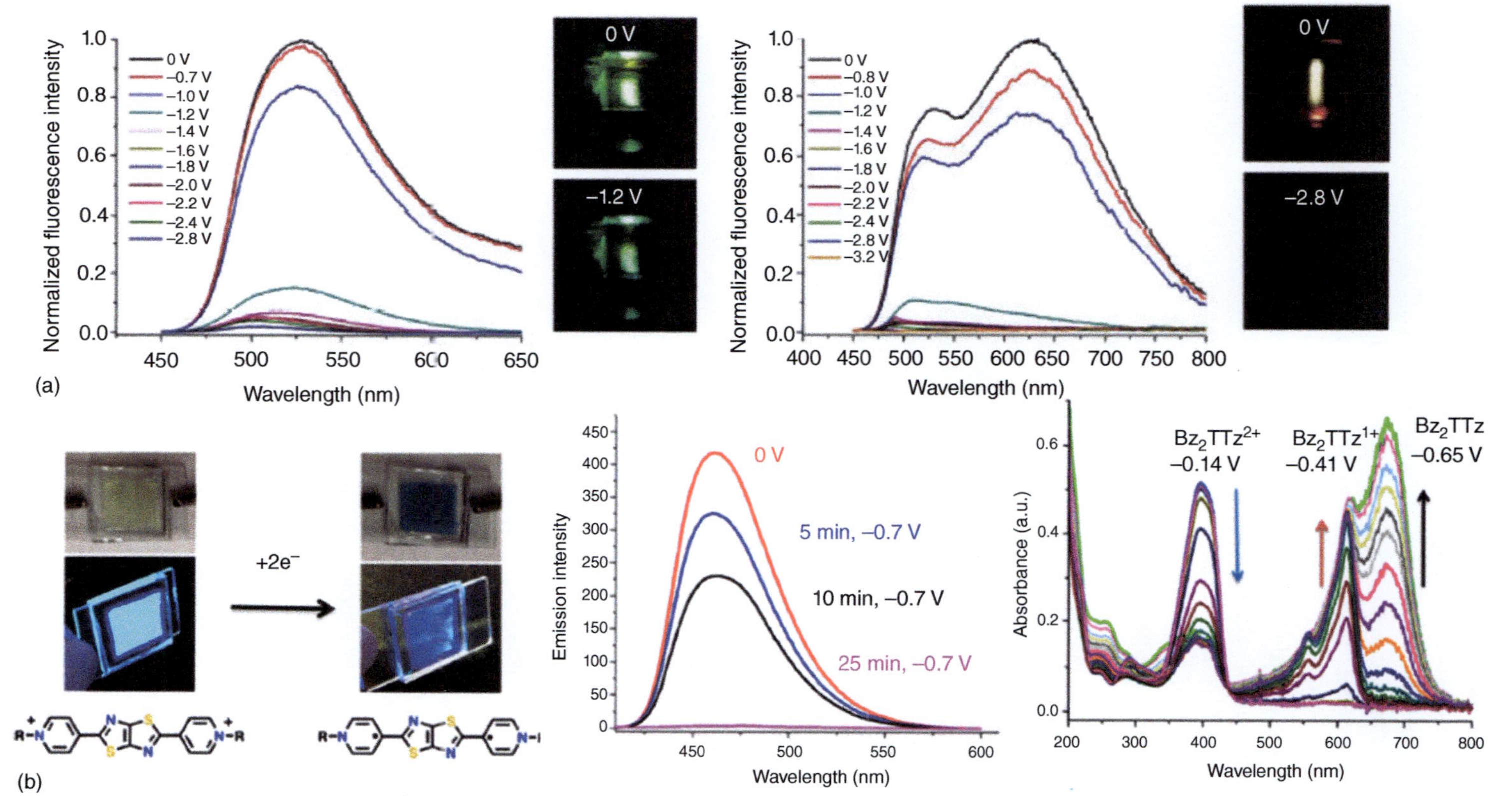

Figure 11.8 (a) EFC performance of M13 [13]. Source: Reproduced with permission from ref. Beneduci et al. [13]. Copyright 2014, Wiley-VCH Verlag GmbH & Co. KGaA, Weinheim. (b) EC and EFC performance of M14 [9]. Source: Reproduced with permission from ref. Woodward et al. [9]. Copyright 2017, American Chemical Society.

Viologens are classical and commercialized electrochromic (EC) materials, which undergo reversible two oxidation processes [2]. Thus, it is reasonable to explore viologen derivatives with high fluorescence for EFC materials. Beneduci et al. firstly reported M13 thienoviologens' EFC performance in different mesophases [26]. Electrical reduction of M13 gave rise to the formation of neutral and radical thienoviologen, leading to increase in red fluorescence. A fluorescence intensity contrast of up to 79% was recorded for M13, and both liquid crystal phases displayed a fast ELC switching time of <3.5 seconds attributed to efficient electron transport facilitated by the liquid crystal packing. One year later, Beneduci et al. exploited M13 at a low concentration of 2–10% w/w incorporating with polyvinyl formal (PVF) polymer matrix to give a highly fluorescent EFC gel [13]. At a higher concentration, another emission at 630 nm occurred due to the aggregation effect, changing the emission color consequently. Figure 11.8 shows ELC performance of M13 in which electroreduction resulted in incomplete quenching of green emission in devices with 2 wt% concentration with a contrast of 87, whereas total fluorescence quenching was observed in device with 10 wt% concentration with a contrast up to 337. Recently, three thiazolo[5,4-*d*]thiazole-based viologens derivatives M14 with different alkyl substituents were reported. M14 compounds exhibited intense blue fluorescence at neutral state with a high quantum yield of 0.8–0.96 due to enhanced planar structure. ELC performance was observed under electrochemical reductions owing to the formation of radical species.

11.4.2 Transition Metal Complexes

Since transition metal complexes have both good photophysical and electrochemical properties, some kinds of transition metal complexes, like ruthenium complexes and iridium complex, are applied in ELC devices where excellent ELC and EC performance are harvested.

(1) Ruthenium complexes
Lehn and coworkers reported a ruthenium–bipyridine complex in 1993, which was considered to be the first example of ELC material [5]. Complex C1 is non-emissive due to the ET process from the ruthenium to the quinone moiety. The phosphorescent emission could be turned on by electrochemical reduction of quinone moiety owing to the generation of C2. Lately, Zhong and coworkers reported a series of $Ru(bpy)_3^{2+}$ complexes, C3 [27], with different redox-active substituents grafted on bipyridine ligands. Results showed that with the increasing number of amine-functionalized bipyridine ligands, the redox potentials were decreased, giving rise to λ_{em} of 655, 720, and 675 nm, which can be reversibly switched under electrical bias due to the redox process of Ru^{2+}.

(2) Europium complexes
Another ELC metal complex is the europium complex [28–30]. Through blending material C4 with diheptyl viologen (HV^{2+}), which can be reduced into HV^{+1} and/or hexyl viologen (HV), the strong red-emission europium complex

was quenched due to EnT process between excited C4 and HV^{+1} or HV [30]. Kobayashis group reported dual-mode digital display devices based on the C4 and HV^{2+} system [28, 29].

(3) Iridium complexes

Zhao, Huang, and coworkers have reported a series of ELC iridium(III) complexes recently [31–35]. Different with most ELC materials, iridium complexes have advantages exhibiting excellent emission color change properties rather than only intensity change with an unconventional mechanism. In 2016, four ionic Ir(III) complexes, C5a–d [31] were firstly reported, which consist the same (2-(2-pyridyl)benzimidazole) ligand and different C^N ligands, as Figure 11.9 shows. Under applied voltage (8 V), a clear boundary between negative and positive electrodes emerged with a color change in emission of the C5a–d solution. The color change of emission can be explained by changes in highest occupied molecular orbital and lowest unoccupied molecular orbital (HOMO-LUMO) energy levels of complexes, which are dependent on the C^N ligands coordinated. Exploiting the mixture of C5 complexes with a non-ELC luminophore as background, multicolor tunable ELC devices for information self-encryption were successfully fabricated (Figure 11.10).

Six ELC materials based on luminescent ion-paired Ir-complexes were reported, which give rise to mixture emission. As shown in Figure 11.11, solutions of C6a–f [34] in acetonitrile show yellow (C7a,b,e,f), red (C7c), and green (C7d) emission at neutral state, and under applying voltage, emission color changes occurred around both the positive and negative electrodes due to ionic migration. In sequent, reduction/oxidation happened to cationic and anionic components of each ion-paired Ir-complexes. After stirring, the original emission color could be recovered.

Most recently, one novel ELC iridium complex has been reported by Zhao and coworkers, which consists of an electron-deficient phosphor and an electron-rich quencher (BH_4^-). Upon applied bias or UV stimulation, the intensity of iridium complex is enhanced due to electrochemical or chemical oxidization (via $^1O_2$) of BH_4^- blocking the ET process between the phosphor and quencher [35].

11.4.3 Polymers

Compared with small molecules, polymers are at advantage of easy processability and good film-forming ability, which is desired for commercialized and large-area device fabrication. Some previous reviews have well summarized the ELC polymer materials. There are considerable examples of ELC polymers, so only some representative and typical advances are highlighted here. Most of ELC polymers are emissive at neutral state, and the luminescence intensity can be modulated upon applied bias. This section classifies ELC polymers into two kinds: non-conjugated and conjugated polymers.

11.4.3.1 Non-Conjugated Polymers

The first ELC polymers were reported in 2007 by Jungmok Yoo et al. The *p*-phenylene vinylene (PPV) polymers P1 and P2 containing a redox-active

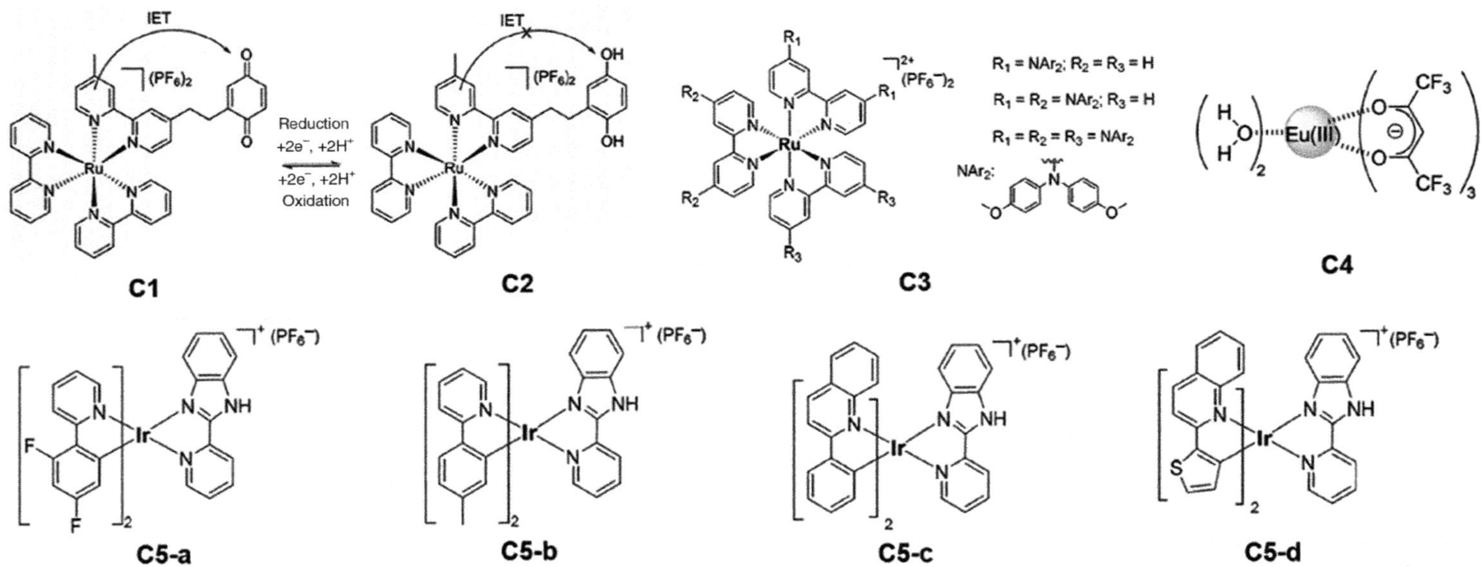

Figure 11.9 Chemical structures of ELC transition metal complexes.

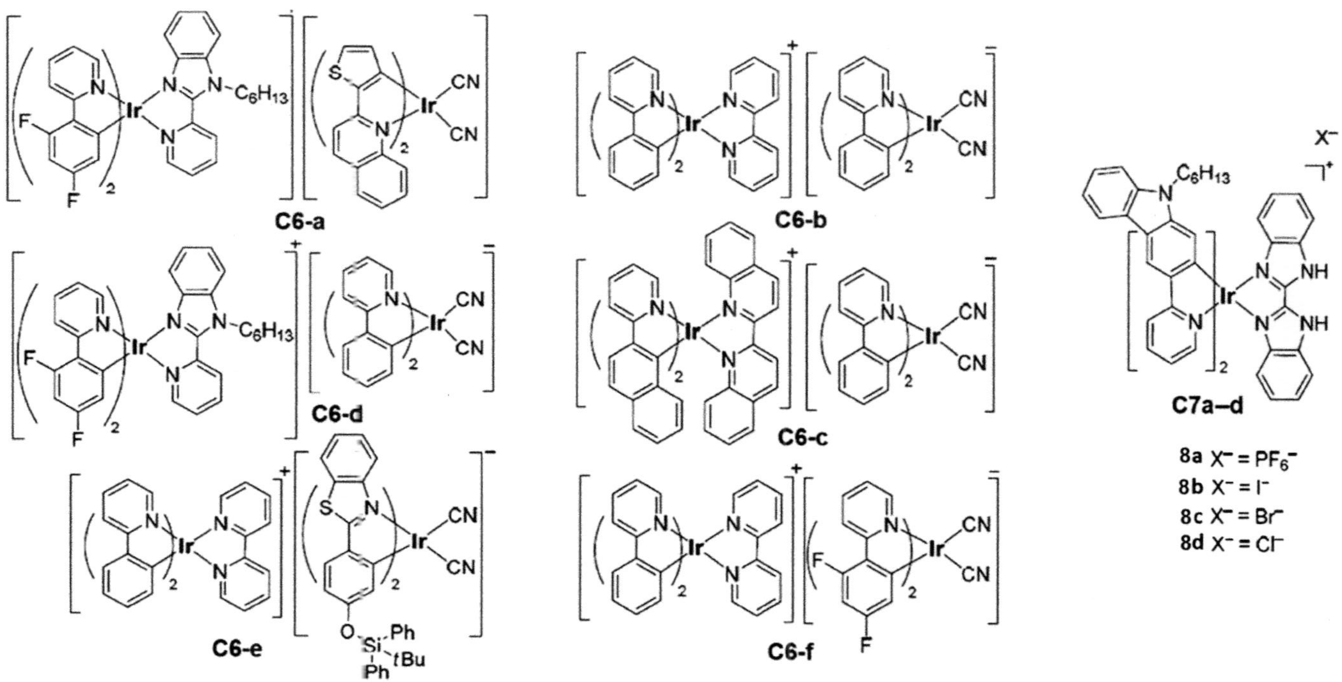

Figure 11.9 (*Continued*)

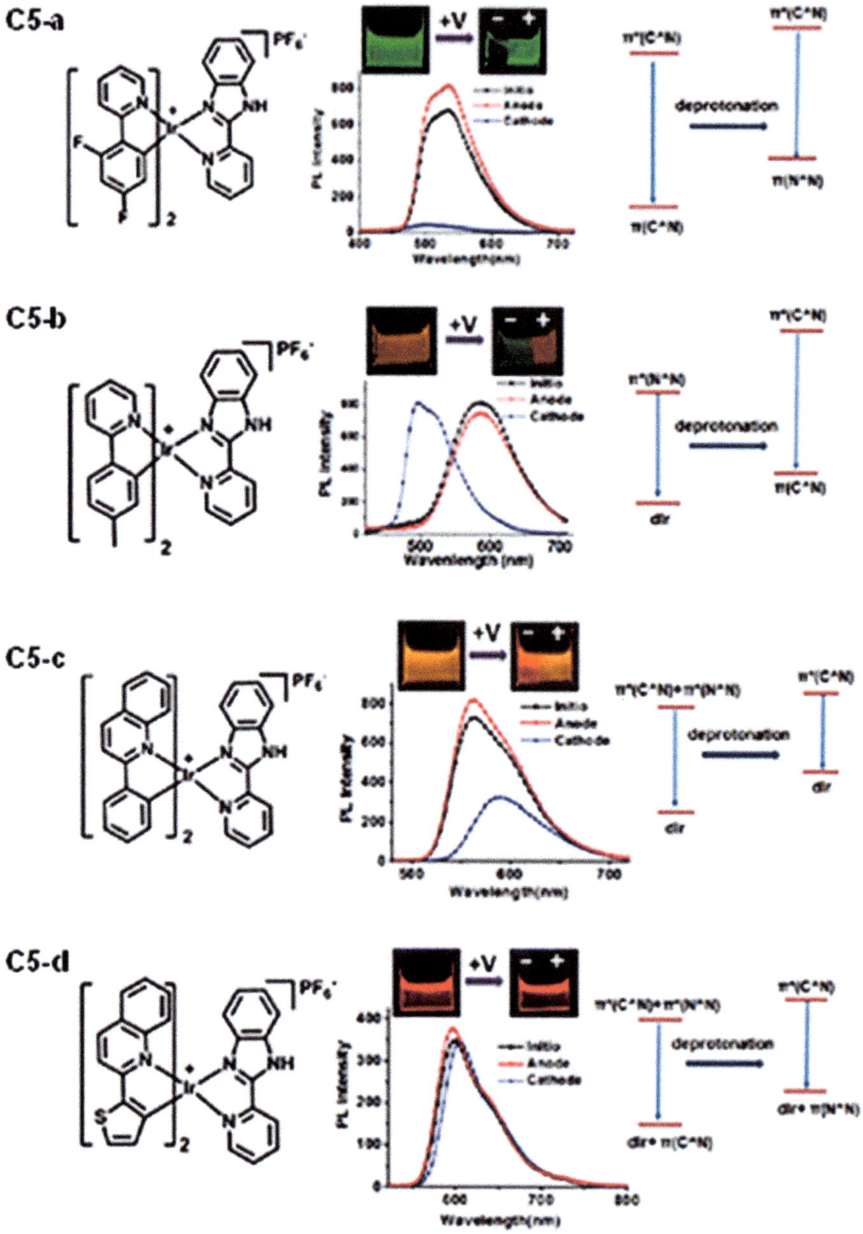

Figure 11.10 ELC properties and mechanism of C5a–d in acetonitrile solution [31]. Source: Reproduced with permission from ref. Zhao et al. [31]. © 2016 John Wiley & Sons.

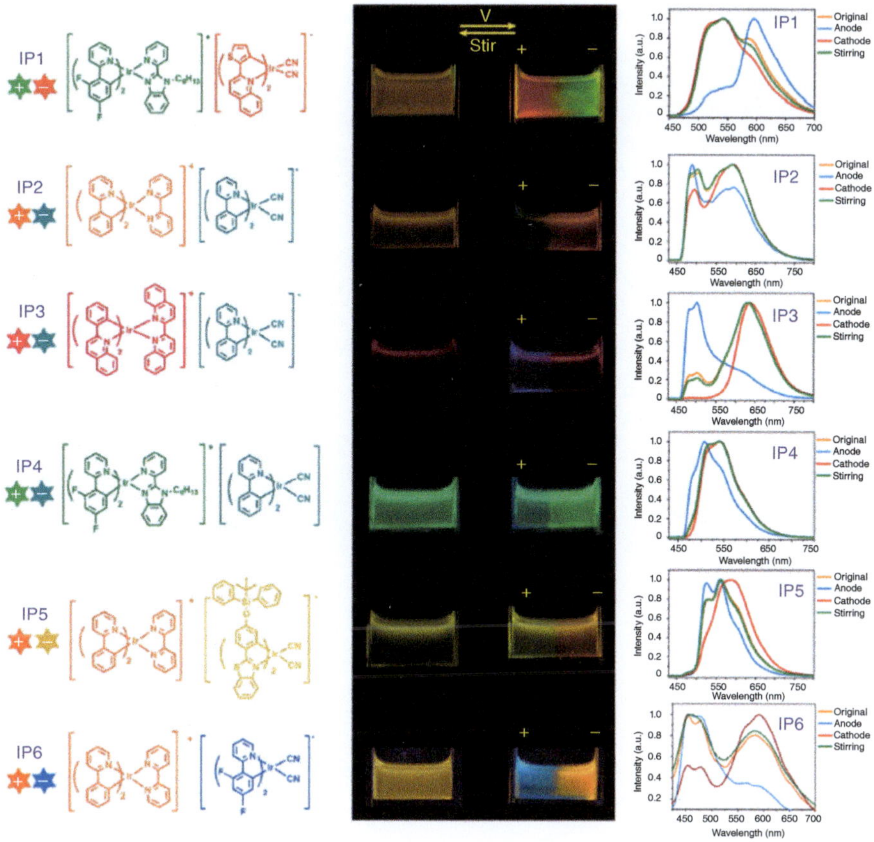

Figure 11.11 ELC properties and mechanism of C6a–d in acetonitrile solution [34]. Source: Reproduced with permission from ref. Guo et al. [34]. Licenced under CC BY 3.0.

s-triazine bridge showed high-performance electroluminochromism as Figure 11.12 shows. The switching fluorescence was achieved in ELC cells containing these polymers and an iodide and iodine couple, which showed on and off reversibly as applied potential changed [36].

Poly(amides)

Fluorene-based polymers are one of the most promising candidates for highly efficient blue luminescence materials in organic light-emitting diodes (OLEDs) due to its rigid plane and large π-conjugated system [37]. Ningwei Sun et al. has reported a series of fluorene-based poly(amides) (Figure 11.13a), showing blue fluorescence modulated by electrochemical oxidation deriving from "electron-transfer" fluorescence quenching mechanism (Figure 11.14). In the first series, P3a–e, five spacers are attached to the carbonyl side (C=O) of amine linkage, demonstrating electrochromic performance with color changing from colorless to green at oxidized states. Moreover, P3e with non-conjugated cyclohexyl exhibits high yield of 47.1% in solution vs P3a–d due to a locally excited HOMO-LUMO transition and the

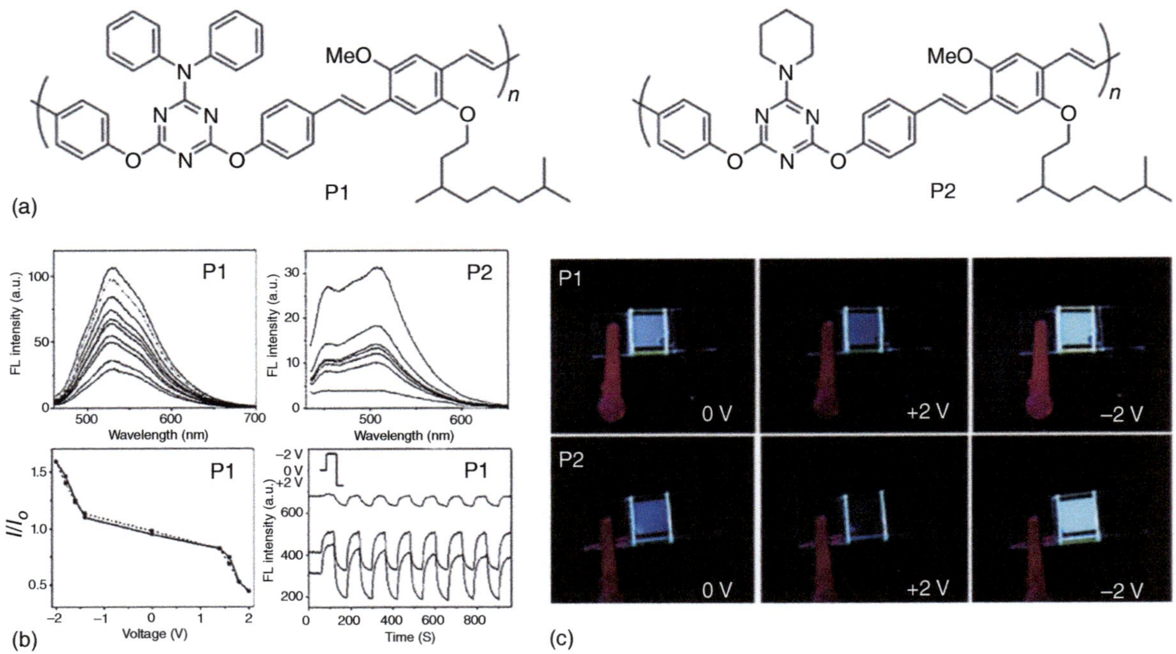

Figure 11.12 (a) Chemical structures of ELC poly(*p*-phenylene vinylene), (b) fluorescence changes of a two-electrode device at different applied potentials, and (c) photographic images of the emission from the device containing. Source: Yoo et al. [36]. Copyright 2007, American Institute of Physics.

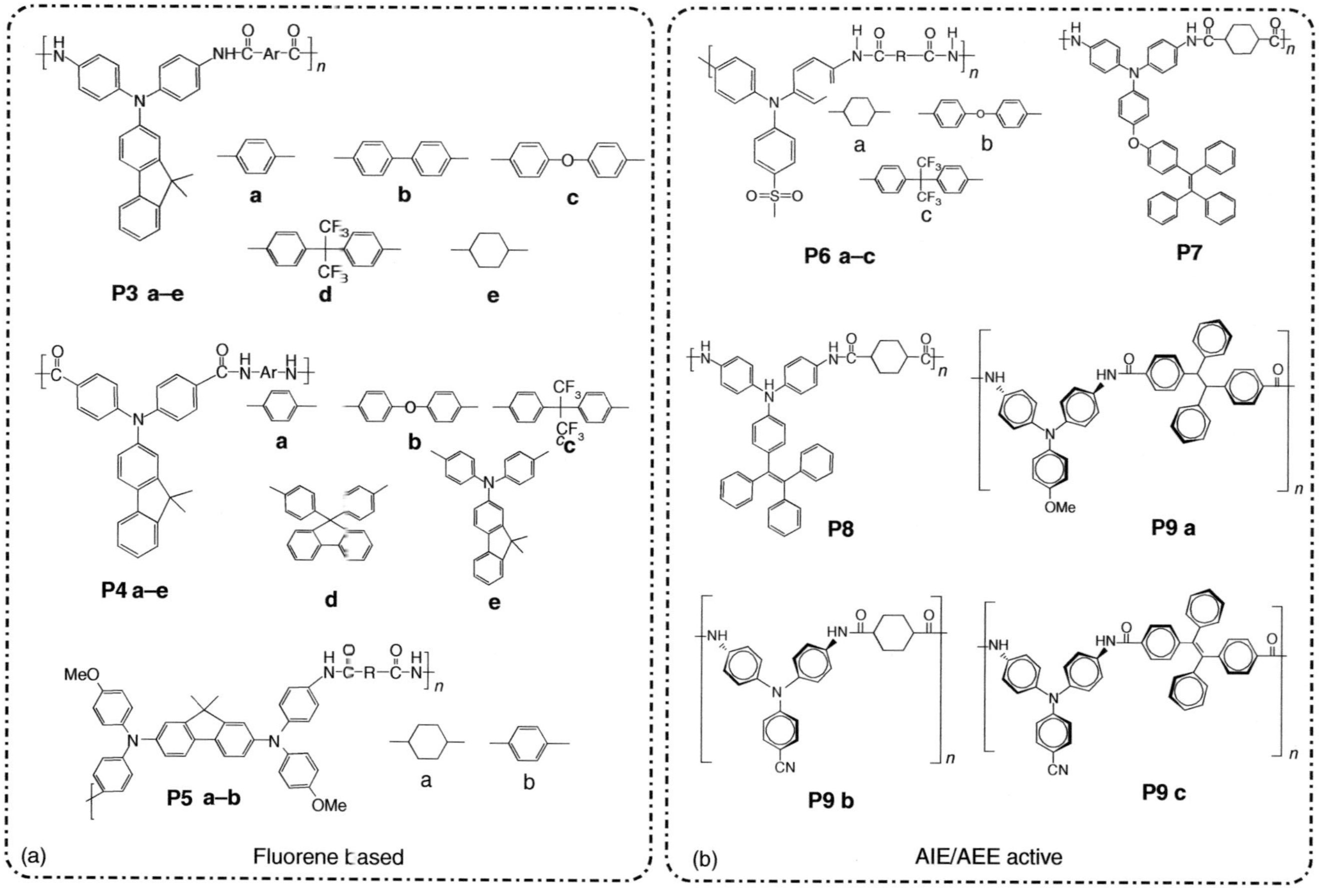

Figure 11.13 (a) Chemical structures of fluorene-based poly(amide) and (b) chemical structures of AIE/AEE active poly(amide).

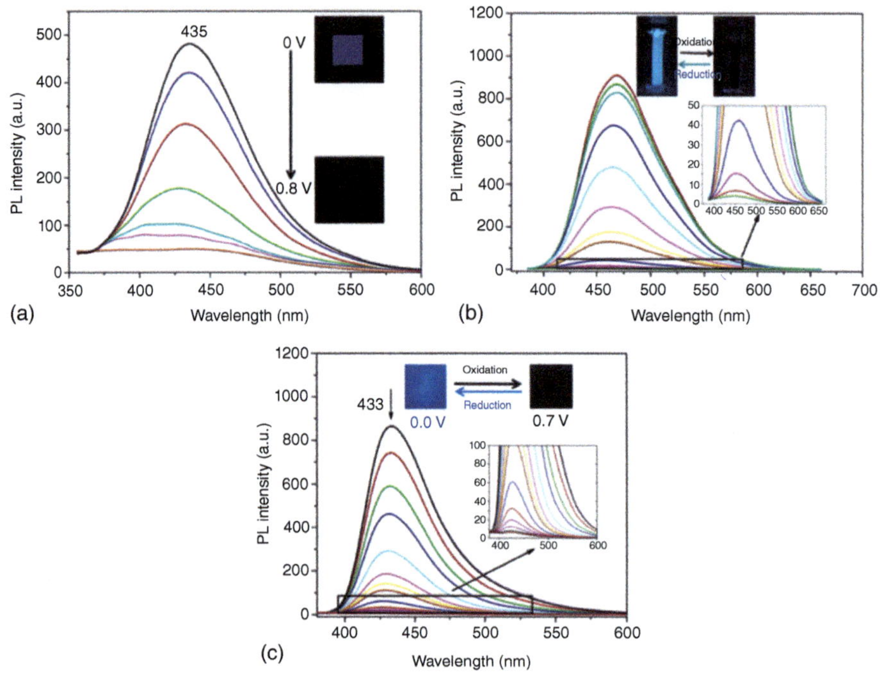

Figure 11.14 Fluorescence switching of polyamide at different voltage (a) P3e [38]. Source: Reproduced with permission from ref. Sun et al. [38]. Copyright 2015, The Royal Society of Chemistry. (b) P4d [39]. Source: Reproduced with permission from ref. Sun et al. [39]. Copyright 2016, Wiley Periodicals, Inc. (c) P5a [40]. Source: Reproduced with permission from ref. Sun et al. [40]. Copyright 2016, The Royal Society of Chemistry.

absence of interchain charge transfer. Thin film device containing P3e exhibits reversible quenching of blue fluorescence with an I_{off}/I_{on} of 12.7 [38]. In the second series, P4e exhibits multicolor performance with color changing from transparent to green at 0.9 V and gray black at 1.5 V arising from two different 2-diphenylamino-9,9-dimethylfluorene groups. Besides, P4d shows the highest yield of 34.1%, and the fluorescence quenched at 1.3 V due to the formation of the triarylamine radical cation, reaching a high contrast of up to 221.4 [39]. In 2016, based on the aforementioned research, bis(diphenylamino)fluorene-based polyamide P5 was reported with non-conjugated aliphatic cyclohexyl as a spacer to reduce the interchain charge transfer effect, leading to stronger blue fluorescence, which reversibly quenched with an optical contrast of 152 upon electrochemical oxidation [40].

Compared with small molecules, polymers have an advantage in better processability, and the film state is desired for solid devices. However, the drawback of polymers is the low quantum efficiency result from the aggregation-caused quenching effect. Recently, some approaches have been proposed by introducing aggregation-induced emission (AIE) units into EC polymers (Figure 11.13b), like tetraphenylethylene (TPE) unit sulfone (SO_2) to achieve high fluorescence quantum yield and hence obtain high fluorescence contrast ratio. In 2017, Chunhai Chen

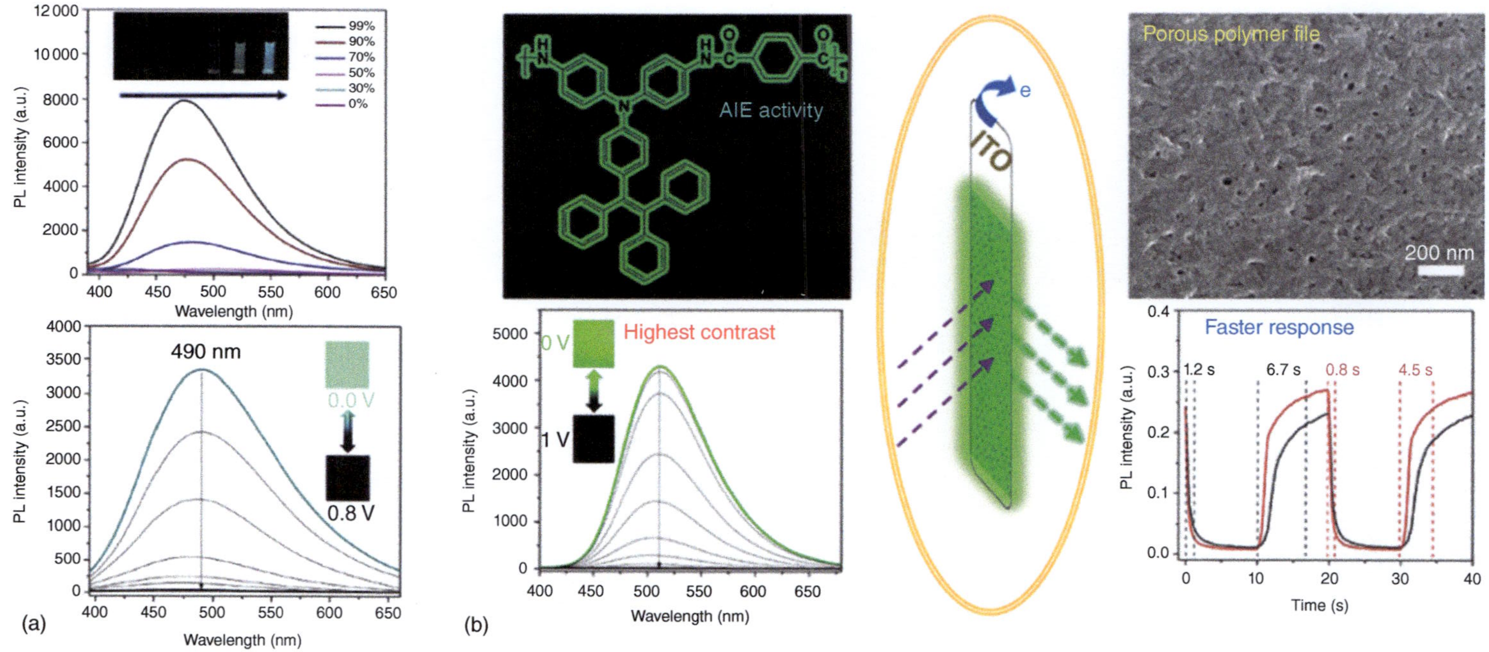

Figure 11.15 Electrofluorochromic performance of (a) P7 [42]. Source: Reproduced with permission from ref. Sun et al. [42]. Copyright 2018, Elsevier Ltd. (b) P8 [43]. Source: Reproduced with permission from ref. Sun et al. [43]. Copyright 2018, American Chemical Society.

has reported a series of AEE-active polyamides with methylsulfonyltriphenylamine units, showing one reversible oxidation along with apparent color changing from colorless to purple. Notably, the polyamide 6a demonstrated an AEE feature in solid state with high fluorescence quantum yield up to 32.1%, and the strong fluorescence could be reversibly modulated by electrochemical switching with a high contrast up to 234 [41]. Another AIE moiety TPE has been introduced into polyamide by attaching to TPA moiety via an ether linkage (P7) [42] or directly attaching to the N-position of diphenylamine (P8) [43] (Figure 11.15). P7 exhibited bright blue fluorescence, which could be quenched to dark under positive voltage due to the energy-transfer effect, giving a high contrast I_{on}/I_{off} up to 206. P8 with AIE activity showed high fluorescence quantum yield of 69% and yellow to brown color changes while offering the highest fluorescence contrast ratio up to 417.

Liou's group also developed a series of EFC polyamides, P9a–c, with TPA as building blocks and investigated the relationship between structure and EFC performance, including effects of different substitutional moieties (OMe and CN) and effects of different spacers (cyclohexyl and TPE) in the polymers. Results showed that P9a and P9c containing TPE moieties exhibited a lower quantum yield in solid state than cyclohexyl-containing P9b, due to a greater PET effect. Figure 11.16 shows the fluorescence of these polymers quenched upon electrochemical oxidation, with a response time of 8.6, 7.1, and 6.5 seconds, respectively. Furthermore, by incorporating HV into the EFC devices, the fluorescence switching contrast ratio of P9a, P9b, and P9c could be increased to 37%, 51%, and 43%, respectively [44].

Poly(imides)

The first AEE-active cyanoarylamine containing EFC polyimide P10a was reported by Guey-Sheng Liou's group (Figure 11.17). The thin film of P10a emits blue fluorescence at the neutral state and changed to a non-fluorescent monocation radical state under −1.6 V, demonstrating high contrast ratio (I_{on}/I_{off}) of 151.9 (Figure 11.18) [45]. In 2016, polyimides P11a–g with fluorene-based TPA unit were prepared by condensation polymerization reactions. The P11g exhibited blue fluorescence at 414 nm, which could be switched with a contrast ratio (I_{on}/I_{off}) up to 75 under applied potentials [46]. Recently, Liou's group developed cyanotriphenylamine-based polyimidothioethers, P12a and P12b. Combining carbonyl moiety and TPA groups, P12a and P12b exhibit multifunctional ambipolar electrochromic and EFC performance. In the neutral state, P12a and P12b display blue emission at 427 and 429 nm, respectively. At 1.7 V, the emissions of thin films were quenched due to the formation of the radical cation of TPA moieties. While at −2.0 V, the quenched emission owes to the formation of radical anion of carbonyl groups (Figure 11.19) [47].

11.4.3.2 Conjugated Polymers

Unlike non-conjugated polymers, the emission and absorption of conjugated copolymers, which consist of electron-deficient and electron-rich units alternatively in the backbone, can be tuned easily by molecular designing, offering a great chance to develop high-performance ELC polymers. Meanwhile, conjugated polymers could

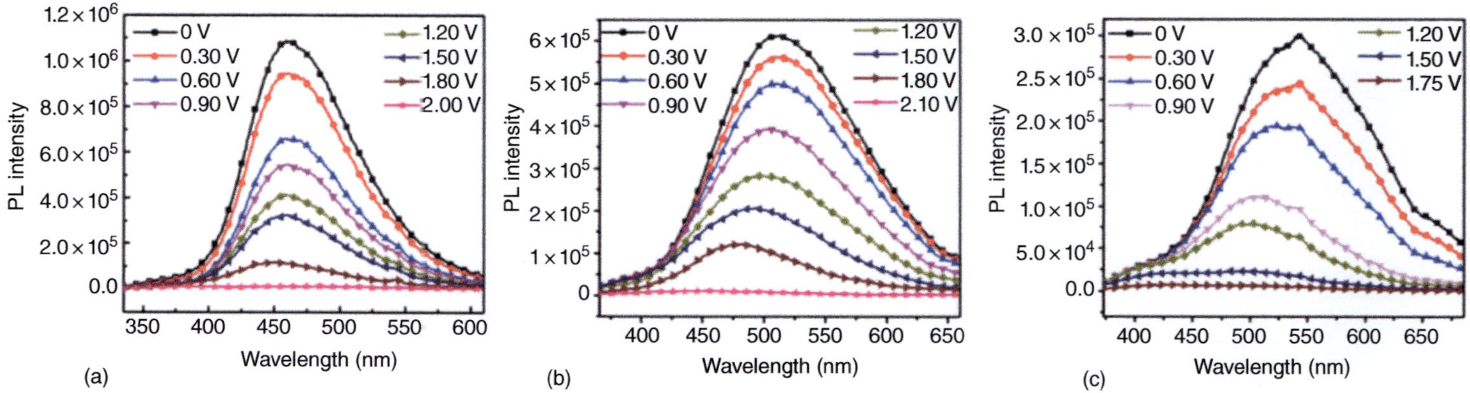

Figure 11.16 Photoluminescence spectra of P9c (a), P9a (b), and P9b (c) thin film devices without HV at different applied voltages. The photoluminescence spectra of P9c, P9a, and P9b thin film devices with HV are as shown in (d–f), respectively. Photos showing the corresponding fluorescence changes of P9c (g), P9a (h), and P9b (i) [44]. Source: Reproduced with permission from ref. Cheng et al. [44]. Copyright 2018, The Royal Society of Chemistry.

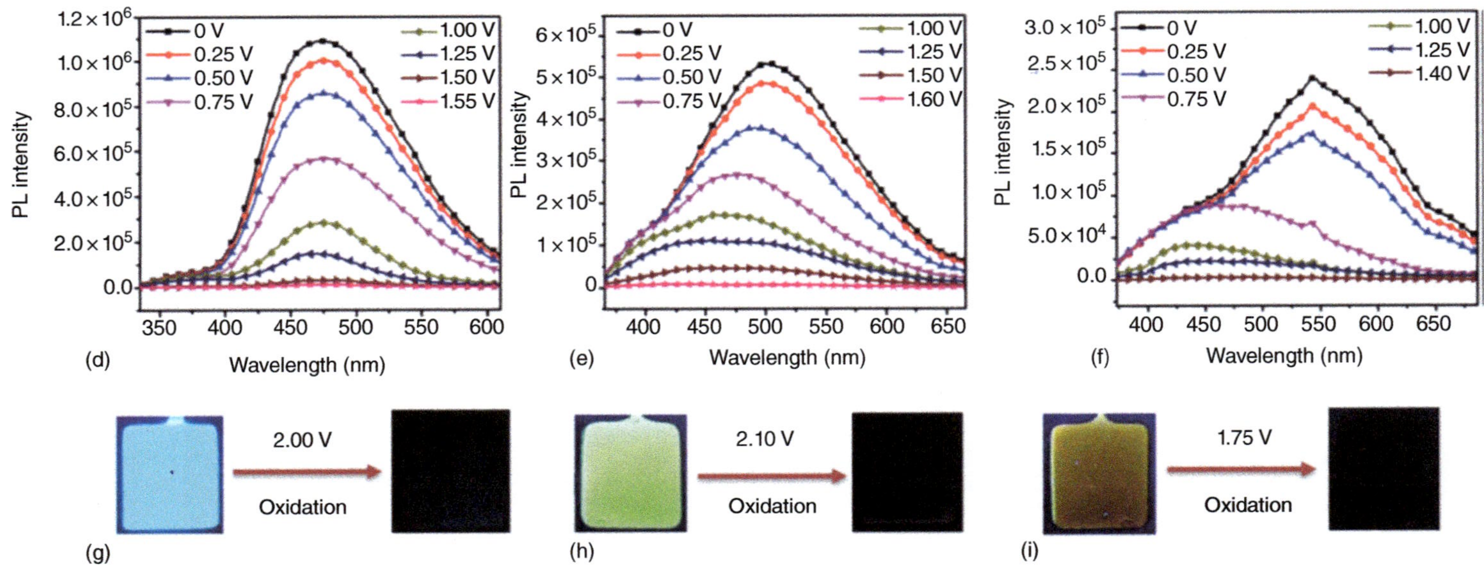

Figure 11.16 *(Continued)*

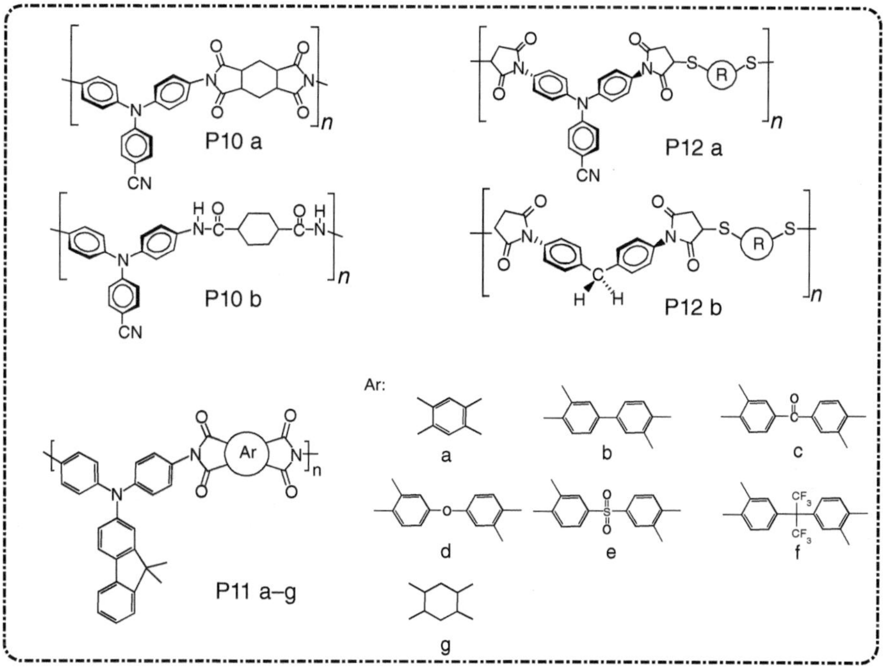

Figure 11.17 Chemical structures of EFC poly(imides).

achieve high sensitivity of fluorescence response due to the rapid excitons transportation [48] and circumvent the problem of crystallization and phase separation, yielding good long-term stabilities [8, 49]. In the part, four key types of conjugated ELC polymers, like TPA-carbazole-, fluorine-, and propylenedioxythiophene (ProDOT)-based ELC conjugated polymers are summarized.

Triphenylamine-based ELC-conjugated polymers

Triphenylamine is easily oxidized to stable radical cations, so it is a classical building block for EC/ELC bifunctional polymers. Liou and coworkers have designed many TPA-based ELC polymers, such as polymer P9a–c and P10 mentioned before [44, 45]. Then they reported a fully conjugated ELC polymer P13 based on cyanated TPA monomers (Figure 11.20) [50]. The EFC devices with P13 and HV exhibited a high photoluminescence (PL) ratio of 242 between blue fluorescent state and quenched state, low operating voltage, and rapid switching time <0.4 seconds, which is attributed to the improved charge transport mobility deriving from the fully conjugated structure of P13. Sun and Liang reported two TPA-based donor–acceptor conjugated polymers, P14a and P14b, which both exhibit simultaneous multicolored EC and ELC behaviors [51]. The stronger ICT in P14b, owing to the electron-withdrawing substituent on the third phenyl ring of TPA, renders higher electrochemical oxidation potential and lower ion diffusion coefficient of P14b. Two dendritic TPA-based polymers P15a and P15b were reported by S. Malik and coworkers. Both of them exhibited good ELC performance, with PL contrast

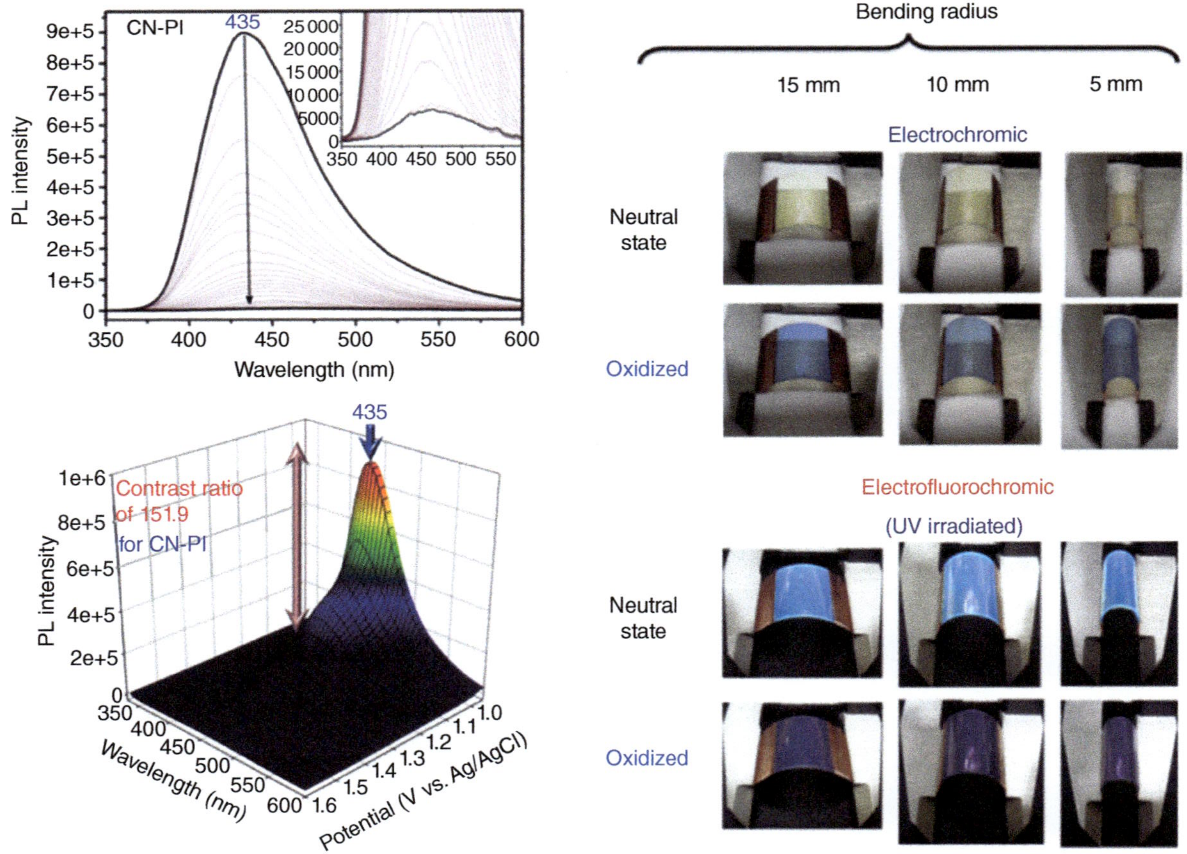

Figure 11.18 EFC behaviors of P10a and bending performance of the single-layer flexible ITO-coated poly(ethylene 2,6-naphthalate) EC and EFC device using polyimide P10a [45]. Source: Reproduced with permission from ref. Yen and Liou [45]. Copyright 2013, The Royal Society of Chemistry.

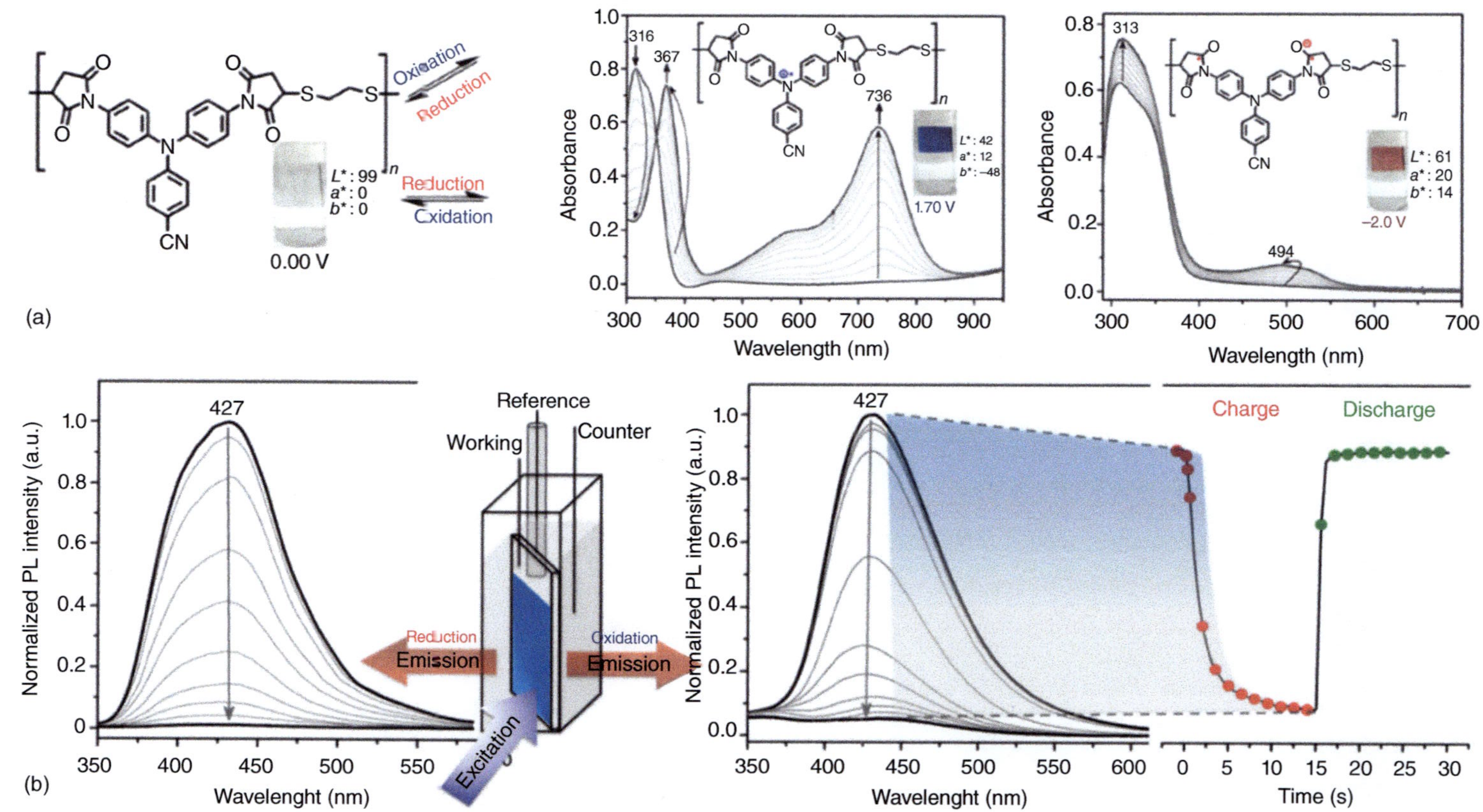

Figure 11.19 EC behaviors (a) and EFC performance (b) of P12a [47]. Source: Reproduced with permission from ref. Yen et al. [47]. Copyright 2018, The Royal Society of Chemistry.

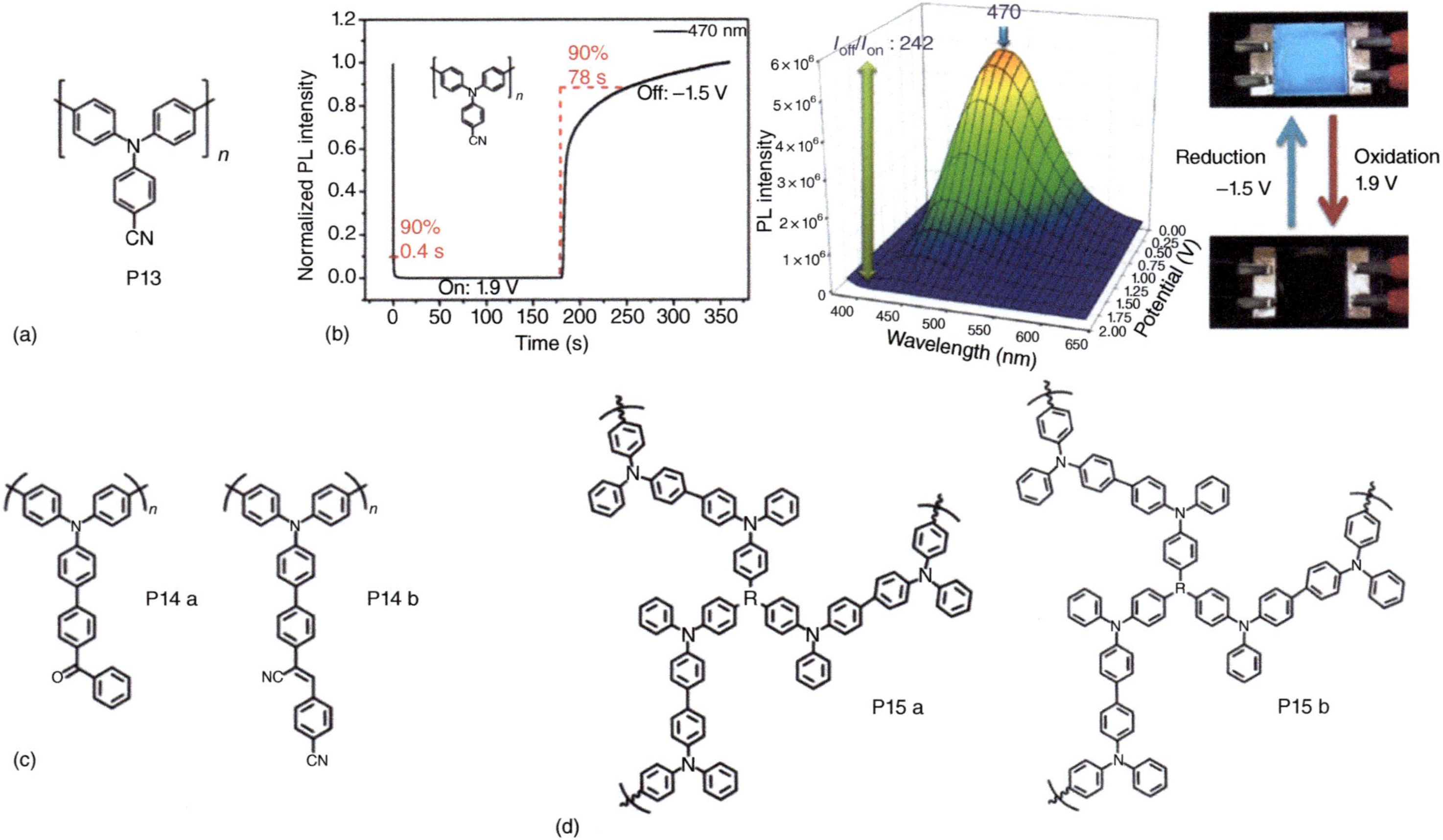

Figure 11.20 (a–d) Structure of TPA-based ELC-conjugated polymers and (b) EFC performance of P13 [50]. Source: Reproduced with permission from ref. Wu et al. [50]. Copyright 2014, Wiley-VCH Verlag GmbH & Co. KGaA, Weinheim.

ratio of 64 and 179, respectively [52]. N. Sun et al. synthesized another polyamide, bearing TPE and TPA pendant based on previous research on P8, which exhibited multicolored EC performance (colorless to green to blue) and fast response speed (0.4 seconds/2.9 seconds for EFC process with a high contrast ratio of 82) [53].

Fluorene-based ELC-conjugated polymers

Fluorene is used as the highly fluorescent moiety to build conjugated ELC polymers (Figure 11.21). In 2012, Cheng-Po Kuo et al. reported a series of high fluorescence fluorene-based ELC-conjugated polymers P16–P18. P16a and P17a showed high quantum efficiencies of 77% and 73%, respectively. And the blue fluorescence can be largely quenched upon electrochemical oxidation due to the formation of TPA radical cations, leading to a contrast ratio of 16.3 for P17a [54]. In contrast, P16a only experienced partial quenching of fluorescence due to the lack of TPA moiety. Compared with P16a and P16b, through introducing benzothiadiazole moiety into copolymers, the emission and absorption of P18a,b could be redshifted, while the quantum yields will be decreased [48]. A white-dark ELC device could be obtained by blending P16b and P18b together in a 1 : 2 ratio, with a contrast ratio of 14.6 (Figure 11.22a) [48]. For further optimization of monomer ratio, blue-emitting

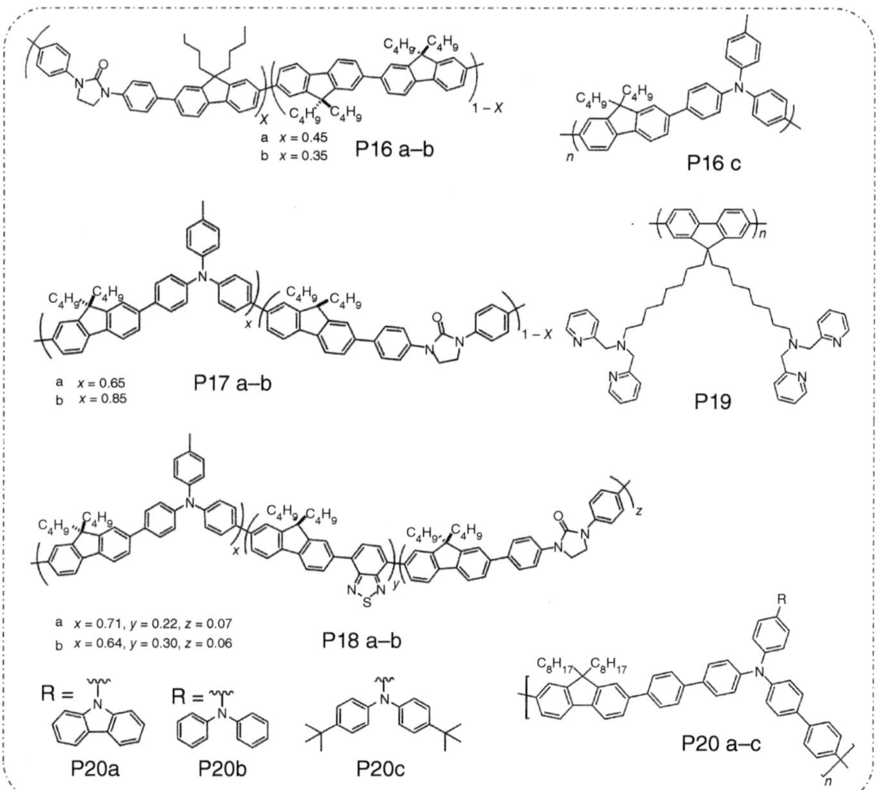

Figure 11.21 Chemical structure of fluorene-based ELC-conjugated polymers.

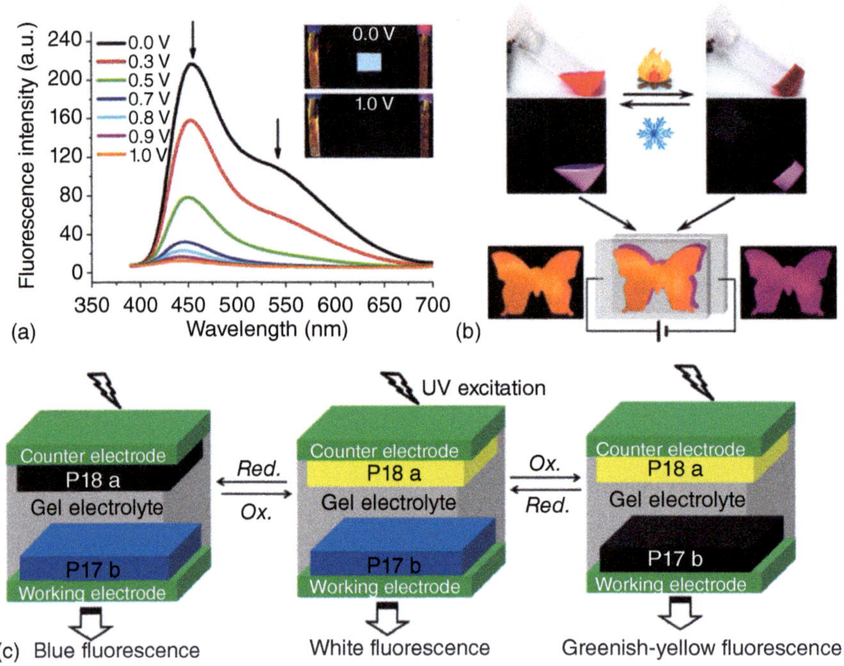

Figure 11.22 (a) White ELC devices using P16b and P18b [48]. Source: Reproduced with permission from ref. Kuo et al. [48]. Copyright 2013, The Royal Society of Chemistry. (b) ELC performance of P19 [55]. Source: Reproduced with permission from ref. Xiang et al. [55]. Copyright 2017, American Chemical Society. (c) White ELC devices using P17b and P18a [56]. Source: Reproduced with permission from ref. Kuo et al. [56]. Copyright 2014, American Chemical Society.

polymer P17b and yellow-emitting polymer P18a were synthesized and utilized in one ELC device by coating onto complementary electrodes separately. This ELC device demonstrated reversibly switchable multi-fluorescence states upon external bias (Figure 11.22c) [56]. Chunlan Xiang et al. reported a thermal and electrochemical responsive fluorescent polymer P19 by cooperating conjugated fluorene with dipicolylamine pedant and small organic dye sulforhodamine B (or its sodium salt) (Figure 11.22b) [55]. A single-layer ELC device without electrolyte could be obtained using P19 polymer gel, giving rise to high fluorescence contrast. More recently, Wang and coworkers reported conjugated polymers P20a–c containing fluorene and TPA moieties, exhibiting blue-to-dark emission switching when applied potentials changed from 0 to 1.5 V [57].

ProDOT-based ELC-conjugated polymers

Phenyl-ProDOT-based conjugated polymers have excellent electrochromic properties, attracting many researchers to develop ProDOT-based ELC polymers (Figure 11.23). In 2014, Eunkyoung Kim et al. reported a chiral conjugated polymer P20 based on ProDOT–phenylene with a chiral alkyl substituent, demonstrating reversible electrochromic and EFC as well as chirality switching, as shown in

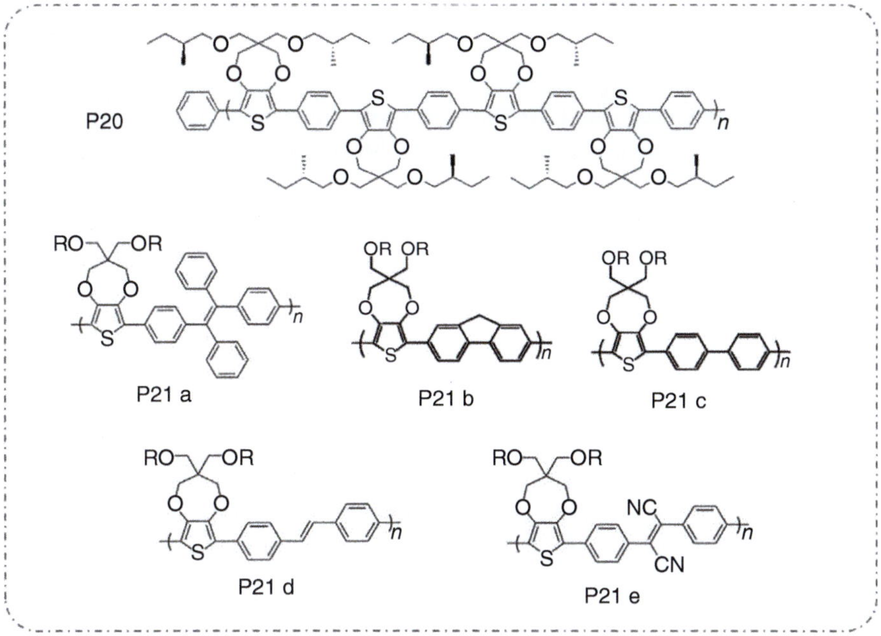

Figure 11.23 Chemical structure of ProDOT-based ELC-conjugated polymers.

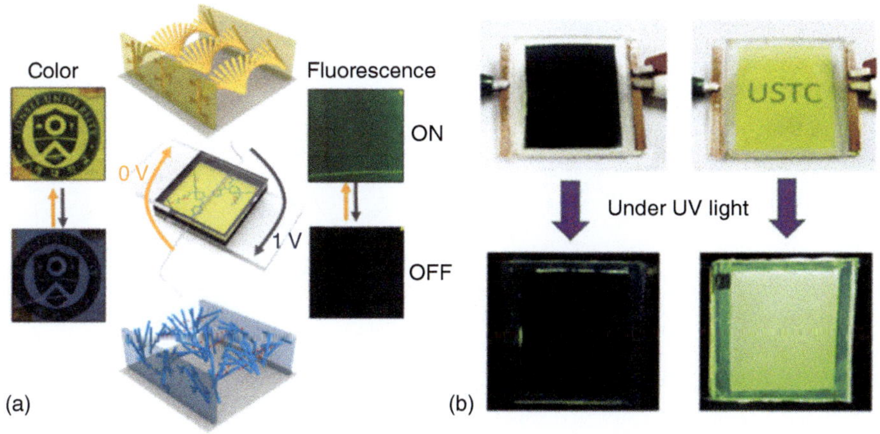

Figure 11.24 ELC performance of (a) P20 [58]. Source: Reproduced with permission from ref. Yang et al. [58]. Copyright 2014, American Chemical Society. (b) P21a [59]. Source: Reproduced with permission from ref. Mi et al. [59]. Copyright 2015, American Chemical Society.

Figure 11.24 [58]. Xu's research group has done a lot of outstanding work in this field. To begin with, Xu et al. exploited AIE moiety TPE and ProDOT to obtain high fluorescence bifunctional polymers P21a. The fluorescence of P21a film was yellow–green at 540 nm, which was quenched under oxidation process owing to the formation of TPE radical cations [59]. After that, a series of ProDOT-based

ELC-conjugated polymers P21b–e have been developed successively. For P21b and P21c, only P21c with biphenyl building blocks demonstrated AIE characteristics, deriving from intramolecular rotation and conformational twisting [60]. Under applied external potential, the P21c film can reversibly be switched from a transparent state (non-fluorescent state) to yellow color (fluorescent state) with optical contrast of 35%. Furtherly, *trans*-stilbene and fumaronitrile groups were incorporated with ProDOT to obtain ELC polymers. Due to the restriction of intramolecular rotation caused by multiple C—H···π bonds, P21d and P21e also emit intense fluorescence and with AIE characteristic [61]. Similar to P21a, yellow–green emission of P21d was observed, which was quenched to dark upon electrochemical oxidation due to the formation of radical anions. However, red-shifted emission at 612 nm was observed in P21e film due to the presence of cyano groups, which also could be quenched under positive potentials.

Carbazole-based ELC-conjugated polymers (Figure 11.25)

Goto et al. reported dithienyl-carbazole-based conjugated ELC polymer P22 in 2012, which exhibited colors changing from yellow in the neutral state to black in the oxidation state, accompanied by fluorescence quenching[62]. The other two bifunctional EC/ELC polymers P23 and P24 were designed, which both mainly consisted of carbazole and benzothiadiazole building blocks and are sensitive with cyanide anions[63, 64]. The emission of nanoporous P23 film is quenched at the oxidized

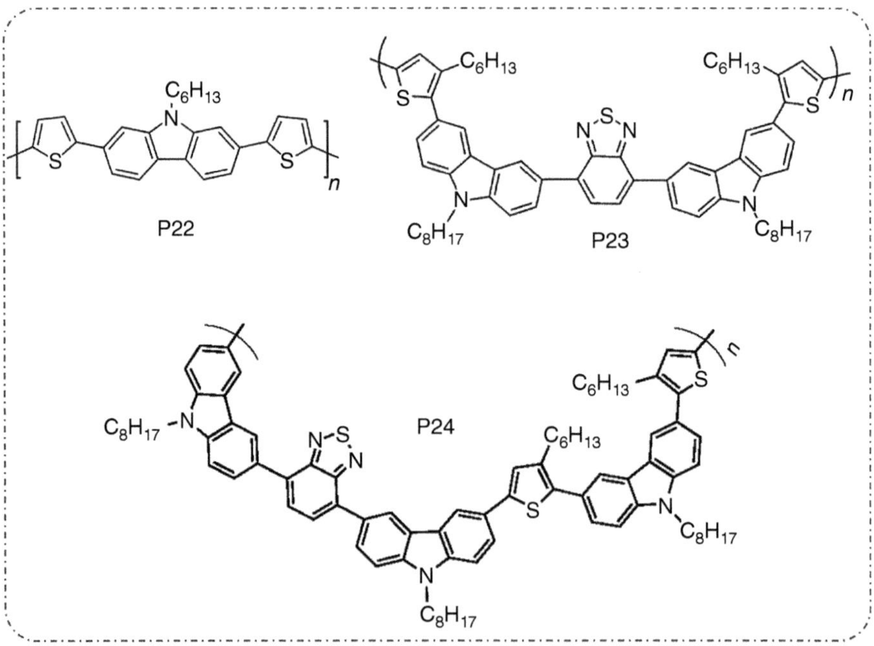

Figure 11.25 Chemical structure of carbazole-based ELC-conjugated polymers.

state with a contrast ratio of 19.2 in the absence of cyanide anions[63]. Similar fluorescence quenching was observed in the thin film device of P24, demonstrating 80% contrast [64]. When the cyanide anions exist, the noncovalent interaction between benzothiadiazole and nucleophilic cyanide anions is energy favorable in the oxidized states, leading to weakened fluorescence quenching, showing potential as detection agents.

11.4.4 Nanocomposite Films

Hybrid organic–inorganic nanocomposite is another strategy to develop multifunctional ELC systems. Dong's group successfully obtained a series of ELC in aqueous solution by using a hybrid film that integrated quantum dots (QDs) and EC polymers. The first hybrid film system is reported in 2011, in which poly(methylene blue) and CdTe QDs were used in aqueous solution. The luminescence of hybrid films could be reversibly switched in a narrow potential (Figure 11.26a) [65]. After that, the authors adopted silica nanoparticles/polyethoxythiophene (PEDOT) and POMs/rhodamine B to fabricate ELC devices. In the former case, pyronin Y-doped silica (PYDS) nanoparticles and PEDOT acted as the luminescent and electroactive components, exhibiting reversibly fluorescence switching under electrochemical oxidation and reduction (Figure 11.26b) [66]. Another hybrid system was fabricated by fluorescent organic dye rhodamine B and electrochromic POMs, whose emission also quenched under applied potentials [67]. These hybrid nanocomposite films not only could be effectively operated in the environment-friendly aqueous solution but also provided another strategy to develop ELC materials. In 2018, Xu and coworkers reported a trifunctional CdSe QDs–ProDOT polymer system, which demonstrated electrochromic, EFC, and light-induced coloration performance. The emission of composite film could be extinguished at 1 V deriving from the quenching of cation radical upon electrochemical oxidation [68].

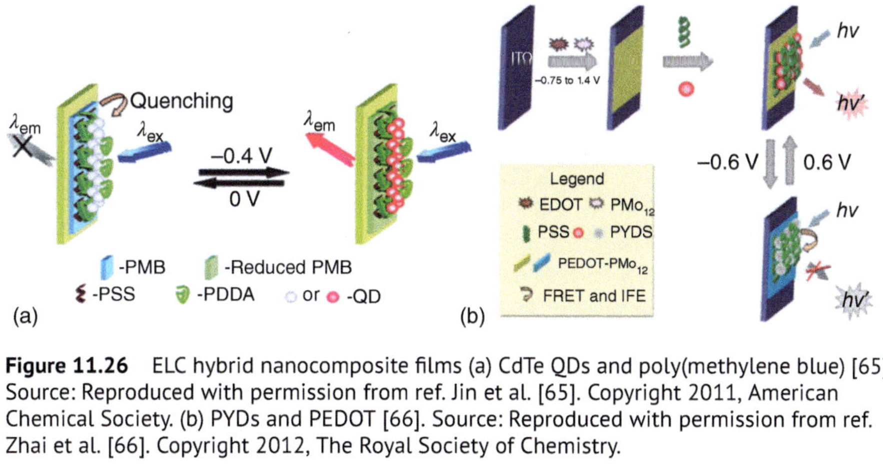

Figure 11.26 ELC hybrid nanocomposite films (a) CdTe QDs and poly(methylene blue) [65]. Source: Reproduced with permission from ref. Jin et al. [65]. Copyright 2011, American Chemical Society. (b) PYDs and PEDOT [66]. Source: Reproduced with permission from ref. Zhai et al. [66]. Copyright 2012, The Royal Society of Chemistry.

11.5 Future Perspectives and Conclusion

In this chapter, we discuss recent progresses in ELC materials including small molecules, transition metal complexes, non-conjugated and conjugated polymers, as well as nanocomposite systems. ELC materials have wide applications in optoelectronic devices, such as dual-mode displays, information encryption devices, chemical and biological sensors, and so on. However, there are still some key issues that require further investigations to meet the needs of practical commercialization, such as long-term stability, PL contrast, fast switching speed, multi-luminochromic, and large-area processability. Besides, the fabrication and optimization of ELC devices are also indispensable, especially for printable optoelectronics or wearable devices. Moreover, in order to accurately design high-performance ELC materials, an in-depth understanding of structure–property relationships should be gained on the molecular levels through synthetic chemistry and theoretical calculation. Interdisciplinary cooperation is also beneficial to establish the correlations among materials, devices, and optoelectronic performance.

References

1 Audebert, P. and Miomandre, F. (2013). Electrofluorochromism: from molecular systems to set-up and display. *Chemical Science* 4 (2): 575–584.
2 Mortimer, R.J. (2011). Electrochromic materials. *Annual Review of Materials Research* 41 (1): 241–268.
3 Gunbas, G. and Toppare, L. (2012). Electrochromic conjugated polyheterocycles and derivatives – highlights from the last decade towards realization of long lived aspirations. *Chemical Communication (Cambridge)* 48 (8): 1083–1101.
4 Chua, M.H., Zhu, Q., Shah, K.W., and Xu, J. (2019). Electroluminochromic materials: from molecules to polymers. *Polymers (Basel)* 11 (1): 98.
5 Goulle, V., Harriman, A., and Lehn, J.M. (1993). An electro-photoswitch: redox switching of the luminescence of a bipyridine metal complex. *Journal of the Chemical Society, Chemical Communications* 12: 1034–1036.
6 Dias, M., Hudhomme, P., Levillain, E. et al. (2004). Electrochemistry coupled to fluorescence spectroscopy: a new versatile approach. *Electrochemistry Communications* 6 (3): 325–330.
7 Kim, Y., Kim, E., Clavier, G., and Audebert, P. (2006). New tetrazine-based fluoroelectrochromic window; modulation of the fluorescence through applied potential. *Chemical Communication (Cambridge)* (34): 3612–3614.
8 Sun, J., Chen, Y., and Liang, Z. (2016). Electroluminochromic materials and devices. *Advanced Functional Materials* 26 (17): 2783–2799.
9 Woodward, A.N., Kolesar, J.M., Hall, S.R. et al. (2017). Thiazolothiazole fluorophores exhibiting strong fluorescence and viologen-like reversible electrochromism. *Journal of the American Chemical Society* 139 (25): 8467–8473.
10 Quinton, C., Alain-Rizzo, V., Dumas-Verdes, C. et al. (2014). Redox-controlled fluorescence modulation (electrofluorochromism) in triphenylamine derivatives. *RSC Advances* 4 (65): 34332–34342.

11 Zhang, Y.M., Li, W., Wang, X. et al. (2014). Highly durable colour/emission switching of fluorescein in a thin film device using "electro-acid/base" as in situ stimuli. *Chemical Communication (Cambridge)* 50 (12): 1420–1422.

12 Seo, S., Pascal, S., Park, C. et al. (2014). NIR electrochemical fluorescence switching from polymethine dyes. *Chemical Science* 5 (4): 1538–1544.

13 Beneduci, A., Cospito, S., Deda, M.L., and Chidichimo, G. (2015). Highly fluorescent thienoviologen-based polymer gels for single layer electrofluorochromic devices. *Advanced Functional Materials* 25 (8): 1240–1247.

14 Su, K., Sun, N., Tian, X. et al. (2019). High-performance blue fluorescent/electroactive polyamide bearing *p*-phenylenediamine and asymmetrical SBF/TPA-based units for electrochromic and electrofluorochromic multifunctional applications. *Journal of Materials Chemistry C* 7 (16): 4644–4652.

15 Corrente, G.A., Fabiano, E., La Deda, M. et al. (2019). High-performance electrofluorochromic switching devices using a novel arylamine-fluorene redox-active fluorophore. *ACS Applied Materials & Interfaces* 11 (13): 12202–12208.

16 Martinez, R., Ratera, I., Tarraga, A. et al. (2006). A simple and robust reversible redox-fluorescence molecular switch based on a 1,4-disubstituted azine with ferrocene and pyrene units. *Chemical Communication (Cambridge)*: 3809–3811.

17 Wu, J.T., Lin, H.T., and Liou, G.S. (2019). Synthesis and characterization of novel triarylamine derivatives with dimethylamino substituents for application in optoelectronic devices. *ACS Applied Materials & Interfaces* 11 (16): 14902–14908.

18 Wang, X., Li, W., Li, W. et al. (2017). An RGB color-tunable turn-on electrofluorochromic device and its potential for information encryption. *Chemical Communication (Cambridge)* 53 (81): 11209–11212.

19 Bill, N.L., Lim, J.M., Davis, C.M. et al. (2014). π-Extended tetrathiafulvalene BODIPY (ex-TTF-BODIPY): a redox switched "on–off–on" electrochromic system with two near-infrared fluorescent outputs. *Chemical Communication (Cambridge)* 50 (51): 6758–6761.

20 Seo, S., Kim, Y., Zhou, Q. et al. (2012). White electrofluorescence switching from electrochemically convertible yellow fluorescent dyad. *Advanced Functional Materials* 22 (17): 3556–3561.

21 Kim, Y., Do, J., Kim, E. et al. (2009). Tetrazine-based electrofluorochromic windows: modulation of the fluorescence through applied potential. *Journal of Electroanalytical Chemistry* 632 (1–2): 201–205.

22 Quinton, C., Alain-Rizzo, V., Dumas-Verdes, C. et al. (2015). Original electroactive and fluorescent bichromophores based on non-conjugated tetrazine and triphenylamine derivatives: towards more efficient fluorescent switches. *RSC Advances* 5 (61): 49728–49738.

23 Lim, H., Seo, S., Pascal, S. et al. (2016). NIR electrofluorochromic properties of aza-boron-dipyrromethene dyes. *Scientific Reports* 6: 18867.

24 Quinton, C., Alain-Rizzo, V., Dumas-Verdes, C. et al. (2015). Redox- and protonation-induced fluorescence switch in a new triphenylamine with six stable active or non-active forms. *Chemistry* 21 (5): 2230–2240.

25 Zhang, J., Chen, Z., Wang, X.-Y. et al. (2017). Redox-modulated near-infrared electrochromism, electroluminochromism, and aggregation-induced fluorescence change in an indolo[3,2-*b*]carbazole-bridged diamine system. *Sensors and Actuators B: Chemical* 246: 570–577.

26 Beneduci, A., Cospito, S., La Deda, M. et al. (2014). Electrofluorochromism in π-conjugated ionic liquid crystals. *Nature Communications* 5: 3105.

27 Nie, H.J., Yang, W.W., Shao, J.Y., and Zhong, Y.W. (2016). Ruthenium-tris(bipyridine) complexes with multiple redox-active amine substituents: tuning of spin density distribution and deep-red to NIR electrochromism and electrofluorochromism. *Dalton Transactions* 45 (25): 10136–10140.

28 Nakamura, K., Kanazawa, K., and Kobayashi, N. (2011). Electrochemically controllable emission and coloration by using europium(III) complex and viologen derivatives. *Chemical Communication (Cambridge)* 47 (36): 10064–10066.

29 Nakamura, K., Kanazawa, K., and Kobayashi, N. (2013). Electrochemically-switchable emission and absorption by using luminescent lanthanide(III) complex and electrochromic molecule toward novel display device with dual emissive and reflective mode. *Displays* 34 (5): 389–395.

30 Kanazawa, K., Nakamura, K., and Kobayashi, N. (2012). Electroswitching of emission and coloration with quick response and high reversibility in an electrochemical cell. *Chemistry - An Asian Journal* 7 (11): 2551–2554.

31 Zhao, Q., Xu, W., Sun, H. et al. (2016). Tunable electrochromic luminescence of iridium(III) complexes for information self-encryption and anti-counterfeiting. *Advanced Optical Materials* 4 (8): 1167–1173.

32 Zhang, K.Y., Chen, X., Sun, G. et al. (2016). Utilization of electrochromically luminescent transition-metal complexes for erasable information recording and temperature-related information protection. *Advanced Materials* 28 (33): 7137–7142.

33 Ma, Y., Yang, J., Liu, S. et al. (2017). Phosphorescent ionic iridium(III) complexes displaying counterion-dependent emission colors for flexible electrochromic luminescence device. *Advanced Optical Materials* 5 (21): 1700587.

34 Guo, S., Huang, T., Liu, S. et al. (2017). Luminescent ion pairs with tunable emission colors for light-emitting devices and electrochromic switches. *Chemical Science* 8 (1): 348–360.

35 Ma, Y., Shen, L., She, P. et al. (2019). Constructing multi-stimuli-responsive luminescent materials through outer sphere electron transfer in ion pairs. *Advanced Optical Materials* 7 (8): 1801657.

36 Yoo, J., Kwon, T., Sarwade, B.D. et al. (2007). Multistate fluorescence switching of *s*-triazine-bridged *p*-phenylene vinylene polymers. *Applied Physics Letters* 91 (24): 241107.

37 Liu, B., Yu, W.L., Lai, Y.H., and Huang, W. (2002). Blue-light-emitting cationic water-soluble polyfluorene derivatives with tunable quaternization degree. *Macromolecules* 35 (13): 4975–4982.

38 Sun, N., Feng, F., Wang, D. et al. (2015). Novel polyamides with fluorene-based triphenylamine: electrofluorescence and electrochromic properties. *RSC Advances* 5 (107): 88181–88190.

39 Sun, N., Zhou, Z., Chao, D. et al. (2017). Novel aromatic polyamides containing 2-diphenylamino-(9,9-dimethylamine) units as multicolored electrochromic and high-contrast electrofluorescent materials. *Journal of Polymer Science Part A: Polymer Chemistry* 55 (2): 213–222.

40 Sun, N., Meng, S., Chao, D. et al. (2016). Highly stable electrochromic and electrofluorescent dual-switching polyamide containing bis(diphenylamino)-fluorene moieties. *Polymer Chemistry* 7 (39): 6055–6063.

41 Sun, N., Zhou, Z., Meng, S. et al. (2017). Aggregation-enhanced emission (AEE)-active polyamides with methylsulfonyltriphenylamine units for electrofluorochromic applications. *Dyes and Pigments* 141: 356–362.

42 Sun, N., Tian, X., Hong, L. et al. (2018). Highly stable and fast blue color/fluorescence dual-switching polymer realized through the introduction of ether linkage between tetraphenylethylene and triphenylamine units. *Electrochimica Acta* 284: 655–661.

43 Sun, N., Su, K., Zhou, Z. et al. (2018). AIE-active polyamide containing diphenylamine-TPE moiety with superior electrofluorochromic performance. *ACS Applied Materials & Interfaces* 10 (18): 16105–16112.

44 Cheng, S.-W., Han, T., Huang, T.-Y. et al. (2018). High-performance electrofluorochromic devices based on aromatic polyamides with AIE-active tetraphenylethene and electro-active triphenylamine moieties. *Polymer Chemistry* 9 (33): 4364–4373.

45 Yen, H.J. and Liou, G.S. (2013). Flexible electrofluorochromic devices with the highest contrast ratio based on aggregation-enhanced emission (AEE)-active cyanotriphenylamine-based polymers. *Chemical Communication (Cambridge)* 49 (84): 9797–9799.

46 Sun, N., Meng, S., Feng, F. et al. (2016). Electrochromic and electrofluorochromic polyimides with fluorene-based triphenylamine. *High Performance Polymers* 29 (10): 1130–1138.

47 Yen, H.-J., Chang, C.-W., Wong, H.Q., and Liou, G.-S. (2018). Cyanotriphenylamine-based polyimidothioethers as multifunctional materials for ambipolar electrochromic and electrofluorochromic devices, and fluorescent electrospun fibers. *Polymer Chemistry* 9 (13): 1693–1700.

48 Kuo, C.-P., Chuang, C.-N., Chang, C.-L. et al. (2013). White-light electrofluorescence switching from electrochemically convertible yellow and blue fluorescent conjugated polymers. *Journal of Materials Chemistry C* 1 (11): 2121.

49 Beaujuge, P.M. and Reynolds, J.R. (2010). Color control in π-conjugated organic polymers for use in electrochromic devices. *Chemical Reviews* 110 (1): 268–320.

50 Wu, J.-H. and Liou, G.-S. (2014). High-performance electrofluorochromic devices based on electrochromism and photoluminescence-active novel poly(4-cyanotriphenylamine). *Advanced Functional Materials* 24 (41): 6422–6429.

51 Sun, J.W. and Liang, Z.Q. (2016). Swift electrofluorochromism of donor–acceptor conjugated polytriphenylamines. *ACS Applied Materials & Interfaces* 8 (28): 18301–18308.

52 Santra, D.C., Nad, S., and Malik, S. (2018). Electrochemical polymerization of triphenylamine end-capped dendron: electrochromic and electrofluorochromic switching behaviors. *Journal of Electroanalytical Chemistry* 823: 203–212.

53 Sun, N., Su, K., Zhou, Z. et al. (2019). High-performance emission/color dual-switchable polymer-bearing pendant tetraphenylethylene (TPE) and triphenylamine (TPA) moieties. *Macromolecules* 52 (14): 5131–5139.

54 Kuo, C.-P., Lin, Y.-S., and Leung, M.-k. (2012). Electrochemical fluorescence switching properties of conjugated polymers composed of triphenylamine, fluorene, and cyclic urea moieties. *Journal of Polymer Science Part A: Polymer Chemistry* 50 (24): 5068–5078.

55 Xiang, C., Wan, H., Zhu, M. et al. (2017). Dipicolylamine functionalized polyfluorene based gel with lower critical solution temperature: preparation, characterization, and application. *ACS Applied Materials & Interfaces* 9 (10): 8872–8879.

56 Kuo, C.P., Chang, C.L., Hu, C.W. et al. (2014). Tunable electrofluorochromic device from electrochemically controlled complementary fluorescent conjugated polymer films. *ACS Applied Materials & Interfaces* 6 (20): 17402–17409.

57 Yang, C., Cai, W., Zhang, X. et al. (2019). Multifunctional conjugated oligomers containing novel triarylamine and fluorene units with electrochromic, electrofluorochromic, photoelectron conversion, explosive detection and memory properties. *Dyes and Pigments* 160: 99–108.

58 Yang, X., Seo, S., Park, C., and Kim, E. (2014). Electrical chiral assembly switching of soluble conjugated polymers from propylenedioxythiophene–phenylene copolymers. *Macromolecules* 47 (20): 7043–7051.

59 Mi, S., Wu, J.C., Liu, J. et al. (2015). AIEE-active and electrochromic bifunctional polymer and a device composed thereof synchronously achieve electrochemical fluorescence switching and electrochromic switching. *ACS Applied Materials & Interfaces* 7 (49): 27511–27517.

60 Liu, J., Li, M., Wu, J.C. et al. (2017). Electrochromic polymer achieving synchronous electrofluorochromic switching for optoelectronic application. *Organic Electronics* 51: 295–303.

61 Wu, J.C., Han, Y.F., Liu, J. et al. (2018). Electrofluorochromic and electrochromic bifunctional polymers with dual-state emission via introducing multiple C–H···π bonds. *Organic Electronics* 62: 481–490.

62 Kawabata, K. and Goto, H. (2012). Dynamically controllable emission of polymer nanofibers: electrofluorescence chromism and polarized emission of polycarbazole derivatives. *Chemistry - A European Journal* 18 (47): 15065–15072.

63 Ding, G.Q., Lin, T.T., Zhou, R. et al. (2014). Electrofluorochromic detection of cyanide anions using a nanoporous polymer electrode and the detection mechanism. *Chemistry - A European Journal* 20 (41): 13226–13233.

64 Ding, G.Q., Zhou, H., Xu, J.W., and Lu, X.H. (2014). Electrofluorochromic detection of cyanide anions using a benzothiadiazole-containing conjugated copolymer. *Chemical Communications* 50 (6): 655–657.

65 Jin, L.H., Fang, Y.X., Wen, D. et al. (2011). Reversibly electroswitched quantum dot luminescence in aqueous solution. *ACS Nano* 5 (6): 5249–5253.

66 Zhai, Y., Jin, L., Zhu, C. et al. (2012). Reversible electroswitchable luminescence in thin films of organic–inorganic hybrid assemblies. *Nanoscale* 4 (24): 7676–7681.

67 Jin, L., Fang, Y., Hu, P. et al. (2012). Polyoxometalate-based inorganic–organic hybrid film structure with reversible electroswitchable fluorescence property. *Chemical Communications* 48 (15): 2101–2103.

68 Liu, J., Luo, G., Mi, S. et al. (2018). Trifunctional CdSe quantum dots-polymer composite film with electrochromic, electrofluorescent and light-induced coloration effects. *Solar Energy Materials & Solar Cells* 177: 82–88.

12

Organic Photoelectrochromic Devices

12.1 Introduction

Currently, most of the energy consumed by human beings comes from the fossil fuels, which contribute to the total energy consumption by 80–85%. But the reserve of fossil fuels is limited, and the use will cause environmental and safety issues. Moreover, the world population of over 7.5 billion people (still increasing) needs a huge amount of energy [1, 2]. Most of the energy consumption is from the heating, cooling, and artificial lighting of residential and commercial buildings. To address these problems, the shifting from fossil fuels to renewables has attracted people's attention, and more and more energy efficient technology appeared. Especially, the smart window is a promising technology, which can adjust a large fraction of energy loss or gains of buildings. The first smart window was investigated by J.S.E.M. Svensson et al. in 1984, which can employ an electrical signal to modulate the solar transmittance to maintain the room temperature [3]. The advantages of smart windows are not only its control of the amount of solar light and lighting input of the building but also the color changes for esthetics [4]. At 2001 Karlsson investigated the smart window under different conditions and obtained the energy saving of 0–50 $kWh/m^2/yr$ in a Stockholm-like climate, 25–75 $kWh/m^2/yr$ in a Denver-like climate, and 50–150 $kWh/m^2/yr$ in a Miami-like climate [5]. While the reduction of energy in hot areas mainly ascribes to the energy demand for cooling, in cold climates savings in lighting energy dominate [6]. Among the various smart window technologies, the electrochromic (EC) windows are recognized as the most suitable and promising technology, which can provide views outside like other conventional building windows regardless of their color states, meanwhile, providing glare control and energy management and saving energy. Hence EC windows are expected to be an important and potential technology of future buildings [7–10].

EC materials and devices have been reported widely during the past decades' years. Major advances have been achieved in electrochromic devices (ECDs). Photoelectrochromic devices (PECDs), first reported by Bechinger et al. at 1996 [11], ensure the good performances in ECD or even have some improvement, without needing the external power supply to change transmittance. A PECD is combining an EC layer in a self-powered battery, such as dye-sensitized solar cells (DSSCs), organic solar cells, perovskite solar cells, or any other possible devices.

Organic Electronics for Electrochromic Materials and Devices, First Edition. Hong Meng.

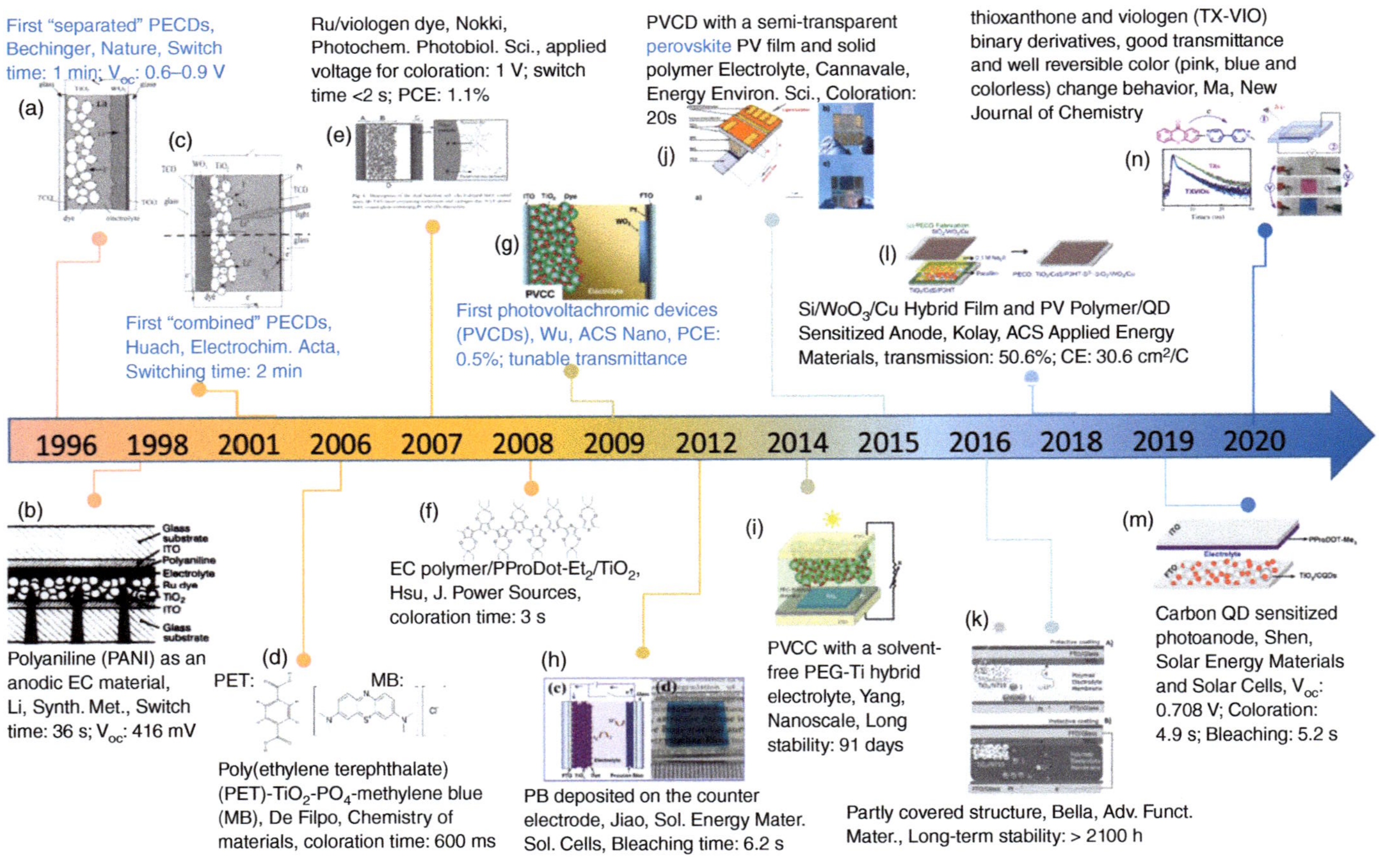

Figure 12.1 PECD technology development. Sources: (g2) Jiao et al. [14], (j(b)) Reproduced with permission of Cannavale et al. [15]. Copyright 2015. © The Royal Society of Chemistry.

The electric power generated by the self-powered photovoltaic part can directly be used by the EC part, which saves a lot of materials and energy comparing with separate components. Also, integrated devices contribute to getting smaller volume and higher energy density and compact devices due to the reduction of the wiring, sharing the electrodes, and encapsulation. In some cases, the PECDs can also be used to supply power to an external load [12, 13]. During the past decades, the PECDs have developed a lot, and it is imperative to review the technique development process as shown in Figure 12.1. In 1996, Bechinger demonstrated the first separated-type PECD; after that many researchers started working on the PECDs. In 2001, Huach proposed the new combined-type PECD, which enhanced the device efficiency. In 2016, the stability of the devices was improved by Bella. During the past period, the PCEDs had developed rapidly and shown great potential to be explored in the near future.

Organic materials play an important role in photoelectric materials and devices. They have advantages of low cost, good electrochemical performance, being flexible, and so on. In PECD, organic materials are also important and have potential. There has been some review on the PECDs, but they are more or less focusing on the inorganic materials. In this review, we will introduce the role of organic materials in PECD and give an outlook of the organic PECDs.

12.2 Structure Design of PECDs

PECDs consist of four main parts: EC materials, electrolyte, power supply, and substrates (Figure 12.2). In this part, the four parts will be introduced separately.

12.2.1 Power Supply for PECD

Specific requirements for the PECD have to be the power supply compared with ECD. The good candidates to be a power supply should be transparent or semi-transparent. Flexibility, robustness, easy processing, low cost, and long-term stability are also necessary properties. In addition, approximately half of the solar energy is infrared radiation; therefore EC smart windows are able to modulate solar heat in near-infrared (NIR) without sacrificing the transmittance for daylighting, and near-ultraviolet (UV) solar cells are promising power source to make full use of

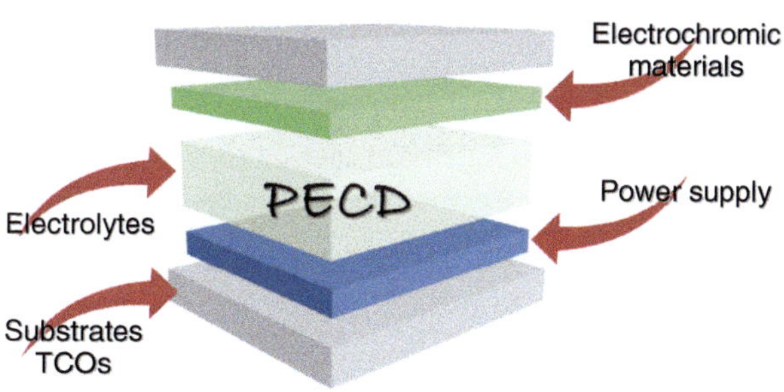

Figure 12.2 Parts of PECDs.

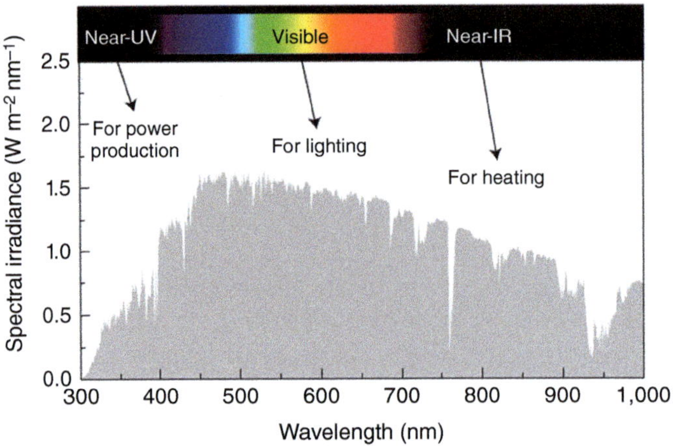

Figure 12.3 Management of the solar spectrum [16]. Source: Davy et al. [16]. © 2017, Springer Nature.

the solar energy (Figure 12.3) [16]. Different cells are used in PECD, such as DSSC [17], organic photovoltaic devices (OPVs) [18], perovskite solar cells [15], silicon solar cells [19], and solid-state heterojunction solar cells [20].

12.2.1.1 DSSC-Based PECD

DSSC is an ideal photovoltaic device for PECD due to its transparency or semi-transparency in visible region, color tunability, robustness, easy processing, and low cost. The structure of DSSC is shown in Figure 12.4. The first PECD was from this design introduced by Bechinger et al. [11]. They introduced an EC layer to replace platinum counter electrode; thus the TiO_2/dyes can offer electrons to promote coloration of the EC layer. Moreover, this device can offer dual functions,

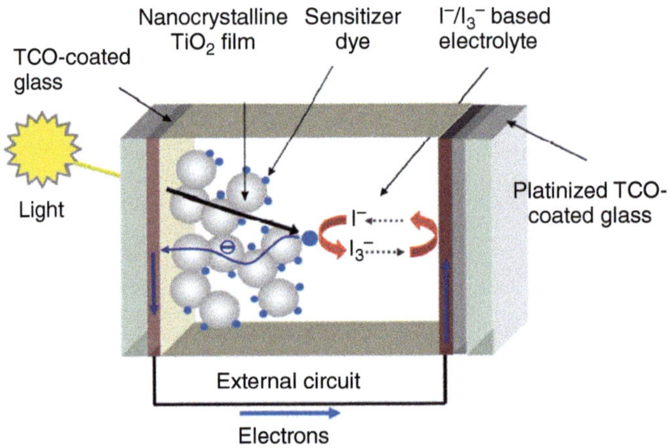

Figure 12.4 Schematic of DSSC [21]. Source: Listorti et al. [21]. © 2011, American Chemical Society.

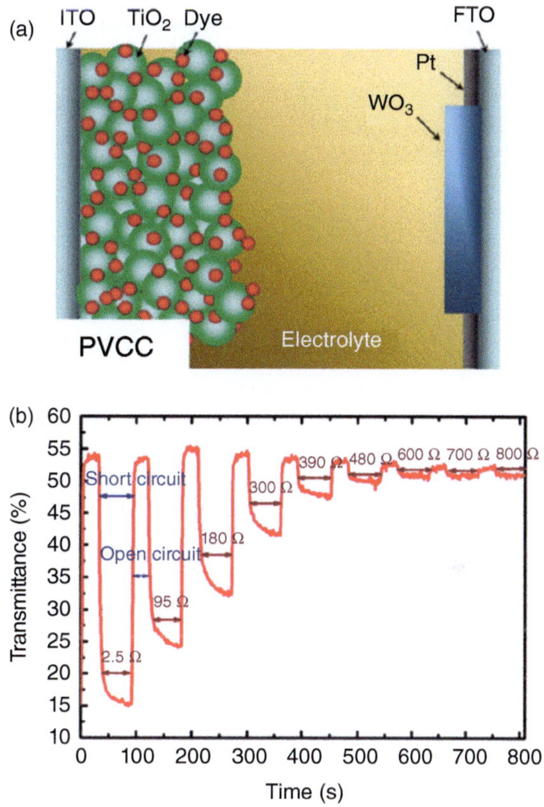

Figure 12.5 (a) Schematic of PVCC and (b) transient optical transmittance at 788 nm of PVCC in series with a variable resistor under illumination [12]. Source: Wu et al. [12]. © 2009 American Chemical Society.

because it can connect an external device to supply power or store energy within the device.

Transparency is an imperative property for power supply aiming at PECD since the high transmittance for daylighting is necessary for saving energy. Films made of oxide nanoparticles offer efficient harvesting of sunlight through the absorbed dyes, since the mesoporous matrix significantly increases the surface area. For the PECD device, the thinner the films are, the more transparent the devices are. Moreover, considering the major function of the DSSC to supply power for coloration of the EC material, very thin photoanodes can reach the ideal results.

According to the structure design, devices can be summarized into four types. The first one is the separated type, which has separated light-sensitive layer and EC layer. The combined type and the partly covered are those with the light-sensitive material and EC layer combined at the same electrode, but fully or partly covered, respectively, by the light-sensitive materials with respect to the EC layer. In addition to consisting all the components on the same substrates, two devices are series connected by external wires achieving an integrated PECD.

Separated-type PECD

Bechinger first introduced the separated-type PECD in 1996 in the Nature journal. Since then researchers started exploring the devices. The dye-sensitized TiO_2 and

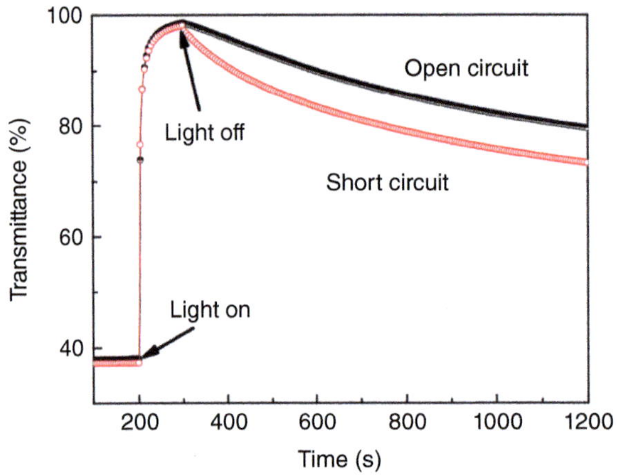

Figure 12.6 Optical transmittance change vs time of the PECD [14]. Source: Jiao et al. [14]. © 2012, Elsevier.

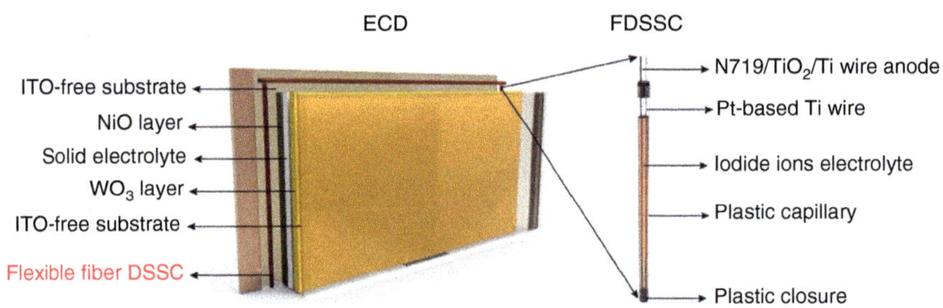

Figure 12.7 Schematic diagram of the large-scale ITO-free PECSW based on ECD–FDSSC system [22]. Source: (a) Xu et al. [22]. © 2019, Elsevier, (b) Reproduced with permission of Xu et al. [22]. Copyright ©2019 Elsevier Ltd.

the EC material are separated by electrolyte, which contains counter ions for internal electron transfer. These devices are colored when light illuminates the PECD at short circuit. Wu et al. demonstrated a photovoltachromic cell (PVCC) (Figure 12.5), of which the photoanode and WO_3 as EC layer are on separated electrodes [12]. The transmittance can be tuned by driving an external load of variable resistance. Similarly, a separated type-PECD based on DSSC power supply was designed by Jiao et al. [14], of which the transmittance can reach 100% when the light is off (Figure 12.6).

Recently, Xu et al. demonstrated new photoelectrochromic smart windows (PECSW) comprising of a large-scale ECD and several integrated fiber-shaped dye-sensitized solar cells (FDSSCs) (Figure 12.7) [22]. The open-circuit voltage (V_{oc}) ranging from 0.742 to 2.968 V can be obtained, and the considerable EC performance of ~80% optical modulation for the PECSW is achieved.

Combined-type PECD

The separated type has a limit that the coloring and bleaching process are competing so that coloring and bleaching cannot reach the highest efficiency simultaneously. The combined-type PECD can promote acceleration of the bleaching by a

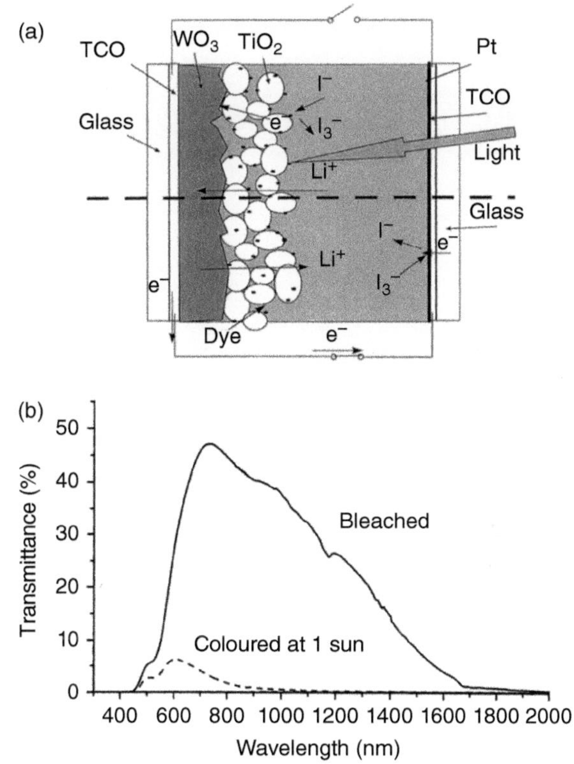

Figure 12.8 (a) Schematic of combined-type PECD and (b) the transmittance of combined PECD [23]. Source: Hauch et al. [23]. © 2001, Elsevier.

catalyst without sacrificing the coloration rate. Hauch et al. demonstrated the first combined-type PECD, combing the dye-sensitized TiO_2 with the EC material at the same electrode (Figure 12.8) [23]. But the highest transmittance is not that ideal, which is less than 50%. Solid electrolyte was introduced in the PECD by Georg et al. [24]. The visible transmittance changes from 62% to 2%. In addition to the structure improvement of combined-type PECD, the technique of the solar cell has also broken through. Typical TiO_2 and WO_3 are prepared from colloids or sol–gel chemistry. Bogati et al. first developed a PECD with sputtered deposited TiO_2 and WO_3 films [25] with visual transmittance change from 61% to 15%.

Partly covered-type PECD

In this type PECD, every layer is designed for maximizing transparency without hampering other properties. For the conventional DSSC-based PECD, thin TiO_2 film cannot absorb sufficient dyes, neither enough photons, causing low V_{oc} and slow switching rate. Moreover, the mesoporous TiO_2 film may cause scattering loss and then low transmittance. Leftheriotis et al. proposed the partly covered-type PECD (Figure 12.9) [26], which consists of a thick and opaque TiO_2 film but only covering a fraction of the EC film. This device shows transparent in visible (with luminous transmittance up to 51%) and exhibits rapid coloration (three minutes) upon solar irradiation in open circuit.

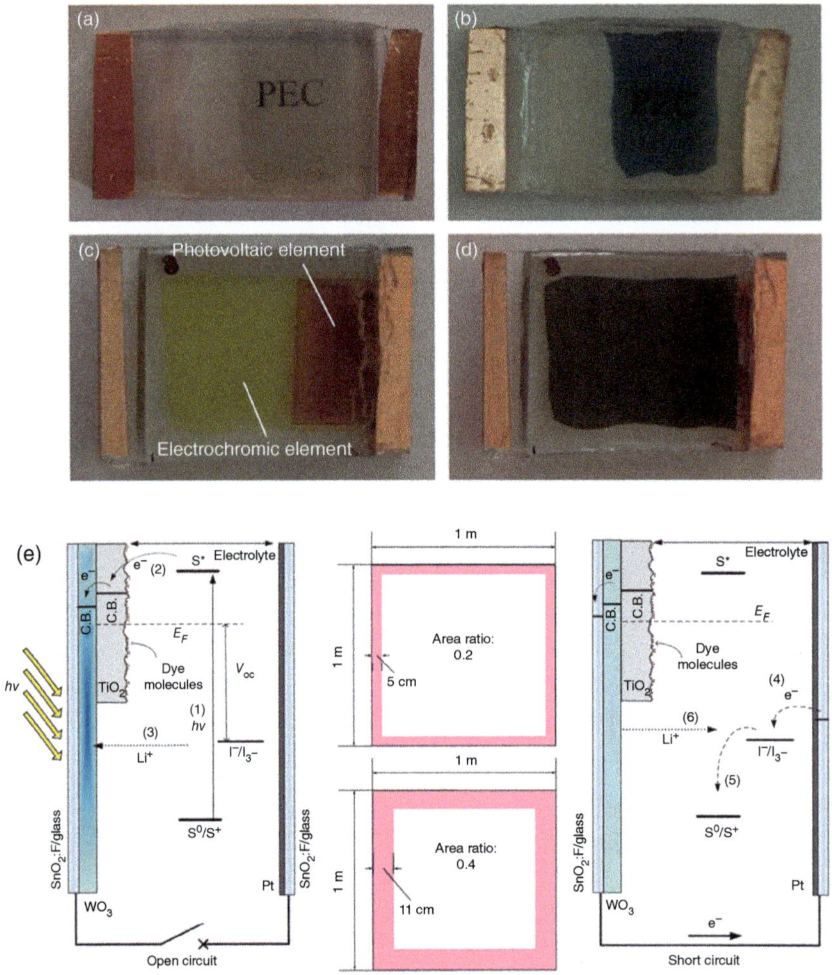

Figure 12.9 Pictures of the bleached and colored state of the PECD (a–d) and the layout and operation of a partly covered PECD (e) [26]. Source: (a) Reproduced with permission of Leftheriotis et al. [26]. Copyright ©2011, Elsevier B.V., (e) Leftheriotis et al. [26]. © 2012, Elsevier.

Other PECD

Xie et al. proposed an integrated smart EC window (Figure 12.10) [27], which is a tandem structure achieved by connecting a DSSC and an EC system to realize self-powered operation. The transmittance reaches 93% in the bleaching state. For this kind of structure, it has a low requirement of the transparency of the power supply, so there are much more choices for the photovoltaic to power the EC system. In addition, when the ECD is extended to be a supercapacitor, the dual function of self-powered and energy storage smart window was formed.

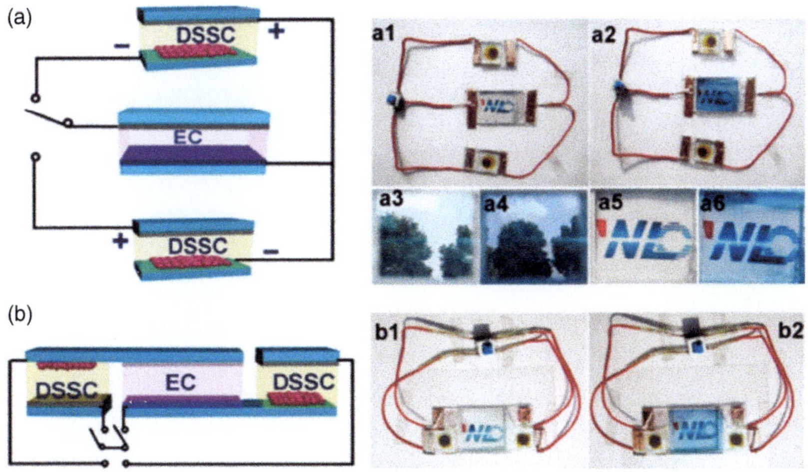

Figure 12.10 Pictures of a DSSC-driven EC smart window under illumination of sunlight: circuits of (a) the independent device and (b) the assembled device [27]. Source: Reproduced with permission of Xie et al. [27]. © 2014 Royal Society of Chemistry.

12.2.1.2 PSC-Based PECD

In recent years, semi-transparent perovskite solar cells have emerged due to their low cost and solution-processable properties. Typically, they form in the ABX3 perovskite crystal structure, where A is Cs, $NH_2CH_2NH_3$ (formamidinium [FA]), or CH_3NH_3 (methylammonium [MA]), B is Sn or Pb, and X is I, Br, or Cl. Different components can achieve bandgap tunability. $MAPbI_3$ and $FAPbI_3$ are the most studied, with a bandgap of 1.57 and 1.48 eV or so. Opaque devices based on these materials have advanced rapidly with power conversion efficiency (PCE) from 3.8% to over 25%. Undoubtedly such technology is ideal to be applied into semi-transparent devices.

Cannavale et al. first reported perovskite solar cell as power supply for PECD[15]. They developed solid-state polymer electrolyte, which significantly increases durability and easy processing. The semi-transparent perovskite was prepared to form micron-sized islands, which are thick enough to absorb incident sunlight. The combination of the fully transparent regions and the fully absorbing islands exhibits homogeneous look as shown in Figure 12.11 [15]. Xia et al. reported a tandem device (Figure 12.12) connecting a perovskite solar cell and an ECD or an LED, which exhibits multifunctionality of solar energy harvesting, electrochemical energy storage, and reutilization [28].

12.2.1.3 OPV-Based PECD

OPVs and other organic devices are attractive owing to their low cost and high-performance characteristics. Here we mainly introduce organic solar cells. The PCEs of such solar cells have reached above 16% [29–31]. Thus OPVs cannot compete with inorganic solar cells with respect to PCE, but they are competitive for the cost, fabrication simplicity, and novel applications. OPV can be classified into

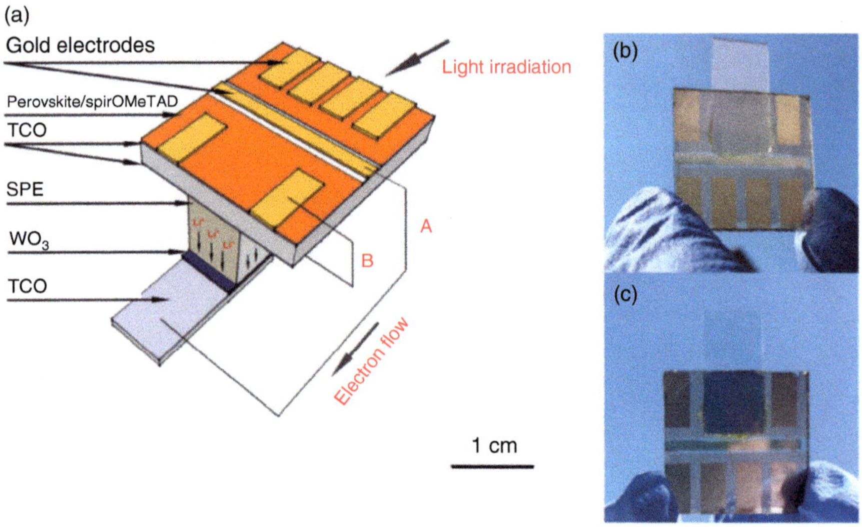

Figure 12.11 (a) Schematic of perovskite-based PECD, (b) bleached state, and (c) colored state [15]. Source: (a) Cannavale et al. [15]. © 2015, The Royal Society of Chemistry, (b) Reproduced with permission of Cannavale et al. [15]. Copyright 2015, ©The Royal Society of Chemistry.

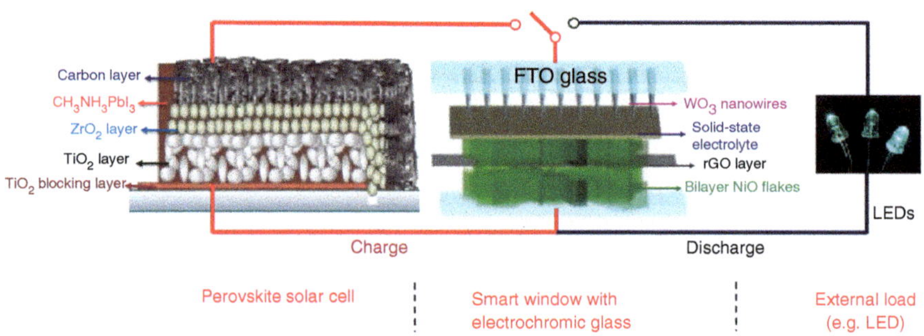

Figure 12.12 Schematics of the device structure and working principle of the tandem device [28]. Source: Reproduced with permission of Xia et al. [28]. © 2012, The Royal Society of Chemistry.

two types according to their molecule constitutes (small molecules or polymers). These two types are different in materials processing, purification, and device fabrication. Polymer solar cells are always processed in solution with organic solvent, while small molecule solar cells are thermal evaporation deposition under high-vacuum environment.

Tang et al. proposed a planar heterojunction of donor–acceptor bilayer in 1979, the PCE of which is about 1% [32]. Later an important breakthrough of OPV was the exploit of C_{60} and its derivatives, such as [6,6]-phenyl-C_{61}-butyric acid methyl ester, [6,6]-phenyl-C61-butyric acid methyl ester (PCBM), to replace n-type molecules in OPVs due to their strong electronegativity and high electron mobility. The bulk heterojunction involving mixing donor–acceptor in the bulk body was

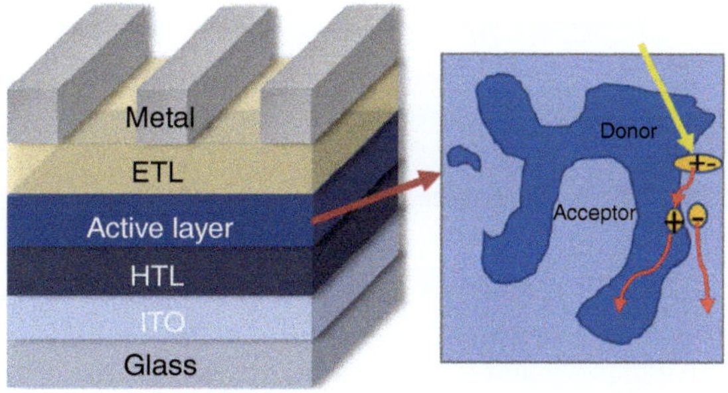

Figure 12.13 Scheme of a bulk-heterojunction organic device. Source: Clarke and Durrant [34]. © 2010, American Chemical Society.

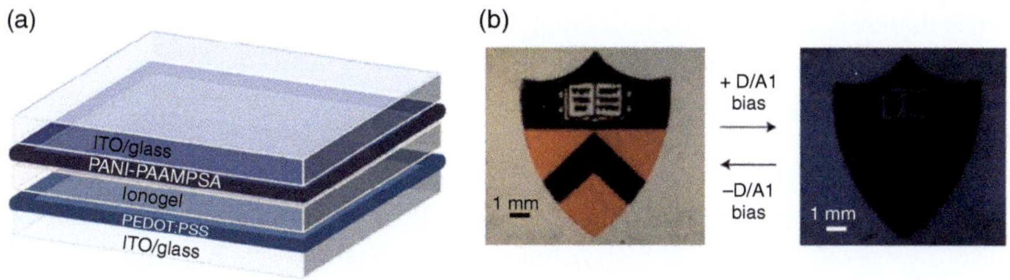

Figure 12.14 (a) Schematic representation of device structure and (b) photograph of the colored and bleached states of the electrochromic window (ECW) powered by a D/A1 [16]. Source: (a) Davy et al. [16]. © 2017, Springer Nature, (b) Reproduced with permission of Davy et al. [16]. Copyright © 2017, Macmillan Publishers, part of Springer Nature.

demonstrated by Hiramoto et al. [33] The typical device structure is shown in Figure 12.13 [34].

Davy et al. reported an EC smart window (shown in Figure 12.14) powered by a near-UV organic single-junction solar cell, which has an open circuit voltage reaching 1.6 V [16]. This organic single-junction solar cell that harvests near-UV photons to achieve regulation of visible and NIR photons for lighting and heating makes full use of the solar spectrum.

12.2.2 Electrochromic Materials in PECD

Electrochromism is the ability of color change under the potential difference, while the EC materials exhibit redox states with distinct electronic (UV–vis) absorption spectra [35]. When the chromophore only absorbs in the UV region, it shows transparent state (bleached); when it absorbs in the visible region, it shows a colored state or multicolored [36] states. Among the enormity of the field of organic chemistry, there are a huge number of organic materials that show electrochromism such as viologens and conducting polymers.

12.2.2.1 Small Molecule

Viologens are salts of quaternized 4,4′-bipyridine [37]; the EC process of which is realized by redox reactions of viologen and its complementary ion in the solution. The prototype viologen is 1,1′-di-methyl-4,4′-bipyridilium, which is often known as methyl viologen (MV), and other simple symmetrical bipyridylium species are

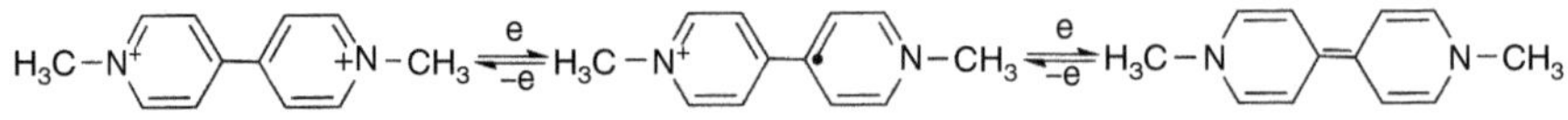

Figure 12.15 The three common viologen redox states.

substituent viologen. They have three common redox states (Figure 12.15). The dication is the most stable and is colorless. The first reduction step is highly reversible and can be cycled many times without significant side reactions. The further reduction is less reversible, because the latter is insoluble and uncharged [38]. The radical cation is intensely colored, which has high molar absorption coefficients, due to the charge transfer between the +1 and zero valent nitrogen. Therefore, the color change can be determined by the nitrogen substituents in viologens to attain the appropriate molecular orbital energy levels.

The change in color of viologens induced by the electric field has utilitarian value to be applied in ECDs. Moreover, viologens have low driving voltage for color change, high contrast, switchable EC property, and easy cell construction [39]. And in addition to the ECD application, viologens are applied as herbicides [40], redox indicators [37], and electron transfer mediators.

At 1933 Michaelis and Hill first reported the electrochemical behavior of a class of compound, which was designated as viologens (1,1′-disubstituted-4,4′-bipyridiniums) [37]. Since then viologens have successive waves of interest. A viologen-modified periodic mesoporous nanocrystalline anatase electrode was performed by Choi et al., which was found to exhibit a fast coloration and bleaching speed and a high color contrast [41]. In another study, an ECD of polysilsesquioxane nanoparticles incorporating a polymeric viologen showed a contrast of 50% at 550 nm and a coloration efficiency of 205 cm^2/C, which validated it to be a good candidate for EC displays [42]. Mortimer and Varley at 2011 reported a novel color-reinforcing ECD in which the anode and cathode reactions simultaneously exhibit reversible colorless to intense purple changes. And it had a very high contrast and fast optical switching [43]. Another study was a new ECD using viologen and compact Prussian blue (PB)/antimony tin oxide (ATO) nano-composites, which showed a high contrast (64.8%, at 600 nm), fast switching time, high coloration efficiency (912 cm^2/C) at 600 nm, and good stability [44]. An all-solid-state ECD using the heptyl viologen ($HV(BF_4)_2$) and 2,2,6,6-tetramethyl-1-piperidinyloxy (TEMPO) exhibited a high optical contrast with coloration efficiencies of c. 65.5 and 342.2 cm^2/C at 610 nm and a high transmittance change of 74% [45]. A great progress was an EC composite developed from the layer-by-layer (LBL) assembly of poly(hexyl viologen) (PXV) and the poly(3,4-ethylenedioxythiophene):poly(styrene sulfonate) (PEDOT:PSS) colloid, the transmittance change of which achieved 82.1% at 525 nm, exceeding most of ECD [46]. At 2015 Moon et al. demonstrated a simple, solid-state, and flexible ECD based on EC ion gel, which showed an excellent coloration efficiency of 105 cm^2/C and good operational stability in air [47]. The viologen derivatives are applied widely in ECDs recently and they have different performance. Recently, Lu et al. fabricated three kinds of viologen-based

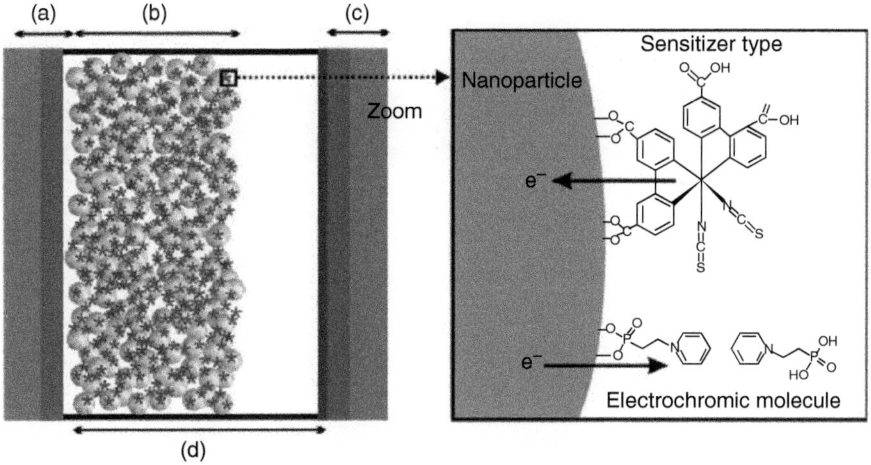

Figure 12.16 Scheme of the PECD. (a) F-doped SnO_2 glass, (b) TiO_2 layer containing ruthenium and viologen dye, (c) F-doped SnO_2-coated glass containing Pt, and (d) electrolyte [49]. Source: Santa-Nokki et al. [49]. © 2007, Royal Society of Chemistry.

ECDs: heptyl viologen ($HV(BF_4)_2$), octyl viologen ($OV(BF_4)_2$), and nonyl viologen ($NV(BF_4)_2$)). Among the three devices, the $NV(BF_4)_2$-based ECD showed the highest coloration efficiency (36.2 cm^2/C) due to the lowest driving energy. Besides, switching between 0 and 1.2 V, the $NV(BF_4)_2$-based ECD shows a desirable initial transmittance change ($\Delta T = 56.7\%$ at 605 nm) and long-term stability ($\Delta T = 45.4\%$ after 4000 cycles) [48]. The stability is always a challenge for ECDs, especially in air. A recent report about an ECD that uses ethyl viologen ($[EV]^{2+}$)- and ferrocene-based redox ionic liquids ($[FcNTf]^-$) as the electroactive species was demonstrated by Gélina et al. It was insensitive to atmospheric O_2 and H_2O in both redox states once dissolved in a typical ionic liquid electrolyte ($[BMIm][NTf_2]$).

When combining the EC materials with the DSSC, in addition to self-power supplied, the electrochemical performance is also improved. As shown in Figure 12.16, Santa-Nokki et al. designed a novel structure using viologen as EC material [49]. The driving voltage for coloration decreased to 1.0 V comparing with bare ECD (2.5 V).

In 2020, Ma et al. reported photophysical and electrochemical properties of newly synthesized thioxanthone and viologen (TX-VIO) binary derivatives. As shown in Figure 12.17, the sandwiched devices displayed a good transmittance and well reversible color (pink, blue, and colorless) change behavior under visible light radiation or an applied bias of −2.4 V [50].

12.2.2.2 Conducting Polymers

Poly(thiophene)s

Poly(thiophene)s are widely used as EC materials since they are easy to synthesize in chemical and electrochemical methods, easy to process, and stable [51]. Poly(thiophene)s thin film is blue (λ_{max} = 730 nm) in doped (oxidized) state and red (λ_{max} = 470 nm) in their undoped state. Color tuning can be achieved by different choice of thiophene monomers, which is also the main advantage of the

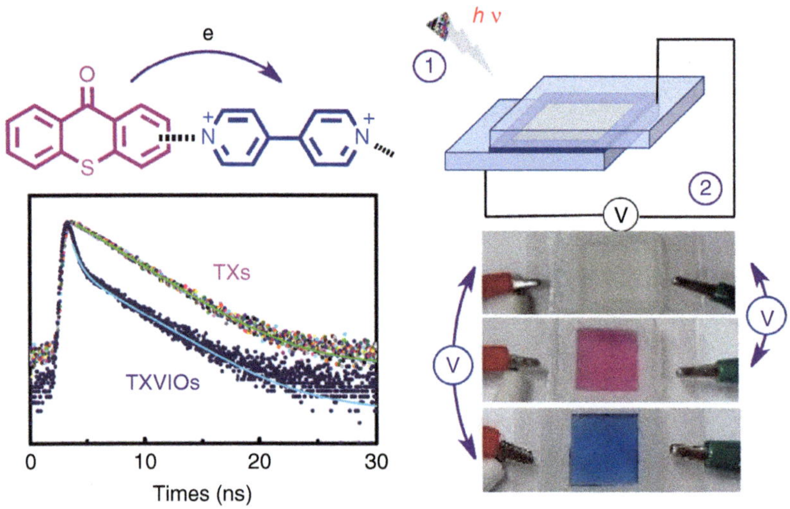

Figure 12.17 TX-VIO molecular structure, time resolved fluorescence decays, and corresponding devices [50]. Source: Ma et al. [50]. © 2020, Royal Society of Chemistry.

conducting polymers for ECD. For example, the colors of films prepared from 3-methylthiophene-based oligomers mainly depend on the relative positions of methyl groups on the polymer backbone [52, 53]. Color changes include pale blue, blue and violet in the oxidized form, and purple, yellow, red, and orange in the reduced state. The color changes are due to the variations of the effective conjugation length of the polymer chain.

A large amount of substituted thiophenes have been synthesized, and therefore numerous novel poly(thiophene)s are studied. Especially poly(3-substituted thiophenes) and poly(3,4-disubstituted thiophenes) attracted a lot of attention [49]. Poly(3,4-ethylenedioxythiophene) (PEDOT) has emerged as the most popular candidate for use in many applications, which was first discovered by Bayer AG research laboratories in the 1980s. Since then, PEDOT has proven to be an extraordinary polymer for its excellent EC properties, high conductivity, and high stability in the oxidized state [54]. The bandgap of PEDOT (1.6–1.7 eV) is 0.5 eV lower than poly(thiophene) owing to the presence of the two electron-donating oxygen atoms adjacent to the thiophene unit, which leads to an absorbance maximum in the red region. PEDOT shows excellent stability in the oxidized state compared with other substituted polythiophenes. But it is insoluble, which is hard for processing and can be overcome by using a water-soluble polyelectrolyte (poly(styrene sulfonate) [PSS]) as the counter ion in the doped form to yield the commercially available product PEDOT:PSS to disperse in water.

In research, improving the color-switching rate, stability, and color contrast of ECDs are most researchers' motivations. The color-switching rate of PEDOT is limited due to slow diffusion rate of counter ions into the film during the redox process. Cho et al. fabricated PEDOT nanotubes (as shown in Figure 12.18) of which thin walls are about 10 nm to reduce diffusion distance of ions, which leads to

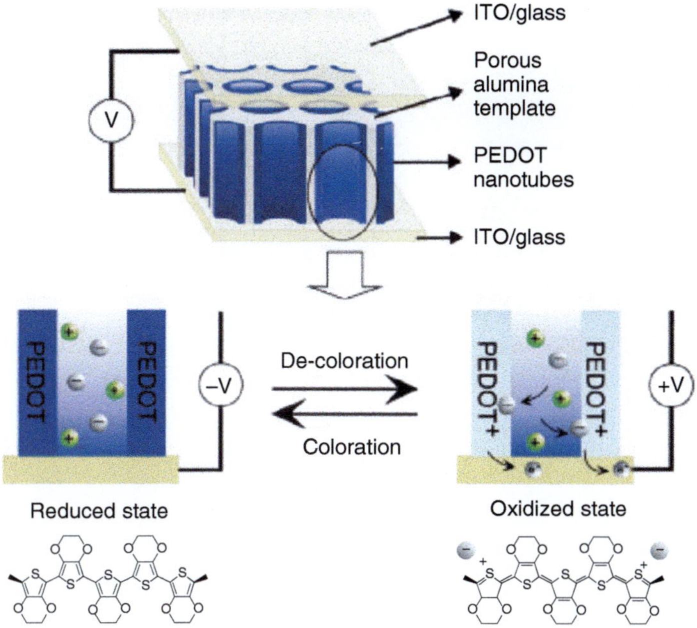

Figure 12.18 Schematics of device structure and redox process [55]. Source: Cho et al. [55]. © 2005, John Wiley & Sons.

ultrafast switching rate, and the long nanotube arrays obtain the strong coloration [55]. Similarly, Xia et al. utilized a highly porous structure of NiO and PEDOT composite film, which can also provide a very short ions diffusion pathway to get fast response time, and this device exhibits multicolor property [56]. The stability of PEDOT is another issue to address because most of the 3,4-ethoxylene dioxy thiophene (EDOT)-containing polymers are oxygen and moisture sensitive in the neutral form. Schwendeman et al. proposed a new polymer, PEDOT-F (as shown in Figure 12.19), introducing highly hydrophobic perfluoroalkyl substituents, and PEDOT doped with PEDOT-F exhibited a water contact angle of 110°, significantly higher than 30° for the PEDOT parent [57]. In addition, Ma et al. proposed the PEDOT/TiO_2 nanocomposites to enhance the interfacial adhesion of the polymers to the substrate and therefore increase the stability of the ECDs [58]. To optimize the color switch contrast of ECDs, the thickness of the pixel electrode and the

Figure 12.19 PEDOT-F.

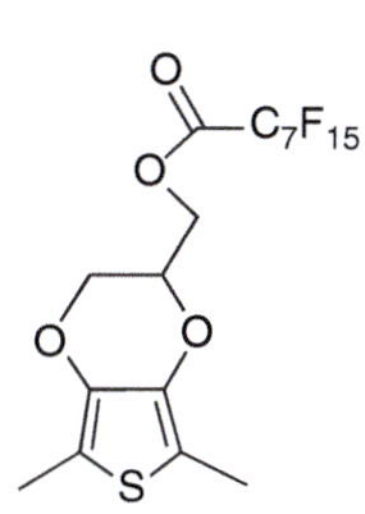

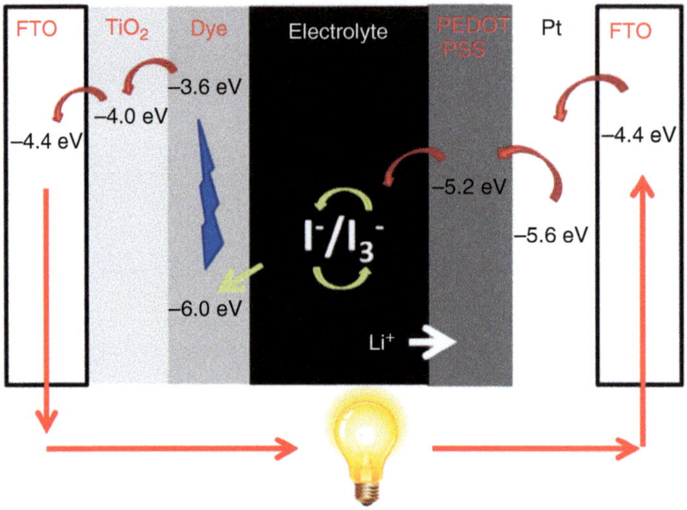

Figure 12.20 Energy diagram and working principle of a dye-sensitized solar cells-electrochromic (DSC-EC) device with the electrochromic material in the counter electrode [60]. Source: Costa et al. [60]. © 2016, Elsevier.

counter electrode was studied by Kawahara et al., and the optimized color switch contrast reached $\Delta E^* \sim 42$ [59].

When combining DSSCs with ECD, the self-powered ECD was obtained known as PECD. Costa et al. studied the PECD performance using PEDOT as EC materials in the counter electrode (shown in Figure 12.20), and the best PECD originates a color contrast of $\Delta E = 47$ at short circuit and an energy conversion efficiency of 1% at V_{oc} of 0.63 V and J_{sc} of 4.5 mA/cm^2 [60]. De Filpo et al. proposed a PECD doped with single wall carbon nanotubes (SWNT) bearing COOH groups to improve electrical conductance using PEDOT:PSS, which has fast coloration time (8 seconds) and fast bleaching time (30 seconds) [61].

Poly(hydroxymethyl-3,4-ethylenedioxythiophene) (PhmEDOT) and PEDOT thin films prepared by electropolymerization method on indium tin oxide (ITO) conducting glass were studied by Lin et al., and they fabricated PECD comprising of the PhmEDOT or the PEDOT thin films and dye-sensitized TiO_2 photoanodes. The PhmEDOT thin film showed shorter photoelectrochromic coloring/bleaching time and better transmittance modulation ability [62].

In addition to PEDOT as popular EC materials, poly(3,4-propylenedioxythiophene) (PProDOT) has enhanced EC properties by increasing the size of the alkylenedioxy ring or the nature of the substituents attached to the ring comparing with PEDOT [63–65]. PProDOT has a transmittance change of 66% at λ_{max} while 54% for PEDOT [64]. Moreover, PProDOT derivatives, PProDOT-Me$_2$ (as shown in Figure 12.21), has a $\Delta\%T$ of close to 78%, which is significantly higher than that of PProDOT or PEDOT. Xu et al. designed a new EC polymer device using PProDOT-Me$_2$ as EC materials, which shows a high contrast ratio of transmittance ($\Delta T > 50\%$) and rapid switching time (0.5–1 second) upon applied voltage of 2.5 V [67]. To achieve

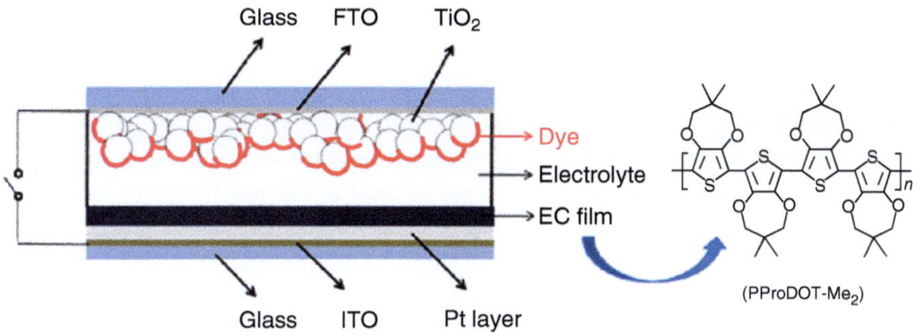

Figure 12.21 Schematic design of the PECD and the chemical structure of PProDOT-Me_2 [66]. Source: Yang et al. [66]. © 2012, Elsevier.

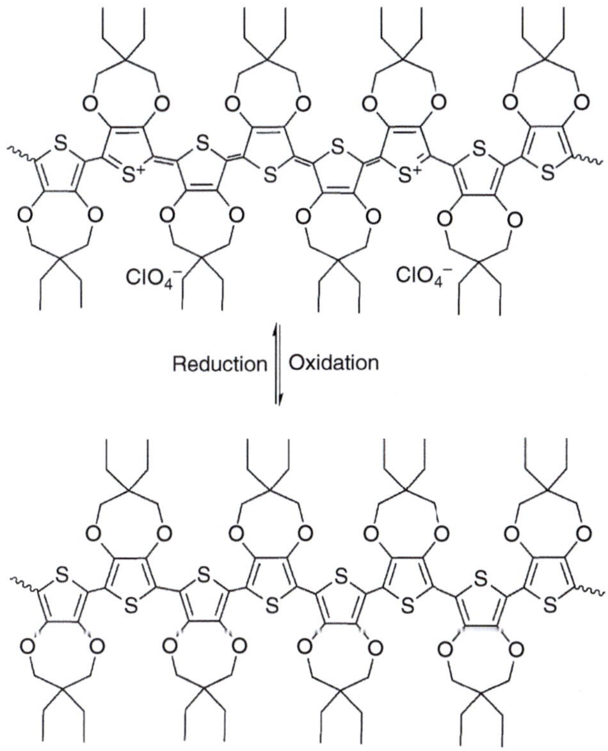

Figure 12.22 The oxidation and reduction of ProDOT-Et_2.

a self-powered ECD, a novel PECD (as shown in Figure 12.21), by adding the PProDOT-Me_2 polymer EC film into a DSSC, was designed with a fast response time (one and four seconds for coloring and bleaching, respectively), and an energy conversion efficiency (ECE) of 1.12%, V_{oc} of 0.66 V and J_{sc} of 2.92 mA/cm^2 in 2012 by Yang et al. [66]. Gaupp et al. studied that as the size and substituents pattern of the alkylenedioxy bridge increases, the properties of the device can be optimized [68]. So they synthesized and studied the EC properties of PProDOT-Et_2 (shown in

Figure 12.22), which has a high contrast with a coloration efficiency higher than that of PProDOT-Me_2 [69].

Polyaniline (PANI)

The EC property of polyaniline (PANI) depends on not only its oxidation state but also its protonation state, and there are several redox mechanisms involving protonation–deprotonation or anion ingress/egress. PANI films show four different poly-EC states, namely, transparent yellow, green, dark blue, and black. Figure 12.23 shows the compositions of different states of PANI. The EC colors of PANI are very close to the three primary colors (green, blue, and red) except for red. If the absorption could be shifted to shorter wavelength up to 500 nm, it could show all three primary colors. To achieve that, introducing a substituent or monomeric unit may get an ECD showing all colors.

Zhang et al. synthesized nanostructured PANI–WO_3 hybrid thin films, which shows multicolor varied from green to royal purple (more positive than 0.25 V), pale yellow to green (between −0.2 and 0.25 V), and finally dark blue (more negative than −0.2 V) [70]. Figure 12.24 shows the CV curves of WO_3 film, sulfuric acid-doped PANI film, and hybrid films. Both curves of hybrid films have the typical redox peaks of PANI (A/A′, B/B′, and C/C′) [70]. The redox peaks of D/D′ of WO_3 are also found in hybrid films. The oxidation potentials of the hybrids are reduced due to the formation of the donor–acceptor (PANI–WO_3) system [71]. It has been studied that in PANI–WO_3 composite materials, electrons in the conduction band of the WO_3 have a negative-enough potential to change the color of PANI from green to yellow [72]. In addition to the coloration change, the switching time and coloration efficiency are also improved due to the porous structure and the electronic structure of the donor–acceptor system. Another study on PANI–WO_3 composite film revealed that the composite films exhibit significant charge storage or discharging capacity after 1000 charge–discharge cycles and enhanced durability, so the composite system has potential for EC and energy storage devices applications [73]. A novel PECD was proposed by Li et al., containing PANI as EC layer and DSSCs, which is

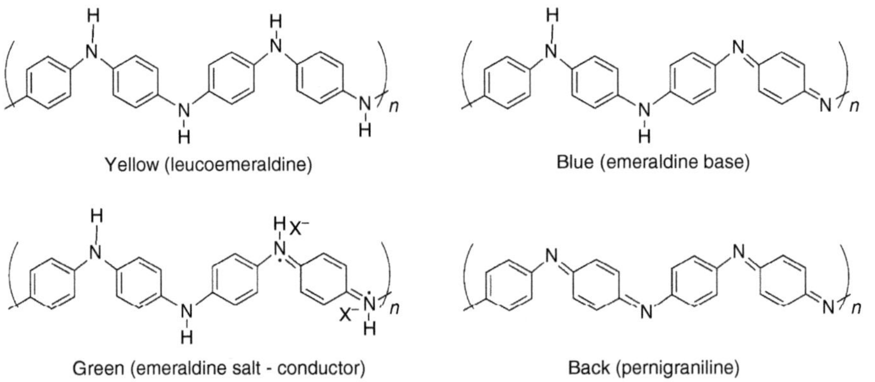

Figure 12.23 Some redox states of polyaniline, from the fully reduced (leucoemeraldine) to the fully oxidized (pernigraniline) forms. X^- is a charge-balancing anion.

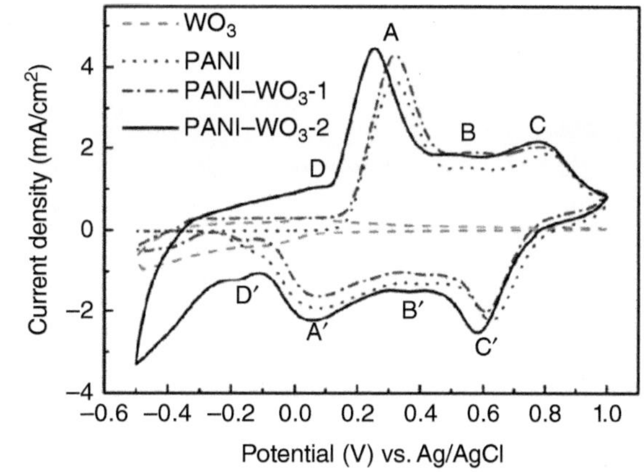

Figure 12.24 CV curves for WO_3 film, sulfuric acid-doped PANI film, and PANI–WO_3 hybrid film [70]. Source: Zhang et al. [70]. © 2011, Royal Society of Chemistry.

self-powered and can modulate the transmission averaged over the whole visible spectral region [74].

Poly(pyrrole)s (Ppys)

Poly(pyrrole)s are also extensively used as EC materials like poly(thiophene)s and can be synthesized easily with a number of optoelectrical properties through different alkyl or alkoxy substitution. The thin film of parent poly(pyrrole)s are yellow/green in the undoped (reduced) insulating form (λ_{max} = 420 nm) and blue/violet in the doped (oxidized) conductive form (λ_{max} = 670 nm), which has a bandgap of 2.7 eV [75].

However, the poly(pyrrole) is not studied widely due to the degradation of the film after repetitive color switching. The dialkoxyl-substituted poly(pyrrole)s, through the addition of the oxygen at the third and fourth positions, have a lower bandgap by raising the highest occupied molecular orbital (HOMO) level. In addition, poly(pyrrole) has relatively low oxidation potential. The aforementioned two facts make the poly(alkylenedioxypyrrole)s the lowest oxidation potential for p-type doping in EC polymers [76].

An interesting design based on electrodeposited poly(pyrrole) film was proposed by Yang et al., which is a bifunctional device exhibiting self-powered ECD and self-rechargeable battery characteristics. The device structure and the mechanism of the Ppy/Al device acting as an ECD and a battery was illustrated in Figure 12.25 [77]. As an ECD, the device can change the color from black to yellow when connecting the two electrodes. As a battery, the capacity of the device is 75.25 mAh/g when it is discharged at a constant current density of 1.0 Ag^{-1}. The open circuit voltage of the device is 0.94 V. When disconnecting the two electrodes, the device can recover the color and capacity [77].

12.2.2.3 Near-Infrared (NIR) Electrochromic Materials

EC materials in NIR region (λ = 780–2500 nm) have attracted a lot of attention due to their advantages applying as variable optical attenuators in optical telecommunications [78]. Current extensively explored NIR EC materials are transition metal oxides

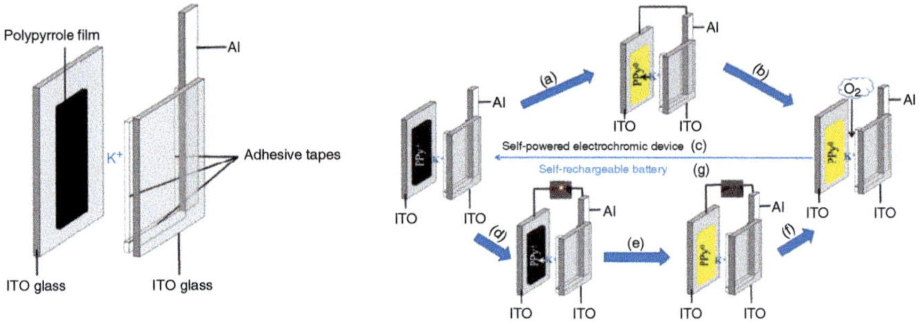

Figure 12.25 (a) Schematic diagram of PPy/Al device, (b) working mechanism for the PPy/Al device (a–c) bleaching and recovery process, and (d–g) the discharging and self-recharging process of the device [77]. Source: Yang et al. [77]. © 2019, Elsevier.

or complexes [79]. Organic EC materials with NIR absorptions are solution processable with short switching time and high color contrast [80], so they are explored as well.

For instance, PEDOT shows chromogenic properties in the NIR region as shown in Figure 12.26 [81]. Different tones of the same color can be obtained by giving different electrical potential; as the potential increases, the absorption in visible region decreases, while the absorption in the NIR region increases. Similarly, aromatic polyamides with triarylamine units were reported by Hung-Ju Yen et al. [82]. The absorption in NIR region increases as the potential increases. The polymer showed electrochromism with high contrast, high coloration efficiency, short switching time, and high stability. In addition, anthraquinone imides [83] and triarylamine derivatives [80] have been reported as well with their NIR absorption.

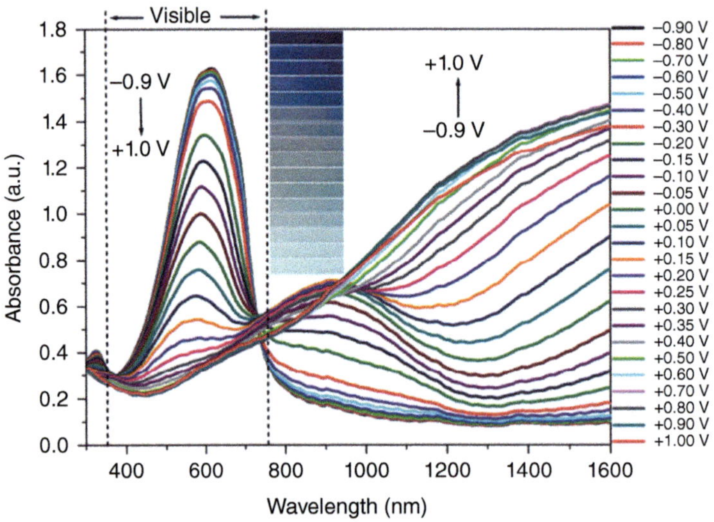

Figure 12.26 Spectroelectrochemistry of PEDOT at different oxidation states with inset pictures of variation in the blue color at different oxidation levels [81]. Source: Sonmez et al. [81]. © 2004, John Wiley & Sons.

12.2.3 Electrolytes in PECD

Electrolyte is one of the main components in ECDs and batteries. In PECDs, electrolyte is responsible for connecting the ECD and the power supply, which has conductive ion media between the cathode and the anode providing the ion flow, preventing short circuit. Generally, electrolyte can be classified into four major categories: liquid, ceramic, solid inorganic, and polymer electrolytes [84–87]. The polymer electrolyte is an ideal candidate for electrochemical system, offering the advantages of easily processable in thin films, long-term durability, high ionic conductivity, and good mechanical and high optical transparency. Also, inexpensive polymers are good for large-scale development.

Polymer electrolytes are generally macromolecular or supramolecular nano-aggregate system characterized by an ionic conductivity (above 10^{-7} S/cm), and they can be classified into five types: solid polymer, gel polymer, plasticized polymer, ionic rubber polymer, and ion conducting polyelectrolytes [88]. Besides polyelectrolyte, most of them are polyethylene oxide (PEO) containing lithium salts. In addition to PEO-based electrolytes, many co-polymer matrices containing polar elements like oxygen or nitrogen along the chain, for example, poly (propylene oxide) (PPO), poly (methyl methacrylate) (PMMA), polyacrylonitrile (PAN), and poly (ethylene imine) (PEI), are effectively used to prepare gel and plasticized polymer electrodes. Recently, poly (vinylidene fluoride) (PVDF) copolymers, particularly poly (vinylidene fluoride co-hexafluoropropylene) (PVDF-HFP) and poly(vinylidene fluoride co-trifluoroethylene) (PVDF-TrFE), are employed to be the plasticized electrolyte in EC and PECDs.

Hechavarría et al. prepared a polymeric electrolyte applying the sol–gel method to a mixture of poly(ethylene glycol) (PEG) and titanium isopropoxide and adding lithium iodide (LiI) in a PECD. The titanium complex in the electrolyte reduced the charge transfer resistance and increased the charge storage capacity of the solar cell [19]. Yang et al. fabricated a PVCC using PEG–Ti hybrid electrolyte, which has a good long-term stability [89]. Wu et al. prepared a gel electrolyte with succinonitrile and poly(vinylidenefluoride-co-hexafluoropropylene) (PVDF-HFP) dissolved in propylene carbonate in the PECD [90]. At 2016, a quasi-solid electrolyte was studied in a PECD by Bella et al., which shows excellent long term stability (>2100 hours) under real outdoor conditions [91–93].

12.2.4 Substrates in PECD

For the standard PECD, the substrates are transparent electrode working as anode, extracting charge carriers. This layer is always covered by a transparent conducting oxide (TCO) film. The most common is ITO on either glass or plastic owing to its excellent optical transparency and good electrical properties. However, the brittle nature and the expensive processing steps of ITO make it an increasing demand, which explores alternative materials.

To make a flexible and truly all-organic device, utilizing PEDOT:PSS as transparent conducting electrode (TCE) to replace ITO brings about the advantages of flexible, stable, and inexpensive ECDs. The first truly all-polymer ECDs were fabricated

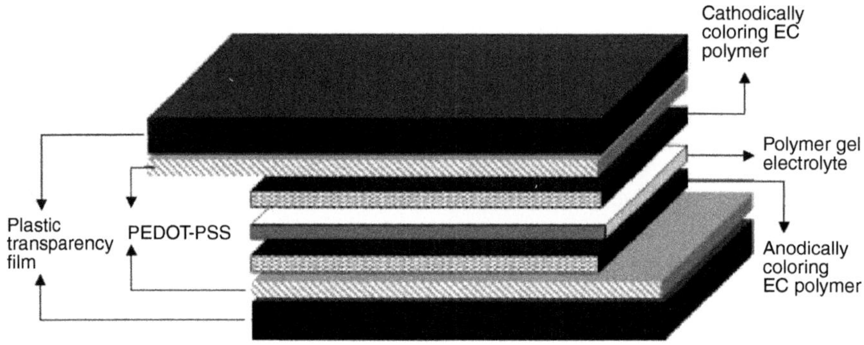

Figure 12.27 ITO-free device structure [94]. Source: Argun et al. [94]. © 2003, John Wiley & Sons.

by Argun et al. at 2003, the device structure of which are shown in Figure 12.27 [94]. It proves that PEDOT:PSS is an excellent TCE to replace for ITO. Moreover, without sacrificing EC contrast, switching time, and coloration efficiency, PEDOT:PSS devices have good mechanical flexibility and superior stability, which has been studied by Singh et al. [95].

12.3 Future Perspectives and Conclusion

PECDs are combining ECD and photovoltaic devices together so no need for external power supply, which is superior to simple ECD. However, more efforts should be taken in this field. Currently, the devices are still focusing on the research, but they have great potential in business and market. The stable large devices are undoubtedly important to fabricate with easy processing techniques for the near future applications in the market. And the development of the new materials is also one of the major ways to improve the performance of PECDs. More importantly, the transparency and the application in UV/NIR will be great research trend for researcher to explore in the PECDs. Organic materials are attractive due to their low cost and good performance in optoelectronics. In particular, conjugated conductive polymers may play a dominant part in the PECDs in the near future. In conclusion, PECDs will be attractive and practical energy-saving devices in future development with long-term usage and good performance.

References

1 Mirandola, A. and Lorenzini, E. (2016). Energy, environment and climate: from the past to the future. *International Journal of Heat and Technology* 34 (2): 159–164.

2 Armaroli, N. and Balzani, V. (2011). Towards an electricity-powered world. *Energy & Environmental Science* 4 (9): 3193–3222.

3 Svensson, J.S.E.M. and Granqvist, C. (1984). Electrochromic tungsten oxide films for energy efficient windows. *Solar Energy Materials* 11 (1–2): 29–34.
4 Khandelwal, H., Schenning, A.P., and Debije, M.G. (2017). Infrared regulating smart window based on organic materials. *Advanced Energy Materials* 7 (14): 1602209.
5 Karlsson, J. (2001). Control system and energy saving potential for switchable windows. In: *Proceedings of Seventh International IBPSA Conference*, 199–206. Rio de Janeiro, Brazil.
6 Pierucci, A., Cannavale, A., Martellotta, F., and Fiorito, F. (2018). Smart windows for carbon neutral buildings: a life cycle approach. *Energy and Buildings* 165: 160–171.
7 Lee, E.S. and DiBartolomeo, D. (2002). Application issues for large-area electrochromic windows in commercial buildings. *Solar Energy Materials and Solar Cells* 71 (4): 465–491.
8 Sbar, N.L., Podbelski, L., Yang, H.M., and Pease, B. (2012). Electrochromic dynamic windows for office buildings. *International Journal of Sustainable Built Environment* 1 (1): 125–139.
9 Clear, R., Inkarojrit, V., and Lee, E. (2006). Subject responses to electrochromic windows. *Energy and Buildings* 38 (7): 758–779.
10 Baetens, R., Jelle, B.P., and Gustavsen, A. (2010). Properties, requirements and possibilities of smart windows for dynamic daylight and solar energy control in buildings: a state-of-the-art review. *Solar Energy Materials and Solar Cells* 94 (2): 87–105.
11 Bechinger, C., Ferrere, S., Zaban, A. et al. (1996). Photoelectrochromic windows and displays. *Nature* 383 (6601): 608–610.
12 Wu, J.-J., Hsieh, M.-D., Liao, W.-P. et al. (2009). Fast-switching photovoltachromic cells with tunable transmittance. *ACS Nano* 3 (8): 2297–2303.
13 Cannavale, A., Manca, M., De Marco, L. et al. (2014). Photovoltachromic device with a micropatterned bifunctional counter electrode. *ACS Applied Materials & Interfaces* 6 (4): 2415–2422.
14 Jiao, Z., Song, J.L., Sun, X.W. et al. (2012). A fast-switching light-writable and electric-erasable negative photoelectrochromic cell based on Prussian blue films. *Solar Energy Materials and Solar Cells* 98: 154–160.
15 Cannavale, A., Eperon, G.E., Cossari, P. et al. (2015). Perovskite photovoltachromic cells for building integration. *Energy & Environmental Science* 8 (5): 1578–1584.
16 Davy, N.C., Sezen-Edmonds, M., Gao, J. et al. (2017). Pairing of near-ultraviolet solar cells with electrochromic windows for smart management of the solar spectrum. *Nature Energy* 2 (8): 1–11.
17 O'regan, B. and Grätzel, M. (1991). A low-cost, high-efficiency solar cell based on dye-sensitized colloidal TiO_2 films. *Nature* 353 (6346): 737.
18 Yu, G., Gao, J., Hummelen, J.C. et al. (1995). Polymer photovoltaic cells: enhanced efficiencies via a network of internal donor–acceptor heterojunctions. *Science* 270 (5243): 1789.

19 Hechavarría, L., Mendoza, N., Rincón, M.E. et al. (2012). Photoelectrochromic performance of tungsten oxide based devices with PEG–titanium complex as solvent-free electrolytes. *Solar Energy Materials and Solar Cells* 100: 27–32.

20 Lee, B., He, J., Chang, R.P., and Kanatzidis, M.G. (2012). All-solid-state dye-sensitized solar cells with high efficiency. *Nature* 485 (7399): 486.

21 Listorti, A., O'Regan, B., and Durrant, J.R. (2011). Electron transfer dynamics in dye-sensitized solar cells. *Chemistry of Materials* 23 (15): 3381–3399.

22 Xu, Z., Li, W., Huang, J. et al. (2019). Flexible, controllable and angle-independent photoelectrochromic display enabled by smart sunlight management. *Nano Energy* 63: 10380.

23 Hauch, A., Georg, A., Baumgärtner, S. et al. (2001). New photoelectrochromic device. *Electrochimica Acta* 46 (13): 2131–2136.

24 Georg, A., Georg, A., and Opara Krašovec, U. (2006). Photoelectrochromic window with Pt catalyst. *Thin Solid Films* 502 (1–2): 246–251.

25 Bogati, S., Georg, A., and Graf, W. (2017). Photoelectrochromic devices based on sputtered WO_3 and TiO_2 films. *Solar Energy Materials and Solar Cells* 163: 170–177.

26 Leftheriotis, G., Syrrokostas, G., and Yianoulis, P. (2012). "Partly covered" photoelectrochromic devices with enhanced coloration speed and efficiency. *Solar Energy Materials and Solar Cells* 96: 86–92.

27 Xie, Z., Jin, X., Chen, G. et al. (2014). Integrated smart electrochromic windows for energy saving and storage applications. *Chemical Communications (Cambridge)* 50 (5): 608–610.

28 Xia, X., Ku, Z., Zhou, D. et al. (2016). Perovskite solar cell powered electrochromic batteries for smart windows. *Materials Horizons* 3 (6): 588–595.

29 Fan, B., Zhang, D., Li, M. et al. (2019). Achieving over 16% efficiency for single-junction organic solar cells. *Science China Chemistry* 62 (6): 746–752.

30 Cui, Y., Yao, H., Zhang, J. et al. (2019). Over 16% efficiency organic photovoltaic cells enabled by a chlorinated acceptor with increased open-circuit voltages. *Nature Communications* 10 (1): 2515.

31 Xu, X., Feng, K., Bi, Z. et al. (2019). Single-junction polymer solar cells with 16.35% efficiency enabled by a platinum(II) complexation strategy. *Advanced Materials* 31 (29): 1901872.

32 Tang, C.W. (1979). Multilayer organic photovoltaic elements. Google Patents No 4 164 431.

33 Hiramoto, M., Fujiwara, H., and Yokoyama, M. (1992). p-i-n like behavior in three-layered organic solar cells having a co-deposited interlayer of pigments. *Journal of Applied Physics* 72 (8): 3781–3787.

34 Clarke, T.M. and Durrant, J.R. (2010). Charge photogeneration in organic solar cells. *Chemical Reviews* 110 (11): 6736–6767.

35 Bamfield, P. (2010). *Chromic Phenomena: Technological Applications of Colour Chemistry*. Royal Society of Chemistry.

36 Xu, F., Wu, X., Zheng, J., and Xu, C. (2019). A new strategy to fabricate multicolor electrochromic device with UV-detecting performance based on TiO_2 and PProDOT-Me_2. *Organic Electronics* 65: 8–14.

37 Michaelis, L. and Hill, E.S. (1933). The viologen indicators. *The Journal of General Physiology* 16 (6): 859.

38 Bird, C. and Kuhn, A. (1981). Electrochemistry of the viologens. *Chemical Society Reviews* 10 (1): 49–82.

39 Chen, P.Y., Chen, C.S., and Yeh, T.H. (2014). Organic multiviologen electrochromic cells for a color electronic display application. *Journal of Applied Polymer Science* 131 (13).

40 Volke, J. (1968). The relationship between herbicidal activity and electrochemical properties of quaternary bipyridylium salts. *Collection of Czechoslovak Chemical Communications* 33 (9): 3044–3048.

41 Choi, S.Y., Mamak, M., Coombs, N. et al. (2004). Electrochromic performance of viologen-modified periodic mesoporous nanocrystalline anatase electrodes. *Nano Letters* 4 (7): 1231–1235.

42 Jain, V., Khiterer, M., Montazami, R. et al. (2009). High-contrast solid-state electrochromic devices of viologen-bridged polysilsesquioxane nanoparticles fabricated by layer-by-layer assembly. *ACS Applied Materials & Interfaces* 1 (1): 83–89.

43 Mortimer, R.J. and Varley, T.S. (2011). Novel color-reinforcing electrochromic device based on surface-confined ruthenium purple and solution-phase methyl viologen. *Chemistry of Materials* 23 (17): 4077–4082.

44 Rong, Y., Kim, S., Su, F. et al. (2011). New effective process to fabricate fast switching and high contrast electrochromic device based on viologen and Prussian blue/antimony tin oxide nano-composites with dark colored state. *Electrochimica Acta* 56 (17): 6230–6236.

45 Hu, C.-W., Lee, K.-M., Chen, K.-C. et al. (2012). High contrast all-solid-state electrochromic device with 2,2,6,6-tetramethyl-1-piperidinyloxy (TEMPO), heptyl viologen, and succinonitrile. *Solar Energy Materials and Solar Cells* 99: 135–140.

46 DeLongchamp, D.M., Kastantin, M., and Hammond, P.T. (2003). High-contrast electrochromism from layer-by-layer polymer films. *Chemistry of Materials* 15 (8): 1575–1586.

47 Moon, H.C., Lodge, T.P., and Frisbie, C.D. (2015). Solution processable, electrochromic ion gels for sub-1 V, flexible displays on plastic. *Chemistry of Materials* 27 (4): 1420–1425.

48 Lu, H.-C., Kao, S.-Y., Yu, H.-F. et al. (2016). Achieving low-energy driven viologens-based electrochromic devices utilizing polymeric ionic liquids. *ACS Applied Materials & Interfaces* 8 (44): 30351–30361.

49 Santa-Nokki, H., Kallioinen, J., and Korppi-Tommola, J. (2007). A dye-sensitized solar cell driven electrochromic device. *Photochemical & Photobiological Sciences* 6 (1): 63–66.

50 Ma, D.-M., Wang, J., Guo, H., and Qian, D.-J. (2020). Photophysical and electrochemical properties of newly synthesized thioxathone–viologen binary derivatives and their photo-/electrochromic displays in ionic liquids and polymer gels. *New Journal of Chemistry* 44 (9): 3654–3663.

51 Roncali, J. (1992). Conjugated poly (thiophenes): synthesis, functionalization, and applications. *Chemical Reviews* 92 (4): 711–738.

52 Mastragostino, M., Arbizzani, C., Bongini, A. et al. (1993). Polymer-based electrochromic devices – I. Poly (3-methylthiophenes). *Electrochimica Acta* 38 (1): 135–140.

53 Mastragostino, M., Arbizzani, C., Ferloni, P., and Marinangeli, A. (1992). Polymer-based electrochromic devices. *Solid State Ionics* 53: 471–478.

54 Groenendaal, L., Jonas, F., Freitag, D. et al. (2000). Poly (3, 4-ethylenedioxythiophene) and its derivatives: past, present, and future. *Advanced Materials (Weinheim, Germany)* 12 (7): 481–494.

55 Cho, S.I., Kwon, W.J., Choi, S.J. et al. (2005). Nanotube-based ultrafast electrochromic display. *Advanced Materials* 17 (2): 171–175.

56 Xia, X., Tu, J., Zhang, J. et al. (2009). Multicolor and fast electrochromism of nanoporous NiO/poly (3,4-ethylenedioxythiophene) composite thin film. *Electrochemistry Communications* 11 (3): 702–705.

57 Schwendeman, I., Gaupp, C.L., Hancock, J.M. et al. (2003). Perfluoroalkanoate-substituted PEDOT for electrochromic device applications. *Advanced Functional Materials* 13 (7): 541–547.

58 Ma, L., Li, Y., Yu, X. et al. (2008). Using room temperature ionic liquid to fabricate PEDOT/TiO_2 nanocomposite electrode-based electrochromic devices with enhanced long-term stability. *Solar Energy Materials and Solar Cells* 92 (10): 1253–1259.

59 Kawahara, J., Ersman, P.A., Engquist, I., and Berggren, M. (2012). Improving the color switch contrast in PEDOT:PSS-based electrochromic displays. *Organic Electronics* 13 (3): 469–474.

60 Costa, C., Mesquita, I., Andrade, L., and Mendes, A. (2016). Photoelectrochromic devices: influence of device architecture and electrolyte composition. *Electrochimica Acta* 219: 99–106.

61 De Filpo, G., Mormile, S., Nicoletta, F.P., and Chidichimo, G. (2010). Fast, self-supplied, all-solid photoelectrochromic film. *Journal of Power Sources* 195 (13): 4365–4369.

62 Lin, C.-L., Chen, C.-Y., Yu, H.-F., and Ho, K.-C. (2019). Comparisons of the electrochromic properties of poly(hydroxymethyl-3,4-ethylenedioxythiophene) and poly(3,4-ethylenedioxythiophene) thin films and the photoelectrochromic devices using these thin films. *Solar Energy Materials and Solar Cells* 202: 110132.

63 Sankaran, B. and Reynolds, J.R. (1997). High-contrast electrochromic polymers from alkyl-derivatized poly (3,4-ethylenedioxythiophenes). *Macromolecules* 30 (9): 2582–2588.

64 Kumar, A., Welsh, D.M., Morvant, M.C. et al. (1998). Conducting poly (3,4-alkylenedioxythiophene) derivatives as fast electrochromics with high-contrast ratios. *Chemistry of Materials* 10 (3): 896–902.

65 Krishnamoorthy, K., Ambade, A.V., Kanungo, M. et al. (2001). Rational design of an electrochromic polymer with high contrast in the visible region: dibenzyl substituted poly (3,4-propylenedioxythiophene) electronic supplementary information (ESI) available: cyclic voltammograms of polymer films in 0.1 M tetrabutylammonium tetrafluoroborate–acetonitrile. See http://www.rsc.org/suppdata/jm/b1/b108654e. *Journal of Materials Chemistry* 11 (12): 2909–2911.

66 Yang, S., Zheng, J., Li, M., and Xu, C. (2012). A novel photoelectrochromic device based on poly (3,4-(2,2-dimethylpropylenedioxy) thiophene) thin film and dye-sensitized solar cell. *Solar Energy Materials and Solar Cells* 97: 186–190.

67 Xu, C., Tamagawa, H., Uchida, M., and Taya, M. (2002). Enhanced contrast ratios and rapid-switching color-changeable devices based on poly (3,4-propylenedioxythiophene) derivative and counterelectrode. In: *Smart Structures and Materials 2002: Electroactive Polymer Actuators and Devices (EAPAD)*, 442–450. International Society for Optics and Photonics.

68 Gaupp, C.L., Welsh, D.M., Rauh, R.D., and Reynolds, J.R. (2002). Composite coloration efficiency measurements of electrochromic polymers based on 3,4-alkylenedioxythiophenes. *Chemistry of Materials* 14 (9): 3964–3970.

69 Gaupp, C.L., Welsh, D.M., and Reynolds, J.R. (2002). Poly (ProDOT-Et_2): a high-contrast, high-coloration efficiency electrochromic polymer. *Macromolecular Rapid Communications* 23 (15): 885–889.

70 Zhang, J., Tu, J.-p., Zhang, D. et al. (2011). Multicolor electrochromic polyaniline–WO_3 hybrid thin films: one-pot molecular assembling synthesis. *Journal of Materials Chemistry* 21 (43): 17316–17324.

71 Meyer, J., Hamwi, S., Schmale, S. et al. (2009). A strategy towards p-type doping of organic materials with HOMO levels beyond 6 eV using tungsten oxide. *Journal of Materials Chemistry* 19 (6): 702–705.

72 Yoneyama, H., Hirao, S., and Kuwabata, S. (1992). Preparation and electrochemical properties of WO_3-incorporated polyaniline films. *Journal of the Electrochemical Society* 139 (11): 3141–3146.

73 Wei, H., Yan, X., Wu, S. et al. (2012). Electropolymerized polyaniline stabilized tungsten oxide nanocomposite films: electrochromic behavior and electrochemical energy storage. *The Journal of Physical Chemistry C* 116 (47): 25052–25064.

74 Li, Y., Hagen, J., and Haarer, D. (1998). Novel photoelectrochromic cells containing a polyaniline layer and a dye-sensitized nanocrystalline TiO_2 photovoltaic cell. *Synthetic Metals* 94 (3): 273–277.

75 Genies, E., Bidan, G., and Diaz, A. (1983). Spectroelectrochemical study of polypyrrole films. *Journal of Electroanalytical Chemistry and Interfacial Electrochemistry* 149 (1–2): 101–113.

76 Schottland, P., Zong, K., Gaupp, C.L. et al. (2000). Poly(3,4-alkylenedioxypyrrole)s: highly stable electronically conducting and electrochromic polymers. *Macromolecules* 33 (19): 7051–7061.

77 Yang, B., Ma, D., Zheng, E., and Wang, J. (2019). A self-rechargeable electrochromic battery based on electrodeposited polypyrrole film. *Solar Energy Materials and Solar Cells* 192: 1–7.

78 Wang, S., Li, X., Xun, S. et al. (2006). Near-infrared electrochromic and electroluminescent polymers containing pendant ruthenium complex groups. *Macromolecules* 39 (22): 7502–7507.

79 Ward, M.D. (2005). Near-infrared electrochromic materials for optical attenuation based on transition-metal coordination complexes. *Journal of Solid State Electrochemistry* 9 (11): 778–787.

80 Choi, K., Yoo, S.J., Sung, Y.-E., and Zentel, R. (2006). High contrast ratio and rapid switching organic polymeric electrochromic thin films based on triarylamine derivatives from layer-by-layer assembly. *Chemistry of Materials* 18: 5823–5825.

81 Sonmez, G., Sonmez, H.B., Shen, C.K.F., and Wudl, F. (2004). Red, green, and blue colors in polymeric electrochromics. *Advanced Materials* 16 (21): 1905–1908.

82 Yen, H.-J., Lin, H.-Y., and Liou, G.-S. (2011). Novel starburst triarylamine-containing electroactive aramids with highly stable electrochromism in near-infrared and visible light regions. *Chemistry of Materials* 23 (7): 1874–1882.

83 Chen, F., Zhang, J., Jiang, H., and Wan, X. (2013). Colorless to purple–red switching electrochromic anthraquinone imides with broad visible/near-IR absorptions in the radical anion state: simulation-aided molecular design. *Chemistry - An Asian Journal* 8 (7): 1497–1503.

84 Vlad, A., Singh, N., Galande, C., and Ajayan, P.M. (2015). Design considerations for unconventional electrochemical energy storage architectures. *Advanced Energy Materials* 5 (19): 1402115.

85 Choudhury, N., Sampath, S., and Shukla, A. (2009). Hydrogel-polymer electrolytes for electrochemical capacitors: an overview. *Energy & Environmental Science* 2 (1): 55–67.

86 Thakur, V.K., Ding, G., Ma, J. et al. (2012). Hybrid materials and polymer electrolytes for electrochromic device applications. *Advanced Materials* 24 (30): 4071–4096.

87 Kalhoff, J., Eshetu, G.G., Bresser, D., and Passerini, S. (2015). Safer electrolytes for lithium-ion batteries: state of the art and perspectives. *ChemSusChem* 8 (13): 2154–2175.

88 Di Noto, V., Lavina, S., Giffin, G.A. et al. (2011). Polymer electrolytes: present, past and future. *Electrochimica Acta* 57: 4–13.

89 Yang, M.-C., Cho, H.-W., and Wu, J.-J. (2014). Fabrication of stable photovoltachromic cells using a solvent-free hybrid polymer electrolyte. *Nanoscale* 6 (16): 9541–9544.

90 Wu, C.-H., Hsu, C.-Y., Huang, K.-C. et al. (2012). A photoelectrochromic device based on gel electrolyte with a fast switching rate. *Solar Energy Materials and Solar Cells* 99: 148–153.

91 Bella, F., Leftheriotis, G., Griffini, G. et al. (2016). A new design paradigm for smart windows: photocurable polymers for quasi-solid photoelectrochromic devices with excellent long-term stability under real outdoor operating conditions. *Advanced Functional Materials* 26 (7): 1127–1137.

92 Lee, M.M., Teuscher, J., Miyasaka, T. et al. (2012). Efficient hybrid solar cells based on meso-superstructured organometal halide perovskites. *Science* 338 (6107): 643.

93 Kim, H.-S., Lee, C.-R., Im, J.-H. et al. (2012). Lead iodide perovskite sensitized all-solid-state submicron thin film mesoscopic solar cell with efficiency exceeding 9%. *Scientific Reports* 2: 591.

94 Argun, A.A., Cirpan, A., and Reynolds, J.R. (2003). The first truly all-polymer electrochromic devices. *Advanced Materials* 15 (16): 1338–1341.

95 Singh, R., Tharion, J., Murugan, S., and Kumar, A. (2016). ITO-free solution-processed flexible electrochromic devices based on PEDOT:PSS as transparent conducting electrode. *ACS Applied Materials & Interfaces* 9 (23): 19427–19435.

13

Application of OEC Devices

With a deep investigation of OEC, the application of the devices has been successfully commercialized in auto-dimming mirror for automobiles and dimmable windows for aircrafts by Gentex Corporation [1–3]. The OEC devices have also shown great potential in many other fields such as sensors, fabrics, glasses, and electronic skins as shown in Figure 13.1 [4–6]. The application of OEC devices mainly follow two principles, light transmission and conduction [7]. Academia and industry have done much work on application of electrochromism, and there are thousands of patents that can be retrieved, indicating a huge commercial prospect in this field. This chapter will introduce the application of electrochromic devices (ECDs) and look forward to the future development in the near future.

13.1 Smart Window

As the demands for energy conservation and emission reductions increase, governments are increasingly demanding building energy consumption. Smart windows can reduce the energy that can be injected into the building from outside in summer by adjusting its color while reducing the outward radiation of the indoor energy in winter. Researchers point out that if the windows of the buildings are replaced by smart windows, the energy consumption of the buildings will be reduced by 65% [8]. It can be seen the smart windows have great potential for development in the future. According to the working mechanism, smart windows can be divided into photochromic, thermochromic, and electrochromic [9]. Photochromic and thermochromic rely on changes of light and heat of environment to change color. Electrochromic can achieve artificially controllable discoloration by applying voltage, so electrochromic window is the best kind for development [10].

13.1.1 The Structure and Working Mechanism of Smart Windows

The structure of smart windows is shown in Figure 13.2. It includes two layers of transparent substrates and an electrochromic layer sandwiched between them [11]. The electrochromic layer can change the color of the window, thus to be the most important part of the window. The composition of the electrochromic part is

Organic Electronics for Electrochromic Materials and Devices, First Edition. Hong Meng.

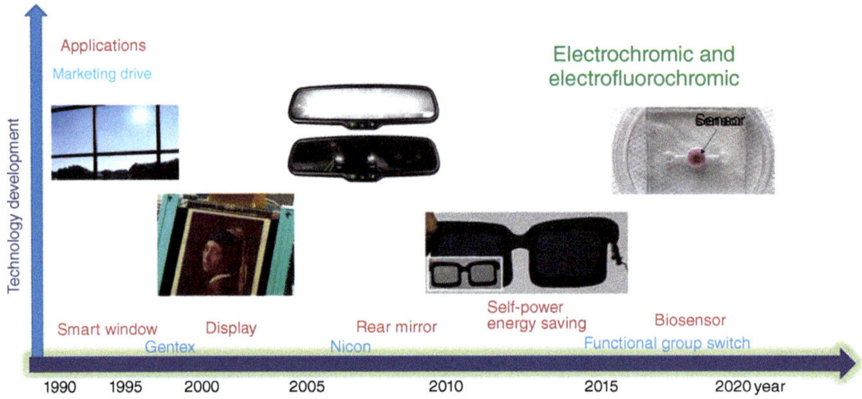

Figure 13.1 Road map of application of OEC technology development. Sources: Yun et al. [4], Lahav et al. [5], Osterholm et al. [6].

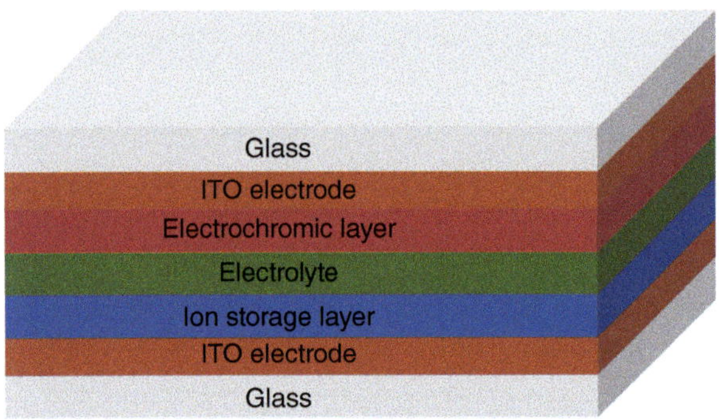

Figure 13.2 Structure of electrochromic smart windows.

composed of a transparent electrode, a light-emitting layer, an electrolyte, an ion storage layer, and another transparent electrode from top to bottom [12].

Taking the cathode electrochromic material WO_3 as an example, if no voltage is applied across the electrodes, the electrochromic layer is colorless or light in color. When a negative voltage is applied on the WO_3 and a positive voltage is applied on the ion storage layer, the lithium ions in the ion storage layer pass through the electrolyte to the color-changing layer under the action of the electric field and enter into the gap of the WO_3 lattice to form $LiWO_{3-x}$. In the process of tungsten ions having positive hexavalent to positive pentad, electrons undergo transitions and absorb photons, causing the color-changing layer to change from a transparent state to a colored state. This is the widely accepted ion/electron injection model, but there are still some phenomena that the model cannot explain [13]. Therefore, the research on the electrochromic mechanism needs devoted more effort.

13.1.2 The Materials for Electrochromic Windows

Electrochromic materials can be divided into organic and inorganic kinds. Many kinds of transition metal oxides and lanthanide metal oxides have electrochromic

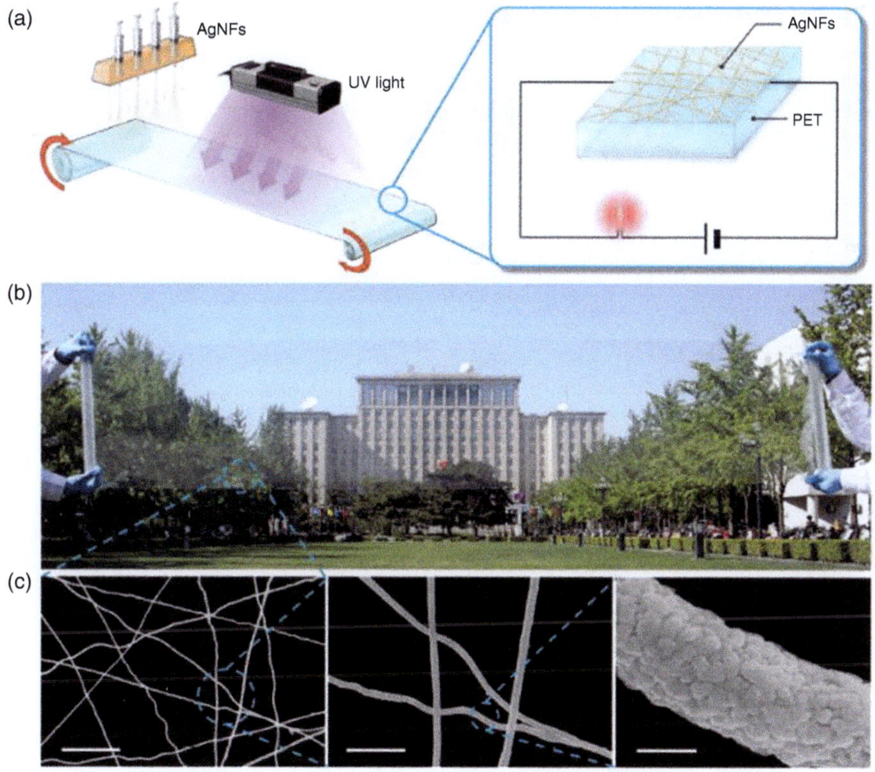

Figure 13.3 The R2R production. (a) Schematic and detail. (b) Photograph of a roll of transparent AgNF electrode material. (c) SEM images of AgNF network (scale bar: 10 and 2 μm and 300 nm). Source: (a) Lin et al. [16]. © 2017, John Wiley & Sons, (b) Reproduced with permission of Lin et al. [16].

effects. Based on the state of the material when it is colored, it can also be divided into cathodic and anodic electrochromic materials. Typical cathodic materials are WO_3, MoO_2, V_2O_5, and TiO_2, and anodic materials include IrO_2, Rh_2O_3, NiO, and Co_3O_4 [14]. WO_3 is a kind of inorganic electrochromic material and has been investigated for a long time and successfully commercialized. It can change from colorless to blue in electric field. Inorganic electrochromic materials exhibit the advantage of chemical stability, but the variety of this kind of material leads to limited variable colors and restrains further development [15].

S. Lin et al. reported a roll-to-roll production method of transparent silver nanofiber network electrodes, which can be used in flexible electrochromic smart windows [16]. Figure 13.3 illustrates the procedure of the roll to roll (R2R) production method. The $AgNO_3$ precursor solution was rapidly stretched into fibers via a jet stream and deposited onto the substrates to form the precursor fiber network. At the same time, UV irradiation was performed on the deposited precursor fiber network for the reduction of Ag^+. The fabricated AgNF network demonstrated excellent conductivity and transmittance. Its sheet resistance vs optical transmittance (550 nm)

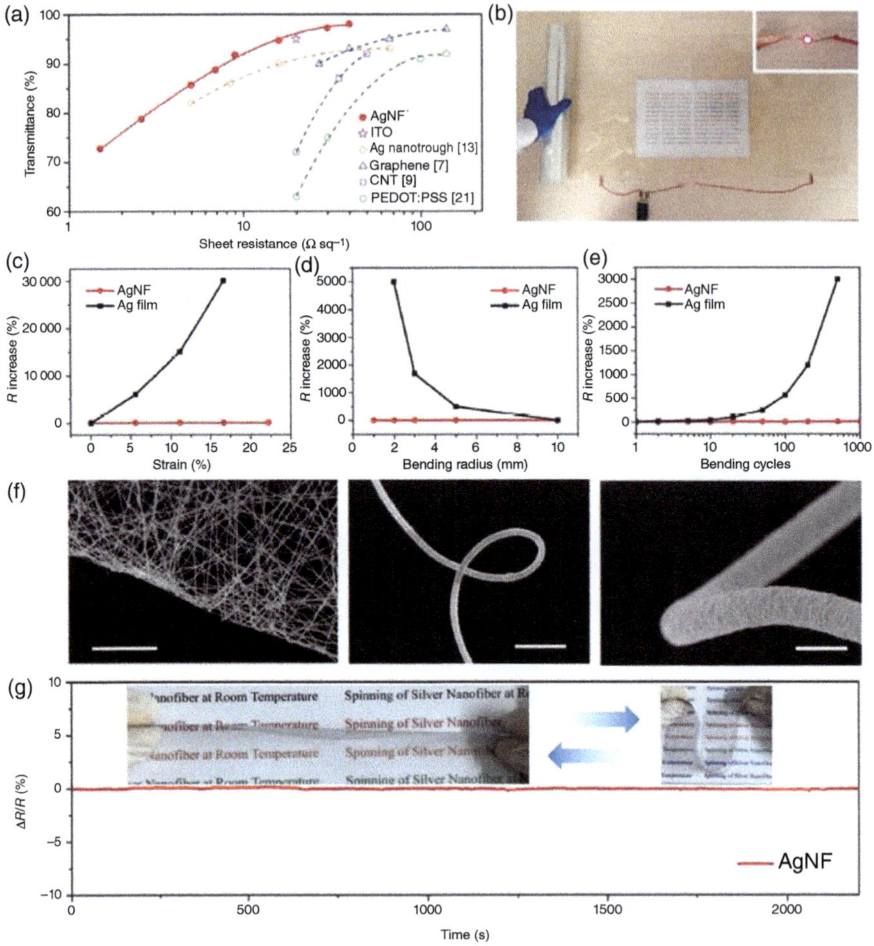

Figure 13.4 Electrical, optical, and mechanical properties of AgNF network. (a) Sheet resistance vs optical transmittance (550 nm) of the AgNF network and other transparent electrodes. (b) Roll of transparent AgNF electrode material used to illuminate an LED. The electrode has a transmittance of 94% (550 nm) and sheet resistance of 15 Ω/sq. (c) Sheet resistance vs uniaxial strain for a stretchable transparent electrode consisting of AgNF network and Ag film on PET substrates. (d) Sheet resistance versus bending radius for bendable transparent electrodes that consisted of AgNF network and Ag film on PET substrates. (e) Variations in the resistance of AgNF network and Ag film on PET substrate, which were both bent to a radius of 4 mm for a 1000 times. (f) SEM images of the bending microstructure of a single AgNF (scale bar: 100 μm, 2 μm, and 500 nm). (g) Stability testing of a PET/AgNF/PET conducting tape during repeated bending. Source: (a, d, e) Lin et al. [16]. © 2017, John Wiley & Sons, (b, f, g) Reproduced with permission of Lin et al. [16].

is comparable with traditional transparent electrodes such as nanotroughs, ITO, graphene, CNTs, and PEDOT:PSS as shown in Figure 13.4a. Figure 13.4b shows a roll of flexible transparent AgNF/polyethylene terephthalate (PET) electrode with a sheet resistance of 15 Ω/sq and 95% transmittance. Figure 13.4c–g shows the porous electrode can maintain its stability and conductivity after it was strained

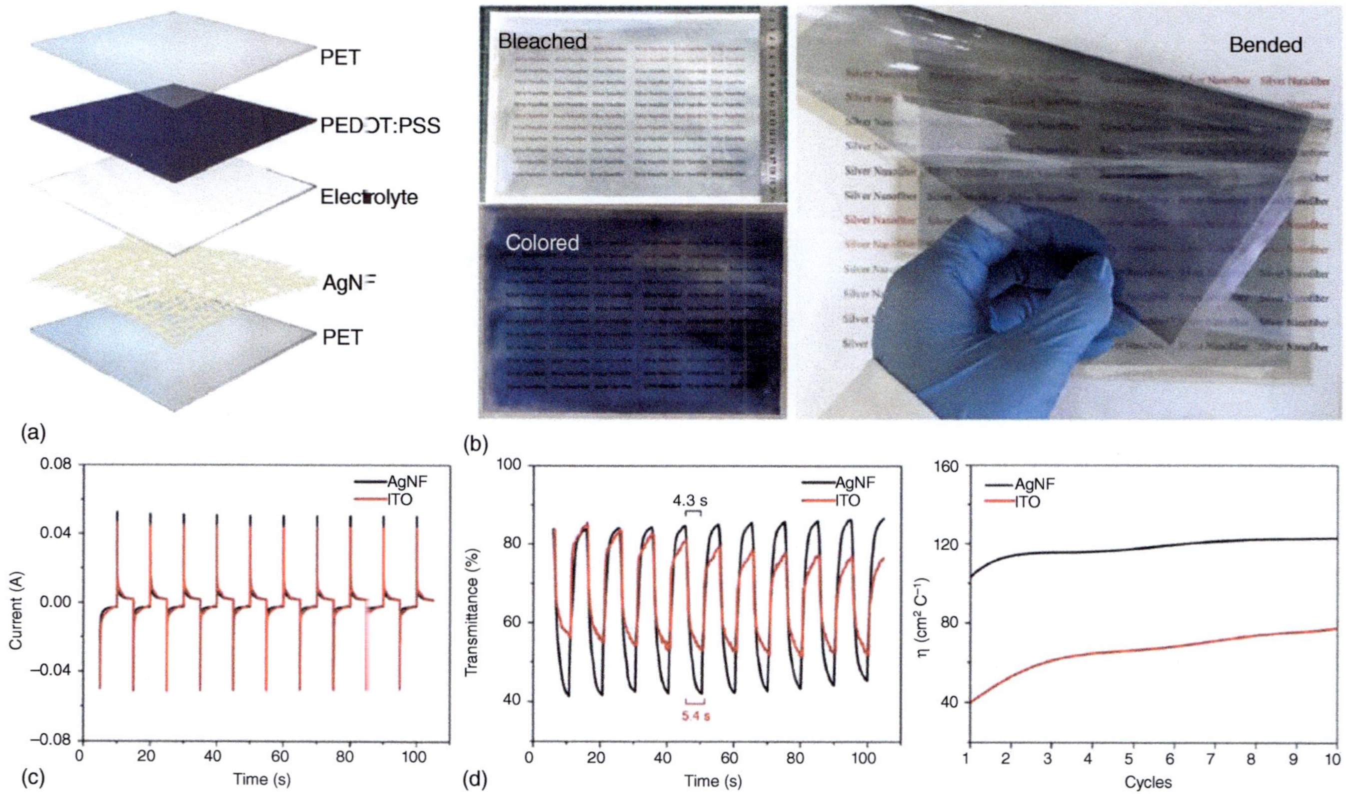

Figure 13.5 Fabrication and properties of AgNF ECSW. Structural schematic (a) and photographs (b) of a flexible A4-sized AgNF ECSW. Current response (c) and transmittance responses (d) during coloring–bleaching cycles of an AgNF ECSW and ITO-based ECSW investigated via multipotential steps (MPS) testing. (e) Coloration efficiency during the first 10 colored–bleached cycles. Source: Reproduced with permission of Lin et al. [16]. © 2017, John Wiley & Sons.

and repeatedly bent for many times. An A4-sized flexible electrochromic smart window with AgNF network is fabricated (Figure 13.5a,b). Figure 13.5c, d show the developed electrochromic smart window (ECSW) exhibits a shorter switching time of coloration compared with that equipped with ITO electrode (4.3 v.s. 5.4 seconds). Figure 13.5e shows the coloration efficiency during the first 10 colored-bleached cycles, indicating the device with AgNF exhibits a smaller variation compared with that equipped with ITO. The decrease of transmittance is only 10% after 1000 coloring–bleaching cycles in 100 ml of 0.2 M $LiClO_4$/DMF electrolyte. Extralarge area, ultrahigh transmittance, good resistance, flexibility as well as the room-temperature R2R production process contribute to the implementation of this material in the fabrication of new-generation, high-quality flexible ECSWs.

13.1.3 Prospects

According to statistics, building energy consumption accounts for 28% of total social energy consumption. Windows are the main way for building energy loss, for 30–50% is lost from the windows [17]. The situation of China is even worse, because 85% of the 44 billion square meters buildings are equipped with ordinary glass. These glasses, with an average heat transfer coefficient of 3.5 W/(m^2 K), often display poor thermal insulation performance and high energy consumption. This directly results of the average building energy consumption in China three times bigger than that of developed countries [18]. In response to this situation, the Ministry of Housing and Urban Rural Development has proposed that by 2020, all new buildings will need to achieve the goal of 65% energy saving. To achieve this goal, the application of energy-saving windows has become an inevitable choice [13].

After decades of development, companies that are preparing to commercialize large-area smart windows include Saga Glass and Soladigm in America, Saint-Gobain in France, E-control-Glass and Gesimat in Germany, ChromoGenics AB in Sweden, and Asahi Glass and Hitachi Chemical in Japan.

13.2 Dimmable Rearview Mirror

Auto-dimming rearview mirror for vehicles is the most successful commercial product of ECDs [1, 3]. It achieves anti-glare effect by adjusting the reflection of light. It works by tuning the voltage applied to the ECD through the optical sensor on the rearview mirror and thus changing the color of the device. In this way, when the sunlight is strong in the daytime or the rear car opening headlights at night, the driver can clearly see the rear situation through the dimmable mirror to ensure safety. The effect of auto-dimming mirror is shown in Figure 13.6.

Gentex is the world's largest rear mirror producer, which has sold millions of devices, accounting for more than 90% of the global market [18]. The Schott company from Germany applied WO_3 and NiO as the electrochromic layer and produced all-solid rear mirrors. This kind of device can be equipped on trucks and vans. At present, many cars have been installed with rearview mirrors, including BMW, Mercedes-Benz, Lexus, Ferrari, Daewoo, and so on.

Figure 13.6 The effect of auto-dimming mirror. Source: Reproduced with permission of Carpi et al. [3]. Copyright, Elsevier.

13.3 Sensors

13.3.1 Application of Electrochromic Sensors on Food Preservation

Food safety is an important issue of human concern. Temperature and oxygen concentration are two important factors, which cause food to deteriorate. There are strict regulations on the temperature and oxygen concentration for food during transportation and storage. For example, the Europe Union (EU) requires that the temperature for pasteurized milk should not exceed 4 °C for storage, and the storage time should not exceed three hours if the environment temperature exceeds 9 °C [19]. In addition, plastic-sealing food, filling inert gas, and adding oxygen adsorbent are ways to isolate food from oxygen and avoid deterioration. Investigators have found that electrochromic materials can change color in different temperatures or oxygen concentrations. Thus, they exhibit great potential to indicate changes in temperature and oxygen concentration of food during storage or transportation.

D. Galliani et al. fabricated a time/temperature indicator through latent pigment approach for the use of handling and storage of perishable goods [19]. The latent pigment approach enables the color of time-temperature indicator (TTI) to be irreversible. As shown in Figure 13.7, they applied a very peculiar class of pigments – 1, 3-squaraines. The change of the optical gap due to the latent pigment process tunes the color of the molecular from blue to colorless. After depositing the derivative 3 on a microscope slide, they characterized the time/temperature response of the

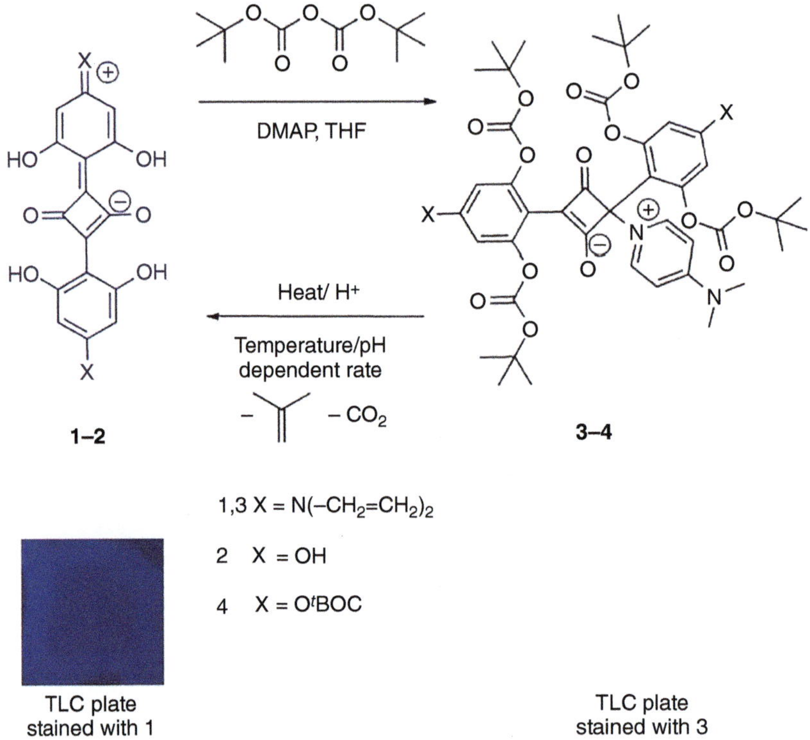

Figure 13.7 Reaction conditions for the preparation of latent pigment 3–4 starting form squaraines 1–2 along with the corresponding cleavage conditions. The two pictures show thin layer chromatography (TLC) plates stained with squaraine 1 and the corresponding latent pigment 3. Source: Reproduced with permission of Galliani et al. [19]. © 2015, John Wiley & Sons.

developed TTI by UV–vis reflectivity. Figure 13.8a display the development of the reflectance spectrum at ambient temperature with different elapsed times. The variation of the reflectance at 638 nm is due to the concentration of the cleaved molecules. Figure 13.8b exhibits if the film is kept in 25 °C for three hours at room temperature (r.t.). The cleavage reaction can be clearly detected by naked eyes as the development of deep blue color. Figure 13.8c–e displays a series of pictures of the same substrate stained with derivative 3 and taken at different times over a period of three hours at r.t. Figure 13.9 displays after changing the substituent group. The obtained derivative 4 will take a much longer time to get a color change. Such label can be applied for goods withstanding longer storage time at r.t. This paper proved the availability of applying squaraines as active material to fabricate very versatile, programmable, cheap, and simple TTI.

Food products are often packaged in modified atmosphere (MA) packaging for the purpose of extending the shelf life. MA typically involves low oxygen (<0.1–2%) and high carbon dioxide to preserve food from decay. L. Roberts et al. evaluated the application of electrochrome-based triggered oxygen indicator in MA packaging [20]. The results showed the polyviologen electrochromes were

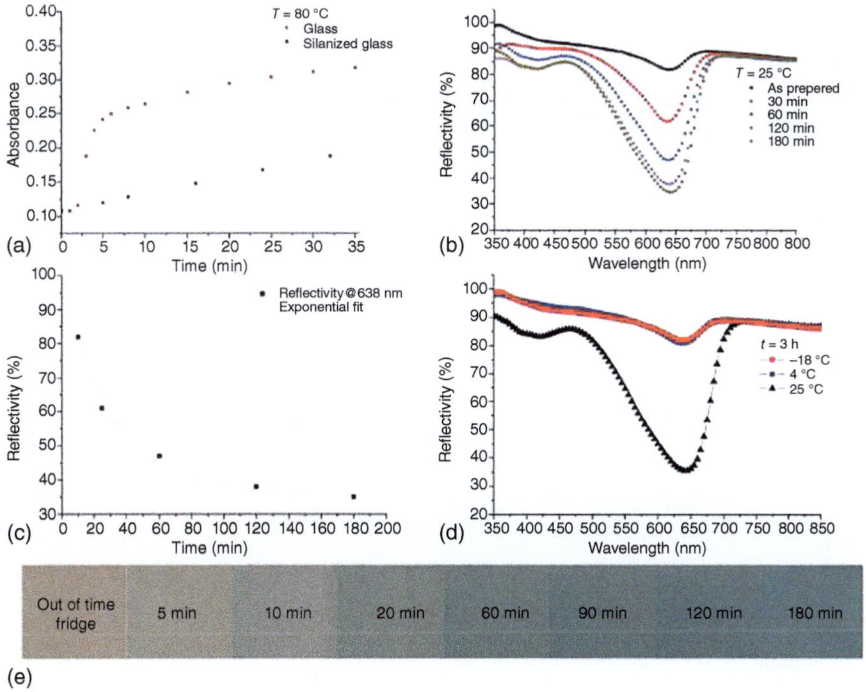

Figure 13.8 Reflectivity spectra of a thin film of 3 adsorbed on a TLC silica plate and stored for (a) three hours at −18 °C (red), 4 °C (blue), and 25 °C (black). (b) Stored at room temperature (r.t.) (25 °C): as-prepared (black), 30 seconds (red), 60 seconds (blue), and 180 seconds (brown). (c) Reflectivity at 638 nm as a function of time for a thin film of 3 on a TLC plate at 25 °C (black) and corresponding exponential fit (red dashed line). (d) Absorbance at 650 nm for a wire bar-coated thin film of 3 on glass (red) and silanized glass (black) at 80 °C. (e) Picture taken at different times of a three-stained TLC plate at 25 °C. Source: (a–d) Galliani et al. [19]. © 2015, John Wiley & Sons, (e) Reproduced with permission of Galliani et al. [19]. Copyright © Wiley.

much more visually impressive than methylene blue after exposure to UV light at low oxygen concentration package (>0.1%) as shown in Figure 13.10. This can be attributed to the lighter color of polyviologen than that of methylene blue in the oxidized state. To evaluate the sensitivity of the inks to oxygen, the initial reflectance transmission spectrum of several selected inks before activation were measured, and the results are listed in Figure 13.11. Figure 13.11 shows that the tolerance of methylene blue, thionine, poly(butylviologen dibromide), and 2,2′-dicyano-1,1′-dimethyl-4,4′-dipyridinium dimesylate to oxygen are 1.16%, 2.45%, 0.03%, and 4%, respectively. 2,2′-Dicyano-1,1′-dimethyl-4,4′-dipyridinium dimesylate shows the highest stability to oxidation, which is due to the higher anodic E° redox potential for this viologen compound. The results showed that thionine and 2,2′-dicyano-1,1′-dimethyl-4,4′-dipyridinium dimesylate can be used to produce oxygen indicators with decreased sensitivity to oxygen.

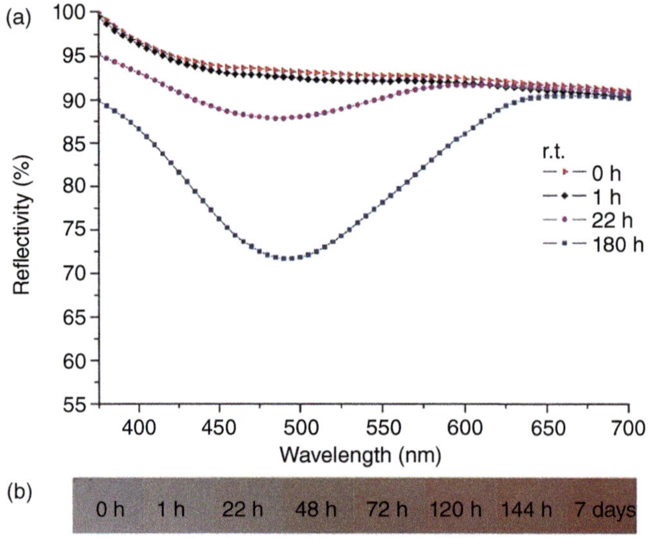

Figure 13.9 (a) Reflectivity spectra of a thin film of 4 on a TLC silica plate as a function of time at r.t. (25 °C). (b) Pictures of the sample taken at different times during a week of exposure to ambient temperature. Source: (a) Galliani et al. [19]. © 2015, John Wiley & Sons, (b) Reproduced with permission of Galliani et al. [19]. Copyright, Wiley.

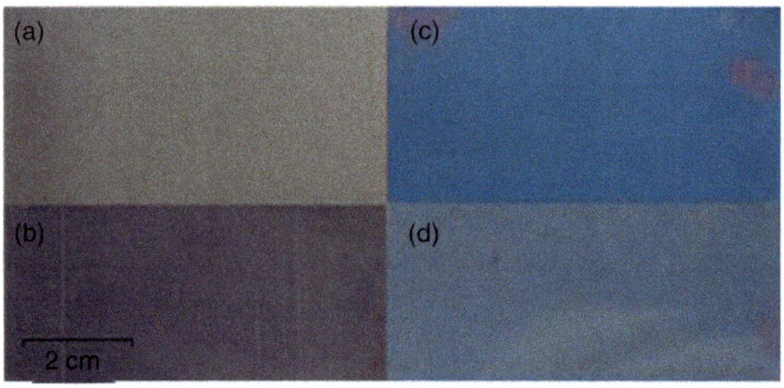

Figure 13.10 Comparison between methylene blue (c,d) and poly(*p*-xylylviologen dibromide) (a,b)-based oxygen indicators, before (a,c) and immediately after (b,d) exposed to UV light. Source: Reproduced with permission of Roberts et al. [20]. © 2011, Elsevier.

13.3.2 Application in Bio-Sensing

Electrochromic devices display a significant application in the field of bio-sensing and cosmetic treatment due to its visible color-change characteristic. This characteristic enables the sensor an omission of electric circuit, which can significantly simplify the device structure.

By integrating dye-sensitized solar cells and ECDs, Q.H. Yang et al. developed a portable self-powered photoelectrochemical device for biological ion detection as shown in Figure 13.12 [21]. Titanium dioxide and nickel-doped tungsten trioxide are separately applied as light-absorbing layer and color-changing layer for

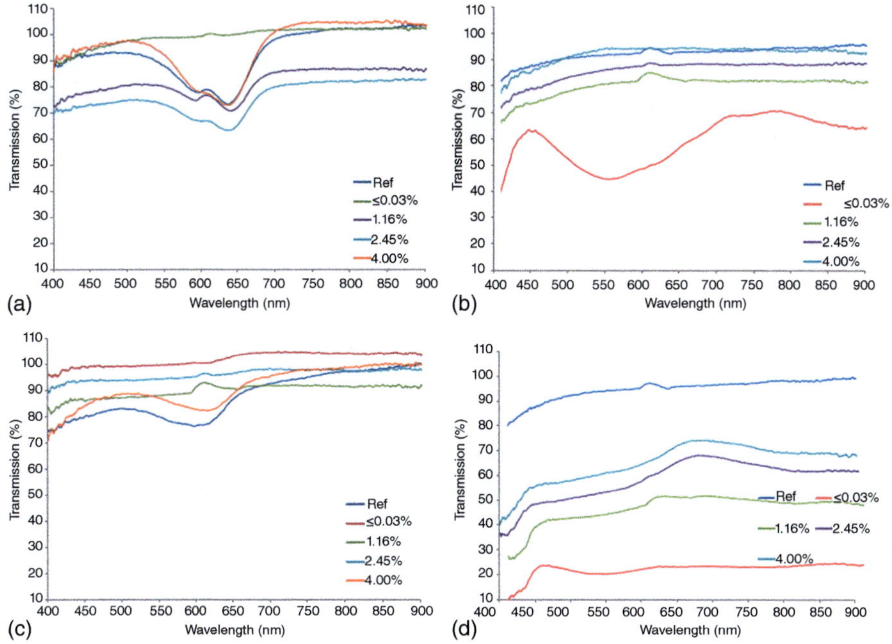

Figure 13.11 UV–vis spectra of methylene blue (a), poly(butylviologen dibromide) (b), thionine (c), and 2,2′-dicyano-1,1′-dimethyl-4,4′-dipyridinium dimesylate (d)- based ink formulations in modified atmosphere packed (MAP) pouch over a range of oxygen concentrations, from ≤0.03% to 4.00%. Source: Roberts et al. [20]. © 2011, Elsevier.

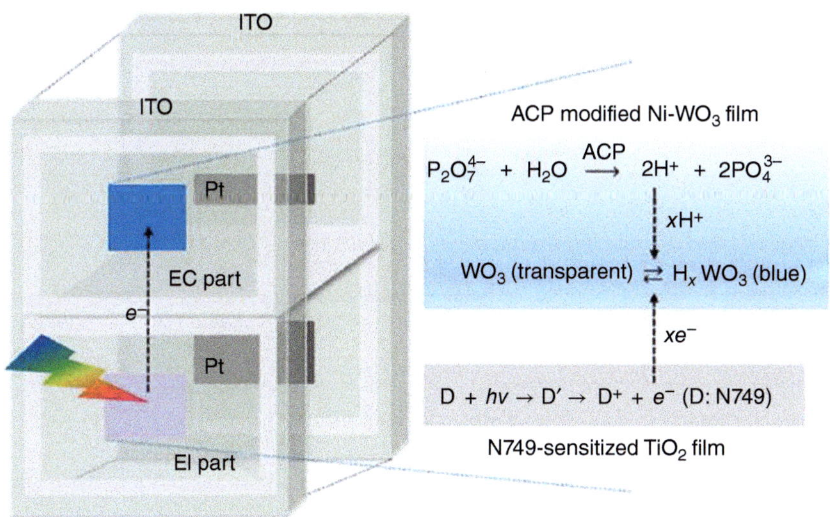

Figure 13.12 The schematic image of the self-powered photoelectrochemical device. Source: Yang et al. [17]. © 2018, American Chemical Society.

Figure 13.13 The color change of WO_3 with an increase of PPi concentration. Source: Yang et al. [17]. © 2018, American Chemical Society.

dye-sensitized solar cells and ECDs. When light illuminates the dye-sensitized solar cell, the voltage generated by the battery will power the ECD. Investigators applied the device to focus on detecting phosphate ions (PPi). The results showed that PPi and tungsten trioxide can undergo hydrolysis reaction to produce blue hydrogen tungsten bronzes as shown in Figure 13.13, which can realize the visual test of PPi. This photoelectrochemical device provides an important reference for practical applications of visualization.

M. Nishizawa and coworkers developed a fully organic disposable electrochromic time indicator as shown in Figure 13.14 [22]. The device applied fructose dehydrogenase (FDH) as anode and realizes the function of self-powered by absorbing the energy of biological cells. The conductive polymer PEDOT:polyurethane (PU) is applied as the cathode material and meanwhile acts as color-changing part. This device can be used in conjunction with a wound stick and acts as a time indicator. With time goes by, the color of PEDOT:PSS cathode will become deeper and deeper

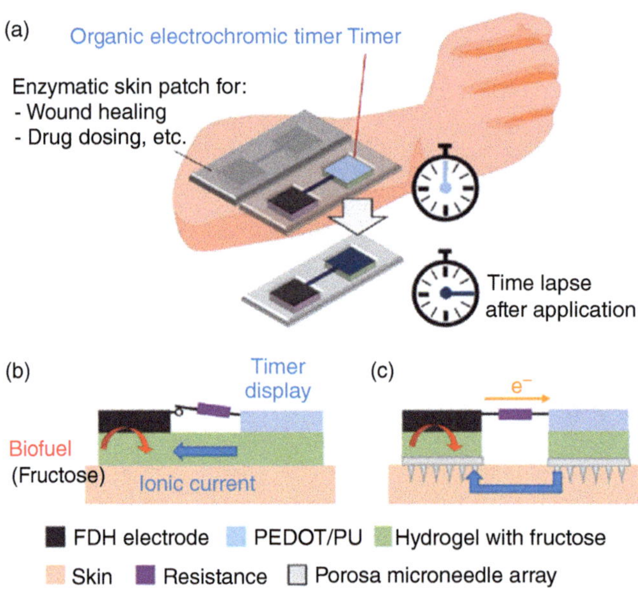

Figure 13.14 Schematic of the organic electrochromic timer for enzymatic skin patches. (a) Concept of the electrochromic timer that displays a time lapse after application of a skin patch. (b) Structure of the electrochromic timer activated by closing an electric circuit with a resistance. (c) Structure of the electrochromic timer with an automatic activation switch after application to the skin. Source: Kai et al. [22]. © 2019, Elsevier.

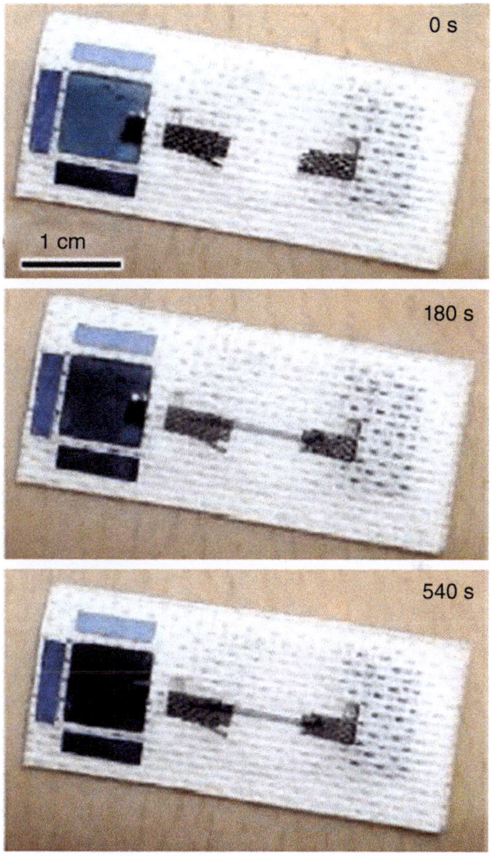

Figure 13.15 The color change of PEDOT:PSS when time goes by. Source: Reproduced with permission of Kai et al. [22]. Copyright, Elsevier.

(Figure 13.15). This device has important application in the prospects of medical and cosmetic fields.

Investigators have also done much work in developing new materials for electrochromic biosensors. H. Azak et al. synthesized a polymer with dithieno(3,2-b: 2′,3′-d) pyrole (DTP)-alkoxy-NH_2 monomer by electropolymerization as shown in Figure 13.16 and studied the electrochemical, photoelectrochemical, and detection behavior to glucose oxidase of this polymer [23]. When applying different voltages across the electrodes, the device with P(DTP-alkoxy-NH2) electrochromic layer can achieve a reversible transition from red-orange to dark blue (Figure 13.17). When the device is exposed to different concentrations of glucose oxidase, it can be seen that the current exhibits a good linear relationship with glucose oxidase in the concentration range of 0.05–0.9 M (Figure 13.18). The developed polymer displays a great potential in detecting glucose oxidase in ECD.

S. Soylemez et al. synthesized the bio-polar material PTTzFr via cyclic voltammetry shown in Figure 13.19 [24]. Applying this polymer as active material, the constructed ECD can detect glucose oxidase with a lower limit of 12.8×10^{-3} mM (Figure 13.20). This developed material also exhibits a great potential in the field of detecting glucose oxidase.

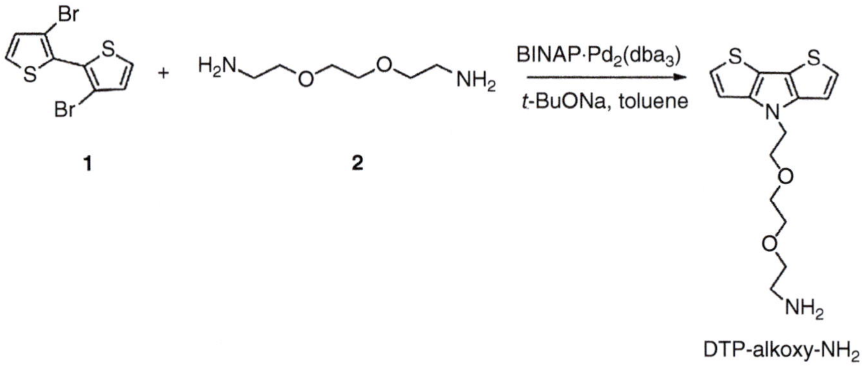

Figure 13.16 Synthesis route of DTP-alkoxy-NH_2 monomer. Source: Azak et al. [23]. © 2018, Elsevier.

Applied potential	0.0 V	0.1 V	0.2 V	0.4 V	0.6 V	0.8 V	1.0 V	1.1 V	1.20 V
colours									

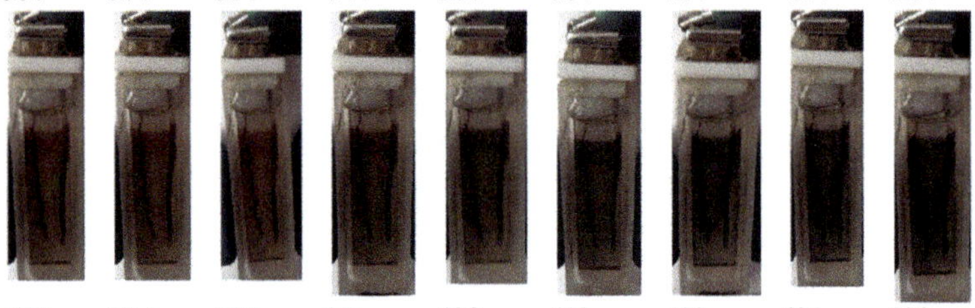

L	108.2	105.8	104.3	87.5	68.8	61.2	51.1	60.1	64.4
a	71.3	68.7	55.8	59.7	55.5	47.5	49.3	49.4	55.1
b	44.5	41.1	34.3	37.2	38.7	34.6	37.4	39.8	43.3

Figure 13.17 Electrochromic colors of P(DTP) films at various potentials and their corresponding data-related color scale. Source: Reproduced with permission of Azak et al. [23]. © 2018, Elsevier.

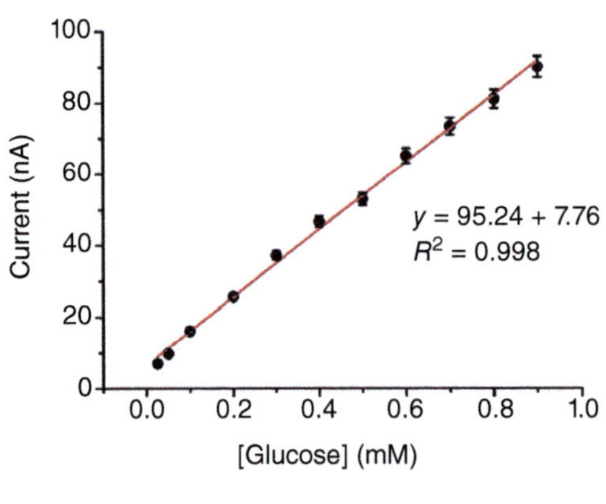

Figure 13.18 Calibration curves for the detection of glucose (in a pH 7 buffer, 0.1–0.7 M). Source: Azak et al. [23]. © 2018, Elsevier.

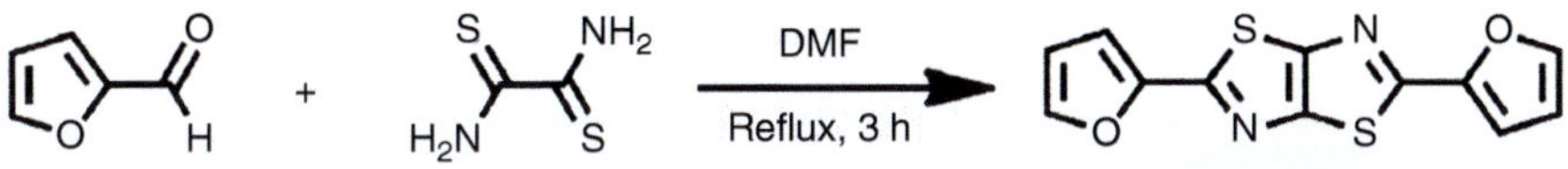

TTzFr

Figure 13.19 Synthetic procedure of the TTzFr. Source: Soylemez et al. [24]. © 2019, Elsevier.

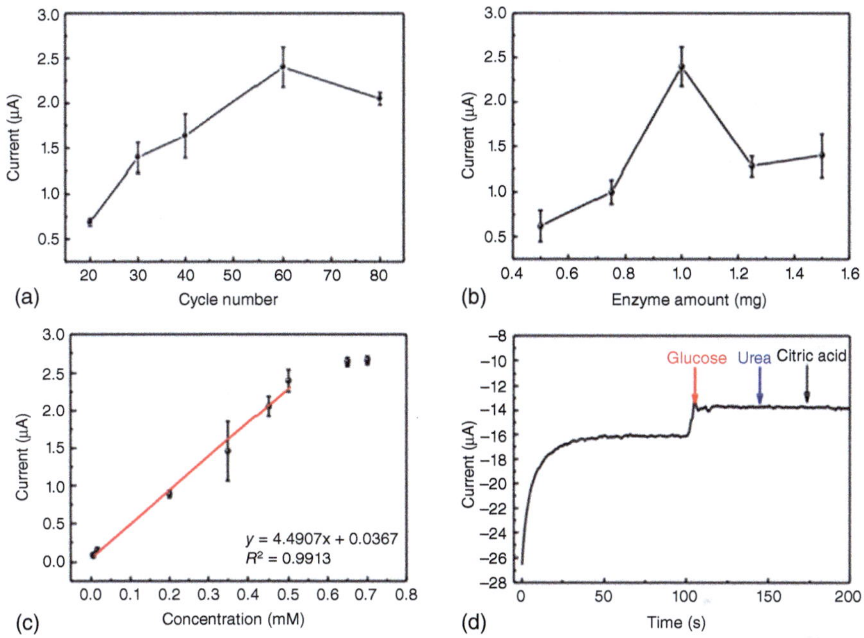

Figure 13.20 Optimization of the (a) cycle number; (b) enzyme amount on biosensor response; (c) calibration curve for glucose in 50 mM phosphate buffer, pH 7.0, 25 °C, and −0.7 V; and (d) amperometric responses of PTTzFr/GOx biosensor to glucose and different studies with citric acid and urea solution. Source: Soylemez et al. [24]. © 2019, Elsevier.

M.P. Pellitero et al. combined a sensor with an electrochromic screen and construct a self-powered visual biosensor [25]. The sensor structure is shown in Figure 13.21. The electrochromic part applied Prussian blue as the color-changing layer. The innovation of the structure is parallel arrangement of the anodic and cathodic electrodes rather than previously reported vertical array. After glucose comes into contact with the biosensor, it provides electrons for the device. Through the path from anode to cathode, electrons will choose the way with the smallest resistance, which will gradually change the color of the electrochromic layer and eventually stabilize. The stable color of the electrochromic layer at different concentrations is shown in Figure 13.22. The device can be self-powered, and the detection results are visualized, which has a huge application prospect in the detection of glucose.

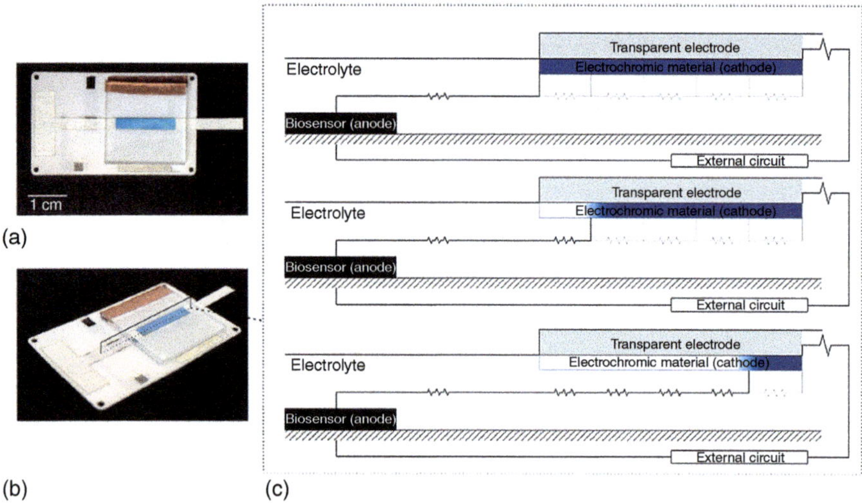

Figure 13.21 Electrochromic biosensor and its operating principle. (a,b) Photographs of the actual electrochromic biosensor device. (c) Schematic representation of the device side view, with the biosensor (anode) placed to the left of the cathode (electrochromic display). As the analyte concentration or reaction time increases, the charge consumed at the cathode increases, resulting in a gradual color change along the electrode length, following the path of least resistance. Source: Reproduced with permission of Pellitero et al. [25]. © 2017, Royal Society of Chemistry.

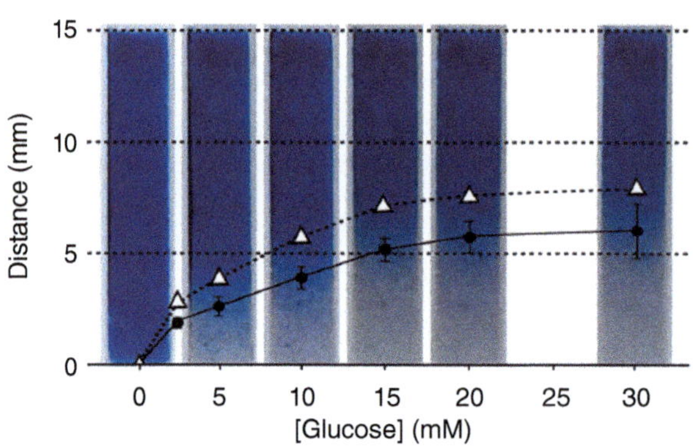

Figure 13.22 Captures of the display 30 seconds after the addition of different glucose solutions. Source: Pellitero et al. [25]. © 2017, Royal Society of Chemistry.

13.4 The Application of Electrochromic Device in Display

Bo-Han Chen et al. developed an inkjet-printing method to fabricate electrochromic thin film patterns with various colors by adjusting digital-printing schemes [26]. Two metallo-supramolecular polymers (MEPEs) separately apply Fe and Ru as centric metallic ligands and are printed on transparent conductive sheets to form ECDs.

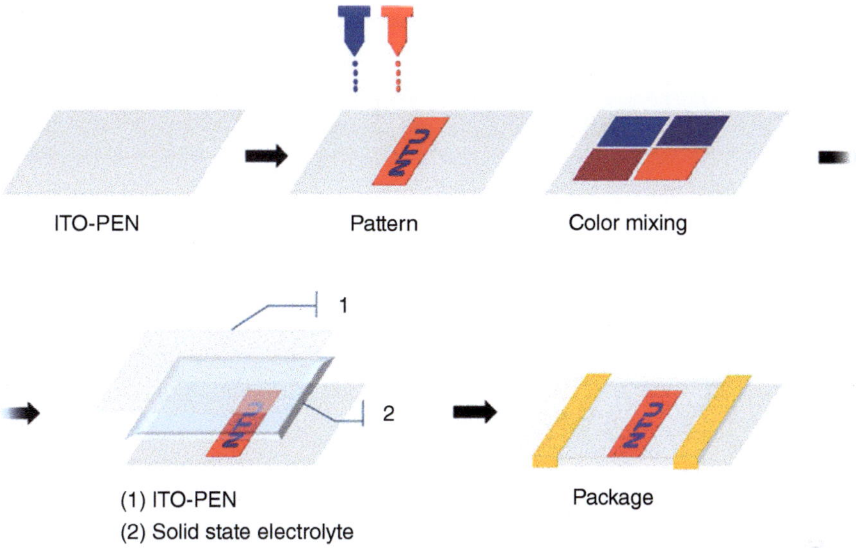

Figure 13.23 Schematic diagram of the experimental steps involved in the study. Source: Chen et al. [26]. © 2015, American Chemical Society.

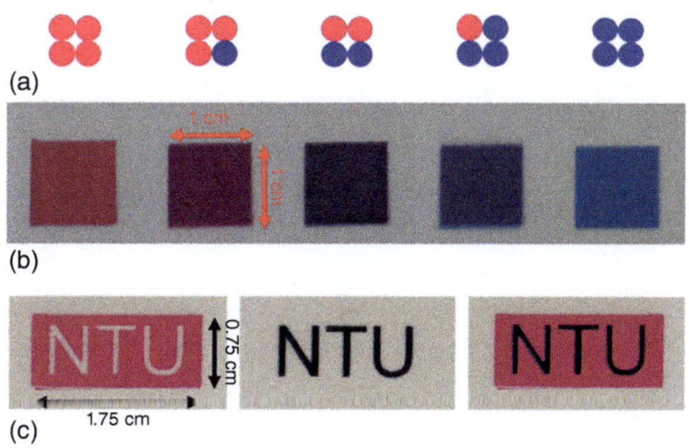

Figure 13.24 (a) Color-mixing scheme and (b) resulting EC material thin films with different mixing ratios of Ru–MEPE to Fe–MEPE. (c) Multicolor patterns of EC material thin films. Source: Reproduced with permission of Chen et al. [26]. © 2015, American Chemical Society.

The fabrication procedure is shown in Figure 13.23. From Figure 13.24, we can see that print patterns with multiple colors with excellent precision can be achieved by tuning the dot ratios between Ru–MEME and Fe–MEPE. The fabricate ECDs show a high contrast with a transmittance difference of 40.1% at the characteristic absorption wavelength of 580 nm with a response time of two seconds (Figures 13.25 and 13.26). The ECDs were also tested under bending condition with radius of curvature of 1 cm. The device shows nearly the same color changes and response time as those flat conditions. This work shows that inkjet-printing method exhibits great potential in fabricating EC thin films with specific colors and patterns. The fabrication process can be extended to fabricated flexible multicolor ECD for display applications.

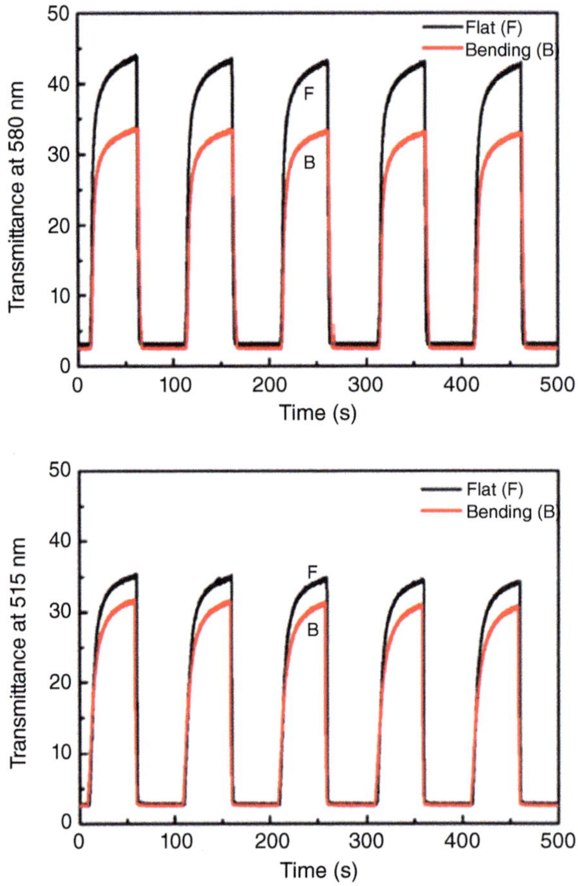

Figure 13.25 Transmittance change (ΔT, %) (a) at 580 nm of the Fe–MEPE ECD and (b) at 515 nm of the Ru–MEPE ECD. Source: Chen et al. [26]. © 2015, American Chemical Society.

S. Takamatsu et al. reported a writing pad combining a light-pen input device with an electrochromic screen as shown in Figure 13.27 [27]. The light-pen is a sensor array, based on a blend of bacterioRhodopsin (bR) and the conducting polymer PEDOT:PSS. The electrochromic display is based on PEDOT:PSS. The application of organic material can ensure the device to be prepared on a flexible substrate. The results showed the successfully recognized various input patterns, even when bent (Figure 13.28). The integration of bR with conductive polymer will lead to large-area flexible optical systems with functions in the future by changing to the other types of bR, which have different optical processing natures.

13.5 Other Applications of OEC

Z.N. Bao et al. was inspired by the color change of chameleon skin and developed a device that can change color according to pressure [28]. The device structure is shown in Figure 13.29. The device is prepared by a full solution method. The

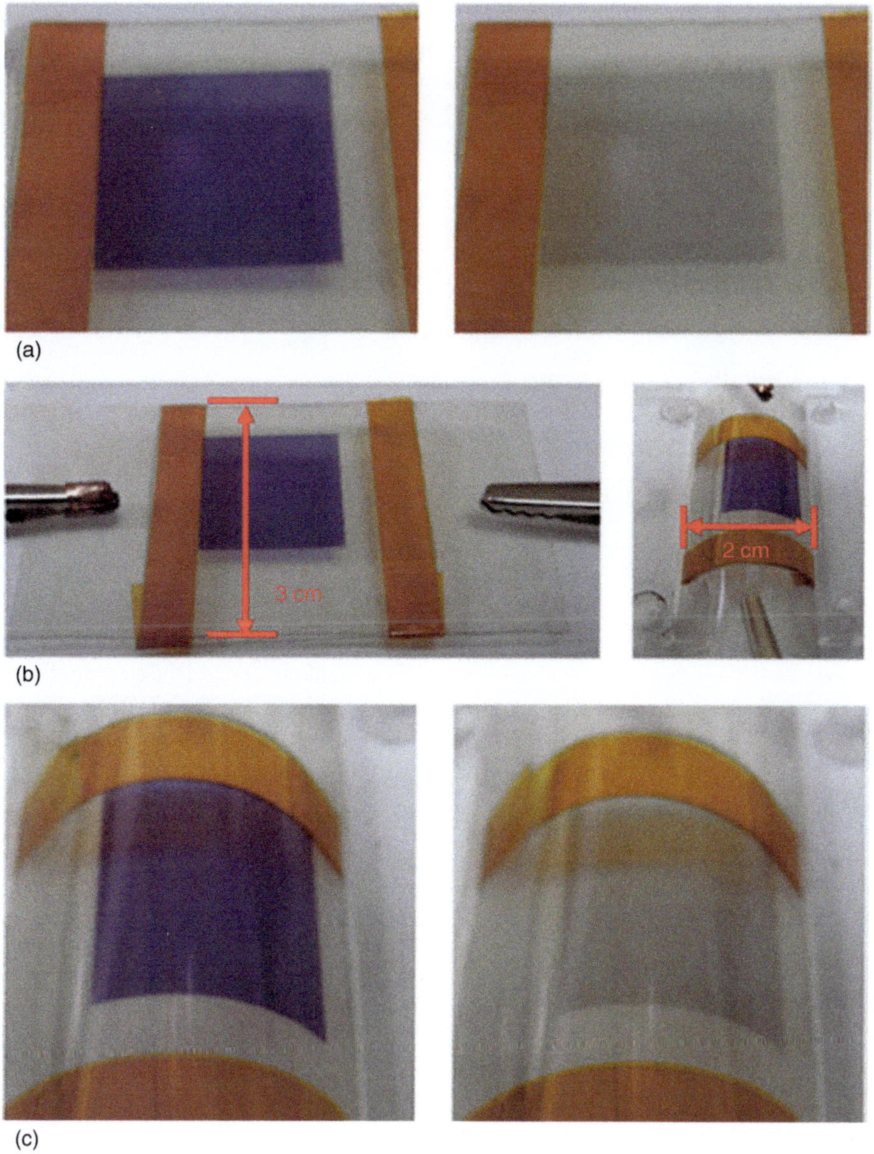

Figure 13.26 (a) Color change between −3 and 3 V of Fe–MEPE ECD under flat condition. (b) The flexible ECD was bent with a radius of curvature of 1 cm. (c) Color change under applied voltage. Source: Reproduced with permission of Chen et al. [26]. © 2015, American Chemical Society.

pressure sensor (PS) part of the device has a wide range of resistance values, which can induce a large range of pressure (Figure 13.30). The device can indicate the pressure according to the color, and displays applications in the fields of wearable devices, artificial limbs, and intelligent robots.

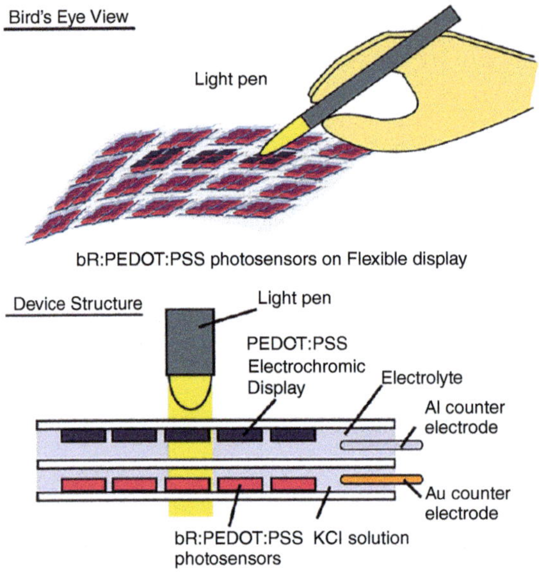

Figure 13.27 (a) Schematic illustration of the device concept. A photosensor array recognizes input from a light pen. The input is displayed on an integrated display. (b) Device layout (not to scale). An electrochromic display element and a photosensor array were assembled together with their pixels in registry. Source: Takamatsu et al. [27]. © 2008, Elsevier.

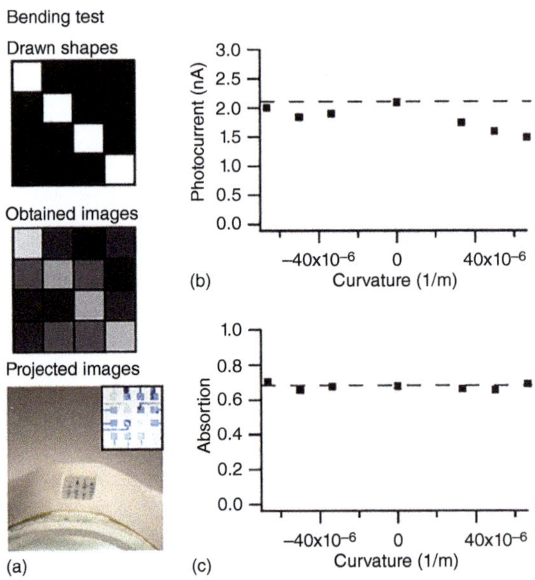

Figure 13.28 The bending test of the device. (a) Demonstration of device operation when bent to the curvature radius of 30 mm, (b) photosensitivity under bending, and (c) light absorption under bending. Source: Reproduced with permission of Takamatsu et al. [27]. © 2008, Elsevier.

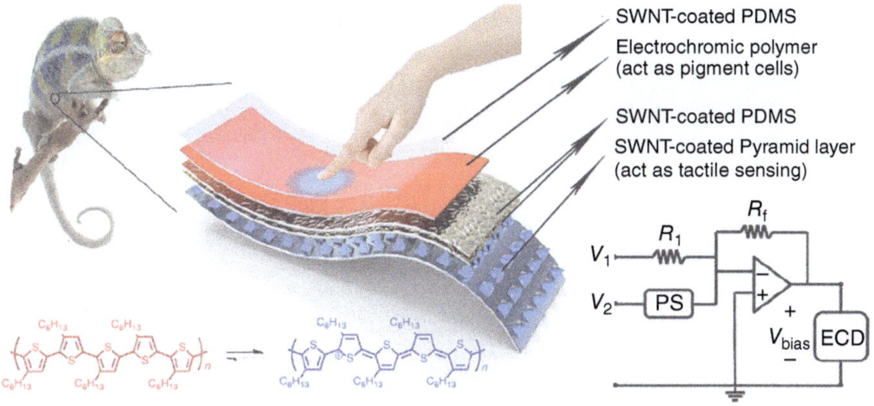

Figure 13.29 Illustration of the concept of a chameleon-inspired e-skin. Also shown are (bottom left) the structures of the neutral and oxidized states of the electrochromic polymer in P3HT and (bottom right) a schematic of the circuit layout (PS, pressure sensor; ECD, electrochromic device). Source: Reproduced with permission of Chou et al. [28]. .Licenced under CC BY 4.0.

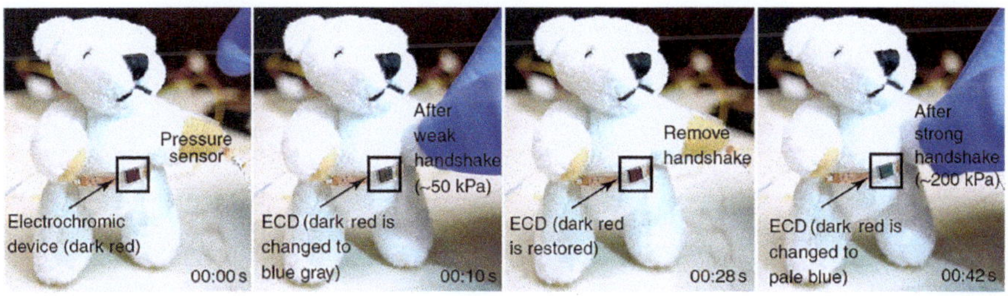

Figure 13.30 An interactive color-changing and tactile-sensing e-skin. Sequential images of a teddy bear show the expression of tactile sensing into visible color changes. In specific, the original color of the ECD is changed from dark red to blue gray upon a weak applied pressure (~50 kPa), reverts back to dark red upon pressure release, and changes to pale blue upon a strong applied pressure (~200 kPa; left to right images). The number at the bottom right corner of each image indicates the elapsed time. Source: Reproduced with permission of Chou et al. [28]. Copyright © Nature Publishing Group.

H.T. Yu et al. synthesized three types of donor–acceptor compounds as shown in Figure 13.31 [29]. Compared with G1 and G2, G3 engineered with the longest side chain exhibits excellent solubility in common solvents such as DCM and toluene (TOL). G3 displays strong absorption in the NIR and IR regions and exhibits unique vegetable green neutral and sand-tone oxidized color (Figure 13.32). Its switching time at fully neutral state and fully oxidized state are only 1 and 1.5 seconds, respectively. At last, a chameleonic fabric was fabricated using G3 as the working electrode. The built device can reversibly change between green and sand-color at an applied potential as a camouflage clothing (Figure 13.33). The G3 shows great potential in the field of military adaptive camouflage.

OEC device has also been successfully applied in mobile phones by One Plus Corporation in 2019[30]. The mobile phone carries a new feature of hidden camera

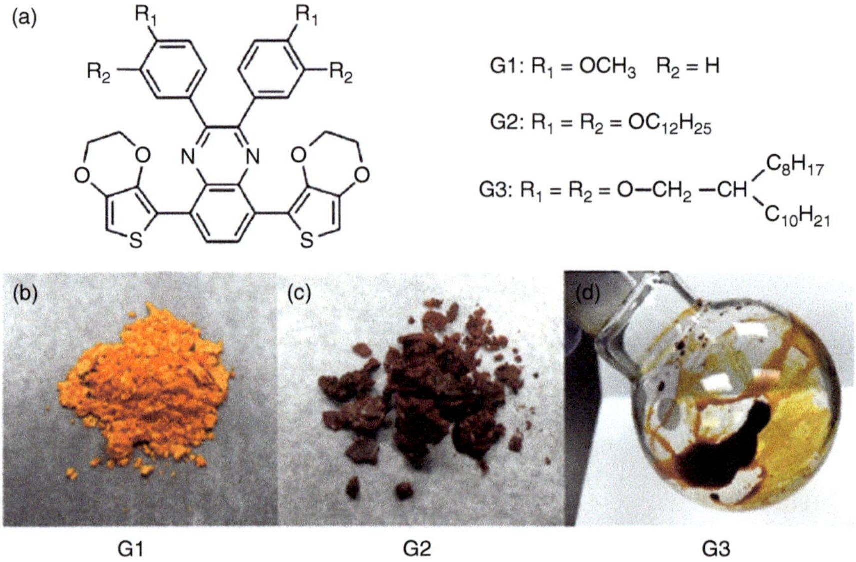

Figure 13.31 (a) The molecular structure of G1, G2, and G3 and (b–d) the appearance of G1, G2, and G3 monomers, respectively. Source: (a) Yu et al. [29]. © 2016, Royal Society of Chemistry, (b) Reproduced with permission of Yu et al. [29]. Copyright, Royal Society of Chemistry.

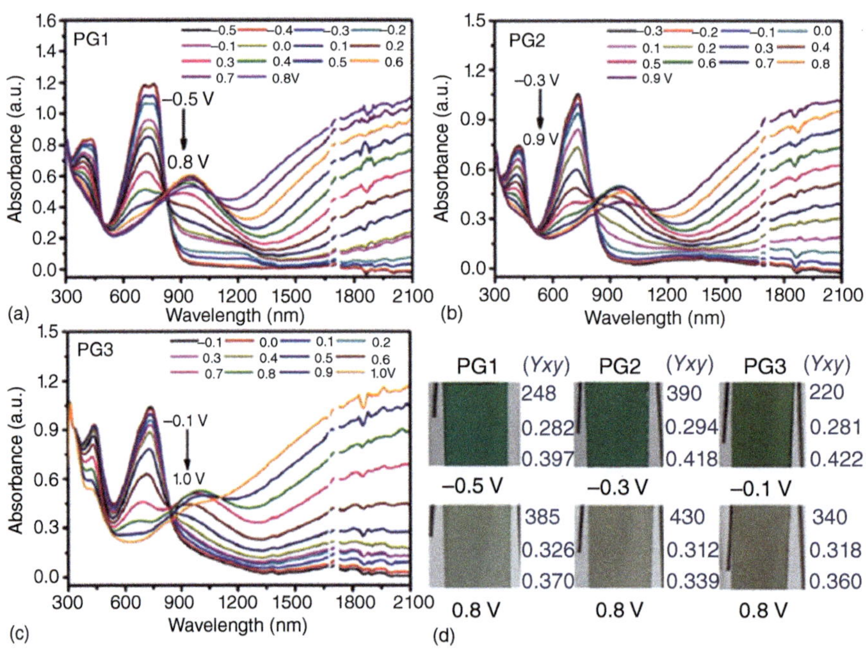

Figure 13.32 Spectroelectrochemistry of PG1, PG2, and PG3 film on an ITO-coated glass slide (0.5 cm width) in a monomer-free 0.1 M TBAPF6/CAN electrolyte–solvent couple at applied potentials: (a) −0.5 to 0.8 V, (b) −0.3 to 0.9 V, (c) −0.1 to 1.0 V. (d) Photographs and corresponding (Y, x, y) values of PG1, PG2, and PG3 film on an ITO-coated glass slide in the fully neutral and oxidized states. Source: (a–c) Yu et al. [29]. © 2016, Royal Society of Chemistry, (d) Reproduced with permission of Yu et al. [29]. Copyright, Royal Society of Chemistry.

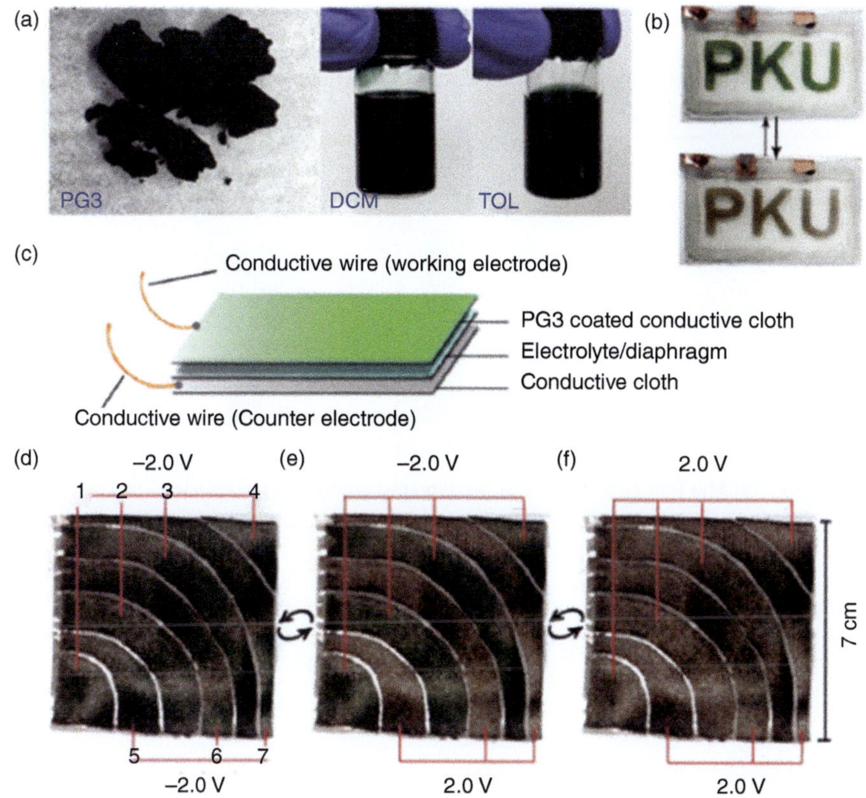

Figure 13.33 (a) The appearance of PG3 and in a DCM or TOL solution. Images of the PG3 camouflage fabric with space between each piece: (b) example of the patterned device in neutral and oxidized states, (c) fabric chameleonic electrochromic device cross-selection for a typical device used in this study, (d) all pieces 1–7 as the working electrode in their neutral, (e) 1–4 in neutral and 5–7 in oxidized, and (f) in oxidized states. Source: Reproduced with permission of Yu et al. [29]. © 2016, Royal Society of Chemistry.

using electrochromic technology, which has never been seen in the mobile phone industry before. The electrochromic technology successfully hides the rear camera module under the back cover of the phone and is difficult to detect when it is not needed as shown in Figure 13.34a. It only appears when the rear camera is used (Figure 13.34b). The time for the ECD transfer from black to transparent needs only 0.7 seconds. So far, this is the fastest transfer speed for ECD. The characteristic of this glass can be attributed to an invisible alloy called indium tin oxide. When ITO is applied to the glass, it can make the surface conductive and the charge can change the composition of the glass molecules and results in less light passing through. Huawei's latest authorized all-in-one mobile phone also applies this hidden rear camera as shown in Figure 13.35. This technology will be largely applied in the market of mobile phones in the future.

(a) (b)

Figure 13.34 (a) The rear camera cannot be detected when it is not needed. (b) The camera appeared when it is used. Source: Hong Meng.

Figure 13.35 The schematic images of the front and rear surfaces of the one-in-all mobile phone. Source: Based on OPPO Guangdong Mobile Communications Co. [30].

References

1 Kim, H.N., Cho, S.M., Ah, C.S. et al. (2016). Electrochromic mirror using viologen-anchored nanoparticles. *Materials Research Bulletin* 82: 16–21.

2 Kim, Y., Han, M., Kim, J., and Kim, E. (2018). Electrochromic capacitive windows based on all conjugated polymers for a dual function smart window. *Energy & Environmental Science* 11 (8): 2124–2133.

3 Carpi, F. and De Rossi, D. (2006). Colours from electroactive polymers: electrochromic, electroluminescent and laser devices based on organic materials. *Optics & Laser Technology* 38 (4-6): 292–305.

4 Yun, T.G., Kim, D., Kim, Y.H. et al. (2017). Photoresponsive smart coloration electrochromic supercapacitor. *Advanced Materials* 29 (32): e1706641–e1706647.

5 Lahav, M. and van der Boom, M.E. (2018). Polypyridyl metallo-organic assemblies for electrochromic applications. *Advanced Materials* 30 (41): e1706641.

6 Osterholm, A.M., Shen, D.E., Kerszulis, J.A. et al. (2015). Four shades of brown: tuning of electrochromic polymer blends toward high-contrast eyewear. *ACS Applied Materials & Interfaces* 7 (3): 1413–1421.

7 Zheng, R., Wang, Y., Jia, C. et al. (2018). Intelligent biomimetic chameleon skin with excellent self-healing and electrochromic properties. *ACS Applied Materials & Interfaces* 10 (41): 35533–35538.

8 Cai, G., Eh, A.L.-S., Ji, L., and Lee, P.S. (2017). Recent advances in electrochromic smart fenestration. *Advanced Sustainable Systems* 1 (12): 1700074–1700096.

9 Granqvist, C.G., Arvizu, M.A., Bayrak, P.I. et al. (2018). Electrochromic materials and devices for energy efficiency and human comfort in buildings: A critical review. *Electrochimica Acta* 259: 1170–1182.

10 Baetens, R., Jelle, B.P., and Gustavsen, A. (2010). Properties, requirements and possibilities of smart windows for dynamic daylight and solar energy control in buildings: a state-of-the-art review. *Solar Energy Materials and Solar Cells* 94 (2): 87–105.

11 Dyer, A.L., Bulloch, R.H., Zhou, Y. et al. (2014). A vertically integrated solar-powered electrochromic window for energy efficient buildings. *Advanced Materials* 26 (28): 4895–4900.

12 Lampert, C.M. (1994). Towards large-area photovoltaic nanocells: experiences learned from smart window technology. *Solar Energy Materials and Solar Cells* 32: 307–321.

13 Granqvist, C.G. (1990). Chromogenic materials for transmittance control of large-area windows. *Critical Reviews in Solid State and Materials Sciences* 16 (5): 291–308.

14 Tian, Y., Zhang, X., Dou, S. et al. (2017). A comprehensive study of electrochromic device with variable infrared emissivity based on polyaniline conducting polymer. *Solar Energy Materials and Solar Cells* 170: 120–126.

15 Xiao, Y., Zhong, X., Guo, J. et al. (2018). The role of interface between LiPON solid electrolyte and electrode in inorganic monolithic electrochromic devices. *Electrochimica Acta* 260: 254–263.

16 Lin, S., Bai, X., Wang, H. et al. (2017). Roll-to-roll production of transparent silver-nanofiber-network electrodes for flexible electrochromic smart windows. *Advanced Materials* 29 (41).

17 Liu, H.-M., Saikia, D., Wu, C.-G. et al. (2017). Solid polymer electrolytes based on coupling of polyetheramine and organosilane for applications in electrochromic devices. *Solid State Ionics* 303: 144–153.

18 Pan, M., Ke, Y., Ma, L. et al. (2018). Single-layer electrochromic device based on hydroxyalkyl viologens with large contrast and high coloration efficiency. *Electrochimica Acta* 266: 395–403.

19 Galliani, D., Mascheroni, L., Sassi, M. et al. (2015). Thermochromic latent-pigment-based time-temperature indicators for perishable goods. *Advanced Optical Materials* 3 (9): 1164–1168.

20 Roberts, L., Lines, R., Reddy, S., and Hay, J. (2011). Investigation of polyviologens as oxygen indicators in food packaging. *Sensors and Actuators B: Chemical* 152 (1): 63–67.

21 Yang, Q.H., Hao, Q., Lei, J., and Ju, H. (2018). Portable photoelectrochemical device integrated with self-powered electrochromic tablet for visual analysis. *Analytical Chemistry* 90 (6): 3703–3707.

22 Kai, H., Suda, W., Yoshida, S., and Nishizawa, M. (2019). Organic electrochromic timer for enzymatic skin patches. *Biosensors & Bioelectronics* 123: 108–113.

23 Azak, H., Yildiz, H.B., and Bezgin Carbas, B. (2018). Synthesis and characterization of a new poly(dithieno (3,2-b:2′, 3′-d) pyrrole) derivative conjugated polymer: its electrochromic and biosensing applications. *Polymer* 134: 44–52.

24 Soylemez, S., Kaya, H.Z., Udum, Y.A., and Toppare, L. (2019). A multipurpose conjugated polymer: electrochromic device and biosensor construction for glucose detection. *Organic Electronics* 65: 327–333.

25 Pellitero, M.A., Guimera, A., Kitsara, M. et al. (2017). Quantitative self-powered electrochromic biosensors. *Chemical Science* 8 (3): 1995–2002.

26 Chen, B.H., Kao, S.Y., Hu, C.W. et al. (2015). Printed multicolor high-contrast electrochromic devices. *ACS Applied Materials & Interfaces* 7 (45): 25069–25076.

27 Takamatsu, S., Nikolou, M., Bernards, D.A. et al. (2008). Flexible, organic light-pen input device with integrated display. *Sensors and Actuators B: Chemical* 135 (1): 122–127.

28 Chou, H.H., Nguyen, A., Chortos, A. et al. (2015). A chameleon-inspired stretchable electronic skin with interactive colour changing controlled by tactile sensing. *Nature Communications* 6: 8011.

29 Yu, H., Shao, S., Yan, L. et al. (2016). Side-chain engineering of green color electrochromic polymer materials: toward adaptive camouflage application. *Journal of Materials Chemistry C* 4 (12): 2269–2273.

30 OPPO Guangdong Mobile Communications Co., Ltd. (2019). Storage medium and method for controlling the electrochromic portion of electrochromic device. CN109116656A[P].

14

Commercialized OEC Materials and Related Analysis of Company Patents

14.1 General Introduction

In this chapter, we will focus on the patents with industrial production of the enterprise. Organic electrochromic industry has risen earlier. Commercialized electrochromic devices were initially reported in the early 1980s by Gentex Corporation. However, in the past few decades, the focus of the industry had changed significantly. More companies are shifting their sweet spot from finding new compounds to designing devices with new structures, which is in line with the current focus of competition in the market by the traditional technology and products to services and application changes. Although the overall trend has been like this, there have been some companies still searching for new materials.

In general, due to a variety of known factors, there are fewer types of organic electrochromic materials available for commercialization, including viologen, aniline derivatives, and thiophene compounds. In particular, viologen is the earliest and most widely used one. Originally electrochromic devices were applied on car rearview mirror to protect against the glare-causing light, but now they are increasingly used in other optical materials.

14.2 Gentex Corporation

Gentex Corporation is a company that's been established for more than 40 years. It could be traced back to a fireproof product manufacturing company founded in 1974 in Michigan. Gentex Corporation, along with its subsidiary, took a colossal 92% of the global auto-dimming interior mirror market in 2018. Behind the scenes, Gentex Corporation possesses 674 patents with the keyword of organic electrochromic. In addition, there are thousands of related patents. Gentex Corporation began to research and develop a new system of automatic rearview mirror system for automotive vehicles in 1982 and applied for the first related patent in 1987 [1]. In these patents, Gentex refers to the earliest commercial materials, viologens. The following will briefly introduce the company's patents from several aspects for learning and communication.

Organic Electronics for Electrochromic Materials and Devices, First Edition. Hong Meng.

The company's main products that Gentex Corporation deals in are drivers safety and information features, side blind zone assistance features, forward driving assistance features, and rear blind zone assistance features.

The early products of Gentex Corporation mainly focused on changing all vehicle rearview mirrors from full reflection mode to partial reflection mode to prevent people from the reflected glare. Generally, electromagnetic radiation reversible transmittance-changing device is realized by electrochromic material [2]. As of today, the anti-glare rearview mirror of the car still uses the same principle.

In previous reports, the usual device structure has four reflectors (shown in Figure 14.1). However, in order to the processing, it is desirable to invent a more simple rearview mirror. After solving some problems such as the electrode being easily corroded by the electrochromic liquid material, electrochromic material cannot get brighter and fainter on the electrode synchronously and so on. In addition to related research on circuits [4–9], Gentex proposed a new simplified structure

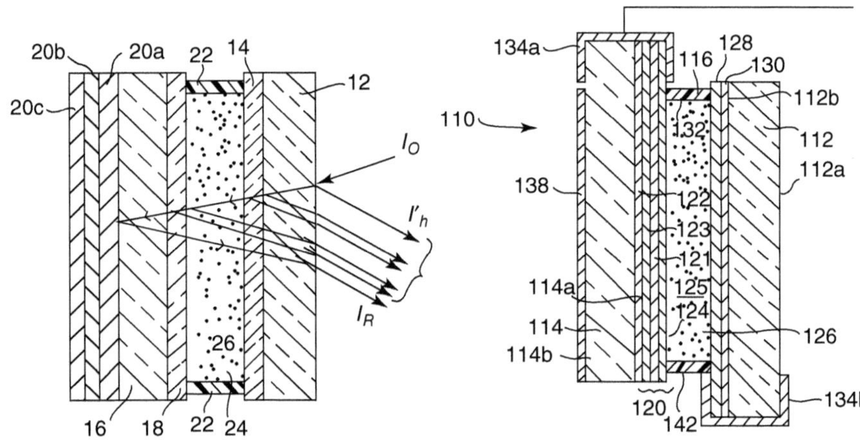

(a)

12-Front plane element
14,18-Transparent conductive coating
16-Rear plane element
22-Sealing element
24-Electrochromic medium
26-Chamber
20a-Silver layer
20b-Copper layer
20c-Protective paint

(b)

110-Motor vehicle mirror device
111a,111b-Rearview mirror element
112-Front transparent element
114-Back element
114a,114b-Element surfaces
116-Sealing element
120-Electrode
121-Reflective material
122-Substrate layer
123-Alloy layer
124,125-Chamber
126-Electrochromic medium
128-Transparent coating
134a,134b-Covering

Figure 14.1 Front view of motor vehicle rearview mirror. (a) Traditional four-reflector structure. (b) Three-reflector structure. Source: Forgette and Anderson [3].

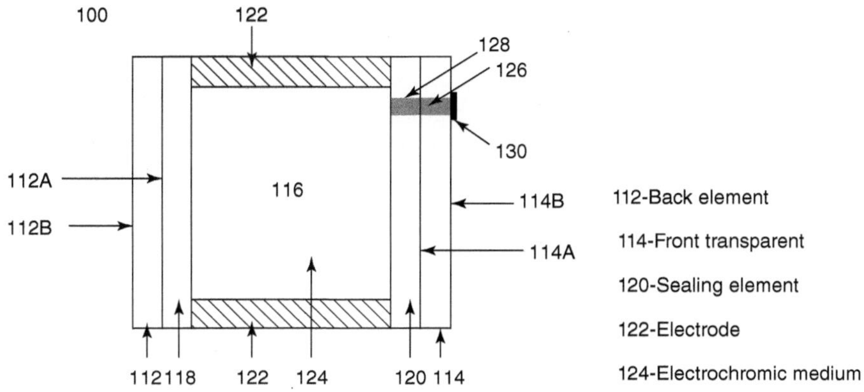

Figure 14.2 General device structure. Source: Baumann Kelvin and Rongguang [17].

[10, 11]. In this device, a layer of transparent conductive material is placed onto the mirror's second surface, and either another layer of transparent conductive material or a combined reflector/electrode is placed onto the mirror's third surface [12].

The selection of electrochromic species is also an important part in commercial production. Electrochromic materials include cathodic materials, anodic materials, and redox regulators. The typical representative of cathodic materials is viologen. They reported a large number of related compounds, expecting to get a high gray scale device [13–16] (see Chapter 4). Anodic materials can be a phenazine, a phenothiazine, a carbazole, a ferrocene with an alkyl chain, and their homologue, which usually serve to reduce the response time or adjust the voltage [14]. Gentex corporation provides simple and stable device preparation process, which is still mainly used till now. These devices introduce cross-linked polymer gel (formed by isocyanate and dibutyltin diacetate) to diminish undesirable visual irregularities substantially and to improve the stability of commercial products (Figure 14.2) [18].

Gentex Corporation also dabbled in other electrochromic-related products, such as variable transmission glass [19], electrochemical energy storage device [20], non-dimer viologen electrochromic device [21], infrared radiation (IR) transmission coating [22, 23] and improved version of the original electrochromic device [24–28] (Space limitations not described in detail here).

As the industrial products are renewed more and more frequently, Gentex continues to synthesize more new functional compounds to meet various needs (shown in Figure 14.3) [29–35].

Overall the company is an absolute leader in the field of motor vehicle rearview mirrors. Gentex Corporation commercialized viologen and a series of anode materials successfully. It is an outstanding company, which pays great attention to innovation. And it is precisely because of its dominance in the automobile rearview mirror industry that other related companies have focused their attention on other industries.

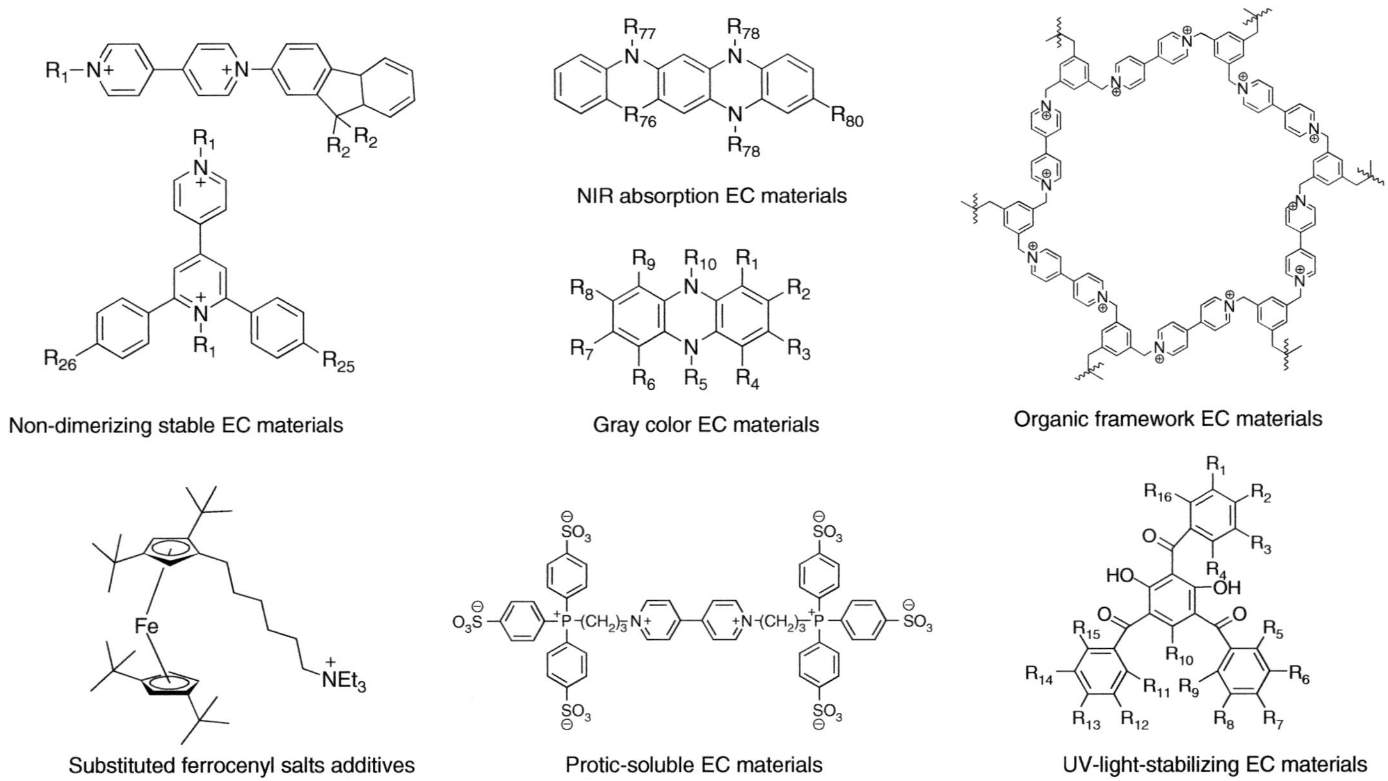

Figure 14.3 The representative EC materials in Gentex. Sources: Dave and Byker Harlan [29], Kloeppner Leroy et al. [30], Rongguang and Baumann Kelvin [31], Baumann Kelvin and Rongguang [32], Erno Zachary and Franz [33], Baumann Kelvin [34], Franz and Baumann Kelvin [35].

14.3 Ricoh Company, Ltd.

Ricoh is a famous manufacturer of office equipments and optical equipments in Japan, and it is also one of the world's top 500 enterprises. The predecessor of Ricoh was the Institute of Physics and Chemistry established by Shi Jingqing in 1936. In 1963, it was officially renamed Ricoh, and the Hong Kong branch was established in the same year.

Ricoh's main business scope is office, optical instruments, and digital cameras. Therefore, Ricoh has a great interest in the electronic paper industry. Traditional electronic paper technology requires light filters to achieve color display, while the filter will reduce the reflectivity and contrast of the electronic paper. So Ricoh noticed the electrochromic device that is non-filter is required when generating colorful patterns.

The problem of electrochromic technology for manufacturing electronic paper is the long response time. So they proposed an invention as shown in Figure 14.4.

The electrochromic material is adsorbed on the surface of the metal oxide, and different electrodes are insulated from each other to achieve independent discoloration [37]. Due to the requirements of electronic paper for color richness, Ricoh synthesized more kinds of viologen electrochromic compounds (as the Figure 14.5 shows).

Compared with the polymerization method used in the Gentex patent, Ricoh tends to use photocurable resins, which can form an electrolyte layer at a lower temperature and in a shorter time [38]. They used ultraviolet radiation polymerization to form a three-dimensional dendrimer so that the electrolyte was dispersed in the electrolyte layer with the form of a continuous phase, which can alleviate the phenomenon that the image gradually becomes blurred after the device is used for a long time (Figure 14.5) [40].

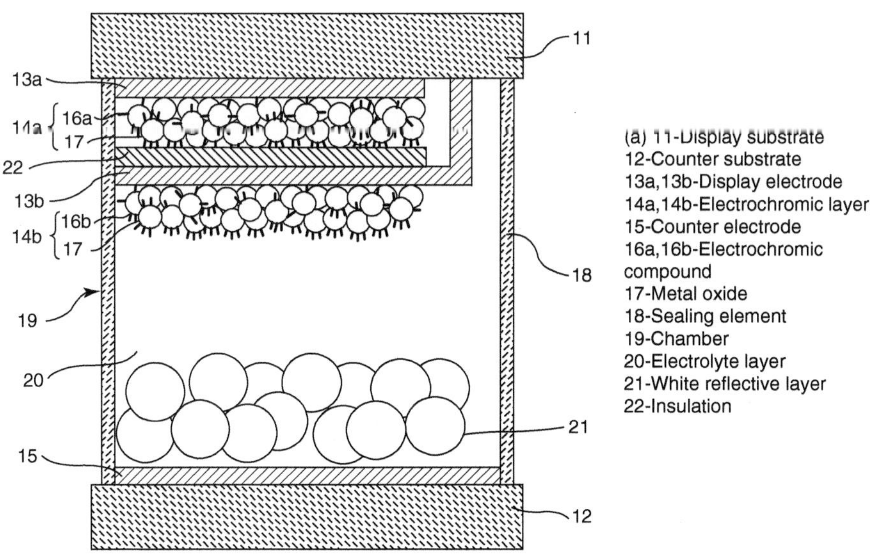

Figure 14.4 Multilayer rapid discoloration device. Source: Hirano and Yashiro [36].

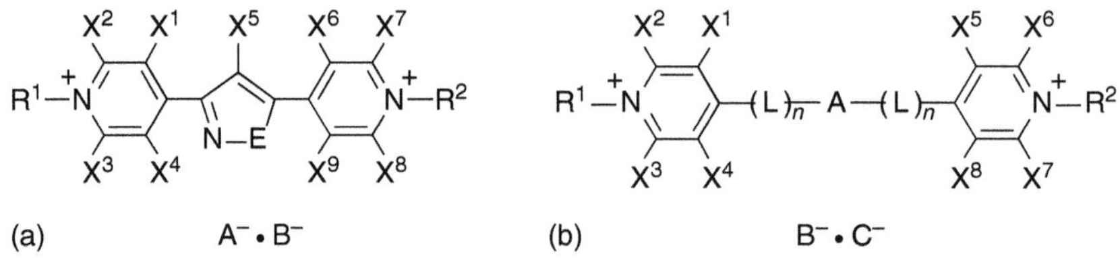

Figure 14.5 Compound structural formula. Source: Yashiro and Hirano [39].

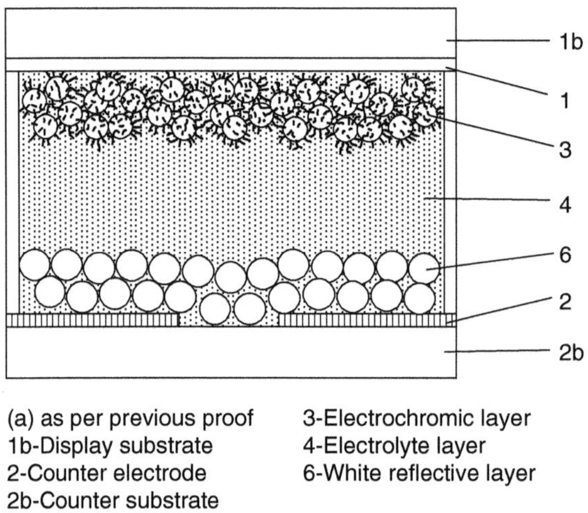

Figure 14.6 Device structure including white reflective layer. Source: Hiroyuki and Tohru [41].

Around the core of electronic paper, Ricoh has also revolutionized the electrode production process. They take simple deposition and inkjet and printing methods to cover the electrode surface with a layer of metal oxide. Moreover, they introduced the opposite oxidation–reduction reaction on the counter electrode to obtain stable color development and erasure (shown in Figure 14.6) [42].

Since then, Ricoh has made many improvements based on the original devices. They added long fibrous conductive particles to the device to get a transparent conductive material with better physical properties than indium tin oxide (ITO) [43]. They also synthesized new electrochromic materials, both non-viologen compounds and viologen compounds [44–54] (shown in Figure 14.7).

Ricoh is a large-scale enterprise, its investment in electrochromic materials is only a small part, and it is basically concentrated on electronic paper. Therefore, the related patent ideas are very unified.

14.4 Canon Inc.

Canon began with the establishment of Precision Optical Instruments Laboratory in 1933, and applied for the registration of the trademark "Canon" in 1935. After years of unremitting efforts, Canon has globalized and expanded its business to various fields. And it has been ranked among the world's top 500 companies.

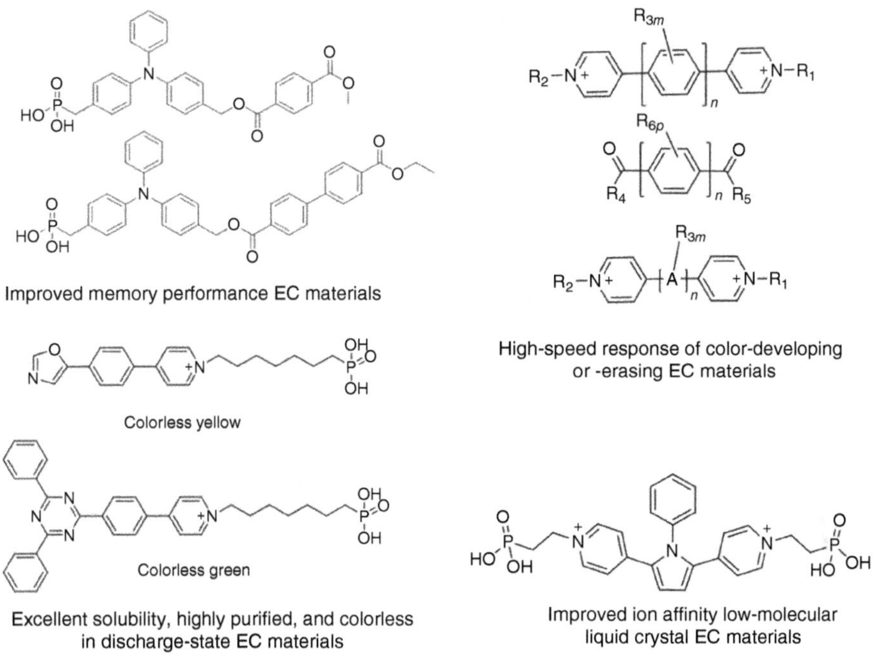

Figure 14.7 The representative EC materials in Ricoh. Sources: Toshiya and Hiroyuki [44], Masato and Satoshi [45], Yoshinori and Keigo [46], Takahiko and Keiichiroh [47], Takashi and Shigenobu [48], Tohru and Kohji [49], Daisuke and Toshiya [50], Tohru and Shigenobu [51], Koh and Shigenobu [52], Shigenobu and Takeshi [53], Koh and Tohru [54].

Canon's product series are distributed in three major areas: personal products, office equipment, and industrial equipment. Canon believes that the oxidation state of free radicals in traditional viologen compounds is unstable. In addition, viologen will irreversibly precipitate substances during the repeated crystallization process.

On the premise of expecting a stable system, Canon pursues higher natural state transparency, higher transmittance change, and more stable performance on optical materials. Canon synthesized a series of compounds (shown in Figure 14.8) [55–58].

Overall if we want to apply electrochromic materials to full-color displays or similar, materials that turn cyan, magenta, and yellow will be needed. Further increasing the range of applications will require increasing the range of spectrum electrochromism (EC) materials can produce.

14.5 BOE Technology Group Co., Ltd. and OPPO Guangdong Mobile Communications Co., Ltd.

BOE and OPPO enterprises are newly established technology companies, and the current number of electrochromic patent applications has entered the top 10 worldwide. Moreover, the two companies are both on research and development of new device structure and performance optimization, and their patent similarity is also very high, so they are introduced together.

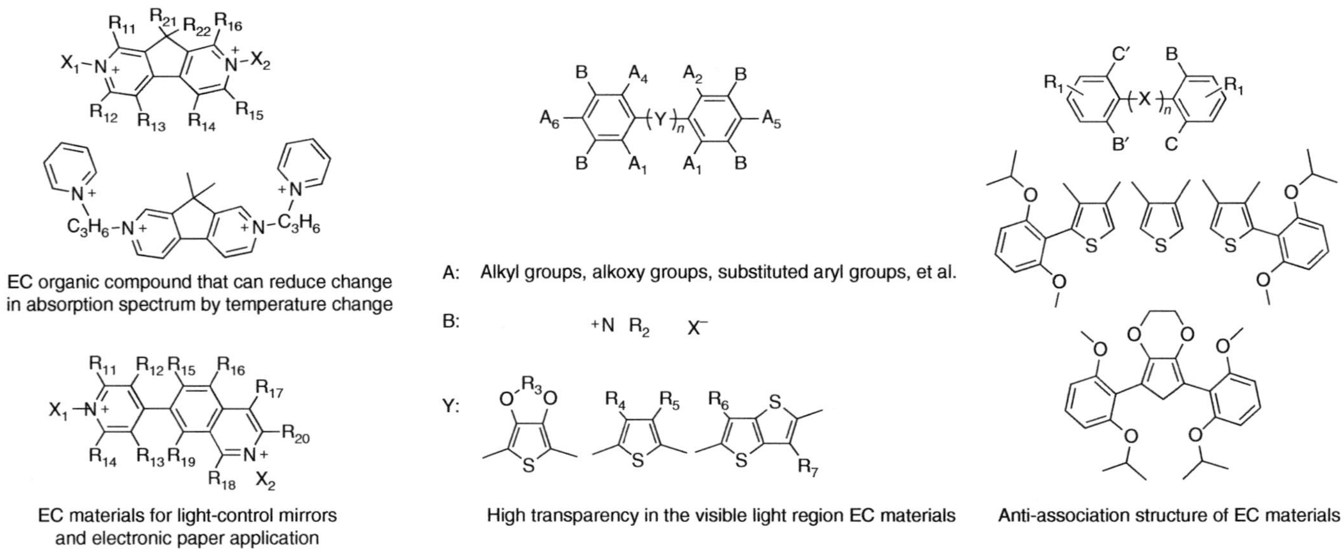

Figure 14.8 Canon's unique EC materials. Source: Kenji and Shinjiro [55], Satoshi and Kenji [56], Jun and Kenji [57], Kenji and Wataru [58].

BOE Technology Group Co., Ltd (hereinafter referred to as BOE), founded in April 1993, is an IoT company providing intelligent interface products and professional services for information interaction and human health. BOE's three core businesses are Interface Devices, Smart IoT Systems, Smart Medicine and Engineering Integration.

Both companies focus on the field of display technology. They pay more attention to the application of electrochromic devices to achieve a series of functions such as color rendering or changing transmittance. Thus the patents of these two companies are fewer studies involved in the changes of electrochromic devices but pay more attention to the overall application.

The method of adopting RGB color photoresist in the traditional liquid crystal display (LCD) screen will cause 2/3 of visible light to be absorbed by the photoresist layer [59], thus reducing the utilization rate of visible light. The researchers designed a new device as shown in Figure 14.9, in which the electrochromic film 5 is attached to the common electrode 3's surface. Then a series of improvements were made on the simple template [61, 62] and improved color filter [63].

In order to achieve the partitioned filling of electrochromic solutions of different colors in two substrates, the researchers used a pixel wall (as the Figure 14.10 shows) formed by photoresist to isolate different types of electrochromic solutions.

In recent years, 3D display technology has become a trend, while grating device can perform 3D display in the glasses-free state (shown in Figure 14.11). The aforementioned electrochromic devices isolated by pixel walls are active gratings. And

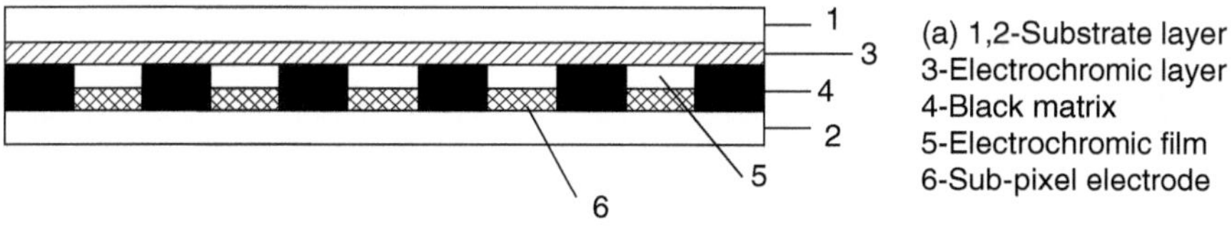

Figure 14.9 Liquid crystal display screen with electrochromic film. Source: Ding [60].

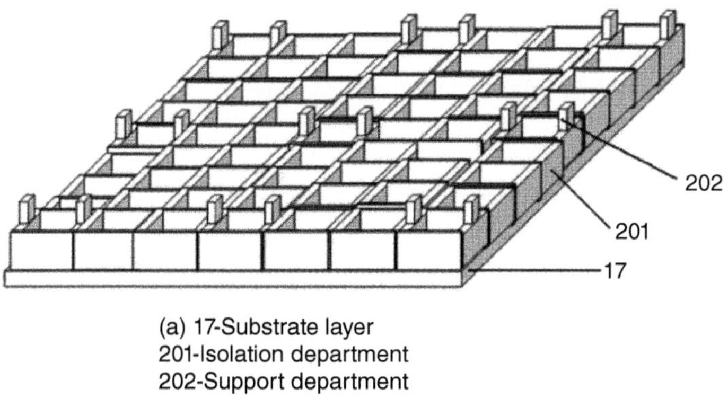

Figure 14.10 Substrate isolated by pixel wall. Source: Wang [64].

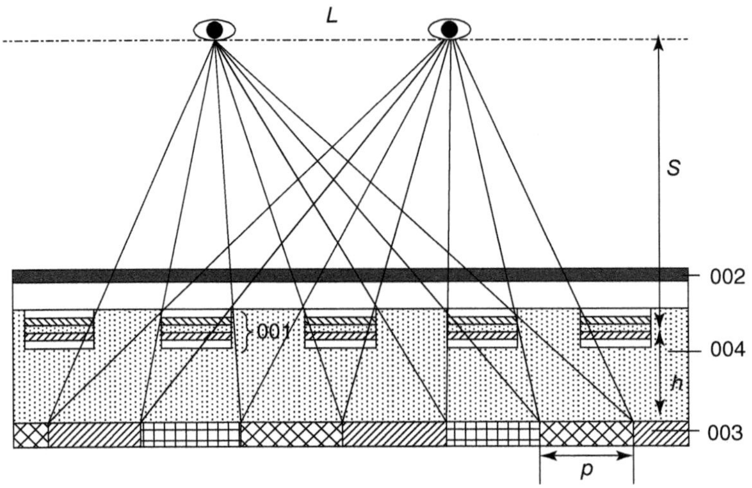

Figure 14.11 Glasses-free 3D grating. Source: Changfeng et al. [65].

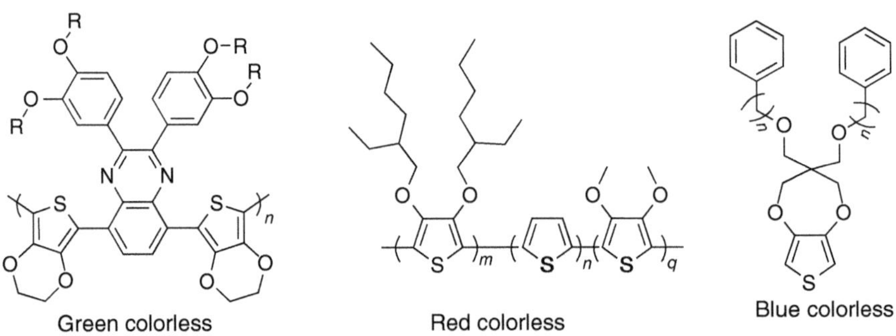

Figure 14.12 Glasses-free 3D grating. Sources: Chunye and Sai [78–80].

these gratings can switch freely between 2D and 3D [66–69]. On this basis, in order to reduce the response time, a gel electrolyte was introduced [70].

As a manufacturer of display devices, electrochromic materials also have the general function of color development and color change. This also spawned a large number of patents and these patents are usually LCD related [71–77]. Although BOE is a company established less than 30 years ago, its number of related patent applications has ranked first in the world, and there is still a growing trend.

Besides, BOE also synthesized electrochromic materials three colors of red, blue, and green [78–80] (as shown in the Figure 14.12).

OPPO Guangdong Mobile Communications Co., Ltd. (hereinafter referred to as OPPO) was founded in 2004. Its core business is electronic products, mobile phones,

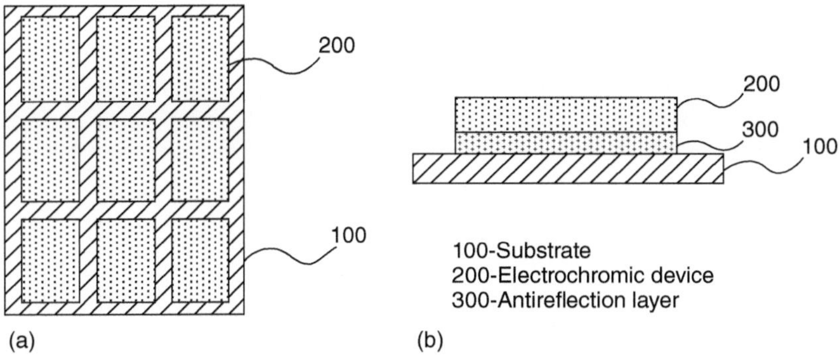

Figure 14.13 (a) Electrochromic mother board. (b) Electrochromic unit. Source: Tianlin et al. [81].

and mobile Internet. In 2018, OPPO became the brand with the largest number of domestic smartphone sales in China. So it is considered to be a promising company.

People in modern society have begun to pursue individuality and fashion, so they put forward higher requirements for the appearance of mobile phones, and the color diversity displayed by electrochromic materials just fits it (shown in Figure 14.13).

By implanting different electrochromic devices on the motherboard, you can get the effect of meeting different needs only with a single product that not only saves raw materials but also increases the fashion of the goods. According to this idea, this electrochromic device can be implanted in the phone case [82–86]. In addition, the color change can also be used as a reminder of information, which means forming different colors in the key area to remind the operators to pay attention [87–89].

OPPO phones are often known for their outstanding camera systems. Furthermore, since the electrochromic film can reversibly change the transmittance, it can also be used in photographic equipment. The application of electrochromic technology to aperture can change the light transmittance of the aperture through voltage, and then the design can flexibly determine the light transmittance area of the aperture to make the camera lens easier to place on mobile devices [90, 91].

14.6 Other Important Enterprises

In addition to the companies mentioned earlier. Although small in scale, their work is also remarkable.

14.6.1 Ninbo Ninuo Electronic Technology Co., Ltd.

Ninbo Ninuo Electronic Technology Co., Ltd. (hereinafter referred to as Ninuo) was registered on 21 June 2010 in Ningbo. Ninuo's main business scope is the research and development; manufacturing and processing of high-performance membrane materials, nanomaterials, and optoelectronic materials; and research and development of electronic products.

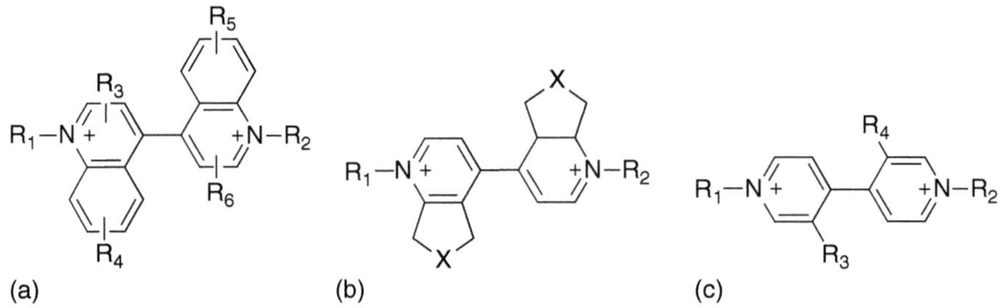

Figure 14.14 Electrochromic compound structure. Source: Zhenhu and Zebin [92].

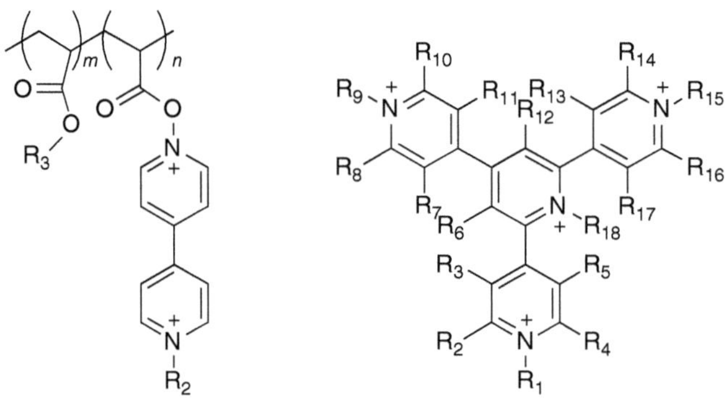

Figure 14.15 Electrochromic compound structure. Sources: Tong and Hu [94], Zhenhu and Shanshan [95].

Ninuo's main products are car rearview mirrors and smart glass. Ninuo pays more attention to the synthesis of new materials. In the company's patent, it is keen to synthesize the viologen compound substituted at the 3,6 position (as shown in Figure 14.14) [92].

The researchers expect to increase the steric hindrance in the planar state to reduce the loss of the viologen molecule in the free radical state [93].

Besides, the company also synthesized polymer branched viologen and viologen compounds containing three quaternized pyridine rings (as shown in Figure 14.15), hoping to increase the diversity of the spectrum [94, 95].

The general ratio of Haoruo's devices is 1 : 1 viologen and methylphenazine (30 mM) dissolved in γ-butyrolactone or polycarbonate (PC) and 4 wt% polyelectrolyte. And the process flow is also quite simple. You just need to inject the electrochromic solution into the prepared substrate and finally seal it with UV glue (as shown in Figure 14.16) [96].

In order to deal with the rupture of the electrochromic device due to physical impact in practice, thereby leaking toxic electrochromic solution and possibly endangering the safety of the driver or other personnel, a repairable gel matrix was achieved by utilizing the reversible radical polymerization of diphenylarylfuranone and the cycloaddition reaction of coumarin groups under light [97, 98].

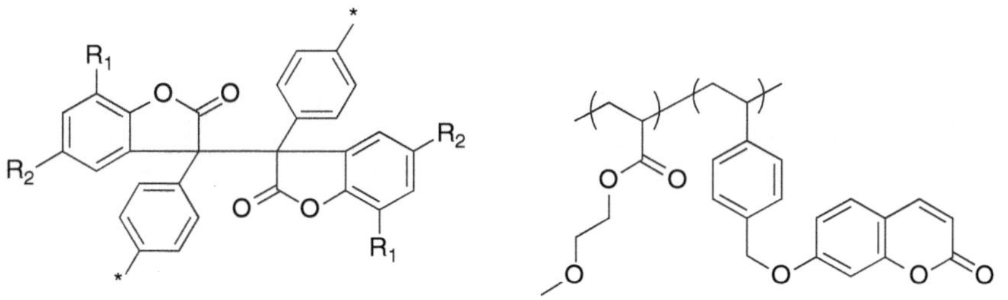

Self-healing EC hyperbranched polymer   Light-healing EC hyperbranched polymer

Figure 14.16 Repairable gel. Source: Modified from Zhenhu et al. [96].

Figure 14.17 Green electrochromic compounds. Source: Liyan [99].

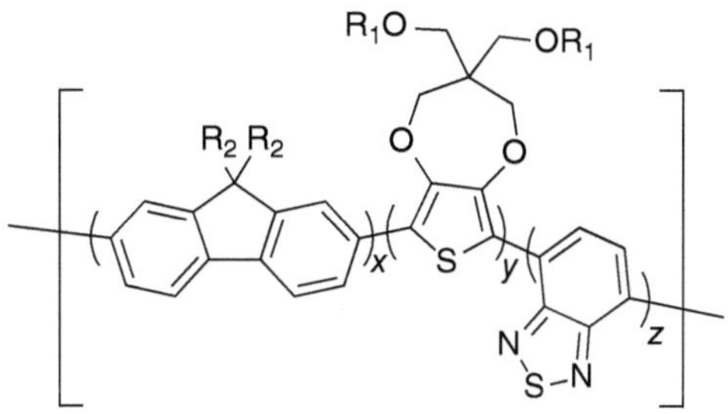

Figure 14.18 Black electrochromic compounds. Source: Liyan [100].

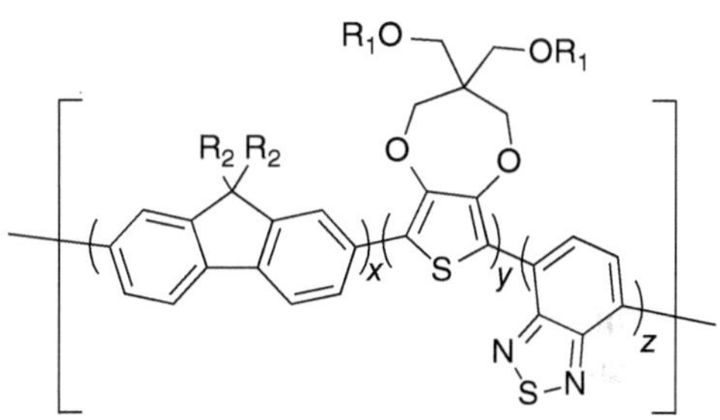

14.6.2 Ambilight Inc.

Ambilight Inc. is a company focused on electrochromism and it was established in Silicon Valley, Shenzhen, and Indiana in 2017. Its main products include auto-dimming mirrors, skylights and windows, and architectural stained glass.

Ambilight synthesized a new compound by itself, which is essentially different from the viologen series compounds. Polymers of the same monomer with different polymerization ratios exhibit different electrochromic colors (Figures 14.17 and 14.18).

Based on these two polymers, they realized electrochromic devices with preset patterns through solvent pretreatment or laser irradiation [101].

14.6.3 Furcifer Inc.

Furcifer Inc was founded in 2015. Its main products are electronic display products and smart windows.

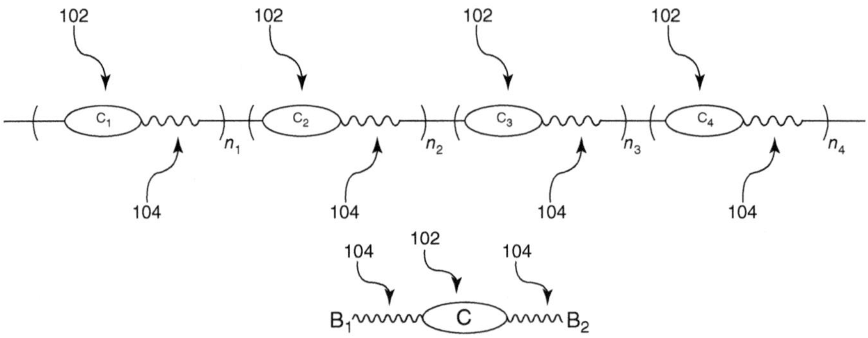

Figure 14.19 Polymer formation process (n_1, n_2, n_3, n_4 are integers between 1 and 50). Source: Jianguo and Yan [102].

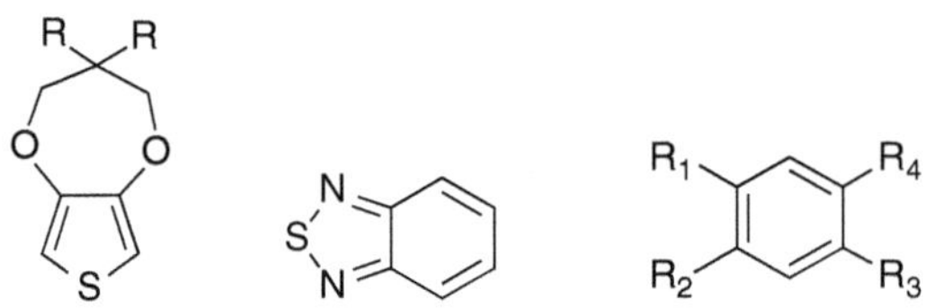

Figure 14.20 Polymerized monomer. Source: Jianguo and Yan [102].

Nowadays black-to-transmissive electrochromic polymers (ECPs) is gradually becoming a hot spot for it has great potential in smart windows and protecting personal privacy. Furcifer disclosed a series of novel semiconducting ECPs, which exhibit high optical contrasts in the visible region. The polymer, in one embodiment, includes a plurality of pi-conjugated chromophores in spaced relation with one another, where each chromophore is colored in a neutral state and is colored or transmissive in an oxidized state (Figures 14.19 and 14.20).

Furcifer adopted the donor–acceptor approach which allows a black-to-transmissive switching polymeric electrochrome.

The composition of the polymer can be controlled, which in turn can affect the physical and chemical properties of the polymer, which makes the processing quite easy [103].

14.6.4 Changzhou Spectrum New Material Co. Ltd.

Changzhou Spectrum New Material Co. Ltd. (hereinafter referred to as YAPU), founded in 2008, is located in Changzhou Wujin district. It specializes in electrically induced discolored lens materials. It claims to be comparable to Gentex in performance of its main products. In addition, the company also has the ability to manufacture large-sized thick EC lenses, which is also one of its core competencies. Furthermore, it has greatly expanded the application field and scope of EC material (Figure 14.21).

YAPU has added optical sensor system and microcontroller unit (MCU) to traditional EC rearview mirror [104, 105]. This product can gradually change color, according to the light intensity, to eliminate glare. In addition, it has developed a

Figure 14.21 YAPU's large-sized thick EC lenses. Source: http://www.yapuglass.com/en/

method of intermediate coating in ECD, which effectively reduced the physical damage caused during packaging [106].

In terms of packaging process, YAPU uses conductive silver paste for encapsulation [107], avoiding the mutual penetration between different sealants. It uses vacuum equipment, which simplifies the packaging process, reduces production costs, and improves the device's conductive performance.

14.7 Conclusion

In general, the pace of the overall development of the current electrochromic industry tends to be gentle, but the recently emerging ECP compounds provide a completely new development prospect. Polymerization of electrochromic compounds may become an important development direction in the future.

References

1 Byker, H.J. (1990) Single-compartment, self-erasing, solution-phase electrochromic devices, solutions for use therein, and uses thereof. USt 4902108 (A).

2 Kmetz, A. (1975). *Nonemissive Electrooptic Displays.*

3 Forgette, J.A.;Anderson, J.S. (2005). Electrochromic review mirror assembly incorporating a display/signal light. CN1643444 (A).

4 Bechtel, J.H., Schmidt, D.J. (1995). Electronic control system. US 5451822 (A).

5 Tonar, W. (1995). Outside automatic rearview mirror for automotive vehicles. US 5448397 (A).

6 Bauer Frederick, T.; Bechtel, J.H. (1995). Automatic rearview mirror incorporating light pipe. US 5434407 (A).

7 Byker, H.J. (1994). Electrochromic devices with bipyridinium salt solutions. US 5336448 (A).

8 Byker, H.J. (1994). Bipyridinium salt solutions. US 5294376 (A).

9 Byker, H.J. (1994). Variable reflectance mirror. US 5282077 (A).

10 Forgette, J.A., Anderson, J.S. (2011). Electrochromic review mirror assembly incorporating a display/signal light. CN102062983 (A).
11 Tonar, W.L., Forgette, J.A. (2003). Electrochromic review mirror assembly incorporating a display/signal light. WO03084780 (A2).
12 Ash Kevin, L., Tonar, W.L. (1999). Electrochromic mirror with two thin glass elements and a gelled electrochromic medium. US 5940201 (A).
13 Rongguang, L., Baumann Kelvin, L. (2016). Electrochromic compounds with improved color stability in their radical states. US2016229803 (A1).
14 Driscoll David, I., Busscher, B.L. (2016). Electro-optic window assembly EMI shield. CN105408203 (A).
15 Baumann Kelvin, L., Guarr, T.F. (2011). Electrochromic compounds and associated media and devices. CN102165032 (A).
16 Ash Kevin, L., Baumann, K.L. (2017). Electrochromic device. CN106662787 (A).
17 Baumann Kelvin, L., Rongguang, L. (2018). Protic-soluble organic electrochromic compounds. CN108699431 (A).
18 Theiste David, A., Kloeppner, L.C. (2015). Electrochemical energy storage devices. US2015346573 (A1).
19 Kloeppner Leroy, J., Guarr, T.F. (2003). Electrochromic medium having a self-healing cross-linked polymer gel and associated electrochromic device. US2003042473 (A1).
20 Theiste David, A., Kloeppner, L.C. (2017). Electrochemical energy storage devices. CN106660977 (A).
21 Rongguang, L.; Baumann Kelvin, L. (2017). Electrochromic compounds with improved color stability in their radical states. CN107406389 (A).
22 Jian, G., Weller Andrew, D. (2018). IR transmitting coating for electro-optic element. CN108369361 (A).
23 Punam, G. (2019). Near-IR absorptive compounds and devices incorporating the same. CN109906219 (A).
24 Kloeppner Leroy, J., Tonar, W.L. (2017). Electrochemical devices with plastic substrates. CN107438793 (A).
25 Ash Kevin, L., Dozeman, G.J. (2018). Flame retardancy of electrochromic devices. CN108885378 (A).
26 Franz Sue, F., Baumann, K.L. (2019). Ultrathin, solution phase electrochromic devices. CN109564372 (A).
27 Ash Kevin, L., Baumann, K.L. (2019). Color excursion mitigation in electrochromic devices. CN109690395 (A).
28 Kambiz, I., Stefan, J. (2018). System and method for electric current leakage detection in land seismic system. CN109073697 (A).
29 Dave, T., Byker Harlan, J. (2001). Near infrared-absorbing electrochromic compounds and devices comprising same. US6193912 (B1).
30 Kloeppner Leroy, J., Guarr, T.F. (2003). Electrochromic medium having a self-healing cross-linked polymer gel and associated electrochromic device. US2003042473 (A1).
31 Rongguang, L.; Baumann Kelvin, L. (2016). Electrochromic compounds with improved color stability in their radical states. US2016229803 (A1).

32 Baumann Kelvin, L., Rongguang, L. (2018). Protic-soluble organic electrochromic compounds. US2018224706 (A1).
33 Erno Zachary, B., Franz, S.F. (2018). Electrochromic organic frameworks. US2018321565 (A1).
34 Baumann, K.L. (2010). Ultraviolet light stabilizing compounds and associated media and devices. US2010073754 (A1).
35 Franz Sue, F., Baumann, K.L. (2018). Electrochromic media and devices with multiple color states. US2018112127 (A1).
36 Hirano, S., Yashiro, T. (2012). Electrochromic display apparatus. CN102414610 (A).
37 Nobe, T., Kotanagi, S. (2011). Drive module and electronic apparatus. CN102193163 (A).
38 Naijo, Y., Yashiro, T. (2011). Electrochromic display apparatus and method of manufacturing the same. CN102193263 (A).
39 Yashiro, T., Hirano, S. (2012). Electrochromic compound, electrochromic composition, and display element. CN102666778 (A).
40 Ishizuka, K., Akazawa, T. (2011). Cleaning device and image-forming device. CN102243467 (A).
41 Hiroyuki, T., Tohru, Y. (2014). Electro-chromic display element, display apparatus, and driving method. CN103562787(A).
42 Yashiro, T., Hirano, S. (2012). Electrochromic display device. CN102540609 (A).
43 Yoshinori, O., Tohru, Y. (2013). Electrochromic display device and method for manufacturing an electrochromic display device. CN103323998(A).
44 Toshiya, S., Hiroyuki, T. (2015). Electrochromic compound, electrochromic composition, and display element. CN104768960(A).
45 Masato, S., Satoshi, Y. (2017). Electrochromic element. CN106886113 (A).
46 Yoshinori, O., Keigo, T. (2019). Electrochromic display device and production method of electrochromic display device. CN110383162 (A).
47 Takahiko, M., Keiichiroh, Y. (2016). Electrochromic apparatus, electrochromic element, and method of manufacturing electrochromic element. US2016209721 (A1).
48 Takashi, O., Shigenobu, H. (2013). Electrochromic compound, electrochromic composition, and display device. US2013135703 (A1).
49 Tohru, Y., Kohji, T. (2017). Electrochromic device and method for producing electrochromic device. US2017010514 (A1).
50 Daisuke, G., Toshiya, S. (2017). Electrochromic compound, electrochromic composition, display element, and dimming element. US2017226413 (A1).
51 Tohru, Y., Shigenobu, H. (2015). Electrochromic device and production method thereof. US2015198857 (A1).
52 Koh, F., Shigenobu, H. (2011). Electrochromic display element. US2011279884 (A1).
53 Shigenobu, H., Takeshi, S. (2009). Electrochromic compound, electrochromic composition and display device. US2009231663 (A1).
54 Koh, F., Tohru, Y. (2015). Electrochromic display element and image display device. US2015378233 (A1).

55 Kenji, Y., Shinjiro, O. (2015). Novel organic compound and electrochromic element including the same. US2015153624 (A1).

56 Satoshi, I., Kenji, Y. (2016). Organic compound, electrochromic element containing the same, optical filter, lens unit, imaging device, and window component. US2016246152 (A1).

57 Jun, Y., Kenji, Y. (2016). Method and apparatus for driving an electrochromic element. US2016041447 (A1).

58 Kenji, Y., Wataru, K. (2018). Electrochromic element, optical filter, lens unit, imaging apparatus, and window member. US2018052375 (A1).

59 Xin, G.; Zaijian, L. (2013). Color film substrate, liquid crystal display panel and method for preparing color film substrate. CN102879946 (A).

60 Ding, Y. (2011). Color film structure, display screen and display device. CN202013465 (U).

61 Xin, G. (2013). Color membrane base plate and display device. CN203250096 (U).

62 Xin, G., Zaijian, L. (2012). Display panel and display device. CN102830526 (A).

63 Yue, S. (2013). Color filter, manufacturing method and displaying device of color filter. CN103217832 (A).

64 Wang, Y. (2015). Electrochromic display panel, manufacturing method thereof and display device. CN104808413 (A).

65 Changfeng, L., Xue, D. (2015). Electrochromic grating, display panel and display device. CN104570536 (A).

66 Changgang, H., Zhenyu, Z., Yanping, L. (2014). Electrochromic grating, preparing method of electrochromic grating and 3D (three-dimensional) display device. CN104238223 (A).

67 Yufeng, W., Xiaoming, W. (2013). Electronic star simulator for APS star sensor. CN103033196 (A).

68 Wenbo, L., Xin, G. (2013). Active grating, display device and active shutter glasses. CN202948233 (U).

69 Shengji, Y., Xue, D. (2015). 3D raster, color film substrate, display device and control method thereof. CN104834103 (A).

70 Changfeng, L., Xue, D. (2015). Electrochromic grating, manufacturing method, display panel and display device. CN104570537 (A).

71 Jiuxia, Y., Feng, B. (2014). Display panel, display device and preparation method of display panel. CN104216177 (A).

72 Yue, S. (2013). Display panel, method for manufacturing same and display device. CN103323999 (A).

73 Xuguang, S., Chenglong, F. (2013). Positive-pressure pneumatic type automatic water sample sampler. CN103439146 (A).

74 Mingchao, L. (2013). Transparent displaying device and manufacturing method thereof. CN 103412434 (A).

75 Wenbo, L. (2016). Display panel and display device. CN205176447 (U).

76 Dongfang, W., Tongshang, S., Yu, W. (2016). Display panel and display apparatus. CN 205539835(U).

77 Bo, S., Shiqi, C. (2015). Display panel, manufacturing method and driving method thereof and display device. CN104965371 (A).
78 Chunye, X., Sai, M. (2014). Green polythiophenes electrochromic material and preparation method thereof as well as component. CN103772664 (A).
79 Chunye, X., Sai, M., Jianming, Z. (2014). Blue electrochromic material as well as preparation method and assembly thereof. CN103666445 (A).
80 Xu, C., Xiaoming, C. (2014). Red electrochromic material and preparation method and assembly thereof. CN103524718 (A).
81 Tianlin, L., Guangming, Y., Yucai, Y. (2018). Electrochromic mother board, electrochromic unit, casing and electronic device. CN108519709 (A).
82 Yang, G. (2018). Housing, preparation method and electronic device. CN108549183 (A).
83 Yang, G. (2018). Housing, preparation method and electronic device. CN108549185 (A).
84 Yang, G. (2018). Shell and preparation method and electronic device. CN108761950 (A).
85 Yang, G. (2019). Casing, electronic equipment. CN208580287 (U).
86 Yang, G. (2018). Information presentation method, device, storage medium, and electronic device. CN109104529 (A).
87 Haiping, Z. (2019). Equipment control method and device, storage medium and electronic equipment. CN109143090 (A).
88 Haiping, Z. (2019). An electronic device, an operating state prompting method, and a storage medium. CN109240416 (A).
89 Yang, X. (2019). Light emitter, time-of-flight assembly and electronic device. CN109842704 (A).
90 Xu, R. (2020). Aperture structure, camera and electronic device. CN110794634 (A).
91 Yanru, Z. (2020). Lens module, imaging device and terminal. CN110727102 (A).
92 Zhenhu, C., Zebin, T. (2013). Electrochromic material and electrochromic device. CN103059831 (A).
93 Zhenhu, C. (2015). Electrochromic material and electrochromic device. CN104910892 (A).
94 Tong, Z., Hu, S. (2012). Electrochromism material and electrochromism device. CN102690646 (A).
95 Zhenhu, C., Shanshan, H. (2017). Cathodic electrochromic compound, and relevant medium and device thereof. CN106916582 (A).
96 Zhenhu, C., Zebin, T., Shanshan, H. (2013). Electrochromic material and electrochromic device. CN103045228 (A).
97 Zhenhu, C., Shanshan, H. (2018). Electrochromic solution capable of being gelatinized and self-repaired at room temperature, and application thereof. CN108251100 (A).
98 Zhenhu, C., Shanshan, H. (2018). Electrochromic solution capable of being gelatinized and self-repaired under illumination, and application thereof. CN108251099 (A).

99 Liyan, Y. (2018). Green electrochromism macromolecule material as well as preparation method and application thereof. CN107674185 (A).

100 Liyan, Y. (2018). Black electrochromic polymer material and preparation method and application thereof. CN107880253 (A).

101 Hong, W., Jiazhi, H. (2019). Patterned electrochromic device, electrochromic glass and application. CN110133935 (A).

102 Jianguo, M., Yan, Z. (2017). Multicolored electrochromic polymer compositions and methods of making and using the same. US2017321002 (A1).

103 Jian, W., Yan, Z. (2017). Integration of electrochromic films on a substrate. US2017298682 (A1).

104 Chunye, X., Hanfeng, L., Jianming, Z. (2014). Front-view intelligent anti-glare device of vehicle and manufacturing method of anti-glare mirror thereof. CN103909868 (A).

105 Chunye, X., Hanfeng, L., Jianming, Z. (2013). Intelligent automotive anti-dazzle control device. CN203047034 (U).

106 Chunye, X., Hanfeng, L., Jianming, Z. (2016). Film plating template and method used for producing ITO glass plated with EC film. CN105271807 (A).

107 Hanfeng, L., Chunye, X., Jianming, Z. (2015). Manufacturing packaging method of electrochromic anti-dazzle device for vehicle. CN104656336 (A).

15

Main Challenges for the Commercialization of OEC

15.1 Introduction

Electrochromic response is an interesting feature of redox-active materials, which exhibit visual color changes during electrochemical reduction or oxidation. There are numerous practical applications for organic electrochromic (OEC) devices, such as smart windows for green buildings and vehicles, low-energy electrochromic (EC) displays, safer anti-glare car rearview mirrors, and smart glasses with better UV radiation protection [1–5]. In particular, high-end electrochromic materials have recently found novel applications in the aerospace industry. The color-changing windows on the latest Boeing 787 Dreamliner, for example, has shocked so many passengers that plastic shades have been scrapped. However, there are still lots of challenges for the prosperous commercialization of OEC such as the long-term stability of OEC materials, the mechanic stability of flexible devices, and the large area process technology.

15.2 The Long-Term Stability of OEC Materials

The basic principle of an electrochromic device (ECD) is the ability to change color by applying an appropriate electrical potential (shown in Figure 15.1), oxidizing the organic film if the electrode potential (positive) is higher than the device's equilibrium potential or reducing/eliminating the film if the electrode potential is lower (negative) [6]. Its working principle determines the structure of the device, which affects the stability of ECDs. Since redox reactions occur in the organic thin films, they need to be in contact with the electrodes, and so the OEC thin films are coated onto separate transparent conductive substrates (usually flexible substrates or glass coated with ITO). For electrical insulation and ion storage, an electrolyte layer is sandwiched between the organic coated electrodes, which are put (facing inwards) on top of each other. Multilayer structures as shown in Figure 15.1b are currently used for most of organic electrochromic devices (OECDs) [7, 8]. For practical application, an OECD should have vivid color and high stability. Generally, the OECD's stability is affected by each of its components, i.e. its transparent electrodes, electrolytes and electrochromic materials as well as their compatibility.

Organic Electronics for Electrochromic Materials and Devices, First Edition. Hong Meng.

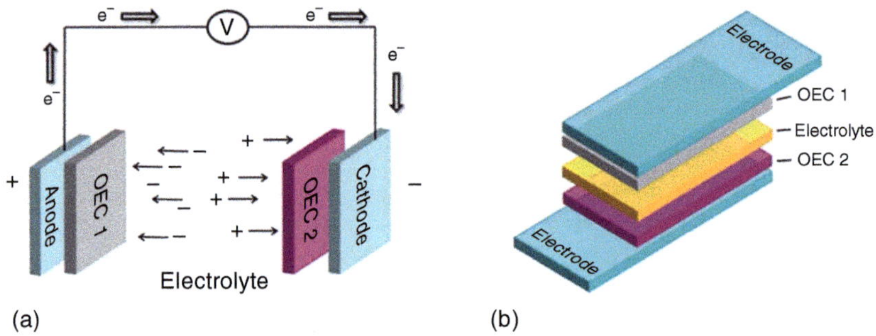

Figure 15.1 Working principle and structure of an OECD. (a) Two OEC materials are coated on electrodes connected to an external power source. (b) The typical structure of OECDs. Source: Jensen et al. [6]. © 2013, Royal Society of Chemistry.

Cyclic voltammetry (CV) is a basic technique to evaluate the electrochemical stability along with spectroelectrochemistry and measure the long-term optical contrast (ΔT). For the sake of electrochemical stability, the LUMO and HOMO levels are of great significance, as they define an electrochemical window in which to operate. Outside of this window the OEC materials are prone to electrochemical degradation by irreversible electrochemical oxidation or reduction. OEC materials with relatively high HOMOs are easily oxidized and as a result "doped" with anions (p-doping) from the electrolyte layer. A high HOMO level is also advantageous when operating an ECD since a low potential is sufficient to drive the device between the two redox states.

A general rule for the photochemical stability of most of donor–acceptor OEC materials is that C—N/C—O bonds, alkyl side chains, and quaternary sites are identified as leading to poor photochemical stability while aromatic polycyclic units stabilized them [9]. Among the different types of electrochromic materials [10, 11], conjugated polymers [12–14] such as polypyrroles, polyanilines, polythiophenes, polyselenophenes, and in particular poly(3,4-ethylenedioxythiophene) (PEDOT) and its derivatives [15, 16] attract increasing interest because of the advantages such as mechanical flexibility, easy color tunability via molecular design, and potential for low-cost processing of large-area devices [17]. Furan, oligofurans, and polyfuran show poor stability [18, 19]. However, oligo-/polyfuran and its derivatives exhibit several unique properties even better than oligo-/polythiophenes, such as higher fluorescence, smaller overlap integrals, lower polarizability, and better solubility [20–22]. Zhen et al. incorporated 3,4-ethylenedioxythiophene (EDOT) into polyfuran backbone and synthesized three electrochromic polymers, namely, 5-(furan-2-yl)-2,3-dihydrothieno[3,4-*b*][1,4]dioxin (EDOT-Fu), 5,7-di(furan-2-yl)-2,3-dihydrothieno[3,4-*b*][1,4]dioxin (Fu-EDOT-Fu), and 2,5-bis(2,3-dihydrothieno[3,4-*b*][1,4]dioxin-5-yl)furan (EDOT-Fu-EDOT). All three polymers exhibit better thermal and long-term redox stability compared to polyfuran. Particularly, P(EDOT-Fu-EDOT) films exhibit better redox stability with the amount of exchange charge still remaining 81% after 3000 cycles [23]. Liou et al. reported on the polyhydrazides and poly(1,3,4-oxadiazole)s bearing main-chain triphenylamine

(TPA) and pendant carbazole units from the dicarboxylic acid monomer 4,4′-Dicarboxy-4″-(carbazol-9-yl)triphenylamine(1′). The polymers could afford amorphous high-Tg films, good thermal stability, and good redox stability [24]. There are also some reports demonstrating that introduction of bulky groups on the electrochemically active sites (C-3 and C-6) of carbazole results in enhanced electrochemical and morphological stability [17, 25]. Hsiao and coworker reported that the incorporation of electron-donating or bulky substituents such as methoxy or *t*-butyl groups at the *para*-position of phenyl rings of the TPA unit leads to stable TPA cationic radicals and decreased oxidation potential, obtaining a significant enhancement on redox and electrochromic stability of the prepared polyamides [26].

The degradation mechanism of polymer-based ECDs has also been studied. Xing et al. investigated the degradation mechanism of ECDs based on poly[2,3-bis-(3-octyloxyphenyl) quinoxaline-5,8-diyl-alt-thiophene-2,5-diyl] (TQ1). They found three main factors involved in the degradation of TQ1 during the electrochromic process. The first factor is anion (ClO_4^-) irreversible trapping, the second one is peroxidation of thiophene group in TQ1. Both factors affect conductivity and electrochromism of TQ1. The third one is structural relaxation resulting in larger conjugated system of TQ1 during CV sweeping (0–1 V) for 400 cycles. However, when a potential range 0–0.7 V is applied, all the three factors prohibited, no electrochromism degradation is observed anymore. Their work suggests that using an appropriate potential to operate electrochromic conjugated polymer (ECPs) is essential for prolonging the lifetime of ECDs [27].

Viologens are particularly prominent small molecular OEC materials because of their high contrast, fast response, and easily obtained rich colors through molecular design [28, 29]. Xiao et al. synthesized bis(dihydroxyalkyl) viologen, 1,1′-bis(2,3-dihydroxypropyl) 4,4′-bipyridine derivative dichloride, $DHPV^{2+}2Cl^-$, and incorporated it as an electrochromic chromophore in an all-in-one ECD on the basis of poly (vinyl butyl) (PVB)-carbonate gel electrolyte. The device shows reversible switching between transparent and deep blue. Spectroelectrochemical studies demonstrated that it only needs a small coloration voltage down to −0.6 V and exhibits high coloration efficiency up to 301 cm^2/C. The optical contrast only reduces by 1% after consecutive operation for 10 000 cycles between −1.0 and 0 V(Figure 15.2) [30]. They also reported two other viologens, 1,10-bis (2-sulfonatoethyl) viologen (SEV) and 1,10-bis (3-sulfonatopropyl) viologen (SPV). High coloration efficiency up to 304 cm^2/C, response times below three seconds, and driving voltage below −0.8 V have been obtained by using the same device structure. SEV-based hydrogel and SPV-based hydrogel ECDs show high cycling stability, with ΔT reducing 2% after 10 000 cycles and reducing 1% after 8000 cycles, respectively [31].

Matching of the charge/discharge capacity of two electrodes determines charge balance and the lifetime of an ECD. Xu and coworkers designed two novel compatible electrochromic materials, cathodic 1-(9-hexyl-9*H*-carbazole)-10-(propylphosphonic acid)-4,40-bipyridilium dichloride (CPD) and anodic (4-(diphenylamino) phenyl)methylphosphonic acid (DPP). They used CPD- and DPP-modified TiO_2 as electrodes and propylene carbonate (PC)/ethylene carbonate/$LiPF_6$ as electrolyte.

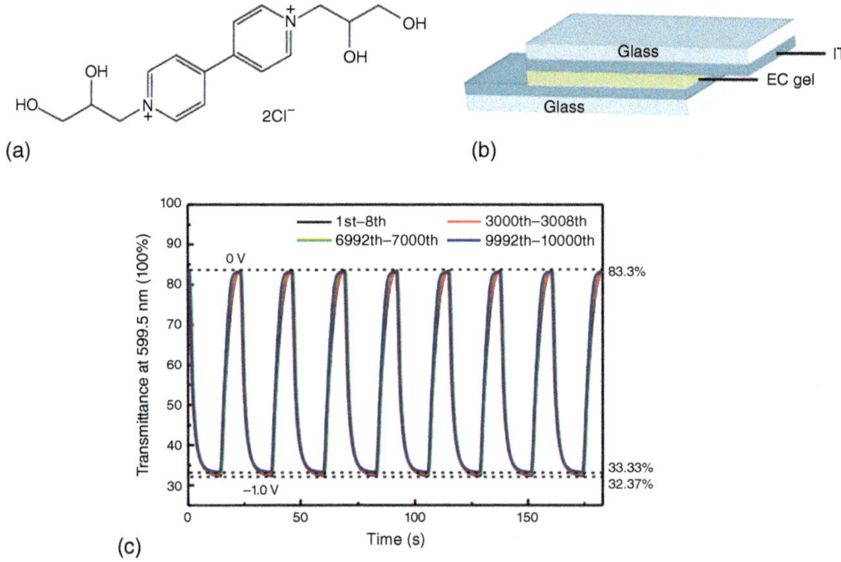

Figure 15.2 (a) Chemical structure of $DHPV^{2+}2Cl^{-}$. (b) Schematic architecture of all-in-one ECD. (c) Comparison of transmittance changes vs time data upon cycling of PVB-based gel ECD between bleached (0 V for 10 seconds) and colored state (−1.0 V for 15 seconds) at 599.5 nm. Source: Zhao et al. [30]. © 2019, Elsevier.

CPD and DPP were fixed on electrodes through chemisorption between phosphonic acid anchored with titanium dioxide on electrodes. The immobilized EC material structures lead to high-speed changing (less than one second) and good electrochromic memory (15 hours). In addition, the similar treatment resulted in the two well matched electrodes as for color degree, color changing voltage, and charge capacity. As a result, the device demonstrated a long lifetime (over 100 K switching cycles) [32].

The thermal stability, chemical stability, and electrochemical stability of electrolytes are also important to improve the long-term stability of ECDs. An appropriate polymer electrolyte should have good thermal stability, as it may release heat during device operation, which may cause degradation of ECDs. Moreover, electrolytes should have long cycle durability at working temperature. In addition, they should have suitable electrochemical stability in the voltage range used for device switching. Ideally, electrolytes should have an electrochemical stability domain extending from 0 V to as high as 4.5 V vs Li/Li^{+} to be compatible with lithium electrode materials. Furthermore, in an ECD configuration, the electrolyte must exhibit good chemical stability so that undesired chemical reactions won't happen when electrodes or cell packaging materials come into contact. The chemical stability must be ensured during both the deposition and the cycling. Moreover, for safety and stability purposes, flammable and toxic products should be preferably avoided. [33].

Electrolytes are classified into liquid electrolytes, gel electrolytes, and solid electrolytes on the basis of their state. Although conventional liquid electrolytes show relatively fast ion diffusion, they suffer from the potential leakage issue of electrical devices. In contrast, the electrical devices with gel or solid electrolytes can be free from the leakage, while they retard the ion diffusion to some extent. Moreover, the operation stability of electrical devices can be improved in specific cases. Ko et al. synthesized gel electrolytes by the formation of microporous organic polymers in the presence of 0.077–0.77 M $LiClO_4$ and *N*-methylpyrrolidone. ECDs fabricated using the gel electrolytes show enhanced operation stability (88% ΔT retained after 1000 cycles) compared with those fabricated using liquid electrolytes, which is attributable to the efficient separation of working viologens from counting molecules [34].

15.3 The Mechanical Stability of OEC Devices (Encapsulation Technology)

The next generation of ideal flexible ECDs (i.e. flexible electrochromic displays and smart clothes) should maintain high mechanical performance without degradation, even upon bending, twisting, or stretching, which is rather difficult to accomplish for conventional rigid ECDs. The mechanical stability of an OECD has two weak points, the electrolyte layer and the conductive substrate [6]. Liquid electrolytes are difficult to process, and the resulting ECDs need to be carefully sealed to prevent potential leakage. To circumvent this, the electrolyte needs to be in gel or solid [35]. The switching time of ECDs is dependent on the ionic mobility of the electrolytes, which drops as a function of increased viscosity of the electrolytes. Using a solid electrolyte compromises the switching time, while gel electrolytes used in ECDs are a better alternative. By using an ionic liquid/polymethyl methacrylate (PMMA)-based polymeric gel, Krebs and coworkers showed that the electrolyte layer can be efficiently deposited onto a polymeric layer using slot die or spray coating, with response time in the order of seconds [36]. Liou and coworkers demonstrated flexible ECDs with a triarylamine-based polyimide, 9Ph-PMPI as electrochromic material, and gel electrolyte based on PMMA and $LiClO_4$. The device exhibits a high contrast ratio and electroactive stability during EC operation with multiple colors at different applied potentials. Reliable and reproducible switching behavior of the flexible ECD also could be obtained under mechanical bending stress [37]. Hsiao and coworker also reported an all-polymeric flexible ECD with bis(triarylamine)-based ambipolar polyimides as electrochromic material. The polyimides were spray coated onto an ITO-coated polyethylene terephthalate (PET) slide and then dried, after which the gel electrolyte was spread on the polymer-coated side of the electrode. The device shows a maximum ΔT of 85% at 940 nm for green coloring and of 91% at 760 nm for deep blue coloring [38]. Liu's group reported flexible OECDs based on a series of conjugated oligomers containing EDOT and aromatic groups. The cycling stability analysis of flexible ECDs based on PET reveals that increasing the number of EDOT groups, the oxidation state of the oligomer is more stable [39].

ITO has been widely used as transparent electrode material so far due to its good optoelectronic properties. This is problematic in large-area devices since ITO has a large sheet resistance. A large sheet resistance can cause an ohmic drop across the substrate, with uneven switching and increased potential near the external contacts [40, 41]. In addition, the mechanical stability (i.e. risk of cracking and scratching) of the ITO layers can be low, and the cost of ITO is too high from a commercial perspective, due to the scarcity of indium and the energy cost of sputtering [42]. Recent studies have reported that several other flexible electrodes such as conducting polymers [14], metal networks [43], and carbon materials [44] are somewhat better alternatives to ITO. However, these materials still suffer from specific shortcomings. In general, conducting polymers have low coloring efficiency and poor cyclic and chemical stability. The optoelectronic performance of carbon materials is relatively poor (the sheet resistance is always higher than 20 Ω/sq and transmittance is less than 90%), and better performing carbon materials are normally fabricated through thermal and plasma chemical vapor deposition methods, both of which are not economically beneficial [45]. Metallic networks are always vulnerable to corrosive electrolytes, especially the most widely used silver and copper networks [46, 47]. An ideal coating material to protect the metallic networks requires stability in EC electrolytes with little impact on the optical and electrical performance of the EC electrode. PEDOT:PSS is widely used for its good conductivity, high light transparency, compatibility with solution deposition, and stability in EC electrolytes [48–52]. Lee's group successfully prepared a stretchable and wearable elastic conductor by embedding a WO_3 ECD and Ag nanowire into polydimethylsiloxane (PDMS). The unique sandwiched structure improved the mechanical robustness and durability [53]. They also developed a stable transparent electrode based on Ag grid/PEDOT:PSS. With the protection of the PEDOT:PSS passivation layer, the atmospheric corrosion resistance and cycling stability in EC and supercapacitor applications can be enhanced [54]. Liu et al. proposed a flexible sandwiched WO_3/Ag nanotrough network/PEDOT:PSS multilayer transparent conducting electrode with prominent conductivity (sheet resistance as low as 1.08 Ω/sq) and transparency (85%). In particular, the flexible transparent electrode exhibits excellent mechanical robustness upon bending and stretching. The hybrid electrode can be further adhered tightly to a layer of WO_3 to form an ECD, which simultaneously displays a rapid switching time (0.75 seconds/1.82 seconds for bleaching/coloring, respectively) and high cyclic performance (86% retention of the initial optical contrast after cycling for 30 000 seconds) [55] (Figure 15.3).

Although the sealant is not a mandatory component of an ECD, sealing the device with an appropriate sealant can greatly enhance the robustness of the ECD, thereby extending its service life. Since the device failure of ECDs may arise from multiple sources, including the degradation of the electrochrome caused by UV radiation, oxygen, and moisture, a sealant can be effective if it acts as a barrier preventing the penetration of the aforementioned damaging species into the ECDs. Much effort has been put into developing barrier films for the encapsulation of these devices. Typical targets for barrier films are ultrahigh such as oxygen transmission rates (OTRs) on the order of 10^5 $cm^3/m^2/d$ and water vapor transmission rates (WVTRs) on the order of 10^{-6} $g/m^2/d$ [56, 57]. These barrier films can be directly deposited onto organic electronic devices by vacuum deposition methods. Another method is to deposit barrier films onto polymer substrates and then seal it around the device with an edge

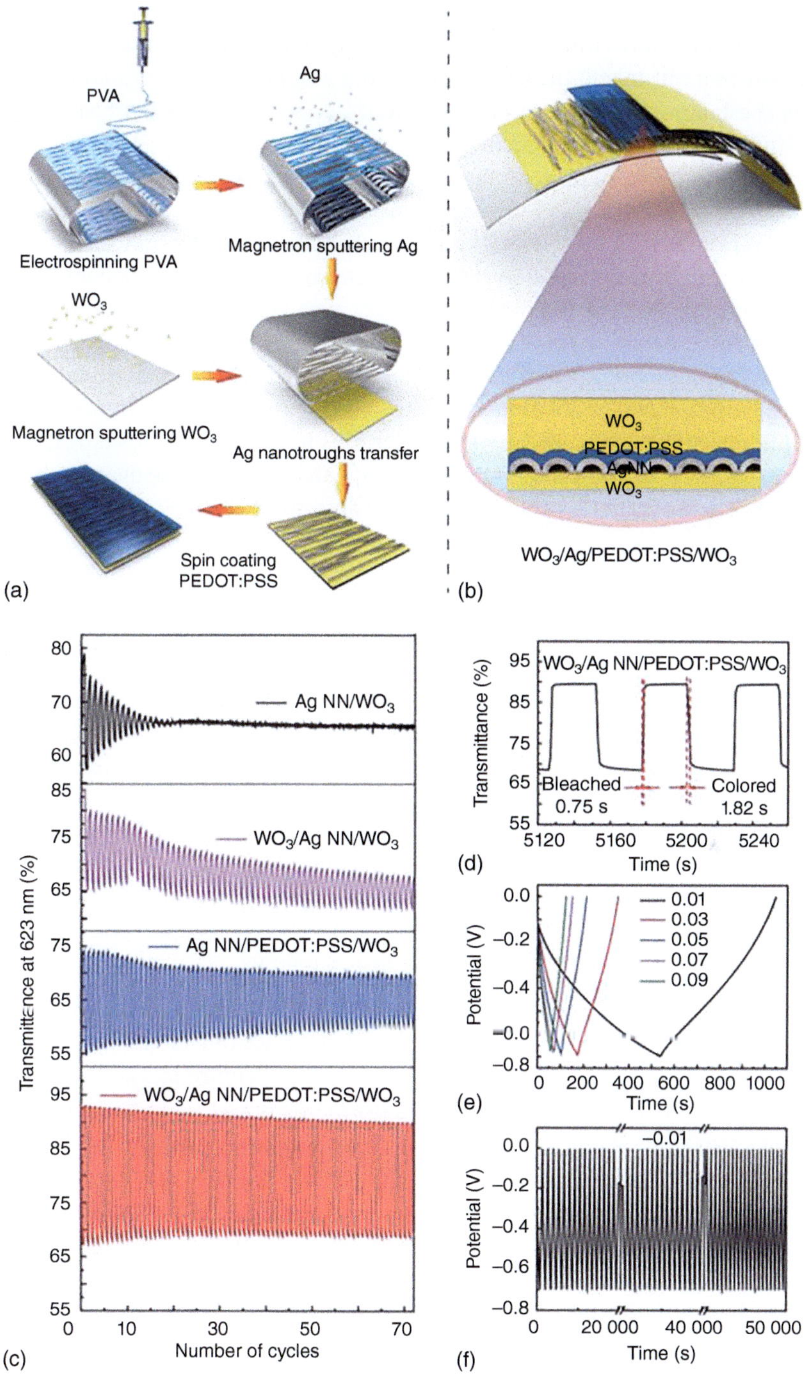

Figure 15.3 (a) Fabrication procedure of the WO_3/Ag NN/PEDOT:PSS transparent electrodes. (b) Schematic illustration of a cross section of the WO_3/Ag NN/PEDOT:PSS/WO_3 hybrid EC film. (c) Characterization of the coloring and bleaching rate by measuring the switching time. (d) Statistical results indicate a bleaching time of 0.75 seconds and coloring time of 1.82 seconds. Galvanostatic charge/discharge profile at (e) current densities from 0.01 to 0.1 mA/cm and (f) 0.01 mA/cm. Source: Liu et al. [55]. © 2018, Royal Society of Chemistry.

sealant or lamination film [58–61]. The latter encapsulation process is of great interest as it offers opportunities for higher manufacturing rates and potential for lower costs especially in a roll-to-roll (R2R) manufacturing process.

Although the lamination and sealing method of device encapsulation has some distinct advantages in device packaging, one of the major weaknesses of this method lies in the performance of the edge sealant itself. Edge sealants used in organic electronics packaging usually consists of thermally or UV-curable epoxies [62, 63] or other adhesives [58–61], which have long diffusion pathways to provide long transient permeation periods to improve sealant performance. A polyisobutylene edge sealant filled with desiccant was also used to encapsulate the ECDs, which revealed improved shelf-life over devices that were sealed with epoxy-based materials as demonstrated by a little loss in optical contrast for devices stored under both ambient and harsh conditions (50 °C and 80% RH). In terms of imparting UV resistance to the ECDs, UV barrier foils have also been investigated. Other secondary but important functions of a sealant are to prevent the evaporation of the electrolyte and also to maintain the physical structures of the ECDs [64].

15.4 Large-Area Process Technology: Spray Coating and Roll-to-Roll Processes

The need for cheap and fast processing of large-area organic thin films has recently become an increasingly important goal within many research fields where the idea is that solution processing of organic electronics has the potential to achieve this goal. In this context, R2R vacuum-free coating and printing is an emerging and ultimate goal processing method within the field of organic/polymer thin film device fabrication. In general, preparing and aligning a small multilayer structure on a solid substrate such as glass is very different from precisely coating and/or printing large areas on a moving flexible substrate [65]. A few R2R or R2R-compatible coating and printing techniques employed in OEC are reviewed as follows.

15.4.1 Inkjet Printing

Inkjet printing is classified as a full-digital, impact-free printing method, which generates images on demand. Two-dimensional patterning is obtained by a pixilated pattern with a defined resolution. Each pixel either receives a droplet of ink or not. The most widely used method, known as drop-on-demand (DOD), uses a piezoelectric material to eject droplets from a nozzle, which is about 1 mm above the substrate. The DOD system requires many nozzles, which limited web speeds and resolution in the early days, but today both high web speeds and high resolution can be achieved with the available commercial systems. The pattern shown in Figure 15.4 was printed on a R2R inkjet printer with a resolution of 600 DPI with web speeds ≤75 m/min. The major advantage of the inkjet printing technique over other 2D techniques is that you can easily change/adjust the printed pattern on a computer without the need to make a physical printing form (i.e. it uses a digital master). Inkjet printing is a relatively new technology on an industrial scale, which has some limitations in processing speed and ink formulation. In particular,

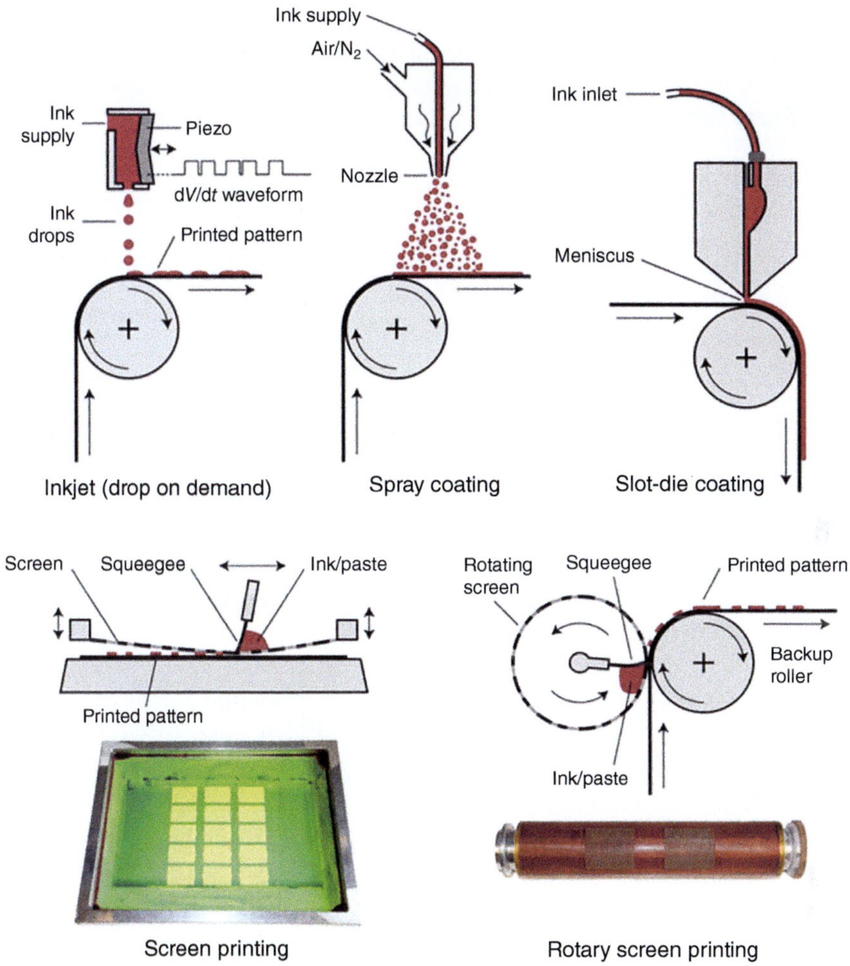

Figure 15.4 Schematic illustrations of coating and printing techniques. Source: Reproduced with permission of Søndergaard et al. [65]. Copyright 2012, Wiley Periodicals, Inc.

the latter has put forward restrictions on the application of organic film forming technology.

Inkjet printing is highly useful if someone wants to prepare patterned EC devices. This is well illustrated in a report by Shim et al. who showed that printing composite dispersions of polyaniline or PEDOT-covered silica nanoparticles onto PET/ITO enabled preparation of ECDs with good resolution [66]. Chen et al. studied electrochemical responses of inkjet-printed multicolored ECDs based on metallo-supramolecular polymers (MEPEs) to evaluate the feasibility of presenting multiple colors in one ECD. The printed ECDs exhibit great contrast with ΔT of 40.1% and a high coloration efficiency up to 445 cm^2/C within a short darkening time of two seconds. The flexible ECDs also show the same darkening time of two seconds and still had a high ΔT of 30.1% under bending condition [67]. Reynolds' group also proposed a series inkjet-printable cyan/magenta/yellow colored-to-colorless electrochromic polymer inks based on alkoxy-functionalized poly(3,4-propylenedioxythiophene)-based P(ProDOT) polymers [68].

15.4.2 Spray Coating

In spray coating, a continuous spray of ink is generated in a nozzle. In pneumatic-based systems, the ink is atomized at the spray nozzle, and pressurized air or gas (e.g. nitrogen) breaks up the liquid bulk into droplets. The liquid properties (surface tension, density, and viscosity), the gas flow properties, and the nozzle design influence the entire atomization process. The kinetic energy of the droplets helps them to spread onto the substrate. The quality of the coated layer is based on several process parameters, such as the distance between the spray nozzle and the substrate, the coating speed, and the number of sprayed layers. Other forms of spray generation are ultrasonication with a directed carrier gas or electrospraying. Pattern formation with spray coating is only possible with the help of shadow masks. The spray method is highly compatible with R2R, but the disadvantage of ink mist should also be considered, as it can cause contamination of the equipment. Ink loss and low edge resolution are also a fundamental disadvantage when using masks. As a coating method for fully covering layers, it can be used in combination with laser ablation or other patterning methods for post-film formation.

In the case of spray coating, the group of Reynolds has contributed to identify a series of polymers of different colors and optimize the conditions for the use of this technique [69–73]. Dyer and coworkers also demonstrated ECDs based on propylenedioxythiophene (ProDOT) polymers using spray coating method [74].

15.4.3 Slot-Die Coating

In slot-die coating, the ink is pumped through a coating head very close to the substrate. The continuous supply of ink forms a meniscus between the moving substrate and the coating head. This creates a continuous coating of fairly uniform thickness over a large area. Stripes may be coated in variable width depending on the shape of the meniscus guide when the slot-die head is fitted with the meniscus guide. Because of the use of pumps, the wet layer thickness can be precisely adjusted according to the pumping speed, meniscus width, and web speed. Slot-die coating is applicable for inks with large varieties in viscosity and solvents, and it has relatively low restriction in the aspect of wetting of the substrate as the ink is "poured" into the substrate compared to other printing techniques where surface tension and surface energy are of great importance in the transfer of the ink. Dewetting will occur if the surface tension of the ink is larger than the surface energy of the substrate. The wetting of inks with a high surface tension can be achieved by pretreatment of the substrate with processes such as corona or plasma treatment, which raises the surface energy of the substrate.

Processing speeds for slot-die coating can be performed at speeds ranging from 0.1 to 200 m/min. Krebs' group reported slot-die-coated ECDs on flexible substrates. The silver electrode structures for $18 \times 18\,cm^2$ devices were processed at a high speed (10 m/min) on PET by flexographic printing, and the EC polymer ECP-Magenta along with a minimal color-changing polymer minimally colour changing polymer (MCCP) was prepared by slot-die coating, showing the potential for fast fabrication of large volumes of low-cost flexible ECDs by using R2R processing techniques [75]. They also demonstrated flexible solid-state ECDs manufactured by sequentially stacking layers in one direction using flexographic printing and slot-die coating methods. The devices were built from the bottom up through in situ photo-cross-linking of the electrolyte layer [76].

15.4.4 Screen Printing

In screen printing, desired pattern is defined by the open area of mesh. A squeegee, which moves relative to the mesh, forces the ink through the open area onto the substrate. Relative thick wet layers are normally obtained (10–500 μm). The technique is generally only useful for rather high-viscosity inks with thixotropic (shear-thinning) behavior, as inks with lower viscosities will simply run through the mesh. There are two types of screen printing – flat bed and rotary screen printing. Flat bed screen printing is a stepwise process, where the screen is put on top of the substrate, then the squeegee swipes over the screen transferring the ink, and finally the screen is lifted resulting in a wet thin film. Stepwise processes are not desirable for large-scale manufacture since they are more time-consuming. However, flat bed screen printing is a powerful tool for small laboratory systems, either in a tabletop configuration or in a small- to medium-scale R2R configuration since the screens used are easily patterned in a variety of sizes and are fairly low in cost. Fully continuous processing can be achieved using rotary screen printing. In rotary screen printing, the web of the screen is folded into a tube with the squeegee and the ink is placed inside. The ink is continuously pushed through the open area of the screen by the fixed squeegee as the screen rotates at the same speed as the substrate, every rotation producing a full print. Much higher processing speed can be achieved using rotary screen printing (>100 m/min) compared to flat bed (0–35 m/min), but the folded screens are quite expensive and are much more difficult to clean due to limited access to the inside of the screen. In addition, the process takes longer time to adjust than flat bed. This results in a higher initial waste, but the process is very reliable once it is running.

Screen-printed flexible ECDs were reported as early as in 1999 by Brotherston et al. [77]. They used PEDOT and V_2O_5 screen printed onto separate ITO-coated Mylar substrates and completed with an electrolyte sandwiched between the two substrates. Zhou and coworkers reported fully screen-printed active matrix electrochromic displays based on PEDOT:PSS employing carbon nanotube thin-film transistors, which have potential applications in disposable tags, medical electronics, and smart home appliances [78]. Santiago et al. proposed a novel screen-printable ink composed of electrochromic polymer, poly(3,4-propylenedioxythiophene)bis(ethylhexyloxy) (ECP-Magenta), antimony-doped tin oxide (ATO/TiO_2) particles, and poly(vinylidene fluoride-co-hexafluoropropene) (P(VDF-co-HFP)) binder. The screen-printed films show significant color change at low potentials (0.3 V vs Ag/AgCl) with higher optical contrast values and coloration efficiency [79].

15.5 Conclusions and Perspective

Though significant progress has been made with OEC materials and OECDs in the past two decades, challenges still exist, such as long-term stability, mechanic stability, and large-area process technology, which are not fully addressed. On one hand, much effort should still be devoted to developing OEC materials and devices with color availability, longer stability, faster switch speed, and higher coloration efficiency. On the other hand, the efficient production of high-quality coatings in an industrial setting and device engineering are equally important. There is a huge difference between preparing and aligning a small multilayer structure on a solid substrate like glass and precise coating and/or printing of large areas with the same accuracy on moving flexible substrates. This requires a high degree of technical skills and the combination of scientific knowhow on how to tune the chemical and physical properties of the ink, with the technical knowhow of different processing procedures and the conditions they apply to. Device engineering needs to address the stability of the electrolytes and the electrical resistivity of the electrodes. Resolving these bottlenecks will certainly facilitate the large-scale production of electrochromic products in the foreseeable future.

References

1 Shin, H., Seo, S., Park, C. et al. (2016). *Energy & Environmental Science* 9: 117.
2 Xu, W., Fu, K., and Bohn, P.W. (2017). *ACS Sensors* 2: 1020.
3 Lv, X., Li, W., Ouyang, M. et al. (2017). *Journal of Materials Chemistry C* 5: 12.
4 Lahav, M. and van der Boom, M.E. (2018). *Advanced Materials* 30: e1706641.
5 Wang, S.-M., Hwang, J., and Kim, E. (2019). *Journal of Materials Chemistry C* 7: 7828.
6 Jensen, J., Madsen, M.V., and Krebs, F.C. (2013). *Journal of Materials Chemistry C* 1: 4826.
7 Mecerreyes, D., Marcilla, R., Ochoteco, E. et al. (2004). *Electrochimica Acta* 49: 3555.
8 Mallouki, M., Aubert, P.-H., Beouch, L. et al. (2012). *Synthetic Metals* 162: 1903.
9 Manceau, M., Bundgaard, E., Carlé, J.E. et al. (2011). *Journal of Materials Chemistry* 21: 4132.
10 Somani, P.R. and Radhakrishnan, S. (2002). *Materials Chemistry and Physics* 77: 117.
11 Mortimer, R.J. (2011). *Annual Review of Materials Research* 41: 241.
12 Sonmez, G. (2005). *Chemical Communications* 42: 5251.
13 Patra, A. and Bendikov, M. (2010). *Journal of Materials Chemistry* 20: 422.
14 Beaujuge, P.M. and Reynolds, J.R. (2010). *Chemical Reviews* 110: 268.
15 Groenendaal, L., Jonas, F., Freitag, D. et al. (2000). *Advanced Materials* 12: 481.
16 Groenendaal, L., Zotti, G., Aubert, P.-H. et al. (2003). *Advanced Materials* 15: 855.
17 Hsiao, S.-H., Wang, H.-M., Guo, W., and Sun, C.-H. (2011). *Macromolecular Chemistry and Physics* 212: 821.
18 Distefano, G., Jones, D., Guerra, M. et al. (1991). *The Journal of Physical Chemistry* 95: 9746.
19 İçli-Özkut, M., İpek, H., Karabay, B. et al. (2013). *Polymer Chemistry* 4: 2457.
20 Takimiya, K., Kunugi, Y., Konda, Y. et al. (2004). *Journal of the American Chemical Society* 126: 5084.

21 Hutchison, G.R., Ratner, M.A., and Marks, T.J. (2005). *Journal of the American Chemical Society* 127: 16866.

22 Gidron, O., Dadvand, A., Sheynin, Y. et al. (2011). *Chemical Communications* 47: 1976.

23 Zhen, S., Xu, J., Lu, B. et al. (2014). *Electrochimica Acta* 146: 666.

24 Liou, G.-S., Huang, N.-K., and Yang, Y.-L. (2007). *Journal of Polymer Science Part A: Polymer Chemistry* 45: 48.

25 Tsai, M.H., Lin, H.W., Su, H.C. et al. (2006). *Advanced Materials* 18: 1216.

26 Wang, H.-M. and Hsiao, S.-H. (2014). *Journal of Polymer Science Part A: Polymer Chemistry* 52: 272.

27 Xing, X., Wang, C., Liu, X. et al. (2017). *Electrochimica Acta* 253: 530.

28 Madasamy, K., Velayutham, D., Suryanarayanan, V. et al. (2019). *Journal of Materials Chemistry C* 7: 4622.

29 Shah, K.W., Wang, S.X., Soo, D.X.Y., and Xu, J. (2019). *Polymers* 11: 1839.

30 Zhao, S., Huang, W., Guan, Z. et al. (2019). *Electrochimica Acta* 298: 533.

31 Xiao, S., Zhang, Y., Ma, L. et al. (2020). *Dyes and Pigments* 174: 108055.

32 Li, M., Wei, Y., Zheng, J. et al. (2014). *Organic Electronics* 15: 428.

33 Thakur, V.K., Ding, G., Ma, J. et al. (2012). *Advanced Materials* 24: 4071.

34 Ko, J.H., Lee, H., Choi, J. et al. (2019). *Polymer Chemistry* 10: 455.

35 Byker, H.J. (2001). *Electrochimica Acta* 46: 2015.

36 Jensen, J., Dam, H.F., Reynolds, J.R. et al. (2012). *Journal of Polymer Science Part B: Polymer Physics* 50: 536.

37 Chen, C.-J., Yen, H.-J., Hu, Y.-C., and Liou, G.-S. (2013). *Journal of Materials Chemistry C* 1: 7623.

38 Wang, H.-M. and Hsiao, S.-H. (2014). *Journal of Materials Chemistry C* 2: 1553.

39 Wan, Z., Zeng, J., Li, H. et al. (2018). *Macromolecular Rapid Communications* 39: e1700886.

40 Kaneko, H. and Miyake, K. (1986). *Applied Physics Letters* 49: 112.

41 Ho, K.-C., Singleton, D.E., and Greenberg, C.B. (1990). *Journal of the Electrochemical Society* 137: 3858.

42 Emmott, C.J.M., Urbina, A., and Nelson, J. (2012). *Solar Energy Materials and Solar Cells* 97: 14.

43 Hsu, P.C., Wang, S., Wu, H. et al. (2013). *Nature Communications* 4: 2522.

44 Qiu, T., Luo, B., Liang, M. et al. (2015). *Carbon* 81: 232.

45 Guo, W., Xu, Z., Zhang, F. et al. (2016). *Advanced Functional Materials* 26: 8855.

46 Yao, S. and Zhu, Y. (2015). *Advanced Materials* 27: 1480.

47 Park, S., Parida, K., and Lee, P.S. (2017). *Advanced Energy Materials* 7: 1701369.

48 Tseng, Y.-C., Mane, A.U., Elam, J.W., and Darling, S.B. (2012). *Solar Energy Materials and Solar Cells* 99: 235.

49 Zhang, W., Zhao, B., He, Z. et al. (2013). *Energy & Environmental Science* 6: 1956.

50 Kim, Y.-S., Chang, M.-H., Lee, E.-J. et al. (2014). *Synthetic Metals* 195: 69.

51 Noh, Y.-J., Kim, S.-S., Kim, T.-W., and Na, S.-I. (2014). *Solar Energy Materials and Solar Cells* 120: 226.

52 Aleksandrova, M., Videkov, V., Ivanova, R. et al. (2016). *Materials Letters* 174: 204.

53 Yan, C., Kang, W., Wang, J. et al. (2014). *ACS Nano* 8: 316.
54 Cai, G., Darmawan, P., Cui, M. et al. (2016). *Advanced Energy Materials* 6: 1501882.
55 Liu, Q., Xu, Z., Qiu, W. et al. (2018). *RSC Advances* 8: 18690.
56 Lewis, J. (2006). *Materials Today* 9: 38.
57 Kim, D.-E., Kang, B.-H., Hong, S.-M. et al. (2010). *Thin Solid Films* 518: 4010.
58 Krebs, F.C. (2009). *Solar Energy Materials and Solar Cells* 93: 1636.
59 Krebs, F.C., Gevorgyan, S.A., Gholamkhass, B. et al. (2009). *Solar Energy Materials and Solar Cells* 93: 1968.
60 Krebs, F.C., Tromholt, T., and Jorgensen, M. (2010). *Nanoscale* 2: 873.
61 Gevorgyan, S.A., Medford, A.J., Bundgaard, E. et al. (2011). *Solar Energy Materials and Solar Cells* 95: 1398.
62 Lewis, J.S. and Weaver, M.S. (2004). *IEEE Journal of Selected Topics in Quantum Electronics* 10: 45.
63 Liao, Y., Yu, F., Long, L. et al. (2011). *Thin Solid Films* 519: 2344.
64 Kim, Y., Kim, H., Graham, S. et al. (2012). *Solar Energy Materials and Solar Cells* 100: 120.
65 Søndergaard, R.R., Hösel, M., and Krebs, F.C. (2013). *Journal of Polymer Science Part B: Polymer Physics* 51: 16.
66 Shim, G.H., Han, M.G., Sharp-Norton, J.C. et al. (2008). *Journal of Materials Chemistry* 18: 594.
67 Chen, B.H., Kao, S.Y., Hu, C.W. et al. (2015). *ACS Applied Materials & Interfaces* 7: 25069.
68 Österholm, A.M., Shen, D.E., Gottfried, D.S., and Reynolds, J.R. (2016). *Advanced Materials Technologies* 1: 1600063.
69 Reeves, B. D.; Grenier, C. R. G.; Argun, A. A.; Cirpan, A.; McCarley, T. D.; Reynolds, J.R.. 2004, 37, 7559.
70 Beaujuge, P.M., Ellinger, S., and Reynolds, J.R. (2008). *Advanced Materials* 20: 2772.
71 Mortimer, R.J., Graham, K.R., Grenier, C.R., and Reynolds, J.R. (2009). *ACS Applied Materials & Interfaces* 1: 2269.
72 Amb, C.M., Beaujuge, P.M., and Reynolds, J.R. (2010). *Advanced Materials* 22: 724.
73 Dyer, A.L., Craig, M.R., Babiarz, J.E. et al. (2010). *Macromolecules* 43: 4460.
74 Jensen, J., Dyer, A.L., Shen, D.E. et al. (2013). *Advanced Functional Materials* 23: 3728.
75 Søndergaard, R.R., Hösel, M., Jørgensen, M., and Krebs, F.C. (2013). *Journal of Polymer Science Part B: Polymer Physics* 51: 132.
76 Jensen, J. and Krebs, F.C. (2014). *Advanced Materials* 26: 7231.
77 Brotherston, I.D., Mudigonda, D.S.K., Osborn, J.M. et al. (1999). *Electrochimica Acta* 44: 2993.
78 Cao, X., Lau, C., Liu, Y. et al. (2016). *ACS Nano* 10: 9816.
79 Santiago, S., Aller, M., Campo, F.J., and Guirado, G. (2019). *Electroanalysis* 31: 1664.

Index

Organic Electronics for Electrochromic Materials and Devices, First Edition. Hong Meng.

e

q

r

s

t